SYMMETRIC FUNCTION
AND
ALLIED TABLES

SYMMETRIC FUNCTION
AND
ALLIED TABLES

BY

F. N. DAVID

M. G. KENDALL

AND

D. E. BARTON

CAMBRIDGE

Published for the Biometrika Trustees

AT THE UNIVERSITY PRESS

1966

PUBLISHED BY

THE SYNDICS OF THE CAMBRIDGE UNIVERSITY PRESS

Bentley House, 200 Euston Road, London, N.W.1
American Branch: 32 East 57th Street, New York, N.Y. 10022
West African Office: P.M.B. 5181, Ibadan, Nigeria

©

F. N. DAVID AND THE BIOMETRIKA TRUSTEES

1966

Printed in Great Britain at the University Printing House, Cambridge
(Brooke Crutchley, University Printer)

LIBRARY OF CONGRESS CATALOGUE
CARD NUMBERS: 65–15310

CONTENTS

TABLES

(The figures in parentheses refer to the pages where the tables are described in the Introduction)

v

Contents

Contents

PREFACE

Combinatorial algebra, in which we may include the algebra of symmetric functions, has been of interest to pure mathematicians from very early days. Many famous scientists, including Newton and Laplace, have made significant contributions, but it is fair to say, as far as the symmetric functions were concerned, that interest lay chiefly in their utility and applicability in the theory of equations. The combinatorial distribution theory to a large extent developed separately of the symmetric function algebra, until the two branches were connected by P. A. McMahon with his books *Combinatory Analysis*, **1** and **2**. Apart from use of rudimentary combinatorial theory in such distributions as the binomial and Poisson, little interest in this field was shown by statisticians until the researches of R. A. Fisher. The idea of randomization was known before him but he was the first to apply such a theory to tests of significance. Similarly, the statisticians of the early nineteen hundreds had had the idea of the calculation of moments of statistical criteria, but Fisher led the way in showing how the elegant methods of McMahon could be used both to short-cut existing procedures and to extend immeasurably the research on these topics. It should be added that the elucidation and further extension of Fisher's work owes much to M. G. Kendall and J. Wishart.

When M. G. Kendall and I started work on what McMahon would have styled the mutual expressibility of symmetric functions, we were interested principally in moment problems and in the possibility of reducing statistical calculations in such problems to a rule-of-thumb procedure. We believe that with the publication of our first table we succeeded in this aim. As a result of this first work we became interested in the general applicability of symmetric function algebra in statistical theory and the four succeeding tables were calculated by us over a further period of time. All tables were calculated *de novo* and were then checked against such previous tables as then existed as we became aware of them. The idea of the augmented symmetric function we believe to be ours.

Tables 1·1–1·5 are the basic tables which have been used in the construction of all the others succeeding. Two further tables have been included with them; tables of bivariate symmetric functions (1·6) and Levine's table of monomial symmetric functions in terms of monomial-monomials (Table 1·7 derived from Table 1·1). It is Tables 1·1–1·5 to which one would turn for the construction of any new table, but 1·6 and 1·7 group themselves naturally with these; 1·6 is a table of bivariate symmetric functions, derived from 1·1, which will be basic for any further tables of multivariate symmetric functions.

The remaining tables fall into two broad groups which reflect the dual applications of symmetric functions in statistics. Tables 2 and 3 will generally be found useful in moment problems, and also Table 5·5 which was constructed for the special purpose of rank-moments, while Tables 4–7 are concerned with problems of distribution. Of these last, Tables 4 deal specifically with the enumeration of partitions, Tables 5 with auxiliary tables concerned with the enumeration of rank-distribution problems, Tables 6 with 'runs' distributions in De Moivre's sense of the word 'run', while Tables 7 give the distributions related to the randomization approach applied to runs of events. Tables 8 are useful for obtaining approximations to distributions which lie outside the scope and range of Tables 6 and 7. Table 9, used for the expansion of the function of a function, has been derived from Tables 1.

Preface

For statisticians who are familiar with the applications of symmetric functions it would be sufficient to preface each table with definitions of the notations used. On the whole, however, in spite of the published researches of the past thirty years, it is clear that many workers do not know of the immense saving of time and labour, not to mention the elegancy introduced into the algebra, which results from using these techniques. This is understandable when it is remembered that McMahon is difficult of comprehension and that Fisher's habit of thought tends to condensation. In the belief therefore that the gathering together of relevant theorems and properties might be useful to the initiated, and in the hope that a certain expansiveness might induce the uninitiated to try out the unfamiliar, we have written at length and, because all the tables are to some extent interconnected, with repetitiveness. In so doing we have summarized all the techniques which we have found useful and have paved the way for further research work.

Our thanks are due to E. S. Pearson, managing editor of *Biometrika*, for suggesting the preparation of this book of tables to us, although it has grown far beyond his first suggestion. His careful reading of the manuscript and subsequent criticisms have done much to clarify our original intent.

UNIVERSITY COLLEGE LONDON   F. N. DAVID

June 1964

INTRODUCTION

PART I. SYMMETRIC FUNCTIONS

PART I A: TABLES 1·1, 1·2, 1·3, 1·4, 1·5, DESCRIPTION AND CONSTRUCTION

Preliminary notations and definitions

Any collection of v non-negative integers whose sum is w is commonly called a v-*partition* of w, the individual integers being spoken of as *parts* of the partition. Usually, unless it is stated otherwise, zero is excluded as a part. The parts are written in descending order of magnitude and the following notation is used:

If there are λ distinct parts, say $p_1, p_2, \ldots, p_\lambda$ with $p_1 > p_2 > p_3 > \ldots > p_\lambda \geqslant 1$ and if p_i is repeated π_i times, $i = 1, 2, \ldots, \lambda$, then we write the partition as $(p_1^{\pi_1} p_2^{\pi_2} \ldots p_\lambda^{\pi_\lambda})$. It is clear that

$$w = \sum_{i=1}^{\lambda} p_i \pi_i$$

and w is often called the *weight* of the partition. The *number* of parts is

$$v = \sum_{i=1}^{\lambda} \pi_i.$$

For example, $(4^2 2 1^3)$ has weight 13, 6 parts and 3 distinct parts. Commonly the order of the parts in the partition is not important. A partition in which the order is of importance is called a *composition*. For example, (414121) and (114241) are two compositions of 13 which if the order were of no importance would both be described by the partition $(4^2 2 1^3)$. Generally the v-partition $(p_1^{\pi_1} \ldots p_\lambda^{\pi_\lambda})$ yields

$$\frac{v!}{\pi_1! \, \pi_2! \ldots \pi_\lambda!}$$

compositions. This partitional notation is useful in the algebra of symmetric functions as will be seen in the succeeding section.

Monomial and augmented symmetric functions

A function of the quantities $x_1, x_2, \ldots, x_n$ is said to be *symmetric* if any number of the quantities can be interchanged and the function remains unaltered. Thus, for example,

$$x_1 + x_2 + \ldots + x_n = \sum_{i=1}^{n} x_i$$

is a symmetric function of the quantities $x_1, x_2, \ldots, x_n$ as is also

$$x_1 x_2 \ldots x_n (e^{x_1} + \ldots + e^{x_n}).$$

The number, n, of the quantities x is not important in the manipulation of the symmetric functions although it will enter into the final enumeration. It is always assumed that $n > w$, the weight. Following McMahon we shall write

$$\sum_{i=1}^{n} x_i = (1), \qquad \sum_{i=1}^{n} x_i^r = (r),$$

and if $r \neq s$,
$$x_1^r x_2^s + x_1^r x_3^s + \ldots + x_1^r x_n^s + x_2^r x_1^s + \ldots = \sum_{i \neq j} x_i^r x_j^s = (rs)$$

and there will be $n(n-1)$ terms in the summation. Or again

$$x_1^r x_2^s x_3^t + x_1^r x_2^s x_4^t + \ldots = \sum_{i \neq j \neq l} x_i^r x_j^s x_l^t = (rst), \quad r \neq s \neq t$$

with $n(n-1)(n-2)$ terms in the summation. When there is equality between the indices the number of terms is less. Thus

$$x_1^r x_2^r + x_1^r x_3^r + x_2^r x_3^r + \ldots = \sum_{i < j} x_i^r x_j^r = (r^2)$$

and there will be $\frac{1}{2}n(n-1)$ terms. Generally, the *monomial symmetric function*, which will be represented by the partition $(p_1^{\pi_1} \ldots p_\lambda^{\pi_\lambda})$, is defined as

$$(p_1^{\pi_1} \ldots p_\lambda^{\pi_\lambda}) = \Sigma(x_{i_1}^{p_1} \ldots x_{i_{\pi_1}}^{p_1})(x_{i_{\pi_1+1}}^{p_2} \ldots x_{i_{\pi_1+\pi_2}}^{p_2})(\ldots)\ldots(\ldots x_{i_v}^{p_\lambda}).$$

The summation is over all recognizably different terms, or, more precisely, the summation is over all selections of v different suffices, $i_1 \ldots i_v$, out of the total of n, and, for each selection, over all distinct arrangements of these into groups of $\pi_1, \pi_2, \ldots \pi_\lambda$. There are, accordingly,

$$^nC_v \frac{v!}{\pi_1! \ldots \pi_\lambda!}$$

terms in the summation. Thus the monomial symmetric function

$$(321) = \sum_{i_1 \neq i_2 \neq i_3} x_{i_1}^3 x_{i_2}^2 x_{i_3}$$

has $p_1 = 3$, $p_2 = 2$, $p_3 = 1$, $\pi_1 = \pi_2 = \pi_3 = 1$, $w = 6$, $v = 3$ and if there were six x variates, there would be $3! \cdot {}^6C_3 = 120$ terms in the summation.

The augmented monomial symmetric functions as defined by David & Kendall (1949) which were termed the *m-functions*, are more convenient for some purposes. We denote these by using square brackets [ ] and define

$$[p_1^{\pi_1} \ldots p_\lambda^{\pi_\lambda}] = \Sigma x_{i_1}^{p_1} \ldots x_{i_v}^{p_\lambda},$$

the summation being now over all selections of v suffices and, for such a selection, over all $v!$ arrangements of these. There will now be $v! \, {}^nC_v$ terms in the summation. We illustrate the difference between the augmented and non-augmented functions by an example. Thus

$$2(2^2) = 2\sum_{i < j} x_i^2 x_j^2 = \sum_{i \neq j} x_i^2 x_j^2 = [2^2],$$

but

$$(32) = \sum_{i < j}(x_i^3 x_j^2 + x_i^2 x_j^3) = \sum_{i \neq j} x_i^3 x_j^2 = [32],$$

and in general
$$[p_1^{\pi_1} \ldots p_\lambda^{\pi_\lambda}] = \pi_1! \pi_2! \ldots \pi_\lambda! (p_1^{\pi_1} \ldots p_\lambda^{\pi_\lambda}).$$

The s, u and h functions

Three special forms of the monomial symmetric functions are of importance:

(i) The *one-part functions* or the *power sums*, called by McMahon the *s-functions*

$$s_r = (r) = [r] = \sum_{i=1}^{n} x_i^r, \quad r = 1, 2, \ldots.$$

(ii) The *unitary* or *a-functions*

$$a^r = (1^r) = \sum_{i_1 < \ldots < i_r} x_{i_1} \ldots x_{i_r}, \quad r = 1, 2, \ldots.$$

The augmented unitary symmetric functions are denoted by

$$u_r = [1^r] = r!(1^r) = r!\,a_r = \Sigma x_{i_1} \dots x_{i_r},$$

where the sum is over all different ordered sets $i_1, \dots, i_r$.

(iii) The *homogeneous product sums*, or *h-functions*. A homogeneous product sum is the sum of all monomial symmetric functions of the same weight. Thus,

$$h_1 = s_1 = a_1,$$
$$h_2 = s_2 + a_2 = (2) + (1^2),$$
$$h_3 = (3) + (21) + (1^3),$$
$$h_4 = (4) + (31) + (2^2) + (21^2) + (1^4),$$

and so on.

The generating functions for s_r, a_r and h_r are easily written down. For any dummy variable t we may write

$$G(t) = \prod_{i=1}^{n} (1 - tx_i) = \sum_{r=0}^{n} a_r(-1)^r t^r = \sum_{r=0}^{n} \frac{u_r}{r!}(-1)^r t^r$$

with conventionally $a_0 = u_0 = 1$. It follows that

$$-\log G(t) = -\sum_{i=1}^{n} \log(1 - tx_i) = \sum_{r=1}^{\infty} \frac{s_r t^r}{r}$$

and

$$\frac{1}{G(t)} = \prod_{i=1}^{n} \frac{1}{1 - tx_i} = \sum_{r=1}^{\infty} h_r t^r.$$

More generally

$$\prod_{i=1}^{n}\left(1 + \sum_{r=1}^{l} \frac{t^r x_i^r}{r!}\right) = 1 + \sum_{m=1}^{n} \sum_{r=m}^{ml} \sum_{P} \frac{[p_1^{\pi_1} \dots p_\lambda^{\pi_\lambda}]}{\pi_1! \dots \pi_\lambda!} \frac{t_{p_1}^{\pi_1} \dots t_{p_\lambda}^{\pi_\lambda}}{(p_1!)^{\pi_1} \dots (p_\lambda!)^{\pi_\lambda}},$$

where the inner sum is over all m-partitions of r.

Mutual expressibility of symmetric functions

The mutual expressibility of symmetric functions has been known for many years (for instance the relations between the power sums and the unitary functions are due to Newton). The relations may be obtained in many different ways. We give two here, the second of which is just a corollary of the first. It is simplest to confine ourselves to the ordinary monomial functions, the relations between the augmented monomials being immediate.

(i) Since

$$-\log G(t) = \sum_{r=1}^{\infty} \frac{s_r t^r}{r}, \quad -G'(t) = G(t) \sum_{r=1}^{\infty} s_r t^{r-1},$$

we may relate the s- and the u-functions by equating coefficients. For example

$$u_1 = s_1,$$
$$u_2 = u_1 s_1 - s_2,$$
$$u_3 = u_2 s_1 - 2u_1 s_2 + 2s_3,$$
$$u_4 = u_3 s_1 - 3u_2 s_2 + 6u_1 s_3 - 6s_4,$$
$$\dots\dots\dots\dots\dots\dots\dots,$$

whence

$$0!\,s_1 = u_1, \qquad\qquad u_1 = s_1,$$
$$1!\,s_2 = u_1^2 - u_2, \qquad\qquad u_2 = s_1^2 - s_2,$$
$$2!\,s_3 = 2u_1^3 - 3u_2 u_1 + u_3, \qquad\qquad u_3 = s_1^3 - 3s_2 s_1 + 2s_3,$$
$$3!\,s_4 = 6u_1^4 - 12u_2 u_1^2 + 3u_2^2 + 4u_3 u_1 - u_4 \qquad u_4 = s_1^4 - 6s_2 s_1^2 + 3s_2^2 + 8s_3 s_1 - 6s_4,$$
$$\dots\dots\dots\dots\dots\dots\dots\dots, \qquad \dots\dots\dots\dots\dots\dots\dots\dots$$

Generally, in determinantal form,

$$s_r = \begin{vmatrix} u_1 & 1 & 0 & 0 & 0 & \dots & 0 \\ u_2 & u_1 & 1 & 0 & 0 & \dots & 0 \\ \dfrac{3u_3}{3!} & \dfrac{u_2}{2!} & u_1 & 1 & 0 & \dots & 0 \\ \dfrac{4u_4}{4!} & \dfrac{u_3}{3!} & \dfrac{u_2}{2!} & u_1 & 1 & \dots & 0 \\ \dots\dots\dots\dots\dots\dots \end{vmatrix}, \qquad u_r = \begin{vmatrix} s_1 & 1 & 0 & 0 & 0 & \dots & 0 \\ s_2 & s_1 & 2 & 0 & 0 & \dots & 0 \\ s_3 & s_2 & s_1 & 3 & 0 & \dots & 0 \\ s_4 & s_3 & s_2 & s_1 & 4 & \dots & 0 \\ \dots\dots\dots\dots\dots\dots \end{vmatrix}.$$

(ii) A slightly different manipulation of the generating functions may prove easier in some cases. Thus s_r/r is the coefficient of t^r in $-\log G$, i.e. it is the coefficient of t^r in

$$\sum_{m=1}^{\infty} \frac{1}{m}(-1)^m \left(\sum_{j=1}^{n} u_j \frac{t^j}{j!}(-1)^j \right)^m.$$

The expression in parentheses may be expanded multinomially and collection of the appropriate coefficients gives the relations

$$s_r = \sum_{m=1} \sum_{P} (-1)^{r+m} \frac{r(m-1)!}{\pi_1! \dots \pi_\lambda!} \frac{u_{p_1}^{\pi_1} \dots u_{p_\lambda}^{\pi_\lambda}}{(p_1!)^{\pi_1} \dots (p_\lambda!)^{\pi_\lambda}}.$$

Conversely, $u_r/r!$ is the coefficient of $t^r(-1)^r$ in $\exp\left\{ -\sum_{j=1}^{\infty} s_j \cdot \frac{t^j}{j} \right\}$ and so

$$u_r = \sum_m \sum_P (-1)^{r+m} \frac{r!}{\pi_1! \dots \pi_\lambda!} \frac{s_{p_1}^{\pi_1} \dots s_{p_\lambda}^{\pi_\lambda}}{p_1^{\pi_1} \dots p_\lambda^{\pi_\lambda}}.$$

The relations between the h- and u-functions follow similarly. From the generating functions we have

$$\frac{1}{1 + \sum\limits_{r=1}^{\infty} h_r t^r} = 1 + \sum_{r=1}^{\infty} a_r(-1)^r t^r = 1 + \sum_{r=1}^{\infty} \frac{u_r}{r!}(-1)^r t^r,$$

or

$$\frac{1}{1 + \sum\limits_{r=1}^{\infty} a_r t^r} = \frac{1}{1 + \sum\limits_{r=1}^{\infty} \dfrac{u_r}{r!} t^r} = 1 + \sum_{r=1}^{\infty} h_r(-1)^r t^r.$$

Consequently any relation between a and h will be the same if a and h are interchanged. We have, by equating coefficients,

$$a_1 = h_1,$$
$$a_2 = h_1^2 - h_2,$$
$$a_3 = h_1^3 - 2h_1 h_2 + h_3,$$
$$\dots\dots\dots\dots\dots$$

and generally

$$a_r = \frac{u_r}{r!} = \sum_m \sum_P (-1)^{r+m} \frac{m!}{\pi_1! \dots \pi_\lambda!} h_{p_1}^{\pi_1} \dots h_{p_\lambda}^{\pi_\lambda},$$

$$h_r = \sum_m \sum_P (-1)^{r+m} \frac{m!}{\pi_1! \dots \pi_\lambda!} a_{p_1}^{\pi_1} \dots a_{p_\lambda}^{\pi_\lambda} = \sum_m \sum_P (-1)^{r+m} \frac{m!}{\pi_1! \dots \pi_\lambda!} \left(\frac{u_{p_1}}{p_1!} \right)^{\pi_1} \dots \left(\frac{u_{p_\lambda}}{p_\lambda!} \right)^{\pi_\lambda}.$$

Similarly, the relations between s_r and h_r may be written down because of the interchangeability of a and h.

It may be shown that any augmented monomial symmetric function of weight w is uniquely expressible as a sum of products of power sums, of weight w, with integer coefficients, and conversely. The same is true in respect of the h- and the u-functions.

The D-operator

The expressions for the monomial symmetric functions (augmented or not) in terms of the three special functions are comparatively easy to calculate. It is enough to give, at this point, one method whereby relations between the monomials and the s-functions (say) may be established, further results following from the mutual expressibility of s, a and h. Generally we shall require to express products of the s-functions in terms of the monomials and vice versa. Consider for example the product $s_3 s_2$. It is clear that this product will be expressible as the sum of certain monomial symmetric functions of weight 5 and that we may write

$$s_3 s_2 = b_1(5) + b_2(41) + b_3(32) + b_4(31^2) + b_5(2^2 1) + b_6(21^3) + b_7(1^5),$$

where the b's are either constants or zero. It remains to determine the b's, which as a routine matter is most easily done by using Hammond's D-operator. The necessary properties of this operator may be summarized by the following four rules with, conventionally, $D_r(r) = 1$, $D_r D_s = D_s D_r$;

$$\text{(i)} \quad D_r s_i s_j = \sum_l D_l s_i D_{r-l} s_j$$

$$\left.\begin{array}{l} \text{(ii)} \quad D_r(rcd) = (cd) \\[4pt] \text{(iii)} \quad D_r(cde) = 0 \\[4pt] \text{(iv)} \quad D_0(rcd) = (rcd) \end{array}\right\} \quad r \neq c \text{ or } d \text{ or } e.$$

It will be recognized that the D-operator is essentially a differential operator. Thus if F is any symmetric function of $x_1, \ldots, x_n$ the operation $D_r F$ may be defined as: (1) increase n to $n+1$, (2) differentiate r times with respect to x_{n+1}, (3) put $x_{n+1} = 0$, (4) divide by $r!$.

Using Rules (i), (ii), (iii) and (iv) above we may, for example, determine the b's of the product $s_3 s_2$. For on the left-hand side we have

$$D_5 s_3 s_2 = 1,$$

and on the right-hand side D_5 obliterates all functions but the first so that $b_1 = 1$. If we operate in turn with $D_5, D_4 D_1, D_3 D_2, D_3 D_1^2, D_2^2 D_1, D_2 D_1^3, D_1^5$ we obtain

$$D_5 s_3 s_2 = 1 = b_1, \qquad D_2^2 D_1 s_3 s_2 = 0 = b_5,$$
$$D_4 D_1 s_3 s_2 = 1 = b_2, \qquad D_2 D_1^3 s_3 s_2 = 0 = b_6,$$
$$D_3 D_2 s_3 s_2 = 1 = b_3, \qquad D_1^5 s_3 s_2 = 0 = b_7.$$
$$D_3 D_1^2 s_3 s_2 = 0 = b_4,$$

giving the relationship

$$s_3 s_2 = (5) + (32) = [5] + [32].$$

Construction of Tables 1·1 $(m-s)$, 1·2 $(m-a)$, 1·3 $(a-h)$, 1·4 $(h-m)$, 1·5 $(a-s)$

From the relations and methods which are given in the previous two sections any of Tables 1 may be built up. There are, however, certain short cuts and checks which are possible and which are worth noticing. Again we illustrate on the first table (Tables 1·1) the others following in like fashion. Possibly for any table a systematic process of building one table from those of lower weight is the simplest. For example we have

$$s_1^2 = (1)^2 = [2] + [1^2].$$

5

Consequently $\qquad s_4 s_1^2 = (4)(1)^2 = [4]\{[2]+[1^2]\} = [4][2]+[4][1^2].$

Remembering the algebraic interpretation of the symbols it is immediate that

$$[4][2] = [6]+[42].$$

For the second product we note that 4 can agree (in suffix) with either of the 1's so that

$$[4][1^2] = 2[51]+[41^2],$$

giving $\qquad\qquad s_4 s_1^2 = [6]+2[51]+[42]+[41^2].$

This building-up process is the obliterating process in reverse. Thus, for example, we may write

$$s_4 s_1^2 = b_1(6)+b_2(51)+b_3(42)+b_4(3^2)+b_5(41^2)$$
$$+b_6(321)+b_7(2^3)+b_8(31^3)+b_9(2^21^2)+b_{10}(21^4)+b_{11}(1^6).$$

Operate on both sides with D_1 to get

$$2s_4 s_1 = b_2(5)+b_5(41)+b_6(32)+b_8(31^2)+b_9(2^21)+b_{10}(21^3)+b_{11}(1^5).$$

From the table of weight 5 we have $\qquad s_4 s_1 = (5)+(41),$

so that $\qquad\qquad\qquad\qquad b_2 = 2 = b_5,$

and $\qquad\qquad\qquad\qquad s_4 s_1^2 = \ldots + 2[51] + \ldots + [41^2] + \ldots.$

The other constants follow using different operators. The power sum-augmented symmetric function tables, shortly denoted by *sm*, were calculated independently using each method. The *ms* tables were obtained by inverting the *sm* tables. These are Tables 1·1. The other tables (1·2, 1·3, 1·4, 1·5) were also built up from those of lower weights in similar fashion.

Previous tables and description of present tables

Sukhatme (1938) gave *ms–sm* tables up to and including weight 8. Kerawala & Hanafi (1941, 1942, 1948) gave *ms* tables for weights 9, 10, 11 and 12. Cayley (1857) gave *ma–am* tables for weights 2–10, Faá di Bruno (1876) for weight 11 and Durfee (1882) for weight 12. McMahon (1884) calculated those (*ma*) for weight 13, Durfee (1887) for weight 14 and Decker (1910) for weight 15. Less extensive tables of the relations between the other functions are listed by Greenwood & Hartley (1962) and Fletcher *et al.* (1962).

The arrangements of our tables differ from any of those previously published and have been made with the object of saving space. In Tables 1·1, reading the tables horizontally or vertically the numbers on the diagonal must perforce be included, as also in Tables 1·2. For the expression of *h* in terms of *a* (and vice versa) it is necessary, as with Tables 1·1 and 1·2 to read right across the table to the diagonal, but because in any *ha* relation the *h*'s and *a*'s may be interchanged only one half of the table need be given. Thus from Table 1·3,

$$h_4 h_1 = -a_4 a_1 + 2a_3 a_1^2 + a_2^2 a_1 - 3a_2 a_1^3 + a_1^5,$$

and consequently $\qquad a_4 a_1 = -h_4 h_1 + 3h_3 h_1^2 + h_2^2 h_1 - 3h_2 h_1^3 + h_1^5.$

In the arrangement of Table 1·4, *h* in terms of *m* and vice versa, we took advantage of Hirsch's Law of Symmetry. Consequently to express *m* in terms of *h* it is necessary to read horizontally until the diagonal in bold type is reached and then continue vertically downwards to the edge of the table. For example, $\qquad (21^2) = -3h_1^4 + 10h_2 h_1^2 - 4h_2^2 - 7h_3 h_1 + 4h_4.$

To express h in terms of m the table should be read vertically downwards to the diagonal and then horizontally to the right edge of the table. For example,

$$h_2 h_1^2 = 12(1^4) + 7(21^2) + 4(2^2) + 3(31) + (4).$$

Table 1·5 gives the unitary functions, a, and therefore the homogeneous product sums, h, in terms of the monomial symmetric functions. In order to express monomial symmetric functions in terms of the a-functions the table is read horizontally up to and including the diagonal. For example,

$$(2)(1)^2 = a_1^4 - 2a_2 a_1^2.$$

To express the a-functions in terms of the monomial symmetric functions the table is to be read vertically downwards as far as and including the bold type diagonal, each coefficient being divided by w! Thus

$$(a_2 a_1^2)\, 4! = 12(1)^4 - 12(2)(1)^2.$$

Because the quantities which border each of the tables may be described in the partitional notation, the coefficients tabled are sometimes referred to as bipartitional functions (see, for example, Sukhatme, 1938). This nomenclature is an unfortunate one and we do not use it because of the distinct possibility of confusion with the bipartite symmetric functions described on pages 13–15 and tabulated in Tables 1·6.

PART I B: TABLES 1·1, 1·2, 1·3, 1·4, 1·5.
SOME APPLICATIONS IN COMBINATORY ANALYSIS

Preliminary notions

It is to McMahon that we owe many of the notions concerning the usefulness of symmetric functions in combinatory analysis, and much of what is written here will be found in his writings. From the statistical point of view the problems of combinatory analysis to which Tables 1·1–1·5 are particularly applicable are those of the enumeration of distributions. But it is first necessary to make clear what is the interpretation of notation and terminology. Suppose that w objects of v different kinds are to be distributed; to fix ideas it may be convenient to think of w balls of v different colours. If there are b_1 balls of one colour, b_2 of another, and so on with

$$b_1 \geqslant b_2 \geqslant \ldots \geqslant b_v \geqslant 1,$$

we may speak of the balls having an *object* specification $B \equiv b_1, b_2, \ldots, b_v$, which may be thought of as a v-partition of w. Now suppose we wish to distribute the objects into certain classes or equivalently to distribute the balls into given boxes. For the sake of example consider eight red, white and blue balls of specification $3^2 2$, and three boxes, each box to contain balls differently coloured each from the other. It is clear that we will need to take account of two cases: (i) the boxes may be distinguishable, in which case the box contents $(RWB)(RWB)(RW)$ can be permuted in three ways to give three distributions, (ii) the boxes may be indistinguishable in which case there can only be one distribution. This leads us to the necessity of defining a *box* specification. Let there be M boxes, c_1 of one kind, c_2 of another, …, c_l of the lth kind. The boxes of different kinds are distinguishable, boxes of the same kind are not. We then say that there is a box specification $C \equiv c_1, c_2, \ldots, c_l$, which again may be written in the partition notation. Thus in (i) above the box specification will be 1^3, while in (ii) it will be 3. The *content* specification within a box follows similarly. A box containing balls of h kinds, d_1 of one kind, d_2 of another, will be said

7

to have content specification $I \equiv d_1, \ldots, d_h$. In the three boxes above the content specifications will be $1^3, 1^3, 1^2$. We should note, however, that although the content specification is most simply written in the partitional notation with the numbers arranged in descending order, a box with content specification 21^2 could, in the example above, have actual contents $(RRWB)$ or $(WWRB)$ or $(BBRW)$ each being reckoned as distinct.

On occasions the number of distinct distributions will depend on the number of times the various parts in a specification are repeated, which, following the partitional notation we would write (for example) as $B \equiv b_1^{\beta_1}, \ldots, b_\lambda^{\beta_\lambda}$ with $b_1 > b_2 > \ldots > b_\lambda \geqslant 1$,

$$\sum_{i=1}^{\lambda} b_i \beta_i = w, \quad \sum_{i=1}^{\lambda} \beta_i = v,$$

where λ is the number of distinct parts. In order to distinguish the specification from the monomial symmetric function we shall write (B) for the latter where (B) denotes $(b_1^{\beta_1} \ldots b_\lambda^{\beta_\lambda})$.

Theorems

The theorems described below are all, in essence at least, given in McMahon's *Combinatory Analysis* (1915) where they are developed from the standpoint of the distribution problem they solve. We develop them here in an order which enables us to treat them more shortly, and it will be noticed that, more or less, they are corollaries of the first. We give a proof of this first, basic, theorem and indicate how the others follow, with some examples.

Theorem 1. The number of ways w balls of object specification B may be distributed into boxes of box specification 1^M so that the boxes have content specifications $I_1, \ldots, I_M$ respectively is the coefficient of (B) in $(I_1) \ldots (I_M)$.

If w_r is the weight of I_r, $r = 1, 2, \ldots, M$, we need only consider the case where

$$\sum_{r=1}^{M} w_r = w,$$

since otherwise the coefficient is zero and the theorem obviously true. Consider how the term $x_1^{b_1} \ldots x_k^{b_k}$ of (B) arises in the product $(I_1) \ldots (I_M)$. Its coefficient is the number of ways a product of one term from each of the monomial symmetric functions (I_r) gives $x_1^{b_1} \ldots x_k^{b_k}$, with b_1 arising as the sum of a set of the exponents of x_1 from one or more of the symmetric functions, and similarly for b_2 to b_k. This amounts to distributing b_1 x_1's into the M boxes, b_2 x_2's, and so on, so that the boxes have their given specification, each distribution corresponding to one product of M terms and only one. Since the coefficient of $x_1^{b_1} \ldots x_k^{b_k}$ is the same as that of (B) the result follows.

Theorem 2. The number of ways w balls of object specification B may be distributed into M different boxes so that Δ_1 of the boxes have content specification I_1, Δ_2 content specification I_2, and so on, no two of $I_1, \ldots, I_\rho$ being the same but otherwise all possible allocations of content specification among the boxes being allowed, is the coefficient of (B) in

$$\frac{M!}{\Delta_1! \ldots \Delta_\rho!} (I_1)^{\Delta_1} \ldots (I_\rho)^{\Delta_\rho}.$$

This is a direct corollary of Theorem 1.

Theorem 3. The number of ways of distributing w balls of specification B into boxes of specification 1^M such that the content specification of each box is one of $I_1, \ldots, I_\rho$, is the coefficient of (B) in

$$((I_1) + (I_2) + \ldots + (I_\rho))^M.$$

This follows by applying Theorem 2 and summing over all compositions $(\Delta_1 \ldots \Delta_\rho)$ of M.

When the balls in each box must be different, or alternatively when no box contains more than one ball of any given colour, a box's content specification depends only on its weight, i.e. $I_i = 1^{w_i}$. It follows that Theorems 1–3 give:

Theorem 4. The number of ways of distributing w balls of specification B into boxes of specification 1^M so that no box contains more than one ball of a given colour is the coefficient of (B) in:

(i) $a_{w_1} \dots a_{w_M}$ *when there are w_1 in the first box, w_2 in the second, and so on;*

(ii) $\dfrac{M!}{\Delta_1! \dots \Delta_\rho!} a_{w_1}^{\Delta_1} \dots a_{w_\rho}^{\Delta_\rho}$

when Δ_i of the boxes must have w_i balls, $i = 1, 2, \dots, \rho$, $w_1 > w_2 > \dots > w_\rho \geqslant 1$;

(iii) $(a_r + a_{r+1} + \dots + a_{r+s})^M$ *when any box may have any number of balls between r and $r+s$.*

Theorem 4(iii), Corollary. It is not difficult to derive from Theorem 4(iii) when $r = 1$ and $s = \infty$ (i.e. the restriction is simply that no box shall be empty) the explicit expression

$$\sum_{i=0}^{M} {}^M C_i (-1)^i \prod_{t=1}^{k} {}^{M-i} C_{b_t}$$

given by McMahon for the coefficient of (B) in $(a_1 + a_2 + \dots)^M$.

When we are interested in specifying the *numbers* in each box (i.e. the weights of the content specifications) without any restrictions regarding the numbers of different colours, it is clearly necessary to sum the coefficients given by Theorems 1 or 2 or 3 over all content specifications of a given weight. The sum of monomial symmetric functions of a given weight w is h_w and so we have:

Theorem 5. The number of ways of distributing w balls of object specification B into boxes of box specification 1^M is the coefficient of (B) in:

(i) $h_{w_1} \dots h_{w_M}$ *when the i-th box must contain w_i balls, $i = 1, 2, \dots, M$;*

(ii) $\dfrac{M!}{\Delta_1! \dots \Delta_\rho!} h_{w_1}^{\Delta_1} \dots h_{w_\rho}^{\Delta_\rho}$

when Δ_i boxes must have w_i balls each, $i = 1, 2, \dots, \rho$; $w_1 > w_2 > \dots > w_\rho \geqslant 1$;

(iii) $(h_r + h_{r+1} + \dots + h_{r+s})^M$ *when boxes may contain between r and $r+s$ balls.*

Theorem 5(iii), Corollary. From Theorem 5(iii), putting $r = 1$, $s = \infty$ we obtain the explicit result of McMahon, that the number of distributions given that no box can be empty is

$$\sum_{l=0}^{M} {}^M C_l (-1)^l \prod_{t=1}^{k} {}^{b_t + M - l} C_{b_t}.$$

So far we have considered the case of distinguishable boxes. If we now have w boxes of specification C and distribute w balls of specification B, no permutation of the contents of the c_i indistinguishable boxes of the ith kind will alter a given arrangement. Thus the set of c_i boxes all alike which must have one ball to a box will give the same distribution as one box which must have all possible content specifications of weight c_i. The corollary to Theorem 5 follows:

Theorem 5(i), Corollary. The number of distinguishable distributions of w objects of object specification B into w boxes of box specification C, one object to a box, is the coefficient of (B) in $h_{c_1} \dots h_{c_\rho}$.

If there is the further restraint that there shall be not more than d_i balls of a given colour among boxes of the ith kind it is plain that h_{c_i}, expressed as a sum of monomial symmetric functions of weight c_i, must be reduced to the sum of all monomial symmetric functions of weight c_i which have no part greater than d_i. Similar conditions may be introduced by the modification of the homogeneous product sums in this way.

Enumeration of patterns

In problems relating to the distribution of numbers into a rectangular array or matrix the theory just developed may be applied if the arrays (say) are thought of as boxes and the elements in a given column as of the same colour. We give two examples of this.

Example 1. Application of Theorem 4 (i). The number of ways in which a $k \times M$ array may be built up of 0's and 1's so that the rows total $w_1, ..., w_M$ respectively and the columns $b_1, ..., b_k$ respectively is the coefficient of (B) in $a_{w_1} ... a_{w_M}$. For instance, if we require the distribution of five ones in a 3×4 lattice so that the sum of the rows is 2, 1, 1, 1, and the sum of the columns 2, 2, 1 we look for the coefficient of $(2^2 1)$ in the expansion of $a_2 a_1^3$. This from Table 1·2 is 12. The actual patterns may be exhibited as follows:

```
1 1 0   1 1 0   1 1 0   1 1 0   1 1 0   1 1 0   1 0 1   1 0 1   1 0 1   0 1 1   0 1 1   0 1 1
1 0 0   0 1 0   1 0 0   0 1 0   0 0 1   0 0 1   0 1 0   0 1 0   1 0 0   1 0 0   0 1 0   1 0 0
0 1 0   0 0 1   0 0 1   1 0 0   0 1 0   1 0 0   1 0 0   0 1 0   0 1 0   0 1 0   1 0 0   1 0 0
0 0 1   1 0 0   0 1 0   0 0 1   1 0 0   0 1 0   0 1 0   1 0 0   0 1 0   1 0 0   1 0 0   0 1 0
```

Example 2. Application of Theorem 5 (i). The number of ways in which a $k \times M$ array may be built up so that the row totals are respectively $w_1, ..., w_M$, and the column totals $b_1, ..., b_k$, is the coefficient of (B) in $h_{w_1} ... h_{w_M}$. For instance, to distribute ten in a 2×3 lattice so that the row totals are 4, 4, 2 and the column totals 5 and 5 we require to find the coefficient of (5^2) in $h_4^2 h_2$ which from Table 1·4 is seen to be 13. These may be exhibited as follows:

```
40   40   31   31   31   22   22   22   13   13   13   04   04
04   13   13   22   04   22   31   13   31   40   22   40   31
11   02   11   02   20   11   02   20   11   02   20   11   20
```

Separations and separates

Separation is an arraying of a partition $B \equiv b_1^{\beta_1} ... b_\lambda^{\beta_\lambda}$ which is apparently similar to but in fact different from those discussed; it consists of dividing B up into a set of *separates* which are themselves partitions, so that in all the parts of all the separates there are β_1 parts of magnitude b_1, β_2 of magnitude b_2, etc. Thus a separation of $3^2 2^3 1$ into partitions is 32, 32, 21. The separates (being partitions) are written in numerical order as a rule. As for partitions, different orderings of the separates among themselves are not counted as distinct. It is easily seen that the number of M-separations of B is the same as the number of distributions of balls of object specifications $\beta_1, ..., \beta_\lambda$ into boxes of specification M (i.e. M indistinguishable boxes). This is a problem which we have not so far considered and is our next concern.

Boxes of general specification

To start with we shall consider M indistinguishable boxes. We will need the new symmetric functions $g_1, g_2, ...$, where

$$g_1 = \sum_{r=1}^{\infty} h_r = (1) + (2) + (1^2) + (3) + (21) + (1^3) + ...$$

and, for any $k, k \geq 1$, $\qquad g_k = (k) + (2k) + (k^2) + (3k) + (2k, k) + (k^3) + ...,$

where, generally, g_k is defined as the result of replacing $(x_1, ..., x_n)$ in g_1 by $(x_1^k, ..., x_n^k)$ respectively. The corresponding symmetric functions

$$H_k = \Sigma \frac{g_{p_1}^{\pi_1} \cdots g_{p_\lambda}^{\pi_\lambda}}{\pi_1! \cdots \pi_\lambda! \, p_1^{\pi_1} \cdots p_\lambda^{\pi_\lambda}}$$

(i.e. H_k is the same function of $g_1, ..., g_k$ as h_k is of $s_1, ..., s_k$) are also needed. We then have

Theorem 6. *The number of ways of distributing balls of object specification $B \equiv b_1, ..., b_k$ into boxes of specification M is the coefficient of $(b_1, ..., b_k)$ in H_M.*

Proof. The key to this theorem is that if we treat the products of $x_1, ..., x_n$ which are summed to give g_1, viz.

$$x_1, ..., x_n; \quad x_1^2, ..., x_n^2; \quad x_1 x_2, ..., x_{n-1} x_n; \quad x_1^3, ...$$

as new elements, $X_1, X_2, ...$, say (for example, $X_i = x_i$, $X_{n+i} = x_i^2, ...$), then,

$$g_r = X_1^r + X_2^r + ...,$$

which is the rth power sum of $X_1, X_2, ...$, and H_M is their Mth homogeneous product sum. Thus each different product $X_{i_1}, ..., X_{i_m}$ occurs just once in H_M. Now if we let $x_1^{c_1} x_2^{c_2} ...$ correspond to the allocation of c_1 balls of the first colour, c_2 of the second, etc., to a box, since this product is just one of $X_1, X_2, ...$, each of these corresponds to such an allocation. Thus the coefficient of $x_1^{b_1} x_2^{b_2} ... x_k^{b_k}$ (or, equally, of $(b_1, ..., b_k)$) in H_M is the number of different ways the balls of specification B may be allocated to the M indistinguishable boxes. It will be noted that if there are more boxes than balls the theorem is trivially true since H_M (being of weight M or more) has no term in $(b_1, ..., b_k)$ in its expansion and so we may say that such a term has coefficient zero corresponding to no ways of leaving no box empty.

Example. Application of Theorem 6. The number of ways of putting three red balls, three blue balls and three black balls into three bags is the coefficient of (3^3) in H_3. Using Table 1·1 we have $6H_3 = g_1^3 + 3g_1 g_2 + 2g_3$. The last term contributes 2 on sight and the second gives 9 (on observing that $g_1 g_2 = (2^3)(1^3) + (2^2)(31^2) + (2)(31^2) +$ terms not giving rise to anything containing (3^3)). The major contribution is from $g_1^3 = (h_1 + h_2 ... + h_7 ...)^3$. This contribution may evidently be written as the coefficient of (3^3) in

$$h_3^3 + 3(h_7 h_1^2 + h_5 h_2^2 + h_1 h_4^2) + 6(h_6 h_2 h_1 + h_5 h_3 h_1 + h_4 h_3 h_2),$$

i.e. it is 811 (from Table 1·5). Hence the required number is $137 = \frac{1}{6}[811 + 9 + 2]$.

Theorem 6 (i), *Corollary. The number of ways of distributing balls of object specification B into M indistinguishable boxes (empty boxes allowed) is the coefficient of (B) in H_M' where H_M' is the same function of $(1 + g_1, ..., 1 + g_M)$ as H_M is of $(g_1, ..., g_M)$.*

Proof. If we introduce a further element X_0 so that

$$1 + g_r = [X_0^r + X_1^r + ...]_{X_0 = 1},$$

it is evident that H_M' would generate all different products of M X's (including X_0) and since $X_0 = 1$, in fact H_M' generates all different products of M or less X's and thus $H_M' = 1 + H_1 + ... + H_M$.

Theorem 6 (ii), *Corollary. The number of ways of distributing balls of object specification B into boxes of specification $C \equiv (c_1, ..., c_l)$ is the coefficient of (B) in $H_{c_1} ... H_{c_l}$.*

Proof. Considering the coefficient of $x_1^{b_1} ... x_k^{b_k}$ in $H_{c_1} ... H_{c_k}$ we see that this is the product of l coefficients (or terms which are products of $x_1, ..., x_k$, each term corresponding to an allocation of the balls of specification B amongst the l sets of boxes) summed over all allocations of the B balls to the l sets of boxes. Now the coefficient of each term enumerates the number of ways the given allocation may be distributed within the corresponding set of like boxes, the product enumerates the total number of distributions corresponding to a given allocation and the total over all allocations is the number required.

PART I C: TABLES 1·1, 1·2, 1·3, 1·4, 1·5.
SOME APPLICATIONS IN MOMENT PROBLEMS

Introductory

The applications of the symmetric function tables in problems concerning moments are manifold, as will be seen from the Tables 2, all of which required the use of Tables 1 for their computation. We give here some slight illustration of the direct uses of Tables 1 in moment problems, reserving more extensive discussion for later tables. We assume throughout that the reader has enough statistical knowledge to be aware of the definitions of both moment and cumulant.

Products of sums of random variables

Assume that there are n independent random variables $x_1, x_2, \ldots, x_n$ with

$$\mathscr{E}(x_i) = \xi, \quad \mathscr{E}(x_i^2) = \sigma^2 + \xi^2, \quad \mathscr{E}(x_i^3) = \mu_3 + 3\sigma^2\xi + 2\xi^3 \quad (i = 1, 2, \ldots, n).$$

Any moment function of these n random variables will be a symmetric function of them. We may use the tables of symmetric functions to express the powers and products of any set of symmetric functions of the n observations in terms of the sum of augmented monomial symmetric functions, whence, using expectation techniques, the expectation of the set may be readily found. A simple example will serve. Consider the product

$$\left(\sum_{\substack{i=1 \\ i \neq j}}^{n} \sum_{j=1}^{n} x_i x_j \right) \left(\sum_{\substack{l=1 \\ l \neq m \neq k}}^{n} \sum_{m=1}^{n} \sum_{k=1}^{n} x_l^2 x_m x_k \right).$$

Each of the two multiple summations here is a symmetric function of $x_1, x_2, \ldots, x_n$, and the product, written in terms of the augmented monomial symmetric functions, is $[1^2][21^2]$. In order to carry out the multiplication it is necessary to express each function in terms of the s (or a) functions; it is immaterial which kind is used. Thus

$$[1^2][21^2] = \{(1)^2 - (2)\}\{2(4) - 2(3)(1) - (2)^2 + (2)(1)^2\}$$

which may be done from Tables 1·1·2 and 1·1·4. Multiplication gives

$$-2(4)(2) + 2(4)(1)^2 + 2(3)(2)(1) + (2)^3 - 2(3)(1)^3 - 2(2)^2(1)^2 + (2)(1)^4,$$

which on referring to Table 1·1·6 becomes

$$4[321] + 2[31^3] + 2[2^3] + 4[2^2 1^2] + [21^4].$$

There is no necessity to write this out in longhand, but if desired this may be done to give

$$4 \sum_{i \neq j \neq k} x_i^3 x_j^2 x_k + 2 \sum_{i \neq j \neq k \neq l} x_i^3 x_j x_k x_l + 2 \sum_{i \neq j \neq l} x_i^2 x_j^2 x_l^2 + 4 \sum_{i \neq j \neq k \neq l} x_i^2 x_j^2 x_k x_l + \sum_{i \neq j \neq k \neq l \neq v} x_i^2 x_j x_k x_l x_v.$$

Remembering the random variables are independent, taking expectations gives

$$4n^{(3)}(\mu_3 + 3\sigma^2\xi + 2\xi^3)(\sigma^2 + \xi^2)\xi + 2n^{(4)}(\mu_3 + 3\sigma^2\xi + 2\xi^3)\xi^3$$
$$+ 2n^{(3)}(\sigma^2 + \xi^2)^3 + 4n^{(4)}(\sigma^2 + \xi^2)^2\xi^2 + n^{(5)}(\sigma^2 + \xi^2)\xi^4.$$

This result follows because if there are ρ parts in the augmented monomial, then the multiplying factor will be $n^{(\rho)}$ since there will be $n^{(\rho)}$ terms in the summation, where $n^{(\rho)}$ is the usual notation for $n!/(n-\rho)!$. This is a simple example and in practice we should not have had resort to symmetric function techniques to write down the required expansion and subsequent expectations. This will not however always be the case and in this connexion Tables 1 will be found to be of great usefulness.

Relations between moments and cumulants

Let μ_r' be the rth crude moment of a random variable x and κ_r, its rth cumulant. As a result of the relation between the moment and the cumulant generating function we have (see Fisher, 1928)

$$\mu_r' = \sum_{\rho=0}^{r} \sum_{P} \frac{\kappa_{p_1}^{\pi_1} \kappa_{p_2}^{\pi_2} \dots \kappa_{p_m}^{\pi_m} \, r!}{\pi_1! \dots \pi_m! (p_1!)^{\pi_1} \dots (p_m!)^{\pi_m}},$$

where the inner sum is over all possible partitions of r and subject to the restriction that

$$\sum_{i=1}^{m} p_i \pi_i = r, \quad \sum_{i=1}^{m} \pi_i = \rho$$

and m is the number of distinct parts. These are the same coefficients as in the expansion of $(1)^r$ of either Tables 1·1 or Tables 1·2 and we may use the last line for arriving at the sum. Thus for example from Table 1·1·4,

$$(1)^4 = [4] + 4[31] + 3[2^2] + 6[21^2] + [1^4]$$

and accordingly

$$\mu_4' = \kappa_4 + 4\kappa_3 \kappa_1 + 3\kappa_2^2 + 6\kappa_2 \kappa_1^2 + \kappa_1^4.$$

To obtain the expressions in terms of moments about the mean, $\mu_1' = \kappa_1$, we ignore the partitions containing unit parts and drop the primes. Thus

$$\mu_4 = \kappa_4 + 3\kappa_2^2.$$

These expressions are important and are reproduced with the converse relations in Tables 2·1·1 and 2·1·2 The converse relations cannot be read directly from Tables 1·1 or 1·2.

PART I D: TABLE 1·6. BIPARTITE SYMMETRIC FUNCTIONS

Notation and definitions

McMahon considers the multipartition of a (multiple) number in the course of his treatment of various problems of distributions. We have not found uses, in statistics, for his distribution theory as yet, but we have found it necessary to study bipartite symmetric functions and the corresponding bipartite s-functions for the development of bivariate moment and cumulant relations. All the essentially new ideas can be treated in this connexion, the extension to multipartition following along similar lines.

Suppose a given bipartite number W, M with partitions $(p_1, \dots, p_v)$ of W and $(q_1, \dots, q_v)$ of M, zero parts being allowed subject to the conditions given below. The set $(p_1 q_1, \dots, p_v q_v)$ is called a bipartition of W, M. Any one, but not both, of a pair $p_i q_i$ may be zero. The partitions $(p_1, \dots, p_v)$ and $(q_1, \dots, q_v)$ are not necessarily in descending order but bipartitions which merely exchange two pairs, say $p_i q_i$ and $p_j q_j$, are counted the same. For example the eight bipartitions of $2, 2$ are

$$(22), (21, 01), (12, 10), (11, 11), (20, 01, 01), (10, 10, 02), (11, 10, 01), (10, 10, 01, 01).$$

We adopt the same exponent convention for repeated biparts as previously given, and would write these eight bipartitions as:

$$(22),\ (21,01),\ (12,10)(11^2),\ (20,01^2),\ (10^2,02),\ (11,10,01),\ (10^2,01^2),$$

the exponent referring to the necessity to repeat both numbers. The commas and parentheses are inserted for clarity. We refer to W, M as biweights and to v as the number of biparts.

Bipartite symmetric functions

Corresponding to bipartitions we have *bi*-nomial (as opposed to monomial) functions symmetric in each of the sets of elements $x_1, ..., x_n$ and $y_1, ..., y_n$. The binomial symmetric function is defined as

$$\Sigma x_{i_1}^{p_1} y_{i_1}^{q_1}, ..., x_{i_v}^{p_v} y_{i_v}^{q_v},$$

where the summation is over all distinguishable terms. These binomial symmetric functions may be augmented, just as in the monomial case. Thus if the pairs $p_1 q_1, ..., p_\lambda q_\lambda$ are all different, and

$$v = \sum_{i=1}^{\lambda} \pi_i,$$

so that λ is the number of different biparts and v is the total number of biparts, then

$$[(p_1 q_1)^{\pi_1}, ...(p_\lambda q_\lambda)^{\pi_\lambda}] = \Sigma x_{i_1}^{p_1} y_{i_1}^{q_1} ... x_{i_v}^{p_\lambda} y_{i_v}^{q_\lambda},$$

where the summation is over all selections of v different integers from the set of integers $1, 2, ..., n$ and over all $v!$ permutations of these. As for the monomials, we have

$$[(p_1 q_1)^{\pi_1} ... (p_\lambda q_\lambda)^{\pi_\lambda}] = \pi_1! ... \pi_\lambda! ((p_1 q_1)^{\pi_1} ... (p_\lambda q_\lambda)^{\pi_\lambda}).$$

Bipartite s and a functions

The special monomial symmetric functions (s, a, h) generalize, but for our purposes we need consider only the s and a functions. The power sums are

$$s_{pq} = \sum_{i=1}^{n} x_i^p y_i^q,$$

with

$$s_{p0} = s_p = \sum_{i=1}^{n} x_i^p.$$

The unitary binomial symmetric functions are defined as having the $\{p_i\}$ and $\{q_i\}$ all zero or unity, so that for a given biweight there is one for each value of w, the sum of the biweights. Thus

$$a_{pq} = ((10)^p, (01)^q) = \Sigma x_{i_1} ... x_{i_p} y_{i_{p+1}} ... y_{i_{p+q}},$$

where the summation is over all sets of $p+q$ different suffices with

$$i < i_2 < ... < i_p, \quad i_{p+1} < ... < i_{p+q},$$

and conventionally

$$a_{p0} = a_p, \quad a_{00} = a_0 = 1.$$

The generating functions of the power-sums and the unitaries, in terms of dummy variables t and z, are

$$G(tz) = \prod_{i=1}^{n} (1 + x_i t + y_i z) = \sum_{p=0}^{n} \sum_{q=0}^{n} a_{pq} t^p z^q, \quad p+q \leqslant n$$

and

$$\log G(tz) = \sum_{r=1}^{\infty} \frac{(-1)^{r-1}}{r} \sum_{p+q=r} {}^r C_p s_{pq} t^p z^q.$$

From these relations we have that

$$\frac{(-1)^{p+q-1}}{p+q}\,^{p+q}C_p\,s_{pq} = \sum_{r=1}^{p+q}\sum_P \frac{(-1)^{r-1}(r-1)!}{\pi_1!\dots\pi_\lambda!}\,a_{p_1 q_1}^{\pi_1}\dots a_{p_\lambda q_\lambda}^{\pi_\lambda}$$

and

$$(-1)^{p+q-1}a_{pq} = \sum_{r=1}^{p+q}\sum_P (^{p_1+q_1}C_{p_1})^{\pi_1}\dots(^{p_\lambda+q_\lambda}C_{p_\lambda})^{\pi_\lambda}\frac{(-1)^{r-1}s_{p_1 q_1}^{\pi_1}\dots s_{p_\lambda q_\lambda}^{\pi_\lambda}}{\pi_1!\dots\pi_\lambda!(p_1+q_1)^{\pi_1}\dots(p_\lambda+q_\lambda)^{\pi_\lambda}},$$

where the inner sum is over all bipartitions of biweight p, q.

Construction of bipartite Tables 1·6

It is possible to extend Hammond's D-operator to the bipartite case but this calls for a multiplicity of notation. The tables are constructed most easily, perhaps, using the process of building up from those of lower weight to those of higher weight. A simple example will serve to illustrate the procedure. Thus for $w = 3$, the (21) partition, we have

$$s_{20}s_{01} = \sum_i x_i^2 \sum_j y_j = \sum_i x_i^2 y_i + \sum_{i\neq j} x_i^2 y_j = [21] + [20, 01].$$

From this we may deduce for a 31-partition, $w = 4$,

$$s_{20}s_{10}s_{01} = s_{10}\{[21] + [20, 01]\} = [31] + [21, 10] + [30, 01] + [20, 11] + [20, 10, 01].$$

The rule being the usual one of adding the multiplier in all possible ways, having regard to the 31-partition, and remembering the inclusion of the extra part as in $[21, 10]$ and $[20, 10, 01]$. The writing out of the algebra in longhand will make this point clear. This process gives products and powers of the power-sums in terms of the binomial augmented symmetric functions. The inverse tables which give the binomial augmented symmetric functions in terms of the products and powers of the power sums were calculated in a similar way. Tables 1·6 were computed by F. N. David and E. Fix and have not been published previously.

Expression of bivariate moments in terms of bivariate cumulants (*Tables* 2·1, 2·2)

The bottom horizontal line in each table gives the numerical coefficients for the expression of moments in terms of cumulants, analogously with the univariate case. Thus, for example, considering $w = 4$ and the 2^2 partition, we read from the bottom horizontal line of $w = 4$, partition (2^2),

$$\mu_{22}' = \kappa_{22} + 2\kappa_{21}\kappa_{01} + \kappa_{20}\kappa_{02} + 2\kappa_{12}\kappa_{10} + 2\kappa_{11}^2 + \kappa_{20}\kappa_{01}^2 + 4\kappa_{11}\kappa_{10}\kappa_{01} + \kappa_{10}^2\kappa_{02} + \kappa_{10}^2\kappa_{01}^2.$$

For moments about the population means, μ_{10}' and μ_{01}', we ignore unit parts and write

$$\mu_{22} = \kappa_{22} + \kappa_{20}\kappa_{02} + 2\kappa_{11}^2.$$

These relations and their converse are given, up to and including weight 8, in Tables 2·1 and 2·2.

Illustration of use of tables

As an illustration of the algebraic use of the tables a simplified form of one of the criteria which originally caused us to calculate them will serve. Assume a bivariate population which has as many moments μ_{rt}' ($r = 0, 1, 2, \dots, t = 0, 1, 2, \dots$), as desired. For brevity write

$$b = \kappa_{11} = \mu_{11}' - \mu_{10}'\mu_{01}'.$$

n independent pairs of observations $\{x_i y_i\}$ $i = 1, 2, \dots, n$, are made and the quantity

$$\theta = \sum_{i=1}^n (x_i^2 - 2bx_i y_i + y_i^2)/n$$

is computed. It is desired to find the sampling mean and variance of θ. We write

$$\theta = (s_{20} - 2bs_{11} + s_{02})/n,$$

whence
$$\mathscr{E}(\theta) = \mu'_{20} - 2b\mu'_{11} + \mu'_{02}.$$

For higher moments we use the tables. Thus

$$\theta^2 = \frac{1}{n^2}\{s_{20}^2 + 4b^2 s_{11}^2 + s_{02}^2 - 4bs_{20}s_{11} + 2s_{20}s_{02} - 4bs_{11}s_{02}\}$$

which, from Tables 1·1·4 and 1·6·4, may be written as

$$\theta^2 = \frac{1}{n^2}(\{[40] + [20, 20]\} + 4b^2\{[22] + [11, 11]\} + \{[04] + [02, 02]\}$$
$$- 4b\{[31] + [20, 11]\} + 2\{[22] + [20, 02]\} - 4b\{[13] + [02, 11]\}).$$

On taking expectations we have

$$\mathscr{E}(\theta^2) = \frac{1}{n^2}\{n(\mu'_{40} + 4b^2\mu'_{22} + \mu'_{04} - 4b\mu'_{31} + 2\mu'_{22} - 4b\mu'_{13})$$
$$+ n^{(2)}(\mu'^2_{20} + 4b^2\mu'^2_{11} + \mu'^2_{02} - 4b\mu'_{20}\mu'_{11} + 2\mu'_{20}\mu'_{02} - 4b\mu'_{02}\mu'_{11})\},$$

which gives on reduction

$$\operatorname{var}(\theta) = \frac{1}{n}\{\mu'_{40} - \mu'^2_{20} + 4b^2(\mu'_{22} - \mu'^2_{11}) + \mu'_{04} - \mu'^2_{02} - 4b(\mu'_{31} - \mu'_{20}\mu'_{11})$$
$$+ 2(\mu'_{22} - \mu'_{20}\mu'_{02}) - 4b(\mu'_{13} - \mu'_{02}\mu'_{11})\}.$$

The third and fourth moments, if required, follow by taking third and fourth powers of θ, expanding the powers and products of the s-functions into the augmented binomial symmetric functions using tables of weight six and weight eight respectively and taking expectations.

PART I E: TABLE 1·7. PRODUCTS OF MONOMIAL SYMMETRIC FUNCTIONS AS LINEAR FUNCTIONS OF MONOMIALS

Uses of tables

In Section C it was noted that one of the direct uses of Tables 1·1 and 1·2 is that one is able to express products of monomial functions as linear functions of monomial functions. The procedure involves expressing each monomial function in the product in terms of the s, a or h functions, carrying out the multiplication, and then using the inverse tables to express the powers and products of the special functions back into monomial functions. Thus, to refresh memory, if we require to express the product $[21][1^2]$ as a linear function of monomials we would use Tables 1·1·2 and 1·1·3 to give

$$[21] = -s_3 + s_2 s_1, \quad [1^2] = -s_2 + s_1^2,$$

so that
$$[21][1^2] = s_3 s_2 - s_3 s_1^2 - s_2^2 s_1 + s_2 s_1^3$$

which, using Table 1·1·5 inversely, gives

$$[21][1^2] = 2[32] + 2[31^2] + 2[2^2 1] + [21^3].$$

It is clear that many of these products of monomial functions, particularly those of low weights, will occur often in statistical research and it is useful therefore to reproduce them as a table ancillary to the first five of this Part I. Functions up to and including those of weight eight only are given. The construction of tables of higher weight presents no difficulties whatever, but we have not computed them chiefly because of the difficulty in presenting them as a concise table. Table 1·7 is not intended in any way as a replacement for the basic Tables 1·1 ... 1·5. It is however useful as an auxiliary, but care must be exercised as far as their introduction into algebra is concerned since it is the ordinary symmetric functions, and not the augmented symmetrics introduced by David & Kendall, which are tabulated. Thus, for example,

$$(21)(1^2) = \tfrac{1}{2}[21][1^2],$$

which from Table 1·1·1 may be written as

$$\tfrac{1}{2}\{-(3)+(2)(1)\}\{-(2)+(1)^2\} = \tfrac{1}{2}\{(3)(2)-(2)^2(1)-(3)(1)^2+(2)(1)^3\}.$$

From the inverse Table 1·1·1 we have, after cancellation of appropriate terms,

$$\tfrac{1}{2}\{2[32]+2[31^2]+2[2^21]+[21^3]\},$$

or

$$\tfrac{1}{2}\{2(32)+4(31^2)+4(2^21)+6(21^3)\},$$

or

$$(21)(1^2) = (32)+2(31^2)+2(2^21)+3(21^3),$$

as given in Table 1·7·5. The beginner will probably find it less confusing to revert to the basic tables.

Tables 1·7 may be read only in one direction, the horizontal. They were originally compiled by Levine (1959).

PART II: MOMENTS, CUMULANTS AND *k*-FUNCTIONS

PART II A: TABLES 2·1·1 AND 2·1·2.   RELATIONS BETWEEN MOMENTS AND CUMULANTS

The univariate case

It has already been noted (Part I C, p. 13) that the coefficients of the monomial symmetric functions of the expansion of $(1)^r$ are the same as those appearing in the expression of the rth crude moment of a random variable in terms of its cumulants, since (see Fisher, 1928)

$$\mu'_r = \sum_{\rho=0}^{r} \sum_{P} \frac{\kappa_{p_1}^{\pi_1} \kappa_{p_2}^{\pi_2} \dots \kappa_{p_\lambda}^{\pi_\lambda} r!}{\pi_1! \dots \pi_\lambda! (p_1!)^{\pi_1} \dots (p_\lambda!)^{\pi_\lambda}},$$

with

$$\sum_{i=1}^{\lambda} p_i \pi_i = r, \quad \sum_{i=1}^{\lambda} \pi_i = \rho.$$

Also we have that

$$\kappa_r = \sum_{\rho=1}^{r} \sum_{P} \frac{(\mu'_{p_1})^{\pi_1} \dots (\mu'_{p_\lambda})^{\pi_\lambda} r!}{(p_1!)^{\pi_1} \dots (p_\lambda!)^{\pi_\lambda} \pi_1! \dots \pi_\lambda!} (-1)^{\rho-1} (\rho-1)!,$$

with the same restrictions on p and π. The coefficients in this last relation are those of $(1)^r$ multiplied by $(-1)^{\rho-1}(\rho-1)!$ where ρ is the number of parts of the partition. For example

$$(1)^4 = [4] + 4[31] + 3[2^2] + 6[21^2] + [1^4],$$

which enables us to write down immediately

$$\mu'_4 = \kappa_4 + 4\kappa_3\kappa_1 + 3\kappa_2^2 + 6\kappa_2\kappa_1^2 + \kappa_1^4$$

and

$$\kappa_4 = \mu'_4 + 4\mu'_3\mu'_1(-1) + 3\mu'^2_2(-1) + 6\mu_2\mu'^2_1(-1)^2(2!) + \mu'^4_1(-1)^3(3!).$$

To obtain the relations in terms of moments about the mean $(\mu'_1 = \kappa_1)$ we ignore the partitions containing unit parts and drop the primes, viz.

$$\mu_4 = \kappa_4 + 3\kappa_2^2, \quad \kappa_4 = \mu_4 - 3\mu_2^2.$$

Multivariate case

The relations between multivariate moments and cumulants may be obtained for the bivariate case by using the bipartite symmetric function Tables 1·6 as described in Part I D, p. 15, the method being entirely analogous to that of the preceding paragraph. Alternatively since we have tabulated only the bivariate sets and multivariate relations may on occasion be required, the following simple operative process may be used. Given random variables $x, y, z, \dots$, with joint cumulative distribution function $F(x, y, z, \dots)$ then $\mu'_{abc\dots}$ is defined as

$$\mu'_{abc\dots} = \int_R x^a y^b z^c \dots dF,$$

the Stieltjes integral being taken over the whole permissible range of variation of the random variables. This definition includes the univariate case (e.g. mean x is $\mu'_{100\dots}$), the bivariate case and so on. By definition, also, the multivariate cumulants are the coefficients in the Taylor expansion of the moment generating function. Now we have, for example, for weight 4 in the univariate case

$$\kappa_4 = \mu'_4 - 4\mu'_3\mu'_1 - 3\mu'^2_2 + 12\mu'_2\mu'^2_1 - 6\mu'^4_1$$

as above. Write this formally as

$$\kappa(r^4) = \mu'(r^4) - 4\mu'(r^3)\mu'(r) - 3(\mu'(r^2))^2 + 12\mu'(r^2)(\mu'(r))^2 - 6(\mu'_1(r))^4,$$

and operate with $s\partial/\partial r$. Cancelling the factor 4 on both sides we have

$$\kappa(r^3 s) = \mu'(r^3 s) - 3\mu'(r^2 s)\mu'(r) - 3\mu'(r^2)\mu'(rs) + 6\mu'(rs)(\mu'(r))^2 - 6(\mu'(r))^3\mu'(s)$$
$$- \mu'(r^3)\mu'(s) \qquad\qquad + 6\mu'(r^2)\mu'(r)\mu'(s)$$

or
$$\kappa_{31} = \mu'_{31} - 3\mu'_{21}\mu'_{10} - 3\mu'_{20}\mu'_{11} + 6\mu'_{11}(\mu'_{10})^2 - 6(\mu'_{10})^3\mu'_{01}$$
$$- \mu'_{30}\mu'_{01} \qquad\qquad + 6\mu'_{20}\mu'_{10}\mu'_{01}$$

κ_{22} follows from the operation on $\kappa(r^3 s)$ by $s\partial/\partial r$. Alternatively if κ_{211} is required it is necessary to operate on $\kappa(r^3 s)$ by $t\partial/\partial r$, to give

$$\kappa(r^2 st) = \mu'(r^2 st) - 2\mu'(rst)\mu'(r) - 2\mu'(rt)\mu'(rs) + 2\mu'(st)(\mu'(r))^2 - 6(\mu'(r))^2\mu'(s)\mu'(t)$$
$$- \mu'(r^2 s)\mu'(t) \quad - \mu'(r^2)\mu'(st) \quad + 4\mu'(rs)\mu'(r)\mu'(t)$$
$$- \mu'(r^2 t)\mu'(s) \qquad\qquad + 4\mu'(rt)\mu'(r)\mu'(s)$$
$$\qquad\qquad\qquad + 2\mu'(r^2)\mu'(s)\mu'(t)$$

or
$$\kappa_{211} = \mu'_{211} - 2\mu'_{111}\mu'_{100} - 2\mu'_{101}\mu'_{110} + 2\mu'_{011}\mu'^2_{100} - 6\mu'^2_{100}\mu'_{010}\mu'_{001}$$
$$- \mu'_{210}\mu'_{001} - \mu'_{200}\mu'_{011} \quad + 4\mu'_{110}\mu'_{100}\mu'_{001}$$
$$- \mu'_{201}\mu'_{010} \qquad\qquad + 4\mu'_{101}\mu'_{100}\mu'_{010}$$
$$\qquad\qquad + 2\mu'_{200}\mu'_{010}\mu'_{001}.$$

Relations of higher weight and the converse expressions for moments in terms of cumulants follow in similar fashion. The validity of this operator technique is found in the fact that it is a shorthand description of the differentiation of the equation expressing the cumulant generating function and the logarithm of the moment generating function (and conversely) with respect to one of the dummy variables.

Tables 2·1·1 *and* 2·1·2

Because the relations between moments and cumulants are important we have compiled them as tables separately from Tables 1. Table 2·1·1 gives the crude univariate and bivariate moments in terms of cumulants up to and including those of weight 8. Table 2·1·2 gives the univariate and bivariate cumulants in terms of the crude moments. In either case conversion to moments about the mean is achieved by omitting all terms containing suffices with a unit part and dropping the primes. For example

$$\mu'_{31} = \kappa_{31} + \kappa_{30}\kappa_{01} + 3\kappa_{21}\kappa_{10} + 3\kappa_{20}\kappa_{11} + 3\kappa_{20}\kappa_{10}\kappa_{01} + 3\kappa_{11}\kappa^2_{10} + \kappa^3_{10}\kappa_{01}$$

and
$$\mu_{31} = \kappa_{31} + 3\kappa_{20}\kappa_{11}.$$

We have not proceeded with trivariate, quadrivariate (etc.) moments since these are not used as frequently as the univariate and the bivariate, and for such cases as they are required they may be derived simply from the bivariate by the operative technique described.

PART II B: TABLE 2·1·3. k-STATISTICS IN TERMS OF PRODUCTS OF THE s-FUNCTIONS

Fisher's k-statistics

A set of independent random variables $x_1, x_2, ..., x_n$ is assumed, with identical probability density function (p.d.f) for the x's. This p.d.f has as many cumulants $\kappa_r, r = 1, 2, ...$ as desired. Fisher (1928) introduced functions, k_r, of the random variables $x_1, x_2, ..., x_n$ such that they are unbiased estimates of κ_r, i.e. so that

$$\mathscr{E}(k_r) = \kappa_r, \quad r = 1, 2, ...,$$

and it was subsequently demonstrated that in addition to their unbiasedness (Halmös, 1946),

$$\mathscr{E}(k_r - \kappa_r)^2$$

is a minimum. Such functions k_r, called by Fisher k-statistics, are necessarily rth order polynomials in $x_1, x_2, ..., x_n$ and the first condition of unbiasedness, together with the restriction that they be symmetric functions of the x's, is enough uniquely to determine them. It is clear that if we denote the rth crude population moment by μ'_r, we have

$$\mu'_r = \mathscr{E}\left(\frac{1}{n}\sum_{i=1}^{n}x_i^r\right) = \frac{1}{n}\mathscr{E}[r]$$

in symmetric function notation. Consequently if we start from any relation between moments and cumulants such as can be taken from Table 2·1·2, the corresponding k-statistics can be written in terms of the augmented monomial symmetric functions. Consider k_3 for example. Table 2·1·2 gives

$$\kappa_3 = \mu'_3 - 3\mu'_2\mu'_1 + 2\mu'^3_1.$$

Now we desire

$$\mathscr{E}(k_3) = \kappa_3.$$

Consequently, remembering the definition of an augmented symmetric function, for example

$$[1^3] = \sum_{i \neq j \neq k} x_i x_j x_k,$$

it is clear that if we write

$$k_3 = \frac{1}{n}[3] - \frac{3}{n^{(2)}}[21] + \frac{2}{n^{(3)}}[1^3],$$

the desired relation is achieved, given the random variables are independent. This expression is enough to define k_3 but it is convenient to express the monomial symmetric functions in terms of the power-sums (Table 1·1·3) and thus obtain

$$k_3 = \frac{1}{n^{(3)}}(n^2 s_3 - 3n s_2 s_1 + 2s_1^3).$$

Generally we have

$$\kappa_r = r! \sum_{\rho=1}^{r}(-1)^{\rho-1}(\rho-1)! \sum_{P}\frac{\mu'^{\pi_1}_{p_1}\mu'^{\pi_2}_{p_2}\cdots\mu'^{\pi_\lambda}_{p_\lambda}}{\pi_1!\ldots\pi_\lambda!(p_1!)^{\pi_1}\ldots(p_\lambda!)^{\pi_\lambda}},$$

giving

$$k_r = r! \sum_{\rho=1}^{r}(-1)^{\rho-1}\frac{(\rho-1)!}{n^{(\rho)}}\sum_{P}\frac{[p_1^{\pi_1}\cdots p_\lambda^{\pi_\lambda}]}{(p_1!)^{\pi_1}\ldots(p_\lambda!)^{\pi_\lambda}\pi_1!\ldots\pi_\lambda!}.$$

It may be shown that the k-statistics found in this way are unique. The first eleven are given in Table 2·1·3. They were constructed in the way described above. The k-statistics of weight 9, 10 and 11 were first calculated by Ziaud-Din (1954, 1959). They have been recalculated by us; we have found that in k_8 the product $s_4 s_3 s_1$ should have a multiplying factor of $-2800n^2$ not $+2800n^2$ as given by him, in k_{10} the product $s_3^2 s_2^2$ should have a multiplying factor $+2{,}457{,}000n^3$ not $-2{,}457{,}000n^3$ and in k_{11} the product $s_7 s_2 s_1^2$ should have a multiplying factor $-242{,}684{,}640n^2$ not $+242{,}684{,}640n^2$.

Multivariate k-statistics

We do not present here tables of bivariate or multivariate k-statistics, but their functional form may be useful in some moment problems and we would call attention to the use of the operative technique, previously described, for their derivation. Illustration for the bivariate case will serve. We recall the definition of the bivariate power sums, given earlier (p. 14), viz.

$$s_{rt} = \sum_{i=1}^{n} x_i^r y_i^t,$$

where (x_i, y_i) $i = 1, 2, \ldots, n$ are a pair of bivariate random variables. Consider the expression for k_3 given above, and write formally

$$k_3 = k(z^3) = \frac{1}{n^{(3)}}\{n^2 s(z^3) - 3ns(z^2)s(z) + 2(s(z))^3\}.$$

Operate on both sides by $w\partial/\partial z$ and cancel the factor 3 to obtain

$$k(z^2 w) = \frac{1}{n^{(3)}}\left\{ \begin{matrix} n^2 s(z^2 w) - 2ns(zw)s(z) + 2(s(z))^2 s(w) \\ -ns(z^2)s(w) \end{matrix} \right\},$$

or

$$k_{21} = \frac{1}{n^{(3)}}\{n^2 s_{21} - 2ns_{11}s_{10} - ns_{20}s_{01} + 2s_{10}^2 s_{01}\}.$$

Thus provided the expressions for the univariate k-statistics are available, the bivariate can easily be deduced. The multivariate k-statistics follow in similar fashion.

PART IIC: TABLE 2·2. POLYKAY FUNCTIONS IN TERMS OF MONOMIAL SYMMETRIC FUNCTIONS AND CONVERSELY

Definition and properties

The k-statistics of R. A. Fisher are based on symmetric functions of random independent variables. When sampling without replacement from a finite population is conceptualized, the random variables are not independent and the process of describing the parameters of the distributions of the various sample moments is not easy. The initial objective remains the same—the definition of a symmetric function of the random variables which will have as its expected value the same function calculated for the population, in this case finite in size. Tukey first discussed the problem in this form (1950). His ideas were clarified and extended by Wishart (1952) and subsequently by Aty (1954). It is Aty's tables for weight 12 which are presented here.

Suppose the sample $x_1, x_2, \ldots, x_n$ to be of size n and the finite population of size N. Then using the general notation of Wishart (1952) we have

$$k_{rs\ldots} = \sum_P \frac{(-1)^{\Sigma(\rho-1)}\Pi(\rho-1)!}{\Pi\{\pi_1!\,\pi_2!\ldots\}} \frac{r!\,s!\ldots}{\Pi\{(p_1!)^{\pi_1}(p_2!)^{\pi_2}\ldots\}} \frac{[\Pi(p_1^{\pi_1}p_2^{\pi_2}\ldots)]}{n^{(\Sigma\rho)}},$$

with [ ] denoting the augmented monomial symmetric function of Part I. For the first suffix of k, i.e. r, there is a partition $(p_1^{\pi_1} p_2^{\pi_2} \dots)$ of r, such that $\Sigma p\pi = r$, having ρ parts, with $\Sigma\pi = \rho$. Similar partitions are supposed for $s, \dots$. The outside summation is over all partitions.

A similar definition is made for a k-statistic for the whole finite population of size N. We write this k-statistic as $K_{rs\dots}$. The structure of $k_{rs\dots}$ and consequently of K_{rs} is so chosen that

$$\mathscr{E}_N(k_{rs\dots}) = K_{rs\dots},$$

where the subscript N to the symbol for expectation denotes the averaging over all possible samples of n from N. Further if the finite population of N is regarded as a sample from a (fictitious) infinite population with cumulants $\kappa_r, \kappa_s, \dots$ the structure of the k's (or polykay functions as we prefer to call them in order to differentiate them from Fisher's k-statistics which are a special case), is such that

$$\mathscr{E}\mathscr{E}_N(k_{rs\dots}) = \mathscr{E}(K_{rs\dots}) = \kappa_r \kappa_s \dots.$$

These ideas were first put forward by Irwin & Kendall (1944), and it is on this last relation that the method depends, for with any symmetric function of weight w we have, for samples of n from an infinite population,

$$\mathscr{E}[p_1^{\pi_1} \dots p_\lambda^{\pi_\lambda}]_n = n^{(w)} \mu_{p_1}'^{\pi_1} \dots \mu_{p_\lambda}'^{\pi_\lambda},$$

and for samples of n from a finite population, N,

$$\mathscr{E}_N[p_1^{\pi_1} \dots p_\lambda^{\pi_\lambda}]_n = \frac{n^{(w)}}{N^{(w)}}[p_1^{\pi_1} \dots p_\lambda^{\pi_\lambda}]_N.$$

The subscript to the right of the brackets denotes the number of random variables on which the symmetric function is based. Since the polykays are symmetric functions of the observations it becomes simple to write them down. As an illustration let us consider k_{31^2}. From Table 2·1·1 we have

$$\kappa_1^2 \kappa_3 = \mu_1'^2 (\mu_3' - 3\mu_2'\mu_1' + 2\mu_1'^3),$$

from which, remembering the definition of a monomial symmetric function and the Irwin–Kendall relation, we have immediately

$$k_{31^2} = \frac{[31^2]_n}{n^{(3)}} - \frac{3[21^3]_n}{n^{(4)}} + \frac{2[1^5]_n}{n^{(5)}}$$

and

$$K_{31^2} = \frac{[31^2]_N}{N^{(3)}} - \frac{3[21^3]_N}{N^{(4)}} + \frac{2[1^5]_N}{N^{(5)}}.$$

One table will serve for both polykay functions if it is remembered that N is associated with K and n with k.

Available tables

Using the basic symmetric function tables (Tables 1) Wishart gave tables of the polykays in terms of the augmented monomial symmetric functions, and conversely, for weights up to and including $w = 6$. Aty produced similar tables up to and including those of weight 12. Because all tables of lower weight may be deduced from those of higher weight simply by cancelling the part of the partition not required, we reproduce here just the weight 12 tables of Aty. For example

$$k_{2^31^6} = -\frac{[1^{12}]}{n^{(12)}} + \frac{3[21^{10}]}{n^{(11)}} - \frac{3[2^21^8]}{n^{(10)}} + \frac{[2^31^6]}{n^{(9)}},$$

whence, using the obliterating operator six times and remembering to lower the power of n,

$$k_{2^3} = -\frac{[1^6]}{n^{(6)}} + \frac{3[21^4]}{n^{(5)}} - \frac{3[2^21^2]}{n^{(4)}} + \frac{[2^3]}{n^{(3)}},$$

as may be checked from Wishart. The inverse tables, in which the monomial symmetric function is expressed as a linear sum of the polykays, may be obtained from those of weight 12 in the same way. Thus Table 2·2, read across the four pages to the diagonal, gives

$$\frac{[61^6]}{n^{(7)}} = k_{1^{12}} + 15k_{21^{10}} + 45k_{2^21^8} + 15k_{2^31^6} + 20k_{31^9} + 60k_{321^7} + 10k_{3^21^6} + 15k_{41^8} + 15k_{421^6} + 6k_{51^7} + k_{61^6},$$

whence again using the obliterating operator we have

$$\frac{[6]}{n} = k_{1^6} + 15k_{21^4} + 45k_{2^21^2} + 15k_{2^3} + 20k_{31^3} + 60k_{321} + 10k_{3^2} + 15k_{41^2} + 15k_{42} + 6k_{51} + k_6$$

which checks with Wishart.

Example of use of tables

The use of the tables in statistical moment problems are manifold, and illustrations will be found in the succeeding pages of this Introduction. Wishart gave the mean value of the square of the second k-statistic of Fisher in repeated samples from a finite population and since the algebra involved is simple we reproduce his example here.

Given a sample of size n, with $\bar{x}$ the sample mean, the second k-statistic is defined as

$$k_2 = \frac{1}{n-1}\left\{\sum_{i=1}^{n}(x_i - \bar{x})^2\right\}.$$

From the Table 2·2 or Table 2·13

$$k_2 = \frac{[2]}{n} - \frac{[1^2]}{n(n-1)} = \frac{1}{n-1}\left\{(2) - \frac{(1)^2}{n}\right\},$$

so that

$$k_2^2 = \frac{1}{(n-1)^2}\left\{(2)^2 - \frac{2}{n}(2)(1)^2 + \frac{(1)^4}{n^2}\right\}.$$

From Table 1·1·4 the right-hand side is put in terms of monomial symmetric functions so that

$$k_2^2 = \frac{1}{(n^{(2)})^2}\{(n-1)^2[4] - 4(n-1)[31] + (n^2 - 2n + 3)[2^2] - 2(n-3)[21^2] + [1^4]\}.$$

Using now the inverse of Table 2·2 to express the monomial symmetric functions in terms of the polykays we obtain, using the obliterating operator 8 times on the expressions for $[41^8]$, $[31^9]$, $[2^21^8]$, $[21^{10}]$ and $[1^{12}]$,

$$k_2^2 = \frac{k_4}{n} + \left(1 + \frac{2}{n-1}\right)k_{22},$$

whence

$$\mathscr{E}_N(k_2^2) = \frac{K_4}{n} + \left(1 + \frac{2}{n-1}\right)K_{22}.$$

K_{22} and k_{22} do not have any easy interpretation in terms of moments. This somewhat devious procedure is rendered unnecessary in a great many cases by Table 2·4 which succeeds these present ones. We cannot however cover all eventualities and from time to time it will be necessary to carry out the process just described.

Polybikays

Following the same argument as in the univariate case it is clear that we may define uniquely symmetric functions of $\{x_i y_i\}$, $i = 1, 2, ..., n$, denoted by $k_{p_1 q_1, ..., p_\lambda q_\lambda}$, so that in sampling from a finite population

$$\mathscr{E}_N(k_{p_1 q_1 ... p_\lambda q_\lambda}) = K_{p_1 q_1 ... p_\lambda q_\lambda},$$

where we call the k's the polybikays of the sample and the K's the polybikays of the finite population. Just as before we may write

$$\mathscr{E}(\mathscr{E}_N(k_{p_1 q_1, ..., p_\lambda q_\lambda})) = \mathscr{E}(K_{p_1 q_1, ..., p_\lambda q_\lambda}) = \kappa_{p_1 q_1, ...} \kappa_{p_\lambda q_\lambda},$$

where the κ's are the bivariate cumulants of the (hypothetical) infinite population. The poly-multikays of a sample of n observations $\{x_i y_i z_i ...\}$, $i = 1, 2, ..., n$, may be similarly defined. (The distinction between the polybikays and those denoted by Hooke (1956) as bipolykays should be noted. Hooke's quantities relate to an essentially different set-up.) Tables whereby the poly-multikays may be utilized have not been computed.

PART II D: TABLES 2·3 AND 2·4. PRODUCTS OF k-STATISTICS AND POLYKAY FUNCTIONS EXPRESSED AS LINEAR SUMS OF POLYKAY FUNCTIONS. POLYKAY FUNCTIONS AS FUNCTIONS OF SINGLE k-STATISTICS

The tables

From the brief discussion of Table 2·2 it will have been noticed that the chief labour in the application of polykays in moment problems is the expression of the quantity studied in terms of the single k-statistics (or their monomial symmetric equivalents), the translation of these single k-statistics into the polykays and, after taking expectations, the conversion back of the polykays of the finite population into single k-statistics.

Wishart (1952) gave formulae up to and including those of weight 8 which expressed powers and products of the single k-statistics in terms of the polykays. We have added to his list various other formulae which we have found useful. These Tables 2·3 were computed using the basic Tables 2·2 and 1·1, which will still be required for the computation of formulae which are not listed by us. The converse procedure, the expression of polykays in terms of single k-statistics, is a considerably more difficult problem and was achieved by Aty (1954) only by back-solving the equations expressed by Table 2·3, the solution of eighth-order determinants being found necessary.

Alternative methods

Because of the labour involved in extending the scope of Wishart's tables we gave some thought to the possibility of using known results for infinite populations. The expectation of powers and products of single k-statistics have been given by Fisher (1928) and by Kendall (1943, chapter 11) and are tabulated by us in Tables 3·1. We indicate here how these known results may be used to build up or to extend the application of Tables 2·3 at present under discussion.

(i) *Moments of k_1 the sample mean.* In finding the moments of k_1, it is necessary to express the appropriate powers of k_1 in terms of the polykays and then to take expectations. Consider the following equivalent procedure. Let there be a sample of n, a finite population $\mathscr{F}$ of size N and an infinite population $\mathscr{I}$ such that

$$\mathscr{E}_{\mathscr{I}}(\mathscr{E}_{\mathscr{F}}(k_1)) = \mathscr{E}_{\mathscr{I}}(K_1) = \kappa_1,$$

the suffices denoting the appropriate population over which expectations are taken. For the infinite population with cumulants κ_r $(r = 1, 2, ...)$

$$\kappa_r(k_1) = \frac{\kappa_r}{n^{r-1}}, \quad \kappa_r(K_1) = \frac{\kappa_r}{N^{r-1}}.$$

Written in terms of crude moments the first relation implies

$$\mathscr{E}_{\mathscr{I}}(k_1) = \kappa_1,$$

$$\mathscr{E}_{\mathscr{I}}(k_1^2) = \frac{\kappa_2}{n} + \kappa_1^2,$$

$$\mathscr{E}_{\mathscr{I}}(k_1^3) = \frac{\kappa_3}{n^2} + \frac{3\kappa_2\kappa_1}{n} + \kappa_1^3,$$

$$\mathscr{E}_{\mathscr{I}}(k_1^4) = \frac{\kappa_4}{n^3} + \frac{3\kappa_2^2}{n^2} + \frac{4\kappa_3\kappa_1}{n^2} + \frac{6\kappa_2\kappa_1^2}{n} + \kappa_1^4,$$

$$\cdots\cdots\cdots\cdots\cdots\cdots\cdots\cdots$$

and so on. Now the raw moments of k_1 in samples from $\mathscr{F}$ are linear sums of the polykays of $\mathscr{F}$ (and it may be shown that these linear sums express them uniquely) so that if we take expectations over $\mathscr{F}$ only

$$\mathscr{E}_{\mathscr{I}}\mathscr{E}_{\mathscr{F}}(k_1) = \mathscr{E}_{\mathscr{I}}(K_1) = \kappa_1,$$

$$\mathscr{E}_{\mathscr{I}}\mathscr{E}_{\mathscr{F}}(k_1^2) = \mathscr{E}_{\mathscr{I}}\left(\frac{K_2}{n} + K_{11}\right) = \frac{\kappa_2}{n} + \kappa_1^2,$$

$$\mathscr{E}_{\mathscr{I}}\mathscr{E}_{\mathscr{F}}(k_1^3) = \mathscr{E}_{\mathscr{I}}\left(\frac{K_3}{n^2} + \frac{3K_{21}}{n} + K_{1^3}\right) = \frac{\kappa_3}{n^2} + \frac{3\kappa_2\kappa_1}{n} + \kappa_1^3,$$

$$\mathscr{E}_{\mathscr{I}}\mathscr{E}_{\mathscr{F}}(k_1^4) = \mathscr{E}_{\mathscr{I}}\left(\frac{K_4}{n^3} + \frac{3K_{22}}{n^2} + \frac{4K_{31}}{n^2} + \frac{6K_{21^2}}{n} + K_{1^4}\right) = \frac{\kappa_4}{n^3} + \frac{3\kappa_2^2}{n^2} + \frac{4\kappa_3\kappa_1}{n^2} + \frac{6\kappa_2\kappa_1^2}{n} + \kappa_1^4.$$

Accordingly, if we consider the first pair of equalities in each case and drop the $\mathscr{E}_{\mathscr{I}}$ we have the relations between the crude moments of k_1 and the polykays of $\mathscr{F}$. The dropping of $\mathscr{E}_{\mathscr{I}}$ is justified by the uniqueness already noted and the functional independence of the cumulants which may be shown.

(ii) *Derivation of general formulae for polykay moments from those for infinite populations.* The process described in (i) immediately above can be used on the formulae listed in Table 3·1 to obtain results for polykay relations. A further illustrative example will serve. Suppose $\mathscr{E}_{\mathscr{F}}(k_3 k_2^2)$ is required. From Table 3·1

$$\kappa(32^2) = \mathscr{E}_{\mathscr{I}}(k_3 - \kappa_3)(k_2 - \kappa_2)^2 = \frac{\kappa_7}{n^2} + \frac{16\kappa_5\kappa_2}{n(n-1)} + \frac{12(2n-3)}{n(n-1)^2}\kappa_4\kappa_3 + \frac{48}{(n-1)^2}\kappa_3\kappa_2^2,$$

whence $\qquad \mathscr{E}_{\mathscr{F}}(k_3 - \kappa_3)(k_2 - \kappa_2)^2 = \frac{K_7}{n^2} + \frac{16K_{52}}{n(n-1)} + \frac{12(2n-3)}{n(n-1)^2}K_{43} + \frac{48}{(n-1)^2}K_{322}.$

Now
$$\mathscr{E}_{\mathscr{I}}(k_3-\kappa_3)(k_2-\kappa_2)^2 = \mathscr{E}_{\mathscr{I}}(k_2^2 k_3) - 2\kappa_2\mathscr{E}_{\mathscr{I}}(k_3 k_2) - \kappa_3\mathscr{E}_{\mathscr{I}}(k_2^2) + 2\kappa_3\kappa_2^2$$

so that on substitution from tables the right-hand side becomes

$$\mathscr{E}_{\mathscr{I}}(k_2^2 k_3) - 2\kappa_2\left(\frac{\kappa_5}{n} + \frac{n+5}{n-1}\kappa_3\kappa_2\right) - \kappa_3\left(\frac{\kappa_4}{n} + \frac{n+1}{n-1}\kappa_2^2\right) + 2\kappa_3\kappa_2^2,$$

whence
$$\mathscr{E}_{\mathscr{I}}(\mathscr{E}_{\mathscr{F}}(k_2^2 k_3)) - \mathscr{E}_{\mathscr{I}}\left(\frac{2K_{52}}{n} + \frac{K_{43}}{n} + \frac{n+13}{n-1}K_{322}\right) = \mathscr{E}_{\mathscr{I}}(\mathscr{E}_{\mathscr{F}}(k_3-\kappa_3)(k_2-\kappa_2))^2.$$

Accordingly, arguing as in the previous section (i)

$$\mathscr{E}_{\mathscr{F}}(k_2^2 k_3) = \frac{K_7}{n^2} + \frac{2(n+7)}{n^{(2)}}K_{52} + \frac{(n^2+22n-35)}{n(n-1)^2}K_{43} + \frac{(n+5)(n+7)}{(n-1)^2}K_{322}$$

as may be verified from Table 2·3. Thus all tables of weight w can be built up from those of lower weights, or from the Fisher–Kendall tables of weight w for the infinite case. An error in one of Wishart's coefficients was found in this way.

(iii) *Illustration of Table* 2·4. On p. 23 above Wishart's example of $\mathscr{E}(k_2^2)$ was used and it was demonstrated that

$$\mathscr{E}_{\mathscr{F}}(k_2^2) = \frac{K_4}{n} + \left(1 + \frac{2}{n-1}\right)K_{22}.$$

It is usually desirable to turn the polykays back into products and powers of the single k-statistic because the single kays are descriptive of some aspect of the shape of the finite population, whereas the interpretation of the polykays in this connexion has not yet been explored. In Wishart's example the conversion is simple. From direct tables we have

$$K_2^2 = \frac{K_4}{N} + \left(1 + \frac{2}{N-1}\right)K_{22},$$

whence by back-solving
$$K_{22} = \frac{N-1}{N+1}\left(K_2^2 - \frac{K_4}{N}\right),$$

as given in Table 2·4 also. Accordingly

$$\mathscr{E}_{\mathscr{F}}(k_2^2) = \frac{K_4}{n} + \frac{n+1}{n-1}\frac{N-1}{N+1}\left[K_2^2 - \frac{K_4}{N}\right].$$

The simplicity of the conversion here lies in the fact that K_{22} may be written down immediately. This will rarely be the case and usually, after a laborious algebraic process, the computations for the polykays in terms of the single kays result in expression of great complexity. An operative technique can possibly be devised whereby expressions of lower order may be deduced from those of higher order, but a building-up process whereby the higher order may be deduced from the lower does not seem practicable. Aty, using Wishart's forward relations, obtained such backward relations as we give, and it is clear that they are of great utility.

PART III. SAMPLING CUMULANTS OF k-STATISTICS AND MOMENTS OF MOMENTS

Introduction

The practical importance of determining the sampling distributions of moments calculated from a sample was early recognized in the development of statistical theory. Further, because the theoretical distributions were found often to have intractable functional forms, it was realized that the moments (or cumulants) of sample moments (or cumulants) provided a means of deriving functions approximating to such theoretical distributions. For it is undeniable the moments of almost any statistical quantity are obtainable (to a greater or less degree of approximation) however difficult it is to derive or manipulate the probability distribution itself. Thus Pearson curves or expansions in terms of Gram Charlier or Edgeworth series with parameters derived from theoretical sampling moments have frequently been used to provide such approximations.

In Part II we have defined the kay and the polykay statistics and have given a brief indication of their uses in statistical moment problems. These Tables 2 were grouped together immediately following the basic Tables 1 because these latter tables were required for their construction. In the construction of the tables of this present section the basic Tables 1 can also be used, but the tables themselves were made by other methods before the basic tables were calculated. Accordingly, we keep them separate from the previous set, as a reminder that Fisher (1928) and Kendall (1943), to whose work the reader is referred, used a different approach from that given here.

The k-statistics are functions of the sample moments and their moments in samples from a defined population may be given algebraic expression. Table 3·1 is concerned with the single k-statistics and an infinite population, Table 3·2, with bivariate k-statistics and an infinite bivariate population, Table 3·3 with single k-statistics and a finite population.

PART III A: TABLE 3·1. FISHER AND KENDALL'S TABLES OF SAMPLING CUMULANTS

Definitions

Let k_r be the rth k-statistic as defined in Part II B, Table 2·1·3, p. 20. Let κ_r, $r = 1, 2, \ldots$, be the cumulants of the infinite population generating a sample of size n of independent elements from which k_r is calculated. Write

$$M'(r^t) = \mathscr{E}(k_r^t),$$

$$M(r^t) = \mathscr{E}(k_r - \kappa_r)^t = \sum_{j=0}^{t} {}^tC_j(-1)^j\kappa_r^{t-j}\mathscr{E}(k_r^j) = \sum_{j=0}^{t} {}^tC_j(-1)^j\kappa_r^{t-j}M'(r^j).$$

Consequently since, by Table 2·1·3, k_r may be expressed in terms of the augmented monomial symmetric functions, M' can be found in terms of population moments about the origin. These can be converted to moments or cumulants about the mean and hence M deduced. Because the structure of cumulant functions (in the infinite population at least) is simpler than the corresponding moment functions, the quantity tabulated is not M but κ, where

$$\kappa(r^t) = t! \sum_{\rho=1}^{t} (-1)^{\rho-1}(\rho-1)! \sum_{P} \frac{(M'(r^{p_1}))^{\pi_1} \ldots (M'(r^{p_\lambda}))^{\pi_\lambda}}{(p_1!)^{\pi_1} \ldots (p_\lambda!)^{\pi_\lambda}\pi_1! \ldots \pi_\lambda!}$$

for the powers of k_r and the corresponding $\kappa(r^t u^v a^b ...)$ for products of powers of $k_r, k_v, k_a, ...$. The expression of these cumulant functions in terms of the M functions has already been outlined in principle in Part II A, pp. 18–19. Thus, for example, from Table 2·1·2 the cumulants and moments of the sixth order of a random variable are related by

$$\kappa_6 = \mu_6 - 15\mu_4\mu_2 - 10\mu_3^2 + 30\mu_2^3.$$

Consequently the sixth-order relation connecting the sixth cumulant and moments of k_r will be

$$\kappa(r^6) = M(r^6) - 15M(r^4)M(r^2) - 10(M(r^3))^2 + 30(M(r^2))^3.$$

Further the cross-cumulants of powers and products of any number of k-statistics may be expressed in terms of the moment-functions of the same weight by a relationship derived as before by the operative process. Thus, for example,

$$\kappa(r^5 u) = M(r^5 u) - 10M(r^3 u)M(r^2) - 5M(r^4)M(ru) - 10M(r^3)M(r^2 u) + 30(M(r^2))^2 M(ru),$$

where

$$M(r^t u^v) = \mathscr{E}\{(k_r - \kappa_r)^t (k_u - \kappa_u)^v\}.$$

Cumulants containing unit parts

It will be noticed that Table 3·1 does not give expressions for the κ's involving any unit parts. In this we have followed Fisher and Kendall who do not tabulate them because of the simplicity with which they may be written down. It is easy to show that

$$\kappa(1^t) = \kappa_t / n^{t-1} \quad (t = 1, 2, ...),$$

where κ_t is the population cumulant and n the sample size. This relation is easily derived from the characteristic function of k_1. It may also be built up from the first relationship

$$\mathscr{E}(k_1) = \kappa(1) = \kappa_1$$

by dividing n each time one is added to the suffix. Thus

$$\kappa(1^2) = \kappa_2 / n$$

and so on. Generally the rule, introduced by Fisher, and rigorized by Kendall, instructs us that in order to introduce unity on the left-hand side it is necessary to add one to the suffices in all possible ways, each time dividing by n. Thus

$$\mathscr{E}(k_2 - \kappa_2)^2 = \kappa(2^2) = \frac{\kappa_4}{n} + \frac{2\kappa_2^2}{n-1},$$

$$\kappa(2^2 1) = \frac{\kappa_5}{n^2} + \frac{4\kappa_3\kappa_2}{n(n-1)},$$

$$\kappa(2^2 1^2) = \frac{\kappa_6}{n^3} + \frac{4\kappa_4\kappa_2}{n^2(n-1)} + \frac{4\kappa_3^2}{n^2(n-1)}$$

and so on.

Illustrative example

The sample estimate of variance is k_2 and the sample mean is k_1. The coefficient of correlation between k_2 and k_1 may be written down immediately, whereas previous to Fisher's tables a great deal of lengthy algebra was called for. If ρ_{12} is this correlation, we have by definition

$$\rho_{12} = \frac{\mathscr{E}\{(k_2 - \kappa_2)(k_1 - \kappa_1)\}}{(\mathscr{E}(k_2 - \kappa_2)^2)^{\frac{1}{2}} (\mathscr{E}(k_1 - \kappa_1)^2)^{\frac{1}{2}}}.$$

Starting from $\mathscr{E}(k_2) = \kappa_2 = \kappa(2)$ and applying Fisher's rule we have

$$\kappa(21) = \mathscr{E}\{(k_2 - \kappa_2)(k_1 - \kappa_1)\} = \frac{\kappa_3}{n},$$

and from Tables 3·1

$$\kappa(2^2) = \frac{\kappa_4}{n} + \frac{2\kappa_2^2}{n-1},$$

while

$$\kappa(1^2) = \frac{\kappa_2}{n}.$$

Consequently

$$\rho_{12} = \kappa_3 \Big/ \left(\kappa_2\left(\kappa_4 + \frac{2n}{n-1}\kappa_2^2\right)\right)^{\frac{1}{2}} = \gamma_1 \Big/ \left(\gamma_2 + \frac{2n}{n-1}\right)^{\frac{1}{2}},$$

where

$$\gamma_1 = \kappa_3/\kappa_2^{\frac{3}{2}} \quad \text{and} \quad \gamma_2 = \kappa_4/\kappa_2^2.$$

For any symmetrical population γ_1 is zero and therefore ρ_{12} is zero also.

PART III B: TABLE 3·2. COOK'S TABLE OF BIVARIATE SAMPLING CUMULANTS

Definitions

The Fisher–Kendall sampling cumulants are used for finding the sampling moments of statistics based on a univariate sample of n independent elements. The obvious extension of these sampling cumulants are those which would be appropriate for quantities based on a multivariate sample. Fisher indicated one method for their computation but it was left to Cook (1951) to devise a satisfactory method for deriving their algebraic expression and systematically to compute them for the bivariate case. The interpretation and expression of the moments and cumulants of a bivariate population was given in Part II A while the definition and expression of the bivariate k-statistics were given in Part II B. Accordingly, the expectation of powers and products of the bivariate k-statistics can be derived by elementary processes, if desired, the real difficulty lying in devising a suitable notation.

If κ_{rs} and κ_{uv} are two bivariate cumulants and k_{rs} and k_{uv} the corresponding bivariate k-statistics such that

$$\mathscr{E}(k_{rs}) = \kappa_{rs}, \quad \mathscr{E}(k_{uv}) = \kappa_{uv},$$

we write

$$\mathscr{E}(k_{rs} - \kappa_{rs})(k_{uv} - \kappa_{uv}) = \kappa \begin{pmatrix} r\ u \\ s\ v \end{pmatrix}.$$

For powers of the k-statistics we rely on repetition within the bracket. Thus we would write, for example,

$$\mathscr{E}(k_{rs} - \kappa_{rs})^2 = \kappa \begin{pmatrix} r\ r \\ s\ s \end{pmatrix},$$

$$\mathscr{E}(k_{rs} - \kappa_{rs})^2 (k_{uv} - \kappa_{uv})^3 = \kappa \begin{pmatrix} r\ r\ u\ u\ u \\ s\ s\ v\ v\ v \end{pmatrix} + 3\kappa \begin{pmatrix} r\ r\ u \\ s\ s\ v \end{pmatrix} K \begin{pmatrix} u\ u \\ v\ v \end{pmatrix}$$

$$+ 6\kappa \begin{pmatrix} r\ u\ u \\ s\ v\ v \end{pmatrix} \kappa \begin{pmatrix} r\ u \\ s\ v \end{pmatrix} + \kappa \begin{pmatrix} r\ r \\ s\ s \end{pmatrix} \kappa \begin{pmatrix} u\ u\ u \\ v\ v\ v \end{pmatrix}.$$

The second example is of weight 5, necessitating the addition of products of cumulants to derive the appropriate moment function. These additional terms follow the application of the technique described in Part III A, p. 28.

Cook's operative technique

The bivariate sampling cumulant tables could be built up from first principles given the definitions of the preceding section. The algebra required is simple—given the bivariate symmetric function Tables 1·6—but it is lengthy and there is much scope for error. Accordingly, Cook put forward an operative process whereby many of the bivariate sampling cumulants may be derived from the known univariate results of Fisher and Kendall and much labour thereby avoided. This operative process, which it is simple to justify, is most easily explained by means of an example.

Consider the derivation of $\kappa\begin{pmatrix} 2 & 2 & 1 \\ 0 & 0 & 1 \end{pmatrix}$. We begin from Table 3·1 which gives

$$\mathscr{E}(k_2 - \kappa_2)^3 = \kappa(2^3) = \frac{\kappa_6}{n^2} + \frac{12}{n^{(2)}}\kappa_4\kappa_2 + \frac{4(n-2)}{n^{(2)}(n-1)}\kappa_3^2 + \frac{8}{(n-1)^2}\kappa_2^3.$$

Write this relationship formally as

$$\kappa\left(r_1^2 r_2^2 r_3^2\right) = \frac{\kappa(r_1^2 r_2^2 r_3^2)}{n^2} + \frac{12}{n^{(2)}}\kappa(r_1 r_2 r_3^2)\kappa(r_1 r_2) + \frac{4(n-2)}{n^{(2)}(n-1)}\left(\kappa(r_1 r_2 r_3)\right)^2 + \frac{8}{(n-1)^2}\kappa(r_1 r_2)\kappa(r_2 r_3)\kappa(r_3 r_1)$$

and operate by

$$\sum_{i=1}^{3} t_i \frac{\partial}{\partial r_i}.$$

On turning indices into suffices we obtain

$$\kappa\begin{pmatrix} 2 & 2 & 1 \\ 0 & 0 & 1 \end{pmatrix} = \frac{\kappa_{51}}{n^2} + \frac{4}{n^{(2)}}\kappa_{40}\kappa_{11} + \frac{8}{n^{(2)}}\kappa_{31}\kappa_{20} + \frac{4(n-2)}{n^{(2)}(n-1)}\kappa_{30}\kappa_{21} + \frac{8}{(n-1)^2}\kappa_{20}^2\kappa_{11}.$$

From symmetry it is apparent that the operation just performed is equivalent to operating with $t\partial/\partial r$ on

$$\kappa\{(r^2)^3\} = \frac{\kappa(r^6)}{n^2} + \frac{12\kappa(r^4)\kappa(r^2)}{n^{(2)}} + \frac{4(n-2)}{n^{(2)}(n-1)}\kappa^2(r^3) + \frac{8}{(n-1)^2}\kappa^3(r^2).$$

Again, since

$$\sum_{i=1}^{3} t^2 \frac{\partial^2}{\partial r_i^2} + \sum_{i \neq j} t_i t_j \frac{\partial^2}{\partial r_i \partial r_j} \equiv s^2 \frac{\partial^2}{\partial r^2},$$

operation with this quantity on $\kappa\{(r^2)^3\}$ will give

$$\kappa\begin{pmatrix} 2 & 2 & 0 \\ 0 & 0 & 2 \end{pmatrix} + 2\kappa\begin{pmatrix} 2 & 1 & 1 \\ 0 & 1 & 1 \end{pmatrix}$$

so that only one of these two quantities need be deduced by another method.

Kaplan's tensor notation

Such quantities as are given in Table 3·2 were worked out by Cook using the methods just described. In this connexion Kendall & Stuart (1958) draw attention to a potentially useful notation technique due to Kaplan (1952), who lists eight fundamental formulae from which, with some concentration, most of the bivariate, trivariate and quadrivariate sampling cumulants may be deduced. The application of these fundamental formulae are not, however, straightforward and until this approach to the problem has been clarified the earlier tried methods are to be preferred.

PART IIIC: TABLE 3·3. SAMPLING MOMENTS FOR
A FINITE POPULATION

In previous sections (Part II) we have discussed the polykay functions through which the moments of statistics based on a sample from a finite population can be derived. From the point of view of algebraic condensation the polykay technique is perhaps the best which can be devised but from a numerical point of view the formulae are cumbersome to manipulate. Consequently we give here the simplest expression of sample moments which we can devise with numerical use in mind.

Definitions

Suppose a population $\mathscr{F}$ of size N composed of elements $u_1, u_2, ..., u_N$. From $\mathscr{F}$ a sample of n elements $x_1, x_2, ..., x_n$ is randomly drawn without replacement. Write

$$\mu_t = \frac{1}{N} \sum_{j=1}^{N} (u_j - \mu_1')^t,$$

where μ_1' is the mean of $\mathscr{F}$ and may without loss of generality (if so desired) be made zero. The two quantities with which we are concerned in this table are

$$m_1' = \frac{1}{n} \sum_{i=1}^{n} x_i, \quad m_2 = \frac{1}{n} \sum_{i=1}^{n} (x_i - m_1')^2.$$

We define
$$M(2^r 1^s) = \mathscr{E}_{\mathscr{F}}\{(m_2 - \mathscr{E}_{\mathscr{F}}(m_2))^r (m_1' - \mathscr{E}_{\mathscr{F}}(m_1'))^s\}.$$

$M(2^r 1^s)$ may be expressed in terms of N, n and μ_t $(t = 2, 3, ...)$ and

$$e_v = \frac{n^{(v)}}{N^{(v)}}.$$

The coefficients required are given in Table 3·3 for quantities up to and including those of weight 8. Sukhatme (1943), who uses the e_v notation, gives a more comprehensive list than ours of the sample moments up to m_8. We have confined ourselves to m_1' and m_2 since these are the ones commonly required.

Illustration

As an illustration of the use of the tables we find the correlation between m_1' and m_2. From Table 3·3 we have

$$M(21) = \frac{N}{n^3} \mu_3 \left\{ (n-1)\frac{n}{N} - \frac{(n-3)n^{(2)}}{N^{(2)}} - \frac{2n^{(3)}}{N^{(3)}} \right\},$$

$$M(2^2) = \frac{N}{n^4} \mu_4 \left\{ (n-1)^2 \frac{n}{N} - \frac{(n^2 - 6n + 7)n^{(2)}}{N^{(2)}} - \frac{(4n-12)n^{(3)}}{N^{(3)}} - \frac{6n^{(4)}}{N^{(4)}} \right\}$$

$$+ \frac{N^2}{n^4} \mu_2^2 \left\{ (n^2 - 2n + 3) \frac{n^{(2)}}{N^{(2)}} + \frac{(2n-6)n^{(3)}}{N^{(3)}} + \frac{3n^{(4)}}{N^{(4)}} \right\},$$

$$M(1^2) = \frac{N}{n^2} \mu_2 \left\{ \frac{n}{N} - \frac{n^{(2)}}{N^{(2)}} \right\}$$

and
$$\rho_{21} = \frac{M(21)}{(M(2^2)\,M(1^2))^{\frac{1}{2}}}.$$

For numerical computation it is usually better to make substitution into $M(21)$, $M(2^2)$, $M(1^2)$ and then compute ρ_{21}. For this particular case the reduction is easy and writing $n = \lambda N$, we have

$$\rho_{21} = \sqrt{\frac{\lambda N - 1}{N(1-\lambda)}}\,\sqrt{\frac{N-2}{N-3}}\left(\frac{N^2(1-\lambda)+2}{N}\right)\frac{\gamma_1}{\sqrt{(A\gamma_2+B)}},$$

where
$$A = -\lambda^2 N^{(2)} + \lambda(N^2+1) - (N+1),$$

$$B = \lambda^2 N(N^2 - 3N + 3) - \lambda(2N^2 - 3) + 3(N+1).$$

PART IV. PARTITIONS

PART IV A: TABLE 4·1. NUMBER OF PARTITIONS OF A UNIPARTITE NUMBER

Introduction

The number of partitions of a unipartite number is needed frequently in the applications of combinatorial theory. To give only one example we may quote Whitworth, Proposition 29: 'To find the number of ways in which n indifferent things can be distributed into r indifferent parcels (no blank lots) or to find the number of different r-partitions of n.' In the algebra of moments and cumulants these numbers of partitions are also useful, particularly in the enumeration of the numbers of different types of terms in summations.

Tables of the number of r-partitions of n must have been calculated many times. Whitworth gives a table for $n = 1(1)27$, and $r = 1(1)20$. The present Table 4·1 gives $n = 1(1)35$, $r = 1(1)35$, and it is easily extendable. Thus if $P_{n,r}$ denotes the number of r-partitions of n, we have the recurrence relation

$$P_{n,r} = P_{n-1,r-1} + P_{n-r,r}.$$

This recurrence was used by us to recompute and extend Whitworth's table. Given this number of r-partitions of n the specification of the actual partitions is straightforward. Space does not permit the reproduction of such a table ancillary to 4·1. We may point out that the partitions for integers up to and including 12 may be read off the borders of any of the symmetric function Tables 1. For integers between 12 and 18 the individual partitions will be found written out in the collected papers of Cayley (1896). The *Royal Society Mathematical Tables*, **4** (1958), give numbers of partitions for n up to 400.

PART IV B: TABLE 4·2. NUMBER OF PARTITIONS OF A BIPARTITE NUMBER

Notions

We are not aware of tables giving the partitions of a bipartite number, although McMahon discusses generating functions for these and for partitions of multipartite numbers. The knowledge of the numbers of partitions is useful in distribution theory, which is why McMahon studied them, but our chief interest has been from the point of view of expansions of bivariate symmetric functions where it is essential to know the different types of term which may occur in cases outside the range of Tables 1·6.

We have written of bipartitions (2-partitions) of a single integer. For example, (21) is a 2-partition of 3, but we may also think of the bipartition as a number itself. Consider for example the product

$$(\textstyle\sum_i x_i)^2 (\textstyle\sum_j y_j)^1,$$

which we may think of as corresponding to the bipartite number (21). It is useful to be certain, in such an expansion, that all the types of terms have been included. Thus

$$(\textstyle\sum_i x_i)^2 (\textstyle\sum_j y_j) = \sum_i x_i^2 y_i^1 + \sum_{i \neq j} x_i^2 y_j^1 + 2\sum_{i \neq j} x_i^1 x_j^1 y_i^1 + \sum_{i \neq j \neq k} x_i^1 x_j^1 y_k^1,$$

which we can write as $\qquad [21] + [20, 01] + 2[11, 10] + [10, 10, 01].$

Table 4·2 for (21) shows that there is 1 one-partition of (21), 2 two-partitions of (21), i.e. [20, 01] and [11, 10] and 1 three-partition, i.e. [10, 10, 01]. The table only gives the numbers, but once these are known the types of term and the numbers of those types may be reasonably easily evaluated. For bipartite numbers $(m, n), (m+n) \leqslant 8$, the actual partitions of the bipartite number may be read off from the borders of 1·6.

Notation

Let $P_{mn,r}$ be the number of bipartitions of (mn) into exactly r biparts. This quantity will have the following properties:

(i) $P_{mn,r} = P_{nm,r}$.

This enables the tables to be given for $m \geqslant n$ only.

(ii) $P_{m0,r} = P_{m,r}$,

where $P_{m,r}$ is, as in A, the number of partitions of m into r parts.

(iii) $P_{mn,r} = 0 \qquad r > m+n,$

$\qquad P_{mn,0} = 0 \qquad m$ and n not both zero,

$\qquad P_{00,0} = 1 \qquad$ conventionally,

$\qquad P_{mn,r} = 1 \qquad$ if $r = m+n.$

The generating function of $P_{mn,r}$ is

(iv) $G(x, y, h) = \Sigma P_{mn,r} x^m y^n h^r = 1 / \prod_{s,t} (1 - x^s y^t h).$

A quantity used for checking the tables is

(v) $S_{wr} = \sum_{m=0}^{w} P_{m(w-m),r},$

where S_{wr} is generated by $\qquad \Sigma S_{wr} x^w h^r = 1 / \prod_{w=1}^{\infty} (1 - x^w h)^{w+1}.$

It may be further noted that

(vi) $G(x, h) = \Sigma P_{mr} x^m h^r = 1 / \prod_{s=1}^{\infty} (1 - x^s h) = G(x, 0, h).$

Construction of tables

1. Table 4·2 was originally built up by direct enumeration. For example it is immediate that

$$(\sum_i x_i)(\sum_j y_j) = s_{10} s_{01} = [11] + [10, 01],$$

from which we may deduce

$$s_{10}^2 s_{01} = [21] + [11, 10] + [20, 01] + [10, 11] + [10, 10, 01].$$

The right-hand side is obtained by adding one in all possible ways. This method is straightforward, although cumbersome.

2. McMahon gives a method depending on the $m-h$ Tables $1\cdot5$ which was thus available for checking only up to $m+n \leqslant 12$. We do not reproduce his method here. It was not less laborious than the previous method but is non-recursive so that errors were isolated.

3. We may write the generating function of $P_{mn,r}$ given above as follows:

$$(1-hx)\,G(x,y,h) = G(x,y,hx)\,G(x,h),$$

so that on equating coefficients we have the recurrence relation

$$P_{mn,r}-P_{(m-1)n,r-1} = \sum_{i=1}\sum_{j=r-i} P_{(m-i)(n-j),\,i}P_{j,r-i}+P_{(m-r)n,r}.$$

The upper limits of summation are $i \leqslant \min(r-1, m-1)$, $j \leqslant n-1$, $2i+j \leqslant m+n$. This recurrence was used to tabulate and check $P_{mn,r}$ up to $m+n = 16$. The table was computed by P. Jackson and K. Solomon.

PART V. QUANTITIES BASED ON THE
FIRST n NATURAL NUMBERS

Introduction

Much of the algebra of symmetric functions is applicable in the so-called non-parametric theory of modern statistical methods, particularly the special case of the algebra of the first n natural numbers. In this Part V and the succeeding Parts VI and VII we have gathered together such tables as have been found useful by us in non-parametric research problems. Many of the tables could not have been computed without the aid of the basic symmetric function tables. Others, such as the tables of binomial coefficients, are well known and are added here for the sake of completeness.

PART VA: TABLE 5·1. COMBINATORIAL COEFFICIENTS

Construction of table and recurrence relations

The coefficients of the binomial $(1 + 1)^n$, n integral and positive, enter universally into statistical theory, but most frequently perhaps in the combinatorial field. The recurrence relation whereby the coefficients may be calculated has been known at least from the fourteenth century. It is obvious, using the common combinatorial notation, that

$$^nC_r + {}^nC_{r-1} = {}^{n+1}C_r.$$

Accordingly, we have the Arithmetic Triangle, commonly attributed to Pascal,

$$
\begin{array}{ccccccc}
 & & & 1 & & & \\
 & & 1 & & 1 & & \\
 & 1 & & 2 & & 1 & \\
1 & & 3 & & 3 & & 1 \\
\end{array}
$$
....

and so on. Table 5·1 was constructed by this method. A comprehensive table extending for $n = 1(1)100$ is given in the *Royal Society Mathematical Tables* (1954). We give enough here for everyday purposes.

For n negative and equal to N/k, i.e. n may not be integral, the recurrence relation becomes

$$^{s+N/k}C_s = {}^{s-1+N/k}C_s - {}^{s-1+N/k}C_{s-1}.$$

A somewhat less simple recurrence is possible for the multinomial coefficients. The tabulation of these coefficients, as a product of binomial coefficients, is a simple process, but we have not presented such tables because of limitations of space. It may be noted from Tables 1·1 that the last line of each table of weight w ($w = 2(1)12$) gives all the multinomial coefficients of weight w, so that up to and including $w = 12$ these may be read off. For example, consider the expansion of $(x+y+z+t)^3$. We have

$$(x+y+z+t)^3 = x^3+y^3+z^3+t^3+3(x^2y+x^2y+x^2t+y^2x+y^2z+y^2t+z^2x$$
$$+z^2y+z^2t+t^2x+t^2y+t^2z)+6(xyz+xyt+xzt+yzt)$$
$$= (3)+3(21)+6(1^3) = [3]+3[21]+[1^3] = (1)^3.$$

For $w > 12$ values of the coefficients could be obtained by building up the expansion of $(1)^w$ from the expansion for $(1)^{12}$ but it is probably simpler to work out the coefficients de novo.

PART VB: TABLES 5·2·1 AND 5·2·2. BERNOULLI POLYNOMIALS

Definitions

Following Nörlund we define the Bernoulli polynomial of nth degree and mth order as the coefficient of $t^n/n!$ in the expansion of $t^m e^{xt}/(e^t - 1)^m$, i.e. we write

$$\frac{t^m e^{xt}}{(e^t - 1)^m} = \sum_{n=0}^{\infty} B_n^{(m)}(x)\frac{t^n}{n!} \quad (|t| < 2\pi).$$

The polynomials $B_n^{(1)}(x)$, (or $B_n(x)$ as usually written), are those equivalent to the classical Bernoulli polynomials and $B_n \equiv B_n(0)$ are equivalent to the classical Bernoulli numbers with, by this definition,

$$B_{2a+1} = 0 \quad (a = 1, 2, \ldots).$$

The generalized polynomials are related to the generalized numbers by the relation

$$B_n^{(m)}(x) = \sum_{r=0}^{n} x^r B_{n-r}^{(m)} \cdot {}^nC_r.$$

Thus, given a table of $B_v^{(m)}$ any generalized polynomial can be constructed. In Table 5·2·1 we give the first twelve polynomials of the first order. The Bernoulli numbers are the coefficient of x^0 in this table divided by the multiplying factor. In Table 5·2·2 we give the numbers $B_v^{(m)}$ as polynomials in m for $v = 1(1)12$. These are all well-known results. For further extensions we would refer to Davis, *Tables of the Higher Mathematical Functions*, **2** (1935).

Uses of the tables

Apart from the uses of Tables 5·2·1 and 5·2·2 in the construction of the Tables of Stirling Numbers (5·3·1 and 5·3·2), the generalized Bernoulli polynomials enter into many branches of combinatory theory, too many to list here. As a brief illustration consider the distribution of the number of 'records' in a randomly arranged sequence of the numbers $1, 2, \ldots, N$. For example if we take the sequence of the first ten natural numbers arranged as follows

$$5 \ \overline{8} \ 7 \ \underline{3} \ \overline{10} \ 4 \ 9 \ \underline{1} \ 6 \ 2$$

there are two upper records and two lower records. Let u be the number of upper records in the random sequence of the first N natural numbers and l be the number of lower records, with

$$R = u + l.$$

Then the joint p.d.f. of u and l (see, for example, David & Barton, 1962, p. 179) may be shown to be

$$p_N(u, l) = \frac{1}{N!} {}^RC_u {}^{N-2}C_{R-1} B_{N-R-1}^{(N-1)}(-1)^{N-R-1},$$

and also

$$p_N(u) = \frac{1}{N!} {}^{N-1}C_{u-1} B_{N-u}^{(N)}(-1)^{N-u}.$$

Either of these p.d.f.'s may be evaluated directly from Table 5·2·2 putting $m = 9$ in the first case, $m = 10$ in the second. The generalized polynomials satisfy the equation

$$B_n^{(m+1)}(x) = \left(1 - \frac{n}{m}\right) B_n^{(m)}(x) + (x - m) \left(\frac{n}{m}\right) B_{n-1}^{(m)}(x),$$

so that for values of m outside the range of the tables the Bernoulli numbers required may be calculated from the recurrence.

Symmetric functions of the natural numbers

The rth power sum, or rth one-part symmetric function, as described in Part I A, is written as s_r. When we consider the rth power sum of the first N natural numbers we may write

$$s_r = \sum_{j=1}^{N} j^r = \frac{1}{r+1} \{B_{r+1}^{(1)}(N+1) - B_{r+1}^{(1)}\}.$$

The Bernoulli polynomials of Table 5·2·1 may therefore be used to express the sum of the rth powers of the first N natural numbers for $r = 1(1)11$, a result which is well known.

The rth unitary symmetric function of the first N natural numbers, a_r, is

$$a_r = (-1)^r {}^N C_r B_r^{(N+1)}, \quad u_r = r! \, a_r, \quad r \leqslant N,$$

while the corresponding homogeneous product sum, h_r, is

$$h_r = {}^{r+N} C_N B_r^{(-N)}.$$

We have used these results in a later section when discussing the polykays of the natural numbers.

PART V C: TABLES 5·3·1 AND 5·3·2. STIRLING'S NUMBERS

Definitions and properties

James Stirling in his *Methodus Differentialis* of 1730 first gave tables of the quantities sometimes referred to as Stirling's numbers of the first and second kinds. Consider the following table of differences of a function $(x+s)^r$ where s is an integer:

$$x^r$$
$$\Delta(x)^r$$
$$(x+1)^r \qquad\qquad \Delta^2(x)^r$$
$$\Delta(x+1)^r \qquad\qquad \Delta^3(x)^r$$
$$(x+2)^r \qquad\qquad \Delta^2(x+1)^r$$
$$\Delta(x+2)^r \qquad\qquad \Delta^3(x+1)^r \quad \dots$$
$$(x+3)^r \qquad\qquad \Delta^2(x+2)^r \qquad\qquad \dots$$

The quantities $\Delta(x)^r$, $\Delta^2(x)^r$, ... are called the leading differences and, on division by $1!$, $2!$..., for the special case $x = 0$ they are often called the difference quotients of zero. These are in fact Stirling's numbers of the second kind. It can be shown that

$$\Delta_{rs} \equiv \frac{\Delta^r(0)^s}{r!} = {}^s C_r B_{s-r}^{(-r)}, \quad r \leqslant s,$$

$$= 0, \qquad r > s.$$

These numbers are connected by the well-known recurrence relation

$$\Delta_{rs} = r\Delta_{r,s-1} + \Delta_{r-1,s-1}, \quad r < s,$$
$$= 1, \qquad\qquad r = s.$$

The Stirling numbers of the first kind, S_m^p, are defined by

$$|S_m^p| = (-1)^{m-p} S_m^p = \frac{1}{p!}\left(\frac{\partial^p}{\partial x^p}(x^{[m]})\right)_{x=0} = {}^{m-1}C_{p-1} B_{m-p}^{(m)}(-1)^{m-p}.$$

They are sometimes called the differentials of zero. It is clear that given Tables 5·2·1 and 5·2·2 of generalized Bernoulli numbers the Stirling numbers may easily be calculated. The recurrence relation for these numbers is

$$|S_{m+1}^p| = m.|S_m^p| + |S_m^{p-1}|,$$

so that again the table may be extended without labour.

Approximate formulae for Stirling numbers of both kinds, for large argument values, involve the use of Tables 8.1 and 8.2 (see p. 57 below).

Illustration of use of tables

The Stirling numbers of both kinds occur in enumeration of many kinds of combinatorial distributions. As an illustration of their use in an everyday problem we would point out an immediate and obvious application in the expression of arithmetic powers in terms of factorial powers and vice versa. It was in this connexion that Stirling himself devised them. Thus writing

$$x^{(m)} = x(x-1)\ldots(x-m+1), \quad x^{[m]} = x(x+1)\ldots(x+m-1),$$

we have

$$x^m = \sum_{r=0}^{m} x^{(r)} \frac{\Delta^r(0)^m}{r!},$$

and

$$x^{[m]} = \sum_{r=0}^{m} \frac{x^r}{r!}\left(\frac{d^r}{dx^r}(x^{[m]})\right)_{x=0}.$$

PART V D: TABLES 5·4·1 AND 5·4·2. DIFFERENCES OF RECIPROCALS OF UNITY

Definitions

A further set of numbers, not previously studied in much detail, arises if we consider the expansion of a negative factorial power in terms of positive arithmetic powers. Thus

$$x^{(-m)} = \frac{1}{(x+1)^{[m]}} = \sum_{r=0}^{\infty} x^r \frac{(-1)^{m-r-1}}{(m-1)!}\left\{\Delta^{m-1}\left(\frac{1}{1^{r+1}}\right)\right\}.$$

These numbers in the brackets { } are styled the differences of reciprocals of unity. In Table 5·4·1 we give the quantity

$$A_{rn} = ((n+1)!)^r \left|\frac{1}{n!}\Delta^n\left(\frac{1}{1^r}\right)\right|$$

for $r = 1(1)10$ and some selected values of n. The function is chosen so that the numbers given are integers, but these numbers for given n increase rapidly with increasing r so that complete tabulation for r is not possible. Table 5·4·2 gives

$$\Delta^n\left(\frac{1}{1^r}\right)$$

computed to ten decimal places for $r = 1(1)20$, $n = 1(1)20$. Laplace in his *Théorie Analytique des Probabilités* gives an expansion for this function (using Tables 8) which is adequate for practical purposes outside the range of Table 5·4·2.

Apart from the usefulness of these differences in expressing a negative factorial power in terms of positive arithmetic powers there are several different combinatorial problems in which the tables have been helpful. For example in discussing sequential occupancy (David & Barton, 1962, p. 274) the mean value of the number of balls required to achieve a given result in what they have termed the 'golliwog' problem may be expressed in terms of these numbers. We have also found them of use in the problem of random intervals on a line.

PART VE: TABLE 5·5. POLYKAYS OF THE NATURAL NUMBERS

Distribution problems connected with rank criteria which are symmetric functions of the observations may be considerably simplified if they are treated as problems of sampling without replacement from a finite population. Thus, for example, the Wilcoxon two-sample test for the identity of two parameters of location (r_1 observations in the first sample and r_2 in the second), can be treated as a distribution problem concerned with the sum of r_1 ranks randomly chosen without replacement from a 'population' consisting of the first $r_1 + r_2$ integers. We have shown (Barton, David & Fix, 1960) that given the first N natural numbers, some of the polykays of this 'population' may be expressed in terms of generalized Bernoulli polynomials (Tables 5·2·1 and 5·2·2), the three simplest being

$$K_r = \frac{B_r}{r}[B_r(N+1) - B_r], \quad r \text{ even},$$

$$= 0, \qquad\qquad r \geqslant 3 \text{ and odd},$$

and
$$K_{1^r} = (-1)^r B_r^{(N+1)} = \frac{u_r}{N^{(r)}}, \quad K_{2^r} = \frac{N^{(r)} r!}{(2r)!}(-1)^r B_{2r}^{(2N+2)}(N+1),$$

where u_r is the augmented unitary symmetric function of weight r. Other relations are given in the paper cited but the list is not complete. Where we have not succeeded in finding an explicit expression, Wishart's formulae and Aty's inversion of them were used to produce the appropriate polynomials by substitution. Table 5·5 gives all polykays of the natural numbers reduced to polynomials in powers of N with the numerical multiplying factor, for all polykays up to and including those of weight 8.

Example

As an example of the use of the tables consider the computation of the correlation between the first and third k-statistics in a sample of n drawn without replacement from the population of the first N integers. The tables reduce such a computation to one of arithmetic. By definition

$$\rho_{31} = \frac{\mathscr{E}_N(k_3 - K_3)(k_1 - K_1)}{\{\mathscr{E}_N(k_3 - K_3)^2 \, \mathscr{E}_N(k_1 - K_1)^2\}^{\frac{1}{2}}}.$$

In the previous section it was noted that $K_3 = 0$ and it is well known that $K_1 = \frac{1}{2}(N+1)$. For the numerator we have

$$\mathscr{E}_N(k_3 - K_3)(k_1 - K_1) = \mathscr{E}_N(k_3 k_1) = \mathscr{E}_N(k_4/n + k_{31})$$

from Table 2·3. Consequently

$$\mathscr{E}_N(k_3 - K_3)(k_1 - K_1) = K_4/n + K_{31}.$$

The values of K_4 and K_{31} may be read off from Table 5·5 and on collecting terms

$$\mathscr{E}_N(k_3 - K_3)(k_1 - K_1) = -\frac{N(N+1)^2(N-n)}{120n}.$$

Using the same procedure we have

$$\mathscr{E}_N(k_3 - K_3)^2 = \frac{K_6}{n} + \frac{9K_{42}}{n-1} + \frac{(n+8)}{n-1}K_{33} + \frac{6n}{(n-1)^{(2)}}K_{222}$$

$$= \frac{N(N+1)(N-n)}{10,080n(n-1)(n-2)}\{N^3(12n^2 + 6n + 80) + N^2(18n^2 - 4n + 160)$$

$$- N(10n^2 + 62n - 40) - (20n^2 + 60n + 40)\}$$

and

$$\mathscr{E}_N(k_1 - K_1)^2 = \frac{(N+1)(N-n)}{120n},$$

giving

$$\rho_{31} = -\frac{(N+1)}{10}\left\{\frac{420N(n-1)(n-2)}{N^3(6n^2 + 3n + 40) + N^2(9n^2 - 2n + 80) - N(5n^2 + 31n - 20) - (10n^2 + 30n + 20)}\right\}^{\frac{1}{2}}.$$

PART V F: TABLE 5·6. ASYMMETRIC SUMS OF PRODUCTS OF NATURAL NUMBERS

Introduction

The technique of using the polykay statistics relies entirely on the fact that the criteria studied are symmetric functions of the observations. There are situations when this will not be so and consequently the polykay technique is inapplicable. In such situations, where the criteria studied are functions of the ranks of the sample (but not symmetric), their moments can often be reduced to asymmetric sums of powers and products of the natural numbers, and Table 5·6 was computed to this end. The table was constructed from Tables 5·2·1 and 5·2·2 (Bernoulli polynomials).

Example

Suppose two sets of observations $x_1, x_2, \ldots, x_n$ and $y_1, y_2, \ldots, y_m$, with $n + m = N$, and rank them together in order of magnitude. Let R_i be the rank of the ith x and consider the criterion

$$X = \sum_{i=1}^{n} iR_i.$$

This is not a symmetric function of the R_i's and to deduce the moments of X it is necessary to return to first principles. The probability that the rank of the ith x is a is

$$P\{R_i = a\} = {}^{a-1}C_{i-1}\,{}^{N-a}C_{n-i}/{}^{N}C_n.$$

The probability that the rank of the ith x is a and that the rank of the jth x is b $(i < j, a < b)$ is

$$P\{R_i = a, R_j = b\} = {}^{a-1}C_{i-1}\,{}^{b-a-1}C_{j-i-1}\,{}^{N-b}C_{n-j}/{}^{N}C_n.$$

Consequently

$$\mathscr{E}(R_i) = i\frac{N+1}{n+1}; \quad \mathscr{E}(R_i^2) = i^2\frac{(N+1)(N+2)}{(n+1)(n+2)} + i\frac{(N+1)(N-n)}{(n+1)(n+2)};$$

$$\mathscr{E}(R_i R_j) = i\frac{(N+1)^2}{n+1} - i(n+1-j)\frac{(N+1)(N+2)}{(n+1)(n+2)}.$$

It follows that

$$\mathscr{E}(X) = \sum_{i=1}^{n} i^2\frac{(N+1)}{n+1} = \frac{n(2n+1)(N+1)}{6},$$

and since

$$\mathscr{E}(X^2) = \mathscr{E}\sum_i i^2 R_i^2 + 2\mathscr{E}\sum_{i<j} ij R_i R_j$$

we have, on substitution for $\mathscr{E}(R_i^2)$ and $\mathscr{E}(R_i R_j)$ and application of Table 5·6 that

$$\mathscr{E}(X^2) = \frac{(N+1)(N+2)}{(n+1)(n+2)}\{\tfrac{3}{2}(n+1)^{(3)} + \tfrac{10}{3}(n+1)^{(4)} + \tfrac{43}{30}(n+1)^{(5)} + \tfrac{13}{90}(n+1)^{(6)}\}$$

$$+ \frac{(N+1)(N-n)}{(n+1)(n+2)}\{\tfrac{5}{6}(n+1)^{(3)} + (n+1)^{(4)} + \tfrac{11}{60}(n+1)^{(5)}\},$$

$$\sigma_x^2 = \frac{n(2n+1)(2n+3)(N+1)(N-n)}{180},$$

since

$$\mathscr{E}(X) = \tfrac{1}{6}n(2n+1)(N+1).$$

The tables were computed for double asymmetric sums of weight eight and for triple sums of weight five, by F. N. David and M. Merrington.

PART V G: TABLE 5·7. WILCOXON'S STATISTIC

Definitions and recurrence relation

Suppose two samples of sizes r_1 and $r_2 (r_1 + r_2 = r)$. It is required to test the hypothesis that the two populations generating these samples are identical, against the alternate hypothesis that the parameters of location of the two populations are different. The procedure is to rank both samples together in order of magnitude (usually ascending), and to calculate the mean of the sum of the ranks of one sample, significance being judged in the usual way. If the ranks of one sample are $R_1, R_2, ..., R_{r_2}$ and

$$m_1 = \frac{1}{r_2}\sum_{i=1}^{r_2} R_i = \frac{1}{r_2}S \quad \text{(say)},$$

the p.d.f. of S is given by the coefficients of the successive powers of h in the expansion of

$$\prod_{i=1}^{r_2}\left(\frac{h^i - h^{r+1}}{1 - h^i}\right)\Big/{}^rC_{r_2}.$$

Table 5·7 may be built up either from this expression or by a recurrence relation. Since it is convenient for tabulation that each distribution shall start from zero, the quantity

$$U = r_1 r_2 + \tfrac{1}{2}r_2(r_2+1) - S$$

is used for entry into the tables instead of S. If $p(U|r_1 r_2)$ denotes the p.d.f. of U for r_1 and r_2 given, then

$$p(U|r_1 r_2) = \frac{r_1}{r}p(U - r_{2+1}|r_1 - 1, r_2) + \frac{r_2}{r}p(U|r_1, r_2 - 1).$$

The tables given here were calculated from the expression for the p.d.f. of S and checked by the recurrence relation, for $r = 4(1)20$, $r_2 = 2(1)[\frac{1}{2}r]$.

Normal approximation

Since the standardized cumulants of order 4 or more tend to zero with increasing sequence length it is clear that the normal distribution is a suitable approximating function to the discrete distribution of U (or of S). The cumulants are

$$\kappa_1(U) = \tfrac{1}{2}(r_1 r_2), \quad \kappa_1(S) = \tfrac{1}{2}r_2(r+1),$$

$$\kappa_2(U) = \kappa_2(S) = \tfrac{1}{12}r_1 r_2(r+1),$$

$$\kappa_3(U) = -\kappa_3(S) = 0,$$

$$\kappa_4(U) = \kappa_4(S) = -\frac{r_1 r_2(r+1)}{120}[r(r+1) - r_1 r_2],$$

whence
$$\gamma_1 = \frac{\kappa_3}{\kappa_2^{\frac{3}{2}}} = 0, \quad \gamma_2 = \frac{\kappa_4}{\kappa_2^2} = -\frac{6}{5}\left[\frac{r}{r_1 r_2} - \frac{1}{r+1}\right],$$

which indicates for a given value of r that γ_2 takes its minimum value when r_1 and r_2 are near equality when r is odd, and are equal when r is even. Accordingly it may be expected that the normal distribution will approximate best to the true distribution in these cases, for r given, and this is the case. For example, for $r = 14$, $r = 8$, $r_2 = 6$ we have

$$P\{U \geqslant 30\} = 0\cdot245 \quad \text{(True)}$$
$$= 0\cdot239 \quad \text{(Normal)}.$$

The more disparate the values of r_1 and r_2 the greater will be the error involved in using the normal approximating function. Thus for $r = 19$, $r_1 = 17$, $r_2 = 2$ we have

$$P\{U \geqslant 25) = 0\cdot175^+ \quad \text{(True)}$$
$$= 0\cdot159 \quad \text{(Normal)}.$$

For the case $r_2 = 2$ however the actual probabilities are so quickly written down that there is no point in resorting to approximation. For this case ($r_2 = 2$) the mean of the discrete distribution is r_1 and, writing [ ] for 'the integral part of', e.g. $[8\frac{1}{2}] = 8$, we have

$$P\{U = U_0 \mid U_0 \leqslant r_1\} = [\tfrac{1}{2}U_0 + 1]/\{\tfrac{1}{2}r(r-1)\},$$

the distribution being symmetrical about r_1.

When we look at $r = 19$, $r_1 = 16$, $r_2 = 3$ the normal approximation is closer than in the previous case. For example
$$P\{U \geqslant 33\} = 0\cdot180 \quad \text{(True)}$$
$$= 0\cdot171 \quad \text{(Normal)},$$

and we can expect this correspondence to become closer as r increases. It is suggested therefore that outside the scope of these tables, knowing the mean and variance already given above, the normal curve can be used to estimate probabilities, care being taken not to put too much reliance on these approximate values when r_1 and r_2 are very different. A continuity correction of $\frac{1}{2}$ should be used. Closer values can be obtained by using the continuity corrections suggested by Fix & Hodges (1955). A further discussion of approximations will be found in Verdooren (1963). The present tables were computed by P. Jackson and K. Solomon.

PART V H: TABLE 5·8. KENDALL'S τ AND MANN'S T

Definitions

Table 5·8 gives the coefficient, C_{nS}, of x^S in the expansion of

$$\prod_{j=1}^{n}\left(\frac{1-x^j}{1-x}\right) = \sum_{S=0}^{M} C_{nS}x^S,$$

where $M = \frac{1}{2}n(n-1)$, for $n = 1(1)16$. The table is symmetrical about $S = \frac{1}{2}M$ so that as the numbers become large the complete distribution has not been written out. The table was constructed by P. Jackson and K. Solomon from the recurrence relation:

$$C_{nS} = C_{n,S-1} + C_{n-1,S} - C_{n-1,S-1}.$$

Kendall's τ

Consider, for example, the sequence of the first five natural numbers written as 3 2 5 1 4. Two numbers can be chosen from these five in 10 ways. We count in the sequence the number of pairs whose order is concordant with their natural order. For instance 3 and 4 form a concordant pair whereas 5 and 1 do not. In the sequence quoted there are five concordant pairs altogether. Generally, assuming all permutations equally likely, if we define for a permutation $i_1, ..., i_n$ of the integers $1, ..., n$, the statistic S as the number of concordances, the p.d.f. of S is

$$p_n(S) = C_{nS}/n!, \quad 0 \leqslant S \leqslant \tfrac{1}{2}n(n-1).$$

It is often convenient, statistically, to use a quantity with fixed range so that we write

$$\tau = -1 + 4S/\{n(n-1)\}, \qquad |\tau| \leqslant 1.$$

τ is commonly referred to as Kendall's τ and it is used as a measure of association between two sets of ranks.

Illustration

The 'resting pulse rate' in beats per minute and the height in inches were each measured for each of 12 individuals (data adapted from A. B. Hill, *Principles of Medical Statistics*). The data were as follows:

Individual	1	2	3	4	5	6	7	8	9	10	11	12
Resting pulse rate	62	74	80	59	65	73	78	86	64	68	75	81
Height	68·5	65·0	73·5	70·5	69·0	66·0	69·5	70·0	72·5	71·5	68·0	67·5

The actual measurements are replaced by ranks and since there is nothing purposeful in the order of the arrangement of the individuals we may, for convenience only, rearrange the two sequences so that one of them, say resting pulse rate, is in its natural order as follows:

Rank (resting pulse rate)	1	2	3	4	5	6	7	8	9	10	11	12
Rank (height)	8	5	11	6	10	2	1	4	7	12	9	3

The number of concordant pairs is 31 and

$$P\{S \leqslant 31 \,|\, n = 12\} = \frac{201,321,589}{12!} = 0\cdot4203$$

from Table 5·8. Kendall's τ is equal to $-0\cdot061$.

44

Mann's T. Definition and illustration

The same basic statistic S, of the number of concordant pairs, is used to test for trend and in this connexion it is referred to as Mann's T. For example, consider the casualties on British roads on successive Sundays in the first quarter of 1958. These were

$$465, 488, 393, 289, 460, 443, 557, 481, 635, 432, 551, 452, 725.$$

Replacing these by their ranks we have

$$7, 9, 2, 1, 6, 4, 11, 8, 12, 3, 10, 5, 13$$

and find that $S = 50$. From Table 5·8

$$P\{S \geqslant 50 \,|\, n = 12\} = 5{,}024{,}811/12! = 0\cdot0105^-.$$

Under the null hypothesis of no trend all permutations of the 12 integers are equally likely. Alternate hypotheses will be either that the numbers of casualties tend to increase with time (and therefore S large) or that the numbers tend to decrease with time (and therefore S small). The result suggests that the former alternative is certainly a possibility.

Normal approximation

For values of n greater than 16 the assumption of a normal distribution for S is not likely to lead to serious error. We have that

$$\mathscr{E}(S) = \frac{n(n-1)}{4}, \quad \mathrm{var}\,(S) = \frac{n(n-1)(2n+5)}{72},$$

and a continuity correction should be applied. For example for $n = 12$ (which is well inside the range of n tabulated)

$$\mathscr{E}(S) = 33, \quad \sigma_S = 7\cdot2915,$$

and

$$P\{S \leqslant 31\cdot5 \,|\, n = 12\} = 0\cdot4185 \quad (\text{error} = -0\cdot0018),$$

$$P\{S \geqslant 49\cdot5 \,|\, n = 12\} = 0\cdot0118 \quad (\text{error} = 0\cdot0013).$$

For larger n it is to be expected that the errors introduced by assuming normality will be less than these.

PART VI. 'RUNS' DISTRIBUTIONS

PART VI A: TABLE 6·1. DISTRIBUTIONS OF NUMBERS OF RUNS IN A RANDOM SEQUENCE

Definitions

The tables in this Section relate to arrangements of the random sequence and the enumeration of the distributions of what we might call 'linear' runs in contradistinction to the runs 'in a ring'. It is supposed that there are r elements of k kinds, the number of the ith kind being r_i, and

$$\sum_{i=1}^{k} r_i = r.$$

It is common in this type of problem to envisage the different kinds of elements as balls of different colours, and this, to fix ideas, is what we shall do here. We suppose the balls mixed together and placed in a line in random order. A succession of balls of like colour is called a run, the length of the run being one, two, …. For example, if there are 10 balls, 2 white (W), 3 gold (G), and 5 black (B) arranged in a line thus

$$B\,B\,G\,W\,W\,G\,B\,B\,G\,B,$$

the sequence is composed of seven runs in all. We shall use T for the total number of runs of all colours and t for the number of runs of one colour. It is clear that T is equal to the number of alternations in the sequence plus one.

Two-colour and one-colour runs

Given r_1 balls of one colour and r_2 of another, it may be shown that

$$P\{T = 2t\} = 2^{r_1-1}C_{t-1}\,{}^{r_2-1}C_{t-1}/{}^rC_{r_1},$$

$$P\{T = 2t+1\} = P\{T = 2t\}\frac{r-2t}{2t}.$$

The numbers ${}^rC_{r_1}P\{T\}$ are given in Table 6·1 for all two-partitions of $r = 6(1)12$.

For the number of runs of one colour (total colour number r_1) we have

$$p(t) = {}^{r_1-1}C_{t-1}\,{}^{r_2+1}C_t/{}^rC_{r_1}.$$

The numerator of this p.d.f. can be read off from Table 6·1 if we take account of the factor 2 and remember to increase r_2 by 2. For example suppose there are $6\,(=r_1)$ black balls and $4\,(=r_2)$ white balls, $r_1+r_2 = 10$. For $r_1 = 6$, $r_2 = 6$ we have from 12 (6^2 partition) that the values of ${}^5C_{t-1}{}^5C_t$ are successively 5, 50, 100, 50, 5 (taking odd values of T), as may be verified from first principles. Extensions of the table may be made quickly using tables of binomial coefficients (Table 5·1).

Multi-coloured runs

Given r_i $(i = 1, 2, …, k)$ balls of different colours

$$P\{T = T_0\} = \frac{\prod_{i=1}^{k} r_i!}{r!} \sum_{T_0} \prod_{i=1}^{k} {}^{r_i-1}C_{t_i-1} \sum_x {}^{T'+1-x}C_{t_k-x} \left\{ \frac{\left(\sum_{j=1}^{k-1} t_j\right)!}{\prod_{j=1}^{k-1} t_j!} P\left\{ T' = \left(\sum_{1}^{k-1} t_i\right) - x\right\} \right\}.$$

Accordingly from the two-colour runs distribution we may build up the distribution for three-colour runs ($k = 3$), from three we may obtain four and so on. Tables 6·1 give the distributions for $k = 3(1)5$ and $r = 6(1)12$, and $k = 6(1)11$ with $r = k(1)12$. They were computed by F. N. David. (Tables 6·1·3 and 6·1·4 were originally published in Barton & David, 1957.)

Approximations to distributions

The factorial moments of the runs distributions are easy to calculate for the one-colour case, and we have

$$\mu_{[s]}^i = s! \, ^{r_2+1}C_s \, ^{r_1}C_s / ^rC_s, \qquad s = 1, 2, \ldots.$$

The moments of the two and more than two runs distributions cannot be expressed quite so succinctly. David & Barton (1962, pp. 124–5) have given the first four moments of $S = r - T$, of which the first two are

$$r\mathscr{E}(S) = F_2,$$

$$r^{(2)}\mathscr{E}(S^{(2)}) = F_2^2 - 2F_3 - 2F_2,$$

where

$$F_w = \sum_{j=1}^{k} r_j^{(w)}.$$

Three approximations to the distribution of T may be adopted, using the first or both the first and second moments:

(i) The range of T is known. Let this range be $n+1$, the same as the range of the positive binomial $(q+p)^n$, and estimate p by making the mean of the distribution of T and the binomial mean agree.

(ii) Use the first two moments of T to estimate both the parameters p and n of the positive binomial, letting n and p be both fractional, or choosing n to the nearest integer and re-estimating p.

(iii) Assume that T is normally distributed with the true mean and variance and use a continuity correction.

Further approximations are possible such as equating three moments of T to the first three moments of a three-parameter hypergeometric series but this does not seem necessary since (ii) and (iii) are adequate for most purposes. (ii) is undoubtedly the best of the three approximations in all circumstances but for distributions of T not too asymmetrical (iii) will serve. None of these suggested approximations are likely to lead to invalidation of tests of significance. Comparisons of the three approximations are given in the tables below. The length of the sequence is 12 in both cases but in the first there are 4 sets of 3 balls and in the second 4 sets, one of 6 balls, one of 4 balls and two of one ball each.

(a) *True distribution of S compared with three approximations.* $r = 12 : [3^4]$

S ...	0	1	2	3	4	5	6	7	8
Approximation (i)	0·100	0·267	0·311	0·208	0·087	0·023	0·004	0·000	0·000
(ii)	·110	·269	·299	·199	·089	·028	·006	·001	·000
(iii)	·120	·227	·304	·227	·095	·022	·003	·000	·000
True value	·112	·267	·297	·200	·090	·028	·006	·001	·000

(b) *True distribution of S compared with three approximations.* $r = 12 : [641^2]$

S ...	0	1	2	3	4	5	6	7	8
Approximation (i)	0·010	0·062	0·170	0·264	0·257	0·160	0·062	0·014	0·001
(ii)	·008	·056	·165	·271	·270	·163	·056	·009	·000
(iii)	·013	·055	·160	·272	·272	·160	·055	·011	·001
True value	·005	·057	·169	·275	·267	·159	·057	·010	·001

PART VIB: TABLE 6·2. MAXIMUM NUMBER OF RUNS

Notions and definitions

The maximum number of runs possible in a sequence has been of interest at least since De Moivre's time. Statistically it is of interest today when considering the possible alternatives to the random hypothesis. The null hypothesis is commonly that of randomness within the sequence with the alternative that there is too great an alternation from one category (represented by balls of one colour) to another (represented by balls of a different colour). We tabulate here the maximum number of runs for sequences up to and including those of length 15 for a great many possible sequences. For brevity only we use the partition notation and a sequence of length 5, for example, composed of 3 sets of balls, one of two, one of two, and one of one we would write as a sequence of composition $2^2 1$.

Because the number of ways of obtaining the maximum number of runs becomes large as r increases it has been found convenient for tabulation purposes to divide by the factorials of the numbers of colour repetitions. Thus if r is composed of λ different colours of frequencies $\rho_1, \dots, \rho_\lambda$, the number of maximum runs divided by $\rho_2! \dots \rho_\lambda!$ is the quantity tabulated. For example if $N(T_{\max.})$ is the maximum number of runs in a sequence of length 14 and of composition $3^3 1^5$, we tabulate

$$N(T_{\max.})/(3!\,5!).$$

These tables were computed by P. Jackson.

PART VIC: TABLES 6·3·1 AND 6·3·2. RING RUNS

Definitions

Two types of ring run appear in combinatorial literature. First consider coloured beads arranged randomly in a line and then the ends of the line bent round and joined together. The number of runs in such a closed circuit will either be equal to the number of runs in the open line or to the number of runs minus one, depending respectively on whether the line begins and finishes with different colours or the same colours. If $N^*(r, T)$ is the number of ways of obtaining T ring runs in this way, and $N(r, T)$ is the number of ways of obtaining T runs in the line (Table 6·1), then

$$N^*(r, T) = {}^r C_T \sum_{t=k}^{T} (-1)^{T-t} N(r, T)/{}^{r-1}C_{t-1}.$$

Tables 6·3·1 were built up in this way from Tables 6·1, for 2, 3 and 4 partitions of r.

Secondly, consider the number of distinguishable ring permutations of balls arranged randomly in a ring. Let this number be $N^{**}(r, T)$ for the case of T runs. When the highest common factor of the numbers, $r_1, r_2, \dots, r_k$, of balls is one, then

$$r.N^{**}(r, T) = N^*(r, T),$$

but when the h.c.f. is greater than one the equality does not hold. We have shown that

$$r.N^{**}(r, T) = \sum_d \phi(d) N^*\left(\frac{r}{d}, \frac{T}{d}\right),$$

where $\phi(d)$ is Euler's function, r/d denotes the composition $r_1/d, ..., r_k/d$ and the summation is over all the divisors d of both T and the h.c.f., h, from 1 to h. Tables 6·3·2 were built up from Tables 6·3·1 (Barton & David, 1958), using this formula. We have styled these distinguishable permutations *permutation runs* in order to distinguish them from the ordinary ring runs. Only those distributions are given which are different from those of Table 6·3·1 divided by r.

PART VI D: TABLE 6·4. NUMBER OF COMPOSITIONS OF A NUMBER

Definitions and notations

Suppose an integer number N. A composition of N, as opposed to a partition, is any collection of integers whose sum is N but in which the order of the parts of the partition is of significance. Table 6·4 gives the number of compositions of N into s parts, the largest part being equal to m. This number of compositions may be denoted by $g^*(N, s, m)$ and is the coefficient of x^N in the expansion of

$$\left(\frac{x}{1-x}\right)^s \{(1-x^m)^s - (1-x^{m-1})^s\}.$$

If

$$g^*(N, s, m) = G(N, s, m) - G(N, s, m-1),$$

then $G(N, s, m)$ is the coefficient of x^N in the expansion of

$$\left(\frac{x(1-x^m)}{1-x}\right)^s,$$

that is, it is

$$G(N, s, m) = \sum_{i=0}^{} {}^sC_i \, {}^{N-mi-1}C_{s-1}(-1)^i.$$

This was the formula used for computation, $g^*(N, s, m)$ following from the differencing of the computed table of $G(N, s, m)$. The table was computed by J. Charles and P. Jackson.

Illustration (longest run of one colour)

The distribution of the 'longest run of luck' dates back to a discussion by de Moivre. Here we discuss the distribution of the longest run of any one fixed colour when r balls of k different colours are arranged in a line in a random order. The result of fixing the colour in advance, in effect, reduces the problem to a two-colour one, since it is only the presence or absence of the fixed colour which need be recorded. If there are r_1 balls of the fixed colour and r $(= r_1 + r_2)$ balls in all there will be ${}^rC_{r_1}$ arrangements of the balls in a line. The number of arrangements with a longest run of the fixed colour of length exactly l is $g^*(r+1, r_2+1, l+1)$. Thus given 4 white balls and 2 black balls arranged in a random order on a line, the number of ways in which the longest run of white is 2 is $g^*(7, 3, 3)$ which from Table 6·4 is 6. These arrangements are easily seen to be

$$WBWWBW, BWWBWW, WWBWWB, WWBBWW, WWBWBW, WBWBWW.$$

For r_1 and r_2 large and outside the tabulated range a useful approximation is given in the next section.

Values outside the tabulated range of r_1 and r_2

It was shown (David & Barton, 1962, chapter 13) that useful approximations can be obtained to the upper tail of the distribution of the longest run of one colour, using the Bonferroni inequality. For r_1 and r_2 large and ϵ equal to 0·05 or less, the $100\epsilon\%$ upper point is given approximately by

$$\frac{-\log[-\log(1-\epsilon)]+\log(r_2+1)}{\log r - \log r_1}.$$

Other approximations are discussed in the reference quoted but the one given here is adequate for most purposes.

PART VIE: TABLE 6·5. LONGEST RUN OF TWO COLOURS

The distribution of the longest run of two colours was computed by Bateman (1948) before the existence of Table 6·4, but in achieving her table she must have begun by computing something like it. The notation used by her is different from ours, but we use such symbols of hers as are necessary to interpret Table 6·5. Let $g^*(N, s, m)$ have the meaning given in D and let g be the length of the longest run of either of the two colours. Write, for brevity,

$$\phi(t_1 t_2 g) = g^*(r_1 t_1 g) \sum_{m \leqslant g} g^*(r_2 t_2 m) + g^*(r_2 t_2 g) \sum_{m \leqslant g-1} g^*(r_1 t_1 m),$$

assuming $r_1 \geqslant r_2$ and $r_1 + r_2 = r$ the sequence length. The number of ways in which the longest run of either colour is of length g, given r_1 and r_2 are arranged randomly in a line together, is

$$A = \sum_{t=1}^{r_1-g+1} \{2\phi(t, t, g) + \phi(t+1, t, g) + \phi(t, t+1, g)\}$$

and the probability of such a longest run is this number divided by ${}^rC_{r_1}$. Table 6·5 gives the values of A for all possible g for $r = 10, (1)20$, for all possible two-partitions of r for r_1 and r_2.

PART VII. RANDOMIZATION DISTRIBUTIONS

General notions

A series of measurements $x_1, x_2, ..., x_N$ consecutive in operational time is supposed. The measurements are replaced by the ranks which are obtained when the x's are ranked in magnitude, and generally these ranks are supposed in a random order. Most of the statistics, the randomization distributions of which are given in Tables 7, are criteria to test for the randomness of the order in time and are based on the pattern of the signs of the successive rank differences. As an illustration consider the page length of each of the first fifteen chapters of *Théorie Analytique des Probabilités* by P-S. Laplace. These are, in successive order:

Page length	53	43	24	17	57	11	96	32	50	16	43⁻	6	12	14	15
Rank	13	11	8	7	14	2	15	9	12	6	10	1	3	4	5
Sign of difference		−	−	−	+	−	+	−	+	−	+	−	+	+	+

There is, possibly, a regularity about the pattern of these signs, which may be brought out by a test for randomness.

The statistics for such kinds of tests fall into three main classes with several types within a class. Most of the statistics have a 'one-way' and a 'both-ways' form.

The tables were computed by P. Jackson.

PART VII A: TABLE 7·1. NUMBER OF PERMUTATIONS WITH *s* POSITIVE SUCCESSIVE DIFFERENCES

Definitions and notations

Let α_i be a random variable which takes the value 1 if $x_i > x_{i-1}$ and is zero if $x_i < x_{i-1}$. Then the number of positive successive differences in a random sequence of the first N natural numbers is

$$s = \sum_{i=1}^{N-1} \alpha_i$$

and the probability distribution of s will be the number of permutations each of which gives rise to a given value of s divided by $N!$. It is this number of permutations which is tabulated in Table 7·1.

Illustration

In the illustration given in the first section $N = 15$, $s = 7$. In Table 7·1 we add the number of permutations for $s = 7(1)14$ and divide by 15! to get

$$P\{s \geqslant 7 \,|\, N = 15\} = 0 \cdot 671.$$

The alternative to the randomness test will be, for this statistic, that there are either too many positive signs or too few, implying that the chapter lengths either tended to increase throughout the book or to diminish. Against such an alternative we should not here reject the hypothesis of randomness.

Construction of table and approximations

The tables was constructed using a recurrence relation derived, for instance, in David & Barton (1962, pp. 150–1). For values of s outside the table it is enough to note that

$$\mathscr{E}(s) = \tfrac{1}{2}(N-1), \quad \operatorname{var} s = \tfrac{1}{12}(N+1),$$

and that with increasing N, s tends to be normally distributed. For the illustration above, since $N = 15$,

$$\mathscr{E}(s) = 7, \quad \operatorname{var} s = \tfrac{4}{3}.$$

Using the continuity correction of $\tfrac{1}{2}$ we have $P\{s > 6\cdot5\} = 0\cdot6675$. (If we take, for illustration, $s = 9$, $N = 16$, $P\{s > 8\cdot5\} = 0\cdot200$, whereas the true value $P\{s \geqslant 9 \mid N = 16\} = 0\cdot195$.) The agreement is clearly close enough for practical purposes.

PART VIIB: TABLES 7·2·1 AND 7·2·2. NUMBER OF PERMUTATIONS OF RUNS: NUMBER OF PERMUTATIONS OF RUNS OF LIKE SIGN

Table 7·2·1. Notation and notions

Given the $N-1$ signs of the successive differences of the random sequence of the first N integers the pattern they form may be investigated by counting the number of runs, t, of both $+$ and $-$ signs. As before the total number of permutations will be $N!$ and the number of permutations containing t runs is found from a recurrence relation the argument for which is set out on pp. 158–9 of David & Barton (1962). Because the numbers are rather large and are always divisible by two, one-half the number of permutations is tabulated in Table 7·2·1.

Illustration

In the Laplace example the number of runs is $t = 10$, and $N = 15$. From Table 7·2·1 we add over the permutations for $t = 10$ and greater, multiply by 2 and divide by 15! to obtain

$$P\{t \geqslant 10\} = 0\cdot548.$$

Approximation

It may be shown that

$$\mathscr{E}(t) = \tfrac{1}{3}(2N-1), \quad \operatorname{var} t = \tfrac{1}{90}(16N - 29),$$

and that with increasing N, t tends to be normally distributed. The normal approximation is adequate for most purposes outside the range of Table 7·2·1, bearing in mind that t is discrete and consequently that care is needed in the interpretation of the nominal significance level. For $t = 10$, $N = 15$, using the normal approximation and a continuity correction

$$P\{t > 9\cdot5\} = 0\cdot543,$$

which agrees closely with the true value. Judged against the hypothesis of randomness the alternate hypotheses will be either that a long chapter tends to be followed by a long chapter or that a long chapter and a short chapter tend to alternate. For the first alternative we calculate $P\{t \leqslant 10\}$ and for the second $P\{t \geqslant 10\}$. In neither case is there any justification for rejecting the null hypothesis.

Table 7·2·2

Usually for sign-run tests it is the total number of runs which is required for a test criterion. Sometimes the number of runs of one sign, commonly called the number of runs up (or down) is used. Let t_u be the number of runs up in a sequence of $(N-1)$ signs. The number of permutations of N which have t_u runs up is given in Table 7·2·2. (The number of permutations with t_d runs down follows from symmetry.) For N, the number of integers in the random sequence, large, the normal approximation may be used. It may be shown that

$$\mathscr{E}(t_u) = \tfrac{1}{6}(2N-1), \quad \operatorname{var}(t_u) = \tfrac{2}{135}(N+1),$$

but since the approach to normality is slow care is needed in the interpretation of a significance level reached by this means.

PART VII C: TABLE 7·3. NUMBER OF PERMUTATIONS WITH p PEAKS

Notation and notions

In Part VII A a characteristic random variable α_i was defined: $\alpha_i = 1$ if in the original sequence $x_{i+1} > x_i$ and is zero if $x_{i+1} < x_i$. It is clear that a 'peak' at i will imply $\alpha_i = 1$ and $\alpha_{i-1} = 0$, while a 'trough' at i will imply $\alpha_i = 0$, $\alpha_{i+1} = 1$. The total number of turning points in the sequence will be the sum of the number of peaks, p, and the number of troughs. The peak-trough effect is perhaps most easily appreciated if the values of the random variable, α, are written down. Thus, in our numerical example these values are, consecutively

$$0\ 0\ 0\ 1\ 0\ 1\ 0\ 1\ 0\ 1\ 0\ 1\ 1\ 1$$

so that there are 4 peaks, 5 troughs and 9 turning points. The number of permutations of N each of which has p peaks is given in Table 7·3.

The normal approximation for p can be assumed for N large, but because of the compactness of the distribution such an assumption will not be adequate unless N *is* large. Thus from Table 7·3

$$P\{p \leqslant 4 \mid N = 15\} = 0 \cdot 578.$$

The mean and variance of p are

$$\mathscr{E}(p) = \tfrac{1}{3}(N-2), \quad \operatorname{var} p = \tfrac{2}{45}(N+1)$$

whence, from normal tables,

$$P\{p < 4 \cdot 5 \mid N = 15\} = 0 \cdot 578,$$

which is clearly good enough.

PART VII D: TABLES 7·4·1 AND 7·4·2. LONGEST RUN UP: LONGEST RUN

Notation and notions

Consider a sequence of N ranks in operational time. Thus if $N = 9$, for example, the rank sequence might be

$$3 \quad 1 \quad 9 \quad 6 \quad 7 \quad 5 \quad 4 \quad 8 \quad 2.$$

In this sequence there are 3 runs up each of length 2 and 4 runs down of lengths 2, 2, 3, 2. If we let l_u denote the length of the longest run up and l denote the length of the longest run up or down, the example gives $l_u = 2$, $l = 3$. In Table 7·4·1 we give the number of permutations of the first N integers in each of which the longest ascending run is l_u and in Table 7·4·2 we give the number of permutations of the first N integers in each of which the length of the longest run up or down is l. Thus for the Laplace chapter illustration we have the sequence of $N = 15$ ranks

$$13 \quad 11 \quad 8 \quad 7 \quad 14 \quad 2 \quad 15 \quad 9 \quad 12 \quad 6 \quad 10 \quad 1 \quad 3 \quad 4 \quad 5.$$

The longest ascending run is of length 4 and this is also the length of the longest run ascending or descending. From Table 7·4·1 $\qquad P\{l_u \geqslant 4 \,|\, N = 15\} = 0.366.$

Table 7·4·2 has been computed only up to $N = 14$ and we cannot therefore compute the exact probability $P\{l \geqslant 4\}$.

Approximations

The distribution of the longest run, either ascending or descending and ascending does not tend to the normal form with increasing N but approaches very slowly to the extreme value double exponential function. Approximating functions useful outside the range of the present tables, but before the double exponential limit is reached, have been discussed elsewhere by David & Barton. They suggest that the cumulative distribution function of l (or l_u) may be taken as

$$\mathscr{P}(l) \doteqdot e^{-\mu},$$

where
$$\mu = \frac{2[(N-l)(l+1)+1]}{(l+2)!}$$

for the longest ascending or descending run and one half this for l_u. Thus for $N = 15$ we have

$$P\{l \leqslant 3 \,|\, N = 15\} = e^{-0.8167} = 0.442,$$

so that
$$P\{l \geqslant 4 \,|\, N = 15\} = 0.558.$$
For the longest ascending run
$$P\{l_u \leqslant 3 \,|\, N = 15\} = e^{-0.4083} = 0.665^-,$$

so that
$$P\{l_u \geqslant 4\} = 0.335^+,$$

which is reasonably in accord with the true value of 0·366. This exponential approximation can be used with confidence for the upper tail of 'longest' distributions; it is useless for the lower tail. It is, however, the upper tail which is the more often required for tests of significance involving this type of criterion, and it is because the approximation is so good that we have not calculated the distribution of the longest run beyond $N = 14$. For closer approximations and approximate significance levels the reader is referred to David & Barton (1962, chapter 13).

PART VII E: TABLES 7·5·1 AND 7·5·2. CONSECUTIVE PAIRS UP: CONSECUTIVE PAIRS UP AND DOWN

Notations and notions

Consider a sequence of the first seven natural numbers, say,

$$3 \quad 2 \quad 1 \quad 4 \quad 5 \quad 7 \quad 6.$$

In this particular sequence there is one consecutive pair of ascending integers and three consecutive pairs of descending integers. If c_u is the number of consecutive pairs up and c_d the number of consecutive pairs down, then c, the number of consecutive pairs up and down is

$$c = c_d + c_u.$$

In the example just given $c_u = 1$, $c_d = 3$ and $c = 4$. We consider a sequence composed of the first N natural numbers arranged in a random order. There are $N!$ permutations of these N. In Table 7·5·1 we give the number of permutations in which there are c_u consecutive pairs up and in Table 7·5·2 one-half the number of permutations in which there are c consecutive pairs up and down. The tables were built up from recurrence relations derived by Barton & David (1962, pp. 168 and 175). For Table 7·5·2 only one-half the number of permutations are given because all numbers of permutations were divisible by two.

Illustration

In the Laplace chapter-length example, $c_u = 2$, $c_d = 2$, $c = 4$ and $N = 15$. From Table 7·5·1 we have $P\{c_u < 2\} = 0.760$ and from Table 7·5·2 we have $P\{c < 4\} = 0.896$ remembering in the latter table that the number of permutations are to be multiplied by 2.

Approximations

c_u tends to be distributed as a Poisson variable with parameter 1 for N large. Consequently since

$$\mathscr{E}(c_u) = 1 - \frac{2}{N} + \frac{1}{N^{(2)}} = \mu \quad \text{(say)}$$

we assume that c_u is distributed as a Poisson variable with parameter μ. For $N = 15$ this gives

$$P\{c_u < 2\} = 0.783,$$

which is reasonably close to the true value. c tends to be distributed as a Poisson variable with parameter 2. It may be shown that

$$\mathscr{E}(c) = 2\left(1 - \frac{1}{N}\right)$$

which, from exponential tables, gives $\quad P\{c < 4\} = 0.880.$

Again the agreement is reasonable although it may be expected to be less so towards the tails of the distribution.

PART VII F: TABLES 7·6·1 AND 7·6·2. RUNS OF CONSECUTIVE PAIRS UP: RUNS OF CONSECUTIVE PAIRS UP AND DOWN

Notation and notions

In the sequence of the first seven natural numbers given in VII F we have

$$3\ \underbrace{2}\ 1\ \ 4\ \ 5\ \overbrace{7}\ 6$$

i.e. one run of consecutive pairs up and two runs of consecutive pairs down, making three runs up or down. In Table 7·6·1 we give the number of permutations with w_u runs of consecutive pairs up, and in Table 7·6·2 we give one-half the number of permutations of the first N integers in which there are w runs of consecutive pairs up or down. The total number of permutations will be $N!$ in each case.

Illustration

In the original illustration in A we have $w_u = 1$, $w = 2$. From Table 7·6·1 we find $P\{w_u \leqslant 1\} = 0·787$. From Table 7·6·2, $P\{w \leqslant 2\} = 0·758$.

Approximations

It may be shown that

$$\mathscr{E}(w) = 2\left(1 - \frac{1}{N} + \frac{1}{N^{(2)}}\right) = 2\mathscr{E}(w_u).$$

Further with increasing N, w tends to be distributed in the Poisson form with parameter 2 and w_u in the Poisson form with parameter 1. We assume therefore that w and w_u each have a Poisson distribution with parameter equal to the appropriate expected value. Under these assumptions the example with $N = 15$ gives

$$P\{w_u \leqslant 1\} = 0·759, \quad P\{w \leqslant 2\} = 0·713.$$

Accordingly, for N greater than 16 the Poisson approximation will probably be adequate.

PART VIII. TABLES FOR THE SOLUTION OF THE EXPONENTIAL EQUATIONS

$$\exp(-a) + ka = 1, \quad \exp a - a/(1-p) = 1$$

Introductory remarks

The equation

$$e^z = 1 + kz$$

has one real root, z, when k is a given positive number. For $0 < k < 1$ the root of the equation is negative. For $k > 1$ the root is positive. These two situations are covered by Tables 8·1 and 8·2 respectively. The use of the tables to evaluate Stirling Numbers outside tabulated values has been given by Barton, David & Merrington (1960, 1963).

PART VIII A: TABLE 8·1. CASE WITH $0 < k < 1$

We consider the solution for a of

$$e^{-a} + ka = 1 \quad (0 < k < 1).$$

For values of $k = 0.050\,(0.001)\,0.190$, a was found approximately by building up a value of $k^{-1} - a$ from third differences, choosing an approximate value for a and solving for a by iteration. For $k = 0.190\,(0.001)\,1.000$ a table of the approximate value of a was built up by assuming constant fifth differences, the exact value of a being then found by iteration. For $k < 0.05$ it is enough to assume $a = k^{-1}$. The table was checked on the Mercury computer of the University of London.

Example

The estimation by maximum likelihood of the parameter of the Poisson distribution when the zero group is missing reduces to the solution of the equation

$$\bar{x} = \frac{\hat{\lambda}}{1 - e^{-\hat{\lambda}}},$$

where $\bar{x}$ is the mean of the N known observations and λ is the Poisson parameter. Van Rest (1937) gives the following table:

Frequency of occurrence of different-sized groups of vessels in the wood of Shorea leprosula

No. of vessels in group	Observed frequency	Theoretical frequencies (i)	Theoretical frequencies (ii)
1	843	838·5	843
2	143	151·9	144
3	18	13·7	17
4	1	0·8	1
Total	1005	1005	1005

He calculated theoretical frequencies under (i) by assuming $x - 1$ was a Poisson variable where x is the number of vessels in a group. The theoretical frequencies (ii) are calculated using the maximum likelihood method of estimation. Thus the mean is $\bar{x} = 1.1811$, whence $k = 1/\bar{x} = 0.8467$ and from Table 8·1 $a = \hat{\lambda} = 0.3426$. The Poisson probabilities were then calculated using this value of $\hat{\lambda}$. The theoretical frequencies (ii) are the individual Poisson probabilities divided by their total of 0·2901 and multiplied by 1005. The agreement with the observed frequency is seen to be very good.

PART VIII B: TABLE 8·2. CASE WITH $k > 1$

For this table we put $b = -a$ and for ease of tabular entry work in terms of $p = 1 - 1/k$, so that instead of the root given in Table 8·1 of

$$e^{-a} + ka = 1,$$

we consider the root of

$$e^b = 1 + b/(1-p).$$

Inverse interpolation into Table 8·1 provided the starting-point for the construction of Table 8·2. Using these initial values obtained by inverse interpolation between a and $a + 1 - 1/k$, the root was obtained by Newtonian iteration.

Example

Fisher (1943) demonstrated a biological interpretation of the logarithmic distribution and gave tables for the solution of the maximum likelihood equation for the estimation of the parameter. He tabulated (in our notation) $\log_{10}[b/(1-p)]$ in terms of $\log_{10}[1/(1-p)]$.

If $x_1, x_2, ..., x_n$ are randomly and independently drawn from the distribution

$$p(x) = \frac{\alpha^x}{x} \frac{1}{\log((1-\alpha)^{-1})} \quad (x = 1, 2, ...),$$

where α, $0 < \alpha < 1$ is the unknown parameter, the mean $\bar{x}$ is sufficient for α and the likelihood equation gives the estimator $\hat{\alpha}$ as the root of

$$\exp\left\{-\left[\bar{x}\left(\frac{1-\hat{\alpha}}{\hat{\alpha}}\right)\right]^{-1}\right\} = 1 - \hat{\alpha}.$$

Accordingly if we write

$$\hat{\alpha} = b\bar{x}/(1 + b\bar{x}),$$

we have the equation

$$e^b = 1 + b\bar{x}.$$

Because

$$p = (\bar{x} - 1)/\bar{x} \quad \text{or} \quad 1 - p = 1/\bar{x}$$

we may use Table 8·2 with this p to obtain b. Having obtained b the value for $\hat{\alpha}$ follows.

PART IX. PARTITION COEFFICENTS FOR THE INVERSION OF FUNCTIONS

Notation and notions

One of the many uses to which the basic Tables 1·1, 1·2, 1·3, 1·4 or 1·5 may be put is in expanding the function of a function. The application of the basic tables is not, however, immediate in this case, and it has been thought fit to calculate, from the basic tables, a further table which is directly applicable to the purposes of the expansion. Assume it is desired to expand the function of a function, say $f(g(x))$, as a Taylor series in x. Purely formally we write

$$f_r = \frac{d^r f}{dg^r}\Big|_{x=0}, \quad g_r = \frac{d^r g}{dx^r}\Big|_{x=0}, \quad c_r = \left[\frac{d^r}{dx^r} f(g(x))\right]_{x=0} \quad (r = 0, 1, 2, \ldots),$$

so that we may express $f(g(x))$ as

$$f(g(x)) = \sum_{r=0}^{\infty} \frac{x^r}{r!} c_r.$$

It is evident that the c_r are expressible in terms of f_r and g_r and in fact

$$c_0 = f_0, \qquad c_1 = f_1 g_1,$$

$$c_2 = f_2 g_1^2 + f_1 g_2, \quad c_3 = f_3 g_1 + 3 f_2 g_1 g_2 + f_1 g_3$$

and so on. From the mode of operation of the differential operator it is evident that c_r is a linear sum of products $g_{p_1}^{\pi_1} \ldots g_{p_\lambda}^{\pi_\lambda}$ over all partitions

$$P \equiv p_1^{\pi_1} \ldots p_\lambda^{\pi_\lambda},$$

each product being multiplied by f_v, where

$$v = \sum_{i=1}^{\lambda} \pi_\lambda,$$

and by a numerical coefficient which we will call $c(P)$. For example if we consider the case where f is the exponential function and $g(x)$ is a cumulant generating function, $c(P)$ is the coefficient in μ'_w of $\kappa_{p_1}^{\pi_1} \ldots \kappa_{p_\lambda}^{\pi_\lambda}$.

Inverse functions

It is often important statistically, as for example in the solution of likelihood equations, to invert a function. Thus if we have two variables x and y and if

$$y = f(x), \quad f(0) = 0,$$

we may put

$$f_r = \left[\frac{d^r}{dx^r} f(x)\right]_{x=0}$$

whence, writing

$$y = f_1 x + f_2 \frac{x^2}{2!} + \ldots, \tag{1}$$

it is also possible to write

$$x = g_1 y + g_2 \frac{y^2}{2!} + \ldots. \tag{2}$$

The analytic conditions for this to be valid are described in Bürmann's Theorem as delineated for example in Whittaker & Watson's *Modern Analysis*. It is essential that $f_1 \neq 0$. We are concerned here with the purely algebraic problem of expressing the coefficients $g_1, g_2, \ldots$ in terms of $f_1, f_2 \ldots$. This may be done by substituting (2) in (1) and equating coefficients. Thus

$$g_1 = \frac{1}{f_1}, \quad g_2 = \frac{1}{f_1}\left(-\frac{f_2}{f_1^2}\right),$$

$$g_3 = \frac{1}{f_1}\left(-\frac{f_3}{f_1^3} + \frac{3f_2^2}{f_1^4}\right),$$

$$\cdots\cdots\cdots\cdots\cdots$$

More conveniently, if we write

$$j_r = -f_{r+1}/f_1^{r+1} \quad (r = 0, 1, 2, \ldots),$$

we have

$$f_1 g_1 = -j_0 = 1, \qquad f_1 g_2 = j_1,$$

$$f_1 g_3 = j_2 + 3j_1^2, \qquad f_1 g_4 = j_3 + 10 j_2 j_1 + 15 j_1^3,$$

$$\cdots\cdots\cdots\cdots\cdots\cdots\cdots$$

We shall also require the coefficients in the like expansions of $x^2, x^3, \ldots$. Now, if $\phi(x)$ is a function (such as $x, x^2, x^3, \ldots$) with $\phi(0) = 0$, Bürmann's theorem gives the Taylor expansion of $\phi(x)$ in terms of y thus:

$$\phi(x) = \sum_{m=1}^{\infty} \frac{y^m}{m!}\left[\frac{d^{m-1}}{dx^{m-1}} \phi'(x) \left(\frac{x}{f(x)}\right)^m\right]_{x=0}.$$

We therefore write

$$x^r = \sum_{m=1}^{\infty} \frac{y^m}{m!} c_{mr},$$

where

$$c_{mr} = r!\,^{m-1}C_{r-1}\left[\frac{d^{m-r}}{dx^{m-r}}\left(\frac{x}{f(x)}\right)^m\right]_{x=0},$$

and where we shall write

$$l(x) = \frac{1}{x}f(x) = f_1 + \frac{f_2}{2}x + \frac{f_3}{3}\frac{x^2}{2!} + \ldots, \quad l_p = \left[\frac{d^p l(x)}{dx^p}\right]_{x=0}, \quad p = 0, 1, \ldots.$$

It is plain that all we now require is the coefficients in the Taylor expansion of l^{-m} which we may easily obtain from our expression for the Taylor expansion of a function (the $-m$th power) of a function. We have

$$\left[\frac{d^n}{dx^n}\left(\frac{1}{l(x)}\right)^m\right]_{x=0} = \sum_P \frac{l_{p_1}^{\pi_1} \ldots l_{p_\lambda}^{\pi_\lambda} m^{[w]}(-1)^w n!}{l_0^{m+w} \pi_1! \ldots \pi_\lambda! (p_1!)^{\pi_1} \ldots (p_\lambda!)^{\pi_\lambda}},$$

where the summation is over all partitions $p_1^{\pi_1} \ldots p_\lambda^{\pi_\lambda}$ of n. Thus when $n = m - r$

$$f_1^r\left[\frac{d^{m-r}}{dx^{m-r}}\left(\frac{1}{l(x)}\right)^m\right]_{x=0} = \sum_P \frac{j_{p_1}^{\pi_1} \ldots j_{p_\lambda}^{\pi_\lambda}}{\pi_1! \ldots \pi_\lambda!} \frac{m^{[w]}(m-r)!}{((p_1+1)!)^{\pi_1} \ldots ((p_\lambda+1)!)^{\pi_\lambda}}.$$

Hence, finally we have

$$f_1^r x^r = r \sum_{n=0}^{\infty} \frac{y^{n+r}}{(n+r)!} \sum_P j_{p_1}^{\pi_1} \ldots j_{p_\lambda}^{\pi_\lambda} M_r(p_1^{\pi_1} \ldots p_\lambda^{\pi_\lambda}),$$

where

$$M_r(p_1^{\pi_1} \ldots p_\lambda^{\pi_\lambda}) = \frac{(n+w+r-1)!}{\pi_1! \ldots \pi_\lambda! ((p_1+1)!)^{\pi_1} \ldots ((p_\lambda+1)!)^{\pi_\lambda}}$$

$$= (n+w+1)^{[r-1]} M_1(p_1^{\pi_1} \ldots p_\lambda^{\pi_\lambda}).$$

For convenience the M-coefficients are tabulated directly for $r = 1$ against partitional arguments, the multipliers $(n+w+1)\ldots(n+w+r-1)$ being tabulated (rather than the M's) to save space. For example, the coefficient of $3y^6/6!$ in x^3 is found from the tables (at $n = 6-3 = 3$) to consist of a sum of three terms, numerical multiples of $j_1^3 f_1^{-3}$, $j_2 j_1 f_1^{-3}$, $j_3 f_1^{-3}$, which are respectively $15 \times 56 = 840$, $42 \times 10 = 420$, $30 \times 1 = 30$, i.e. the coefficient is

$$-840 f_2^3 f_1^{-9} + 420 f_2 f_3 f_1^{-8} - 30 f_4 f_1^{-7}.$$

This form of tabulation was devised by P. G. Marks and the coefficients are distinct from those arising when a series

$$y = a_1 x + a_2 x^2 + \ldots$$

is inverted to give

$$x = b_1 y + b_2 y^2 + \ldots$$

(and similar series for x^2, x^3, x^4 ...). These latter coefficients were first given by J. McMahon (1894) for $r = 1$ and we will call them McMahon's form. They are evidently simple multiples of Marks's coefficients and tend to have smaller values and so are tabulated directly. For example

$$a_1 b_1 = 1, \quad a_1 b_2 = -a_2 a_1^{-2}, \quad a_1 b_3 = 2 a_2^2 a_1^{-4} - a_3 a_1^{-3}, \ldots$$

or, putting

$$A_r = -a_{r+1}/a_1^{r+1},$$

we have

$$a_1 b_1 = 1, \qquad a_1 b_2 = A_1,$$

$$a_1 b_3 = A_2 + 2 A_1^2, \quad a_1 b_4 = A_3 + 5 A_1 A_2 + 5 A_1^3.$$

Since the relation to the previous notation is given by putting $a_r = f_r/r!$ and $b_r = g_r/r!$, which implies $A_r = j_r/(r+1)!$, we have that the coefficient of $y^{n+r} A_{p_1}^{\pi_1} \ldots A_{p_\lambda}^{\pi_\lambda}$ in $f_1^r x^r$ is

$$M_r'(P) = \frac{r(n+r+w-1)!}{(n+r)! \, \pi_1! \ldots \pi_\lambda!}.$$

Illustration (1)

An illustration of the application of the tables, which is typical of a field where they are of great utility, is in the problem of estimating the two parameters of the Pearson type III distribution with origin at zero. Suppose a sample of n independent observations of the random variable X with p.d.f.

$$p(X) = \alpha^\beta X^{\beta-1} e^{-\alpha X}/\Gamma(\beta) \quad (0 < X < \infty).$$

The arithmetic and geometric means, m and g say, are jointly sufficient for the unspecified parameters α and β. The maximum likelihood estimators of α and β are the roots of

$$\hat{\beta} = m\hat{\alpha}, \quad \log g + \log \hat{\alpha} - \Gamma_1(\hat{\beta}-1) = 0$$

where $\Gamma_j(z)$ denotes the jth derivative of the logarithm of $\Gamma(z+1)$. Thus $\hat{\beta}$ is the root of

$$\log \hat{\beta} - \Gamma_1(\hat{\beta}-1) = \log m/g \quad \text{and} \quad \hat{\alpha} = \hat{\beta}/m.$$

Concerning ourselves purely with $\hat{\beta}$, for the purposes of illustration, if we put

$$f(\hat{\beta}-\beta) \equiv \log \hat{\beta} - \Gamma_1(\hat{\beta}-1) - \{\log \beta - \Gamma_1(\beta-1)\},$$

$$y \equiv \log m/g - \log \beta + \Gamma_1(\beta-1) \quad (x \equiv \hat{\beta}-\beta),$$

we have to study the statistical behaviour of the root of $f(x) = y$ where, for n large at least, x and y are both small except with negligible probability. The moment generating function of y is available, viz. $M(t) = \mathscr{E}(e^{ty})$ and we have

$$\log M(t) = n \log \left\{ \frac{\Gamma(\beta - t/n)}{\Gamma(\beta)} \right\} - \log \left\{ \frac{\Gamma(\beta n - t)}{\Gamma(n\beta)} \right\} - t \log n\beta + t \Gamma_1(\beta - 1),$$

so that the cumulants of y are

$$\kappa_1 = \Gamma_1(\beta n - 1) - \log \beta n \sim -\frac{1}{2n\beta} - \frac{1}{12n^2\beta^2} + \frac{1}{120\beta^4 n^4} + O\left(\frac{1}{n^6}\right)$$

and, for $r \geqslant 2$,

$$\kappa_r = (-1)^{r-1}\left\{\Gamma_r(\beta n - 1) - \frac{\Gamma_r(\beta - 1)}{n^{r-1}}\right\}$$

$$\sim \frac{1}{n^{r-1}}\left\{(-1)^r \Gamma_r(\beta - 1) - \frac{(r-2)!}{\beta^{r-1}} - \frac{(r-1)!}{2\beta^r n} - \frac{r!}{12\beta^{r+1}n^2} + \frac{(r+2)!}{720\beta^{r+3}n^4} + O\left(\frac{1}{n^6}\right)\right\}.$$

Now, returning to the solution (in the notation of the tables) of

$$y = f(x) = a_1 x + a_2 x^2 + \dots,$$

we have

$$a_r = -\frac{\Gamma_{r+1}(\beta - 1)}{r!} + \frac{(-1)^{r+1}}{r\beta^r}.$$

Taking the case $\beta = 2$ for illustration, and using Peters's tables for the gamma derivatives (1922, Table 10), we have, using the McMahon form of coefficients, the inverse series

$$x = b_1 y + b_2 y^2 + \dots,$$

where the numerical values of $A_r = -a_{r+1}/a_1^{r+1}$ and b_r are given ($r = 1, 2 \dots 8$), in the table below. They are scaled by $(0\cdot3)^r$ for convenience of tabulation.

r	$-(0\cdot3)^r A_r$	$(0\cdot3)^r b_r$	c_r
1	0·0990454	− 0·0990454	− 0·0600602
2	·1081694	− ·0885494	− ·709525
3	·1173168	− ·0686067	+ ·108997
4	·1264523	− ·0425689	+ 51·1819
5	·1355578	− ·0153009	—
6	·1446295	·0079475	—
7	·1536749	·0228698	—
8	·1627095	·0272727	—

The series for $x = \hat{\beta} - \beta$ may now be averaged, termwise, over the distribution of y, and terms of the same order in $1/n$ collected to give

$$\mathscr{E}(\hat{\beta}) = 2 + \frac{c_1}{n} + \frac{c_2}{n^2} + \dots,$$

where the numbers $\{c_r\}$, are given in the third column of the table. It is worth remarking that what has been done in this illustration is to evaluate, more expeditiously, the coefficients which would have been obtained by use of the saddle-point method of approximation which, however, provides the rigorous justification for the validity of the expansion for $\mathscr{E}(\hat{\beta})$. The equivalent method of Laplace specifically requires use of the tables given here.

Illustration (2)

An illustration of an algebraic use of the series inversion method is provided by its use in obtaining an asymptotic series for the variance of the median in a sample of n independent observations from a normal population whose p.d.f. is

$$\phi(x) \equiv e^{-\frac{1}{2}x^2}/\sqrt{(2\pi)} \quad (-\infty < x < +\infty),$$

when n is odd. If X is the median and $n = 2m + 1$ (m an integer) then the required variance is

$$\sigma^2 = \int_{-\infty}^{+\infty} \frac{n!}{(m!)^2} X^2 \phi(X) \Phi^m(X) (1 - \Phi(X))^m \, dX,$$

where

$$\Phi(x) = \int_{-\infty}^{x} \phi(t) \, dt, \quad \text{since} \quad \mathscr{E}(X) = 0.$$

The rth moment of $2\pi(\Phi(X) - \tfrac{1}{2})^2$ is easily seen to be

$$\left(\frac{\pi}{2}\right)^r \frac{(2r - 1)(2r - 3) \dots 3.1}{(n + 2)(n + 4) \dots (n + 2r)}$$

and

$$2\pi(\Phi(X) - \tfrac{1}{2})^2 = X^2 - \frac{X^4}{3} + \frac{7X^6}{90} - \frac{X^8}{70} + \frac{83X^{10}}{37{,}800} - \frac{73X^{12}}{249{,}480} + \dots.$$

If then we invert this series, using the tables, and take expectations termwise we obtain the series

$$\sigma^2 = \frac{\pi}{2(n + 2)} + \frac{\pi^2}{4(n + 2)(n + 4)} + \frac{13}{48} \frac{\pi^3}{(n + 2)(n + 4)(n + 6)}$$

$$+ \frac{11}{24} \frac{\pi^4}{(n + 2)(n + 4)(n + 6)(n + 8)} + \frac{4069}{3840} \frac{\pi^5}{(n + 2)(n + 4)(n + 6)(n + 8)(n + 10)}$$

$$+ \frac{8963}{2880} \frac{\pi^6}{(n + 2)(n + 4)(n + 6)(n + 8)(n + 10)(n + 12)} + O\left(\frac{1}{n^7}\right).$$

REFERENCES

ABDEL-ATY, S. H. (1954). Tables of generalised k-statistics. *Biometrika*, **41**, 253–60.

BARTON, D. E. & DAVID, F. N. (1957). Multiple runs. *Biometrika*, **44**, 168–78.

BARTON, D. E. & DAVID, F. N. (1958). Runs in a ring. *Biometrika*, **45**, 572–8.

BARTON, D. E., DAVID, F. N. & FIX, E. (1960). Polykays of the natural numbers. *Biometrika*, **47**, 53–9.

BARTON, D. E., DAVID, F. N. & MERRINGTON, M. (1960). Tables for the solution of the exponential equation $e^{-a} + ka = 1$. *Biometrika*, **47**, 439–45. (1963). Tables for the solution of the exponential equation $e^b - b/(1 - p) = 1$. *Biometrika*, **50**, 169–76.

BATEMAN, G. I. (1948). Power function of the longest-run test. *Biometrika*, **35**, 97–112.

CAYLEY, A. (1857). Memoir on the symmetric functions of the roots of an equation. *Philos. Trans.* **147**, 489–96.

CAYLEY, A. (1896). *Collected Mathematical Papers*, **11**, 357–64. Cambridge University Press. (Alternatively see *Amer. J. Math.* (1881), **4**, 248–55.)

COOK, M. B. (1951). Bivariate k-statistics and cumulants of their joint sampling distribution. *Biometrika*, **38**, 179–95.

DAVID, F. N. & BARTON, D. E. (1962). *Combinatorial Chance*. London: Charles Griffin and Co.

DAVID, F. N. & KENDALL, M. G. (1949, 1951, 1953, 1955). Tables of symmetric functions. I, II, III, IV, V. *Biometrika*, **36**, 431–49; **38**, 435–62; **40**, 427–46, **42**, 223–42.

DAVIS, H. T. (1935). *Tables of the Higher Mathematical Functions*, **2**. Bloomington, Indiana: Principia Press.

DECKER, F. F. (1910). *Symmetric Function Tables of the Fifteenthic*. Washington: Carnegie Institution.

DURFEE, W. P. (1882). Tables of symmetric functions of the twelfthic. *Amer. J. Math.* **5**, 46–61.

DURFEE, W. P. (1887). Symmetric functions of the fourteenthic. *Amer. J. Math.* **9**, 278–96.

FAÀ DI BRUNO, F. (1876). *Théorie des Formes Binaires*. Turin: Brero.

FISHER, R. A. (1928). Moments and product-moments of sampling distributions. *Proc. Lond. Math. Soc.* **30**, 199–238.

FISHER, R. A. (1943). The relation between the number of species and the number of individuals in a random sample from an animal population. *J. Anim. Ecol.* **12**, 54–7.

References

Fix, E. & Hodges, J. L. (1955). Significance probabilities of the Wilcoxon test. *Ann. Math. Statist.* **26**, 301–12.

Fletcher, A., Miller, J. C. P., Rosenhead, L. & Comrie, L. J. (1962). *An Index of Mathematical Tables.* (Second edition). Oxford: Blackwell Scientific Publications Ltd.

Greenwood, J. Arthur & Hartley, H. O. (1962). *Guide to Tables in Mathematical Statistics.* Princeton University Press.

Halmös, P. R. (1946). The theory of unbiased estimation. *Ann. Math. Statist.* **17**, 34–43.

Hill, A. B. (1950). *Principles of Medical Statistics* (5th edition). The Lancet.

Hooke, R. (1956). Symmetric functions of a two way array. *Ann. Math. Statist.* **27**, 55–79.

Irwin, J. O. & Kendall, M. G. (1944). Sampling moments of moments from a finite population. *Ann. Eugen.* **12**, 138–42.

Kaplan, E. (1952). Tensor notation and the sampling cumulants of k-statistics. *Biometrika*, **39**, 319–23.

Kendall, M. G. (1943). *The Advanced Theory of Statistics*, **1**. London: Charles Griffin and Co.

Kendall, M. G. & Stuart, A. (1958). *The Advanced Theory of Statistics*, **1**. London: Charles Griffin and Co.

Kerawala, S. M. & Hanafi, A. R. (1941, 1942). Tables of monomial symmetric functions. *Proc. Nat. Acad. Sci. Ind.* **11**, 51–63; **2**, 81–96.

Kerawala, S. M. & Hanafi, A. R. (1948). Tables of monomial symmetric functions. *Sankhyā*, **8**, 345–59.

Laplace, P-S. (1812). *Théorie Analytique des Probabilités.* Paris: Courcier.

Levine, J. (1959). Monomial-monomial symmetric function tables. *Biometrika*, **46**, 205–13.

McMahon, J. (1894). On the general term in the reversion of series. *Bull. Amer. Math. Soc.* **3**, 170–2.

McMahon, P. A. (1884). Symmetric functions of the thirteenthic. *Amer. J. Math.* **6**, 289–300.

McMahon, P. A. (1915–16). *Combinatory Analysis*, **1** and **2**. Cambridge University Press. **1**, **2** (1960). New York: Chelsea Publishing Co.

Peters, J. (1922). *Zehnstellige Logarithmen der Zahlen von 1 bis 100,000 nebst einem Anhang mathematischer Tafeln.* Berlin: Reichsamt für Landesaufnahme.

Royal Society Mathematical Tables.

   Miller, J. C. P. (1954). **3.** *Tables of Binomial Coefficients.*

   Gupta, H., Gwyther, C. E. & Miller, J. C. P. (1958). **4.** *Tables of Partitions.*

Stirling, J. (1730). *Methodus Differentialis.* London: G. Strahan.

Sukhatme, P. V. (1938). On bipartitional functions. *Phil. Trans.* A, **237**, 375–409.

Sukhatme, P. V. (1943). Moments and product-moments of moment-statistics for samples of the finite and infinite populations. *Sankhyā*, **6**, 363–82.

Tukey, J. W. (1950). Some sampling simplified. *J. Amer. Statist. Ass.* **45**, 501–19.

Van Rest, E. D. (1937). Examples of statistical methods in forest products research. *J. R. Statist. Soc.* (Suppl.), **4**, 184–203.

Verdooren, L. R. (1963). Extended tables of critical values for Wilcoxon's test statistic. *Biometrika*, **50**, 177–86.

Whittaker, E. T. & Watson, G. N. (1946). *Modern Analysis* (4th edition). Cambridge University Press.

Whitworth, W. A. (1878). *Choice and Chance* (3rd edition). Cambridge.

Wishart, J. (1952). Moment coefficients of the k-statistics in samples from a finite population. *Biometrika*, **39**, 1–16.

Ziaud-Din, M. (1954, 1959). Expression of the k-statistics in terms of power sums and sample moments. *Ann. Math. Statist.* **25**, 800–3, **30**, 825–7.

An additional bibliography will be found in the Appendix on pp. 276–8 below.

TABLES

Table 1·1. *Augmented monomial symmetric functions in terms of the power sums (s-functions) and conversely (up to and including weight 12)*

Note that the final digit (or digits) in the subheadings indicates the weight, e.g. the section for weight 5 is described as Table 1·1·5.

To express an augmented monomial symmetric in terms of s-functions the appropriate coefficients are found by reading vertically downwards up to and including the diagonal, the unit entry in which is shown in bold type, e.g. from Table 1·1·5

$$[31^2] = 2(5) - 2(4)(1) - (3)(2) + (3)(1)^2.$$

Conversely, to express a product of s-functions in terms of augmented monomial symmetrics the appropriate coefficients are found by reading horizontally up to and including the bold-faced diagonal unit. Thus again from Table 1·1·5

$$(2)^2(1) = [5] + [41] + 2[32] + [2^2 1].$$

Table 1·1·2

$w=2$	$[2]$	$[1^2]$
(2)	1	−1
$(1)^2$	1	1

Table 1·1·3

$w=3$	$[3]$	$[21]$	$[1^3]$
(3)	1	−1	2
$(2)(1)$	1	1	−3
$(1)^3$	1	3	1

Table 1·1·4

$w=4$	$[4]$	$[31]$	$[2^2]$	$[21^2]$	$[1^4]$
(4)	1	−1	−1	2	−6
$(3)(1)$	1	1	.	−2	8
$(2)^2$	1	.	1	−1	3
$(2)(1)^2$	1	2	1	1	−6
$(1)^4$	1	4	3	6	1

Table 1·1·5

$w=5$	$[5]$	$[41]$	$[32]$	$[31^2]$	$[2^2 1]$	$[21^3]$	$[1^5]$
(5)	1	−1	−1	2	2	−6	24
$(4)(1)$	1	1	.	−2	−1	6	−30
$(3)(2)$	1	.	1	−1	−2	5	−20
$(3)(1)^2$	1	2	1	1	.	−3	20
$(2)^2(1)$	1	1	2	.	1	−3	15
$(2)(1)^3$	1	3	4	3	3	1	−10
$(1)^5$	1	5	10	10	15	10	1

Table 1·1·6

$w=6$	$[6]$	$[51]$	$[42]$	$[41^2]$	$[3^2]$	$[321]$	$[31^3]$	$[2^3]$	$[2^2 1^2]$	$[21^4]$	$[1^6]$
(6)	1	−1	−1	2	−1	2	−6	2	−6	24	−120
$(5)(1)$	1	1	.	−2	.	−1	6	.	4	−24	144
$(4)(2)$	1	.	1	−1	.	−1	3	−3	5	−18	90
$(4)(1)^2$	1	2	1	1	.	.	−3	.	−1	12	−90
$(3)^2$	1	.	.	.	1	−1	2	.	2	−8	40
$(3)(2)(1)$	1	1	1	.	1	1	−3	.	−4	20	−120
$(3)(1)^3$	1	3	3	3	1	3	1	.	.	−4	40
$(2)^3$	1	.	3	.	.	.	.	1	−1	3	−15
$(2)^2(1)^2$	1	2	3	1	2	4	.	1	1	−6	45
$(2)(1)^4$	1	4	7	6	4	16	4	3	6	1	−15
$(1)^6$	1	6	15	15	10	60	20	15	45	15	1

Table 1·1·7

$w=7$	[7]	[61]	[52]	$[51^2]$	[43]	[421]	$[41^3]$	$[3^21]$	$[32^2]$	$[321^2]$	$[31^4]$	$[2^31]$	$[2^21^3]$	$[21^5]$	$[1^7]$
(7)	1	−1	−1	2	−1	2	−6	2	2	−6	24	−6	24	−120	720
(6)(1)	1	1	.	−2	.	−1	6	−1	.	4	−24	2	−18	120	−840
(5)(2)	1	.	1	−1	.	−1	3	.	−2	3	−12	6	−18	84	−504
(5)(1)^2	1	2	1	1	.	.	−3	.	.	−1	12	.	6	−60	504
(4)(3)	1	.	.	.	1	−1	2	−2	−1	4	−14	3	−14	70	−420
(4)(2)(1)	1	1	1	.	1	1	−3	.	.	−2	12	−3	15	−90	630
(4)(1)^3	1	3	3	3	1	3	1	.	.	.	−4	.	−1	20	−210
(3)^2(1)	1	1	.	.	2	.	.	1	.	.	−2	8	6	−40	280
(3)(2)^2	1	.	2	.	1	.	.	.	1	−1	3	−3	8	−35	210
(3)(2)(1)^2	1	2	2	1	3	2	.	.	2	1	−6	.	−6	50	−420
(3)(1)^4	1	4	6	6	5	12	4	4	3	6	1	.	.	−5	70
(2)^3(1)	1	3	3	.	3	3	.	3	3	.	.	1	−3	15	−105
(2)^2(1)^3	1	3	5	3	7	9	1	6	7	6	.	3	1	−10	105
(2)(1)^5	1	5	11	10	15	35	10	20	25	40	5	15	10	1	−21
(1)^7	1	7	21	21	35	105	35	70	105	210	35	105	105	21	1

Table 1·1·8

$w=8$ (i)	[8]	[71]	[62]	$[61^2]$	[53]	[521]	$[51^3]$	$[4^2]$	[431]	$[42^2]$	$[421^2]$	$[41^4]$
(8)	1	−1	−1	2	−1	2	−6	−1	2	2	−6	24
(7)(1)	1	1	.	−2	.	−1	6	.	−1	.	4	−24
(6)(2)	1	.	1	−1	.	−1	3	.	.	−2	3	−12
(6)(1)^2	1	2	1	1	.	.	−3	.	.	.	−1	12
(5)(3)	1	.	.	.	1	−1	2	.	−1	.	2	−8
(5)(2)(1)	1	1	1	.	1	1	−3	.	.	.	−2	12
(5)(1)^3	1	3	3	3	1	3	1	.	.	.	.	−4
(4)^2	1	.	.	.	.	.	.	1	−1	−1	2	−6
(4)(3)(1)	1	1	.	.	1	.	.	1	1	.	−2	8
(4)(2)^2	1	.	2	.	.	.	.	1	.	1	−1	3
(4)(2)(1)^2	1	2	2	1	2	2	.	1	2	1	1	−6
(4)(1)^4	1	4	6	6	4	12	4	1	4	3	6	1
(3)^2(2)	1	.	1	.	2	.	.	.	.	.	.	.
(3)^2(1)^2	1	2	1	1	2	.	.	2	4	.	.	.
(3)(2)^2(1)	1	1	2	.	3	2	.	1	1	1	.	.
(3)(2)(1)^3	1	3	4	3	5	6	1	3	9	3	3	.
(3)(1)^5	1	5	10	10	11	30	10	5	25	15	30	5
(2)^4	1	.	4	.	.	.	.	3	.	6	.	.
(2)^3(1)^2	1	2	4	1	6	6	.	3	6	6	3	.
(2)^2(1)^4	1	4	8	6	12	20	4	7	28	16	18	1
(2)(1)^6	1	6	16	15	26	66	20	15	90	60	105	15
(1)^8	1	8	28	28	56	168	56	35	280	210	420	70

$w=8$ (ii)	$[3^22]$	$[3^21^2]$	$[32^21]$	$[321^3]$	$[31^5]$	$[2^4]$	$[2^31^2]$	$[2^21^4]$	$[21^6]$	$[1^8]$
(8)	2	−6	−6	24	−120	−6	24	−120	720	−5040
(7)(1)	.	4	2	−18	120	.	−12	96	−720	5760
(6)(2)	−1	1	4	−12	60	8	−20	84	−480	3360
(6)(1)^2	.	−1	.	6	−60	.	2	−36	360	−3360
(5)(3)	−2	4	4	−14	64	.	−12	64	−384	2688
(5)(2)(1)	.	.	−2	9	−60	.	12	−72	504	−4032
(5)(1)^3	.	.	.	−1	20	.	.	8	−120	1344
(4)^2	.	.	1	−6	30	3	−6	30	−180	1260
(4)(3)(1)	.	−4	−1	12	−70	.	6	−56	420	−3360
(4)(2)^2	.	.	−1	3	−15	−6	9	−33	180	−1260
(4)(2)(1)^2	.	.	.	−3	30	.	−3	30	−270	2520
(4)(1)^4	.	.	.	.	−5	.	.	−1	30	−420
(3)^2(2)	1	−1	−2	5	−20	.	6	−28	160	−1120
(3)^2(1)^2	1	1	1	−3	20	.	.	12	−120	1120
(3)(2)^2(1)	2	.	1	−3	15	.	−6	32	−210	1680
(3)(2)(1)^3	4	3	3	1	−10	.	.	−8	100	−1120
(3)(1)^5	10	10	15	10	1	.	.	.	−6	112
(2)^4	.	.	.	.	.	1	−1	3	−15	105
(2)^3(1)^2	6	.	6	.	.	1	1	−6	45	−420
(2)^2(1)^4	20	12	28	8	.	3	6	1	−15	210
(2)(1)^6	70	60	150	80	6	15	45	15	1	−28
(1)^8	280	280	840	560	56	105	420	210	28	1

Table 1·1·9

$w = 9$ (i)	$[9]$	$[81]$	$[72]$	$[71^2]$	$[63]$	$[621]$	$[61^3]$	$[54]$	$[531]$	$[52^2]$	$[521^2]$	$[51^4]$	$[4^21]$	$[432]$	$[431^2]$	$[42^21]$	$[421^3]$
(9)	*1*	−1	−1	2	−1	2	−6	−1	2	2	−6	24	2	2	−6	−6	24
$(8)(1)$	1	*1*	·	−2	·	−1	6	·	−1	·	4	−24	−1	·	4	2	−18
$(7)(2)$	1	·	*1*	−1	·	−1	3	·	·	−2	3	−12	·	−1	1	4	−12
$(7)(1)^2$	1	2	1	*1*	·	·	−3	·	·	·	−1	12	·	·	−1	·	6
$(6)(3)$	1	·	·	·	*1*	−1	2	·	−1	·	2	−8	·	−1	2	·	−8
$(6)(2)(1)$	1	1	1	·	1	*1*	−3	·	·	·	−2	12	·	·	·	2	9
$(6)(1)^3$	1	3	3	3	1	3	*1*	·	·	·	·	−4	·	·	·	·	−1
$(5)(4)$	1	·	·	·	·	·	·	*1*	−1	−1	·	−6	−2	−1	4	3	−12
$(5)(3)(1)$	1	1	·	·	1	·	·	·	*1*	1	−2	8	·	·	−2	·	6
$(5)(2)^2$	1	·	2	·	·	·	·	·	1	*1*	1	−1	3	·	·	−1	3
$(5)(2)(1)^2$	1	2	2	1	2	2	·	1	2	1	*1*	−6	·	·	·	·	−3
$(5)(1)^4$	1	4	6	6	4	12	4	1	1	4	3	6	*1*	·	·	·	−3
$(4)^2(1)$	1	1	·	·	·	·	·	·	2	·	·	·	2	·	·	−2	6
$(4)(3)(2)$	1	·	1	·	1	·	·	·	1	·	·	·	·	*1*	·	−1	5
$(4)(3)(1)^2$	1	2	1	2	1	·	·	·	3	2	·	·	·	2	*1*	−2	−5
$(4)(2)^2(1)$	1	1	2	·	2	·	·	·	2	·	1	·	·	1	2	*1*	−3
$(4)(2)(1)^3$	1	3	4	3	4	6	1	·	4	6	3	3	·	3	4	3	*1*
$(4)(1)^5$	1	5	10	10	10	30	10	6	20	20	15	30	5	5	10	10	10
$(3)^3$	1	·	·	·	3	·	·	·	·	·	·	·	·	·	·	·	·
$(3)^2(2)(1)$	1	1	1	·	3	1	·	·	2	2	·	·	·	2	·	·	·
$(3)^2(1)^3$	1	3	3	3	3	3	1	6	6	·	·	·	6	6	6	6	·
$(3)(2)^3$	1	·	3	·	·	·	·	3	3	3	·	·	·	3	3	·	·
$(3)(2)^2(1)^2$	1	2	3	1	5	4	·	5	6	3	2	·	·	2	7	1	4
$(3)(2)(1)^4$	1	4	7	6	9	16	4	11	20	9	12	1	·	23	18	12	4
$(3)(1)^6$	1	6	15	15	21	60	20	21	66	45	90	15	30	75	75	90	60
$(2)^4(1)$	1	1	4	·	4	4	·	6	·	6	·	·	·	3	12	6	·
$(2)^3(1)^3$	1	3	6	3	10	12	1	12	18	12	9	·	·	9	30	18	3
$(2)^2(1)^5$	1	5	12	10	20	40	10	26	60	36	50	5	35	100	70	80	30
$(2)(1)^7$	1	7	22	21	42	112	35	56	182	126	231	35	105	350	315	420	245
$(1)^9$	1	9	36	36	84	252	84	126	504	378	756	126	315	1260	1260	1890	1260

$w = 9$ (ii)	$[41^5]$	$[3^3]$	$[3^221]$	$[3^21^3]$	$[32^3]$	$[32^21^2]$	$[321^4]$	$[31^6]$	$[2^41]$	$[2^31^3]$	$[2^21^5]$	$[21^7]$	$[1^9]$
(9)	−120	2	−6	24	−6	24	−120	720	24	−120	720	−5040	40320
$(8)(1)$	120	·	2	−18	·	−12	96	−720	−6	72	−600	5040	−45360
$(7)(2)$	60	·	2	−6	6	−14	60	−360	−24	90	−480	3240	−25920
$(7)(1)^2$	−60	·	·	6	·	2	−36	360	·	−18	240	−2520	25920
$(6)(3)$	40	−3	5	−14	2	−14	64	−360	−8	58	−360	2520	−20160
$(6)(2)(1)$	−60	·	−1	3	·	8	−48	360	8	−60	420	−3360	30240
$(6)(1)^3$	20	·	·	−1	·	·	8	−120	·	2	−60	840	−10080
$(5)(4)$	54	·	2	−12	3	−10	54	−324	−12	54	−324	2268	−18144
$(5)(3)(1)$	−40	·	−2	12	·	8	−56	384	·	−36	320	−2688	24192
$(5)(2)^2$	−15	·	·	·	−3	4	−15	90	12	−36	174	−1134	9072
$(5)(2)(1)^2$	30	·	·	·	·	−2	18	−180	·	18	−180	1764	−18144
$(5)(1)^4$	−5	·	·	·	·	·	−1	30	·	·	10	−210	3024
$(4)^2(1)$	−30	·	·	6	·	2	−24	180	3	−18	150	−1260	11340
$(4)(3)(2)$	−20	·	−2	6	−3	9	−38	210	12	−51	280	−1890	15120
$(4)(3)(1)^2$	20	·	·	−6	·	−1	24	−210	·	9	−140	1470	−15120
$(4)(2)^2(1)$	15	·	·	·	·	−2	12	−90	−6	27	−165	1260	−11340
$(4)(2)(1)^3$	−10	·	·	·	·	·	−4	60	·	−3	50	−630	7560
$(4)(1)^5$	*1*	·	·	·	·	·	·	−6	·	·	−1	42	−756
$(3)^3$	·	*1*	−1	2	·	2	−8	40	·	−6	40	−280	2240
$(3)^2(2)(1)$	·	1	*1*	−3	·	−4	20	−120	·	18	−140	1120	−10080
$(3)^2(1)^3$	·	1	3	*1*	·	·	−4	40	·	·	20	−280	3360
$(3)(2)^3$	·	·	·	·	*1*	−1	3	−15	−4	11	−50	315	−2520
$(3)(2)^2(1)^2$	·	2	4	1	1	*1*	−6	45	−9	·	80	−735	7560
$(3)(2)(1)^4$	·	4	16	4	3	6	*1*	−15	−15	·	−10	175	−2520
$(3)(1)^6$	6	10	60	20	15	45	15	*1*	·	−3	15	−7	168
$(2)^4(1)$	·	·	·	·	4	·	·	·	*1*	−3	15	−105	945
$(2)^3(1)^3$	·	6	18	10	10	9	9	·	3	*1*	−10	105	−1260
$(2)^2(1)^5$	1	20	100	20	40	70	10	10	15	10	*1*	−21	378
$(2)(1)^7$	21	70	490	140	210	525	140	7	105	105	21	*1*	−36
$(1)^9$	126	280	2520	840	1260	3780	1260	84	945	1260	378	36	*1*

Table 1·1·10

$w = 10$ (i)	$[10]$	$[91]$	$[82]$	$[81^2]$	$[73]$	$[721]$	$[71^3]$	$[64]$	$[631]$	$[62^2]$	$[621^2]$	$[61^4]$
(10)	1	−1	−1	2	−1	2	−6	−1	2	2	−6	24
$(9)(1)$	1	1	.	−2	.	−1	6	.	−1	.	4	−24
$(8)(2)$	1	.	1	−1	.	−1	3	.	.	−2	3	−12
$(8)(1)^2$	1	2	1	1	.	.	−3	.	.	.	−1	12
$(7)(3)$	1	.	.	.	1	−1	2	.	−1	.	2	−8
$(7)(2)(1)$	1	1	1	.	1	1	−3	.	.	.	−2	12
$(7)(1)^3$	1	3	3	3	1	3	1	.	.	.	.	−4
$(6)(4)$	1	.	.	.	.	.	.	1	−1	−1	2	−6
$(6)(3)(1)$	1	1	.	2	.	1	.	.	1	1	−2	8
$(6)(2)^2$	1	.	2	.	.	.	.	.	1	.	1	3
$(6)(2)(1)^2$	1	2	2	1	2	2	.	1	2	1	1	−6
$(6)(1)^4$	1	4	6	6	4	12	4	1	4	3	6	1
$(5)^2$	1	.	.	.	.	.	.	.	.	.	.	.
$(5)(4)(1)$	1	1	.	.	.	.	.	1	.	.	.	.
$(5)(3)(2)$	1	.	1	.	1	.	.	.	2	.	.	.
$(5)(3)(1)^2$	1	2	1	1	1	.	.	2	2	.	.	.
$(5)(2)^2(1)$	1	1	2	.	2	2	.	1	.	1	.	.
$(5)(2)(1)^3$	1	3	4	3	4	6	1	3	6	3	3	.
$(5)(1)^5$	1	5	10	10	10	30	10	5	20	15	30	5
$(4)^2(2)$	1	.	1	.	.	.	.	2	.	.	.	.
$(4)^2(1)^2$	1	2	1	1	.	.	.	2	.	.	.	.
$(4)(3)^2$	1	.	.	.	2	.	.	1	1	.	.	.
$(4)(3)(2)(1)$	1	1	1	.	2	1	.	2	2	.	.	.
$(4)(3)(1)^3$	1	3	3	3	2	3	1	4	3	.	.	.
$(4)(2)^3$	1	.	3	.	.	.	.	4	.	3	.	.
$(4)(2)^2(1)^2$	1	2	3	1	4	4	.	4	4	3	2	.
$(4)(2)(1)^4$	1	4	7	6	8	16	4	8	16	9	12	1
$(4)(1)^6$	1	6	15	15	20	60	20	16	60	45	90	15
$(3)^3(1)$	1	1	.	.	3	.	.	3	3	.	.	.
$(3)^2(2)^2$	1	.	2	.	2	.	.	1	.	1	.	.
$(3)^2(2)(1)^2$	1	2	2	1	4	2	.	5	6	1	1	.
$(3)^2(1)^4$	1	4	6	6	6	12	4	9	12	3	6	1
$(3)(2)^3(1)$	1	1	3	.	4	3	.	4	1	3	.	.
$(3)(2)^2(1)^3$	1	3	5	3	8	9	1	10	15	7	6	.
$(3)(2)(1)^5$	1	5	11	10	16	35	10	20	45	25	40	5
$(3)(1)^7$	1	7	21	21	36	105	35	42	147	105	210	35
$(2)^5$	1	.	5	.	.	.	.	10	.	10	.	.
$(2)^4(1)^2$	1	2	5	1	8	8	.	10	8	10	4	.
$(2)^3(1)^4$	1	4	9	6	16	24	4	22	40	24	24	1
$(2)^2(1)^6$	1	6	17	15	32	72	20	46	120	76	120	15
$(2)(1)^8$	1	8	29	28	64	176	56	98	336	238	448	70
$(1)^{10}$	1	10	45	45	120	360	120	210	840	630	1260	210

$w = 10$ (ii)	$[5^2]$	$[541]$	$[532]$	$[531^2]$	$[52^21]$	$[521^3]$	$[51^5]$	$[4^22]$	$[4^21^2]$	$[43^2]$	$[4321]$
(10)	−1	2	2	−6	−6	24	−120	2	−6	2	−6
$(9)(1)$	.	−1	.	4	2	−18	120	.	4	.	2
$(8)(2)$	.	.	−1	1	4	−12	60	−1	1	.	2
$(8)(1)^2$	.	.	.	−1	−1	6	−60	.	−1	.	.
$(7)(3)$	.	.	−1	2	2	−8	40	.	.	−2	3
$(7)(2)(1)$	.	.	.	.	−2	9	−60	.	.	.	−1
$(7)(1)^3$	.	.	.	.	.	−1	20	.	.	.	.
$(6)(4)$	.	−1	.	2	1	−6	30	−2	4	−1	3
$(6)(3)(1)$	.	.	.	−2	.	6	−40	.	.	.	−1
$(6)(2)^2$	.	.	.	.	−1	3	−15	.	.	.	.
$(6)(2)(1)^2$	.	.	.	.	.	−3	30	.	.	.	.
$(6)(1)^4$	.	.	.	.	.	.	−5	.	.	.	.
$(5)^2$	1	−1	−1	2	2	−6	24	.	2	.	1
$(5)(4)(1)$	1	1	.	−2	−1	6	−30	.	−4	.	−1
$(5)(3)(2)$	1	.	1	−1	−2	5	−20	.	.	.	−1
$(5)(3)(1)^2$	1	2	1	1	.	−3	20	.	.	.	.
$(5)(2)^2(1)$	1	1	2	.	1	−3	15	.	.	.	.
$(5)(2)(1)^3$	1	3	4	3	3	1	−10	.	.	.	.
$(5)(1)^5$	1	5	10	10	15	10	1	.	.	.	.
$(4)^2(2)$	.	.	.	.	.	.	.	1	−1	.	−1
$(4)^2(1)^2$	2	4	.	.	.	.	.	1	1	.	−1
$(4)(3)^2$	1	1	.	.	.	.	.	1	.	1	1
$(4)(3)(2)(1)$	1	1	1	.	.	.	.	1	.	1	1
$(4)(3)(1)^3$	3	9	3	3	.	.	.	3	3	1	3
$(4)(2)^3$	.	.	.	.	.	.	.	3	.	.	.
$(4)(2)^2(1)^2$	2	4	4	.	2	.	.	3	1	2	4
$(4)(2)(1)^4$	4	16	16	12	12	4	.	7	6	4	16
$(4)(1)^6$	6	36	60	60	90	60	6	15	15	10	60
$(3)^3(1)$	.	.	.	.	.	.	.	.	.	3	.
$(3)^2(2)^2$	2	.	4	.	.	.	.	.	.	1	.
$(3)^2(2)(1)^2$	2	4	4	2	.	.	.	2	.	5	4
$(3)^2(1)^4$	6	24	12	12	.	.	.	12	12	9	24
$(3)(2)^3(1)$	3	3	9	.	3	.	.	9	3	3	3
$(3)(2)^2(1)^3$	5	15	19	9	9	2	.	9	3	13	21
$(3)(2)(1)^5$	11	55	61	50	45	20	1	35	30	35	115
$(3)(1)^7$	21	147	231	231	315	210	21	105	105	105	525
$(2)^5$	.	.	.	.	.	.	.	15	.	.	.
$(2)^4(1)^2$	6	12	24	.	12	.	.	15	3	12	24
$(2)^3(1)^4$	12	48	72	36	48	12	.	39	18	48	120
$(2)^2(1)^6$	26	156	232	180	216	100	6	135	105	160	600
$(2)(1)^8$	56	448	784	728	1008	616	56	455	420	560	2800
$(1)^{10}$	126	1260	2520	2520	3780	2520	252	1575	1575	2100	12600

Table 1·1·10 (cont.)

$w = 10$ (iii)	$[431^3]$	$[42^3]$	$[42^21^2]$	$[421^4]$	$[41^6]$	$[3^31]$	$[3^22^2]$	$[3^221^2]$	$[3^21^4]$	$[32^31]$
(10)	24	−6	24	−120	720	−6	−6	24	−120	24
$(9)(1)$	−18	.	−12	96	−720	2	.	−12	96	−6
$(8)(2)$	−6	6	−14	60	−360	.	4	−8	36	−18
$(8)(1)^2$	6	.	2	−36	360	.	.	2	−36	.
$(7)(3)$	−8	.	−8	40	−240	6	4	−16	64	−12
$(7)(2)(1)$	3	.	8	−48	360	.	.	4	−24	6
$(7)(1)^3$	−1	.	.	8	−120	.	.	.	8	.
$(6)(4)$	−12	5	−12	54	−300	3	1	−10	54	−8
$(6)(3)(1)$	6	.	4	−32	240	−3	.	10	−56	2
$(6)(2)^2$	.	−3	4	−15	90	.	−1	1	−3	6
$(6)(2)(1)^2$	.	.	−2	18	−180	.	.	−1	6	.
$(6)(1)^4$	.	.	.	−1	30	.	.	.	−1	.
$(5)^2$	−6	.	−4	24	−144	.	2	−4	24	−6
$(5)(4)(1)$	12	.	6	−48	324	.	.	4	−48	3
$(5)(3)(2)$	3	.	4	−20	120	.	−4	6	−24	12
$(5)(3)(1)^2$	−3	.	.	12	−120	.	.	−2	24	.
$(5)(2)^2(1)$	.	.	−2	12	−90	.	.	.	.	−3
$(5)(2)(1)^3$	.	.	.	−4	60	.	.	.	.	.
$(5)(1)^5$	.	.	.	.	−6	.	.	.	.	.
$(4)^2(2)$	3	−3	5	−18	90	.	.	2	−12	3
$(4)^2(1)^2$	−3	.	−1	12	−90	.	.	.	12	.
$(4)(3)^2$	2	.	2	−8	40	−3	−1	6	−22	3
$(4)(3)(2)(1)$	−3	.	−4	20	−120	.	.	−4	24	−3
$(4)(3)(1)^3$	1	.	.	−4	40	.	.	.	−8	.
$(4)(2)^3$	.	1	−1	3	−15	.	.	.	.	−1
$(4)(2)^2(1)^2$	.	1	1	−6	45	.	.	.	.	.
$(4)(2)(1)^4$	4	3	6	1	−15	.	.	.	.	.
$(4)(1)^6$	20	15	45	15	1	.	.	.	.	.
$(3)^3(1)$	.	.	.	.	.	1	.	−2	8	.
$(3)^2(2)^2$	.	.	.	.	.	.	1	−1	3	−3
$(3)^2(2)(1)^2$	8	.	.	.	.	2	1	1	−6	.
$(3)^2(1)^4$	.	.	.	.	.	4	3	6	1	.
$(3)(2)^3(1)$	.	1	3	.	.	.	3	.	.	1
$(3)(2)^2(1)^3$	1	3	3	.	.	6	7	6	.	3
$(3)(2)(1)^5$	30	15	30	5	.	20	25	40	5	15
$(3)(1)^7$	175	105	315	105	7	70	105	210	35	105
$(2)^5$	.	10	.	.	.	.	.	.	.	.
$(2)^4(1)^2$	.	10	6	.	.	.	12	.	.	8
$(2)^3(1)^4$	12	28	36	3	.	24	48	36	.	40
$(2)^2(1)^6$	140	120	240	45	1	120	220	300	30	240
$(2)(1)^8$	840	630	1680	490	28	560	1120	1960	280	1680
$(1)^{10}$	4200	3150	9450	3150	210	2800	6300	12600	2100	12600

$w = 10$ (iv)	$[32^21^3]$	$[321^5]$	$[31^7]$	$[2^5]$	$[2^41^2]$	$[2^31^4]$	$[2^21^6]$	$[21^8]$	$[1^{10}]$
(10)	−120	720	−5040	24	−120	720	−5040	40320	−362880
$(9)(1)$	72	−600	5040	.	48	−480	4320	−40320	403200
$(8)(2)$	66	−360	2520	−30	102	−504	3240	−25200	226800
$(8)(1)^2$	−18	240	−2520	.	−6	144	−1800	20160	−226800
$(7)(3)$	64	−360	2400	.	48	−336	2400	−19200	172800
$(7)(2)(1)$	−42	300	−2520	.	−48	360	−2880	25920	−259200
$(7)(1)^3$	2	−60	840	.	.	−24	480	−6720	86400
$(6)(4)$	48	−300	2100	−20	52	−300	2100	−16800	151200
$(6)(3)(1)$	−42	320	−2520	.	−16	232	−2160	20160	−201600
$(6)(2)^2$	−18	90	−630	20	−44	186	−1110	8400	−75600
$(6)(2)(1)^2$	12	−120	1260	.	8	−120	1260	−13440	151200
$(6)(1)^4$	.	10	−210	.	.	2	−90	1680	−25200
$(5)^2$	24	−144	1008	.	24	−144	1008	−8064	72576
$(5)(4)(1)$	−30	270	−2268	.	−24	216	−1944	18144	−181440
$(5)(3)(2)$	−40	204	−1344	.	−48	264	−1728	13440	−120960
$(5)(3)(1)^2$	12	−140	1344	.	.	−72	960	−10752	120960
$(5)(2)^2(1)$	12	−75	630	.	24	−144	1044	−9072	90720
$(5)(2)(1)^3$	−2	30	−420	.	.	24	−360	4704	−60480
$(5)(1)^5$	.	−1	42	.	.	.	12	−336	6048
$(4)^2(2)$	−15	90	−630	15	−27	126	−810	6300	−56700
$(4)^2(1)^2$	3	−60	630	.	3	−36	450	−5040	56700
$(4)(3)^2$	−20	110	−700	.	−12	96	−700	5600	−50400
$(4)(3)(2)(1)$	27	−190	1470	.	24	−204	1680	−15120	151200
$(4)(3)(1)^3$	−1	40	−490	.	.	12	−280	3920	−50400
$(4)(2)^3$	3	−15	105	−10	14	−51	285	−2100	18900
$(4)(2)^2(1)^2$	−3	30	−315	.	−6	54	−495	5040	−56700
$(4)(2)(1)^4$	.	−5	105	.	.	−3	75	−1260	18900
$(4)(1)^6$	.	.	−7	.	.	.	−1	56	−1260
$(3)^3(1)$	6	−40	280	.	.	−24	240	−2240	22400
$(3)^2(2)^2$	8	−35	210	.	12	−60	370	−2800	25200
$(3)^2(2)(1)^2$	−6	50	−420	.	.	36	−420	4480	−50400
$(3)^2(1)^4$	.	−5	70	.	.	.	30	−560	8400
$(3)(2)^3(1)$	−3	15	−105	.	−8	44	−300	2520	−25200
$(3)(2)^2(1)^3$	1	−10	105	.	.	−12	160	−1960	25200
$(3)(2)(1)^5$	10	1	−21	.	.	.	−12	280	−5040
$(3)(1)^7$	105	21	1	.	.	.	.	−8	240
$(2)^5$	.	.	.	1	−1	3	−15	105	−945
$(2)^4(1)^2$	.	.	.	1	1	−6	45	−420	4725
$(2)^3(1)^4$	12	.	.	3	6	1	−15	210	−3150
$(2)^2(1)^6$	140	12	.	15	45	15	1	−28	630
$(2)(1)^8$	1400	224	8	105	420	210	28	1	−45
$(1)^{10}$	12600	2520	120	945	4725	3150	630	45	1

Table 1·1·11

$w=11$ (i)	$[11]$	$[10,1]$	$[92]$	$[91^2]$	$[83]$	$[821]$	$[81^3]$	$[74]$	$[731]$	$[72^2]$	$[721^2]$	$[71^4]$	$[65]$	$[641]$	$[632]$
(11)	1	-1	-1	2	-1	2	-6	-1	2	2	-6	24	-1	2	2
$(10)(1)$	1	1	.	-2	.	-1	6	.	-1	.	4	-24	.	-1	.
$(9)(2)$	1	.	1	-1	.	-1	3	.	.	-2	3	-12	.	.	-1
$(9)(1)^2$	1	2	1	1	.	.	-3	.	.	.	-1	12	.	.	.
$(8)(3)$	1	.	.	.	1	.	-1	2	.	-1	2	-8	.	.	.
$(8)(2)(1)$	1	1	1	.	1	1	-3	.	-1	-1	-2	12	.	.	.
$(8)(1)^3$	1	3	3	3	1	3	1	.	-1	-1	-6	-4	.	-1	.
$(7)(4)$	1	.	.	.	.	.	.	1	1	-1	-2	-6	.	.	.
$(7)(3)(1)$	1	1	.	.	1	.	.	1	1	.	-2	8	.	-1	.
$(7)(2)^2$	1	.	2	.	.	.	.	1	.	1	-1	3	.	.	.
$(7)(2)(1)^3$	1	2	2	1	2	2	.	1	2	1	1	-6	.	.	.
$(7)(1)^4$	1	4	6	6	4	12	4	1	4	3	6	1	.	-1	-1
$(6)(5)$	1	.	.	.	.	.	.	1	.	.	.	.	1	1	-1
$(6)(4)(1)$	1	1	.	.	.	.	.	1	.	.	.	.	1	1	.
$(6)(3)(2)$	1	.	1	.	1	.	.	.	.	.	.	.	1	.	1
$(6)(3)(1)^2$	1	2	1	1	1	.	.	2	2	1	.	.	1	2	2
$(6)(2)^2(1)$	1	1	2	.	2	2	.	1	.	.	.	.	1	3	4
$(6)(2)(1)^3$	1	3	4	3	4	6	1	3	6	3	3	.	2	5	10
$(6)(1)^5$	1	5	10	10	10	30	10	5	20	15	30	5	2	.	.
$(5)^2(1)$	1	1	.	.	.	.	.	.	.	.	.	.	2	.	.
$(5)(4)(2)$	1	.	1	.	.	.	.	1	.	.	.	.	1	.	.
$(5)(4)(1)^2$	1	2	1	1	.	.	.	1	.	.	.	.	3	2	.
$(5)(3)^2$	1	.	.	.	2	.	.	.	.	.	.	.	1	.	1
$(5)(3)(2)(1)$	1	1	1	.	2	1	.	1	1	.	.	.	2	6	3
$(5)(3)(1)^3$	1	3	3	3	2	3	1	3	3	.	.	.	4	6	4
$(5)(2)^3$	1	.	3	.	.	.	.	3	.	3	.	.	1	.	4
$(5)(2)^2(1)^2$	1	2	4	1	4	4	.	3	4	3	2	.	3	2	16
$(5)(2)(1)^4$	1	4	7	6	8	16	4	7	16	9	12	1	5	12	50
$(5)(1)^6$	1	6	15	15	20	60	20	15	60	45	90	15	7	30	60
$(4)^2(3)$	1	.	.	.	1	.	.	2	.	.	.	.	2	.	.
$(4)^2(2)(1)$	1	1	1	.	1	1	.	2	.	.	.	.	2	2	.
$(4)^2(1)^3$	1	3	3	3	2	3	1	2	.	.	.	.	6	6	.
$(4)(3)^2(1)$	1	1	.	.	2	.	.	3	2	1	.	.	1	1	2
$(4)(3)(2)^2$	1	.	2	.	1	.	.	2	.	1	.	.	2	.	2
$(4)(3)(2)(1)^2$	1	2	2	1	3	2	.	4	4	1	1	.	4	4	6
$(4)(3)(1)^4$	1	4	6	6	5	12	4	6	8	3	6	1	10	16	6
$(4)(2)^3(1)$	1	1	3	.	3	3	.	4	.	3	.	.	4	12	14
$(4)(2)^2(1)^3$	1	3	5	3	7	9	1	8	12	7	6	.	10	40	50
$(4)(2)(1)^5$	1	5	11	10	15	35	10	16	40	25	40	5	16	40	50
$(4)(1)^7$	1	7	21	21	35	105	35	36	140	105	210	35	28	112	210
$(3)^3(2)$	1	.	1	.	3	.	.	6	6	.	.	.	3	6	3
$(3)^3(1)^2$	1	2	1	1	3	.	.	3	2	1	.	.	3	1	3
$(3)^2(2)^2(1)$	1	1	2	.	4	2	.	6	12	3	3	.	5	15	6
$(3)^2(2)(1)^3$	1	3	4	3	6	6	1	9	12	3	3	.	9	45	30
$(3)^2(1)^5$	1	5	10	10	12	30	10	15	30	15	30	5	21	45	30
$(3)(2)^4$	1	.	4	.	1	.	.	6	6	6	.	.	4	8	4
$(3)(2)^3(1)^2$	1	2	4	1	7	6	.	8	8	6	3	.	10	8	16
$(3)(2)^2(1)^4$	1	4	8	6	13	20	4	18	32	16	18	1	20	40	48
$(3)(2)(1)^6$	1	6	16	15	27	66	20	36	96	60	105	15	42	120	156
$(3)(1)^8$	1	8	28	28	57	168	56	78	288	210	420	70	84	336	588
$(2)^5(1)$	1	1	5	.	5	5	.	10	.	10	.	.	10	10	20
$(2)^4(1)^3$	1	3	7	3	13	15	.	18	24	18	12	.	22	30	52
$(2)^3(1)^5$	1	5	13	10	25	45	10	38	80	48	60	5	46	110	160
$(2)^2(1)^7$	1	7	23	21	49	119	35	78	224	148	252	35	98	322	504
$(2)(1)^9$	1	9	37	36	93	261	84	162	576	414	792	126	210	882	1596
$(1)^{11}$	1	11	55	55	165	495	165	330	1320	990	1980	330	462	2310	4620

Table 1·1·11 (cont.)

$w=11$ (ii)	$[631^2]$	$[62^21]$	$[621^3]$	$[61^6]$	$[5^21]$	$[542]$	$[541^2]$	$[53^2]$	$[5321]$	$[531^3]$	$[52^3]$	$[52^21^2]$	$[521^4]$	$[51^6]$	$[4^23]$
(11)	−6	−6	24	−120	2	2	−6	2	−6	24	−6	24	−120	720	2
$(10)(1)$	4	2	−18	120	−1	·	4	·	2	−18	−6	−12	96	−720	·
$(9)(2)$	1	4	−12	60	·	−1	1	·	2	−6	6	−14	60	−360	·
$(9)(1)^2$	−1	·	6	−60	·	·	−1	·	6	·	2	−36	360	·	·
$(8)(3)$	2	2	−8	40	·	·	·	−2	3	−8	·	2	−36	360	−1
$(8)(2)(1)$	·	−2	9	−60	·	·	·	·	−1	3	·	−8	40	−240	·
$(8)(1)^3$	·	·	·	20	·	·	·	·	−1	3	−1	8	−48	360	·
$(7)(4)$	2	1	−6	30	·	·	·	2	1	−6	·	−6	30	−180	−2
$(7)(3)(1)$	−2	·	6	−40	·	·	·	·	−1	6	3	4	−32	240	·
$(7)(2)^2$	·	−1	3	−15	·	·	·	·	·	·	−3	4	−15	90	·
$(7)(2)(1)^2$	·	·	−3	30	·	·	·	·	·	·	·	−2	18	−180	·
$(7)(1)^4$	·	·	−5	−5	·	·	·	·	·	·	·	−1	18	30	·
$(6)(5)$	2	2	−6	24	−2	−1	4	−1	3	−12	2	−10	48	−264	·
$(6)(4)(1)$	−2	−1	6	−30	·	·	−2	·	6	6	·	2	−24	180	·
$(6)(3)(2)$	−1	−2	5	−20	·	·	·	·	3	3	·	4	−20	120	·
$(6)(3)(1)^2$	1	·	−3	20	·	·	·	·	−3	−3	·	·	12	−120	·
$(6)(2)^2(1)$	·	1	−3	15	·	·	·	·	·	·	·	−2	12	−90	·
$(6)(2)(1)^3$	·	·	1	−10	·	·	·	·	·	·	·	·	−4	60	·
$(6)(1)^5$	10	15	10	1	·	·	·	·	·	·	·	·	·	−6	·
$(5)^2(1)$	·	·	·	·	1	·	−2	·	−1	6	6	4	−24	144	·
$(5)(4)(2)$	·	·	·	·	·	1	−1	·	−1	3	−3	5	−18	90	·
$(5)(4)(1)^2$	·	·	·	·	2	1	1	·	−3	−3	·	−1	12	−90	·
$(5)(3)^2$	·	·	·	·	·	·	·	1	−1	−1	2	·	−8	40	·
$(5)(3)(2)(1)$	·	·	·	·	1	1	·	1	−3	−3	−3	2	20	−120	·
$(5)(3)(1)^3$	·	·	·	·	1	3	3	1	3	1	·	−3	−4	40	·
$(5)(2)^3$	·	·	·	·	·	3	·	1	·	·	·	−1	3	−15	·
$(5)(2)^2(1)^2$	·	2	·	·	2	3	·	2	4	·	1	1	−6	45	·
$(5)(2)(1)^4$	12	12	4	·	4	7	6	4	16	4	3	6	6	−15	·
$(5)(1)^6$	60	90	60	6	6	15	15	10	60	20	15	45	15	1	·
$(4)^2(3)$	·	·	·	·	·	·	·	·	·	·	·	·	·	·	1
$(4)^2(2)(1)$	·	·	·	·	·	2	·	·	·	·	·	·	·	·	1
$(4)^2(1)^3$	·	·	·	·	6	6	·	·	·	·	·	·	·	·	1
$(4)(3)^2(1)$	·	·	·	·	·	6	6	1	·	·	·	·	·	·	2
$(4)(3)(2)^2$	·	·	·	·	·	2	·	·	·	·	·	·	·	·	1
$(4)(3)(2)(1)^2$	·	·	·	·	2	4	1	2	2	2	·	·	·	·	3
$(4)(3)(1)^4$	6	3	·	·	12	18	18	4	12	12	4	·	·	·	5
$(4)(2)^3(1)$	·	·	·	·	·	6	·	·	·	·	1	·	·	·	3
$(4)(2)^2(1)^3$	6	9	2	·	6	14	6	6	12	·	·	3	·	·	7
$(4)(2)(1)^5$	40	45	20	1	20	46	40	20	80	20	15	30	5	·	15
$(4)(1)^7$	210	315	210	21	42	126	126	70	420	140	105	315	105	7	35
$(3)^3(2)$	·	·	·	·	·	·	·	3	·	·	·	·	·	·	6
$(3)^3(1)^2$	3	·	·	·	·	·	·	3	·	·	·	·	·	·	2
$(3)^2(2)^2(1)$	·	·	·	·	2	·	·	5	4	·	·	·	·	·	12
$(3)^2(2)(1)^3$	9	3	1	·	6	12	6	9	12	2	2	·	·	·	30
$(3)^2(1)^5$	30	15	10	1	30	60	60	21	60	20	20	·	·	·	12
$(3)(2)^4$	·	·	·	·	·	12	·	·	·	·	4	·	·	·	3
$(3)(2)^3(1)^2$	1	6	8	·	6	18	3	12	18	·	4	18	3	·	9
$(3)(2)^2(1)^4$	30	28	80	·	20	52	30	32	76	12	12	18	18	·	35
$(3)(2)(1)^6$	135	150	80	6	66	186	165	96	366	100	60	1260	135	·	105
$(3)(1)^8$	588	840	560	56	168	588	588	336	1848	616	420	1260	420	28	315
$(2)^5(1)$	·	10	·	·	·	30	18	·	·	·	10	·	·	·	15
$(2)^4(1)^3$	12	30	4	·	18	66	18	36	72	·	22	18	18	·	39
$(2)^3(1)^5$	100	120	40	1	60	198	120	120	360	60	76	120	15	·	135
$(2)^2(1)^7$	420	532	280	21	182	658	546	392	1624	420	336	756	175	·	455
$(2)(1)^9$	1512	2142	1344	126	504	2142	2016	1344	7056	2184	1638	4536	1386	84	1575
$(1)^{11}$	4620	6930	4620	462	1386	6930	6930	4620	27720	9240	6930	20790	6930	462	5775

Table 1·1·11 (cont.)

w = 11 (iii)	[4²21]	[4²1³]	[43²1]	[432²]	[4321²]	[431⁴]	[42³1]	[42²1³]	[421⁵]	[41⁷]	[3³2]	[3³1²]	[3²2²1]	[3²21³]	[3²1⁵]
(11)	−6	24	−6	−6	24	−120	24	−120	720	−5040	−6	24	24	−120	720
(10)(1)	2	−18	2	.	−12	96	−6	72	−600	5040	.	−12	−6	72	−600
(9)(2)	2	−6	.	4	−8	36	−18	66	−360	2520	2	−2	−12	42	−240
(9)(1)²	.	6	.	.	2	−36	.	−18	240	−2520	.	2	.	−18	240
(8)(3)	1	−2	4	2	−10	40	−6	40	−240	1680	6	−18	−16	70	−360
(8)(2)(1)	−1	3	.	.	4	−24	6	−42	300	−2520	.	.	4	−24	180
(8)(1)³	.	−1	.	.	.	8	.	2	−60	840	.	.	2	.	−60
(7)(4)	4	−12	4	3	−12	54	−12	54	−300	1980	.	−12	−6	48	−300
(7)(3)(1)	.	.	−2	.	6	−32	.	−24	200	−1680	.	12	4	−48	320
(7)(2)²	.	.	.	−1	1	−3	6	−18	90	−630	.	.	2	−6	30
(7)(2)(1)²	.	.	.	.	−1	6	.	12	−120	1260	.	.	.	6	−60
(7)(1)⁴	.	.	.	.	.	−1	.	.	10	−210	.	.	.	.	10
(6)(5)	2	−12	1	2	−8	48	−8	42	−264	1848	3	−6	−10	42	−264
(6)(4)(1)	−2	12	−1	.	6	−48	5	−36	270	−2100	.	6	1	−30	270
(6)(3)(2)	.	.	.	−2	3	−12	6	−22	120	−840	−3	3	10	−29	140
(6)(3)(1)²	.	.	.	.	−1	12	.	6	−80	840	.	−3	.	15	−140
(6)(2)²(1)	.	.	.	.	.	.	−3	12	−75	630	.	.	−1	3	−15
(6)(2)(1)³	.	.	.	.	.	.	.	−2	30	−420	.	.	.	−1	10
(6)(1)⁵	.	.	.	.	.	.	.	.	−1	42	.	.	.	.	−1
(5)²(1)	.	6	.	.	2	−24	.	−12	120	−1008	.	.	2	−12	120
(5)(4)(2)	−2	6	.	−2	5	−24	9	−33	174	−1134	.	.	4	−18	120
(5)(4)(1)²	.	−6	.	.	−1	24	.	9	−120	1134	.	.	6	−120	−120
(5)(3)²	.	.	−1	.	2	−8	.	−6	40	−280	−3	6	6	−22	104
(5)(3)(2)(1)	.	.	.	.	−2	12	.	12	−100	840	.	.	−4	18	−120
(5)(3)(1)³	.	.	.	.	.	−4	.	.	20	−280	.	.	.	−2	40
(5)(2)³	.	.	.	.	.	.	−1	3	−15	105	.	.	.	.	.
(5)(2)²(1)²	.	.	.	.	.	.	.	−3	30	−315	.	.	.	.	.
(5)(2)(1)⁴	.	.	.	.	.	.	.	.	−5	105	.	.	.	.	.
(5)(1)⁶	.	.	.	.	.	.	.	.	.	−7	.	.	.	.	.
(4)²(3)	−1	2	−2	−1	4	−14	3	−14	70	−420	.	6	2	−18	100
(4)²(2)(1)	1	−3	.	.	−2	12	−3	15	−90	630	.	.	.	6	−60
(4)²(1)³	3	1	.	.	−4	.	.	−1	20	−210	.	.	.	.	20
(4)(3)²(1)	.	.	1	.	−2	8	.	6	−40	280	.	−6	−1	18	−110
(4)(3)(2)²	.	.	.	1	−1	3	−3	8	−35	210	.	.	−2	6	−30
(4)(3)(2)(1)²	2	.	2	1	1	−6	1	−6	50	−420	.	.	.	−6	60
(4)(3)(1)⁴	12	4	4	3	6	.	.	.	−5	70	.	.	.	.	−10
(4)(2)³(1)	3	.	.	3	.	.	1	−3	15	−105	.	.	.	.	.
(4)(2)²(1)³	9	1	6	7	6	.	3	1	−10	105	.	.	.	.	.
(4)(2)(1)⁵	35	10	20	25	40	5	15	10	1	−21	.	.	.	.	.
(4)(1)⁷	105	35	70	105	210	35	105	105	21	1	.	.	.	.	.
(3)³(2)	.	.	6	.	.	.	.	.	.	.	1	−1	−2	5	−20
(3)³(1)²	.	.	.	.	.	.	.	.	.	.	1	1	.	−3	20
(3)²(2)²(1)	.	.	1	2	.	.	.	.	.	.	2	.	1	−3	15
(3)²(2)(1)³	6	.	15	6	6	.	.	.	.	.	4	3	3	1	−10
(3)²(1)⁵	60	20	45	30	60	10	.	.	.	.	10	10	15	10	1
(3)(2)⁴	.	.	.	6	.	.	.	.	.	.	.	.	.	.	.
(3)(2)³(1)²	6	4	6	12	3	.	2	.	.	.	6	.	6	8	.
(3)(2)²(1)⁴	36	4	52	44	42	1	12	4	.	.	20	12	28	8	.
(3)(2)(1)⁶	210	60	210	210	345	45	90	60	6	.	70	60	150	80	6
(3)(1)⁸	840	280	840	1050	2100	350	840	840	168	8	280	280	840	560	56
(2)⁵(1)	15	.	.	30	.	.	10	.	.	.	.	.	.	.	.
(2)⁴(1)³	45	3	36	78	36	.	30	6	.	.	24	.	36	.	.
(2)³(1)⁵	195	30	240	300	300	15	140	60	.	.	120	60	240	60	.
(2)²(1)⁷	945	245	1120	1400	2100	245	840	560	63	1	560	420	1540	700	42
(2)(1)⁹	4095	1260	5040	6930	12600	1890	5670	5040	882	36	2800	2520	10080	5880	504
(1)¹¹	17325	5775	23100	34650	69300	11550	34650	34650	6930	330	15400	15400	69300	46200	4620

Table 1·1·11 (cont.)

$w = 11$ (iv)	$[32^4]$	$[32^3 1^2]$	$[32^2 1^4]$	$[321^6]$	$[31^8]$	$[2^5 1]$	$[2^4 1^3]$	$[2^3 1^5]$	$[2^2 1^7]$	$[21^9]$	$[1^{11}]$
(11)	24	−120	720	−5040	40320	−120	720	−5040	40320	−362880	3628800
$(10)(1)$	·	48	−480	4320	−40320	24	−360	3600	−35280	362880	−3991680
$(9)(2)$	−24	78	−384	2520	−20160	120	−552	3360	−25200	221760	−2217600
$(9)(1)^2$	·	−6	144	−1800	20160	·	72	−1200	15120	−181440	2217600
$(8)(3)$	−6	60	−360	2400	−18480	30	−300	2280	−18480	166320	−1663200
$(8)(2)(1)$	·	−36	264	−2160	20160	−30	306	−2520	22680	−226800	2494800
$(8)(1)^3$	·	·	−24	480	−6720	·	−6	240	−4200	60480	−831600
$(7)(4)$	−12	42	−276	1980	−15840	60	−288	1980	−15840	142560	−1425600
$(7)(3)(1)$	·	−24	256	−2160	19200	·	144	−1680	16800	−172800	1900800
$(7)(2)^2$	12	−24	102	−630	5040	−60	216	−1170	8280	−71280	712800
$(7)(2)(1)^2$	·	6	−84	900	−10080	·	−72	900	−10080	116640	−1425600
$(7)(1)^4$	·	·	2	−90	1680	·	·	−30	840	−15120	237600
$(6)(5)$	−8	46	−264	1848	−14784	40	−264	1848	−14784	133056	−1330560
$(6)(4)(1)$	·	−16	192	−1800	16800	−20	150	−1500	14700	−151200	1663200
$(6)(3)(2)$	8	−44	212	−1320	10080	−40	256	−1660	12600	−110880	1108800
$(6)(3)(1)^2$	·	2	−84	960	−10080	·	−24	580	−7560	90720	−1108800
$(6)(2)^2(1)$	·	12	−72	540	−5040	20	−132	930	−7770	75600	−831600
$(6)(2)(1)^3$	·	·	16	−240	3360	·	8	−200	2940	−40320	554400
$(6)(1)^5$	·	·	·	12	−336	·	·	2	−126	3024	−55440
$(5)^2(1)$	·	−12	96	−864	8064	·	72	−720	7056	−72576	798336
$(5)(4)(2)$	12	−33	168	−1134	9072	−60	252	−1512	11340	−99792	997920
$(5)(4)(1)^2$	·	3	−60	810	−9072	·	−36	540	−6804	81648	−997920
$(5)(3)^2$	·	−18	104	−664	4928	·	72	−600	4928	−44352	443520
$(5)(3)(2)(1)$	·	24	−160	1224	−10752	·	−144	1320	−12096	120960	−1330560
$(5)(3)(1)^3$	·	·	16	−280	3584	·	·	−120	2240	−32256	443520
$(5)(2)^3$	−4	5	−18	105	−840	20	−60	294	−1974	16632	−166320
$(5)(2)^2(1)^2$	·	−3	24	−225	2520	·	36	−360	3654	−40824	498960
$(5)(2)(1)^4$	·	·	−2	45	−840	·	·	30	−630	10584	−166320
$(5)(1)^6$	·	·	·	−1	56	·	·	·	14	−504	11088
$(4)^2(3)$	3	−12	86	−600	4620	−15	78	−570	4620	−41580	415800
$(4)^2(2)(1)$	·	6	−60	540	−5040	15	−81	630	−5670	56700	−623700
$(4)^2(1)^3$	·	·	4	−120	1680	·	3	−60	1050	−15120	207900
$(4)(3)^2(1)$	·	6	−80	660	−5600	·	−36	480	−4900	50400	−554400
$(4)(3)(2)^2$	−6	15	−65	390	−2940	30	−120	675	−4830	41580	−415800
$(4)(3)(2)(1)^2$	·	−3	54	−570	5880	·	36	−510	5880	−68040	831600
$(4)(3)(1)^4$	·	·	−1	60	−980	·	·	15	−490	8820	−138600
$(4)(2)^3(1)$	·	−2	12	−90	840	−10	42	−255	1995	−18900	207900
$(4)(2)^2(1)^3$	·	·	−4	60	−840	·	−6	90	−1155	15120	−207900
$(4)(2)(1)^5$	·	·	·	−6	168	·	·	−3	105	−2268	41580
$(4)(1)^7$	·	·	·	·	−8	·	·	·	−1	72	−1980
$(3)^3(2)$	·	6	−28	160	−1120	·	−24	180	−1400	12320	−123200
$(3)^3(1)^2$	·	·	12	−120	1120	·	·	−60	840	−10080	123200
$(3)^2(2)^2(1)$	·	−6	32	−210	1680	·	36	−300	2590	−25200	277200
$(3)^2(2)(1)^3$	·	·	−8	100	−1120	·	·	60	−980	13440	−184800
$(3)^2(1)^5$	·	·	·	−6	112	·	·	·	42	−1008	18480
$(3)(2)^4$	1	−1	3	−15	105	−5	14	−65	420	−3465	34650
$(3)(2)^3(1)^2$	1	1	−6	45	−420	·	−12	110	−1050	11340	−138600
$(3)(2)^2(1)^4$	3	6	1	−15	210	·	·	−15	280	−4410	69300
$(3)(2)(1)^6$	15	45	15	1	−28	·	·	·	−14	420	−9240
$(3)(1)^8$	105	420	210	28	1	·	·	·	·	−9	330
$(2)^5(1)$	5	·	·	·	·	1	−3	15	−105	945	−10395
$(2)^4(1)^3$	13	12	·	·	·	3	1	−10	105	−1260	17325
$(2)^3(1)^5$	55	100	15	·	·	15	10	1	−21	378	−6930
$(2)^2(1)^7$	315	840	245	14	·	105	105	21	1	−36	990
$(2)(1)^9$	2205	7560	3150	336	9	945	1260	378	36	1	−55
$(1)^{11}$	17325	69300	34650	4620	165	10395	17325	6930	990	55	1

Table 1·1·12

$w = 12$ (i)	[12]	[11, 1]	[10, 2]	[10, 1²]	[93]	[921]	[91³]	[84]	[831]	[82²]	[821²]	[81⁴]	[75]	[741]
(12)	1	−1	−1	2	−1	2	−6	−1	2	2	−6	24	−1	2
(11)(1)	1	1	.	−2	.	−1	6	.	−1	.	4	−24	.	−1
(10)(2)	1	.	1	−1	.	−1	3	.	.	−2	3	−12	.	.
(10)(1)²	1	2	1	1	.	.	−3	.	.	.	3	12	.	.
(9)(3)	1	.	.	.	1	−1	2	.	−1	.	2	−8	.	.
(9)(2)(1)	1	1	1	.	1	1	−3	.	.	.	−2	12	.	.
(9)(1)³	1	3	3	3	1	3	1	.	.	.	.	−4	.	.
(8)(4)	1	.	.	.	.	.	.	1	−1	−1	2	−6	.	−1
(8)(3)(1)	1	1	.	.	1	.	.	1	1	.	−2	8	.	.
(8)(2)²	1	.	2	.	.	.	.	1	.	1	−1	3	.	.
(8)(2)(1)²	1	2	2	1	2	2	.	1	2	1	1	−6	.	.
(8)(1)⁴	1	4	6	6	4	12	4	1	4	3	6	1	.	.
(7)(5)	1	.	.	.	.	.	.	.	.	.	.	.	1	−1
(7)(4)(1)	1	1	.	.	.	.	.	1	.	.	.	.	1	1
(7)(3)(2)	1	.	1	.	1	.	.	.	.	.	.	.	1	.
(7)(3)(1)²	1	2	1	1	1	.	.	2	2	.	.	.	1	2
(7)(2)²(1)	1	1	2	.	2	2	.	1	.	1	.	.	1	1
(7)(2)(1)³	1	3	4	3	4	6	1	3	6	3	3	.	1	3
(7)(1)⁵	1	5	10	10	10	10	10	5	20	15	30	5	1	5
(6)²	1	.	.	.	.	.	.	.	.	.	.	.	.	.
(6)(5)(1)	1	1	.	.	.	.	.	.	.	.	.	.	1	.
(6)(4)(2)	1	.	.	1	.	.	.	1	.	.	.	.	1	.
(6)(4)(1)²	1	2	1	1	.	.	.	1	.	.	.	.	2	2
(6)(3)²	1	.	.	.	2	.	.	.	.	.	.	.	.	.
(6)(3)(2)(1)	1	1	1	.	2	1	.	1	1	.	.	.	1	.
(6)(3)(1)³	1	3	3	3	2	3	1	3	3	.	.	.	3	6
(6)(2)³	1	.	3	.	.	.	.	3	.	3	.	.	.	.
(6)(2)²(1)²	1	2	3	1	4	4	.	3	4	3	2	.	2	2
(6)(2)(1)⁴	1	4	7	6	8	16	4	7	16	9	12	1	4	12
(6)(1)⁶	1	6	15	15	20	60	20	15	60	45	90	15	6	30
(5)²(2)	1	.	1	.	.	.	.	.	.	.	.	.	2	.
(5)²(1)²	1	2	1	1	.	.	.	.	.	.	.	.	2	.
(5)(4)(3)	1	.	.	.	1	.	.	1	.	.	.	.	1	.
(5)(4)(2)(1)	1	1	1	.	1	1	.	1	.	.	.	.	2	1
(5)(4)(1)³	1	3	3	3	1	3	1	1	.	.	.	.	4	3
(5)(3)²(1)	1	1	.	.	2	.	.	2	2	.	.	.	1	.
(5)(3)(2)²	1	.	2	.	1	.	.	1	.	1	.	.	3	.
(5)(3)(2)(1)²	1	2	2	1	3	2	.	3	4	1	1	.	3	2
(5)(3)(1)⁴	1	4	6	6	5	12	4	5	8	3	6	1	7	12
(5)(2)³(1)	1	1	3	.	3	3	.	3	.	3	.	.	4	3
(5)(2)²(1)³	1	3	5	3	7	9	1	7	12	7	6	.	6	9
(5)(2)(1)⁵	1	5	11	10	15	35	10	15	40	25	40	5	12	35
(5)(1)⁷	1	7	21	21	35	105	35	35	140	105	210	35	22	105
(4)³	1	.	.	.	.	.	.	3	.	.	.	.	.	.
(4)²(3)(1)	1	1	.	.	1	.	.	3	1	.	.	.	2	2
(4)²(2)²	1	.	2	.	.	.	.	3	.	1	.	.	.	.
(4)²(2)(1)²	1	2	2	1	2	2	.	3	2	1	1	.	4	4
(4)²(1)⁴	1	4	6	6	4	12	4	3	4	3	6	1	8	8
(4)(3)²(2)	1	.	.	.	2	.	.	1	.	.	.	.	2	.
(4)(3)²(1)²	1	2	1	1	2	.	.	5	4	.	.	.	4	6
(4)(3)(2)²(1)	1	1	2	.	3	2	.	3	1	1	.	.	4	2
(4)(3)(2)(1)³	1	3	4	3	5	6	1	7	9	3	3	.	8	12
(4)(3)(1)⁵	1	5	10	10	11	30	10	11	25	15	30	5	16	30
(4)(2)⁴	1	.	4	.	.	.	.	7	.	6	.	.	.	.
(4)(2)³(1)²	1	2	4	1	6	6	.	7	6	6	3	.	8	8
(4)(2)²(1)⁴	1	4	8	6	12	20	4	15	28	16	18	1	16	32
(4)(2)(1)⁶	1	6	16	15	26	66	20	31	90	60	105	15	32	96
(4)(1)⁸	1	8	28	28	56	168	56	71	280	210	420	70	64	288
(3)⁴	1	.	.	.	4	.	.	3	.	.	.	.	.	.
(3)³(2)(1)	1	1	1	.	4	1	.	3	3	.	.	.	3	.
(3)³(1)³	1	3	3	3	4	3	1	9	9	.	.	.	9	18
(3)²(2)³	1	.	3	.	2	.	.	3	.	3	.	.	6	.
(3)²(2)²(1)²	1	2	3	1	6	4	.	7	8	3	2	.	8	6
(3)²(2)(1)⁴	1	4	7	6	10	16	4	15	24	9	12	1	18	36
(3)²(1)⁶	1	6	15	15	22	60	20	27	72	45	90	15	36	90
(3)(2)⁴(1)	1	1	4	.	5	4	.	7	1	6	.	.	10	6
(3)(2)³(1)³	1	3	6	3	11	12	1	15	21	12	9	.	18	24
(3)(2)²(1)⁵	1	5	12	10	21	40	10	31	65	36	50	5	38	90
(3)(2)(1)⁷	1	7	22	21	43	112	35	63	189	126	231	35	78	252
(3)(1)⁹	1	9	36	36	85	252	84	135	513	378	756	126	162	702
(2)⁶	1	.	6	.	.	.	.	15	.	15	.	.	.	.
(2)⁵(1)²	1	2	6	1	10	10	.	15	10	15	5	.	20	20
(2)⁴(1)⁴	1	4	10	6	20	28	4	31	52	33	30	1	40	72
(2)³(1)⁶	1	6	18	15	38	78	20	63	150	93	135	15	84	228
(2)²(1)⁸	1	8	30	28	72	184	56	127	392	267	476	70	176	624
(2)(1)¹⁰	1	10	46	45	130	370	120	255	930	675	1305	210	372	1620
(1)¹²	1	12	66	66	220	660	220	495	1980	1485	2970	495	792	3960

Table 1·1·12 (cont.)

$w=12$ (ii)	$[732]$	$[731^2]$	$[72^21]$	$[721^3]$	$[71^5]$	$[6^2]$	$[651]$	$[642]$	$[641^2]$	$[63^2]$	$[6321]$	$[631^3]$	$[62^3]$	$[62^21^2]$
(12)	2	−6	−6	24	−120	−1	2	2	−6	2	−6	24	−6	24
(11)(1)	·	4	2	−18	120	·	−1	·	4	·	2	−18	·	−12
(10)(2)	−1	1	4	−12	60	·	·	−1	1	·	2	−6	6	−14
$(10)(1)^2$	·	−1	·	6	−60	·	·	·	−1	·	·	6	·	2
(9)(3)	−1	2	2	−8	40	·	·	·	·	−2	3	−8	·	−8
(9)(2)(1)	·	·	−2	9	−60	·	·	·	·	·	−1	3	·	8
$(9)(1)^3$	·	·	·	−1	20	·	·	·	·	·	·	−1	·	·
(8)(4)	·	2	1	−6	30	·	·	−1	2	·	1	−6	3	−6
(8)(3)(1)	·	−2	·	6	−40	·	·	·	·	·	−1	6	·	4
$(8)(2)^2$	·	·	−1	3	−15	·	·	·	·	·	·	·	−3	4
$(8)(2)(1)^2$	·	·	·	−3	30	·	·	·	·	·	·	·	·	−2
$(8)(1)^4$	·	·	·	·	−5	·	·	·	·	·	·	·	·	·
(7)(5)	−1	2	2	−6	24	·	−1	·	2	·	1	−6	·	−4
(7)(4)(1)	·	−2	−1	6	−30	·	·	·	−2	·	·	6	·	2
(7)(3)(2)	1	−1	−2	5	−20	·	·	·	·	·	·	−1	·	4
$(7)(3)(1)^2$	1	1	·	−3	20	·	·	·	·	·	·	−3	·	4
$(7)(2)^2(1)$	2	·	1	−3	15	·	·	·	·	·	·	·	·	−2
$(7)(2)(1)^3$	4	3	3	1	−10	·	·	·	·	·	·	·	·	·
$(7)(1)^5$	10	10	15	10	1	·	·	·	·	·	·	·	·	·
$(6)^2$	·	·	·	·	·	1	−1	−1	2	−1	2	−6	2	−6
(6)(5)(1)	·	·	·	·	·	1	1	·	−2	·	−1	6	·	4
(6)(4)(2)	·	·	·	·	·	1	1	1	·	−1	·	3	−3	5
$(6)(4)(1)^2$	·	·	·	·	·	1	2	1	1	·	−1	−3	·	−1
$(6)(3)^2$	·	·	·	·	·	1	·	·	·	1	−1	2	·	2
(6)(3)(2)(1)	1	·	·	·	·	1	1	1	·	·	1	−3	·	−4
$(6)(3)(1)^3$	3	3	·	·	·	1	3	3	3	1	3	1	·	−1
$(6)(2)^3$	·	·	2	·	·	1	·	3	·	·	·	·	1	−1
$(6)(2)^2(1)^2$	4	·	·	·	·	1	2	3	1	2	4	·	1	1
$(6)(2)(1)^4$	16	12	12	4	·	1	4	7	6	4	16	4	3	6
$(6)(1)^6$	60	60	90	60	6	1	6	15	15	10	60	20	15	45
$(5)^2(2)$	·	·	·	·	·	2	4	·	·	·	·	·	·	·
$(5)^2(1)^2$	·	·	·	·	·	2	·	·	·	·	·	·	·	·
(5)(4)(3)	·	·	·	·	·	·	·	·	·	·	·	·	·	·
(5)(4)(2)(1)	·	·	·	·	·	1	1	1	·	·	·	·	·	·
$(5)(4)(1)^3$	·	·	·	·	·	3	9	3	3	·	·	·	·	·
$(5)(3)^2(1)$	·	·	·	·	·	1	1	·	·	1	·	·	·	·
$(5)(3)(2)^2$	2	·	·	·	·	·	·	·	·	·	·	·	·	·
$(5)(3)(2)(1)^2$	2	1	·	·	·	2	4	2	·	·	2	2	·	·
$(5)(3)(1)^4$	6	6	·	·	·	4	16	12	12	4	12	4	·	·
$(5)(2)^3(1)$	6	·	3	·	·	1	1	3	·	·	·	·	1	·
$(5)(2)^2(1)^3$	14	6	9	2	·	3	9	9	3	6	12	·	3	3
$(5)(2)(1)^5$	50	40	45	20	1	5	25	35	30	20	80	20	15	30
$(5)(1)^7$	210	210	315	210	21	7	49	105	105	70	420	140	105	315
$(4)^3$	·	·	·	·	·	·	·	·	·	·	·	·	·	·
$(4)^2(3)(1)$	·	·	·	·	·	·	·	·	·	·	·	·	·	·
$(4)^2(2)^2$	·	·	·	·	·	2	·	4	·	·	·	·	·	·
$(4)^2(2)(1)^2$	·	·	·	·	·	2	4	4	2	·	·	·	·	·
$(4)^2(1)^4$	·	·	·	·	·	6	24	12	12	·	·	·	·	·
$(4)(3)^2(2)$	·	·	·	·	·	1	·	1	·	1	1	·	·	·
$(4)(3)^2(1)^2$	2	2	·	·	·	1	2	1	1	1	1	·	·	·
$(4)(3)(2)^2(1)$	4	·	1	·	·	2	2	4	·	2	2	·	·	·
$(4)(3)(2)(1)^3$	8	6	3	1	·	4	12	10	6	4	6	1	·	·
$(4)(3)(1)^5$	20	20	15	10	1	10	50	40	40	10	30	10	·	·
$(4)(2)^4$	·	·	·	·	·	4	·	16	·	·	·	·	4	·
$(4)(2)^3(1)^2$	12	·	6	·	·	4	8	16	4	·	12	·	4	3
$(4)(2)^2(1)^4$	40	24	28	8	·	8	32	40	24	20	56	8	12	18
$(4)(2)(1)^6$	140	120	150	80	6	16	96	136	120	70	300	80	60	135
$(4)(1)^8$	560	560	840	560	56	28	224	448	448	280	1680	560	420	1200
$(3)^4$	·	·	·	·	·	3	·	·	·	6	·	·	·	·
$(3)^3(2)(1)$	3	·	·	·	·	3	3	3	·	6	3	·	·	·
$(3)^3(1)^3$	9	9	·	·	·	3	9	9	9	6	9	3	·	·
$(3)^2(2)^3$	6	·	·	·	·	1	·	3	·	1	·	·	1	·
$(3)^2(2)^2(1)^2$	10	2	·	·	·	5	10	11	1	11	12	·	1	1
$(3)^2(2)(1)^4$	30	24	12	4	·	9	36	39	30	21	48	12	3	6
$(3)^2(1)^6$	90	90	90	60	6	21	126	135	135	51	180	60	15	45
$(3)(2)^4(1)$	16	·	6	·	·	4	4	16	·	4	4	·	4	·
$(3)(2)^3(1)^3$	36	12	18	3	·	10	30	42	12	28	48	1	10	9
$(3)(2)^2(1)^5$	112	80	80	30	1	20	100	140	100	80	240	50	40	70
$(3)(2)(1)^7$	372	336	420	245	21	42	294	462	420	252	1092	315	210	525
$(3)(1)^9$	1296	1296	1890	1260	126	84	756	1512	1512	924	5292	1764	1260	3780
$(2)^6$	·	·	·	·	·	10	·	60	·	·	·	·	20	10
$(2)^5(1)^2$	40	·	20	16	·	10	20	60	10	20	40	·	20	60
$(2)^4(1)^4$	112	48	72	16	·	22	88	148	60	88	208	16	52	60
$(2)^3(1)^6$	336	240	288	120	6	46	276	468	330	280	960	200	196	360
$(2)^2(1)^8$	1024	896	1184	672	56	98	784	1484	1288	896	4032	1120	868	2128
$(2)(1)^{10}$	3000	2880	4140	2640	252	210	2100	4620	4410	2940	15960	5040	3780	10710
$(1)^{12}$	7920	7920	11880	7920	792	462	5544	13860	13860	9240	55440	18480	13860	41580

Table 1·1·12 (*cont.*)

$w = 12$ (iii)	$[621^4]$	$[61^6]$	$[5^22]$	$[5^21^2]$	$[543]$	$[5421]$	$[541^3]$	$[53^21]$	$[532^2]$	$[5321^2]$	$[531^4]$	$[52^31]$	$[52^21^3]$	$[521^5]$
(12)	−120	720	2	−6	2	−6	24	−6	−6	24	−120	24	−120	720
$(11)(1)$	96	−720	·	4	·	2	−18	2	·	−12	96	−6	72	−600
$(10)(2)$	60	−360	−1	1	·	2	−6	·	4	−8	36	−18	66	−360
$(10)(1)^2$	−36	360	·	−1	·	·	6	·	·	2	−36	·	−18	240
$(9)(3)$	40	−240	·	·	−1	1	−2	4	2	−10	40	−6	40	−240
$(9)(2)(1)$	−48	360	·	·	·	−1	3	·	·	4	−24	6	−42	300
$(9)(1)^3$	8	−120	·	·	·	·	−1	·	·	·	8	·	2	−60
$(8)(4)$	30	−180	·	·	−1	2	−6	2	1	−6	30	−6	30	−180
$(8)(3)(1)$	−32	240	·	·	·	·	·	−2	·	6	−32	·	−24	200
$(8)(2)^2$	−15	90	·	·	·	·	·	·	−1	1	−3	6	−18	90
$(8)(2)(1)^2$	18	−180	·	·	·	·	·	·	·	−1	6	·	12	−120
$(8)(1)^4$	−1	30	·	·	·	·	·	·	·	·	−1	·	·	10
$(7)(5)$	24	−144	−2	4	−1	3	−12	2	4	−10	48	−12	48	−264
$(7)(4)(1)$	−24	180	·	·	·	−1	6	·	·	2	−24	3	−18	150
$(7)(3)(2)$	−20	120	·	·	·	·	·	·	−2	3	−12	6	−22	120
$(7)(3)(1)^2$	12	−120	·	·	·	·	·	·	·	−1	12	·	6	−80
$(7)(2)^2(1)$	12	−90	·	·	·	·	·	·	·	·	·	−3	12	−75
$(7)(2)(1)^3$	−4	60	·	·	·	·	·	·	·	·	·	·	−2	30
$(7)(1)^5$	·	−6	·	·	·	·	·	·	·	·	·	·	·	−1
$(6)^2$	24	−120	·	2	·	1	−6	1	·	−4	24	−2	18	−120
$(6)(5)(1)$	−24	144	·	−4	·	−1	12	−1	·	6	−48	2	−30	240
$(6)(4)(2)$	−18	90	·	·	·	−1	3	·	·	2	−12	3	−15	90
$(6)(4)(1)^2$	12	−90	·	·	·	·	−3	·	·	·	12	·	3	−60
$(6)(3)^2$	−8	40	·	·	·	·	·	−1	·	·	−8	·	−6	40
$(6)(3)(2)(1)$	20	−120	·	·	·	·	·	·	·	−2	12	·	12	−100
$(6)(3)(1)^3$	−4	40	·	·	·	·	·	·	·	·	−4	·	·	20
$(6)(2)^3$	3	−15	·	·	·	·	·	·	·	·	·	−1	3	−15
$(6)(2)^2(1)^2$	−6	45	·	·	·	·	·	·	·	·	·	·	−3	30
$(6)(2)(1)^4$	1	−15	·	·	·	·	·	·	·	·	·	·	·	−5
$(6)(1)^6$	15	1	·	·	·	·	·	·	·	·	·	·	·	·
$(5)^3(2)$	·	·	1	−1	·	−1	3	·	−2	3	−12	6	−18	84
$(5)^2(1)^2$	·	·	1	1	·	−1	−3	·	·	−1	12	·	6	−60
$(5)(4)(3)$	·	·	·	·	1	−1	2	−2	−1	4	−14	3	−14	70
$(5)(4)(2)(1)$	·	·	1	·	1	1	−3	·	·	−2	12	−3	15	−90
$(5)(4)(1)^3$	·	·	3	3	1	3	1	·	·	·	−4	·	−1	20
$(5)(3)^2(1)$	·	·	·	·	2	1	·	1	1	−1	3	·	6	−40
$(5)(3)(2)^2$	·	·	2	·	1	·	·	·	1	1	−6	−3	8	−35
$(5)(3)(2)(1)^2$	·	·	2	1	3	2	·	2	1	3	6	·	−6	50
$(5)(3)(1)^4$	·	·	6	6	5	12	4	4	·	3	·	·	·	−5
$(5)(2)^3(1)$	·	·	3	·	3	3	·	·	·	3	·	1	−3	15
$(5)(2)^2(1)^3$	·	·	5	3	7	9	10	6	7	6	·	3	1	−10
$(5)(2)(1)^5$	5	·	11	10	15	35	10	20	25	40	5	15	10	1
$(5)(1)^7$	105	7	21	21	35	105	35	70	105	210	35	105	105	21
$(4)^3$	·	·	·	·	2	·	·	·	·	·	·	·	·	·
$(4)^2(3)(1)$	·	·	·	·	·	·	·	·	·	·	·	·	·	·
$(4)^2(2)^2$	·	·	·	·	·	·	·	·	·	·	·	·	·	·
$(4)^2(2)(1)^2$	·	·	2	·	4	4	8	·	·	·	·	·	·	·
$(4)^2(1)^4$	·	·	12	12	8	24	8	·	·	·	·	·	·	·
$(4)(3)^2(2)$	·	·	·	·	2	·	·	2	·	·	·	·	·	·
$(4)(3)^2(1)^2$	·	·	·	·	6	·	·	·	·	·	·	·	·	·
$(4)(3)(2)^2(1)$	·	·	2	·	4	2	1	6	1	3	3	·	·	·
$(4)(3)(2)(1)^3$	·	·	6	3	14	12	·	6	3	30	·	·	·	·
$(4)(3)(1)^5$	·	·	30	30	36	90	30	20	15	30	5	·	·	·
$(4)(2)^4$	·	·	·	·	6	·	·	·	·	6	·	·	·	·
$(4)(2)^3(1)^2$	·	·	6	·	12	12	·	24	·	6	·	2	4	·
$(4)(2)^2(1)^4$	2	·	20	12	40	56	8	24	28	24	·	12	60	·
$(4)(2)(1)^6$	30	1	66	60	116	276	80	120	150	240	30	90	60	6
$(4)(1)^8$	420	28	168	168	336	1008	336	560	840	1680	280	840	840	168
$(3)^4$	·	·	·	·	·	·	·	·	·	·	·	·	·	·
$(3)^3(2)(1)$	·	·	·	·	6	·	·	3	·	·	·	·	·	·
$(3)^3(1)^3$	·	·	6	·	18	6	·	9	6	·	·	·	·	·
$(3)^2(2)^3$	·	·	6	·	6	6	·	6	6	·	·	·	·	·
$(3)^2(2)^2(1)^2$	·	·	6	2	18	8	8	10	6	4	2	·	·	·
$(3)^2(2)(1)^4$	·	·	18	12	54	48	8	36	18	24	30	·	·	·
$(3)^2(1)^6$	·	·	90	90	162	360	120	126	90	180	30	·	·	·
$(3)(2)^4(1)$	·	·	12	·	18	12	·	36	18	·	·	4	·	·
$(3)(2)^3(1)^3$	·	·	24	9	60	54	3	36	42	27	·	12	·	·
$(3)(2)^2(1)^5$	10	·	72	50	186	260	50	160	156	190	15	60	30	2
$(3)(2)(1)^7$	140	7	252	231	588	1302	385	672	756	1281	175	420	315	42
$(3)(1)^9$	1260	84	756	756	1890	5292	1764	3024	4158	8316	1386	3780	3780	756
$(2)^6$	·	·	·	·	60	60	·	·	·	60	·	·	·	·
$(2)^5(1)^2$	·	·	30	·	84	·	·	·	·	·	·	20	·	·
$(2)^4(1)^4$	4	·	84	36	216	264	24	144	216	144	·	88	24	·
$(2)^3(1)^6$	60	1	258	180	708	1188	240	720	888	1080	90	456	240	18
$(2)^2(1)^8$	560	28	840	728	2352	5264	1456	3136	4032	6496	840	2688	2016	280
$(2)(1)^{10}$	3360	210	2646	2520	7980	21420	6720	13440	18900	35280	5460	16380	15120	2772
$(1)^{12}$	13860	924	8316	8316	27720	83160	27720	55440	83160	166320	27720	83160	83160	16632

$w=12$ (iv)	$[51^7]$	$[4^3]$	$[4^231]$	$[4^22^2]$	$[4^221^2]$	$[4^21^4]$	$[43^2]$	$[43^21^2]$	$[432^21]$	$[4321^3]$	$[431^5]$	$[42^4]$	$[42^31^2]$
(12)	-5040	2	-6	-6	24	-120	-6	24	24	-120	720	24	-120
$(11)(1)$	5040	·	2	·	-12	96	·	-12	-6	72	-600	·	48
$(10)(2)$	2520	·	·	4	-8	36	2	-2	-12	42	-240	-24	78
$(10)(1)^2$	-2520	·	·	·	2	-36	·	2	·	-18	240	·	-6
$(9)(3)$	1680	·	2	·	-4	16	4	-12	-10	46	-240	·	36
$(9)(2)(1)$	-2520	·	·	·	4	-24	·	·	4	-24	180	·	-36
$(9)(1)^3$	840	·	·	·	·	8	·	·	·	2	-60	·	·
$(8)(4)$	1260	-3	5	5	-14	54	2	-14	-10	54	-300	-18	54
$(8)(3)(1)$	-1680	·	-1	·	2	-8	·	8	2	-30	200	·	-12
$(8)(2)^2$	-630	·	·	-1	1	-3	·	·	2	-6	30	12	-24
$(8)(2)(1)^2$	1260	·	·	·	-1	6	·	·	·	6	-60	·	6
$(8)(1)^4$	-210	·	·	·	·	-1	·	·	·	·	10	·	·
$(7)(5)$	1728	·	2	·	-8	48	2	-8	-8	42	-264	·	36
$(7)(4)(1)$	-1260	·	-2	·	8	-48	·	8	3	-36	270	·	-24
$(7)(3)(2)$	-840	·	·	·	·	·	-2	2	6	-17	80	·	-24
$(7)(3)(1)^2$	840	·	·	·	·	·	·	-2	·	9	-80	·	·
$(7)(2)^2(1)$	630	·	·	·	·	·	·	·	-1	3	-15	·	12
$(7)(2)(1)^3$	-420	·	·	·	·	·	·	·	·	-1	10	·	·
$(7)(1)^5$	42	·	·	·	·	·	·	·	·	·	-1	·	·
$(6)^2$	840	·	·	2	-4	24	1	-2	-4	18	-120	-8	22
$(6)(5)(1)$	-1848	·	·	·	4	-48	·	2	2	-24	240	·	-16
$(6)(4)(2)$	-630	·	·	-4	6	-24	-1	1	6	-21	120	20	-41
$(6)(4)(1)^2$	630	·	·	·	-2	24	·	-1	·	9	-120	·	5
$(6)(3)^2$	-280	·	·	·	·	·	·	-1	2	2	-120	·	5
$(6)(3)(2)(1)$	840	·	·	·	·	·	·	·	2	-8	40	·	-6
$(6)(3)(1)^3$	-280	·	·	·	·	·	·	·	·	-1	20	·	12
$(6)(2)^3$	105	·	·	·	·	·	·	·	·	·	·	·	·
$(6)(2)^2(1)^2$	-315	·	·	·	·	·	·	·	·	·	·	-4	5
$(6)(2)(1)^4$	105	·	·	·	·	·	·	·	·	·	·	·	-3
$(6)(1)^6$	-7	·	·	·	·	·	·	·	·	·	·	·	·
$(5)^2(2)$	-504	·	·	·	2	-12	·	·	·	2	-9	·	-12
$(5)^2(1)^2$	504	·	·	·	·	12	·	·	·	3	-60	·	·
$(5)(4)(3)$	-420	·	-2	·	-4	-16	·	8	5	-26	134	·	-18
$(5)(4)(2)(1)$	630	·	·	·	-4	24	·	·	-2	15	-120	·	18
$(5)(4)(1)^3$	-210	·	·	·	·	-8	·	·	·	-1	40	·	·
$(5)(3)^2(1)$	280	·	·	·	·	·	·	-2	·	6	-40	·	·
$(5)(3)(2)^2$	210	·	·	·	·	·	·	·	·	-1	·	·	6
$(5)(3)(2)(1)^2$	-420	·	·	·	·	·	·	·	·	3	-15	·	·
$(5)(3)(1)^4$	70	·	·	·	·	·	·	·	·	-3	30	·	·
$(5)(2)^3(1)$	-105	·	·	·	·	·	·	·	·	·	-5	·	·
$(5)(2)^2(1)^3$	105	·	·	·	·	·	·	·	·	·	·	·	-2
$(5)(2)(1)^5$	-21	·	·	·	·	·	·	·	·	·	·	·	·
$(5)(1)^7$	1	·	·	·	·	·	·	·	·	·	·	·	·
$(4)^3$	·	1	-1	-1	2	-6	·	·	2	1	-6	3	-6
$(4)^2(3)(1)$	·	1	1	·	-2	8	·	-4	-1	12	-70	3	6
$(4)^2(2)^2$	·	1	1	·	-1	3	·	·	-1	3	-15	-6	9
$(4)^2(2)(1)^2$	·	1	2	1	1	-6	·	·	·	3	30	-6	-3
$(4)^2(1)^4$	·	1	4	3	6	1	·	·	·	·	-5	-6	-3
$(4)(3)^2(2)$	·	·	·	·	·	·	1	-1	·	-2	5	-20	6
$(4)(3)^2(1)^2$	·	2	4	·	·	·	1	1	·	-3	20	·	·
$(4)(3)(2)^2(1)$	·	1	1	1	·	·	2	·	1	-3	15	·	-6
$(4)(3)(2)(1)^3$	·	3	9	3	3	·	4	3	3	·	-10	·	·
$(4)(3)(1)^5$	·	5	25	15	30	5	10	10	15	10	1	·	-1
$(4)(2)^4$	·	3	6	6	·	·	·	·	·	·	·	1	-1
$(4)(2)^3(1)^2$	·	3	6	6	3	·	6	·	6	·	·	1	1
$(4)(2)^2(1)^4$	·	7	28	16	·	3	20	12	28	8	·	3	6
$(4)(2)(1)^6$	·	15	90	60	105	15	70	60	150	80	6	15	45
$(4)(1)^8$	8	35	280	210	420	70	280	280	840	560	56	105	420
$(3)^4$	·	·	·	·	·	·	·	·	·	·	·	·	·
$(3)^3(2)(1)$	·	·	·	·	·	·	3	·	·	·	·	·	·
$(3)^3(1)^3$	·	6	18	·	·	·	9	9	·	·	·	·	·
$(3)^2(2)^3$	·	·	·	·	·	·	3	·	·	·	·	·	·
$(3)^2(2)^2(1)^2$	·	2	4	2	·	·	11	1	4	·	·	·	·
$(3)^2(2)(1)^4$	·	12	48	12	12	·	39	30	24	8	·	·	·
$(3)^2(1)^6$	·	30	180	90	180	30	135	135	180	120	12	·	·
$(3)(2)^4(1)$	·	3	3	6	·	·	12	·	6	·	·	1	·
$(3)(2)^3(1)^3$	·	9	27	18	9	·	48	9	36	3	·	3	3
$(3)(2)^2(1)^5$	·	35	175	80	90	5	200	130	220	70	1	15	30
$(3)(2)(1)^7$	1	105	735	420	735	105	840	735	1470	805	63	105	315
$(3)(1)^9$	36	315	2835	1890	3780	630	3780	3780	9450	6300	630	945	3780
$(2)^6$	·	15	·	45	·	·	60	·	60	·	·	15	·
$(2)^5(1)^2$	·	15	30	45	15	·	60	·	60	·	·	15	10
$(2)^4(1)^4$	·	39	156	123	90	3	264	72	312	48	·	43	60
$(2)^3(1)^6$	·	135	810	495	585	45	1200	720	1800	600	18	195	420
$(2)^2(1)^8$	8	455	3640	2345	3780	490	5600	4480	11200	5600	392	1155	3360
$(2)(1)^{10}$	120	1575	15750	11025	20475	3150	27300	25200	69300	42000	3780	7875	28350
$(1)^{12}$	792	5775	69300	51975	103950	17325	138600	138600	415800	277200	27720	51975	207900

Table 1·1·12 (cont.)

$w=12$ (v)	$[42^2 1^4]$	$[421^6]$	$[41^8]$	$[3^4]$	$[3^3 21]$	$[3^3 1^3]$	$[3^2 2^3]$	$[3^2 2^2 1^2]$	$[3^2 21^4]$	$[3^2 1^6]$	$[32^4 1]$	$[32^3 1^3]$	$[32^2 1^5]$
(12)	720	−5040	40320	−6	24	−120	24	−120	720	−5040	−120	720	−5040
$(11)(1)$	−480	4320	−40320	.	−6	72	.	48	−480	4320	24	−360	3600
$(10)(2)$	−384	2520	−20160	.	−6	18	−18	54	−264	1800	96	−432	2640
$(10)(1)^2$	144	−1800	20160	.	.	−18	.	−6	144	−1800	.	72	−1200
$(9)(3)$	−240	1680	−13440	8	−20	76	−12	72	−384	2400	48	−348	2400
$(9)(2)(1)$	264	−2160	20160	.	2	−6	.	−24	168	−1440	−24	234	−1920
$(9)(1)^3$	−24	480	−6720	.	.	2	.	.	−24	480	.	−6	240
$(8)(4)$	−300	1980	−15120	.	−6	54	−6	38	−276	1980	42	−252	1860
$(8)(3)(1)$	160	−1440	13440	.	6	−54	.	−32	280	−2160	−6	180	−1800
$(8)(2)^2$	102	−630	5040	.	.	.	6	−10	42	−270	−36	126	−690
$(8)(2)(1)^2$	−84	900	−10080	.	.	.	.	4	−48	540	.	−54	660
$(8)(1)^4$	2	−90	1680	.	.	.	.	.	2	−90	.	.	−30
$(7)(5)$	−240	1728	−13824	.	−6	36	−12	40	−240	1728	48	−252	1728
$(7)(4)(1)$	216	−1800	15840	.	.	−36	.	−12	192	−1800	−12	126	−1380
$(7)(3)(2)$	128	−840	6720	.	6	−18	12	−36	160	−960	−48	228	−1360
$(7)(3)(1)^2$	−48	600	−6720	.	.	18	.	4	−96	960	.	−36	640
$(7)(2)^2(1)$	−72	540	−5040	.	.	.	.	4	−24	180	12	−72	510
$(7)(2)(1)^3$	16	−240	3360	.	.	.	.	.	8	−120	.	6	−140
$(7)(1)^5$	.	12	−336	.	.	.	.	.	.	12	.	.	2
$(6)^2$	−120	840	−6720	3	−6	18	−2	22	−120	840	16	−120	840
$(6)(5)(1)$	168	−1584	14784	.	3	−18	.	−20	168	−1584	−8	138	−1320
$(6)(4)(2)$	180	−1110	8400	.	3	−9	3	−21	114	−810	−32	168	−1080
$(6)(4)(1)^2$	−72	810	−8400	.	.	9	.	1	−60	810	.	−24	480
$(6)(3)^2$	40	−280	2240	−6	9	−24	2	−24	112	−640	−8	94	−640
$(6)(3)(2)(1)$	−88	720	−6720	.	−3	9	.	20	−116	840	8	−132	1060
$(6)(3)(1)^3$	8	−160	2240	.	.	−3	.	.	20	−280	.	2	−140
$(6)(2)^3$	−18	105	−840	.	.	.	−1	1	−3	15	8	−24	120
$(6)(2)^2(1)^2$	24	−225	2520	.	.	.	.	−1	6	−45	.	18	−180
$(6)(2)(1)^4$	−2	45	−840	.	.	.	.	.	−1	15	.	.	20
$(6)(1)^6$	.	−1	56	.	.	.	.	.	.	−1	.	.	.
$(5)^2(2)$	72	−504	4032	.	.	.	6	−10	48	−360	−24	90	−528
$(5)^2(1)^2$	−24	360	−4032	.	.	.	6	2	−24	360	.	−18	240
$(5)(4)(3)$	120	−804	6048	.	6	−36	6	−28	172	−1128	−24	150	−1064
$(5)(4)(2)(1)$	−132	1044	−9072	.	.	.	.	8	−72	720	12	−99	840
$(5)(4)(1)^3$	12	−240	3024	.	.	.	.	.	8	−240	.	3	−100
$(5)(3)^2(1)$	−24	240	−2240	.	−3	18	.	12	−88	624	.	−54	520
$(5)(3)(2)^2$	−32	210	−1680	.	.	.	−6	8	−30	180	24	−78	404
$(5)(3)(2)(1)^2$	24	−300	3360	.	.	.	.	−4	36	−360	.	36	−400
$(5)(3)(1)^4$	.	30	−560	.	.	.	.	−2	60	.	.	.	20
$(5)(2)^3(1)$	12	−90	840	.	.	.	.	.	.	.	−4	15	−90
$(5)(2)^2(1)^3$	−4	60	−840	.	.	.	.	.	.	.	.	−3	40
$(5)(2)(1)^5$	.	−6	168	.	.	.	.	.	.	.	.	.	−2
$(5)(1)^7$	.	.	−8	.	.	.	.	.	.	.	.	.	.
$(4)^3$	30	−180	1260	.	.	−6	.	−2	24	−180	−3	18	−150
$(4)^2(3)(1)$	−56	420	−3360	.	.	18	.	4	−72	600	3	−36	430
$(4)^2(2)^2$	−33	180	−1260	.	.	.	.	2	−12	90	6	−27	165
$(4)^2(2)(1)^2$	30	−270	2520	.	.	.	.	.	12	−180	.	9	−150
$(4)^2(1)^4$	−1	30	−420	.	.	.	.	.	.	30	.	.	5
$(4)(3)^2(2)$	−28	160	−1120	.	−3	9	−3	13	−58	330	12	−69	420
$(4)(3)^2(1)^2$	12	−120	1120	.	.	−9	.	−1	36	−330	.	9	−200
$(4)(3)(2)^2(1)$	32	−210	1680	.	.	.	.	−4	24	−180	−6	45	−325
$(4)(3)(2)(1)^3$	−8	100	−1120	.	.	.	.	.	−8	120	.	−3	90
$(4)(3)(1)^5$	.	−6	112	.	.	.	.	.	.	−12	.	.	−1
$(4)(2)^4$	3	−15	105	.	.	.	.	.	.	.	−1	3	−15
$(4)(2)^3(1)^2$	−6	45	−420	.	.	.	.	.	.	.	.	−3	30
$(4)(2)^2(1)^4$	1	−15	210	.	.	.	.	.	.	.	.	.	−5
$(4)(2)(1)^6$	15	1	−28	.	.	.	.	.	.	.	.	.	.
$(4)(1)^8$	210	28	1	.	.	.	.	.	.	.	.	.	.
$(3)^4$	.	.	.	1	−1	2	.	2	−8	40	.	−6	40
$(3)^3(2)(1)$	.	.	.	1	1	−3	.	−4	20	−120	.	18	−140
$(3)^3(1)^3$	.	.	.	1	3	1	1	.	−4	40	.	11	20
$(3)^2(2)^3$	.	.	.	.	.	.	1	−1	3	−15	−4	−9	−50
$(3)^2(2)^2(1)^2$	.	.	.	2	4	.	1	1	−6	45	.	.	80
$(3)^2(2)(1)^4$	.	.	.	4	16	4	3	6	1	−15	.	.	−10
$(3)^2(1)^6$	.	.	.	10	60	20	15	45	15	1	.	.	.
$(3)(2)^4(1)$	.	.	.	.	.	.	4	.	.	.	1	−3	15
$(3)(2)^3(1)^3$	.	.	.	6	18	.	10	9	.	.	3	1	−10
$(3)(2)^2(1)^5$	.	.	.	20	100	20	40	70	10	.	15	10	1
$(3)(2)(1)^7$	5	7	.	70	490	140	210	525	140	7	105	105	21
$(3)(1)^9$	1890	252	9	280	2520	840	1260	3780	1260	84	945	1260	378
$(2)^6$	.	.	.	.	.	.	20	.	.	.	10	.	.
$(2)^5(1)^2$	6	.	.	24	96	.	88	72	.	.	52	16	.
$(2)^4(1)^4$	90	3	1	120	720	120	460	720	90	56	330	200	18
$(2)^3(1)^6$	1120	84	.	560	4480	1120	2800	6160	1400	.	2520	2240	392
$(2)^2(1)^8$	12600	1470	45	2800	28000	8400	18900	50400	14700	840	22050	25200	6300
$(1)^{12}$	103950	13860	495	15400	184800	61600	138600	415800	138600	9240	207900	277200	83160

Table 1·1·12 (*cont.*)

$w = 12$ (vi)	[321⁷]	[31⁸]	[2⁶]	[2⁵1²]	[2⁴1⁴]	[2³1⁶]	[2²1⁸]	[21¹⁰]	[1¹²]
(12)	40320	−362880	−120	720	−5040	40320	−362880	3628800	−39916800
(11) (1)	−35280	36288o	.	−240	2880	−30240	322560	−3628800	43545600
(10) (2)	−20160	181440	144	−624	3600	−25920	221760	−2177280	23950080
(10) (1)²	15120	−181440	.	24	−720	10800	−141120	1814400	−23950080
(9) (3)	−18480	161280	.	−240	2112	−17760	161280	−1612800	17740800
(9) (2) (1)	17640	−181440	.	240	−2208	20160	−201600	2217600	−26611200
(9) (1)³	−4200	60480	.	.	96	−2400	40320	−604800	8870400
(8) (4)	−15120	136080	90	−300	1908	−15120	136080	−1360800	14968800
(8) (3) (1)	16800	−166320	.	60	−1200	13680	−147840	1663200	−19958400
(8) (2)²	5040	−45360	−90	270	−1314	8640	−70560	680400	−7484400
(8) (2) (1)²	−7560	90720	.	−30	612	−7560	90720	−1134000	14968800
(8) (1)⁴	840	−15120	.	.	−6	360	−8400	151200	−2494800
(7) (5)	−13824	124416	.	−240	1728	−13824	124416	−1244160	13685760
(7) (4) (1)	13860	−142560	.	120	−1152	11880	−126720	1425600	−17107200
(7) (3) (2)	9960	−86400	.	240	−1632	12240	−105600	1036800	−11404800
(7) (3) (1)²	−7560	86400	.	.	288	−5040	67200	−864000	11404800
(7) (2)² (1)	−4410	45360	.	−120	864	−7020	66240	−712800	8553600
(7) (2) (1)³	2100	−30240	.	.	−96	1800	−26880	388800	−5702400
(7) (1)⁵	−126	3024	.	.	.	−36	1344	−30240	570240
(6)²	−6720	60480	40	−120	840	−6720	60480	−604800	6652800
(6) (5) (1)	12936	−133056	.	80	−1056	11088	−118272	1330560	−15966720
(6) (4) (2)	8400	−75600	−120	280	−1512	10800	−92400	907200	−9979200
(6) (4) (1)²	−6300	75600	.	−20	312	−4500	58800	−756000	9979200
(6) (3)²	4760	−40320	.	40	−496	4400	−40320	403200	−4435200
(6) (3) (2) (1)	−9240	90720	.	−80	1024	−9960	100800	−1108800	13305600
(6) (3) (1)³	2240	−30240	.	.	−32	1160	−20160	302400	−4435200
(6) (2)³	−840	7560	40	−80	336	−2040	15960	−151200	1663200
(6) (2)² (1)²	1890	−22680	.	20	−264	2790	−31080	378000	−4989600
(6) (2) (1)⁴	−420	7560	.	.	8	−300	5880	−100800	1663200
(6) (1)⁶	14	−504	.	.	.	2	−168	5040	−110880
(5)² (2)	4032	−36288	.	120	−720	5184	−44352	435456	−4790016
(5)² (1)²	−3024	36288	.	.	144	−2160	28224	−362880	4790016
(5) (4) (3)	8316	−72576	.	120	−960	7992	−72576	725760	−7983360
(5) (4) (2) (1)	−7938	81648	.	−120	1008	−9072	90720	−997920	11975040
(5) (4) (1)³	1890	−27216	.	.	−48	1080	−18144	272160	−3991680
(5) (3)² (1)	−4648	44352	.	.	288	−3600	39424	−443520	5322240
(5) (3) (2)²	−2814	24192	.	−120	672	−4572	37632	−362880	3991680
(5) (3) (2) (1)²	4284	−48384	.	.	−288	3960	−48384	604800	−7983360
(5) (3) (1)⁴	−490	8064	.	.	.	−180	4480	−80640	1330560
(5) (2)³ (1)	735	−7560	.	40	−240	1764	−15792	166320	−1995840
(5) (2)² (1)³	−525	7560	.	.	48	−720	9744	−136080	1995840
(5) (2) (1)⁵	63	−1512	.	.	.	36	−1008	21168	−399168
(5) (1)⁷	−1	72	.	.	.	.	16	−720	19008
(4)³	1260	−11340	−15	30	−162	1260	−11340	113400	−1247400
(4)² (3) (1)	−4200	41580	.	−30	312	−3420	36960	−415800	4989600
(4)² (2)²	−1260	11340	45	−75	333	−2160	17640	−170100	1871100
(4)² (2) (1)²	1890	−22680	.	15	−162	1890	−22680	283500	−3742200
(4)² (1)⁴	−210	3780	.	.	3	−90	2100	−37800	623700
(4) (3)² (2)	−3010	25200	.	−60	456	−3540	30800	−302400	3326400
(4) (3)² (1)²	2310	−25200	.	.	−72	1440	−19600	252000	−3326400
(4) (3) (2)² (1)	2730	−26460	.	60	−480	4050	−38640	415800	−4989600
(4) (3) (2) (1)³	−1330	17640	.	.	48	−1020	15680	−226800	3326400
(4) (3) (1)⁵	84	−1764	.	.	.	18	−784	17640	−332640
(4) (2)⁴	105	−945	−15	20	−72	405	−3045	28350	−311850
(4) (2)³ (1)²	−315	3780	.	−10	84	−765	7980	−94500	1247400
(4) (2)² (1)⁴	105	−1890	.	.	−6	135	−2310	37800	−623700
(4) (2) (1)⁶	−7	252	.	.	.	−3	140	−3780	83160
(4) (1)⁸	.	−9	.	.	.	.	−1	90	−2970
(3)⁴	−280	2240	.	.	24	−240	2240	−22400	246400
(3)³ (2) (1)	1120	−10080	.	.	−96	1080	−11200	123200	−1478400
(3)³ (1)³	−280	3360	.	.	.	−120	2240	−33600	492800
(3)² (2)³	315	−2520	.	20	−104	670	−5320	50400	−554400
(3)² (2)² (1)²	−735	7560	.	.	72	−900	10360	−126000	1663200
(3)² (2) (1)⁴	175	−2520	.	.	.	90	−1960	33600	−554400
(3)² (1)⁶	−7	168	.	.	.	.	56	−1680	36960
(3) (2)⁴ (1)	−105	945	.	−10	56	−390	3360	−34650	415800
(3) (2)³ (1)³	105	−1260	.	.	−16	220	−2800	37800	−554400
(3) (2)² (1)⁵	−21	378	.	.	.	−18	448	−8820	166320
(3) (2) (1)⁷	1	−36	.	.	.	.	−16	600	−15840
(3) (1)⁹	36	1	.	.	.	.	.	−10	440
(2)⁶	.	.	1	−1	3	−15	105	−945	10395
(2)⁵ (1)²	.	.	1	1	−6	45	−420	4725	−62370
(2)⁴ (1)⁴	.	.	3	6	1	−15	210	−3150	51975
(2)³ (1)⁶	.	.	15	45	15	1	−28	630	−13860
(2)² (1)⁸	16	.	105	420	210	28	1	−45	1485
(2) (1)¹⁰	480	10	945	4725	3150	630	45	1	−66
(1)¹²	7920	220	10395	62370	51975	13860	1485	66	1

Table 1·2. *Monomial symmetric functions in terms of the unitary functions (a-functions) and conversely (up to and including weight 12)*

Note that the final digit (or digits) in the subheadings indicates the weight, w.

To express a monomial symmetric in terms of a-functions the appropriate coefficients are found by reading vertically downwards up to and including the diagonal, the unit entry in which is shown in bold type, e.g. from Table 1·2·11

$$(2^3 1^5) = -77a_{11} + 27a_{10}a_1 - 7a_9a_2 + a_8a_3.$$

Conversely, to express a product of a-functions in terms of monomial symmetrics, the appropriate coefficients are found by reading horizontally up to and including the bold-faced diagonal unit. Thus again from Table 1·2·11

$$a_8 a_3 = 165(1^{11}) + 36(21^9) + 7(2^2 1^7) + (2^3 1^5).$$

Table 1·2·2

$w=2$	(1^2)	(2)
a_2	**1**	-2
a_1^2	2	**1**

Table 1·2·3

$w=3$	(1^3)	(21)	(3)
a_3	**1**	-3	3
$a_2 a_1$	3	**1**	-3
a_1^3	6	3	**1**

Table 1·2·4

$w=4$	(1^4)	(21^2)	(2^2)	(31)	(4)
a_4	**1**	-4	2	4	-4
$a_3 a_1$	4	**1**	-2	-1	4
a_2^2	6	2	**1**	-2	2
$a_2 a_1^2$	12	5	2	**1**	-4
a_1^4	24	12	6	4	**1**

Table 1·2·5

$w=5$	(1^5)	(21^3)	$(2^2 1)$	(31^2)	(32)	(41)	(5)
a_5	**1**	-5	5	5	-5	-5	5
$a_4 a_1$	5	**1**	-3	-1	5	1	-5
$a_3 a_2$	10	3	**1**	-2	-1	5	-5
$a_3 a_1^2$	20	7	2	**1**	-2	-1	5
$a_2^2 a_1$	30	12	5	2	**1**	-3	5
$a_2 a_1^3$	60	27	12	7	3	**1**	-5
a_1^5	120	60	30	20	10	5	**1**

Table 1·2·6

$w=6$	(1^6)	(21^4)	$(2^2 1^2)$	(31^3)	(2^3)	(321)	(41^2)	(3^2)	(42)	(51)	(6)
a_6	**1**	-6	9	6	-2	-12	-6	3	6	6	-6
$a_5 a_1$	6	**1**	-4	-1	2	7	1	-3	-6	-1	6
$a_4 a_2$	15	4	**1**	-2	-2	4	2	-3	2	-6	6
$a_4 a_1^2$	30	9	2	**1**	·	-3	-1	3	2	1	-6
a_3^2	20	6	2	·	**1**	-3	3	3	-3	-3	3
$a_3 a_2 a_1$	60	22	8	3	3	**1**	-3	3	4	7	-12
$a_3 a_1^3$	120	48	18	10	6	3	**1**	·	-2	-1	6
a_2^3	90	36	15	6	6	3	·	**1**	-2	2	-2
$a_2^2 a_1^2$	180	78	34	18	15	8	2	2	**1**	-4	9
$a_2 a_1^4$	360	168	78	48	36	22	9	6	4	**1**	-6
a_1^6	720	360	180	120	90	60	30	20	15	6	**1**

Table 1·2·7

$w=7$	(1^7)	(21^5)	$(2^2 1^3)$	(31^4)	$(2^3 1)$	(321^2)	(41^3)	(32^2)	$(3^2 1)$	(421)	(51^2)	(43)	(52)	(61)	(7)
a_7	**1**	-7	14	7	-7	-21	-7	7	7	14	7	-7	-7	-7	7
$a_6 a_1$	7	**1**	-5	-1	5	9	1	-7	-4	-8	-1	7	7	**1**	-7
$a_5 a_2$	21	5	**1**	-2	-3	6	2	3	-7	-4	-2	7	-3	7	-7
$a_5 a_1^2$	42	11	2	**1**	·	-4	-1	2	4	3	1	-7	2	-1	7
$a_4 a_3$	35	10	3	·	**1**	·	1	3	-1	5	-2	-5	7	7	-7
$a_4 a_2 a_1$	105	35	11	4	3	**1**	-3	3	-2	8	3	-2	-4	-8	14
$a_4 a_1^3$	210	75	24	13	6	3	**1**	·	·	-3	-1	3	2	1	-7
$a_3^2 a_1$	140	50	18	6	7	2	·	**1**	-2	-1	4	5	-7	-4	7
$a_3 a_2^2$	210	80	31	12	12	5	·	2	**1**	-2	2	-1	3	-7	7
$a_3 a_2 a_1^2$	420	170	68	34	27	13	3	5	2	**1**	-4	-3	6	9	-21
$a_3 a_1^4$	840	360	150	88	60	34	13	12	6	4	**1**	·	-2	-1	7
$a_2^3 a_1$	630	270	117	60	51	27	6	12	7	3	·	**1**	-3	5	-7
$a_2^2 a_1^3$	1260	570	258	150	117	68	24	31	18	11	2	3	**1**	-5	14
$a_2 a_1^5$	2520	1200	570	360	270	170	75	80	50	35	11	10	5	**1**	-7
a_1^7	5040	2520	1260	840	630	420	210	210	140	105	42	35	21	7	**1**

Table 1·2·8

$w = 8$ (i)	(1^8)	(21^6)	(2^21^4)	(31^5)	(2^31^2)	(321^3)	(41^4)	(2^4)	(32^21)	(3^21^2)
a_8	1	−8	20	8	−16	−32	−8	2	24	12
a_7a_1	8	1	−6	−1	9	11	1	−2	−17	−5
a_6a_2	28	6	1	−2	−4	8	2	2	·	−9
$a_6a_1^2$	56	13	2	1	·	−5	−1	2	5	5
a_5a_3	56	15	4	·	1	−3	3	−2	6	3
$a_5a_2a_1$	168	51	14	5	3	1	−3	·	−3	−1
$a_5a_1^3$	336	108	30	16	6	6	1	·	·	·
a_4^2	70	20	6	·	2	·	·	1	−4	2
$a_4a_3a_1$	280	95	32	10	11	3	·	4	1	−2
$a_4a_2^2$	420	150	53	20	18	7	·	6	2	1
$a_4a_2a_1^2$	840	315	114	55	39	18	4	12	5	2
$a_4a_1^4$	1680	660	246	140	84	46	17	24	12	6
$a_3^2a_2$	560	210	80	30	31	12	·	12	5	2
$a_3^2a_1^2$	1120	440	172	80	68	30	6	28	12	4
$a_3a_2^2a_1$	1680	690	284	140	117	58	12	48	24	12
$a_3a_2a_1^3$	3360	1440	612	340	258	141	46	108	58	30
$a_3a_1^5$	6720	3000	1320	800	570	340	140	240	140	80
a_2^4	2520	1080	468	240	204	108	24	90	48	28
$a_2^3a_1^2$	5040	2250	1008	570	453	258	84	204	117	68
$a_2^2a_1^4$	10080	4680	2172	1320	1008	612	246	468	284	172
$a_2a_1^6$	20160	9720	4680	3000	2250	1440	660	1080	690	440
a_1^8	40320	20160	10080	6720	5040	3360	1680	2520	1680	1120

$w = 8$ (ii)	(421^2)	(51^3)	(3^22)	(42^2)	(431)	(521)	(61^2)	(4^2)	(53)	(62)	(71)	(8)
a_8	24	8	−8	−8	−16	−16	−8	4	8	8	8	−8
a_7a_1	−10	−1	8	+8	9	9	1	−4	−8	−8	−1	8
a_6a_2	−6	−2	2	−4	16	4	2	−4	−8	4	−8	8
$a_6a_1^2$	4	1	−5	−2	−9	−3	−1	4	8	2	1	−8
a_5a_3	−9	−3	−7	8	1	1	8	−4	7	−8	−8	8
$a_5a_2a_1$	11	3	5	−4	−10	−8	−3	8	1	4	9	−16
$a_5a_1^3$	−4	−1	·	2	4	3	1	−4	−3	−2	−1	8
a_4^2	4	−4	4	−4	−8	8	4	6	−4	−4	−4	4
$a_4a_3a_1$	−1	4	−1	·	10	−10	−9	−8	1	16	9	−16
$a_4a_2^2$	−2	2	−2	4	·	−4	−2	−4	8	−4	8	−8
$a_4a_2a_1^2$	1	−4	·	−2	−1	11	4	4	−9	−6	−10	24
$a_4a_1^4$	4	1	·	·	−3	−1	·	·	3	2	1	−8
$a_3^2a_2$	·	·	1	−2	−1	5	−5	·	−7	2	8	−8
$a_3^2a_1^2$	2	·	2	1	−2	−1	5	4	3	−9	−5	12
$a_3a_2^2a_1$	5	·	5	2	1	−3	5	−4	6	·	−17	24
$a_3a_2a_1^3$	18	3	12	7	3	1	5	·	−3	8	11	−32
$a_3a_1^5$	55	16	30	20	10	5	1	·	−2	−1	·	8
a_2^4	12	·	12	6	4	·	·	1	−2	2	−2	2
$a_2^3a_1^2$	39	6	31	18	11	3	·	2	1	−4	9	−16
$a_2^2a_1^4$	114	30	80	53	32	14	2	6	4	1	−6	20
$a_2a_1^6$	315	108	210	150	95	51	13	20	15	6	1	−8
a_1^8	840	336	560	420	280	168	56	70	56	28	8	1

Table 1·2·9

$w=9$ (i)	(1^9)	(21^7)	$(2^2 1^5)$	(31^6)	$(2^3 1^3)$	(321^4)	(41^5)	$(2^4 1)$	$(32^2 1^2)$	$(3^2 1^3)$	(421^3)	(51^4)	(32^3)
a_9	1	-9	27	9	-30	-45	-9	9	54	18	36	9	-9
$a_8 a_1$	9	1	-7	-1	14	13	1	-7	-30	-6	-12	-1	9
$a_7 a_2$	36	7	1	-2	-5	10	2	5	-5	-11	-8	-2	-5
$a_7 a_1^2$	72	15	2	1	·	-6	-1	·	9	6	5	1	-2
$a_6 a_3$	84	21	5	·	1	-3	3	-3	9	3	-12	-3	3
$a_6 a_2 a_1$	252	70	17	6	3	1	-3	·	-4	-1	14	3	2
$a_6 a_1^3$	504	147	36	19	6	3	3	·	·	·	-5	-1	·
$a_5 a_4$	126	35	10	·	3	·	·	1	-4	2	4	-4	-1
$a_5 a_3 a_1$	504	161	50	15	15	4	4	4	1	-2	-1	4	-2
$a_5 a_2^2$	756	252	81	30	24	9	·	6	2	1	-2	2	·
$a_5 a_2 a_1^2$	1512	525	172	81	51	23	5	12	5	2	1	-4	·
$a_5 a_1^4$	3024	1092	366	204	108	58	21	24	12	6	4	1	·
$a_4^2 a_1$	630	210	70	20	24	6	·	9	2	·	·	·	1
$a_4 a_3 a_2$	1260	455	165	60	60	22	·	22	8	3	·	·	3
$a_4 a_3 a_1^2$	2520	945	350	155	129	54	10	48	19	6	3	·	7
$a_4 a_2^2 a_1$	3780	1470	565	270	213	101	20	78	36	17	7	·	12
$a_4 a_2 a_1^3$	7560	3045	1200	645	459	241	75	168	85	42	25	4	27
$a_4 a_1^5$	15120	6300	2550	1500	990	570	225	360	200	110	75	21	60
a_3^3	1680	630	240	90	93	36	·	36	15	6	·	·	6
$a_3^2 a_2 a_1$	5040	2030	820	390	333	158	30	136	65	30	12	·	27
$a_3^2 a_1^3$	10080	4200	1740	920	720	372	110	300	152	72	42	6	64
$a_3 a_2^3$	7560	3150	1320	660	555	282	60	234	120	64	27	12	51
$a_3 a_2^2 a_1^2$	15120	6510	2800	1530	1203	656	200	516	281	152	85	12	120
$a_3 a_2 a_1^4$	30240	13440	5940	3480	2610	1516	570	1140	656	372	241	58	282
$a_3 a_1^6$	60480	27720	12600	7800	5670	3480	1500	2520	1530	920	645	204	660
$a_2^4 a_1$	22680	10080	4500	2520	2016	1140	360	906	516	300	168	24	234
$a_2^3 a_1^3$	45360	20790	9540	5670	4383	2610	990	2016	1203	720	459	108	555
$a_2^2 a_1^5$	90720	42840	20220	12600	9540	5940	2550	4500	2800	1740	1200	366	1320
$a_2 a_1^7$	181440	88200	42840	27720	20790	13440	6300	10080	6510	4200	3045	1092	3150
a_1^9	362880	181440	90720	60480	45360	30240	15120	22680	15120	10080	7560	3024	7560

$w=9$ (ii)	$(3^2 21)$	$(42^2 1)$	(431^2)	(521^2)	(61^3)	(3^3)	(432)	(52^2)	$(4^2 1)$	(531)	(621)	(71^2)	(54)	(63)	(72)	(81)	(9)
a_9	-27	-27	-27	-27	-9	3	18	9	9	18	18	9	-9	-9	-9	-9	9
$a_8 a_1$	19	19	11	11	1	-3	-18	-9	-5	-10	-10	-1	9	9	9	1	-9
$a_7 a_2$	13	-1	20	6	2	-3	-4	5	-9	-18	-4	-2	9	9	-5	9	-9
$a_7 a_1^2$	-12	-5	-11	-4	-1	3	11	2	5	10	3	1	-9	-9	-2	-1	9
$a_6 a_3$	-18	9	·	9	3	6	-9	-9	-9	·	8	-9	9	-9	9	9	-9
$a_6 a_2 a_1$	7	-12	-13	-11	-3	-3	4	4	14	10	8	3	-18	·	-4	-10	18
$a_6 a_1^3$	·	5	4	4	1	·	-5	-2	-5	-4	-3	-1	9	3	2	1	-9
$a_5 a_4$	7	-3	-13	7	9	-3	2	1	11	2	-18	-9	-11	9	9	9	-9
$a_5 a_3 a_1$	4	2	-15	-15	-3	-3	-2	8	-6	5	10	10	2	·	-18	-10	18
$a_5 a_2^2$	-3	6	·	-6	-2	3	-6	·	-1	8	4	2	1	-9	5	-9	9
$a_5 a_2 a_1^2$	·	-3	-1	15	4	·	5	-6	1	-15	-11	-4	7	9	6	11	-27
$a_5 a_1^4$	·	·	·	-4	-1	·	·	2	·	4	3	1	-4	-3	-2	-1	9
$a_4^2 a_1$	-3	-1	5	1	-5	3	-2	-1	-5	-6	14	5	11	-9	-9	-5	18
$a_4 a_3 a_2$	1	-2	-1	5	-5	-3	8	-6	-2	-2	4	11	·	·	-4	-18	18
$a_4 a_3 a_1^2$	2	1	-2	-1	5	·	-1	·	·	5	-11	-13	-13	9	20	11	-27
$a_4 a_2^2 a_1$	5	2	1	-3	5	·	-2	6	-1	2	-12	-5	-3	9	-1	19	-27
$a_4 a_2 a_1^3$	12	7	3	1	-5	·	·	-2	·	-1	14	5	4	-12	-8	-12	36
$a_4 a_1^5$	30	20	10	5	1	·	-3	·	·	-3	-3	-1	·	3	2	1	-9
a_3^3	3	·	·	2	·	1	3	3	3	-3	-3	3	-3	6	-3	-3	3
$a_3^2 a_2 a_1$	13	5	2	·	·	3	1	-3	-3	4	7	-12	7	-18	19	19	-27
$a_3^2 a_1^3$	30	17	6	2	·	6	3	1	·	-2	-1	6	2	3	-11	-6	18
$a_3 a_2^3$	27	12	7	·	·	6	3	·	1	-2	2	-2	-1	3	-5	9	-9
$a_3 a_2^2 a_1^2$	65	36	19	5	15	8	2	2	2	1	-4	9	-4	9	-5	-30	54
$a_3 a_2 a_1^4$	158	101	54	23	36	36	22	9	6	4	1	-6	·	-3	10	13	-45
$a_3 a_1^6$	390	270	155	81	19	90	60	30	20	15	6	·	1	·	-2	-1	9
$a_2^4 a_1$	136	78	48	12	6	36	22	6	9	4	·	1	·	-3	5	-7	9
$a_2^3 a_1^3$	333	213	129	51	6	93	60	24	24	15	3	2	3	1	-5	14	-30
$a_2^2 a_1^5$	820	565	350	172	36	240	165	81	70	50	17	10	5	·	5	-7	+27
$a_2 a_1^7$	2030	1470	945	525	147	630	455	252	210	161	70	15	35	21	7	1	-9
a_1^9	5040	3780	2520	1512	504	1680	1260	756	630	504	252	72	126	84	36	9	1

Table 1·2·10

$w = 10$ (i)	(1^{10})	(21^8)	$(2^2 1^6)$	(31^7)	$(2^3 1^4)$	(321^5)	(41^6)	$(2^4 1^2)$	$(32^2 1^3)$
a_{10}	1	−10	35	10	−50	−60	−10	25	100
$a_9 a_1$	10	1	−8	−1	20	15	1	−16	−40
$a_8 a_2$	45	8	1	−2	−6	12	2	9	−12
$a_8 a_1^2$	90	17	2	1	.	−7	−1	.	14
$a_7 a_3$	120	28	6	.	1	−3	3	−4	12
$a_7 a_2 a_1$	360	92	20	7	3	1	−3	.	−5
$a_7 a_1^3$	720	192	42	22	6	3	1	.	.
$a_6 a_4$	210	56	15	.	4	.	.	1	−4
$a_6 a_3 a_1$	840	252	72	21	19	5	.	4	1
$a_6 a_2^2$	1260	392	115	42	30	11	.	6	2
$a_6 a_2 a_1^2$	2520	812	242	112	63	28	6	12	5
$a_6 a_1^4$	5040	1680	510	280	132	70	25	24	12
a_5^2	252	70	20	.	6	.	.	2	.
$a_5 a_4 a_1$	1260	406	130	35	42	10	.	14	3
$a_5 a_3 a_2$	2520	868	296	105	99	35	.	32	11
$a_5 a_3 a_1^2$	5040	1792	622	266	210	85	15	68	26
$a_5 a_2^2 a_1$	7560	2772	990	462	339	156	30	108	48
$a_5 a_2 a_1^3$	15120	5712	2082	1092	720	368	111	228	112
$a_5 a_1^5$	30240	11760	4380	2520	1530	860	330	480	260
$a_4^2 a_2$	3150	1120	400	140	144	50	.	53	18
$a_4^2 a_1^2$	6300	2310	840	350	306	120	20	114	42
$a_4 a_3^2$	4200	1540	570	210	213	80	.	80	31
$a_4 a_3 a_2 a_1$	12600	4900	1900	875	735	335	60	284	128
$a_4 a_3 a_1^3$	25200	10080	3990	2030	1566	775	215	612	294
$a_4 a_2^3$	18900	7560	3015	1470	1194	585	120	468	228
$a_4 a_2^2 a_1^2$	37800	15540	6330	3360	2547	1340	390	1008	523
$a_4 a_2 a_1^4$	75600	31920	13290	7560	5436	3050	1095	2172	1196
$a_4 a_1^6$	151200	65520	27900	16800	11610	6900	2850	4680	2730
$a_3^3 a_1$	16800	6720	2700	1260	1092	510	90	444	210
$a_3^2 a_2^2$	25200	10360	4280	2100	1776	880	180	740	370
$a_3^2 a_2 a_1^2$	50400	21280	8980	4760	3792	2000	570	1604	844
$a_3^2 a_1^4$	100800	43680	18840	10640	8100	4520	1580	3480	1920
$a_3 a_2^3 a_1$	75600	32760	14220	7770	6180	3390	1020	2688	1479
$a_3 a_2^2 a_1^3$	151200	67200	29820	17220	13212	7610	2730	5844	3358
$a_3 a_2 a_1^5$	302400	137760	62520	37800	28260	17000	6900	12720	7610
$a_3 a_1^7$	604800	282240	131040	82320	60480	37800	16800	27720	17220
a_2^5	113400	50400	22500	12600	10080	5700	1800	4530	2580
$a_2^4 a_1^2$	226800	103320	47160	27720	21564	12720	4680	9876	5844
$a_2^3 a_1^4$	453600	211680	98820	60480	46152	28260	11610	21564	13212
$a_2^2 a_1^6$	907200	433440	207000	131040	98820	62520	27900	47160	29820
$a_2 a_1^8$	1814400	887040	433440	282240	211680	137760	65520	103320	67200
a_1^{10}	3628800	1814400	907200	604800	453600	302400	151200	226800	151200

$w = 10$ (ii)	$(3^2 1^4)$	(421^4)	(51^5)	(2^5)	$(32^3 1)$	$(3^2 2 1^2)$	$(42^2 1^2)$	(431^3)	(521^3)	(61^4)
a_{10}	25	50	10	−2	−40	−60	−60	−40	−40	−10
$a_9 a_1$	−7	−14	−1	2	31	33	33	13	13	1
$a_8 a_2$	−13	−10	−2	−2	−8	28	4	24	8	2
$a_8 a_1^2$	7	6	1	.	−7	−21	−9	−13	−5	−1
$a_7 a_3$	3	−15	−3	2	−2	−24	18	12	12	3
$a_7 a_2 a_1$	−1	17	3	.	5	9	−23	−16	−14	−3
$a_7 a_1^3$	.	−6	−1	.	.	9	9	6	5	1
$a_6 a_4$	2	4	−4	−2	8	.	−12	−8	16	4
$a_6 a_3 a_1$	−2	−1	4	.	−3	6	3	5	−19	−4
$a_6 a_2^2$	1	−2	2	.	.	−4	8	.	−8	−2
$a_6 a_2 a_1^2$	2	1	−4	.	.	.	−4	−1	19	4
$a_6 a_1^4$	6	4	1	.	.	.	.	.	−5	−1
a_5^2	.	.	.	1	−5	5	5	−5	−5	5
$a_5 a_4 a_1$	.	.	.	5	1	−3	−1	5	1	−5
$a_5 a_3 a_2$	4	.	.	10	3	1	−2	−1	5	−5
$a_5 a_3 a_1^2$	8	4	.	20	7	2	1	−2	−1	+5
$a_5 a_2^2 a_1$	22	9	.	30	12	5	2	1	−3	5
$a_5 a_2 a_1^3$	54	32	5	60	27	12	7	3	1	−5
$a_5 a_1^5$	140	95	26	120	60	30	20	10	5	1
$a_4^2 a_2$	6	.	.	20	7	2	.	.	.	.
$a_4^2 a_1^2$	12	6	.	45	16	4	2	.	.	.
$a_4 a_3^2$	12	.	.	30	12	5	.	.	.	.
$a_4 a_3 a_2 a_1$	56	22	.	110	49	21	8	3	.	.
$a_4 a_3 a_1^3$	132	76	10	240	112	48	27	9	3	.
$a_4 a_2^3$	115	48	.	180	87	42	18	10	.	.
$a_4 a_2^2 a_1^2$	270	149	20	390	198	99	54	27	7	.
$a_4 a_2 a_1^4$	650	416	95	840	450	236	149	76	32	4
$a_4 a_1^6$	1580	1095	330	1800	1020	570	390	215	111	25
$a_3^3 a_1$	96	36	.	180	87	42	15	6	.	.
$a_3^2 a_2^2$	188	78	.	310	156	80	34	18	.	.
$a_3^2 a_2 a_1^2$	436	236	30	680	358	186	99	48	12	.
$a_3^2 a_1^4$	1032	650	140	1500	820	436	270	132	54	6
$a_3 a_2^3 a_1$	820	450	60	1170	645	358	198	112	27	.
$a_3 a_2^2 a_1^3$	1920	1196	260	2580	1479	844	523	294	112	12
$a_3 a_2 a_1^5$	4520	3050	860	5700	3390	2000	1340	775	368	70
$a_3 a_1^7$	10640	7560	2520	12600	7770	4760	3360	2030	1092	280
a_2^5	1500	840	120	2040	1170	680	390	240	60	.
$a_2^4 a_1^2$	3480	2172	480	4530	2688	1604	1008	612	228	24
$a_2^3 a_1^4$	8100	5436	1530	10080	6180	3792	2547	1566	720	132
$a_2^2 a_1^6$	18840	13290	4380	22500	14220	8980	6330	3990	2082	510
$a_2 a_1^8$	43680	31920	11760	50400	32760	21280	15540	10080	5712	1680
a_1^{10}	100800	75600	30240	113400	75600	50400	37800	25200	15120	5040

Table 1·2·10 (*cont.*)

$w=10$ (iii)	$(3^2 2^2)$	$(4 2^3)$	$(3^3 1)$	(4321)	$(5 2^2 1)$	$(4^2 1^2)$	$(5 3 1^2)$	$(6 2 1^2)$	$(7 1^3)$	$(4 3^2)$	$(4^2 2)$
a_{10}	15	10	10	60	30	−6	30	30	10	−10	−10
$a_9 a_1$	−15	−10	−7	−42	−21	−11	−12	−12	−1	10	10
$a_8 a_2$	1	6	−10	−28	2	−11	−22	−6	−2	10	2
$a_8 a_1^2$	7	2	7	26	5	6	12	4	1	−10	−6
$a_7 a_3$	6	−10	11	3	−9	−15	−9	−9	−3	−11	10
$a_7 a_2 a_1$	−7	4	−4	21	12	17	13	11	3	1	−12
$a_7 a_1^3$		−2		12	−12	−6	−5	−6	−1	3	6
$a_6 a_4$	−9	10	2	−15	−18	9	−6	15	4	13	−14
$a_6 a_3 a_1$	3		−7	−15	18	−3	15	15	2	4	4
$a_6 a_2^2$	2	−4	4	−8	4	−1	10	6	2	−4	10
$a_6 a_2 a_1^2$		2		7	−17	1	−19	−15	−4	−3	−6
$a_6 a_1^4$					5		5	4	1		
a_5^2	5	−5	−5	−5	10	5	10	−15	−5	5	5
$a_5 a_4 a_1$	−1		5	5	−1	−8	−12	18	11	−8	4
$a_5 a_3 a_2$	−2	4	−1	5	−13	1	−4	15	5	1	−12
$a_5 a_3 a_1^2$		−2		4	2	2	3	−19	−5	−3	2
$a_5 a_2^2 a_1$				−3	9	−1	2	−17	−5	3	2
$a_5 a_2 a_1^3$					−3		−1	19	5		
$a_5 a_1^5$								−4	−1		
$a_4^2 a_2$	1	−2	−2	4	2	−3	2	−6	6	2	2
$a_4^2 a_1^2$	2	1		−3	−1	3	2	1	−6	3	−3
$a_4 a_3^2$	2		1	−3	3	3	−3	−3	3	−1	2
$a_4 a_3 a_2 a_1$	8	3	3	1	−3	−3	4	7	−12	−3	4
$a_4 a_3 a_1^3$	18	10	6	3	1		−2	−1	6		
$a_4 a_2^3$	15	6	6	3		1	−2	−2	−2		−2
$a_4 a_2^2 a_1^2$	34	18	15	8	2	2	1	−4	9		
$a_4 a_2 a_1^4$	78	48	36	22	9	6	4	4	−6		
$a_4 a_1^6$	180	120	90	60	30	20	15	6	1		
$a_3^3 a_1$	18	6	10	3		2				1	−2
$a_3^2 a_2^2$	34	15	18	8		2				2	1
$a_3^2 a_2 a_1^2$	80	42	42	21	5	4	2			5	2
$a_3^2 a_1^4$	188	115	96	56	22	12	8	2		12	6
$a_3 a_2^3 a_1$	156	87	87	49	12	16	7			12	7
$a_3 a_2^2 a_1^3$	370	228	210	128	48	42	26	5		31	18
$a_3 a_2 a_1^5$	880	585	510	335	156	120	85	28	3	80	50
$a_3 a_1^7$	2100	1470	1260	875	462	350	266	112	22	210	140
a_2^5	310	180	180	110	30	45	20			30	20
$a_2^4 a_1^2$	740	468	444	284	108	114	68	12		80	53
$a_2^3 a_1^4$	1776	1194	1092	735	339	306	210	63	6	213	144
$a_2^2 a_1^6$	4280	3015	2700	1900	990	840	622	242	42	570	400
$a_2 a_1^8$	10360	7560	6720	4900	2772	2310	1792	812	192	1540	1120
a_1^{10}	25200	18900	16800	12600	7560	6300	5040	2520	720	4200	3150

$w=10$ (iv)	(532)	$(6 2^2)$	(541)	(631)	(721)	$(8 1^2)$	(5^2)	(64)	(73)	(82)	(91)	(10)
a_{10}	−20	−10	−20	−20	−20	−10	5	10	10	10	10	−10
$a_9 a_1$	20	11	20	20	11	1	−5	−10	−10	−10	−1	10
$a_8 a_2$	4	−6	20	20	−3	−1	−5	−10	10	2	1	−10
$a_8 a_1^2$	−12	−2	−11	−11	−1	10	5	10	11	−10	−10	−10
$a_7 a_3$	−1	10	20	−1	−8	−3	10	20	−1	4	11	−20
$a_7 a_2 a_1$	−3	−4	−31	−10	−3	−3	10	−10	4	−2	−1	10
$a_7 a_1^3$	5	2	11	4	1	1	−5	−10	−3	−10	−10	10
$a_6 a_4$	20	−2	−4	−4	20	10	−5	14	−10	20	−10	10
$a_6 a_3 a_1$	−19	−8	−7	−4	−10	−11	10	−4	−1	−6	11	−20
$a_6 a_2^2$	−4		−8	−8	−4	−2	5	−2	10		10	−10
$a_6 a_2 a_1^2$	15	6	18	15	11	4	−15	−6	−9	−6	−12	30
$a_6 a_1^4$	−5	−2	−5	−4	−3	−1	5	4	3	2	1	−10
a_5^2	−15	5	−15	10	10	5	10	−5	−5	−5	−5	5
$a_5 a_4 a_1$	10	−8	23	−7	−31	−11	−15	−4	20	20	11	−20
$a_5 a_3 a_2$	17	−4	10	−19	−3	−12	−15	20	−1	4	20	−20
$a_5 a_3 a_1^2$	−4	10	−12	15	13	12	10	−6	−9	−22	−12	30
$a_5 a_2^2 a_1$	−13	4	−1	18	12	5	10	−9	12	2	−21	30
$a_5 a_2 a_1^3$	5	−8	1	−19	−14	−5	−5	16	−3	8	13	−40
$a_5 a_1^5$		2		4	3	1		−4	2	−2	−1	10
$a_4^2 a_2$	−12	10	4	4	−12	−6	5	−14	10	2	10	−10
$a_4^2 a_1^2$	1	−1	−8	−3	17	6	5	9	−15	−11	−6	15
$a_4 a_3^2$	1	−4	−8	13	1	−10	5	−2	−11	10	10	−10
$a_4 a_3 a_2 a_1$	5	−8		−15	21	26	−5	12	3	−28	−42	60
$a_4 a_3 a_1^3$	−1		5	5	−16	−13	−5	−8	12	24	13	−40
$a_4 a_2^3$	4	−4			4	2	−5	10	−10	6	−10	10
$a_4 a_2^2 a_1^2$	−2	8	−1	3	−23	−9	5	−12	18	4	33	−60
$a_4 a_2 a_1^4$		−2		−1	17	6		4	−15	−10	−14	50
$a_4 a_1^6$					−3	−1			3	2	1	−10
$a_3^3 a_1$	−1	4	5	−7	−4	7	−5	2	11	−10	−7	10
$a_3^2 a_2^2$	−2	2	−1	3	−7	7	5	−9	6	1	−15	15
$a_3^2 a_2 a_1^2$	1	−4	−3	6	9	−21	5		−24	28	33	−60
$a_3^2 a_1^4$	4	1		−2	−1	7		2	3	−13	−7	25
$a_3 a_2^3 a_1$	3		1	−3	−5	−7	−5	8	−2	−8	31	−40
$a_3 a_2^2 a_1^3$	11	2	3	1	−5	14		−4	12	−12	−46	100
$a_3 a_2 a_1^5$	35	11	10	5	1	−7			−3	12	15	−60
$a_3 a_1^7$	105	42	35	21	7	1		−2	2	−2	−1	10
a_2^5	10		5				1		2	−2	2	−2
$a_2^4 a_1^2$	32	6	14	4			2	1	−4	9	−16	25
$a_2^3 a_1^4$	99	30	42	19	3		6	4	1	−6	20	−50
$a_2^2 a_1^6$	296	115	130	72	20	2	20	15		1	−8	35
$a_2 a_1^8$	868	392	406	252	92	17	70	56	28	8	1	−10
a_1^{10}	2520	1200	1260	840	360	90	252	210	120	45	10	1

Table 1·2·11

$w = 11$ (i)	(1^{11})	(21^9)	(2^21^7)	(31^8)	(2^31^5)	(321^6)	(41^7)	(2^41^3)	(32^21^4)	(3^21^5)
a_{11}	1	−11	44	11	−77	−77	−11	55	165	33
$a_{10}a_1$	11	1	−9	−1	27	17	1	−30	−65	−8
a_9a_2	55	9	1	−2	−7	14	2	14	−21	−15
$a_9a_1^2$	110	19	2	1	.	−8	−1	.	20	8
a_8a_3	165	36	7	.	1	−3	−3	−5	15	3
$a_8a_2a_1$	495	117	23	8	3	1	−3	.	−6	−1
$a_8a_1^3$	990	243	48	25	6	3	1	.	.	.
a_7a_4	330	84	21	.	5	.	.	1	−4	2
$a_7a_3a_1$	1320	372	98	28	23	6	.	4	1	−2
$a_7a_2^2$	1980	576	155	56	36	13	.	6	2	1
$a_7a_2a_1^2$	3960	1188	324	148	75	33	7	12	5	2
$a_7a_1^4$	7920	2448	678	368	156	82	29	24	12	6
a_6a_5	462	126	35	.	10	.	.	3	.	.
$a_6a_4a_1$	2310	714	217	56	65	15	.	19	4	.
$a_6a_3a_2$	4620	1512	483	168	148	51	.	42	14	5
$a_6a_3a_1^2$	9240	3108	1008	420	311	123	21	88	33	10
$a_6a_2^2a_1$	13860	4788	1589	728	495	223	42	138	60	27
$a_6a_2a_1^3$	27720	9828	3318	1708	1041	522	154	288	139	66
$a_6a_1^5$	55440	20160	6930	3920	2190	1210	455	600	320	170
$a_5^2a_1$	2772	882	280	70	90	20	.	30	6	.
$a_5a_4a_2$	6930	2394	826	280	285	95	.	99	32	10
$a_5a_4a_1^2$	13860	4914	1722	686	600	225	35	210	74	20
$a_5a_3^2$	9240	3276	1162	420	411	150	.	144	53	20
$a_5a_3a_2a_1$	27720	10332	3808	1708	1383	611	105	492	213	90
$a_5a_3a_1^3$	55440	21168	7938	3920	2916	1398	371	1044	484	210
$a_5a_2^3$	41580	15876	5985	2856	2211	1053	210	792	372	181
$a_5a_2^2a_1^2$	83160	32508	12474	6468	4665	2388	672	1680	843	422
$a_5a_2a_1^4$	166320	66528	25998	14448	9846	5382	1869	3564	1904	1006
$a_5a_1^6$	332640	136080	54180	31920	20790	12060	4830	7560	4290	2420
$a_4^2a_3$	11550	4200	1540	560	570	210		213	80	30
$a_4^2a_2a_1$	34650	13230	5040	2240	1920	840	140	735	316	130
$a_4^2a_1^3$	69300	27090	10500	5110	4050	1910	490	1566	714	300
$a_4a_3^2a_1$	46200	18060	7070	3220	2775	1260	210	1092	497	220
$a_4a_3a_2^2$	69300	27720	11095	5320	4440	2135	420	1776	854	415
$a_4a_3a_2a_1^2$	138600	56700	23100	11900	9375	4785	1295	3792	1917	950
$a_4a_3a_1^4$	277200	115920	48090	26320	19800	10670	3535	8100	4292	2210
$a_4a_2^3a_1$	207900	86940	36225	19320	15015	7995	2310	6180	3276	1745
$a_4a_2^2a_1^3$	415800	177660	75390	42420	31725	17730	6090	13212	7319	4030
$a_4a_2a_1^5$	831600	362880	156870	92400	67050	39150	15225	28260	16320	9350
$a_4a_1^7$	1663200	740880	326340	199920	141750	86100	36750	60480	36330	21700
$a_3^3a_2$	92400	37800	15540	7560	6420	3150	630	2664	1320	660
$a_3^3a_1^2$	184800	77280	32340	16800	13560	7020	1890	5700	2952	1500
$a_3^2a_2^2a_1$	277200	118440	50680	27160	21720	11660	3360	9324	5016	2700
$a_3^2a_2a_1^3$	554400	241920	105420	59360	45900	25740	8750	19980	11168	6180
$a_3^2a_1^5$	1108800	493920	219240	128800	97020	56600	21700	42840	24820	14200
$a_3a_2^4$	415800	181440	79380	43680	34800	19260	5880	15282	8496	4780
$a_3a_2^3a_1^2$	831600	370440	165060	94920	73560	42330	14910	32784	18876	10880
$a_3a_2^2a_1^4$	1663200	756000	343140	204960	155520	92700	36330	70380	41864	24820
$a_3a_2a_1^6$	3326400	1542240	713160	440160	328860	202320	86100	151200	92700	56600
$a_3a_1^8$	6652800	3144960	1481760	940800	695520	440160	199920	325080	204960	128800
$a_2^5a_1$	1247400	567000	258300	151200	117900	69300	25200	53910	31800	18900
$a_2^4a_1^3$	2494800	1156680	536760	325080	249300	151200	60480	115884	70380	42840
$a_2^3a_1^5$	4989600	2358720	1115100	695520	527220	328860	141750	249300	155520	97020
$a_2^2a_1^7$	9979200	4808160	2315880	1481760	1115100	713160	326340	536760	343140	219240
$a_2a_1^9$	19958400	9797760	4808160	3144960	2358720	1542240	740880	1156680	756000	493920
a_1^{11}	39916800	19958400	9979200	6652800	4989600	3326400	1663200	2494800	1663200	1108800

Table 1·2·11 (*cont.*)

$w = 11$ (ii)	(421^5)	(51^6)	(2^51)	(32^31^2)	(3^221^3)	(42^21^3)	(431^4)	(521^4)	(61^5)	(32^4)
a_{11}	66	11	−11	−110	−110	−110	−55	−55	−11	11
$a_{10}a_1$	−16	−1	9	70	50	50	15	15	1	−11
a_9a_2	−12	−2	−7	−7	47	11	28	10	2	7
$a_9a_1^2$	7	1	.	−16	−32	−14	−15	−6	−1	2
a_8a_3	−18	−3	5	−10	−30	30	15	15	3	−5
$a_8a_2a_1$	20	3	.	9	11	−37	−19	−17	−3	−2
$a_8a_1^3$	−7	−1	.	.	.	14	7	6	1	.
a_7a_4	4	−4	−3	12	−2	−16	−8	20	4	3
$a_7a_3a_1$	−1	4	.	−4	8	4	5	−23	−4	2
$a_7a_2^2$	−2	2	.	.	−5	10	.	−10	−2	.
$a_7a_2a_1^2$	1	−4	.	.	.	−5	−1	23	4	.
$a_7a_1^4$	4	1	.	.	.	.	.	−6	−1	.
a_6a_5	.	.	1	−5	5	5	−5	−5	5	−1
$a_6a_4a_1$	.	.	5	1	−3	−1	5	1	−5	−2
$a_6a_3a_2$	.	.	10	3	1	−2	−1	5	−5	.
$a_6a_3a_1^2$	5	.	20	7	2	1	−2	−1	5	.
$a_6a_2^2a_1$	11	.	30	12	5	2	1	−3	5	.
$a_6a_2a_1^3$	39	6	60	27	12	7	3	1	−5	.
$a_6a_1^5$	115	31	120	60	30	20	10	5	1	.
$a_5^2a_1$	.	.	11	2	.	.	.	.	.	1
$a_5a_4a_2$	.	.	35	11	3	.	.	.	.	4
$a_5a_4a_1^2$	10	.	75	25	6	3	.	.	.	9
$a_5a_3^2$	.	.	50	18	7	.	.	.	.	6
$a_5a_3a_2a_1$	35	.	170	71	29	11	4	.	.	22
$a_5a_3a_1^3$	120	15	360	160	66	37	12	4	.	48
$a_5a_2^3$	75	.	270	123	57	24	13	.	.	36
$a_5a_2^2a_1^2$	231	30	570	276	133	72	35	9	.	78
$a_5a_2a_1^4$	639	141	1200	618	314	197	98	41	5	168
$a_5a_1^6$	1665	486	2520	1380	750	510	275	141	31	360
$a_4^2a_3$	.	.	80	31	12	.	.	.	.	12
$a_4^2a_2a_1$	50	.	285	121	48	18	6	6	.	48
$a_4^2a_1^3$	170	20	615	272	108	60	18	6	.	109
$a_4a_3^2a_1$	80	.	430	197	89	31	12	.	.	78
$a_4a_3a_2^2$	170	.	710	341	164	68	34	.	.	136
$a_4a_3a_2a_1^2$	505	60	1530	766	375	196	90	22	.	306
$a_4a_3a_1^4$	1370	275	3300	1718	864	527	244	98	10	688
$a_4a_2^3a_1$	945	120	2520	1326	697	378	203	48	.	528
$a_4a_2^2a_1^3$	2475	510	5430	2973	1613	983	527	197	20	1182
$a_4a_2a_1^5$	6225	1665	11700	6660	3750	2475	1370	639	115	2640
$a_4a_1^7$	15225	4830	25200	14910	8750	6090	3535	1869	455	5880
$a_3^3a_2$	270	.	1110	555	282	117	60	.	.	234
$a_3^3a_1^2$	780	90	2400	1248	642	327	156	36	.	528
$a_3^2a_2^2a_1$	1450	180	4010	2162	1168	628	338	78	.	934
$a_3^2a_2a_1^3$	3750	750	8700	4852	2682	1613	864	314	30	2112
$a_3^2a_1^5$	9350	2420	18900	10880	6180	4030	2210	1006	170	4780
$a_3a_2^4$	2640	360	6720	3750	2112	1182	688	168	.	1656
$a_3a_2^3a_1^2$	6660	1380	14610	8415	4852	2973	1718	618	60	3750
$a_3a_2^2a_1^4$	16320	4290	31800	18876	11168	7319	4292	1904	320	8496
$a_3a_2a_1^6$	39150	12060	69300	42330	25740	17730	10670	5382	1210	19260
$a_3a_1^8$	92400	31920	151200	94920	59360	42420	26320	14448	3920	43680
$a_2^5a_1$	11700	2520	24690	14610	8700	5430	3300	1200	120	6720
$a_2^4a_1^3$	28260	7560	53910	32784	19980	13212	8100	3564	600	15282
$a_2^3a_1^5$	67050	20790	117900	73560	45900	31725	19800	9846	2190	34800
$a_2^2a_1^7$	156870	54180	258300	165060	105420	75390	48090	25998	6930	79380
$a_2a_1^9$	362880	136080	567000	370440	241920	177660	115920	66528	20160	181440
a_1^{11}	831600	332640	1247400	831600	554400	415800	277200	166320	55440	415800

Table 1·2·11 (*cont.*)

$w=11$ (iii)	(3^22^21)	(42^31)	(3^31^2)	(4321^2)	(52^21^2)	(4^21^3)	(531^3)	(621^3)	(71^4)	(3^32)	(432^2)	(52^3)
a_{11}	66	44	22	132	66	22	44	44	11	−11	−33	−11
$a_{10}a_1$	−51	−34	−12	−72	−36	−7	−14	−14	−1	11	33	11
a_9a_2	−12	10	−19	−60	−3	−13	−26	−8	−2	5	−3	−7
$a_9a_1^2$	24	7	12	45	9	7	14	5	1	−8	−15	−2
a_8a_3	30	−20	14	−12	−18	−18	−12	−12	−3	−13	9	11
$a_8a_2a_1$	−17	16	−5	44	23	20	16	14	3	8	−6	−4
$a_8a_1^3$	.	−7	.	−21	−9	−7	−6	−5	−1	.	7	2
a_7a_4	−24	12	6	36	−24	6	−16	−16	−4	11	−9	−3
$a_7a_3a_1$	.	−2	−9	−21	33	−3	19	19	4	2	.	−8
$a_7a_2^2$	5	−10	5	−10	10	10	−1	12	8	−5	10	.
$a_7a_2a_1^2$	.	5	.	9	−32	1	−23	−19	−4	.	−7	6
$a_7a_1^4$	.	.	.	.	9	.	6	5	1	.	.	−2
a_6a_5	9	−4	−7	−12	9	8	16	−14	−11	−4	3	1
$a_6a_4a_1$	6	2	3	−12	−3	−5	−6	24	5	−7	6	2
$a_6a_3a_2$	−3	6	−1	6	−18	1	−4	20	5	5	−12	6
$a_6a_3a_1^2$	.	−3	.	6	3	2	3	−24	−5	5	.	.
$a_6a_2^2a_1$	.	.	.	−4	12	−1	2	−22	−5	.	3	−6
$a_6a_2a_1^3$	.	.	.	.	−4	.	−1	24	5	.	.	2
$a_6a_1^5$	.	.	.	.	.	.	.	−5	−1	.	.	2
$a_5^2a_1$	−4	−1	2	7	1	−3	−6	−1	6	4	−3	−1
$a_5a_4a_2$	1	−2	−2	4	2	−3	2	−6	6	−1	2	.
$a_5a_4a_1^2$	2	1	.	−3	−1	3	2	1	−6	.	−1	.
$a_5a_3^2$	2	.	1	−3	3	3	−3	−3	3	−2	6	−6
$a_5a_3a_2a_1$	8	3	3	1	−3	−3	4	7	−12	.	.	6
$a_5a_3a_1^3$	18	10	6	3	1	.	−2	−1	6	.	.	−2
$a_5a_2^3$	15	6	6	3	.	1	−2	2	−2	.	.	.
$a_5a_2^2a_1^2$	34	18	15	8	2	2	1	−4	9	.	.	.
$a_5a_2a_1^4$	78	48	36	22	9	6	4	1	−6	.	.	.
$a_5a_1^6$	180	120	90	60	30	20	15	6	1	.	.	.
$a_4^2a_3$	5	.	2	.	.	.	.	.	.	1	−3	3
$a_4^2a_2a_1$	19	7	6	2	.	.	.	.	.	3	1	−3
$a_4^2a_1^3$	42	23	12	6	2	.	.	.	.	6	3	1
$a_4a_3^2a_1$	36	12	17	5	.	.	.	.	.	7	2	.
$a_4a_3a_2^2$	65	27	30	13	.	3	.	.	.	12	5	.
$a_4a_3a_2a_1^2$	148	76	69	34	8	6	3	.	.	27	13	3
$a_4a_3a_1^4$	338	203	156	90	35	18	12	3	.	60	34	13
$a_4a_2^3a_1$	273	147	138	76	18	23	10	.	.	51	27	6
$a_4a_2^2a_1^3$	628	378	327	196	72	60	37	7	.	117	68	24
$a_4a_2a_1^5$	1450	945	780	505	231	170	120	39	4	270	170	75
$a_4a_1^7$	3360	2310	1890	1295	672	490	371	154	29	630	420	210
$a_3^3a_2$	120	51	64	27	.	6	.	.	.	27	12	.
$a_3^3a_1^2$	276	138	146	69	15	12	6	.	.	64	30	6
$a_3^2a_2^2a_1$	506	273	276	148	34	42	18	.	.	120	65	15
$a_3^2a_2a_1^3$	1168	697	642	375	133	108	66	12	.	282	164	57
$a_3^2a_1^5$	2700	1745	1500	950	422	300	210	66	6	660	415	181
$a_3a_2^4$	934	528	528	306	78	109	48	.	.	234	136	36
$a_3a_2^3a_1^2$	2162	1326	1248	766	276	272	160	27	.	555	341	123
$a_3a_2^2a_1^4$	5016	3276	2952	1917	843	714	484	139	12	1320	854	372
$a_3a_2a_1^6$	11660	7995	7020	4785	2388	1910	1398	522	82	3150	2135	1053
$a_3a_1^8$	27160	19320	16800	11900	6468	5110	3920	1708	368	7560	5320	2856
$a_2^5a_1$	4010	2520	2400	1530	570	615	360	60	.	1110	710	270
$a_2^4a_1^3$	9324	6180	5700	3792	1680	1566	1044	288	24	2664	1776	792
$a_2^3a_1^5$	21720	15015	13560	9375	4665	4050	2916	1041	156	6420	4440	2211
$a_2^2a_1^7$	50680	36225	32340	23100	12474	10500	7938	3318	678	15540	11095	5985
$a_2a_1^9$	118440	86940	77280	56700	32508	27090	21168	9828	2448	37800	27720	15876
a_1^{11}	277200	207900	184800	138600	83160	69300	55440	27720	7920	92400	69300	41580

Table 1·2·11 (*cont.*)

$w=11$ (iv)	(43^21)	(4^221)	(5321)	(62^21)	(541^2)	(631^2)	(721^2)	(81^2)	(4^23)	(53^2)	(542)	(632)
a_{11}	−33	−33	−66	−33	−33	−33	−33	−11	11	11	22	22
$a_{10}a_1$	23	23	46	23	13	13	13	1	−11	−11	−22	−22
a_9a_2	33	15	30	−3	24	24	6	2	−11	−11	−4	−4
$a_9a_1^2$	−23	−14	−28	−5	−13	−13	−4	−1	11	11	13	13
a_8a_3	−15	21	−6	9	33	9	9	3	1	13	−22	2
$a_8a_2a_1$	−8	−34	−20	−12	−37	−13	−11	−3	10	−2	26	2
$a_8a_1^3$	7	14	12	5	13	5	4	1	−7	−3	−13	−5
a_7a_4	−23	−23	38	19	5	5	5	11	17	−11	6	−22
$a_7a_3a_1$	27	12	−33	−18	−18	−15	−15	−4	−18	−2	16	20
$a_7a_2^2$	−5	13	−16	−4	−10	−10	−6	−2	−3	11	−10	4
$a_7a_2a_1^2$	−4	−8	41	17	22	19	15	4	7	−9	−11	−15
$a_7a_1^4$	·	−12	−5	−6	−5	−4	−1	·	3	6	5	8
a_6a_5	18	3	−24	3	−27	3	33	11	−11	4	8	8
$a_6a_4a_1$	3	6	6	−24	18	−12	−18	−12	−6	7	−14	14
$a_6a_3a_2$	−3	−6	30	−12	3	−21	−15	−5	10	−17	−4	8
$a_6a_3a_1^2$	−7	−4	−11	23	−5	22	19	5	7	6	1	−21
$a_6a_2^2a_1$	4	3	−18	16	−1	23	17	5	−7	6	14	−12
$a_6a_2a_1^3$	·	·	7	−22	1	−24	−19	−5	·	−3	−6	20
$a_6a_1^5$	·	·	·	5	·	5	4	1	·	·	·	−5
$a_5^2a_1$	−13	2	9	2	12	7	−23	−6	11	−4	−8	−8
$a_5a_4a_2$	5	−2	−18	14	11	1	−11	−13	−6	7	8	−4
$a_5a_4a_1^2$	5	−2	2	−1	−15	−5	22	13	−5	−3	11	3
$a_5a_3^2$	·	−6	6	6	−3	6	−9	−3	−1	2	7	−17
$a_5a_3a_2a_1$	−1	8	−7	−18	2	−11	41	12	−2	6	−18	30
$a_5a_3a_1^3$	·	·	4	2	2	3	−23	−6	·	−3	2	−4
$a_5a_2^3$	·	−3	6	−6	·	·	6	2	3	−6	·	6
$a_5a_2^2a_1^2$	·	·	−3	12	−1	3	−32	−9	·	3	2	−18
$a_5a_2a_1^4$	·	·	·	−3	·	−1	23	6	·	·	·	5
$a_5a_1^6$	·	·	·	·	·	·	−4	−1	·	·	·	·
$a_4^2a_3$	−1	5	−2	−7	−5	7	7	−7	1	−1	−6	10
$a_4^2a_2a_1$	−2	−1	8	3	−2	−4	−8	14	5	−6	−2	−6
$a_4^2a_1^3$	·	·	−3	−1	3	2	1	−7	·	3	−3	1
$a_4a_3^2a_1$	·	−2	−1	4	5	−7	−4	7	−1	·	5	−3
$a_4a_3a_2^2$	2	1	−2	2	−1	3	−7	7	−3	6	2	−12
$a_4a_3a_2a_1^2$	5	2	1	−4	−3	6	9	−21	·	−3	4	6
$a_4a_3a_1^4$	12	6	4	1	·	−2	−1	7	·	·	·	−1
$a_4a_2^3a_1$	12	7	3	·	1	−3	5	−7	·	·	−2	6
$a_4a_2^2a_1^3$	31	18	11	2	3	1	−5	14	·	·	·	−2
$a_4a_2a_1^5$	80	50	35	11	10	5	1	−7	·	·	·	·
$a_4a_1^7$	210	140	105	42	35	21	7	1	·	·	·	·
$a_3^3a_2$	7	3	·	·	·	·	·	·	1	−2	−1	5
$a_3^3a_1^2$	17	6	3	·	·	·	·	·	2	1	−2	−1
$a_3^2a_2^2a_1$	36	19	8	·	2	·	·	·	5	2	1	−3
$a_3^2a_2a_1^3$	89	48	29	5	6	2	·	·	12	7	3	1
$a_3^2a_1^5$	220	130	90	27	20	10	2	·	30	20	10	5
$a_3a_2^4$	78	48	22	·	9	·	·	·	12	6	4	·
$a_3a_2^3a_1^2$	197	121	71	12	25	7	·	·	31	18	11	3
$a_3a_2^2a_1^4$	497	316	213	60	74	33	5	·	80	53	32	14
$a_3a_2a_1^6$	1260	840	611	223	225	123	33	3	210	150	95	51
$a_3a_1^8$	3220	2240	1708	728	686	420	148	25	560	420	280	168
$a_2^5a_1$	430	285	170	30	75	20	·	·	80	50	35	10
$a_2^4a_1^3$	1092	735	492	138	210	88	12	·	213	144	99	42
$a_2^3a_1^5$	2775	1920	1383	495	600	311	75	6	570	411	285	148
$a_2^2a_1^7$	7070	5040	3808	1589	1722	1008	324	48	1540	1162	826	483
$a_2a_1^9$	18060	13230	10332	4788	4914	3108	1188	243	4200	3276	2394	1512
a_1^{11}	46200	34650	27720	13860	13860	9240	3960	990	11550	9240	6930	4620

Table 1·2·11 (cont.)

$w = 11$ (v)	(72^2)	(5^21)	(641)	(731)	(821)	(91^2)	(65)	(74)	(83)	(92)	$(10,1)$	(11)
a_{11}	11	11	22	22	22	11	-11	-11	-11	-11	-11	11
$a_{10}a_1$	-11	-6	-12	-12	-12	-1	11	11	11	11	1	-11
a_9a_2	7	-11	-22	-22	-4	-2	11	11	11	-7	11	-11
$a_9a_1^2$	2	6	12	12	3	1	-11	-11	-11	-2	-1	11
a_8a_3	-11	-11	-22	2	2	-11	11	11	-13	11	11	-11
$a_8a_2a_1$	4	17	34	10	8	3	-22	-22	2	-4	-12	22
$a_8a_1^3$	-2	-6	-12	-4	-3	-1	11	11	3	2	1	-11
a_7a_4	3	-11	6	6	-22	-11	11	-17	11	11	11	-11
$a_7a_3a_1$	8	17	6	3	10	12	-22	6	2	-22	-12	22
$a_7a_2^2$	.	11	8	8	4	2	-11	3	-11	7	-11	11
$a_7a_2a_1^2$	-6	-23	-18	-15	-11	-4	33	5	9	6	13	-33
$a_7a_1^4$	2	6	5	4	3	1	-11	-4	-3	-2	-1	11
a_6a_5	-11	19	8	-22	-22	-11	-19	11	11	11	11	-11
$a_6a_4a_1$	8	-13	.	6	34	12	8	6	-22	-22	-12	22
$a_6a_3a_2$	4	-8	14	20	2	13	8	-22	2	-4	-22	22
$a_6a_3a_1^2$	-10	7	-12	-15	-13	-13	3	5	9	24	13	-33
$a_6a_2^2a_1$	-4	2	-24	-18	-12	-5	3	19	-3	23	23	-33
$a_6a_2a_1^3$	8	-1	24	19	14	5	-14	-16	-12	-8	-14	44
$a_6a_1^5$	-2	.	-5	-4	-3	-1	5	4	3	2	1	-11
$a_5^2a_1$	11	-9	-13	17	17	6	19	-11	-11	-11	-6	11
$a_5a_4a_2$	-10	-8	-14	16	26	13	8	6	-22	-4	-22	22
$a_5a_4a_1^2$	-10	12	18	-18	-37	-13	-27	5	33	24	13	-33
$a_5a_3^2$	11	-4	7	-2	-2	11	4	-11	13	-11	-11	11
$a_5a_3a_2a_1$	-16	9	6	-33	-20	-28	-24	38	-6	30	46	-66
$a_5a_3a_1^3$	12	-6	-6	19	16	14	16	-16	-12	-26	-14	44
$a_5a_2^3$	.	-1	2	-8	-4	-2	1	-3	11	-7	11	-11
$a_5a_2^2a_1^2$	10	1	-3	33	23	9	9	-24	-18	-3	-36	66
$a_5a_2a_1^4$	-10	.	1	-23	-17	-6	-5	20	15	10	15	-55
$a_5a_1^6$	2	.	4	.	3	1	.	-4	-3	-2	-1	11
$a_4^2a_3$	-3	11	-6	-18	10	11	-11	17	1	-11	-11	11
$a_4^2a_2a_1$	13	2	6	12	-34	-14	3	-23	21	15	23	-33
$a_4^2a_1^3$	-1	-3	-5	-3	20	7	8	6	-18	-13	-7	22
$a_4a_3^2a_1$	-5	-13	3	27	-8	-23	18	-23	-15	33	23	-33
$a_4a_3a_2^2$	10	-3	6	.	-6	-15	3	-9	9	-3	33	-33
$a_4a_3a_2a_1^2$	-10	7	-12	-21	44	45	-12	36	-12	-60	-72	132
$a_4a_3a_1^4$	.	.	5	5	-19	-15	-5	-8	15	28	15	-55
$a_4a_2^3a_1$	-10	-1	2	-2	16	7	-4	12	-20	10	-34	44
$a_4a_2^2a_1^3$	10	.	-1	4	-37	-14	5	-16	30	11	50	-110
$a_4a_2a_1^5$	-2	.	.	.	-1	20	7	.	4	-18	-16	66
$a_4a_1^7$	.	.	.	.	-3	-1	.	.	3	2	1	-11
$a_3^3a_2$	-5	4	-7	2	8	-8	-4	11	-13	5	11	-11
$a_3^3a_1^2$	5	2	3	-9	-5	12	-7	6	14	-19	-12	22
$a_3^2a_2^2a_1$	5	-4	6	.	-17	24	9	-24	30	-12	-51	66
$a_3^2a_2a_1^3$	-5	.	-3	8	11	-32	5	-2	-30	47	50	-110
$a_3^2a_1^5$	1	.	.	-2	-1	8	.	2	3	-15	-8	33
$a_3a_2^4$	.	1	-2	2	-2	2	-1	3	-5	7	-11	11
$a_3a_2^3a_1^2$	.	2	1	-4	9	-16	-5	12	-10	-7	70	-110
$a_3a_2^2a_1^4$	2	6	4	1	-6	20	.	-4	15	-21	-65	165
$a_3a_2a_1^6$	13	20	15	6	1	-8	.	.	-3	14	17	-77
$a_3a_1^8$	56	70	56	28	8	1	.	.	.	-2	-1	11
$a_2^5a_1$	.	11	5	.	.	.	1	-3	5	-7	9	-11
$a_2^4a_1^3$	6	30	19	4	.	.	3	1	-5	14	-30	55
$a_2^3a_1^5$	36	90	65	23	3	.	10	5	1	-7	27	-77
$a_2^2a_1^7$	155	280	217	98	23	2	35	21	7	1	-9	44
$a_2a_1^9$	576	882	714	372	117	19	126	84	36	9	1	-11
a_1^{11}	1980	2772	2310	1320	495	110	462	330	165	55	11	1

Table 1·2·12

$w=12$ (i)	(1^{12})	(21^{10})	$(2^2 1^8)$	(31^9)	$(2^3 1^6)$	(321^7)	(41^8)	$(2^4 1^4)$	$(32^2 1^5)$	$(3^2 1^6)$
a_{12}	1	−12	54	12	−112	−96	−12	105	252	42
$a_{11}a_1$	12	1	−10	−1	35	19	1	−50	−87	−9
$a_{10}a_2$	66	10	1	−2	−8	16	2	20	−32	−17
$a_{10}a_1^2$	132	21	2	1	·	−9	−1	·	27	9
a_9a_3	220	45	8	·	1	−3	3	−6	18	3
$a_9a_2a_1$	660	145	26	9	3	1	−3	·	−7	−1
$a_9a_1^3$	1320	300	54	28	6	3	1	·	·	·
a_8a_4	495	120	28	·	6	·	·	1	−4	2
$a_8a_3a_1$	1980	525	128	36	27	7	·	4	1	−2
$a_8a_2^2$	2970	810	201	72	42	15	·	6	2	1
$a_8a_2a_1^2$	5940	1665	418	189	87	38	8	12	5	2
$a_8a_1^4$	11880	3420	870	468	180	94	33	24	12	6
a_7a_5	792	210	56	·	15	·	·	4	·	·
$a_7a_4a_1$	3960	1170	336	84	93	21	·	24	5	·
$a_7a_3a_2$	7920	2460	736	252	207	70	·	52	17	6
$a_7a_3a_1^2$	15840	5040	1528	624	432	168	28	108	40	12
$a_7a_2^2a_1$	23760	7740	2392	1080	681	302	56	168	72	32
$a_7a_2a_1^3$	47520	15840	4968	2520	1422	703	204	348	166	78
$a_7a_1^5$	95040	32400	10320	5760	2970	1620	600	720	380	200
a_6^2	924	252	70	·	20	·	·	6	·	·
$a_6a_5a_1$	5544	1722	532	126	165	35	·	52	10	·
$a_6a_4a_2$	13860	4620	1526	504	498	161	·	160	50	15
$a_6a_4a_1^2$	27720	9450	3164	1218	1041	378	56	336	115	30
$a_6a_3^2$	18480	6300	2128	756	707	252	·	228	81	30
$a_6a_3a_2a_1$	55440	19740	6888	3024	2337	1008	168	760	320	132
$a_6a_3a_1^3$	110880	40320	14280	6888	4890	2289	588	1596	722	306
$a_6a_2^3$	83160	30240	10752	5040	3696	1722	336	1206	552	262
$a_6a_2^2a_1^2$	166320	61740	22288	11340	7737	3878	1064	2532	1241	608
$a_6a_2a_1^4$	332640	126000	46200	25200	16200	8680	2940	5316	2780	1440
$a_6a_1^6$	665280	257040	95760	55440	33930	19320	7560	11160	6210	3440
$a_5^2a_2$	16632	5670	1932	630	660	210	·	228	70	20
$a_5^2a_1^2$	33264	11592	4004	1512	1380	490	70	480	160	40
$a_5a_4a_3$	27720	9870	3528	1260	1266	455	·	456	165	60
$a_5a_4a_2a_1$	83160	30870	11396	4914	4188	1771	280	1536	635	250
$a_5a_4a_1^3$	166320	63000	23604	11088	8766	3983	966	3240	1420	570
$a_5a_3^2a_1$	110880	42000	15848	7056	5952	2632	420	2220	976	420
$a_5a_3a_2^2$	166320	64260	24696	11592	9417	4410	840	3552	1652	776
$a_5a_3a_2a_1^2$	332640	131040	51128	25704	19722	9793	2548	7500	3670	1762
$a_5a_3a_1^4$	665280	267120	105840	56448	41310	21644	6888	15840	8132	4056
$a_5a_2^3a_1$	498960	200340	79632	41580	31221	16212	4536	12012	6177	3192
$a_5a_2^2a_1^3$	997920	408240	164808	90720	65412	35658	11844	25380	13656	7308
$a_5a_2a_1^5$	1995840	831600	341040	196560	137070	78120	29400	53640	30130	16800
$a_5a_1^7$	3991680	1693440	705600	423360	287280	170520	70560	113400	66360	38640
a_4^3	34650	12600	4620	1680	1710	630	·	639	240	90
$a_4^2a_3a_1$	138600	53550	20720	9240	8040	3570	560	3132	1390	600
$a_4^2a_2^2$	207900	81900	32270	15120	12720	5950	1120	5022	2340	1090
$a_4^2a_2a_1^2$	415800	166950	66780	33390	26640	13160	3360	10620	5180	2460
$a_4^2a_1^4$	831600	340200	138180	73080	55800	28980	9030	22464	11440	5620
$a_4a_3^2a_2$	277200	111300	44800	21420	18075	8680	1680	7308	3525	1710
$a_4a_3^2a_1^2$	554400	226800	92680	47040	37860	19110	4900	15468	7780	3840
$a_4a_3a_2^2a_1$	831600	346500	144200	75600	59925	31360	8680	24864	12980	6760
$a_4a_3a_2a_1^3$	1663200	705600	298200	163800	125550	68495	22260	52668	28530	15270
$a_4a_3a_1^5$	3326400	1436400	616560	352800	263070	149100	54600	111600	62600	34600
$a_4a_2^4$	1247400	529200	224280	120960	94860	51240	15120	40002	21600	11760
$a_4a_2^3a_1^2$	2494800	1077300	463680	260820	198765	111510	37800	84780	47385	26460
$a_4a_2^2a_1^4$	4989600	2192400	958440	559440	416520	241920	91140	179748	103780	59640
$a_4a_2a_1^6$	9979200	4460400	1980720	1194480	872910	523320	214200	381240	226950	134400
$a_4a_1^8$	19958400	9072000	4092480	2540160	1829520	1128960	493920	808920	495600	302400
a_3^4	369600	151200	62160	30240	25800	12600	2520	10656	5280	2640
$a_3^3a_2a_1$	1108800	470400	199920	105840	85140	45150	12600	36336	19320	10260
$a_3^3a_1^3$	2217600	957600	413280	228480	178380	98280	31920	77040	42360	23040
$a_3^2a_2^3$	1663200	718200	310800	168840	134760	73500	21840	58536	32040	17660
$a_3^2a_2^2a_1^2$	3326400	1461600	642320	362880	282360	159460	54040	124176	70120	39520
$a_3^2a_2a_1^4$	6652800	2973600	1327200	776160	591060	344960	129360	263520	153220	88560
$a_3^2a_1^6$	13305600	6048000	2741760	1653120	1239840	744240	302400	559440	334320	198400
$a_3a_2^4a_1$	4989600	2230200	997920	574560	446940	257880	90720	200328	115800	67200
$a_3a_2^3a_1^3$	9979200	4536000	2061360	1224720	936540	556290	214200	425376	252600	149940
$a_3a_2^2a_1^5$	19958400	9223200	4257120	2600640	1962540	1197000	495600	903600	550240	334320
$a_3a_2a_1^7$	39916800	18748800	8789760	5503680	4112640	2569560	1128960	1920240	1197000	744240
$a_3a_1^9$	79833600	38102400	18144000	11612160	8618400	5503680	2540160	4082400	2600640	1653120
a_2^6	7484400	3402000	1549800	907200	707400	415800	151200	323460	190800	113400
$a_2^5a_1^2$	14968800	6917400	3200400	1927800	1482300	894600	352800	687240	415500	252000
$a_2^4a_1^4$	29937600	14061600	6607440	4082400	3106080	1920240	808920	1460736	903600	559440
$a_2^3a_1^6$	59875200	28576800	13638240	8618400	6508620	4112640	1829520	3106080	1962540	1239840
$a_2^2a_1^8$	119750400	58060800	28143360	18144000	13638240	8789760	4092480	6607440	4257120	2741760
$a_2a_1^{10}$	239500800	117936000	58060800	38102400	28576800	18748800	9072000	14061600	9223200	6048000
a_1^{12}	479001600	239500800	119750400	79833600	59875200	39916800	19958400	29937600	19958400	13305600

Table 1·2·12 (*cont.*)

$w = 12$ (ii)	(421^6)	(51^7)	(2^51^2)	(32^31^3)	(3^221^4)	(42^21^4)	(431^5)	(521^5)	(61^6)	(2^6)	(32^41)
a_{12}	84	12	−36	−240	−180	−180	−72	−72	−12	2	60
$a_{11}a_1$	−18	−1	25	130	70	70	17	17	1	−2	−49
$a_{10}a_2$	−14	−2	−16	.	70	20	32	12	2	2	20
$a_{10}a_1^2$	8	1	.	−30	−45	−20	−17	−7	−1	.	9
a_9a_3	−21	−3	9	−21	−36	45	18	18	3	−2	−6
$a_9a_2a_1$	23	3	.	14	13	−54	−22	−20	−3	.	−7
$a_9a_1^3$	−8	−1	.	.	.	20	8	7	1	.	.
a_8a_4	4	−4	−4	16	−4	−20	−8	24	4	2	−4
$a_8a_3a_1$	−1	4	.	−5	10	5	5	−27	−4	.	5
$a_8a_2^2$	−2	2	.	.	−6	12	.	−12	−2	.	.
$a_8a_2a_1^2$	1	−4	.	.	.	−6	−1	27	4	.	.
$a_8a_1^4$	4	1	.	.	.	.	.	−7	−1	.	.
a_7a_5	.	.	1	−5	5	5	−5	−5	5	−2	10
$a_7a_4a_1$	.	.	5	1	−3	−1	5	1	−5	.	−3
$a_7a_3a_2$	.	.	10	3	1	−2	−1	5	−5	.	.
$a_7a_3a_1^2$	6	.	20	7	2	1	−2	−1	5	.	.
$a_7a_2^2a_1$	13	.	30	12	5	2	1	−3	5	.	.
$a_7a_2a_1^3$	46	7	60	27	12	7	3	1	−5	.	.
$a_7a_1^5$	135	36	120	60	30	20	10	5	1	.	.
a_6^2	.	.	2	.	.	.	.	.	.	1	−6
$a_6a_5a_1$	.	.	17	3	.	.	.	.	.	6	1
$a_6a_4a_2$	.	.	50	15	4	4	.	.	.	15	4
$a_6a_4a_1^2$	15	.	105	34	8	4	.	.	.	30	9
$a_6a_3^2$	.	.	70	24	9	.	.	.	.	20	6
$a_6a_3a_2a_1$	51	.	230	93	37	14	5	.	.	60	22
$a_6a_3a_1^3$	174	21	480	208	84	47	15	5	.	120	48
$a_6a_2^3$	108	.	360	159	72	30	16	.	.	90	36
$a_6a_2^2a_1^2$	331	42	750	354	167	90	43	11	.	180	78
$a_6a_2a_1^4$	910	196	1560	786	392	245	120	50	6	360	168
$a_6a_1^6$	2355	672	3240	1740	930	630	335	171	37	720	360
$a_5^2a_2$	.	.	81	24	6	6	.	.	.	30	9
$a_5^2a_1^2$	20	.	172	54	12	6	.	.	.	66	20
$a_5a_4a_3$	.	.	165	60	22	.	.	.	.	60	22
$a_5a_4a_2a_1$	95	.	565	228	86	32	10	.	.	210	83
$a_5a_4a_1^3$	320	35	1200	507	192	106	30	10	.	450	184
$a_5a_3^2a_1$	150	.	820	357	154	53	20	.	.	300	128
$a_5a_3a_2^2$	315	.	1320	606	279	114	55	.	.	480	216
$a_5a_3a_2a_1^2$	926	105	2800	1344	632	327	145	35	.	1020	476
$a_5a_3a_1^4$	2489	476	5940	2976	1442	872	390	155	15	2160	1048
$a_5a_2^3a_1$	1713	210	4500	2277	1153	618	321	75	.	1620	798
$a_5a_2^2a_1^3$	4446	882	9540	5037	2640	1593	828	306	30	3420	1752
$a_5a_2a_1^5$	11085	2856	20220	11130	6070	3970	2135	985	171	7200	3840
$a_5a_1^7$	26880	8232	42840	24570	14000	9660	5460	2856	672	15120	8400
a_4^3	.	.	240	93	36	.	.	.	.	90	36
$a_4^2a_3a_1$	210	.	1225	546	238	80	30	.	.	480	216
$a_4^2a_2^2$	440	.	1990	924	424	172	80	.	.	795	368
$a_4^2a_2a_1^2$	1280	140	4245	2046	954	488	210	50	.	1710	817
$a_4^2a_1^4$	3420	630	9060	4524	2160	1294	560	220	20	3690	1812
$a_4a_3^2a_2$	690	.	2960	1434	699	284	140	.	.	1200	584
$a_4a_3^2a_1^2$	1950	210	6320	3174	1570	781	360	80	.	2580	1296
$a_4a_3a_2^2a_1$	3575	420	10300	5361	2779	1466	755	170	60	4260	2210
$a_4a_3a_2a_1^3$	9110	1715	22020	11841	6284	3711	1905	675	60	9180	4900
$a_4a_3a_1^5$	22425	5460	47100	26130	14250	9140	4800	2135	335	19800	10860
$a_4a_2^4$	6420	840	16800	9054	4900	2694	1500	360	.	7020	3768
$a_4a_2^3a_1^2$	15975	3150	35940	19986	11085	6678	3705	1305	120	15120	8352
$a_4a_2^2a_1^4$	38670	9660	76920	44088	25120	16201	9140	3970	630	32580	18504
$a_4a_2a_1^6$	91725	26880	164700	97200	56990	38670	22425	11085	2355	70200	40980
$a_4a_1^8$	214200	70560	352800	214200	129360	91140	54600	29400	7560	151200	90720
a_3^4	1080	.	4440	2220	1128	468	240	.	.	1860	936
$a_3^3a_2a_1$	5400	630	15540	8289	4428	2328	1230	270	.	6660	3564
$a_3^3a_1^3$	13590	2520	33300	18300	9972	5820	3060	1050	90	14400	7920
$a_3^2a_2^3$	9630	1260	25470	13986	7728	4248	2390	570	.	11100	6114
$a_3^2a_2^2a_1^2$	23720	4620	54040	30858	17404	10428	5840	2020	180	24060	13592
$a_3^2a_2a_1^4$	56990	14000	117300	68040	39244	25120	14250	6070	930	52200	30220
$a_3^2a_1^6$	134440	38640	252000	149940	88560	59640	34600	16800	3440	113400	67200
$a_3a_2^4a_1$	40980	8400	89850	52020	30220	18504	10860	3840	360	40320	23376
$a_3a_2^3a_1^3$	97200	24570	193140	114657	68040	44088	26130	11130	1740	87660	52020
$a_3a_2^2a_1^5$	226950	66360	415500	252600	153220	103780	62600	30130	6210	190800	115800
$a_3a_2a_1^7$	523320	170520	894600	556290	344960	241920	149100	78120	19320	415800	257880
$a_3a_1^9$	1194480	423360	1927800	1224720	776160	559440	352800	196560	55440	907200	574560
a_2^6	70200	15120	148140	87660	52200	32580	19800	7200	720	67950	40320
$a_2^5a_1^2$	164700	42840	318930	193140	117300	76920	47100	20220	3240	148140	89850
$a_2^4a_1^4$	381240	113400	687240	425376	263520	179748	111600	53640	11160	323460	200328
$a_2^3a_1^6$	872910	287280	1482300	936540	591660	416520	263070	137070	33930	707400	446940
$a_2^2a_1^8$	1980720	705600	3200400	2061360	1327200	958440	610560	341040	95760	1549800	997920
$a_2a_1^{10}$	4460400	1693440	6917400	4536000	2973600	2192400	1436400	831600	257040	3402000	2230200
a_1^{12}	9979200	3991680	14968800	9979200	6652800	4989600	3326400	1995840	665280	7484400	4989600

Table 1·2·12 (*cont.*)

$w=12$ (iii)	$(3^22^21^2)$	(42^31^2)	(3^31^3)	(4321^3)	(52^21^3)	(4^21^4)	(531^4)	(621^4)	(71^5)	(3^22^3)	(42^4)	(3^221)
a_{12}	180	120	40	240	120	30	60	60	12	-24	-12	-48
$a_{11}a_1$	-114	-76	-18	-108	-54	-8	-16	-16	-1	24	12	37
$a_{10}a_2$	-45	10	-30	-100	-10	-15	-30	-10	-2	-6	-8	28
$a_{10}a_1^2$	54	16	18	68	14	8	16	6	1	-9	-2	-27
a_9a_3	63	-39	17	-33	-30	-21	-15	-15	-3	-3	12	-42
$a_9a_2a_1$	-30	39	-6	73	37	23	19	17	3	9	-4	19
$a_9a_1^3$	·	-16	·	-32	-14	-8	-7	-6	-1	·	2	·
a_8a_4	-28	24	8	48	-40	6	-20	-20	-4	8	-12	16
$a_8a_3a_1$	-5	-5	-11	-27	52	-3	23	23	4	-5	·	13
$a_8a_2^2$	9	-18	6	-12	18	-1	14	10	2	-2	4	-12
$a_8a_2a_1^2$	·	9	·	11	-51	1	-27	-23	-4	·	-2	·
$a_8a_1^4$	·	·	·	·	14	·	7	6	1	·	·	·
a_7a_5	-5	-15	-5	5	20	5	10	-25	-5	-11	12	13
$a_7a_4a_1$	9	3	3	-17	-4	-5	-6	29	5	3	·	-18
$a_7a_3a_2$	-4	8	-1	7	-23	1	-4	25	5	2	-4	7
$a_7a_3a_1^2$	·	-4	·	8	4	2	3	-29	-5	·	2	·
$a_7a_2^2a_1$	·	·	·	-5	15	-1	2	-27	-5	·	·	·
$a_7a_2a_1^3$	·	·	·	·	-5	·	-1	29	5	·	·	·
$a_7a_1^5$	·	·	·	·	·	·	·	-6	-1	·	·	·
a_6^2	9	6	-2	-12	-6	3	6	6	6	-6	-6	-12
$a_6a_5a_1$	-4	-1	2	7	1	-3	-6	-1	6	-1	·	7
$a_6a_4a_2$	1	-2	-2	4	2	-3	2	-6	6	-2	4	4
$a_6a_4a_1^2$	2	1	·	-3	-1	3	2	1	-6	·	-2	·
$a_6a_3^2$	2	·	1	-3	3	3	-3	-3	3	·	·	-3
$a_6a_3a_2a_1$	8	3	3	1	-3	-3	4	7	-12	·	·	·
$a_6a_3a_1^3$	18	10	6	3	1	·	-2	-1	6	·	·	·
$a_6a_2^3$	15	6	6	3	·	1	-2	2	-2	·	·	·
$a_6a_2^2a_1^2$	34	18	15	8	2	2	1	-4	9	·	·	·
$a_6a_2a_1^4$	78	48	36	22	9	6	4	1	-6	·	·	·
$a_6a_1^6$	180	120	90	60	30	20	15	6	1	·	·	·
$a_5^2a_2$	2	·	·	·	·	·	·	·	·	1	-2	-3
$a_5^2a_1^2$	4	2	·	·	·	·	·	·	·	2	1	·
$a_5a_4a_3$	8	·	3	·	·	·	·	·	·	3	·	1
$a_5a_4a_2a_1$	30	11	9	3	·	·	·	·	·	11	4	3
$a_5a_4a_1^3$	66	36	18	9	3	·	·	·	·	24	13	6
$a_5a_3^2a_1$	54	18	24	7	·	·	·	·	·	18	6	7
$a_5a_3a_2^2$	96	39	42	18	·	4	·	·	·	31	12	12
$a_5a_3a_2a_1^2$	216	110	96	47	11	8	4	·	·	68	34	27
$a_5a_3a_1^4$	488	291	216	124	48	24	16	4	·	150	88	60
$a_5a_2^3a_1$	390	207	189	103	24	30	13	·	·	117	60	51
$a_5a_2^2a_1^3$	886	528	444	264	96	78	48	9	·	258	150	117
$a_5a_2a_1^5$	2020	1305	1050	675	306	220	155	50	5	570	360	270
$a_5a_1^7$	4620	3150	2520	1715	882	630	476	196	36	1260	840	630
a_4^3	15	·	6	·	·	·	·	·	·	6	·	3
$a_4^2a_3a_1$	96	31	42	12	·	·	·	·	·	39	12	18
$a_4^2a_2^2$	168	68	72	30	·	6	·	·	·	68	28	30
$a_4^2a_2a_1^2$	376	189	162	78	18	12	6	·	·	153	76	66
$a_4^2a_1^4$	844	498	360	204	78	36	24	6	·	344	201	144
$a_4a_3^2a_2$	286	117	141	58	·	12	·	·	·	117	48	58
$a_4a_3^2a_1^2$	644	314	318	147	31	24	12	·	·	264	126	132
$a_4a_3a_2^2a_1$	1140	599	582	305	68	80	34	22	·	467	244	237
$a_4a_3a_2a_1^3$	2578	1506	1332	763	264	204	124	120	10	1056	610	540
$a_4a_3a_1^5$	5840	3705	3060	1905	828	560	390	120	10	2390	1500	1230
$a_4a_2^4$	2026	1116	1080	610	150	201	88	·	·	828	456	438
$a_4a_2^3a_1^2$	4592	2754	2505	1506	528	498	291	48	·	1875	1116	1008
$a_4a_2^2a_1^4$	10428	6678	5820	3711	1593	1294	872	245	20	4248	2694	2328
$a_4a_2a_1^6$	23720	15975	13590	9110	4446	3420	2489	910	135	9630	6420	5400
$a_4a_1^8$	54040	37800	31920	22260	11844	9030	6888	2940	600	21840	15120	12600
a_3^4	480	204	256	108	·	24	·	·	·	204	90	108
$a_3^3a_2a_1$	1914	1008	1038	540	117	144	60	·	·	828	438	451
$a_3^3a_1^3$	4332	2505	2364	1332	444	360	216	36	·	1884	1080	1038
$a_3^2a_2^3$	3386	1875	1884	1056	258	344	150	·	·	1486	828	828
$a_3^2a_2^2a_1^2$	7672	4592	4332	2578	886	844	488	78	·	3386	2026	1914
$a_3^2a_2a_1^4$	17404	11085	9972	6284	2640	2100	1442	392	30	7728	4900	4428
$a_3^2a_1^6$	39520	26460	23040	15270	7308	5620	4056	1440	200	17660	11760	10260
$a_3a_2^4a_1$	13592	8352	7920	4900	1752	1812	1048	168	60	6114	3768	3564
$a_3a_2^3a_1^3$	30858	19986	18300	11841	5037	4524	2976	786	60	13986	9054	8289
$a_3a_2^2a_1^5$	70120	47385	42360	28530	13656	11440	8132	2780	380	32040	21600	19320
$a_3a_2a_1^7$	159460	111510	98280	68495	35658	28980	21644	8680	1620	73500	51240	45150
$a_3a_1^9$	362880	260820	228480	163800	90720	73080	56448	25200	5760	168840	120960	105840
a_2^6	24060	15120	14400	9180	3420	3690	2160	360	·	11100	7020	6660
$a_2^5a_1^2$	54640	35940	33300	22020	9540	9060	5940	1560	120	25470	16800	15540
$a_2^4a_1^4$	124176	84780	77040	52668	25380	22464	15840	5316	720	58536	40002	36336
$a_2^3a_1^6$	282360	198765	178380	125550	65412	55800	41310	16200	2970	134760	94860	85140
$a_2^2a_1^8$	642320	463680	413280	298200	164808	138180	105840	46200	10320	310800	224280	199920
$a_2a_1^{10}$	1461600	1077300	957600	705600	408240	340200	267120	126000	32400	718200	529200	470400
a_1^{12}	3326400	2494800	2217600	1663200	997920	831600	665280	332640	95040	1663200	1247400	1108800

Table 1·2·12 (*cont.*)

$w=12$ (iv)	(432^21)	(52^31)	(43^21^2)	(4^221^2)	(5321^2)	(62^21^2)	(541^3)	(631^3)	(721^3)	(81^4)	(3^4)	(43^22)	(4^22^3)
a_{12}	−144	−48	−72	−72	−144	−72	−48	−48	−48	−12	3	36	18
$a_{11}a_1$	111	37	39	39	78	39	15	15	15	1	−3	−36	−18
$a_{10}a_2$	24	−12	62	32	64	2	28	28	8	2	−3	−16	2
$a_{10}a_1^2$	−51	−7	−39	−24	−48	−9	−15	−15	−5	−1	3	26	8
a_9a_3	9	21	−9	45	9	18	39	12	12	3	6	18	−18
$a_9a_2a_1$	−36	−16	−20	−62	−43	−23	−43	−16	−14	−3	−3	−2	16
$a_9a_1^3$	24	7	12	24	21	9	15	6	5	1	.	−8	−8
a_8a_4	−16	16	−40	−40	48	24	16	16	16	4	−3	−4	22
$a_8a_3a_1$	16	−26	34	16	−63	−33	−22	−19	−19	−4	−3	−14	−4
$a_8a_2^2$	24	−4	−6	16	−32	−10	−12	−12	−8	−2	3	.	−14
$a_8a_2a_1^2$	−17	23	−5	−10	75	32	26	23	19	4	.	8	8
$a_8a_1^4$	.	−7	.	.	−21	−9	−7	−6	−5	−1	.	.	.
a_7a_5	4	−22	2	2	−31	37	−22	13	13	12	−3	−1	−18
$a_7a_4a_1$	3	3	15	15	21	−42	11	−24	−24	−5	6	5	−4
$a_7a_3a_2$	−19	19	−4	−7	39	−27	3	−26	−20	−5	−3	5	16
$a_7a_3a_1^2$	.	−2	−9	−6	−15	42	−5	27	24	5	.	2	−4
$a_7a_2^2a_1$	5	−15	5	4	−23	33	−1	28	22	5	.	−5	−2
$a_7a_2a_1^3$	.	5	.	.	9	−41	1	−29	−24	−5	.	.	.
$a_7a_1^5$	.	.	.	.	.	9	.	6	5	1	.	.	.
a_6^2	.	12	18	.	.	−18	−12	−12	24	6	3	.	9
$a_6a_5a_1$	−1	−1	−20	1	7	2	19	14	−28	−13	−3	1	.
$a_6a_4a_2$	−8	−4	2	8	−16	22	4	4	−24	−6	−3	8	−12
$a_6a_4a_1^2$	6	2	3	−9	−3	−3	−8	−3	29	6	3	−7	8
$a_6a_3^2$	9	−9	.	−9	9	9	−3	6	−12	−3	3	−9	.
$a_6a_3a_2a_1$	−3	9	−1	11	−11	−25	2	−11	53	12	.	5	−4
$a_6a_3a_1^3$	.	−3	.	.	.	6	3	2	−29	−6	.	.	.
$a_6a_2^3$	.	.	.	−4	.	−8	.	.	8	2	.	.	2
$a_6a_2^2a_1^2$	.	.	.	.	.	−4	16	−1	−41	−9	.	.	.
$a_6a_2a_1^4$	.	.	.	.	.	−4	.	−1	29	6	.	.	.
$a_6a_1^6$	.	.	.	.	.	.	.	.	−5	−1	.	.	.
$a_5^2a_2$	6	2	3	−7	−4	−2	7	−3	7	−7	3	−9	8
$a_5^2a_1^2$	−4	−1	2	4	3	1	−7	−2	−1	7	.	4	−4
$a_5a_4a_3$	−3	3	−1	5	−2	−7	−5	7	7	−7	−3	11	−4
$a_5a_4a_2a_1$	1	−3	−2	−1	8	3	−2	−4	−8	14	.	−1	.
$a_5a_4a_1^3$	3	1	.	.	−3	−1	3	2	1	−7	.	.	.
$a_5a_3^2a_1$	2	.	1	−2	−1	4	5	−7	−4	7	.	−2	4
$a_5a_3a_2^2$	5	.	2	1	−2	2	−1	3	−7	7	.	.	−2
$a_5a_3a_2a_1^2$	13	3	5	2	1	−4	−3	6	9	−21	.	.	.
$a_5a_3a_1^4$	34	13	12	6	4	1	.	−2	−1	7	.	.	.
$a_5a_2^3a_1$	27	6	12	7	3	.	1	−3	5	−7	.	.	.
$a_5a_2^2a_1^3$	68	24	31	18	11	2	3	1	−5	14	.	.	.
$a_5a_2a_1^5$	170	75	80	50	35	11	10	5	1	−7	.	.	.
$a_5a_1^7$	420	210	210	140	105	42	35	21	7	1	.	.	.
a_4^3	.	.	.	.	.	.	.	.	.	.	1	−4	2
$a_4^2a_3a_1$	5	.	2	.	.	.	.	.	.	.	4	1	−2
$a_4^2a_2^2$	12	.	4	2	.	.	.	.	.	.	6	2	1
$a_4^2a_2a_1^2$	31	7	10	4	2	.	2	.	.	.	12	5	2
$a_4^2a_1^4$	80	30	24	12	8	2	.	.	.	.	24	12	6
$a_4a_3^2a_2$	24	.	12	5	.	.	.	.	.	.	12	5	2
$a_4a_3^2a_1^2$	60	12	29	10	5	.	.	.	.	.	28	12	4
$a_4a_3a_2^2a_1$	123	27	60	31	13	.	3	.	.	.	48	24	12
$a_4a_3a_2a_1^3$	305	103	147	78	47	8	9	3	.	.	108	58	30
$a_4a_3a_1^5$	755	321	360	210	145	43	30	15	3	.	240	140	80
$a_4a_2^4$	244	60	126	76	34	.	13	.	.	.	90	48	28
$a_4a_2^3a_1^2$	599	207	314	189	110	18	36	10	.	.	204	117	68
$a_4a_2^2a_1^4$	1466	618	781	488	327	90	106	47	7	.	468	284	172
$a_4a_2a_1^6$	3575	1713	1950	1280	926	331	320	174	46	4	1080	690	440
$a_4a_1^8$	8680	4536	4900	3360	2548	1064	966	588	204	33	2520	1680	1120
a_3^4	48	.	28	12	.	.	.	.	.	.	24	12	6
$a_3^3a_2a_1$	237	51	132	66	27	.	6	.	.	.	108	58	30
$a_3^3a_1^3$	582	189	318	162	96	15	18	6	.	.	256	141	72
$a_3^2a_2^3$	467	117	264	153	68	.	24	.	.	.	204	117	68
$a_3^2a_2^2a_1^2$	1140	390	644	376	216	34	66	18	.	.	480	286	168
$a_3^2a_2a_1^4$	2779	1153	1570	954	632	167	192	84	12	.	1128	699	424
$a_3^2a_1^6$	6760	3192	3840	2460	1762	608	570	306	78	6	2640	1710	1090
$a_3a_2^3a_1$	2210	798	1296	817	476	78	184	48	.	.	936	584	368
$a_3a_2^2a_1^3$	5361	2277	3174	2046	1344	354	507	208	27	.	2220	1434	924
$a_3a_2a_1^5$	12980	6177	7780	5180	3670	1241	1420	722	166	12	5280	3525	2340
$a_3a_2a_1^7$	31360	16212	19110	13100	9793	3878	3983	2289	703	94	12600	8680	5950
$a_3a_1^9$	75600	41580	47040	33390	25704	11340	11088	6888	2520	468	30240	21420	15120
a_2^6	4260	1620	2580	1710	1020	180	450	120	.	.	1860	1200	795
$a_2^5a_1^2$	10300	4500	6320	4245	2800	750	1200	480	60	.	4440	2960	1990
$a_2^4a_1^4$	24864	12012	15468	10620	7500	2532	3240	1596	348	24	10656	7308	5022
$a_2^3a_1^6$	59925	31221	37860	26640	19722	7737	8766	4890	1422	180	25680	18075	12720
$a_2^2a_1^8$	144200	79632	92680	66780	51128	22288	23604	14280	4968	870	62160	44800	32270
$a_2a_1^{10}$	346500	200340	226800	166950	131040	61740	63000	40320	15840	3420	151200	111300	81900
a_1^{12}	831600	498960	554400	415800	332640	166320	166320	110880	47520	11880	369600	277200	207900

Table 1·2·12 (cont.)

$w=12$ (v)	(532²)	(62³)	(4²31)	(53²1)	(5421)	(6321)	(72²1)	(5²1²)	(641²)	(731²)	(821²)	(91³)	(4³)	(543)
a_{12}	36	12	36	36	72	72	36	18	36	36	36	12	-4	-24
$a_{11}a_1$	-36	-12	-25	-25	-50	-50	-25	-7	-14	-14	-14	-1	4	24
$a_{10}a_2$	4	8	-36	-36	-32	-32	4	-13	-26	-26	-6	-2	4	24
$a_{10}a_1^2$	16	2	25	25	30	30	5	7	14	14	4	1	-4	-24
a_9a_3	-9	-12	-9	18	-45	9	-9	-18	-36	-9	-9	-3	4	-3
$a_9a_2a_1$	5	4	34	7	73	19	12	20	40	13	11	3	-8	-21
$a_9a_1^3$	-7	-2	-16	-7	-30	-12	-5	-7	-14	-5	-4	-1	4	15
a_8a_4	-20	4	44	-4	-8	-40	-20	-18	-4	-4	-4	-12	-12	-8
$a_8a_3a_1$	29	8	-34	-1	31	33	18	25	18	15	15	4	8	11
$a_8a_2^2$	4	·	-4	20	·	16	4	13	10	10	6	2	4	-8
$a_8a_2a_1^2$	-21	-6	9	-24	-39	-41	-17	-27	-22	-19	-15	-4	-4	13
$a_8a_1^4$	7	2	·	7	14	12	5	7	6	5	4	1	·	-7
a_7a_5	34	-12	-1	-1	33	-37	-1	17	-1	-1	-36	-12	4	-11
$a_7a_4a_1$	-14	8	-26	-6	-19	55	24	-10	11	11	18	13	8	19
$a_7a_3a_2$	-23	4	10	-17	-28	30	12	-4	27	21	15	5	-8	14
$a_7a_3a_1^2$	4	-10	16	11	16	-49	-23	3	-25	-22	-19	-5	-4	-22
$a_7a_2^2a_1$	19	-4	-9	8	18	-42	-16	2	-29	-23	-17	-5	4	-6
$a_7a_2a_1^3$	-7	8	·	-4	-8	53	22	-1	29	24	19	5	·	7
$a_7a_1^5$	·	-2	·	·	·	·	-12	·	-6	-5	-4	-1	·	·
a_6^2	-18	6	-18	·	·	·	36	-18	18	-18	-18	-6	2	12
$a_6a_5a_1$	2	·	26	5	-25	-27	26	-16	-27	15	50	13	-8	-13
$a_6a_4a_2$	16	-12	-8	4	16	-24	16	-5	-18	30	10	14	8	-16
$a_6a_4a_1^2$	-2	2	1	2	7	5	-29	9	17	-25	-22	-14	-4	5
$a_6a_3^2$	9	·	9	-9	9	-27	9	·	·	9	9	3	-4	3
$a_6a_3a_2a_1$	-11	8	-10	14	-10	56	-42	1	5	-49	-41	-12	8	-4
$a_6a_3a_1^3$	3	·	·	-7	-4	-11	28	-2	-3	27	23	6	·	7
$a_6a_2^3$	-4	4	4	·	-8	8	-4	-1	2	-10	-6	-2	-4	8
$a_6a_2^2a_1^2$	2	-8	·	4	3	-25	33	1	-3	42	32	9	·	-7
$a_6a_2a_1^4$	·	2	·	·	·	·	7	·	1	-29	-23	-6	·	·
$a_6a_1^6$	·	·	·	·	·	·	·	·	·	5	4	1	·	·
$a_5^2a_2$	-4	2	1	1	-3	17	-19	-7	-4	-4	21	7	-4	11
$a_5^2a_1^2$	1	-1	-8	-5	10	1	2	7	9	3	-27	-7	6	1
$a_5a_4a_3$	-11	8	-10	11	8	-4	-6	1	5	-22	13	15	8	-14
$a_5a_4a_2a_1$	3	·	10	-10	-10	-10	18	10	7	16	-39	-30	-8	8
$a_5a_4a_1^3$	-1	·	·	5	-2	2	-1	-7	-8	-5	26	15	·	-5
$a_5a_3^2a_1$	2	-8	·	·	-10	14	8	-5	2	11	-24	-7	-4	11
$a_5a_3a_2^2$	4	-4	-1	2	3	-11	19	1	-2	4	-21	-7	4	-11
$a_5a_3a_2a_1^2$	-2	8	·	-1	8	-11	-23	3	-3	-15	75	21	·	-2
$a_5a_3a_1^4$	·	-2	·	·	·	·	4	·	2	3	-27	-7	·	·
$a_5a_2^3a_1$	·	·	·	·	-3	9	-15	-1	2	-2	23	7	·	3
$a_5a_2^2a_1^3$	·	·	·	·	·	-3	15	·	-1	4	-51	-14	·	·
$a_5a_2a_1^5$	·	·	·	·	·	·	-3	·	·	-1	27	7	·	·
$a_5a_1^7$	·	·	·	·	·	·	·	·	·	·	-4	-1	·	·
a_4^3	4	-4	4	-4	-8	8	4	6	-4	-4	-4	4	-4	8
$a_4^2a_3a_1$	-1	4	-1	·	10	-10	-9	-8	1	16	9	-16	4	-10
$a_4^2a_2^2$	-2	2	-2	4	-1	-4	-2	-4	8	-4	8	-8	2	-4
$a_4^2a_2a_1^2$	1	-4	·	-2	-1	11	4	4	-9	-6	-10	24	·	5
$a_4^2a_1^4$	4	1	·	·	·	-3	-1	·	3	2	1	-8	·	·
$a_4a_3^2a_2$	·	·	1	-2	-1	5	-5	4	-7	2	8	-8	-4	11
$a_4a_3^2a_1^2$	2	·	2	1	-2	-1	5	2	3	-9	-5	12	·	-1
$a_4a_3a_2^2a_1$	5	·	5	2	1	-3	-5	-4	6	·	-17	24	·	-3
$a_4a_3a_2a_1^3$	18	3	12	7	3	1	1	·	-3	8	11	-32	·	·
$a_4a_3a_1^5$	55	16	30	20	10	5	5	·	·	-2	-1	8	·	·
$a_4a_2^4$	12	·	12	6	4	·	·	1	·	2	-2	2	·	·
$a_4a_2^3a_1^2$	39	6	31	18	11	3	·	2	1	-4	9	-16	·	·
$a_4a_2^2a_1^4$	114	30	80	53	32	14	2	6	4	1	-6	20	·	·
$a_4a_2a_1^6$	315	108	210	150	95	51	13	20	15	6	1	-8	·	·
$a_4a_1^8$	840	336	560	420	280	168	56	70	56	28	8	1	·	·
a_3^4	·	·	4	·	·	·	·	·	·	·	·	·	1	-3
$a_3^3a_2a_1$	12	·	18	7	3	·	·	·	·	·	·	·	3	1
$a_3^3a_1^3$	42	6	42	24	9	3	·	·	·	·	·	·	6	3
$a_3^2a_2^3$	31	·	39	18	11	·	·	2	·	·	·	·	6	3
$a_3^2a_2^2a_1^2$	96	15	96	54	30	8	5	4	2	·	·	·	15	8
$a_3^2a_2a_1^4$	279	72	238	154	86	37	12	12	8	2	·	·	36	22
$a_3^2a_1^6$	776	262	600	420	250	132	32	40	30	12	2	·	90	60
$a_3a_2^4a_1$	216	36	216	128	83	22	·	20	9	·	·	·	36	22
$a_3a_2^3a_1^3$	606	159	546	357	228	93	12	54	34	7	·	·	93	60
$a_3a_2^2a_1^5$	1652	552	1390	976	635	320	72	160	115	40	5	·	240	165
$a_3a_2a_1^7$	4410	1722	3570	2632	1771	1008	302	490	378	168	38	3	630	455
$a_3a_1^9$	11592	5040	9240	7056	4914	3024	1080	1512	1218	624	189	28	1680	1260
a_2^6	480	90	480	300	210	60	·	66	30	20	·	·	90	60
$a_2^5a_1^2$	1320	360	1225	820	565	230	30	172	105	30	·	·	240	165
$a_2^4a_1^4$	3552	1206	3132	2220	1536	760	168	480	336	108	12	·	639	456
$a_2^3a_1^6$	9417	3696	8040	5952	4188	2337	681	1380	1041	432	87	6	1710	1266
$a_2^2a_1^8$	24696	10752	20720	15848	11396	6888	2392	4004	3164	1528	418	54	3528	3528
$a_2a_1^{10}$	64260	30240	53550	42000	30870	19740	7740	11592	9450	5040	1665	300	12600	9870
a_1^{12}	166320	83160	138600	110880	83160	55440	23760	33264	27720	15840	5940	1320	34650	27720

Table 1·2·12 (cont.)

$w=12$ (vi)	(63^2)	(5^22)	(642)	(732)	(82^2)	(651)	(741)	(831)	(921)	$(10,1^2)$	(6^3)	(75)	(84)	(93)	$(10,2)$	$(11,1)$	(12)
a_{12}	-12	-12	-24	-24	-12	-24	-24	-24	-24	-12	6	12	12	12	12	12	-12
$a_{11}a_1$	12	12	24	24	12	13	13	13	13	1	-6	-12	-12	-12	-12	-1	12
$a_{10}a_2$	12	2	4	4	-8	24	24	24	4	2	-6	-12	-12	-12	8	-12	12
$a_{10}a_1^2$	-12	-7	-14	-14	-2	-13	-13	-13	-3	-1	6	12	12	12	2	1	-12
a_9a_3	-15	12	24	-3	12	24	24	-3	-3	12	-6	-12	-12	15	-12	-12	-12
$a_9a_2a_1$	3	-14	-28	-1	-4	-37	-37	-10	-8	-3	12	24	24	-3	4	13	-24
$a_9a_1^3$	3	7	14	5	2	13	13	4	3	1	-6	-12	-12	-3	-2	-1	12
a_8a_4	12	12	-8	24	-4	24	24	-8	-8	24	-6	-12	20	-12	-12	-12	-24
$a_8a_3a_1$	3	-24	-16	-21	-8	-37	-5	-2	-10	-13	12	24	-8	-3	24	13	-24
$a_8a_2^2$	-12	-2	12	-4	$\cdot$	-24	-8	-8	-4	-2	6	12	-4	12	-8	12	-12
$a_8a_2a_1^2$	9	21	10	15	6	50	18	15	11	4	-18	-36	-4	-9	-6	-14	36
$a_8a_1^4$	-3	-7	-6	-5	-2	-13	-5	-4	-3	-1	6	12	4	3	2	1	-12
a_7a_5	12	-23	24	-11	12	-11	-11	24	24	12	-6	23	-12	-12	-12	-12	12
$a_7a_4a_1$	-24	11	-16	-13	-8	-2	2	-5	-37	-13	12	-11	-8	24	24	13	-24
$a_7a_3a_2$	3	21	-28	-8	-4	-13	-13	-21	-1	-14	12	-11	24	-3	4	24	-24
$a_7a_3a_1^2$	9	-4	30	21	10	15	11	15	13	14	-18	-1	-4	-9	-26	-14	36
$a_7a_2^2a_1$	9	-19	16	12	4	26	24	18	12	5	-18	-1	-20	-9	4	-25	36
$a_7a_2a_1^3$	-12	7	-24	-20	-8	-28	-24	-19	-14	-5	24	13	16	12	8	15	-48
$a_7a_1^5$	3	$\cdot$	6	5	2	6	5	4	3	1	-6	-5	-4	-3	-2	-1	12
a_6^2	-12	6	-24	12	6	-24	12	12	12	6	15	-6	-6	-6	-6	-6	6
$a_6a_5a_1$	12	11	24	-13	-24	40	-2	-37	-37	-13	-24	-11	24	24	24	13	-24
$a_6a_4a_2$	12	-14	28	-28	12	24	-16	-16	-28	-14	-24	-24	-8	24	4	24	-24
$a_6a_4a_1^2$	$\cdot$	-4	-18	27	10	-27	11	18	40	14	18	-1	-4	-36	-26	-14	36
$a_6a_3^2$	6	-12	12	3	-12	12	-24	3	3	-12	-12	12	12	-15	12	12	-12
$a_6a_3a_2a_1$	-27	17	-24	30	16	-27	55	33	19	30	36	-37	-40	9	-32	-50	72
$a_6a_3a_1^3$	6	-3	4	-26	-12	14	-24	-19	-16	-15	-12	13	16	12	28	15	-48
$a_6a_2^3$	$\cdot$	2	-12	4	$\cdot$	$\cdot$	8	8	4	2	6	-12	4	-12	8	-12	12
$a_6a_2^2a_1^2$	9	-2	22	-27	-10	2	-42	-33	-23	-9	-18	37	24	18	2	39	-72
$a_6a_2a_1^4$	-3	$\cdot$	-6	25	10	-1	29	23	17	6	6	-25	-20	-15	-10	-16	60
$a_6a_1^6$	$\cdot$	$\cdot$	$\cdot$	-5	-2	$\cdot$	-5	-4	-3	-1	$\cdot$	5	4	3	2	1	-12
$a_5^2a_2$	-12	3	-14	21	-2	11	11	-24	-14	-7	6	-23	12	12	2	12	-12
$a_5^2a_1^2$	$\cdot$	-7	-5	-4	13	-16	-10	25	20	7	9	17	-18	-18	-13	-7	18
$a_5a_4a_3$	3	11	-16	14	-8	-13	19	19	11	-21	12	-11	-8	-3	24	24	-24
$a_5a_4a_2a_1$	9	-3	16	-28	$\cdot$	-25	-19	31	73	30	$\cdot$	33	-8	-45	-32	-50	72
$a_5a_4a_1^3$	-3	7	4	3	12	19	11	-22	-43	-15	-12	-22	16	39	28	15	-48
$a_5a_3^2a_1$	-9	1	4	-17	20	5	-6	-1	7	25	$\cdot$	-1	-4	18	-36	-25	36
$a_5a_3a_2^2$	9	-4	16	-23	4	2	-14	29	5	16	-18	34	-20	-9	4	-36	36
$a_5a_3a_2a_1^2$	9	-4	-16	39	-32	7	21	-63	-43	-48	$\cdot$	-31	48	9	64	78	-144
$a_5a_3a_1^4$	-3	$\cdot$	2	-4	14	-6	-6	23	19	16	6	10	-20	-15	-30	-16	60
$a_5a_2^3a_1$	-9	2	-4	19	-4	-1	3	-26	-16	-7	12	-22	16	21	-12	37	-48
$a_5a_2^2a_1^3$	3	$\cdot$	2	-23	18	1	-4	52	37	14	-6	20	-40	-30	-10	-54	120
$a_5a_2a_1^5$	$\cdot$	$\cdot$	$\cdot$	5	-12	$\cdot$	1	-27	-20	-7	$\cdot$	-5	24	18	12	17	-72
$a_5a_1^7$	$\cdot$	$\cdot$	$\cdot$	$\cdot$	2	$\cdot$	$\cdot$	4	3	1	$\cdot$	$\cdot$	-4	-3	-2	-1	12
a_4^3	-4	-4	8	-8	4	-8	8	8	-8	-4	2	4	-12	4	4	4	-4
$a_4^2a_3a_1$	9	1	10	-4	26	$\cdot$	-26	-34	34	25	-18	-1	44	-9	-36	-25	36
$a_4^2a_2^2$	$\cdot$	8	-12	16	-14	$\cdot$	-4	-4	16	8	9	-18	22	-18	2	-18	18
$a_4^2a_2a_1^2$	-9	-7	8	-7	16	1	15	16	-62	-24	$\cdot$	2	-40	45	32	39	-72
$a_4^2a_1^4$	3	$\cdot$	-3	1	-1	-3	-5	-3	23	8	3	5	6	-21	-15	-8	30
$a_4a_3^2a_2$	-9	-9	8	5	$\cdot$	1	5	-14	-2	26	$\cdot$	-1	-4	18	-16	-36	36
$a_4a_3^2a_1^2$	$\cdot$	3	2	-4	-6	-20	15	34	-20	-39	18	2	-40	-9	62	39	-72
$a_4a_3a_2^2a_1$	9	6	-8	-19	24	-1	3	16	-36	-51	-12	4	-16	9	24	111	-144
$a_4a_3a_2a_1^3$	-3	$\cdot$	4	7	-12	7	-17	-27	73	68	$\cdot$	5	48	-33	-100	-108	240
$a_4a_3a_1^5$	$\cdot$	$\cdot$	$\cdot$	-1	$\cdot$	$\cdot$	5	5	-22	-17	-6	-5	-8	18	32	17	-72
$a_4a_2^4$	$\cdot$	-2	4	-4	4	$\cdot$	$\cdot$	$\cdot$	-4	-2	-6	12	-12	12	-8	12	-12
$a_4a_2^3a_1^2$	$\cdot$	$\cdot$	-2	8	-18	-1	3	-5	39	16	6	-15	24	-39	10	-76	120
$a_4a_2^2a_1^4$	$\cdot$	$\cdot$	$\cdot$	-2	12	$\cdot$	-1	5	-54	-20	$\cdot$	5	-20	45	20	70	-180
$a_4a_2a_1^6$	$\cdot$	$\cdot$	$\cdot$	$\cdot$	-2	$\cdot$	$\cdot$	-1	5	23	8	$\cdot$	4	-21	-14	-18	84
$a_4a_1^8$	$\cdot$	$\cdot$	$\cdot$	$\cdot$	$\cdot$	$\cdot$	$\cdot$	$\cdot$	-3	-1	$\cdot$	$\cdot$	$\cdot$	3	2	1	-12
a_3^4	3	3	-3	-3	3	-3	6	-3	-3	3	3	-3	-3	6	-3	-3	3
$a_3^2a_2a_1$	-3	-3	4	7	-12	7	-18	13	19	-27	-12	13	16	-42	28	37	-48
$a_3^3a_1^3$	1	$\cdot$	-2	-1	6	2	2	-11	-6	18	-2	-5	8	17	-30	-18	40
$a_3^2a_2^3$	$\cdot$	1	-2	2	-2	-1	3	-5	-9	-9	6	-11	8	-3	-6	24	-24
$a_3^2a_2^2a_1^2$	2	2	1	-4	9	-4	9	-5	-30	54	9	-5	-28	63	-45	-114	180
$a_3^2a_2a_1^4$	9	6	4	1	-6	$\cdot$	-3	10	13	-45	$\cdot$	5	-4	-36	70	70	-180
$a_3^2a_1^6$	30	20	15	6	1	$\cdot$	$\cdot$	-2	-1	9	$\cdot$	$\cdot$	2	3	-17	-9	42
$a_3a_2^4a_1$	6	9	4	$\cdot$	$\cdot$	1	-3	5	-7	9	-6	10	-4	-6	20	-49	60
$a_3a_2^3a_1^3$	24	24	15	3	3	3	1	-5	14	-30	$\cdot$	-5	16	-21	$\cdot$	130	-240
$a_3a_2^2a_1^5$	81	70	50	17	2	10	5	1	-7	27	$\cdot$	$\cdot$	-4	18	-32	-87	252
$a_3a_2a_1^7$	252	210	161	70	15	35	21	7	1	$\cdot$	$\cdot$	$\cdot$	$\cdot$	-3	16	19	-96
$a_3a_1^9$	756	630	504	252	72	126	84	36	9	1	$\cdot$	$\cdot$	$\cdot$	$\cdot$	-2	-1	12
a_2^6	20	30	15	$\cdot$	$\cdot$	6	$\cdot$	$\cdot$	$\cdot$	$\cdot$	1	-2	2	-2	2	-2	2
$a_2^5a_1^2$	70	81	50	10	$\cdot$	17	5	$\cdot$	$\cdot$	$\cdot$	2	1	-4	9	-16	25	-36
$a_2^4a_1^4$	228	228	160	52	6	52	24	4	$\cdot$	$\cdot$	6	4	1	-6	20	-50	105
$a_2^3a_1^6$	707	660	498	207	42	165	93	27	3	$\cdot$	20	15	6	1	-8	35	-112
$a_2^2a_1^8$	2128	1932	1526	736	201	532	336	128	26	2	70	56	28	8	1	-10	54
$a_2a_1^{10}$	6300	5670	4620	2460	810	1722	1170	525	145	21	252	210	120	45	10	1	-12
a_1^{12}	18480	16632	13860	7920	2970	5544	3960	1980	660	132	924	792	495	220	66	12	1

Table 1·3. *Unitary functions (a-functions) in terms of homogeneous product sums (h-functions) and conversely (up to and including weight 12)*

Note that the final digit (or digits) in the subheadings indicates the weight, w.

To express a product of h-functions in terms of a-functions, the appropriate coefficients are found by reading horizontally *right across the table*, e.g. from Table 1·3·5

$$h_4h_1 = -a_4a_1 + 2a_3a_1^2 + a_2^2a_1 - 3a_2a_1^3 + a_1^5.$$

To express a-functions in terms of h-functions, we note that any relation expressing h's in terms of a's is equally valid if h and a are interchanged in it.

Table 1·3·2

$w=2$	a_2	a_1^2
h_2	-1	1
h_1^2		1

Table 1·3·3

$w=1.\quad h_1=a_1.$

$w=3$	a_3	a_2a_1	a_1^3
h_3	1	-2	1
h_2h_1		-1	1
h_1^3			1

Table 1·3·4

$w=4$	a_4	a_3a_1	a_2^2	$a_2a_1^2$	a_1^4
h_4	-1	2	$\cdot$	-3	1
h_3h_1		1		-2	1
h_2^2			1	-2	1
$h_2h_1^2$				-1	1
h_1^4					1

Table 1·3·5

$w=5$	a_5	a_4a_1	a_3a_2	$a_3a_1^2$	$a_2^2a_1$	$a_2a_1^3$	a_1^5
h_5	1	-2	-2	3	3	-4	1
h_4h_1		-1	$\cdot$	2	1	-3	1
h_3h_2			-1	1	2	-3	1
$h_3h_1^2$				1	$\cdot$	-2	1
$h_2^2h_1$					1	-2	1
$h_2h_1^3$						-1	1
h_1^5							1

Table 1·3·6

$w=6$	a_6	a_5a_1	a_4a_2	$a_4a_1^2$	a_3^2	$a_3a_2a_1$	$a_3a_1^3$	a_2^3	$a_2^2a_1^2$	$a_2a_1^4$	a_1^6
h_6	-1	2	2	-3	1	-6	4	-1	6	-5	1
h_5h_1		1	$\cdot$	-2	$\cdot$	-2	3	$\cdot$	3	-4	1
h_4h_2			1	-1	$\cdot$	-2	2	-1	4	-4	1
$h_4h_1^2$				-1	$\cdot$	$\cdot$	2	$\cdot$	1	-3	1
h_3^2					1	-4	2	$\cdot$	4	-4	1
$h_3h_2h_1$						-1	1	$\cdot$	2	-3	1
$h_3h_1^3$							1	$\cdot$	$\cdot$	-2	1
h_2^3								-1	3	-3	1
$h_2^2h_1^2$									1	-2	1
$h_2h_1^4$										-1	1
h_1^6											1

Table 1·3·7

$w=7$	a_7	a_6a_1	a_5a_2	$a_5a_1^2$	a_4a_3	$a_4a_2a_1$	$a_4a_1^3$	$a_3^2a_1$	$a_3a_2^2$	$a_3a_2a_1^2$	$a_3a_1^4$	$a_2^3a_1$	$a_2^2a_1^3$	$a_2a_1^5$	a_1^7
h_7	1	-2	-2	3	-2	6	-4	3	3	-12	5	-4	10	-6	1
h_6h_1		-1	$\cdot$	2	$\cdot$	2	-3	1	$\cdot$	-6	4	-1	6	-5	1
h_5h_2			-1	1	$\cdot$	2	-2	$\cdot$	2	-5	3	-3	7	-5	1
$h_5h_1^2$				1	$\cdot$	$\cdot$	-2	$\cdot$	$\cdot$	-2	3	$\cdot$	3	-4	1
h_4h_3					-1	2	-1	2	1	-7	3	-2	7	-5	1
$h_4h_2h_1$						1	-1	$\cdot$	$\cdot$	-2	2	-1	4	-4	1
$h_4h_1^3$							-1	$\cdot$	$\cdot$	$\cdot$	2	$\cdot$	1	-3	1
$h_3^2h_1$								1	$\cdot$	-4	2	$\cdot$	4	-4	1
$h_3h_2^2$									1	-2	1	-2	5	-4	1
$h_3h_2h_1^2$										-1	1	$\cdot$	2	-3	1
$h_3h_1^4$											1	$\cdot$	$\cdot$	-2	1
$h_2^3h_1$												-1	3	-3	1
$h_2^2h_1^3$													1	-2	1
$h_2h_1^5$														-1	1
h_1^7															1

Table 1·3·8

$w=8$ (i)	a_8	a_7a_1	a_6a_2	$a_6a_1^2$	a_5a_3	$a_5a_2a_1$	$a_5a_1^3$	a_4^2	$a_4a_3a_1$	$a_4a_2^2$	$a_4a_2a_1^2$
h_8	-1	2	2	-3	2	-6	4	1	-6	-3	12
h_7h_1		1	$\cdot$	-2	$\cdot$	-2	3	$\cdot$	-2	$\cdot$	6
h_6h_2			1	-1	$\cdot$	-2	2	$\cdot$	$\cdot$	-2	5
$h_6h_1^2$				-1	$\cdot$	$\cdot$	2	$\cdot$	$\cdot$	$\cdot$	2
h_5h_3					1	$\cdot$	$\cdot$				4
$h_5h_2h_1$						-2	1	$\cdot$	-2	$\cdot$	2
$h_5h_1^3$						-1	1	$\cdot$	$\cdot$	$\cdot$	$\cdot$
h_4^2								1	$\cdot$	$\cdot$	6
$h_4h_3h_1$									-4	-2	2
$h_4h_2^2$									-1	-1	2
$h_4h_2h_1^2$											1

$w=8$ (ii)	$a_4a_1^4$	$a_3^2a_2$	$a_3^2a_1^2$	$a_3a_2^2a_1$	$a_3a_2a_1^3$	$a_3a_1^5$	a_2^4	$a_2^3a_1^2$	$a_2^2a_1^4$	$a_2a_1^6$	a_1^8
h_8	-5	-3	6	12	-20	6	1	-10	15	-7	1
h_7h_1	-4	$\cdot$	3	3	-12	5	$\cdot$	-4	10	-6	1
h_6h_2	-3	-1	1	6	-10	4	1	-7	11	-6	1
$h_6h_1^2$	-3	$\cdot$	1	$\cdot$	-6	4	$\cdot$	-1	6	-5	1
h_5h_3	-2	-2	3	7	-12	4	$\cdot$	-6	11	-6	1
$h_5h_2h_1$	-2	$\cdot$	$\cdot$	2	-5	3	$\cdot$	-3	7	-5	1
$h_5h_1^3$	-2	$\cdot$	$\cdot$	$\cdot$	-2	3	$\cdot$	$\cdot$	3	-4	1
h_4^2	-2	$\cdot$	4	4	-12	4	1	-6	11	-6	1
$h_4h_3h_1$	-1	$\cdot$	2	1	-7	3	$\cdot$	-2	7	-5	1
$h_4h_2^2$	-1	$\cdot$	$\cdot$	2	-4	2	1	-5	8	-5	1
$h_4h_2h_1^2$	-1	$\cdot$	$\cdot$	$\cdot$	-2	2	$\cdot$	-1	4	-4	1
$h_4h_1^4$	-1	$\cdot$	$\cdot$	$\cdot$	$\cdot$	2	$\cdot$	$\cdot$	1	-3	1
$h_3^2h_2$	$\cdot$	-1	1	4	-6	2	$\cdot$	-4	8	-5	1
$h_3^2h_1^2$	$\cdot$	$\cdot$	1	$\cdot$	-4	2	$\cdot$	$\cdot$	4	-4	1
$h_3h_2^2h_1$	$\cdot$	$\cdot$	$\cdot$	1	-2	1	$\cdot$	-2	5	-4	1
$h_3h_2h_1^3$	$\cdot$	$\cdot$	$\cdot$	$\cdot$	-1	1	$\cdot$	$\cdot$	2	-3	1
$h_3h_1^5$	$\cdot$	$\cdot$	$\cdot$	$\cdot$	$\cdot$	1	$\cdot$	$\cdot$	$\cdot$	-2	1
h_2^4	$\cdot$	$\cdot$	$\cdot$	$\cdot$	$\cdot$	$\cdot$	1	-4	6	-4	1
$h_2^3h_1^2$	$\cdot$	$\cdot$	$\cdot$	$\cdot$	$\cdot$	$\cdot$	$\cdot$	-1	3	-3	1
$h_2^2h_1^4$	$\cdot$	$\cdot$	$\cdot$	$\cdot$	$\cdot$	$\cdot$	$\cdot$	$\cdot$	1	-2	1
$h_2h_1^6$	$\cdot$	$\cdot$	$\cdot$	$\cdot$	$\cdot$	$\cdot$	$\cdot$	$\cdot$	$\cdot$	-1	1
h_1^8	$\cdot$	$\cdot$	$\cdot$	$\cdot$	$\cdot$	$\cdot$	$\cdot$	$\cdot$	$\cdot$	$\cdot$	1

Table 1·3·9

$w=9$ (i)	a_9	a_8a_1	a_7a_2	$a_7a_1^2$	a_6a_3	$a_6a_2a_1$	$a_6a_1^3$	a_5a_4	$a_5a_3a_1$	$a_5a_2^2$	$a_5a_2a_1^2$	$a_5a_1^4$	$a_4^2a_1$	$a_4a_3a_2$	$a_4a_3a_1^2$
h_9	1	-2	-2	3	-2	6	-4	-2	6	3	-12	5	3	6	-12
h_8h_1		-1	$\cdot$	2	$\cdot$	2	-3	$\cdot$	2	$\cdot$	-6	4	1	$\cdot$	-6
h_7h_2			-1	1	$\cdot$	2	-2	$\cdot$	$\cdot$	2	-5	3	$\cdot$	2	-2
$h_7h_1^2$				1	$\cdot$	$\cdot$	-2	$\cdot$	$\cdot$	$\cdot$	-2	3	$\cdot$	$\cdot$	-2
h_6h_3					-1	2	-1	$\cdot$	$\cdot$	$\cdot$	-4	2	$\cdot$	2	-3
$h_6h_2h_1$						1	-1	$\cdot$	$\cdot$	$\cdot$	-2	2	$\cdot$	$\cdot$	$\cdot$
$h_6h_1^3$							-1	$\cdot$	$\cdot$	$\cdot$	$\cdot$	2	$\cdot$	$\cdot$	$\cdot$
h_5h_4								-1	2	$\cdot$	-3	1	2	2	-7
$h_5h_3h_1$									1	$\cdot$	-2	1	$\cdot$	$\cdot$	-2
$h_5h_2^2$										1	-2	1	$\cdot$	$\cdot$	$\cdot$
$h_5h_2h_1^2$											-1	1	$\cdot$	$\cdot$	$\cdot$
$h_5h_1^4$												1	$\cdot$	$\cdot$	$\cdot$
$h_4^2h_1$													1	$\cdot$	-4
$h_4h_3h_2$														1	-1
$h_4h_3h_1^2$															-1

$w=9$ (ii)	$a_4a_2^2a_1$	$a_4a_2a_1^3$	$a_4a_1^5$	a_3^3	$a_3^2a_2a_1$	$a_3^2a_1^3$	$a_3a_2^3$	$a_3a_2^2a_1^2$	$a_3a_2a_1^4$	$a_3a_1^6$	$a_2^4a_1$	$a_2^3a_1^3$	$a_2^2a_1^5$	$a_2a_1^7$	a_1^9
h_9	-12	20	-6	1	-12	10	-4	30	-30	7	5	-20	21	-8	1
h_8h_1	-3	12	-5	$\cdot$	-3	6	$\cdot$	12	-20	6	1	-10	15	-7	1
h_7h_2	-6	10	-4	$\cdot$	-3	3	-3	15	-17	5	4	-14	16	-7	1
$h_7h_1^2$	$\cdot$	6	-4	$\cdot$	$\cdot$	3	$\cdot$	3	-12	5	$\cdot$	-4	10	-6	1
h_6h_3	-4	8	-3	1	-8	5	-1	18	-19	5	2	-13	16	-7	1
$h_6h_2h_1$	-2	5	-3	$\cdot$	-1	1	$\cdot$	6	-10	4	1	-7	11	-6	1
$h_6h_1^3$	$\cdot$	2	-3	$\cdot$	$\cdot$	1	$\cdot$	$\cdot$	-6	4	$\cdot$	-1	6	-5	1
h_5h_4	-5	10	-3	$\cdot$	-4	6	-2	15	-19	5	3	-13	16	-7	1
$h_5h_3h_1$	$\cdot$	4	-2	$\cdot$	-2	3	$\cdot$	7	-12	4	$\cdot$	-6	11	-6	1
$h_5h_2^2$	-2	4	-2	$\cdot$	$\cdot$	$\cdot$	-2	7	-8	3	3	-10	12	-6	1
$h_5h_2h_1^2$	$\cdot$	2	-2	$\cdot$	$\cdot$	$\cdot$	$\cdot$	2	-5	3	$\cdot$	-3	7	-5	1
$h_5h_1^4$	$\cdot$	$\cdot$	-2	$\cdot$	$\cdot$	$\cdot$	$\cdot$	$\cdot$	-2	3	$\cdot$	$\cdot$	3	-4	1
$h_4^2h_1$	-2	6	-2	$\cdot$	$\cdot$	4	$\cdot$	4	-12	4	1	-6	11	-6	1
$h_4h_3h_2$	-2	3	-1	$\cdot$	-2	2	-1	8	-10	3	2	-9	12	-6	1
$h_4h_3h_1^2$	$\cdot$	2	-1	$\cdot$	$\cdot$	2	$\cdot$	1	-7	3	$\cdot$	-2	7	-5	1
$h_4h_2^2h_1$	-1	2	-1	$\cdot$	$\cdot$	$\cdot$	$\cdot$	2	-4	2	1	-5	8	-5	1
$h_4h_2h_1^3$	$\cdot$	1	-1	$\cdot$	$\cdot$	$\cdot$	$\cdot$	$\cdot$	-2	2	$\cdot$	-1	4	-4	1
$h_4h_1^5$	$\cdot$	$\cdot$	-1	$\cdot$	$\cdot$	$\cdot$	$\cdot$	$\cdot$	$\cdot$	2	$\cdot$	$\cdot$	1	-3	1
h_3^3	$\cdot$	$\cdot$	$\cdot$	1	-6	3	$\cdot$	12	-12	3	$\cdot$	-8	12	-6	1
$h_3^2h_2h_1$	$\cdot$	$\cdot$	$\cdot$	$\cdot$	-1	1	$\cdot$	4	-6	2	2	-4	8	-5	1
$h_3^2h_1^3$	$\cdot$	$\cdot$	$\cdot$	$\cdot$	$\cdot$	1	$\cdot$	$\cdot$	-4	2	$\cdot$	$\cdot$	4	-4	1
$h_3h_2^3$	$\cdot$	$\cdot$	$\cdot$	$\cdot$	$\cdot$	$\cdot$	1	$\cdot$	-3	1	$\cdot$	$\cdot$	9	-5	1
$h_3h_2^2h_1^2$	$\cdot$	$\cdot$	$\cdot$	$\cdot$	$\cdot$	$\cdot$	$\cdot$	1	-2	1	$\cdot$	-7	5	-4	1
$h_3h_2h_1^4$	$\cdot$	$\cdot$	$\cdot$	$\cdot$	$\cdot$	$\cdot$	$\cdot$	$\cdot$	-1	1	$\cdot$	-2	4	-3	1
$h_3h_1^6$	$\cdot$	$\cdot$	$\cdot$	$\cdot$	$\cdot$	$\cdot$	$\cdot$	$\cdot$	$\cdot$	1	$\cdot$	$\cdot$	3	-2	1
$h_2^4h_1$	$\cdot$	$\cdot$	$\cdot$	$\cdot$	$\cdot$	$\cdot$	$\cdot$	$\cdot$	$\cdot$	$\cdot$	1	-4	6	-4	1
$h_2^3h_1^3$	$\cdot$	$\cdot$	$\cdot$	$\cdot$	$\cdot$	$\cdot$	$\cdot$	$\cdot$	$\cdot$	$\cdot$	$\cdot$	-1	3	-3	1
$h_2^2h_1^5$	$\cdot$	$\cdot$	$\cdot$	$\cdot$	$\cdot$	$\cdot$	$\cdot$	$\cdot$	$\cdot$	$\cdot$	$\cdot$	$\cdot$	1	-2	1
$h_2h_1^7$	$\cdot$	$\cdot$	$\cdot$	$\cdot$	$\cdot$	$\cdot$	$\cdot$	$\cdot$	$\cdot$	$\cdot$	$\cdot$	$\cdot$	$\cdot$	-1	1
h_1^9	$\cdot$	$\cdot$	$\cdot$	$\cdot$	$\cdot$	$\cdot$	$\cdot$	$\cdot$	$\cdot$	$\cdot$	$\cdot$	$\cdot$	$\cdot$	$\cdot$	1

Table 1·3·10

$w=10$ (i)	a_{10}	a_9a_1	a_8a_2	$a_8a_1^2$	a_7a_3	$a_7a_2a_1$	$a_7a_1^3$	a_6a_4	$a_6a_3a_1$	$a_6a_2^2$	$a_6a_2a_1^2$	$a_6a_1^4$	a_5^2	$a_5a_4a_1$
h_{10}	-1	2	2	-3	2	-6	4	2	-6	-3	12	-5	1	-6
h_9h_1		1	·	-2	·	-2	3	·	-2	·	6	-4	·	-2
h_8h_2			1	-1	·	-2	2	·	·	-2	5	-3	·	·
$h_8h_1^2$				-1	·	·	2	·	·	·	2	-3	·	·
h_7h_3					1	-2	1	·	-2	·	4	-2	·	·
$h_7h_2h_1$						-1	1	·	·	·	2	-2	·	·
$h_7h_1^3$							1	·	·	·	·	-2	·	·
h_6h_4								1	-2	-1	3	-1	·	-2
$h_6h_3h_1$									-1	·	2	-1	·	·
$h_6h_2^2$										-1	2	-1	·	·
$h_6h_2h_1^2$											1	-1	·	·
$h_6h_1^4$												-1	·	·
h_5^2													1	-4
$h_5h_4h_1$														-1

$w=10$ (ii)	$a_5a_3a_2$	$a_5a_3a_1^2$	$a_5a_2^2a_1$	$a_5a_2a_1^3$	$a_5a_1^5$	$a_4^2a_2$	$a_4^2a_1^2$	$a_4a_3^2$	$a_4a_3a_2a_1$	$a_4a_3a_1^3$	$a_4a_2^3$	$a_4a_2^2a_1^2$	$a_4a_2a_1^4$	$a_4a_1^6$
h_{10}	-6	12	12	-20	6	-3	6	-3	24	-20	4	-30	30	-7
h_9h_1	·	6	3	-12	5	·	3	·	6	-12	·	-12	20	-6
h_8h_2	-2	2	6	-10	4	-1	1	·	6	-6	3	-15	17	-5
$h_8h_1^2$	·	2	·	-6	4	·	1	·	·	-6	·	-3	12	-5
h_7h_3	-2	3	4	-8	3	·	·	-2	10	-6	·	-12	14	-4
$h_7h_2h_1$	·	·	2	-5	3	·	·	·	2	-2	·	-6	10	-4
$h_7h_1^3$	·	·	·	-2	3	·	·	·	·	-2	·	·	6	-4
h_6h_4	·	4	2	-6	2	-2	3	-1	10	-10	3	-15	16	-4
$h_6h_3h_1$	·	2	·	-4	2	·	·	·	2	-3	·	-4	8	-3
$h_6h_2^2$	·	·	2	-4	2	·	·	·	·	·	2	-7	8	-3
$h_6h_2h_1^2$	·	·	·	-2	2	·	·	·	·	·	·	-2	5	-3
$h_6h_1^4$	·	·	·	·	2	·	·	·	·	·	·	·	2	-3
h_5^2	-4	6	6	-8	2	·	4	·	8	-12	·	-12	16	-4
$h_5h_4h_1$	·	2	1	-3	1	·	2	·	2	-7	·	-5	10	-3
$h_5h_3h_2$	-1	1	2	-3	1	·	·	·	2	-2	·	-4	6	-2
$h_5h_3h_1^2$	·	1	·	-2	1	·	·	·	·	-2	·	·	4	-2
$h_5h_2^2h_1$	·	·	1	-2	1	·	·	·	·	·	·	-2	4	-2
$h_5h_2h_1^3$	·	·	·	-1	1	·	·	·	·	·	·	·	2	-2
$h_5h_1^5$	·	·	·	·	1	·	·	·	·	·	·	·	·	-2
$h_4^2h_2$	·	·	·	·	·	-1	1	·	4	-4	2	-8	8	-2
$h_4^2h_1^2$	·	·	·	·	·	·	1	·	·	-4	·	-2	6	-2
$h_4h_3^2$	·	·	·	·	·	·	·	-1	4	-2	·	-4	4	-1
$h_4h_3h_2h_1$	·	·	·	·	·	·	·	·	1	-1	·	-2	3	-1
$h_4h_3h_1^3$	·	·	·	·	·	·	·	·	·	-1	·	·	2	-1
$h_4h_2^3$	·	·	·	·	·	·	·	·	·	·	1	-3	3	-1
$h_4h_2^2h_1^2$	·	·	·	·	·	·	·	·	·	·	·	-1	2	-1
$h_4h_2h_1^4$	·	·	·	·	·	·	·	·	·	·	·	·	1	-1
$h_4h_1^6$	·	·	·	·	·	·	·	·	·	·	·	·	·	-1

$w=10$ (iii)	$a_3^3a_1$	$a_3^2a_2^2$	$a_3^2a_2a_1^2$	$a_3^2a_1^4$	$a_3a_2^3a_1$	$a_3a_2^2a_1^3$	$a_3a_2a_1^5$	$a_3a_1^7$	a_2^5	$a_2^4a_1^2$	$a_2^3a_1^4$	$a_2^2a_1^6$	$a_2a_1^8$	a_1^{10}
h_{10}	4	6	-30	15	-20	60	-42	8	-1	15	-35	28	-9	1
h_9h_1	1	·	-12	10	-4	30	-30	7	·	5	-20	21	-8	1
h_8h_2	·	3	-9	6	-12	32	-26	6	-1	11	-25	22	-8	1
$h_8h_1^2$	·	·	-3	6	·	12	-20	6	·	1	-10	15	-7	1
h_7h_3	3	3	-18	8	-10	37	-28	6	·	8	-24	22	-8	1
$h_7h_2h_1$	·	·	-3	3	-3	15	-17	5	·	4	-14	16	-7	1
$h_7h_1^3$	·	·	·	3	·	3	-12	5	·	·	-4	10	-6	1
h_6h_4	2	1	-15	9	-8	34	-28	6	-1	9	-24	22	-8	1
$h_6h_3h_1$	1	·	-8	5	-1	18	-19	5	·	2	-13	16	-7	1
$h_6h_2^2$	·	1	-2	1	-6	16	-14	4	-1	8	-18	17	-7	1
$h_6h_2h_1^2$	·	·	-1	1	·	6	-10	4	·	1	-7	11	-6	1
$h_6h_1^4$	·	·	·	1	·	·	-6	4	·	·	-1	6	-5	1
h_5^2	·	4	-12	9	-12	34	-28	6	·	9	-24	22	-8	1
$h_5h_4h_1$	·	·	-4	6	-2	15	-19	5	·	3	-13	16	-7	1
$h_5h_3h_2$	·	2	-5	3	-7	19	-16	4	·	6	-17	17	-7	1
$h_5h_3h_1^2$	·	·	-2	3	·	7	-12	4	·	·	-6	11	-6	1
$h_5h_2^2h_1$	·	·	·	·	-2	7	-8	3	·	3	-10	12	-6	1
$h_5h_2h_1^3$	·	·	·	·	·	2	-5	3	·	·	-3	7	-5	1
$h_5h_1^5$	·	·	·	·	·	·	-2	3	·	·	·	3	-4	1
$h_4^2h_2$	·	·	-4	4	-4	16	-16	4	-1	7	-17	17	-7	1
$h_4^2h_1^2$	·	·	·	4	·	4	-12	4	·	1	-6	11	-6	1
$h_4h_3^2$	2	1	-11	5	-4	22	-18	4	·	4	-16	17	-7	1
$h_4h_3h_2h_1$	·	·	-2	2	-1	8	-10	3	·	2	-9	12	-6	1
$h_4h_3h_1^3$	·	·	·	2	·	1	-7	3	·	·	-2	7	-5	1
$h_4h_2^3$	·	·	·	·	-2	6	-6	2	-1	6	-13	13	-6	1
$h_4h_2^2h_1^2$	·	·	·	·	·	2	-4	2	·	1	-5	8	-5	1
$h_4h_2h_1^4$	·	·	·	·	·	·	-2	2	·	·	-1	4	-4	1
$h_4h_1^6$	·	·	·	·	·	·	·	2	·	·	·	1	-3	1
$h_3^3h_1$	1	·	-6	3	-4	12	-12	3	·	·	-8	12	-6	1
$h_3^2h_2^2$	·	1	-2	1	-4	10	-8	2	·	4	-12	13	-6	1
$h_3^2h_2h_1^2$	·	·	-1	1	·	4	-6	2	·	·	-4	8	-5	1
$h_3^2h_1^4$	·	·	1	1	·	·	-4	2	·	·	·	4	-4	1
$h_3h_2^3h_1$	·	·	·	·	-1	3	-3	1	·	2	-7	9	-5	1
$h_3h_2^2h_1^3$	·	·	·	·	·	1	-2	1	·	·	-2	5	-4	1
$h_3h_2h_1^5$	·	·	·	·	·	·	-1	1	·	·	·	2	-3	1
$h_3h_1^7$	·	·	·	·	·	·	·	1	·	·	·	·	-2	1
h_2^5	·	·	·	·	·	·	·	·	-1	5	-10	10	-5	1
$h_2^4h_1^2$	·	·	·	·	·	·	·	·	·	1	-4	6	-4	1
$h_2^3h_1^4$	·	·	·	·	·	·	·	·	·	·	-1	3	-3	1
$h_2^2h_1^6$	·	·	·	·	·	·	·	·	·	·	·	1	-2	1
$h_2h_1^8$	·	·	·	·	·	·	·	·	·	·	·	·	-1	1
h_1^{10}	·	·	·	·	·	·	·	·	·	·	·	·	·	1

Table 1.3.11

$w = 11$(i)

$w=11$(i)	a_{11}	$a_{10}a_1$	a_9a_2	$a_9a_1^2$	a_8a_3	$a_8a_2a_1$	$a_8a_1^3$	a_7a_4	$a_7a_3a_1$	$a_7a_2^2$	$a_7a_2a_1^2$	$a_7a_1^4$
h_{11}	1	-2	-2	3	-2	6	-4	-2	6	3	-12	5
$h_{10}h_1$		-1	·	2		2	-3		2	·	-6	4
h_9h_2			-1	1		2	-2		·	2	-5	3
$h_9h_1^2$				1		·	-2		·		-2	3
h_8h_3					-1	2	-1		2		-4	2
$h_8h_2h_1$						1	-1		·		-2	2
$h_8h_1^3$							-1		·			2
h_7h_4								-1	2	1	-3	1
$h_7h_3h_1$									1	·	-2	1
$h_7h_2^2$										1	-2	1
$h_7h_2h_1^2$											-1	1
$h_7h_1^4$												1

$w = 11$(ii)

$w=11$(ii)	a_6a_5	$a_6a_4a_1$	$a_6a_3a_2$	$a_6a_3a_1^2$	$a_6a_2^2a_1$	$a_6a_2a_1^3$	$a_6a_1^5$	$a_5^2a_1$	$a_5a_4a_2$	$a_5a_4a_1^2$	$a_5a_3^2$	$a_5a_3a_2a_1$	$a_5a_3a_1^3$
h_{11}	-2	6	6	-12	-12	20	-6	3	6	-12	3	-24	20
$h_{10}h_1$		2		-6	-3	12	-5	1		-6		-6	12
h_9h_2				-2	-6	10	-4		2	-2		-6	6
$h_9h_1^2$				-2		6	-4			-2			6
h_8h_3			2	-3	-4	8	-3				2	-10	6
$h_8h_2h_1$					-2	5	-3					-2	2
$h_8h_1^3$						2	-3						2
h_7h_4		2		-4	-2	6	-2		2	-3		-4	6
$h_7h_3h_1$				-2	-2	4	-2					-2	3
$h_7h_2^2$					-2	4	-2						
$h_7h_2h_1^2$						2	-2						
$h_7h_1^4$							-2						
h_6h_5	-1	2	2	-3	-3	4	-1	2	2	-7	1	-10	6
$h_6h_4h_1$		1		-2	-1	3	-1			-2		-2	4
$h_6h_3h_2$			1	-1	-2	3	-1					-2	2
$h_6h_3h_1^2$				-1		2	-1						2
$h_6h_2^2h_1$					-1	2	-1						
$h_6h_2h_1^3$						1	-1						
$h_6h_1^5$							-1						
$h_5^2h_1$								1		-4		-4	6
$h_5h_4h_2$									1	-1		-2	2
$h_5h_4h_1^2$										-1			2
$h_5h_3^2$											1	-4	2
$h_5h_3h_2h_1$												-1	1
$h_5h_3h_1^3$													1

$w = 11$(iii)

$w=11$(iii)	$a_5a_2^3$	$a_5a_2^2a_1^2$	$a_5a_2a_1^4$	$a_5a_1^6$	$a_4^2a_3$	$a_4^2a_2a_1$	$a_4^2a_1^3$	$a_4a_3^2a_1$	$a_4a_3a_2^2$	$a_4a_3a_2a_1^2$	$a_4a_3a_1^4$	$a_4a_2^3a_1$	$a_4a_2^2a_1^3$	$a_4a_2a_1^5$	$a_4a_1^7$
h_{11}	-4	30	-30	7	3	-12	10	-12	-12	60	-30	20	-60	42	-8
$h_{10}h_1$		12	-20	6		-3	6	-3		24	-20	4	-30	30	-7
h_9h_2	-3	15	-17	5		-3	3		-6	18	-12	12	-32	26	-6
$h_9h_1^2$		3	-12	5			3			6	-12		-12	20	-6
h_8h_3		12	-14	4	1	-2	1	-6	-3	24	-11	6	-27	22	-5
$h_8h_2h_1$		6	-10	4		-1	1			6	-6	3	-15	17	-5
$h_8h_1^3$			-6	4			1				-6		-3	12	-5
h_7h_4	-2	9	-11	3	2	-6	4	-7	-5	30	-15	10	-32	24	-5
$h_7h_3h_1$		4	-8	3				-2		10	-6		-12	14	-4
$h_7h_2^2$	-2	7	-8	3					-2	4	-2	6	-16	14	-4
$h_7h_2h_1^2$		2	-5	3						2	-2		-6	10	-4
$h_7h_1^4$			-2	3							-2			6	-4
h_6h_5	-1	12	-13	3		-4	6	-2	-4	24	-17	8	-29	24	-5
$h_6h_4h_1$		2	-6	2		-2	3	-1		10	-10	3	-15	16	-4
$h_6h_3h_2$		4	-6	2					-2	5	-3	4	-12	11	-3
$h_6h_3h_1^2$			-4	2						2	-3		-4	8	-3
$h_6h_2^2h_1$		2	-4	2								2	-7	8	-3
$h_6h_2h_1^3$			-2	2									-2	5	-3
$h_6h_1^5$				2										2	-3
$h_5^2h_1$		6	-8	2			4			8	-12		-12	16	-4
$h_5h_4h_2$	-1	4	-4	1		-2	2		-2	9	-7	5	-15	13	-3
$h_5h_4h_1^2$		1	-3	1			2			2	-7		-5	10	-3
$h_5h_3^2$		4	-4	1				-2		8	-4		-8	8	-2
$h_5h_3h_2h_1$		2	-3	1						2	-2		-4	6	-2
$h_5h_3h_1^3$			-2	1							-2			4	-2
$h_5h_2^3$	-1	3	-3	1								2	-6	6	-2
$h_5h_2^2h_1^2$		1	-2	1									-2	4	-2
$h_5h_2h_1^4$			-1	1										2	-2
$h_5h_1^6$				1											-2
$h_4^2h_3$					1	-2	1	-4	-2	14	-6	4	-14	10	-2
$h_4^2h_2h_1$						-1	1			4	-4	2	-8	8	-2
$h_4^2h_1^3$							1				-4		-2	6	-2
$h_4h_3^2h_1$								-1		4	-2		-4	4	-1
$h_4h_3h_2^2$									-1	2	-1	2	-5	4	-1
$h_4h_3h_2h_1^2$										1	-1		-2	3	-1
$h_4h_3h_1^4$											-1			2	-1
$h_4h_2^3h_1$												1	-3	3	-1
$h_4h_2^2h_1^3$													-1	2	-1
$h_4h_2h_1^5$														1	-1
$h_4h_1^7$															-1

Table 1·3·11 (cont.)

$w = 11$ (iv)	$a_3^3 a_2$	$a_3^3 a_1^2$	$a_3^2 a_2^2 a_1$	$a_3^2 a_2 a_1^3$	$a_3^2 a_1^5$	$a_3 a_2^4$	$a_3 a_2^3 a_1^2$	$a_3 a_2^2 a_1^4$	$a_3 a_2 a_1^6$	$a_3 a_1^8$	$a_2^5 a_1$	$a_2^4 a_1^3$	$a_2^3 a_1^5$	$a_2^2 a_1^7$	$a_2 a_1^9$	a_1^{11}
h_{11}	−4	10	30	−60	21	5	−60	105	−56	9	−6	35	−56	36	−10	1
$h_{10}h_1$	·	4	6	−30	15	·	−20	60	−42	8	−1	15	−35	28	−9	1
h_9h_2	−1	1	12	−22	10	4	−34	60	−37	7	−5	25	−41	29	−9	1
$h_9h_1^2$	·	1	·	−12	10	·	−4	30	−30	7	·	5	−20	21	−8	1
h_8h_3	−3	6	18	−35	12	1	−34	67	−39	7	−2	21	−40	29	−9	1
$h_8h_2h_1$	·	·	3	−9	6	·	−12	32	−26	6	−1	11	−25	22	−8	1
$h_8h_1^3$	·	·	·	−3	6	·	·	12	−20	6	·	1	−10	15	−7	1
h_7h_4	·	6	9	−33	13	3	−29	64	−39	7	−4	22	−40	29	−9	1
$h_7h_3h_1$	·	3	3	−18	8	·	−10	37	−28	6	·	8	−24	22	−8	1
$h_7h_2^2$	·	·	3	−6	3	3	−18	32	−22	5	−4	18	−30	23	−8	1
$h_7h_2h_1^2$	·	·	·	−3	3	·	−3	15	−17	5	·	4	−14	16	−7	1
$h_7h_1^4$	·	·	·	·	3	·	·	3	−12	5	·	·	−4	10	−6	1
h_6h_5	−2	3	15	−30	13	2	−33	64	−39	7	−3	22	−40	29	−9	1
$h_6h_4h_1$	·	2	1	−15	9	·	−8	34	−28	6	−1	9	−24	22	−8	1
$h_6h_3h_2$	−1	1	8	−13	5	1	−19	37	−24	5	−2	15	−29	23	−8	1
$h_6h_3h_1^2$	·	1	·	−8	5	·	−1	18	−19	5	·	2	−13	16	−7	1
$h_6h_2^2h_1$	·	·	1	−2	1	·	−6	16	−14	4	−1	8	−18	17	−7	1
$h_6h_2h_1^3$	·	·	·	−1	1	·	·	6	−10	4	·	1	−7	11	−6	1
$h_6h_1^5$	·	·	·	·	·	·	·	·	−6	4	·	·	−1	6	−5	1
$h_5^2h_1$	·	·	4	−12	9	·	−12	34	−28	6	·	9	−24	22	−8	1
$h_5h_4h_2$	·	·	4	−10	6	2	−17	34	−24	5	−3	16	−29	23	−8	1
$h_5h_4h_1^2$	·	·	·	−4	6	·	−2	15	−19	5	·	3	−13	16	−7	1
$h_5h_3^2$	−2	3	11	−20	7	·	−20	42	−26	5	·	12	−28	23	−8	1
$h_5h_3h_2h_1$	·	·	2	−5	3	·	−7	19	−16	4	·	6	−17	17	−7	1
$h_5h_3h_1^3$	·	·	·	−2	3	·	·	7	−12	4	·	·	−6	11	−6	1
$h_5h_2^3$	·	·	·	·	·	2	−9	15	−11	3	−3	13	−22	18	−7	1
$h_5h_2^2h_1^2$	·	·	·	·	·	·	−2	7	−8	3	·	3	−10	12	−6	1
$h_5h_2h_1^4$	·	·	·	·	·	·	·	2	−5	3	·	·	−3	7	−5	1
$h_5h_1^6$	·	·	·	·	·	·	·	·	−2	3	·	·	·	3	−4	1
$h_4^2h_3$	·	4	4	−20	8	1	−14	39	−26	5	−2	13	−28	23	−8	1
$h_4^2h_2h_1$	·	·	·	−4	4	·	−4	16	−16	4	−1	7	−17	17	−7	1
$h_4^2h_1^3$	·	·	·	·	4	·	·	4	−12	4	·	1	−6	11	−6	1
$h_4h_3^2h_1$	·	2	1	−11	5	·	−4	22	−18	4	·	4	−16	17	−7	1
$h_4h_3h_2^2$	·	·	2	−4	2	1	−9	18	−13	3	−2	11	−21	18	−7	1
$h_4h_3h_2h_1^2$	·	·	·	−2	2	·	−1	8	−10	3	·	2	−9	12	−6	1
$h_4h_3h_1^4$	·	·	·	·	2	·	·	1	−7	3	·	·	−2	7	−5	1
$h_4h_2^3h_1$	·	·	·	·	·	·	−2	6	−6	2	−1	6	−13	13	−6	1
$h_4h_2^2h_1^3$	·	·	·	·	·	·	·	2	−4	2	·	1	−5	8	−5	1
$h_4h_2h_1^5$	·	·	·	·	·	·	·	·	−2	2	·	·	−1	4	−4	1
$h_4h_1^7$	·	·	·	·	·	·	·	·	·	2	·	·	·	1	−3	1
$h_3^3h_2$	−1	1	6	−9	3	·	−12	24	−15	3	·	8	−20	18	−7	1
$h_3^3h_1^2$	·	1	·	−6	3	·	·	12	−12	3	·	·	−8	12	−6	1
$h_3^2h_2^2h_1$	·	·	1	−2	1	·	−4	10	−8	2	·	4	−12	13	−6	1
$h_3^2h_2h_1^3$	·	·	·	−1	1	·	·	4	−6	2	·	·	−4	8	−5	1
$h_3^2h_1^5$	·	·	·	·	1	·	·	·	−4	2	·	·	·	4	−4	1
$h_3h_2^4$	·	·	·	·	·	1	−4	6	−4	1	−2	9	−16	14	−6	1
$h_3h_2^3h_1^2$	·	·	·	·	·	·	−1	3	−3	1	·	2	−7	9	−5	1
$h_3h_2^2h_1^4$	·	·	·	·	·	·	·	1	−2	1	·	·	−2	5	−4	1
$h_3h_2h_1^6$	·	·	·	·	·	·	·	·	−1	1	·	·	·	2	−3	1
$h_3h_1^8$	·	·	·	·	·	·	·	·	·	1	·	·	·	·	−2	1
$h_2^5h_1$	·	·	·	·	·	·	·	·	·	·	−1	5	−10	10	−5	1
$h_2^4h_1^3$	·	·	·	·	·	·	·	·	·	·	·	1	−4	6	−4	1
$h_2^3h_1^5$	·	·	·	·	·	·	·	·	·	·	·	·	−1	3	−3	1
$h_2^2h_1^7$	·	·	·	·	·	·	·	·	·	·	·	·	·	1	−2	1
$h_2h_1^9$	·	·	·	·	·	·	·	·	·	·	·	·	·	·	−1	1
h_1^{11}	·	·	·	·	·	·	·	·	·	·	·	·	·	·	·	1

Table 1·3·12

w = 12 (i)

$w=12$ (i)	a_{12}	$a_{11}a_1$	$a_{10}a_2$	$a_{10}a_1^2$	a_9a_3	$a_9a_2a_1$	$a_9a_1^3$	a_8a_4	$a_8a_3a_1$	$a_8a_2^2$	$a_8a_2a_1^2$	$a_8a_1^4$	a_7a_5	$a_7a_4a_1$	$a_7a_3a_2$
h_{12}	−1	2	2	−3	2	−6	4	2	−6	−3	12	−5	2	−6	−6
$h_{11}h_1$		1	.	−2	.	−2	3	.	−2	.	6	−4	.	−2	.
$h_{10}h_2$			1	−1	.	−2	2	.	.	−2	5	−3	.	.	.
$h_{10}h_1^2$				−1	.	.	2	.	.	.	2	−3	.	.	−2
h_9h_3					1	−2	2	.	.	−2	4	−2	.	.	.
$h_9h_2h_1$						.	1	.	.	.	2	−2	.	.	.
$h_9h_1^3$							1	.	.	.	.	−2	.	.	−2
h_8h_4								1	−2	.	3	−1	.	−2	.
$h_8h_3h_1$									−1	.	2	−1	.	.	−2
$h_8h_2^2$										−1	2	−1	.	.	.
$h_8h_2h_1^2$											1	−1	.	.	.
$h_8h_1^4$												−1	.	.	.
h_7h_5													1	−2	−2
$h_7h_4h_1$														−1	.
$h_7h_3h_2$															−1

w = 12 (ii)

$w=12$ (ii)	$a_7a_3a_1^2$	$a_7a_2^2a_1$	$a_7a_2a_1^3$	$a_7a_1^5$	a_6^2	$a_6a_5a_1$	$a_6a_4a_2$	$a_6a_4a_1^2$	$a_6a_3^2$	$a_6a_3a_2a_1$	$a_6a_3a_1^3$	$a_6a_2^3$	$a_6a_2^2a_1^2$	$a_6a_2a_1^4$	$a_6a_1^6$
h_{12}	12	12	−20	6	1	−6	−6	12	−3	24	−20	4	−30	30	−7
$h_{11}h_1$	6	3	−12	5	.	−2	.	6	.	6	−12	.	−12	20	−6
$h_{10}h_2$	2	6	−10	4	.	.	−2	2	.	6	−6	3	−15	17	−5
$h_{10}h_1^2$	2	.	−6	4	.	.	.	2	.	.	−6	.	−3	12	−5
h_9h_3	3	4	−8	3	.	.	.	.	−2	10	−6	.	−12	14	−4
$h_9h_2h_1$	.	2	−5	3	.	.	.	.	.	2	−2	.	−6	10	−4
$h_9h_1^3$	.	.	−2	3	.	.	.	.	.	.	−2	.	.	6	−4
h_8h_4	4	2	−6	2	.	.	.	−2	3	4	−6	−6	−9	11	−3
$h_8h_3h_1$	2	.	−4	2	.	.	.	.	.	2	−3	−3	−4	8	−3
$h_8h_2^2$	.	2	−4	2	.	.	.	.	.	.	.	.	−7	8	−3
$h_8h_2h_1^2$	.	.	−2	2	.	.	.	.	.	.	.	.	−2	5	−3
$h_8h_1^4$	.	.	.	2	.	.	.	.	.	.	.	.	.	2	−3
h_7h_5	3	3	−4	1	.	−2	.	.	4	4	−6	−6	−6	8	−2
$h_7h_4h_1$	2	1	−3	1	.	.	.	.	2	.	−4	−4	−2	6	−2
$h_7h_3h_2$	1	2	−3	1	.	.	.	.	.	.	−2	−2	−4	6	−2
$h_7h_3h_1^2$	1	.	−2	1	.	.	.	.	.	.	.	.	.	4	−2
$h_7h_2^2h_1$	.	1	−2	1	.	.	.	.	.	.	.	.	−2	4	−2
$h_7h_2h_1^3$	.	.	−1	1	.	.	.	.	.	.	.	.	.	2	−2
$h_7h_1^5$	.	.	.	1	.	.	.	.	.	.	.	.	.	.	−2
h_6^2	.	.	.	.	1	−4	−4	6	−2	12	−8	2	−12	10	−2
$h_6h_5h_1$	.	.	.	.	.	−1	.	.	2	2	−3	−3	−3	4	−1
$h_6h_4h_2$	.	.	.	.	.	.	−1	1	.	2	−2	−2	−4	4	−1
$h_6h_4h_1^2$	.	.	.	.	.	.	.	1	.	.	−2	−2	−1	3	−1
$h_6h_3^2$	.	.	.	.	.	.	.	.	1	4	−2	−2	−4	4	−1
$h_6h_3h_2h_1$	.	.	.	.	.	.	.	.	.	−1	1	−1	−2	3	−1
$h_6h_3h_1^3$	.	.	.	.	.	.	.	.	.	.	.	−1	.	2	−1
$h_6h_2^3$	.	.	.	.	.	.	.	.	.	.	.	1	−3	3	−1
$h_6h_2^2h_1^2$	.	.	.	.	.	.	.	.	.	.	.	.	−1	2	−1
$h_6h_2h_1^4$	.	.	.	.	.	.	.	.	.	.	.	.	.	1	−1
$h_6h_1^6$	.	.	.	.	.	.	.	.	.	.	.	.	.	.	−1

w = 12 (iii)

$w=12$ (iii)	$a_5^2a_2$	$a_5^2a_1^2$	$a_5a_4a_3$	$a_5a_4a_2a_1$	$a_5a_4a_1^3$	$a_5a_3^2a_1$	$a_5a_3a_2^2$	$a_5a_3a_2a_1^2$	$a_5a_3a_1^4$	$a_5a_2^3a_1$	$a_5a_2^2a_1^3$	$a_5a_2a_1^5$	$a_5a_1^7$	a_4^3	$a_4^2a_3a_1$
h_{12}	−3	6	−6	24	−20	12	12	−60	30	−20	60	−42	8	−1	12
$h_{11}h_1$	.	3	.	6	−12	3	.	−24	20	−4	30	−30	7	.	3
$h_{10}h_2$	−1	1	.	6	−6	.	6	−18	12	−12	32	−26	6	.	.
$h_{10}h_1^2$	.	1	.	.	−6	.	.	−6	12	.	12	−20	6	.	.
h_9h_3	.	.	−2	4	−2	6	3	−24	11	−6	27	−22	5	.	3
$h_9h_2h_1$	.	.	.	2	−2	.	.	−6	6	−3	15	−17	5	.	.
$h_9h_1^3$	.	.	.	.	−2	.	.	.	6	.	3	−12	5	.	.
h_8h_4	.	.	−2	6	−4	4	2	−18	10	−6	22	−18	4	−1	8
$h_8h_3h_1$	.	.	.	2	−2	.	.	−10	6	.	12	−14	4	.	1
$h_8h_2^2$	.	.	.	.	.	.	2	−4	2	−6	16	−14	4	.	.
$h_8h_2h_1^2$	.	.	.	.	.	.	.	−2	2	.	6	−10	4	.	.
$h_8h_1^4$	.	.	.	.	.	.	.	.	2	.	.	−6	4	.	.
h_7h_5	−2	3	−2	10	−10	3	7	−24	14	−10	27	−20	4	.	4
$h_7h_4h_1$	.	.	.	2	−3	.	.	−4	6	−2	9	−11	3	.	2
$h_7h_3h_2$	.	.	.	.	.	.	2	−5	3	−4	12	−11	3	.	.
$h_7h_3h_1^2$	.	.	.	.	.	.	.	−2	3	.	4	−8	3	.	.
$h_7h_2^2h_1$	.	.	.	.	.	.	.	.	.	−2	7	−8	3	.	.
$h_7h_2h_1^3$	.	.	.	.	.	.	.	.	.	.	2	−5	3	.	.
$h_7h_1^5$	.	.	.	.	.	.	.	.	.	.	.	−2	3	.	.
h_6^2	.	4	.	8	−12	4	.	−24	16	−4	24	−20	4	.	.
$h_6h_5h_1$	.	2	.	2	−7	1	.	−10	10	−1	12	−13	3	.	.
$h_6h_4h_2$	.	.	.	2	−2	.	.	−4	4	−2	8	−8	2	.	.
$h_6h_4h_1^2$	.	.	.	.	−2	.	.	.	4	.	2	−6	2	.	.
$h_6h_3^2$	.	.	.	.	.	2	.	−8	4	.	8	−8	2	.	.
$h_6h_3h_2h_1$	.	.	.	.	.	.	.	−2	2	.	4	−6	2	.	.
$h_6h_3h_1^3$	.	.	.	.	.	.	.	.	2	.	.	−4	2	.	.
$h_6h_2^3$	.	.	.	.	.	.	.	.	.	−2	6	−6	2	.	.
$h_6h_2^2h_1^2$	.	.	.	.	.	.	.	.	.	.	2	−4	2	.	.
$h_6h_2h_1^4$	.	.	.	.	.	.	.	.	.	.	.	−2	2	.	.
$h_6h_1^6$	.	.	.	.	.	.	.	.	.	.	.	.	2	.	.
$h_5^2h_2$	−1	1	.	4	−4	.	4	−10	6	−6	14	−10	2	.	.
$h_5^2h_1^2$	.	1	.	.	−4	.	.	−4	6	.	6	−8	2	.	.
$h_5h_4h_3$	.	.	−1	2	−1	2	1	−7	3	−2	7	−5	1	.	2
$h_5h_4h_2h_1$	.	.	.	1	−1	.	.	−2	2	−1	4	−4	1	.	.
$h_5h_4h_1^3$	.	.	.	.	−1	.	.	.	2	.	1	−3	1	.	.
$h_5h_3^2h_1$	.	.	.	.	.	1	.	−4	2	.	4	−4	1	.	.
$h_5h_3h_2^2$	.	.	.	.	.	.	1	−2	1	−2	5	−4	1	.	.
$h_5h_3h_2h_1^2$	.	.	.	.	.	.	.	−1	1	.	2	−3	1	.	.
$h_5h_3h_1^4$	.	.	.	.	.	.	.	.	1	.	.	−2	1	.	.
$h_5h_2^3h_1$	.	.	.	.	.	.	.	.	.	−1	3	−3	1	.	.
$h_6h_2^2h_1^3$	.	.	.	.	.	.	.	.	.	.	1	−2	1	.	.
$h_6h_2h_1^5$	.	.	.	.	.	.	.	.	.	.	.	−1	1	.	.
$h_5h_1^7$	.	.	.	.	.	.	.	.	.	.	.	.	1	.	.
h_4^3	.	.	.	.	.	.	.	.	.	.	.	.	.	−1	6
$h_4^2h_3h_1$	.	.	.	.	.	.	.	.	.	.	.	.	.	.	1

Table 1·3·12 (*cont.*)

$w=12$ (iv)	$a_4^2a_2^2$	$a_4^2a_2a_1^2$	$a_4^2a_1^4$	$a_4a_3^2a_2$	$a_4a_3^2a_1^2$	$a_4a_3a_2^2a_1$	$a_4a_3a_2a_1^3$	$a_4a_3a_1^5$	$a_4a_2^4$	$a_4a_2^3a_1^2$	$a_4a_2^2a_1^4$	$a_4a_2a_1^6$	$a_4a_1^8$	a_3^4	$a_3^3a_2a_1$	$a_3^3a_1^3$
h_{12}	6	−30	15	12	−30	−60	120	−42	−5	60	−105	56	−9	1	−20	20
$h_{11}h_1$	·	−12	10	·	−12	−12	60	−30	·	20	−60	42	−8	·	−4	10
$h_{10}h_2$	3	−9	6	3	−3	−24	44	−20	−4	34	−60	37	−7	·	−4	4
$h_{10}h_1^2$	·	−3	6	·	−3	·	24	−20	·	4	−30	30	−7	·	·	4
h_9h_3	·	−6	3	6	−12	−24	50	−18	·	24	−52	32	−6	1	−14	11
$h_9h_2h_1$	·	−3	3	·	·	−6	18	−12	·	12	−32	26	−6	·	−1	1
$h_9h_1^3$	·	·	3	·	·	·	6	−12	·	·	−12	20	−6	·	·	1
h_8h_4	4	−15	6	3	−18	−24	62	−22	−4	31	−59	34	−6	·	−6	12
$h_8h_3h_1$	·	−2	1	·	−6	−3	24	−11	·	6	−27	22	−5	·	−3	6
$h_8h_2^2$	1	−2	1	·	·	−6	12	−6	−3	18	−32	22	−5	·	·	·
$h_8h_2h_1^2$	·	−1	1	·	·	·	6	−6	·	3	−15	17	−5	·	·	·
$h_8h_1^4$	·	·	1	·	·	·	·	−6	·	·	−3	12	−5	·	·	·
h_7h_5	·	−12	8	4	−12	−24	58	−24	·	26	−56	34	−6	·	−6	9
$h_7h_4h_1$	·	−6	4	·	−7	−5	30	−15	·	10	−32	24	−5	·	·	6
$h_7h_3h_2$	·	·	·	2	−2	−10	16	−6	·	12	−26	18	−4	·	−3	3
$h_7h_3h_1^2$	·	·	·	·	−2	·	10	−6	·	·	−12	14	−4	·	·	3
$h_7h_2^2h_1$	·	·	·	·	·	−2	4	−2	·	6	−16	14	−4	·	·	·
$h_7h_2h_1^3$	·	·	·	·	·	·	2	−2	·	·	−6	10	−4	·	·	·
$h_7h_1^5$	·	·	·	·	·	·	·	−2	·	·	·	6	−4	·	·	·
h_6^2	4	−12	9	4	−6	−24	52	−24	−4	30	−56	34	−6	1	−12	8
$h_6h_5h_1$	·	−4	6	·	−2	−4	24	−17	·	8	−29	24	−5	·	−2	3
$h_6h_4h_2$	2	−5	3	1	−1	−10	20	−10	−3	18	−31	20	−4	·	−2	2
$h_6h_4h_1^2$	·	−2	3	·	−1	·	10	−10	·	3	−15	16	−4	·	·	2
$h_6h_3^2$	·	·	·	2	−3	−8	16	−6	·	8	−20	14	−3	1	−10	6
$h_6h_3h_2h_1$	·	·	·	·	·	−2	5	−3	·	4	−12	11	−3	·	−1	1
$h_6h_3h_1^3$	·	·	·	·	·	·	2	−3	·	·	−4	8	−3	·	·	1
$h_6h_2^3$	·	·	·	·	·	·	·	·	−2	9	−15	11	−3	·	·	·
$h_6h_2^2h_1^2$	·	·	·	·	·	·	·	·	·	2	−7	8	−3	·	·	·
$h_6h_2h_1^4$	·	·	·	·	·	·	·	·	·	·	−2	5	−3	·	·	·
$h_6h_1^6$	·	·	·	·	·	·	·	·	·	·	·	2	−3	·	·	·
$h_5^2h_2$	·	−4	4	·	·	−8	20	−12	·	12	−28	20	−4	·	·	·
$h_5^2h_1^2$	·	·	4	·	·	·	8	−12	·	·	−12	16	−4	·	·	·
$h_5h_4h_3$	·	−4	2	2	−7	−9	26	−10	·	10	−25	16	−3	·	−4	6
$h_5h_4h_2h_1$	·	−2	2	·	·	−2	9	−7	·	5	−15	13	−3	·	·	·
$h_5h_4h_1^3$	·	·	2	·	·	·	2	−7	·	·	−5	10	−3	·	·	·
$h_5h_3^2h_1$	·	·	·	·	−2	·	8	−4	·	·	−8	8	−2	·	−2	3
$h_5h_3h_2^2$	·	·	·	·	·	−2	4	−2	·	4	−10	8	−2	·	·	·
$h_5h_3h_2h_1^2$	·	·	·	·	·	·	2	−2	·	·	−4	6	−2	·	·	·
$h_5h_3h_1^4$	·	·	·	·	·	·	·	−2	·	·	·	4	−2	·	·	·
$h_5h_2^3h_1$	·	·	·	·	·	·	·	·	·	2	−6	6	−2	·	·	·
$h_5h_2^2h_1^3$	·	·	·	·	·	·	·	·	·	·	−2	4	−2	·	·	·
$h_5h_2h_1^5$	·	·	·	·	·	·	·	·	·	·	·	2	−2	·	·	·
$h_5h_1^7$	·	·	·	·	·	·	·	·	·	·	·	·	−2	·	·	·
h_4^3	3	−9	3	·	−12	−12	36	−12	−3	18	−33	18	−3	·	·	8
$h_4^2h_3h_1$	·	−2	1	·	−4	−2	14	−6	−2	4	−14	10	−2	·	·	4
$h_4^2h_2^2$	1	−2	1	·	·	−4	8	−4	−2	10	−16	10	−2	·	·	·
$h_4^2h_2h_1^2$	·	−1	1	·	·	·	4	−4	·	2	−8	8	−2	·	·	·
$h_4^2h_1^4$	·	·	1	·	·	·	·	−4	·	·	−2	6	−2	·	·	·
$h_4h_3^2h_2$	·	·	·	1	−1	−4	6	−2	·	4	−8	5	−1	·	−2	2
$h_4h_3^2h_1^2$	·	·	·	·	−1	·	4	−2	·	·	−4	4	−1	·	·	2
$h_4h_3h_2^2h_1$	·	·	·	·	·	−1	2	−1	·	2	−5	4	−1	·	·	·
$h_4h_3h_2h_1^3$	·	·	·	·	·	·	1	−1	·	·	−2	3	−1	·	·	·
$h_4h_3h_1^5$	·	·	·	·	·	·	·	−1	·	·	·	2	−1	·	·	·
$h_4h_2^4$	·	·	·	·	·	·	·	·	−1	4	−6	4	−1	·	·	·
$h_4h_2^3h_1^2$	·	·	·	·	·	·	·	·	·	1	−3	3	−1	·	·	·
$h_4h_2^2h_1^4$	·	·	·	·	·	·	·	·	·	·	−1	2	−1	·	·	·
$h_4h_2h_1^6$	·	·	·	·	·	·	·	·	·	·	·	1	−1	·	·	·
$h_4h_1^8$	·	·	·	·	·	·	·	·	·	·	·	·	−1	·	·	·
h_3^4	·	·	·	·	·	·	·	·	·	·	·	·	·	1	−8	4
$h_3^3h_2h_1$	·	·	·	·	·	·	·	·	·	·	·	·	·	·	−1	1
$h_3^3h_1^3$	·	·	·	·	·	·	·	·	·	·	·	·	·	·	·	1

Table 1·3·12 (cont.)

$w = 12$ (v)	$a_3^2a_2^3$	$a_3^2a_2^2a_1^2$	$a_3^2a_2a_1^4$	$a_3^2a_1^6$	$a_3a_2^4a_1$	$a_3a_2^3a_1^3$	$a_3a_2^2a_1^5$	$a_3a_2a_1^7$	$a_3a_1^9$	a_2^6	$a_2^5a_1^2$	$a_2^4a_1^4$	$a_2^3a_1^6$	$a_2^2a_1^8$	$a_2a_1^{10}$	a_1^{12}
h_{12}	−10	90	−105	28	30	−140	168	−72	10	1	−21	70	−84	45	−11	1
$h_{11}h_1$	·	30	−60	21	5	−60	105	−56	9	·	−6	35	−56	36	−10	1
$h_{10}h_2$	−6	36	−45	15	20	−80	102	−50	8	1	−16	50	−63	37	−10	1
$h_{10}h_1^2$	·	6	−30	15	·	−20	60	−42	8	·	−1	15	−35	28	−9	1
h_9h_3	−4	54	−62	17	13	−84	111	−52	8	·	−10	45	−62	37	−10	1
$h_9h_2h_1$	·	12	−22	10	4	−34	60	−37	7	·	−5	25	−41	29	−9	1
$h_9h_1^3$	·	·	−12	10	·	−4	30	−30	7	·	·	5	−20	21	−8	1
h_8h_4	−3	39	−61	18	14	−76	108	−52	8	1	−13	46	−62	37	−10	1
$h_8h_3h_1$	·	18	−35	12	1	−34	67	−39	7	·	−2	21	−40	29	−9	1
$h_8h_2^2$	−3	12	−15	6	12	−44	58	−32	6	1	−12	36	−47	30	−9	1
$h_8h_2h_1^2$	·	3	−9	6	·	−12	32	−26	6	·	−1	11	−25	22	−8	1
$h_8h_1^4$	·	·	−3	6	·	·	12	−20	6	·	·	1	−10	15	−7	1
h_7h_5	−6	42	−58	18	17	−80	108	−52	8	·	−12	46	−62	37	−10	1
$h_7h_4h_1$	·	9	−33	13	3	−29	64	−39	7	·	−4	22	−40	29	−9	1
$h_7h_3h_2$	−3	21	−26	8	10	−47	65	−34	6	·	−8	32	−46	30	−9	1
$h_7h_3h_1^2$	·	3	−18	8	·	−10	37	−28	6	·	·	8	−24	22	−8	1
$h_7h_2^2h_1$	·	3	−6	3	3	−18	32	−22	5	·	−4	18	−30	23	−8	1
$h_7h_2h_1^3$	·	·	−3	3	·	−3	15	−17	5	·	·	4	−14	16	−7	1
$h_7h_1^5$	·	·	·	3	·	·	3	−12	5	·	·	·	−4	10	−6	1
h_6^2	−2	48	−58	18	12	−80	108	−52	8	1	−12	46	−62	37	−10	1
$h_6h_5h_1$	·	15	−30	13	2	−33	64	−39	7	·	−3	22	−40	29	−9	1
$h_6h_4h_2$	−1	16	−24	9	8	−42	62	−34	6	1	−10	33	−46	30	−9	1
$h_6h_4h_1^2$	·	1	−15	9	·	−8	34	−28	6	·	−1	9	−24	22	−8	1
$h_6h_3^2$	−1	34	−37	10	4	−50	72	−36	6	·	−4	28	−45	30	−9	1
$h_6h_3h_2h_1$	·	8	−13	5	1	−19	37	−24	5	·	−2	15	−29	23	−8	1
$h_6h_3h_1^3$	·	·	−8	5	·	−1	18	−19	5	·	·	2	−13	16	−7	1
$h_6h_2^3$	−1	3	−3	1	6	−22	30	−18	4	1	−9	26	−35	24	−8	1
$h_6h_2^2h_1^2$	·	1	−2	1	·	−6	16	−14	4	·	−1	8	−18	17	−7	1
$h_6h_2h_1^4$	·	·	−1	1	·	·	6	−10	4	·	·	1	−7	11	−6	1
$h_6h_1^6$	·	·	·	1	·	·	·	−6	4	·	·	·	−1	6	−5	1
$h_5^2h_2$	−4	16	−21	9	12	−46	62	−34	6	·	−9	33	−46	30	−9	1
$h_5^2h_1^2$	·	4	−12	9	·	−12	34	−28	6	·	·	9	−24	22	−8	1
$h_5h_4h_3$	−2	23	−35	11	7	−45	69	−36	6	·	−6	29	−45	30	−9	1
$h_5h_4h_2h_1$	·	4	−10	6	2	−17	34	−24	5	·	−3	16	−29	23	−8	1
$h_5h_4h_1^3$	·	·	−4	6	·	−2	15	−19	5	·	·	3	−13	16	−7	1
$h_5h_3^2h_1$	·	11	−20	7	·	−20	42	−26	5	·	·	12	−28	23	−8	1
$h_5h_3h_2^2$	−2	7	−8	3	7	−26	35	−20	4	·	−6	23	−34	24	−8	1
$h_5h_3h_2h_1^2$	·	2	−5	3	·	−7	19	−16	4	·	·	6	−17	17	−7	1
$h_5h_3h_1^4$	·	·	−2	3	·	·	7	−12	4	·	·	·	−6	11	−6	1
$h_5h_2^3h_1$	·	·	·	·	2	−9	15	−11	3	·	−3	13	−22	18	−7	1
$h_5h_2^2h_1^3$	·	·	·	·	·	−2	7	−8	3	·	·	3	−10	12	−6	1
$h_5h_2h_1^5$	·	·	·	·	·	·	2	−5	3	·	·	·	−3	7	−5	1
$h_5h_1^7$	·	·	·	·	·	·	·	−2	3	·	·	·	·	3	−4	1
h_4^3	·	12	−36	12	6	−36	66	−36	6	1	−9	30	−45	30	−9	1
$h_4^2h_3h_1$	·	4	−20	8	1	−14	39	−26	5	·	−2	13	−28	23	−8	1
$h_4^2h_2^2$	·	4	−8	4	4	−20	32	−20	4	1	−8	24	−34	24	−8	1
$h_4^2h_2h_1^2$	·	·	−4	4	·	−4	16	−16	4	·	−1	7	−17	17	−7	1
$h_4^2h_1^4$	·	·	·	4	·	·	4	−12	4	·	·	1	−6	11	−6	1
$h_4h_3^2h_2$	−1	12	−16	5	4	−26	40	−22	4	·	−4	20	−33	24	−8	1
$h_4h_3^2h_1^2$	·	1	−11	5	·	−4	22	−18	4	·	·	4	−16	17	−7	1
$h_4h_3h_2^2h_1$	·	2	−4	2	1	−9	18	−13	3	·	−2	11	−21	18	−7	1
$h_4h_3h_2h_1^3$	·	·	−2	2	·	−1	8	−10	3	·	·	2	−9	12	−6	1
$h_4h_3h_1^5$	·	·	·	2	·	·	1	−7	3	·	·	·	−2	7	−5	1
$h_4h_2^4$	·	·	·	·	2	−8	12	−8	2	1	−7	19	−26	19	−7	1
$h_4h_2^3h_1^2$	·	·	·	·	·	−2	6	−6	2	·	−1	6	−13	13	−6	1
$h_4h_2^2h_1^4$	·	·	·	·	·	·	2	−4	2	·	·	1	−5	8	−5	1
$h_4h_2h_1^6$	·	·	·	·	·	·	·	−2	2	·	·	·	−1	4	−4	1
$h_4h_1^8$	·	·	·	·	·	·	·	·	2	·	·	·	·	1	−3	1
h_3^4	·	24	−24	6	·	−32	48	−24	4	·	·	16	−32	24	−8	1
$h_3^3h_2h_1$	·	6	−9	3	·	−12	24	−15	3	·	·	8	−20	18	−7	1
$h_3^3h_1^3$	·	·	−6	3	·	·	12	−12	3	·	·	·	−8	12	−6	1
$h_3^2h_2^3$	−1	3	−3	1	·	−4	10	−8	3	·	·	4	−12	12	−6	1
$h_3^2h_2^2h_1^2$	·	1	−2	1	·	−4	10	−8	2	·	·	4	−12	13	−6	1
$h_3^2h_2h_1^4$	·	·	−1	1	·	·	4	−6	2	·	·	·	−4	8	−5	1
$h_3^2h_1^6$	·	·	·	1	·	·	·	−4	2	·	·	·	·	4	−4	1
$h_3h_2^4h_1$	·	·	·	·	·	−4	6	−4	1	·	−2	9	−16	14	−6	1
$h_3h_2^3h_1^3$	·	·	·	·	·	−1	3	−3	1	·	·	2	−7	9	−5	1
$h_3h_2^2h_1^5$	·	·	·	·	·	·	1	−2	1	·	·	·	−2	5	−4	1
$h_3h_2h_1^7$	·	·	·	·	·	·	·	−1	1	·	·	·	·	2	−3	1
$h_3h_1^9$	·	·	·	·	·	·	·	·	1	·	·	·	·	·	−2	1
h_2^6	·	·	·	·	·	·	·	·	·	1	−6	15	−20	15	−6	1
$h_2^5h_1^2$	·	·	·	·	·	·	·	·	·	·	−1	5	−10	10	−5	1
$h_2^4h_1^4$	·	·	·	·	·	·	·	·	·	·	·	1	−4	6	−4	1
$h_2^3h_1^6$	·	·	·	·	·	·	·	·	·	·	·	·	−1	3	−3	1
$h_2^2h_1^8$	·	·	·	·	·	·	·	·	·	·	·	·	·	1	−2	1
$h_2h_1^{10}$	·	·	·	·	·	·	·	·	·	·	·	·	·	·	−1	1
h_1^{12}	·	·	·	·	·	·	·	·	·	·	·	·	·	·	·	1

Table 1·4. *Monomial symmetric functions in terms of homogeneous product sums (h-functions) and conversely (up to and including weight 12)*

Note that the final digit (or digits) in the subheadings indicates the weight, w.

In order to save space use has been made of Meyer Hirsch's law of symmetry which states that the coefficient of the monomial symmetric function $(r_1 r_2 \dots r_l)$ in the expansion of $h_{l_1} h_{l_2} \dots h_{l_m}$ is equal to the coefficient of the monomial symmetric function $(l_1 l_2 \dots l_m)$ in the expansion of $h_{r_1} h_{r_2} \dots h_{r_l}$. This enables each table to be written compactly and read, using the following rules:

In order to express a monomial symmetric in terms of h-functions, read horizontally until the diagonal figure in bold type is reached and then continue vertically downwards to the edge of the table. E.g. from Table 1·4·5

$$(31^2) = 3h_1^5 - 13h_2 h_1^3 + 10h_2^2 h_1 + 12h_3 h_1^2 - 8h_3 h_2 - 9h_4 h_1 + 5h_5.$$

Conversely, to express a product of h-functions in terms of monomial symmetrics, read down vertically until the diagonal figure in bold type is reached and then continue horizontally to the right-hand edge of the table. Thus

$$(h_3 h_1^2) = 20(1^5) + 13(21^3) + 8(2^2 1) + 7(31^2) + 4(32) + 3(41) + (5).$$

The two diagonal figures are separated by a bold horizontal rule.

Table 1·4·2

$w=2$	h_1^2	h_2
(1^2)	$\frac{2}{1}$	1
	$\cdot$	
(2)	$\cdot$ / -1	$\frac{1}{2}$

Table 1·4·3

$w=3$	h_1^3	$h_2 h_1$	h_3
(1^3)	$\frac{6}{1}$	3	1
(21)	$\cdot$ / -2	$\frac{2}{5}$	1
(3)	1	-3	$\frac{1}{3}$

Table 1·4·4

$w=4$	h_1^4	$h_2 h_1^2$	h_2^2	$h_3 h_1$	h_4
(1^4)	$\frac{24}{1}$	12	6	4	1
(21^2)	-3	$\frac{7}{10}$	4	3	1
(2^2)	1	-4	$\frac{3}{3}$	2	1
(31)	2	-7	2	$\frac{2}{7}$	1
(4)	-1	4	-2	-4	$\frac{1}{4}$

Table 1·4·5

$w=5$	h_1^5	$h_2 h_1^3$	$h_2^2 h_1$	$h_3 h_1^2$	$h_3 h_2$	$h_4 h_1$	h_5
(1^5)	$\frac{120}{1}$	60	30	20	10	5	1
(21^3)	-4	$\frac{33}{17}$	18	13	7	4	1
$(2^2 1)$	3	-14	$\frac{11}{14}$	8	5	3	1
(31^2)	3	-13	10	$\frac{7}{12}$	4	3	1
(32)	-2	10	-11	-8	$\frac{3}{11}$	2	1
(41)	-2	9	-7	-9	5	$\frac{2}{9}$	1
(5)	1	-5	5	5	-5	-5	$\frac{1}{5}$

Table 1·4·6

$w=6$	h_1^6	$h_2h_1^4$	$h_2^2h_1^2$	h_2^3	$h_3h_1^3$	$h_3h_2h_1$	$h_4h_1^2$	h_3^2	h_4h_2	h_5h_1	h_6
(1^6)	$\frac{720}{1}$	360	180	90	120	60	30	20	15	6	1
(21^4)	-5	$\frac{192}{26}$	102	54	72	38	21	14	11	5	1
(2^21^2)	6	-33	$\frac{58}{46}$	33	42	24	14	10	8	4	1
(2^3)	-1	6	-10	$\frac{21}{4}$	24	15	9	7	6	3	1
(31^3)	4	-21	26	-4	$\frac{34}{19}$	19	13	8	7	4	1
(321)	-6	34	-48	8	-30	$\frac{12}{61}$	8	6	5	3	1
(41^2)	-3	16	-20	4	-15	21	$\frac{7}{15}$	4	4	3	1
(3^2)	1	-6	9	-1	6	-15	-3	$\frac{4}{0}$	3	2	1
(42)	2	-12	19	-6	10	-20	-10	3	$\frac{3}{14}$	2	1
(51)	2	-11	14	-2	11	-17	-11	3	6	$\frac{2}{11}$	1
(6)	-1	6	-9	2	-6	12	6	-3	-6	-6	$\frac{1}{6}$

Table 1·4·7

$w=7$	h_1^7	$h_2h_1^5$	$h_2^2h_1^3$	$h_2^3h_1$	$h_3h_1^4$	$h_3h_2h_1^2$	$h_3h_2^2$	$h_3^2h_1$	$h_4h_1^3$	$h_4h_2h_1$	h_4h_3	$h_5h_1^2$	h_5h_2	h_6h_1	h_7
(1^7)	$\frac{5040}{1}$	2520	1260	630	840	420	210	140	210	105	35	42	21	7	1
(21^5)	-6	$\frac{1320}{37}$	690	360	480	250	130	90	135	70	25	31	16	6	1
(2^21^3)	10	-64	$\frac{378}{117}$	207	270	148	81	58	84	46	18	22	12	5	1
(2^31)	-4	27	-54	$\frac{120}{30}$	150	87	51	37	51	30	13	15	9	4	1
(31^4)	5	-31	53	-21	$\frac{208}{28}$	114	62	46	73	39	15	21	11	5	1
(321^2)	-12	78	-143	62	-70	$\frac{67}{195}$	39	30	43	25	11	14	8	4	1
(32^2)	3	-21	44	-26	17	-54	$\frac{25}{26}$	19	25	16	8	9	6	3	1
(3^21)	3	-20	37	-14	20	-60	13	$\frac{16}{25}$	20	13	7	8	5	3	1
(41^3)	-4	25	-43	18	-23	55	-14	-14	$\frac{34}{22}$	19	8	13	7	4	1
(421)	6	-40	76	-36	36	-101	30	29	-33	$\frac{12}{64}$	6	8	5	3	1
(43)	-2	14	-28	13	-14	45	-13	-19	11	-26	$\frac{4}{19}$	4	3	2	1
(51^2)	3	-19	33	-14	18	-43	12	10	-18	25	-7	$\frac{7}{18}$	4	3	1
(52)	-2	14	-29	17	-12	36	-17	-7	12	-24	7	-12	$\frac{3}{17}$	2	1
(61)	-2	13	-23	9	-13	33	-7	-10	13	-20	7	-13	7	$\frac{2}{13}$	1
(7)	1	-7	14	-7	7	-21	7	7	-7	14	-7	7	-7	-7	$\frac{1}{7}$

Table 1·4·8

$w=8$ (i)	h_1^8	$h_2h_1^6$	$h_2^2h_1^4$	$h_2^3h_1^2$	h_2^4	$h_3h_1^5$	$h_3h_2h_1^3$	$h_3h_2^2h_1$	$h_3^2h_1^2$	$h_3^2h_2$
(1^8)	40320 / 1	20160	10080	5040	2520	6720	3360	1680	1120	560
(21^6)	−7	10440 / 50	5400	2790	1440	3720	1920	990	680	350
(2^21^4)	15	−110	2892 / 251	1548	828	2040	1092	584	412	220
(2^31^2)	−10	76	−183	861 / 146	480	1110	618	345	248	139
(2^4)	1	−8	21	−20	282 / 5	600	348	204	148	88
(31^5)	6	−43	94	−63	6	1520 / 39	820	440	320	170
(321^3)	−20	148	−338	240	−24	−134	463 / 489	260	194	108
(32^21)	12	−93	228	−182	20	81	−322	154 / 258	116	69
(3^21^2)	6	−45	103	−70	6	44	−165	102	92 / 66	54
(3^22)	−3	24	−61	50	−4	−22	94	−84	−33	35 / 35
(41^4)	−5	36	−79	54	−6	−33	111	−66	−35	16
(421^2)	12	−90	209	−154	20	82	−298	192	99	−48
(42^2)	−3	24	−62	56	−12	−20	80	−64	−23	14
(431)	−6	46	−108	76	−8	−46	176	−107	−74	31
(4^2)	1	−8	20	−16	3	8	−32	20	14	−4
(51^3)	4	−29	64	−44	4	27	−91	58	27	−16
(521)	−6	46	−110	84	−8	−42	159	−119	−49	37
(53)	2	−16	40	−31	2	16	−67	54	27	−23
(61^2)	−3	22	−49	34	−4	−21	71	−43	−22	11
(62)	2	−16	41	−36	6	14	−56	48	15	−14
(71)	2	−15	34	−23	2	15	−53	31	19	−8
(8)	−1	8	−20	16	−2	−8	32	−24	−12	8

$w=8$ (ii)	$h_4h_1^4$	$h_4h_2h_1^2$	$h_4h_2^2$	$h_4h_3h_1$	h_4^2	$h_5h_1^3$	$h_5h_2h_1$	h_5h_3	$h_6h_1^2$	h_6h_2	h_7h_1	h_8
(1^8)	1680	840	420	280	70	336	168	56	56	28	8	1
(21^6)	1020	525	270	185	50	228	117	41	43	22	7	1
(2^21^4)	606	324	173	122	36	150	80	30	32	17	6	1
(2^31^2)	354	198	111	80	26	96	54	22	23	13	5	1
(2^4)	204	120	72	52	19	60	36	16	16	10	4	1
(31^5)	500	265	140	100	30	136	71	26	31	16	6	1
(321^3)	286	160	89	66	22	85	47	19	22	12	5	1
(32^21)	162	96	57	43	16	52	31	14	15	9	4	1
(3^21^2)	126	76	45	36	14	44	26	12	14	8	4	1
(3^22)	70	45	29	23	10	26	17	9	9	6	3	1
(41^4)	209 / 31	115	63	47	16	73	39	15	21	11	5	1
(421^2)	−76	68 / 211	40	31	12	43	25	11	14	8	4	1
(42^2)	20	−66	26 / 36	20	9	25	16	8	9	6	3	1
(431)	40	−121	32	17 / 98	8	20	13	7	8	5	3	1
(4^2)	−8	28	−12	−24	5 / 10	8	6	4	4	3	2	1
(51^3)	−26	60	−14	−28	4	34 / 26	19	8	13	7	4	1
(521)	39	−105	28	54	−8	−39	12 / 76	6	8	5	3	1
(53)	−13	39	−8	−31	4	13	−31	4 / 23	4	3	2	1
(61^2)	21	−50	14	23	−4	−21	29	−8	7 / 21	4	3	1
(62)	−14	42	−20	−16	4	14	−28	8	−14	3 / 20	2	1
(71)	−15	38	−8	−23	4	15	−23	8	−15	8	2 / 15	1
(8)	8	−24	8	16	−4	−8	16	−8	8	−8	−8	1 / 8

Table 1·4·9

$w=9$ (i)	h_1^9	$h_2h_1^7$	$h_2^2h_1^5$	$h_2^3h_1^3$	$h_2^4h_1$	$h_3h_1^6$	$h_3h_2h_1^4$	$h_3h_2^2h_1^2$	$h_3h_2^3$	$h_3^2h_1^3$	$h_3^2h_2h_1$	h_3^3	$h_4h_1^5$
(1^9)	362880/1	181440	90720	45360	22680	60480	30240	15120	7560	10080	5040	1680	15120
(21^7)	-8	93240/65	47880	24570	12600	32760	16800	8610	4410	5880	3010	1050	8820
(2^21^5)	21	-174	25260/478	13320	7020	17640	9300	4900	2580	3420	1800	660	5070
(2^31^3)	-20	170	-484	7227/517	3924	9450	5130	2787	1515	1980	1077	417	2880
(2^41)	5	-44	132	-154	2202/55	5040	2820	1584	894	1140	644	264	1620
(31^6)	7	-57	152	-146	36	12840/52	6840	3630	1920	2600	1370	510	4020
(321^4)	-30	250	-687	686	-177	-228	3764/1039	2064	1128	1508	822	324	2250
(32^21^2)	30	-258	741	-789	222	231	-1110	1173/1297	666	868	493	207	1250
(32^3)	-4	36	-111	134	-50	-30	153	-202	396/50	496	295	132	690
(3^21^3)	10	-84	231	-226	54	81	-374	392	-48	664/149	378	162	950
(3^221)	-12	105	-306	325	-82	-99	494	-597	78	-192	227/327	105	520
(3^3)	1	-9	27	-29	6	9	-48	63	-6	21	-45	55/10	210
(41^5)	-6	49	-131	127	-33	-45	195	-195	27	-68	78	-6	1545/42
(421^3)	20	-168	466	-474	132	154	-699	740	-112	249	-304	24	-143
(42^21)	-12	105	-309	344	-110	-93	451	-536	98	-154	222	-18	87
(431^2)	-12	102	-284	283	-74	-100	463	-477	66	-189	216	-18	91
(432)	6	-54	164	-188	60	50	-256	324	-60	96	-157	15	-44
(4^21)	3	-26	74	-76	22	26	-122	124	-19	52	-51	3	-26
(51^4)	5	-41	110	-107	27	38	-165	169	-23	56	-70	6	-36
(521^2)	-12	102	-287	297	-82	-94	434	-479	74	-151	204	-18	88
(52^2)	3	-27	83	-100	38	23	-115	148	-38	35	-51	3	-23
(531)	6	-52	148	-150	36	52	-250	275	-34	104	-142	15	-46
(54)	-2	18	-54	60	-19	-18	90	-104	19	-38	47	-3	18
(61^3)	-4	33	-89	87	-22	-31	135	-139	18	-46	60	-6	30
(621)	6	-52	150	-160	44	48	-229	268	-38	79	-127	15	-45
(63)	-2	18	-54	59	-15	-18	93	-117	15	-39	72	-12	15
(71^2)	3	-25	68	-67	18	24	-105	106	-16	37	-42	3	-24
(72)	-2	18	-55	65	-23	-16	80	-103	23	-25	41	-3	16
(81)	-2	17	-47	46	-11	-17	77	-78	9	-30	35	-3	17
(9)	1	-9	27	-30	9	9	-45	54	-9	18	-27	3	-9

$w=9$ (ii)	$h_4h_2h_1^3$	$h_4h_2^2h_1$	$h_4h_3h_1^2$	$h_4h_3h_2$	$h_4^2h_1$	$h_5h_1^4$	$h_5h_2h_1^2$	$h_5h_2^2$	$h_5h_3h_1$	h_5h_4	$h_6h_1^3$	$h_6h_2h_1$	h_6h_3	$h_7h_1^2$	h_7h_2	h_8h_1	h_9
(1^9)	7560	3780	2520	1260	630	3024	1512	756	504	126	504	252	84	72	36	9	1
(21^7)	4515	2310	1575	805	420	1932	987	504	343	91	357	182	63	57	29	8	1
(2^21^5)	2670	1405	980	515	280	1206	634	333	232	66	246	129	47	44	23	7	1
(2^31^3)	1566	852	606	330	186	738	402	219	156	48	165	90	35	33	18	6	1
(2^41)	912	516	372	212	123	444	252	144	104	35	108	62	26	24	14	5	1
(31^6)	2115	1110	785	410	230	1044	543	282	197	56	229	118	42	43	22	7	1
(321^4)	1229	669	486	263	154	626	339	183	132	41	151	81	31	32	17	6	1
(32^21^2)	710	403	298	169	102	370	210	119	88	30	97	55	23	23	13	5	1
(32^3)	408	243	181	109	67	216	129	78	58	22	61	37	17	16	10	4	1
(3^21^3)	548	313	240	135	86	302	172	97	74	26	86	48	20	22	12	5	1
(3^221)	313	188	145	87	56	174	105	63	49	19	53	32	15	15	9	4	1
(3^3)	135	87	69	45	30	78	51	33	27	12	27	18	10	9	6	3	1
(41^5)	840	455	335	180	110	501	266	141	101	31	136	71	26	31	16	6	1
(421^3)	479/520	272	206	116	74	287	161	90	67	23	85	47	19	22	12	5	1
(42^21)	-344	163/270	125	75	49	163	97	58	44	17	52	31	14	15	9	4	1
(431^2)	-344	217	101/277	60	42	127	77	46	37	15	44	26	12	14	8	4	1
(432)	184	-154	-131	39/112	27	71	46	30	24	11	26	17	9	9	6	3	1
(4^21)	104	-69	-91	38	21/39	48	32	21	18	9	20	13	7	8	5	3	1
(51^4)	119	-71	-71	36	18	209/35	115	63	47	16	73	39	15	21	11	5	1
(521^2)	-317	209	200	-113	-55	-84	68/228	40	31	12	43	25	11	14	8	4	1
(52^2)	94	-80	-54	42	19	21	-68	26/38	20	9	25	16	8	9	6	3	1
(531)	175	-110	-143	74	42	42	-123	28	17/97	8	20	13	7	8	5	3	1
(54)	-76	57	67	-38	-29	-14	47	-19	-38	5/29	8	6	4	4	3	2	1
(61^3)	-99	61	56	-31	-13	-30	69	-16	-32	9	34/30	19	8	13	7	4	1
(621)	166	-120	-95	68	22	45	-121	32	62	-18	-45	12/88	6	8	5	3	1
(63)	-60	45	45	-36	-9	-15	45	-9	-36	9	15	-36	4/27	4	3	2	1
(71^2)	81	-49	-50	25	13	24	-57	16	26	-9	-24	33	-9	7/24	4	3	1
(72)	-64	55	34	-32	-9	-16	48	-23	-18	9	16	-32	9	-16	3/23	2	1
(81)	-60	35	43	-18	-13	-17	43	-9	-26	9	17	-26	9	-17	9	2/17	1
(9)	36	-27	-27	18	9	9	-27	9	18	-9	-9	18	-9	9	-9	-9	1/9

Table 1·4·10

$w = 10$ (i)	h_1^{10}	$h_2 h_1^8$	$h_2^2 h_1^6$	$h_2^3 h_1^4$	$h_2^4 h_1^2$	h_2^5	$h_3 h_1^7$	$h_3 h_2 h_1^5$	$h_3 h_2^2 h_1^3$
(1^{10})	3628800/1	1814400	907200	453600	226800	113400	604800	302400	151200
(21^8)	-9	927360/82	473760	241920	123480	63000	322560	164640	84000
$(2^2 1^6)$	28	-259	247320/834	129060	67320	35100	171360	89400	46620
$(2^3 1^4)$	-35	330	-1090	68868/1476	36756	19620	90720	48420	25848
$(2^4 1^2)$	15	-145	496	-708	20100/371	11010	47880	26160	14316
(2^5)	-1	10	-36	56	-35	6210/6	25200	14100	7920
(31^7)	8	-73	230	-290	124	-8	122640/67	64680	34020
(321^5)	-42	390	-1256	1628	-720	48	-358	34960/1965	18850
$(32^2 1^3)$	60	-570	1892	-2555	1194	-84	518	-2946	10434/4655
$(32^2 1)$	-20	196	-680	982	-512	40	-172	1023	-1746
$(3^2 1^4)$	15	-140	451	-578	246	-15	134	-741	1102
$(3^2 21^2)$	-30	288	-963	1296	-582	36	-273	1584	-2531
$(3^2 2^2)$	6	-60	213	-315	166	-10	54	-333	596
$(3^3 1)$	4	-39	132	-177	74	-4	39	-234	383
(41^6)	-7	64	-202	256	-112	8	-59	313	-449
(421^4)	30	-280	907	-1188	543	-42	258	-1412	2106
$(42^2 1^2)$	-30	288	-969	1338	-660	60	-261	1491	-2368
(42^3)	4	-40	143	-218	130	-20	34	-204	352
(431^3)	-20	188	-610	790	-348	24	-182	1006	-1485
(4321)	24	-234	798	-1106	528	-40	222	-1306	2124
(43^2)	-3	30	-105	148	-68	4	-30	186	-320
$(4^2 1^2)$	6	-57	187	-246	114	-10	56	-310	452
$(4^2 2)$	-3	30	-106	156	-86	12	-28	168	-280
(51^5)	6	-55	174	-221	96	-6	51	-271	393
(521^3)	-20	188	-614	812	-372	24	-174	961	-1463
$(52^2 1)$	12	-117	402	-572	292	-20	105	-613	1020
(531^2)	12	-114	374	-489	214	-12	112	-629	950
(532)	-6	60	-212	310	-160	8	-56	342	-600
(541)	-6	58	-194	260	-120	8	-58	330	-500
(5^2)	1	-10	35	-50	25	-1	10	-60	100
(61^4)	-5	46	-146	186	-81	6	-43	229	-333
(621^2)	12	-114	377	-506	236	-20	106	-594	920
(62^2)	-3	30	-107	162	-94	12	-26	156	-272
(631)	-6	58	-194	258	-112	8	-58	336	-525
(64)	2	-20	70	-100	51	-6	20	-120	196
(71^3)	4	-37	118	-151	66	-4	35	-187	273
(721)	-6	58	-196	270	-128	8	-54	311	-503
(73)	2	-20	70	-99	46	-2	20	-123	212
(81^2)	-3	28	-90	116	-52	4	-27	145	-210
(82)	2	-20	71	-106	59	-6	18	-108	188
(91)	2	-19	62	-80	34	-2	19	-105	154
(10)	-1	10	-35	50	-25	2	-10	60	-100

Table 1·4·10 (cont.)

$w=10$ (ii)	$h_3 h_2^3 h_1$	$h_3^2 h_1^4$	$h_3^2 h_2 h_1^2$	$h_3^2 h_2^2$	$h_3^3 h_1$	$h_4 h_1^6$	$h_4 h_2 h_1^4$	$h_4 h_2^2 h_1^2$	$h_4 h_2^3$	$h_4 h_3 h_1^3$
(1^{10})	75600	100800	50400	25200	16800	151200	75600	37800	18900	25200
(21^8)	42840	57120	29120	14840	10080	85680	43680	22260	11340	15120
$(2^2 1^6)$	24300	32280	16820	8760	6060	48060	25050	13050	6795	9030
$(2^3 1^4)$	13800	18180	9708	5184	3648	26730	14274	7623	4071	5364
$(2^4 1^2)$	7848	10200	5596	3076	2196	14760	8088	4440	2442	3168
(2^5)	4470	5700	3220	1830	1320	8100	4560	2580	1470	1860
(31^7)	17850	24080	12600	6580	4620	36960	19320	10080	5250	7070
(321^5)	10140	13560	7280	3900	2790	20340	10930	5860	3135	4195
$(32^2 1^3)$	5760	7600	4200	2318	1686	11130	6154	3399	1875	2472
$(32^3 1)$	__3288__	4240	2418	1382	1017	6060	3450	1968	1125	1447
	774									
$(3^2 1^4)$	−363	__5672__	3156	1748	1296	8300	4650	2590	1435	1932
		299								
$(3^2 2 1^2)$	902	−637	__1818__	1044	786	4490	2594	1495	859	1126
			1527							
$(3^2 2^2)$	−282	123	−330	__626__	474	2420	1442	862	517	652
				120						
$(3^3 1)$	−122	104	−273	48	__370__	1770	1074	651	393	504
					65					
(41^6)	151	−116	228	−45	−30	__13290__	7155	3840	2055	2795
						55				
(421^4)	−744	529	−1090	227	148	−240	__3985__	2212	1224	1645
							1091			
$(42^2 1^2)$	910	−551	1236	−282	−174	243	−1166	__1273__	732	960
								1366		
(42^3)	−160	69	−164	42	20	−34	178	−248	__441__	556
									76	
(431^3)	502	−410	829	−162	−122	168	−777	814	−112	__749__
										622
(4321)	−796	524	−1248	276	210	−198	978	−1164	184	−752
(43^2)	116	−83	228	−48	−55	24	−124	156	−20	104
$(4^2 1^2)$	−156	129	−240	45	30	−56	266	−284	47	−224
$(4^2 2)$	112	−66	152	−29	−22	28	−148	200	−54	112
(51^5)	−135	99	−200	43	26	−48	206	−205	26	−140
(521^3)	542	−355	755	−182	−98	163	−735	774	−104	514
$(52^2 1)$	−446	217	−510	162	60	−99	476	−554	84	−323
(531^2)	−332	259	−543	120	78	−103	477	−493	58	−389
(532)	280	−130	331	−122	−41	50	−250	298	−36	189
(541)	181	−140	277	−61	−35	58	−280	299	−40	245
(5^2)	−45	25	−55	20	5	−10	50	−55	5	−45
(61^4)	110	−84	175	−33	−26	41	−176	180	−26	115
(621^2)	−328	221	−498	102	78	−100	459	−512	86	−307
(62^2)	128	−52	128	−40	−16	26	−132	176	−48	80
(631)	169	−144	336	−57	−65	52	−249	279	−40	201
(64)	−72	52	−120	21	22	−20	104	−132	30	−88
(71^3)	−94	69	−142	30	20	−34	146	−147	18	−97
(721)	193	−115	277	−67	−44	51	−239	273	−36	158
(73)	−82	53	−144	36	31	−17	85	−102	10	−68
(81^2)	73	−55	107	−23	−13	27	−118	119	−18	83
(82)	−88	37	−92	31	10	−18	90	−116	26	−56
(91)	−49	43	−87	15	13	−19	86	−87	10	−67
(10)	40	−25	60	−15	−10	10	−50	60	−10	40

Table 1·4·10 (*cont.*)

$w=10$ (iii)	$h_4h_3h_2h_1$	$h_4h_3^2$	$h_4^2h_1^2$	$h_4^2h_2$	$h_5h_1^5$	$h_5h_2h_1^3$	$h_5h_2^2h_1$	$h_5h_3h_1^2$	$h_5h_3h_2$	$h_5h_4h_1$	h_5^2
(1^{10})	12600	4200	6300	3150	30240	15120	7560	5040	2520	1260	252
(21^8)	7700	2660	3990	2030	18480	9408	4788	3248	1652	854	182
(2^21^6)	4700	1690	2520	1310	11100	5778	3006	2078	1080	578	132
(2^31^4)	2865	1077	1584	846	6570	3510	1875	1320	705	390	96
(2^41^2)	1744	688	990	547	3840	2112	1164	832	460	262	70
(2^5)	1060	440	615	355	2220	1260	720	520	300	175	51
(31^7)	3675	1330	2030	1050	9240	4788	2478	1722	889	483	112
(321^5)	2240	850	1280	680	5380	2872	1530	1089	578	327	82
(32^21^3)	1363	545	800	440	3100	1710	942	684	376	220	60
(32^31)	828	350	496	285	1770	1011	579	426	245	147	44
(3^21^4)	1066	432	652	356	2460	1366	754	560	306	186	52
(3^221^2)	647	279	402	230	1390	802	461	348	199	124	38
(3^22^2)	392	180	246	149	780	468	282	214	130	82	28
(3^31)	306	145	198	120	600	366	222	174	105	69	24
(41^6)	1490	570	890	470	4050	2139	1128	803	422	242	62
(421^4)	907	367	560	306	2275	1249	684	501	273	164	46
(42^21^2)	551	237	348	199	1270	726	415	310	177	110	34
(42^3)	334	153	214	130	705	420	252	190	115	73	25
(431^3)	431	189	286	162	970	564	325	252	143	94	30
(4321)	261 / 1157	123	174	105	535	325	197	154	93	62	22
(43^2)	−195	65 / 55	84	55	220	143	93	75	49	34	14
(4^21^2)	249	−27	126 / 98	75	350	218	134	110	66	48	18
(4^22)	−180	22	−49	49 / 54	190	124	81	66	43	31	13
(51^6)	166	−20	44	−20	1546 / 46	841	456	336	181	111	32
(521^3)	−648	80	−168	80	−155	480 / 561	273	207	117	75	24
(52^21)	465	−57	109	−58	95	−379	164 / 311	126	76	50	18
(531^2)	470	−63	136	−58	95	−358	232	102 / 282	61	43	16
(532)	−295	41	−59	28	−50	215	−193	−154	40 / 157	28	12
(541)	−280	32	−108	44	−50	201	−141	−172	90	22 / 143	10
(5^2)	55	−5	20	−5	10	−45	40	40	−35	−35	6 / 15
(61^4)	−146	20	−34	20	−40	130	−75	−75	35	35	−5
(621^2)	431	−63	95	−66	96	−339	215	205	−105	−102	15
(62^2)	−128	16	−31	30	−24	96	−80	−50	36	32	−5
(631)	−291	53	−63	44	−48	177	−102	−141	61	73	−10
(64)	132	−22	39	−34	16	−64	42	54	−20	−44	5
(71^3)	122	−17	28	−14	34	−112	69	63	−35	−29	5
(721)	−247	41	−43	28	−51	188	−136	−107	77	49	−10
(73)	123	−31	15	−10	17	−68	51	51	−41	−20	5
(81^2)	−94	10	−28	14	−27	91	−55	−56	28	29	−5
(82)	92	−10	19	−18	18	−72	62	38	−36	−20	5
(91)	78	−10	24	−10	19	−67	39	48	−20	−29	5
(10)	−60	10	−15	10	−10	40	−30	−30	20	20	−5

Table 1·4·10 (cont.)

$w = 10$ (iv)	$h_6h_1^4$	$h_6h_2h_1^2$	$h_6h_2^2$	$h_6h_3h_1$	h_6h_4	$h_7h_1^3$	$h_7h_2h_1$	h_7h_3	$h_8h_1^2$	h_8h_2	h_9h_1	h_{10}
(1^{10})	5040	2520	1260	840	210	720	360	120	90	45	10	1
(21^8)	3360	1708	868	588	154	528	268	92	73	37	9	1
(2^21^6)	2190	1138	591	408	113	378	196	70	58	30	8	1
(2^31^4)	1398	747	399	281	83	264	141	53	45	24	7	1
(2^41^2)	876	484	268	192	61	180	100	40	34	19	6	1
(2^5)	540	310	180	130	45	120	70	30	25	15	5	1
(31^7)	1960	1008	518	357	98	358	183	64	57	29	8	1
(321^5)	1230	652	345	244	72	247	130	48	44	23	7	1
(32^21^3)	758	417	229	166	53	166	91	36	33	18	6	1
(32^31)	460	264	152	112	39	109	63	27	24	14	5	1
(3^21^4)	646	354	193	142	46	152	82	32	32	17	6	1
(3^221^2)	386	222	127	96	34	98	56	24	23	13	5	1
(3^22^2)	228	138	84	64	25	62	38	18	16	10	4	1
(3^31)	186	114	69	55	22	54	33	16	15	9	4	1
(41^6)	1045	544	283	198	57	229	118	42	43	22	7	1
(421^4)	627	340	184	133	42	151	81	31	32	17	6	1
(42^21^2)	371	211	120	89	31	97	55	23	23	13	5	1
(42^3)	217	130	79	59	23	61	37	17	16	10	4	1
(431^3)	303	173	98	75	27	86	48	20	22	12	5	1
(4321)	175	106	64	50	20	53	32	15	15	9	4	1
(43^2)	79	52	34	28	13	27	18	10	9	6	3	1
(4^21^2)	128	78	47	38	16	44	26	12	14	8	4	1
(4^22)	72	47	31	25	12	26	17	9	9	6	3	1
(51^5)	501	266	141	101	31	136	71	26	31	16	6	1
(521^3)	287	161	90	67	23	85	47	19	22	12	5	1
(52^21)	163	97	58	44	17	52	31	14	15	9	4	1
(531^2)	127	77	46	37	15	44	26	12	14	8	4	1
(532)	71	46	30	24	11	26	17	9	9	6	3	1
(541)	48	32	21	18	9	20	13	7	8	5	3	1
(5^2)	16	12	9	8	5	8	6	4	4	3	2	1
(61^4)	209 40	115	63	47	16	73	39	15	21	11	5	1
(621^2)	−96	68 261	40	31	12	43	25	11	14	8	4	1
(62^2)	24	−78	26 44	20	9	25	16	8	9	6	3	1
(631)	48	−141	32	17 112	8	20	13	7	8	5	3	1
(64)	−16	54	−22	−44	5 34	8	6	4	4	3	2	1
(71^3)	−34	78	−18	−36	10	34 34	19	8	13	7	4	1
(721)	51	−137	36	70	−20	−51	12 100	6	8	5	3	1
(73)	−17	51	−10	−41	10	17	−41	4 31	4	3	2	1
(81^2)	27	−64	18	29	−10	−27	37	−10	7 27	4	3	1
(82)	−18	54	−26	−20	10	18	−36	10	−18	3 26	2	1
(91)	−19	48	−10	−29	10	19	−29	10	−19	10	2 19	1
(10)	10	−30	10	20	−10	−10	20	−10	10	−10	−10	1 10

Table 1·4·11

$w=11$ (i)	$h_1{}^{11}$	$h_2 h_1{}^9$	$h_2{}^2 h_1{}^7$	$h_2{}^3 h_1{}^5$	$h_2{}^4 h_1{}^3$	$h_2{}^5 h_1$	$h_3 h_1{}^8$	$h_3 h_2 h_1{}^6$	$h_3 h_2{}^2 h_1{}^4$	$h_3 h_2{}^3 h_1{}^2$
(1^{11})	39916800/1	19958400	9979200	4989600	2494800	1247400	6652800	3326400	1663200	831600
(21^9)	−10	10160640/101	5171040	2630880	1338120	680400	3507840	1784160	907200	461160
$(2^2 1^7)$	36	−368	2678760/1361	1387260	718200	371700	1844640	955080	494340	255780
$(2^3 1^5)$	−56	581	−2190	731520/3614	385740	203400	967680	510300	269100	141900
$(2^4 1^3)$	35	−370	1430	−2444	207324/1742	111510	506520	272160	146340	78744
$(2^5 1)$	−6	65	−260	468	−364	61260/91	264600	144900	79500	43710
(31^8)	9	−91	331	−519	325	−55	1303680/84	682080	356160	185640
(321^6)	−56	574	−2123	3398	−2182	380	−530	363840/3411	193740	102990
$(32^2 1^4)$	105	−1095	4140	−6818	4544	−828	1005	−6634	105284/13359	57156
$(32^3 1^2)$	−60	640	−2494	4284	−3035	602	−576	3926	−8298	31731/5583
(32^4)	5	−55	224	−412	329	−85	47	−332	744	−562
$(3^2 1^5)$	21	−216	799	−1270	798	−132	206	−1332	2580	−1488
$(3^2 2 1^3)$	−60	630	−2392	3929	−2566	438	−598	3994	−8085	4926
$(3^2 2^2 1)$	30	−324	1278	−2214	1557	−282	300	−2091	4527	−3095
$(3^3 1^2)$	10	−106	405	−663	420	−66	105	−714	1458	−856
$(3^3 2)$	−4	44	−177	312	−219	34	−42	303	−684	487
(41^7)	−8	81	−205	464	−294	52	−75	471	−888	510
(421^5)	42	−432	1604	−2584	1688	−312	400	−2569	4980	−2961
$(42^2 1^3)$	−60	630	−2402	4007	−2745	546	−578	3824	−7717	4845
$(42^3 1)$	20	−216	856	−1510	1128	−260	192	−1319	2820	−1954
(431^4)	−30	310	−1152	1843	−1178	207	−298	1925	−3713	2151
(4321^2)	60	−636	2442	−4074	2742	−516	606	−4074	8296	−5118
(432^2)	−12	132	−534	963	−734	170	−120	849	−1879	1352
$(43^2 1)$	−12	129	−501	837	−547	94	−129	891	−1849	1103
$(4^2 1^3)$	10	−104	389	−628	411	−78	101	−652	1250	−729
$(4^2 21)$	−12	129	−504	862	−608	130	−123	836	−1722	1088
$(4^2 3)$	3	−33	132	−230	162	−34	33	−234	502	−316
(51^6)	7	−71	259	−408	258	−44	66	−415	787	−458
(521^4)	−30	310	−1157	1875	−1229	219	−288	1860	−3647	2220
$(52^2 1^2)$	30	−318	1227	−2079	1449	−282	291	−1950	4027	−2645
(52^3)	−4	44	−179	329	−264	70	−38	266	−589	442
(531^3)	20	−208	778	−1252	800	−132	202	−1316	2571	−1526
(5321)	−24	258	−1008	1718	−1178	204	−246	1690	−3568	2354
(53^2)	3	−33	132	−229	155	−22	33	−237	525	−355
(541^2)	−12	126	−476	776	−511	94	−124	810	−1576	943
(542)	6	−66	266	−476	360	−84	62	−434	940	−660
$(5^2 1)$	3	−32	123	−204	135	−22	32	−214	430	−270
(61^5)	−6	61	−223	352	−223	39	−57	359	−682	393
(621^3)	20	−208	782	−1278	846	−156	194	−1263	2500	−1522
$(62^2 1)$	−12	129	−507	881	−636	130	−117	799	−1699	1164
(631^2)	−12	126	−476	773	−497	86	−124	819	−1619	949
(632)	6	−66	266	−474	348	−68	62	−440	978	−704
(641)	6	−64	246	−408	272	−52	64	−428	852	−505
(65)	−2	22	−88	154	−110	21	−22	154	−330	225
(71^4)	5	−51	187	−296	188	−33	48	−303	577	−333
(721^2)	−12	126	−479	793	−533	102	−118	778	−1563	961
(72^2)	3	−33	134	−245	194	−50	29	−203	449	−332
(731)	6	−64	246	−406	262	−44	64	−434	883	−520
(74)	−2	22	−88	154	−111	25	−22	154	−326	208
(81^3)	−4	41	−151	240	−153	26	−39	247	−472	277
(821)	6	−64	248	−420	288	−52	60	−405	842	−545
(8_3)	−2	22	−88	153	−105	17	−22	157	−345	230
(91^2)	3	−31	115	−184	119	−22	30	−191	364	−213
(92)	−2	22	−89	161	−124	29	−20	140	−309	227
$(10, 1)$	−2	21	−79	127	−80	13	−21	137	−265	150
(11)	1	−11	44	−77	55	−11	11	−77	165	−110

Table 1·4·11 (*cont.*)

$w = 11$ (ii)	$h_3h_2^4$	$h_3^2h_1^5$	$h_3^2h_2h_1^3$	$h_3^2h_2^2h_1$	$h_3^3h_1^2$	$h_3^3h_2$	$h_4h_1^7$	$h_4h_2h_1^5$	$h_4h_2^2h_1^3$	$h_4h_2^3h_1$
(1^{11})	415800	1108800	554400	277200	184800	92400	1663200	831600	415800	207900
(21^9)	234360	614880	312480	158760	107520	54600	922320	468720	238140	120960
(2^21^7)	132300	340200	175980	91000	62580	32340	507780	262710	135870	70245
(2^31^5)	74820	187740	99000	52200	36420	19200	277830	146520	77265	40740
(2^41^3)	42402	103320	55620	29964	21180	11424	151200	81360	43812	23610
(2^51)	24090	56700	31200	17210	12300	6810	81900	45000	24780	13680
(31^8)	96600	249760	129920	67480	47040	24360	381360	198240	102900	53340
(321^6)	54660	137720	73100	38740	27420	14490	207060	109950	58290	30855
(32^21^4)	30996	75700	41068	22256	15972	8640	111930	60750	32939	17841
(32^31^2)	17622	41480	23032	12794	9288	5163	60270	33450	18573	10317
(32^4)	**10050**	22660	12892	7358	5388	3090	32340	18360	10452	5970
	85									
(3^21^5)		**55480**	30340	16540	12060	6540	82180	45110	24670	13445
	117	**546**								
(3^221^3)			**17010**	9516	7026	3918	44030	24740	13873	7762
	-402	-1638	**5235**							
(3^22^21)				**5478**	4080	2352	23520	13530	7788	4485
	272	828	-2872	**1923**						
(3^31^2)					**3098**	1792	17010	9930	5787	3363
	62	315	-1043	540	**245**					
(3^32)						**1081**	9030	5400	3237	1944
	-34	-126	472	-357	-98	**85**				
(41^7)							**129990**	69405	36960	19635
	-44	-182	517	-255	-87	33	**70**			
(421^5)								**38010**	20760	11310
	270	999	-2934	1503	504	-198	-373	**2044**		
(42^21^3)									**11642**	6519
	-486	-1478	4561	-2483	-795	331	539	-3063	**4843**	
(42^31)										**3762**
	240	489	-1598	974	262	-124	-183	1092	-1870	**840**
(431^4)	-183	-793	2318	-1145	-428	158	277	-1532	2278	-778
(4321^2)	476	1676	-5305	2820	1041	-408	-552	3194	-5096	1896
(432^2)	-170	-327	1126	-732	-198	102	111	-682	1208	-560
(43^21)	-90	-400	1341	-675	-327	111	111	-662	1085	-380
(4^21^3)	68	271	-761	365	131	-44	-100	560	-839	300
(4^221)	-118	-344	1076	-552	-204	66	120	-716	1192	-488
(4^23)	34	106	-364	180	92	-23	-30	188	-332	136
(51^6)	38	159	-457	237	75	-33	-62	327	-467	155
(521^4)	-195	-717	2140	-1177	-354	170	270	-1471	2185	-764
(52^21^2)	262	740	-2348	1425	384	-213	-273	1553	-2457	946
(52^3)	-70	-93	308	-200	-44	22	38	-232	414	-204
(531^3)	120	544	-1616	868	287	-136	-188	1037	-1524	504
(5321)	-196	-686	2276	-1446	-414	258	222	-1290	2058	-752
(53^2)	22	105	-385	267	88	-65	-27	162	-265	88
(541^2)	-86	-340	967	-498	-159	66	124	-698	1041	-365
(542)	84	172	-560	343	90	-43	-62	380	-660	298
(5^21)	21	91	-270	161	42	-26	-32	182	-270	89
(61^5)	-33	-138	402	-202	-70	28	54	-285	412	-140
(621^3)	136	489	-1493	808	263	-118	-183	1006	-1528	556
(62^21)	-118	-298	986	-642	-162	102	111	-644	1060	-444
(631^2)	-78	-347	1062	-540	-210	84	115	-646	980	-337
(632)	68	174	-615	447	107	-85	-56	336	-570	242
(641)	46	186	-557	270	109	-37	-64	372	-583	214
(65)	-21	-66	215	-141	-37	26	22	-132	215	-84
(71^4)	29	117	-342	169	61	-22	-46	243	-352	118
(721^2)	-94	-303	947	-498	-177	66	112	-625	972	-357
(72^2)	50	72	-243	155	39	-17	-29	176	-314	154
(731)	42	190	-616	306	139	-46	-58	335	-528	178
(74)	-25	-68	222	-108	-50	11	22	-136	236	-100
(81^3)	-22	-96	279	-148	-47	22	38	-201	289	-97
(821)	46	157	-515	313	93	-52	-57	324	-515	192
(83)	-17	-69	250	-162	-58	35	19	-114	190	-68
(91^2)	20	76	-215	108	35	-14	-30	161	-233	81
(92)	-29	-51	173	-120	-25	17	20	-120	209	-98
$(10, 1)$	-11	-58	170	-81	-32	11	21	-116	170	-54
(11)	11	33	-110	66	22	-11	-11	66	-110	44

Table 1·4·11 (*cont.*)

$w=11$ (iii)	$h_4h_3h_1^4$	$h_4h_3h_2h_1^2$	$h_4h_3h_2^2$	$h_4h_3^2h_1$	$h_4^2h_1^3$	$h_4^2h_2h_1$	$h_4^2h_3$	$h_5h_1^6$	$h_5h_2h_1^4$	$h_5h_2^2h_1^2$	$h_5h_2^3$	$h_5h_3h_1^3$
(1^{11})	277200	138600	69300	46200	69300	34650	11550	332640	166320	83160	41580	55440
(21^9)	161280	81900	41580	28140	42210	21420	7350	196560	99792	50652	25704	34272
(2^21^7)	93450	48300	24955	17150	25620	13230	4690	114660	59262	30618	15813	21042
(2^31^5)	53910	28425	14905	10455	15480	8160	3000	66150	34884	18393	9696	12834
(2^41^3)	30960	16692	9006	6372	9306	5025	1923	37800	20376	10992	5934	7776
(2^51)	17700	9780	5420	3880	5565	3090	1235	21420	11820	6540	3630	4680
(31^8)	71680	37100	19180	13300	20230	10430	3710	92400	47712	24612	12684	17024
(321^6)	41270	21815	11515	8120	12230	6440	2380	52620	27798	14664	7725	10334
(32^21^4)	23642	12797	6919	4957	7344	3966	1530	29730	16094	8703	4701	6234
(32^31^2)	13478	7489	4163	3023	4382	2437	985	16680	9264	5148	2862	3736
(32^4)	7648	4372	2510	1840	2599	1495	635	9300	5304	3036	1746	2224
(3^21^5)	18050	9810	5315	3860	5820	3140	1220	23060	12574	6830	3697	4978
(3^221^3)	10254	5733	3197	2359	3458	1926	788	12830	7192	4021	2242	2972
(3^22^21)	5798	3342	1927	1438	2042	1179	509	7100	4094	2362	1363	1762
(3^31^2)	4386	2553	1479	1127	1602	930	410	5370	3138	1827	1059	1392
(3^32)	2460	1482	864	685	936	567	265	2940	1770	1068	645	816
(41^7)	26635	14105	7455	5320	8260	4340	1610	37590	19761	10374	5439	7343
(421^5)	15215	8265	4480	3255	4970	2680	1040	20805	11256	6078	3276	4397
(42^21^3)	8642	4830	2696	1990	2964	1649	673	11460	6386	3555	1977	2616
(42^31)	4883	2815	1626	1214	1754	1012	436	6285	3609	2076	1197	1546
(431^4)	6573 / 1237	3697	2070	1555	2356	1310	540	8615	4871	2738	1531	2069
(4321^2)	-2551	2151 / 5904	1248	951	1384	801	351	4700	2742	1595	925	1218
(432^2)	509	-1320	755 / 426	579	808	489	228	2555	1538	928	562	712
(43^21)	576	-1464	294	456 / 488	630	385	185	1890	1159	709	432	557
(4^21^3)	-465	920	-188	-188	950 / 200	549	243	2990	1782	1052	616	834
(4^221)	578	-1360	325	320	-240	334 / 383	159	1610	992	609	373	484
(4^23)	-166	456	-111	-161	60	-119	85 / 69	630	410	267	174	216
(51^6)	-239	472	-95	-91	84	-96	22	13326 / 59	7185	3864	2073	2819
(521^4)	1090	-2249	485	440	-391	470	-110	-256	4010 / 1160	2232	1239	1665
(52^21^2)	-1135	2532	-624	-495	413	-552	132	259	-1240	1289 / 1447	744	976
(52^3)	157	-392	134	66	-65	109	-25	-34	175	-238	450 / 70	568
(531^3)	-845	1708	-340	-360	316	-360	88	176	-815	859	-106	765 / 651
(5321)	1024	-2363	550	529	-359	484	-130	-214	1060	-1257	166	-826
(53^2)	-140	357	-72	-111	41	-60	23	27	-140	174	-16	121
(541^2)	593	-1151	239	217	-262	288	-61	-112	535	-578	86	-452
(542)	-300	724	-214	-137	131	-210	50	54	-280	362	-84	214
(5^21)	-160	307	-63	-53	72	-68	11	32	-160	181	-21	144
(61^5)	204	-417	85	86	-69	86	-22	-52	221	-219	29	-146
(621^3)	-731	1576	-352	-330	254	-336	88	175	-784	828	-118	533
(62^21)	452	-1060	296	206	-163	235	-59	-107	516	-612	100	-342
(631^2)	529	-1140	231	271	-191	250	-73	-107	492	-510	66	-303
(632)	-255	630	-192	-129	87	-126	34	56	-285	354	-50	216
(641)	-321	684	-138	-159	141	-186	50	56	-209	291	-46	230
(65)	115	-252	63	48	-52	63	-11	-22	115	-141	21	-104
(71^4)	-174	363	-73	-80	57	-74	22	45	-191	192	-27	122
(721^2)	453	-1037	233	244	-150	222	-73	-108	489	-535	88	-317
(72^2)	-117	302	-104	-61	45	-79	25	27	-135	178	-50	76
(731)	-281	675	-132	-201	91	-144	62	54	-253	273	-36	199
(74)	118	-300	75	89	-50	89	-39	-18	90	-108	25	-72
(81^3)	146	-293	59	59	-50	60	-15	-38	163	-164	20	-108
(821)	-233	548	-126	-124	76	-114	34	57	-267	305	-40	176
(83)	95	-252	57	81	-26	45	-23	-19	95	-114	11	-76
(91^2)	-122	237	-51	-43	46	-52	11	30	-131	132	-20	92
(92)	82	-204	69	33	-31	51	-11	-20	100	-129	29	-62
$(10,1)$	95	-192	33	43	-37	43	-11	-21	95	-96	11	-74
(11)	-55	132	-33	-33	22	-33	11	11	-55	66	-11	44

Table 1·4·11 (*cont.*)

$w = 11$ (iv)	$h_5h_3h_2h_1$	$h_5h_3^2$	$h_5h_4h_1^2$	$h_5h_4h_2$	$h_5^2h_1$	$h_6h_1^5$	$h_6h_2h_1^3$	$h_6h_2^2h_1$	$h_6h_3h_1^2$	$h_6h_3h_2$	$h_6h_4h_1$	h_6h_5
(1^{11})	27720	9240	13860	6930	2772	55440	27720	13860	9240	4620	2310	462
(21^9)	17388	5964	8946	4536	1890	35280	17892	9072	6132	3108	1596	336
$(2^2 1^7)$	10864	3850	5754	2968	1288	22050	11382	5873	4032	2079	1099	245
$(2^3 1^5)$	6765	2487	3684	1941	876	13560	7149	3768	2629	1385	754	179
$(2^4 1^3)$	4200	1608	2346	1269	594	8220	4440	2400	1700	920	515	131
$(2^5 1)$	2600	1040	1485	830	401	4920	2730	1520	1090	610	350	96
(31^8)	8764	3108	4718	2422	1078	19040	9772	5012	3444	1764	938	210
(321^6)	5437	2006	3023	1585	736	11530	6058	3179	2229	1167	643	154
$(32^2 1^4)$	3365	1297	1924	1036	500	6890	3719	2005	1433	771	439	113
$(32^3 1^2)$	2078	840	1216	677	338	4070	2263	1259	914	509	298	83
(32^4)	1280	544	763	443	227	2380	1366	788	578	336	201	61
$(3^2 1^5)$	2686	1044	1584	848	424	5690	3074	1655	1198	641	374	98
$(3^2 21^3)$	1655	677	998	553	286	3320	1854	1032	762	422	254	72
$(3^2 2^2 1)$	1018	440	624	361	192	1920	1110	642	480	278	171	53
$(3^3 1^2)$	807	355	510	294	162	1530	894	519	398	229	146	46
$(3^3 2)$	495	232	315	192	108	870	528	321	247	151	97	34
(41^7)	3843	1414	2177	1134	546	9275	4816	2499	1743	903	497	119
(421^5)	2365	913	1389	743	374	5410	2896	1548	1107	590	339	88
$(42^2 1^3)$	1453	591	878	486	254	3125	1730	957	699	386	230	65
$(42^3 1)$	891	383	550	318	171	1790	1027	591	438	253	155	48
(431^4)	1151	475	726	399	218	2485	1386	769	575	316	196	57
(4321^2)	705	309	452	260	146	1410	818	473	360	207	132	42
(432^2)	431	201	279	170	97	795	480	291	223	136	88	31
$(43^2 1)$	340	163	227	138	82	615	378	231	183	111	75	27
$(4^2 1^3)$	484	216	332	189	114	990	580	337	264	151	102	34
$(4^2 21)$	295	141	203	123	75	550	337	206	163	99	68	25
$(4^2 3)$	141	75	99	65	41	230	151	99	81	53	38	16
(51^6)	1508	582	908	482	254	4051	2140	1129	804	423	243	63
(521^4)	922	377	575	316	174	2276	1250	685	502	274	165	47
$(52^2 1^2)$	563	245	360	207	118	1271	727	416	311	178	111	35
(52^3)	343	159	223	136	79	706	421	253	191	116	74	26
(531^3)	443	197	298	170	102	971	565	326	253	144	95	31
(5321)	$\overline{270}$	129	183	111	68	536	326	198	155	94	63	23
	1305											
(53^2)	−228	$\overline{69}$	90	59	38	221	144	94	76	50	35	15
		65										
(541^2)	534	−63	$\overline{135}$	81	54	351	219	135	111	67	49	19
			390									
(542)	−326	37	−183	$\overline{53}$	35	191	125	82	67	44	32	14
				160								
$(5^2 1)$	−191	26	−128	52	$\overline{26}$	112	76	51	44	29	23	11
					56							
(61^5)	174	−22	87	−44	−22	$\overline{1546}$	841	456	336	181	111	32
						51						
(621^3)	−685	91	−331	182	89	−170	$\overline{480}$	273	207	117	75	24
							608					
$(62^2 1)$	518	−72	223	−146	−68	102	−402	$\overline{164}$	126	76	50	18
								326				
(631^2)	485	−72	260	−133	−73	102	−378	235	$\overline{102}$	61	43	16
									293			
(632)	−366	61	−135	92	52	−51	216	−192	−147	$\overline{40}$	28	12
										152		
(641)	−270	37	−198	102	57	−51	200	−132	−168	74	$\overline{22}$	10
											136	
(65)	156	−26	93	−52	−41	17	−74	63	63	−52	−52	$\overline{6}$
												41
(71^4)	−150	19	−69	38	16	−45	146	−84	−84	39	39	−11
(721^2)	429	−57	185	−121	−43	108	−381	241	230	−117	−114	33
(72^2)	−116	11	−56	54	11	−27	108	−90	−56	40	36	−11
(731)	−273	46	−114	72	27	−54	199	−114	−159	68	82	−22
(74)	94	−11	61	−50	−11	18	−72	47	61	−22	−50	11
(81^3)	136	−19	62	−31	−16	38	−125	77	70	−39	−32	11
(821)	−276	46	−95	62	27	−57	210	−152	−119	86	54	−22
(83)	138	−35	33	−22	−11	19	−76	57	57	−46	−22	11
(91^2)	−104	11	−62	31	16	−30	101	−61	−62	31	32	−11
(92)	102	−11	42	−40	−11	20	−80	69	42	−40	−22	11
$(10, 1)$	86	−11	53	−22	−16	21	−74	43	53	−22	−32	11
(11)	−66	11	−33	22	11	−11	44	−33	−33	22	22	−11

Table 1·4·11 (*cont.*)

$w=11$ (v)	$h_7h_1^4$	$h_7h_2h_1^2$	$h_7h_2^2$	$h_7h_3h_1$	h_7h_4	$h_8h_1^3$	$h_8h_2h_1$	h_8h_3	$h_9h_1^2$	h_9h_2	$h_{10}h_1$	h_{11}
(1^{11})	7920	3960	1980	1320	330	990	495	165	110	55	11	1
(21^9)	5472	2772	1404	948	246	747	378	129	91	46	10	1
(2^21^7)	3702	1908	983	674	183	552	284	100	74	38	9	1
(2^31^5)	2454	1293	681	475	136	399	210	77	59	31	8	1
(2^41^3)	1596	864	468	332	101	282	153	59	46	25	7	1
(2^51)	1020	570	320	230	75	195	110	45	35	20	6	1
(31^8)	3392	1732	884	604	162	529	269	93	73	37	9	1
(321^6)	2218	1159	605	422	120	379	197	71	58	30	8	1
(32^21^4)	1422	765	411	293	89	265	142	54	45	24	7	1
(32^31^2)	896	499	278	202	66	181	101	41	34	19	6	1
(32^4)	556	322	188	138	49	121	71	31	25	15	5	1
(3^21^5)	1254	670	357	256	78	248	131	49	44	23	7	1
(3^221^3)	778	432	239	176	58	167	92	37	33	18	6	1
(3^22^21)	476	276	160	120	43	110	64	28	24	14	5	1
(3^31^2)	402	234	135	104	38	99	57	25	23	13	5	1
(3^32)	240	147	90	70	28	63	39	19	16	10	4	1
(41^7)	1961	1009	519	358	99	358	183	64	57	29	8	1
(421^5)	1231	653	346	245	73	247	130	48	44	23	7	1
(42^21^3)	759	418	230	167	54	166	91	36	33	18	6	1
(42^31)	461	265	153	113	40	109	63	27	24	14	5	1
(431^4)	647	355	194	143	47	152	82	32	32	17	6	1
(4321^2)	387	223	128	97	35	98	56	24	23	13	5	1
(432^2)	229	139	85	65	26	62	38	18	16	10	4	1
(43^21)	187	115	70	56	23	54	33	16	15	9	4	1
(4^21^3)	304	174	99	76	28	86	48	20	22	12	5	1
(4^221)	176	107	65	51	21	53	32	15	15	9	4	1
(4^23)	80	53	35	29	14	27	18	10	9	6	3	1
(51^6)	1045	544	283	198	57	229	118	42	43	22	7	1
(521^4)	627	340	184	133	42	151	81	31	32	17	6	1
(52^21^2)	371	211	120	89	31	97	55	23	23	13	5	1
(52^3)	217	130	79	59	23	61	37	17	16	10	4	1
(531^3)	303	173	98	75	27	86	48	20	22	12	5	1
(5321)	175	106	64	50	20	53	32	15	15	9	4	1
(53^2)	79	52	34	28	13	27	18	10	9	6	3	1
(541^2)	128	78	47	38	16	44	26	12	14	8	4	1
(542)	72	47	31	25	12	26	17	9	9	6	3	1
(5^21)	48	32	21	18	9	20	13	7	8	5	3	1
(61^5)	501	266	141	101	31	136	71	26	31	16	6	1
(621^3)	287	161	90	67	23	85	47	19	22	12	5	1
(62^21)	163	97	58	44	17	52	31	14	15	9	4	1
(631^2)	127	77	46	37	15	44	26	12	14	8	4	1
(632)	71	46	30	24	11	26	17	9	9	6	3	1
(641)	48	32	21	18	9	20	13	7	8	5	3	1
(65)	16	12	9	8	5	8	6	4	4	3	2	1
(71^4)	$\frac{209}{45}$	115	63	47	16	73	39	15	21	11	5	1
(721^2)	−108	$\frac{68}{294}$	40	31	12	43	25	11	14	8	4	1
(72^2)	27	−88	$\frac{26}{50}$	20	9	25	16	8	9	6	3	1
(731)	54	−159	36	$\frac{17}{127}$	8	8	6	4	4	3	2	1
(74)	−18	61	−25	−50	$\frac{5}{39}$	8	6	4	4	3	2	1
(81^3)	−38	87	−20	−40	11	$\frac{34}{38}$	19	8	13	7	4	1
(821)	57	−153	40	78	−22	−57	$\frac{12}{112}$	6	8	5	3	1
(83)	−19	57	−11	−46	11	19	−46	$\frac{4}{35}$	4	3	2	1
(91^2)	30	−71	20	32	−11	−30	41	−11	$\frac{7}{30}$	4	3	1
(92)	−20	60	−29	−22	11	20	−40	11	−20	$\frac{3}{29}$	2	1
$(10,1)$	−21	53	−11	−32	11	21	−32	11	−21	11	$\frac{2}{21}$	1
(11)	11	−33	11	22	−11	−11	22	−11	11	−11	−11	$\frac{1}{11}$

Table 1·4·12

$w=12$ (i)	h_1^{12}	$h_2 h_1^{10}$	$h_2^2 h_1^8$	$h_2^3 h_1^6$	$h_2^4 h_1^4$	$h_2^5 h_1^2$	h_2^6	$h_3 h_1^9$	$h_3 h_2 h_1^7$	$h_3 h_2^2 h_1^5$
(1^{12})	**479001600** 1	239500800	119750400	59875200	29937600	14968800	7484400	79833600	39916800	19958400
(21^{10})	−11	**121564800** 122	61689600	31298400	15876000	8051400	4082400	41731200	21168000	10735200
($2^2 1^8$)	45	−504	**31772160** 2107	16359840	8421840	4334400	2230200	21772800	11208960	5769120
($2^3 1^6$)	−84	952	−4039	**8550900** 7890	4469040	2335500	1220400	11340000	5927040	3097620
($2^4 1^4$)	70	−805	3480	−6970	**2372256** 6376	1259640	669060	5896800	3129840	1661760
($2^5 1^2$)	−21	246	−1090	2260	−2178	**680070** 812	367560	3061800	1650600	890700
(2^6)	1	−12	55	−120	126	−56	**202410** 7	1587600	869400	477000
(31^9)	10	−111	458	−861	720	−215	10	**15240960** 103	7922880	4112640
(321^7)	−72	808	−3378	6452	−5500	1680	−80	−750	**4184040** 5545	2206680
($32^2 1^5$)	168	−1911	8124	−15846	13872	−4380	216	1767	−13314	**1183000** 32770
($32^3 1^3$)	−140	1620	−7040	14136	−12872	4288	−224	−1480	11418	−29044
($32^4 1$)	30	−355	1590	−3332	3236	−1197	70	315	−2500	6636
($3^2 1^6$)	28	−315	1317	−2504	2106	−624	28	300	−2225	5330
($3^2 2 1^4$)	−105	1200	−5115	9958	−8614	2628	−120	−1140	8654	−21363
($3^2 2^2 1^2$)	90	−1050	4596	−9264	8385	−2694	126	984	−7698	19830
($3^2 2^3$)	−10	120	−546	1164	−1149	426	−20	−108	876	−2388
($3^3 1^3$)	20	−230	984	−1911	1626	−474	20	226	−1734	4299
($3^3 2 1$)	−20	236	−1044	2118	−1902	576	−24	−228	1821	−4788
(3^4)	1	−12	54	−111	99	−27	1	12	−99	270
(41^8)	−9	100	−413	778	−655	200	−10	−93	675	−1584
(421^6)	56	−630	2641	−5066	4362	−1372	72	586	−4326	10364
($42^2 1^4$)	−105	1200	−5130	10088	−8972	2952	−168	−1110	8374	−20633
($42^3 1^2$)	60	−700	3074	−6268	5867	−2086	140	636	−4932	12630
(42^4)	−5	60	−274	592	−609	260	−30	−52	416	−1116
(431^5)	−42	474	−1988	3796	−3224	984	−48	−454	3364	−8037
(4321^3)	120	−1380	5924	−11642	10242	−3264	168	1316	−10024	24810
($432^2 1$)	−60	708	−3144	6468	−6000	2100	−120	−660	5214	−13605
($43^2 1^2$)	−30	348	−1503	2952	−2556	780	−36	−345	2667	−6651
($43^2 2$)	12	−144	651	−1362	1287	−434	20	138	−1122	3024
($4^2 1^4$)	15	−170	716	−1375	1183	−375	21	164	−1214	2890
($4^2 21^2$)	−30	348	−1509	3006	−2709	918	−60	−333	2550	−6338
($4^2 2^2$)	6	−72	327	−696	694	−280	30	66	−528	1400
($4^2 31$)	12	−141	618	−1236	1100	−358	20	141	−1104	2786
(4^3)	−1	12	−54	112	−106	40	−4	−12	96	−248
(51^7)	8	−89	368	−694	584	−176	8	83	−603	1420
(521^5)	−42	474	−1994	3840	−3316	1032	−48	−442	3275	−7901
($52^2 1^3$)	60	−690	2972	−5899	5300	−1740	84	638	−4852	12115
($52^3 1$)	−20	236	−1052	2190	−2110	772	−40	−212	1667	−4371
(531^4)	30	−340	1432	−2745	2334	−699	30	328	−2443	5882
(5321^2)	−60	696	−3018	6000	−5334	1680	−72	−666	5136	−12946
(532^2)	12	−144	654	−1389	1364	−502	20	132	−1065	2878
($53^2 1$)	12	−141	618	−1233	1078	−316	12	141	−1113	2857
(541^3)	−20	228	−966	1866	−1612	504	−24	−222	1654	−3972
(5421)	24	−282	1242	−2522	2322	−788	40	270	−2104	5368
(543)	−6	72	−324	670	−620	204	−8	−72	582	−1540
($5^2 1^2$)	6	−69	295	−575	500	−154	6	68	−512	1246
($5^2 2$)	−3	36	−163	344	−335	122	−4	−34	272	−724
(61^6)	−7	78	−323	610	−514	156	−8	−73	531	−1252
(621^4)	30	−340	1437	−2782	2417	−762	42	318	−2368	5746
($62^2 1^2$)	−30	348	−1515	3048	−2787	942	−60	−321	2469	−6259
(62^3)	4	−48	219	−472	482	−200	20	42	−336	904
(631^3)	−20	228	−966	1862	−1590	480	−24	−222	1666	−4039
(6321)	24	−282	1242	−2516	2284	−732	40	270	−2122	5486
(63^2)	−3	36	−162	334	−303	90	−4	−36	294	−792
(641^2)	12	−138	590	−1150	1003	−322	20	136	−1024	2480
(642)	−6	72	−326	688	−672	256	−24	−68	544	−1440
(651)	−6	70	−304	602	−530	164	−8	−70	538	−1338
(6^2)	1	−12	54	−112	105	−36	3	12	−96	252
(71^5)	6	−67	278	−526	444	−135	6	63	−459	1084
(721^3)	−20	228	−970	1892	−1658	528	−24	−214	1605	−3927
($72^2 1$)	12	−141	624	−1283	1208	−422	20	129	−1009	2621
(731^2)	12	−138	590	−1147	986	−300	12	136	−1033	2532
(732)	−6	72	−326	686	−658	228	−8	−68	550	−1484
(741)	−6	70	−304	602	−532	172	−8	−70	538	−1330
(75)	2	−24	108	−224	210	−71	2	24	−192	504
(81^4)	−5	56	−233	442	−374	114	−6	−53	387	−916
(821^2)	12	−138	593	−1170	1039	−338	20	130	−986	2444
(82^2)	−3	36	−164	352	−356	144	−12	−32	256	−688
(831)	−6	70	−304	600	−520	156	−8	−70	544	−1367
(84)	2	−24	108	−224	211	−76	6	24	−192	500
(91^3)	4	−45	188	−358	304	−92	4	43	−315	748
(921)	−6	70	−306	616	−558	180	−8	−66	511	−1303
(93)	2	−24	108	−223	204	−63	2	24	−195	522
($10, 1^2$)	−3	34	−143	274	−235	74	−4	−33	243	−577
($10, 2$)	2	−24	109	−232	230	−88	6	22	−176	472
($11, 1$)	2	−23	98	−189	160	−47	2	23	−173	417
(12)	−1	12	−54	112	−105	36	−2	−12	96	−252

Table 1·4·12 (*cont.*)

w = 12 (ii)	$h_3h_2^3h_1^3$	$h_3h_2^4h_1$	$h_3^2h_1^6$	$h_3^2h_2h_1^4$	$h_3^2h_2^2h_1^2$	$h_3^2h_2^3$	$h_3^3h_1^3$	$h_3^3h_2h_1$	h_3^4	$h_4h_1^8$	$h_4h_2h_1^6$
(1^{12})	9979200	4989600	13305600	6652800	3326400	1663200	2217600	1108800	369600	19958400	9979200
(21^{10})	5443200	2759400	7257600	3679200	1864800	945000	1260000	638400	218400	10886400	5518800
(2^21^8)	2968560	1527120	3951360	2032800	1045520	537600	715680	367920	129360	5906880	3039120
(2^31^6)	1618740	845820	2147040	1112940	586200	306240	406260	212220	76800	3190320	1667250
(2^41^4)	882576	468888	1164240	618480	328656	174696	230400	122496	45696	1716120	911520
(2^51^2)	481140	260190	630000	340500	184240	99810	130500	70740	27240	919800	496800
(2^6)	262260	144540	340200	187200	103260	57120	73800	40860	16200	491400	270000
(31^9)	2131920	1103760	2862720	1481760	766080	395640	530880	273840	97440	4354560	2252880
(321^7)	1162350	611520	1554000	817600	429660	225540	301500	158130	57960	2338560	1230600
(32^21^5)	633660	339120	841680	450540	240960	128760	171120	91380	34560	1251600	670170
(32^31^3)	**345405** 27013	188256	454860	247920	135114	73626	96960	52833	20652	667800	363960
(32^41)	−6630	**104628** 1889	245280	136220	75744	42174	54840	30552	12360	355320	197160
(3^21^6)	−4506	948	**608320** 926	328240	176720	94940	127200	68220	26160	907200	490560
(3^221^4)	18750	−4098	−3606	**180516** 14613	99116	54332	72108	39492	15672	482160	265570
$(3^22^21^2)$	−18513	4350	3156	−13482	**55576** 13627	31150	40788	22866	9408	255640	143440
(3^22^3)	2466	−732	−339	1521	−1702	**17896** 316	23016	13236	5652	135240	77310
(3^31^3)	−3714	762	762	−3141	2852	−292	**30372** 745	17130	7168	183120	104010
(3^321)	4509	−974	−783	3492	−3693	402	−808	**9929** 1183	4324	96600	55920
(3^4)	−261	48	45	−216	252	−21	56	−105	**2008** 15	36120	21600
(41^8)	1326	−288	−269	1008	−858	96	−195	189	−9	**1421280** 87	753480
(421^6)	−8900	2000	1731	−6646	5823	−678	1302	−1302	63	−548	**407415** 3523
(42^21^4)	18383	−4350	−3340	13254	−12120	1500	−2613	2745	−135	1038	−6848
(42^31^2)	−11910	3072	1926	−7968	7790	−1066	1534	−1758	87	−600	4093
(42^4)	1136	−340	−153	654	−674	100	−116	136	−6	52	−374
(431^5)	6798	−1473	−1404	5385	−4626	513	−1116	1080	−54	424	−2741
(4321^3)	−21906	5004	4188	−16830	15246	−1804	3582	−3686	192	−1214	8098
(432^21)	13006	−3310	−2118	9053	−9146	1250	−1864	2274	−126	606	−4203
(43^21^2)	5777	−1250	−1188	4888	−4323	474	−1182	1143	−63	309	−2100
(43^22)	−2978	748	477	−2156	2328	−318	486	−678	45	−120	852
(4^21^4)	−2446	546	509	−1911	1605	−181	384	−348	15	−161	1048
(4^221^2)	5638	−1346	−1069	4252	−3756	450	−894	834	−36	324	−2200
(4^22^2)	−1376	380	210	−892	880	−112	176	−192	9	−66	476
(4^231)	−2448	562	500	−2066	1776	−204	516	−432	18	−132	922
(4^3)	224	−56	−44	184	−152	16	−48	32	−1	12	−88
(51^7)	−1198	260	239	−901	786	−90	171	−177	9	−78	488
(521^5)	6875	−1557	−1303	5049	−4569	555	−966	1044	−54	415	−2057
(52^21^3)	−11051	2694	1916	−7721	7381	−998	1473	−1689	87	−599	3952
(52^31)	4316	−1234	−633	2669	−2764	482	−478	574	−24	203	−1395
(531^4)	−5041	1086	1021	−3955	3529	−399	802	−860	48	−307	1979
(5321^2)	11758	−2724	−2138	8799	−8468	1064	−1825	2148	−126	612	−4094
(532^2)	−2946	884	417	−1831	1998	−390	342	−432	18	−123	865
(53^21)	−2577	538	499	−2147	2091	−222	519	−633	45	−123	842
(541^3)	3417	−770	−696	2631	−2299	274	−511	514	−24	220	−1432
(5421)	−4988	1252	874	−3578	3424	−472	716	−786	36	−264	1810
(543)	1460	−360	−262	1150	−1140	156	−280	313	−15	66	−470
(5^21^2)	−1095	250	217	−830	760	−97	155	−174	9	−68	444
(5^22)	730	−220	−109	460	−485	99	−80	93	−3	34	−238
(61^6)	1053	−225	−211	801	−696	75	−156	162	−9	69	−432
(621^4)	−5012	1116	945	−3708	3377	−380	728	−812	48	−300	1931
(62^21^2)	5802	−1394	−971	4009	−3958	484	−776	984	−63	303	−2024
(62^3)	−928	280	123	−530	566	−92	92	−120	6	−42	298
(631^3)	3463	−730	−706	2774	−2481	258	−585	642	−42	208	−1353
(6321)	−5128	1156	884	−3788	3888	−432	798	−1122	90	−246	1674
(63^2)	762	−156	−132	612	−684	66	−151	255	−30	30	−210
(641^2)	−2130	472	440	−1688	1455	−156	348	−342	18	−136	900
(642)	1408	−368	−222	948	−971	118	−186	244	−15	68	−484
(651)	1190	−259	−234	930	−874	95	−188	229	−15	70	−468
(6^2)	−240	54	42	−180	189	−18	38	−60	6	−12	84
(71^5)	−913	201	183	−696	597	−70	138	−132	6	−60	376
(721^3)	3459	−810	−643	2547	−2303	294	−512	520	−24	203	−1317
(72^21)	−2528	674	391	−1654	1688	−260	316	−384	18	−123	836
(731^2)	−2182	484	447	−1784	1560	−186	399	−378	18	−127	839
(732)	1492	−408	−224	1001	−1102	182	−203	283	−15	62	−434
(741)	1157	−271	−238	941	−795	99	−213	174	−6	70	−476
(75)	−485	130	84	−355	355	−59	75	−83	3	−24	168
(81^4)	773	−166	−155	591	−513	56	−117	118	−6	51	−320
(821^2)	−2178	504	397	−1602	1475	−176	328	−348	18	−124	815
(82^2)	704	−216	−95	410	−431	78	−74	84	−3	32	−226
(831)	1195	−247	−242	1010	−917	91	−243	253	−15	64	−433
(84)	−464	116	86	−364	332	−40	88	−80	3	−24	172
(91^3)	−637	138	127	−483	429	−48	93	−102	6	−42	264
(921)	1208	−283	−205	859	−858	105	−172	223	−15	63	−421
(93)	−501	114	87	−396	423	−51	97	−138	12	−21	147
$(10,1^2)$	400	−111	−100	375	−321	39	−72	69	−3	33	−210
$(10,2)$	−480	140	67	−290	315	−54	50	−68	3	−22	154
$(11,1)$	−350	71	75	−290	246	−24	62	−59	3	−23	150
(12)	240	−60	−42	180	−180	24	−40	48	−3	12	−84

Table 1·4·12 (cont.)

$w=12$ (iii)	$h_4 h_2^2 h_1^4$	$h_4 h_2^3 h_1^2$	$h_4 h_2^4$	$h_4 h_3 h_1^5$	$h_4 h_3 h_2 h_1^3$	$h_4 h_3 h_2^2 h_1$	$h_4 h_3^2 h_1^2$	$h_4 h_3^2 h_2$	$h_4^2 h_1^4$	$h_4^2 h_2 h_1^2$	$h_4^2 h_2^2$	$h_4^2 h_3 h_1$
(1^{12})	4989600	2494800	1247400	3326400	1663200	831600	554400	277200	831600	415800	207900	138600
(21^{10})	2797200	1417500	718200	1890000	957600	485100	327600	165900	491400	248850	126000	85050
(2^21^8)	1563240	803880	413280	1070160	550200	282800	193480	99400	289380	148680	76370	52220
(2^31^6)	871200	455175	237780	603810	315450	164775	114180	59625	169740	88650	46290	32070
(2^41^4)	484308	257400	136842	339480	180468	95964	67308	35808	99144	52740	28062	19692
(2^51^2)	268620	145410	78810	190200	103020	55870	39020	21530	57660	31305	17020	12085
(2^6)	148680	82080	45450	106200	58680	32520	23280	12960	33390	18540	10335	7410
(31^9)	1164240	601020	309960	806400	415800	214200	147840	76020	224280	115290	59220	40740
(321^7)	646800	339570	178080	454020	238170	124740	87290	45640	131460	68740	35910	25060
(32^21^5)	358540	191655	102360	254660	136040	72610	51480	27435	76660	40870	21770	15410
(32^31^3)	198348	108084	58890	142320	77550	42252	30312	16512	44484	24234	13200	9468
(32^41)	109524	60918	33930	79260	44108	24582	17812	9950	25692	14333	8010	5810
(3^21^6)	264600	142380	76440	190840	102490	54920	39400	21030	59380	31720	16910	12080
(3^221^4)	146040	80175	43940	106350	58336	31941	23210	12671	34340	18776	10246	7432
$(3^22^21^2)$	80468	45128	25298	59060	33128	18574	13640	7644	19764	11084	6212	4504
(3^22^3)	44268	25395	14598	32690	18770	10801	7994	4617	11324	6527	3772	2797
(3^31^3)	58980	33375	18840	43920	24858	14028	10458	5877	15180	8568	4818	3594
(3^321)	32388	18762	10866	24240	14061	8154	6126	3554	8664	5034	2922	2202
(3^4)	12948	7776	4674	9840	5928	3576	2740	1660	3744	2268	1374	1060
(41^8)	398580	210420	110880	284760	149940	77820	55800	29260	86310	45150	23590	16590
(421^6)	219810	118350	63600	159195	85510	45850	32970	17615	50310	26860	14320	10230
(42^21^4)	**121027** / 13791	66537	36528	88640	48647	26663	19427	10019	29140	15922	8690	6302
(42^31^2)	−8664	**37398** / 5908	21018	49170	27609	15503	11421	6410	16782	9408	5276	3876
(42^4)	864	−700	**12126** / 140	27180	15632	9014	6696	3874	9615	5543	3209	2379
(431^5)	5308	−3099	264	**66130** / 2246	36560	20145	14890	8155	22570	12370	6760	4960
(4321^3)	−16376	10092	−920	−6622	**20713** / 20777	11707	8757	4929	12930	7288	4098	3054
(432^21)	9022	−6134	640	3328	−11202	**6804** / 6988	5133	2983	7372	4282	2488	1875
(43^21^2)	4299	−2582	212	1840	−6019	3120	**3941** / 2076	2299	5630	3302	1928	1482
(43^22)	−1874	1286	−112	−706	2510	−1668	−780	**1395** / 498	3182	1930	1172	906
(4^21^4)	−2040	1215	−116	−872	2521	−1258	−673	244	**8512** / 372	4816	2714	2052
(4^221^2)	4550	−2930	328	1818	−5738	3108	1644	−624	−764	**2819** / 1808	1646	1263
(4^22^2)	−1088	848	−156	−368	1280	−848	−336	156	164	−464	**1002** / 207	774
(4^231)	−1952	1228	−120	−828	2800	−1464	−1024	337	328	−868	202	**614** / 595
(4^3)	200	−144	24	80	−288	160	112	−28	−36	112	−42	−84
(51^7)	−917	524	−42	−374	1062	−534	−261	108	140	−272	52	104
(521^5)	5135	−3039	252	2052	−6020	3164	1500	−658	−776	1566	−312	−612
(52^21^3)	−7953	4970	−436	−3035	9355	−5276	−2346	1128	1155	−2480	528	992
(52^31)	2980	−2042	200	1036	−3406	2160	816	−456	−408	952	−224	−376
(531^4)	−3806	2195	−168	−1627	4740	−2413	−1259	532	630	−1246	232	518
(5321^2)	8290	−5088	400	3343	−10509	5854	2913	−1386	−1253	2734	−544	−1232
(532^2)	−1890	1306	−104	−671	2302	−1530	−594	360	254	−607	122	275
(53^21)	−1735	1042	−72	−742	2464	−1350	−836	388	258	−608	112	360
(541^3)	2771	−1630	136	1205	−3426	1736	860	−352	−524	1008	−192	−384
(5421)	−3780	2448	−224	−1494	4756	−2731	−1272	599	632	−1437	312	618
(543)	1020	−660	48	422	−1480	873	527	−253	−160	413	−76	−274
(5^21^2)	−855	495	−35	−381	1065	−546	−249	112	170	−302	50	100
(5^22)	510	−340	22	194	−630	394	147	−81	−85	187	−28	−71
(61^6)	817	−474	42	327	−942	479	236	−98	−120	242	−52	−94
(621^4)	−3778	2296	−220	−1473	4401	−2370	−1109	502	549	−1154	272	448
(62^21^2)	4172	−2750	308	1519	−4806	2870	1182	−630	−575	1298	−368	−484
(62^3)	−676	532	−96	−208	712	−496	−156	96	89	−236	102	76
(631^3)	2639	−1558	136	1115	−3317	1722	903	−390	−427	882	−192	−368
(6321)	−3492	2280	−224	−1342	4367	−2667	−1165	677	491	−1117	308	446
(63^2)	450	−294	24	180	−627	405	198	−135	−57	135	−36	−63
(641^2)	−1782	1087	−114	−766	2241	−1146	−593	233	338	−695	164	273
(642)	1080	−810	132	384	−1292	856	314	−184	−171	440	−180	−152
(651)	930	−561	48	408	−1193	635	298	−143	−183	349	−72	−118
(6^2)	−180	126	−18	−72	228	−144	−54	36	33	−72	27	18
(71^5)	−712	408	−32	−285	834	−419	−221	86	102	−212	40	94
(721^3)	2601	−1570	128	1008	−3099	1666	839	−358	−365	802	−160	−368
(72^21)	−1764	1180	−104	−620	2049	−1289	−525	295	229	−536	122	235
(731^2)	−1663	970	−70	−708	2203	−1122	−680	260	261	−584	104	312
(732)	942	−640	44	343	−1191	809	326	−223	−119	281	−56	−134
(741)	971	−589	48	419	−1313	663	415	−139	−181	415	−76	−226
(75)	−355	225	−12	−149	485	−284	−142	71	65	−142	18	71
(81^4)	607	−352	32	243	−708	356	186	−72	−88	184	−40	−80
(821^2)	−1640	1023	−106	−623	1959	−1065	−543	224	226	−528	132	249
(82^2)	508	−386	60	160	−556	376	138	−72	−65	176	−66	−76
(831)	885	−533	48	373	−1227	640	418	−158	−131	328	−76	−202
(84)	−380	264	−36	−152	528	−304	−184	68	66	−184	58	116
(91^3)	−499	288	−22	−203	580	−300	−142	64	76	−148	28	56
(921)	864	−549	44	320	−1031	612	268	−146	−115	262	−56	−110
(93)	−315	201	−12	−126	447	−279	−153	90	39	−99	18	63
$(10,1^2)$	400	−234	22	167	−472	237	115	−46	−67	130	−28	−47
$(10,2)$	−340	250	−32	−112	380	−264	−82	56	45	−112	38	36
$(11,1)$	−290	164	−12	−127	372	−177	−105	36	52	−105	18	47
(12)	180	−120	12	72	−240	144	72	−36	−30	72	−18	−36

Table 1·4·12 (cont.)

$w=12$ (iv)	h_4^3	$h_5h_1^7$	$h_5h_2h_1^5$	$h_5h_2^2h_1^3$	$h_5h_2^3h_1$	$h_5h_3h_1^4$	$h_5h_3h_2h_1^2$	$h_5h_3h_2^2$	$h_5h_3^2h_1$	$h_5h_4h_1^3$	$h_5h_4h_2h_1$	$h_5h_4h_3$	$h_5^2h_1^2$
(1^{12})	34650	3991680	1995840	997920	498960	665280	332640	166320	110880	166320	83160	27720	33264
$(2 1^{10})$	22050	2298240	1164240	589680	298620	398160	201600	102060	68880	103320	52290	17850	21672
$(2^2 1^8)$	14070	1310400	673680	346248	177912	236880	121688	62496	42728	63924	32816	11508	14084
$(2^3 1^6)$	9000	740880	387090	202212	105615	140130	73182	38211	26472	39366	20550	7428	9120
$(2^4 1^4)$	5769	415800	221040	117540	62520	82440	43860	23340	16380	24120	12840	4800	5880
$(2^5 1^2)$	3705	231840	125520	68040	36930	48240	26200	14250	10120	14700	8005	3105	3772
(2^6)	2385	128520	70920	39240	21780	28080	15600	8700	6240	8910	4980	2010	2406
$(3 1^9)$	11130	1028160	529200	272100	139860	187488	96264	49392	33936	51408	26334	9240	11592
$(3 2 1^7)$	7140	575400	301560	157878	82572	110404	57687	30114	21000	31033	16485	5971	7518
$(3 2^2 1^5)$	4590	320040	170950	91236	48651	64672	34464	18350	12984	19350	10291	3863	4848
$(3 2^3 1^3)$	2955	177030	96450	52545	28623	37686	20529	11181	8019	11766	6408	2502	3108
$(3 2^4 1)$	1905	97440	54180	30168	16824	21848	12192	6816	4944	7112	3081	1622	1980
$(3^2 1^6)$	3660	243000	131040	70308	37632	50536	26999	14384	10276	15550	8258	3112	4016
$(3^2 2 1^4)$	2364	133840	73530	40320	22067	29334	16036	8749	6346	9424	5132	2018	2568
$(3^2 2^2 1^2)$	1527	73220	41100	23062	12934	16940	9502	5326	3914	5676	3182	1310	1632
$(3^2 2^3)$	987	39900	22890	13158	7581	9734	5614	3247	2408	3398	1969	851	1030
$(3^3 1^3)$	1230	54600	30990	17544	9903	13092	7386	4152	3096	4524	2541	1059	1344
$(3^3 2 1)$	795	29610	17190	9981	5793	7494	4356	2529	1907	2697	1569	689	846
(3^4)	415	11760	7080	4272	2580	3264	1980	1200	928	1260	768	364	432
$(4 1^8)$	4830	393120	205800	107604	56196	75768	39508	20580	14420	22134	11494	4158	5446
$(4 2 1^6)$	3120	215040	115095	61506	32820	44059	23470	12486	8898	13534	7177	2696	3526
$(4 2^2 1^4)$	2019	117180	64150	35079	19161	25492	13907	7579	5487	8214	4466	1750	2264
$(4 2^3 1^2)$	1308	63630	35640	19965	11187	14676	8219	4605	3379	4950	2771	1137	1442
$(4 2^4)$	849	34440	19740	11340	6534	8408	4844	2802	2076	2963	1715	739	911
$(4 3 1^5)$	1620	86940	48185	26598	14628	19720	10806	5903	4332	6603	3586	1414	1888
$(4 3 2 1^3)$	1053	47005	26675	15099	8525	11311	6374	3583	2670	3958	2218	920	1196
$(4 3 2^2 1)$	684	25340	14725	8556	4972	6456	3750	2179	1642	2357	1369	599	752
$(4 3^2 1^2)$	555	18550	10940	6435	3773	4942	2897	1691	1299	1868	1090	486	618
$(4 3^2 2)$	360	9940	6005	3633	2201	2797	1697	1031	797	1100	670	317	384
$(4^2 1^4)$	729	29190	16870	9684	5526	7474	4238	2392	1814	2780	1554	648	892
$(4^2 2 1^2)$	477	15610	9250	5460	3212	4238	2484	1452	1116	1642	955	423	556
$(4^2 2^2)$	312	8330	5060	3074	1870	2392	1452	884	684	964	586	276	344
$(4^2 3 1)$	255	6020	3720	2294	1411	1814	1116	684	542	756	464	225	280
(4^3)	120 / _20_	1890	1230	801	522	648	423	276	225	297	195	105	123
$(5 1^7)$	−8	_130872_ / _74_	70056	37422	19950	27076	14399	7644	5488	8533	4508	1694	2324
$(5 2 1^5)$	48	−393	_38491_ / _2149_	21102	11544	15541	8483	4621	3380	5172	2805	1103	1498
$(5 2^2 1^3)$	−80	567	−3218	_11886_ / _5090_	6687	8874	4986	2798	2080	3108	1739	719	956
$(5 2^3 1)$	32	−191	1142	−1974	_3879_ / _920_	5042	2923	1698	1277	1853	1075	469	604
$(5 3 1^4)$	−40	289	−1598	2375	−801	_6794_ / _1290_	3845	2166	1640	2493	1395	583	800
$(5 3 2 1^2)$	96	−580	3351	−5329	1970	−2676	_6162_ / _2251_	1314	1009	1476	859	381	502
$(5 3 2^2)$	−20	119	−739	1340	−666	545	−1444	_566_ / _800_	618	868	528	249	312
$(5 3^2 1)$	−28	115	−678	1085	−358	590	−1447	278	_490_ / _453_	683	419	203	256
$(5 4 1^3)$	32	−204	1147	−1727	615	−958	1919	−409	−383	_1035_ / _826_	602	270	378
$(5 4 2 1)$	−56	240	−1426	2360	−971	1150	−2664	695	572	−960	_1422_ / _368_	177	232
$(5 4 3)$	24	−58	358	−620	255	−310	814	−215	−253	227	−400	_202_ / _95_	114
$(5^2 1^2)$	−6	68	−391	600	−215	340	−694	159	133	−306	336	−71	_160_ / 136
$(5^2 2)$	4	−34	214	−390	198	−170	434	−176	−71	153	−247	59	−68
$(6 1^6)$	8	−66	346	−492	161	−249	495	−94	−100	169	−200	48	−53
$(6 2 1^4)$	−40	286	−1551	2295	−786	1135	−2361	470	490	−788	980	−240	255
$(6 2^2 1^2)$	48	−289	1637	−2572	954	−1187	2658	−578	−560	839	−1141	281	−280
$(6 2^3)$	−12	38	−228	396	−184	150	−360	100	64	−120	192	−40	35
$(6 3 1^3)$	32	−196	1079	−1585	513	−875	1797	−327	−409	635	−762	199	−216
$(6 3 2 1)$	−40	238	−1382	2191	−755	1106	−2573	505	632	−760	1046	−292	277
$(6 3^2)$	4	−30	180	−291	87	−159	405	−63	−135	93	−135	51	−36
$(6 4 1^2)$	−28	124	−698	1043	−358	592	−1171	214	248	−516	603	−139	182
$(6 4 2)$	24	−60	360	−598	236	−286	656	−128	−140	244	−368	80	−77
$(6 5 1)$	8	−70	408	−629	211	−366	763	−142	−169	331	−373	83	−148
(6^2)	−2	12	−72	114	−36	66	−144	18	36	−60	72	−12	27
$(7 1^5)$	−8	58	−304	438	−151	213	−435	94	88	−139	176	−48	41
$(7 2 1^3)$	32	−195	1067	−1623	605	−758	1646	−397	−340	508	−694	199	−155
$(7 2^2 1)$	−20	119	−692	1153	−501	476	−1145	351	224	−335	514	−150	110
$(7 3 1^2)$	−28	119	−665	1011	−360	534	−1162	262	275	−369	504	−166	116
$(7 3 2)$	8	−62	377	−657	295	−286	741	−251	−161	195	−316	110	−76
$(7 4 1)$	24	−62	359	−568	223	−302	653	−158	−150	259	−363	115	−82
$(7 5)$	−4	24	−149	260	−118	130	−319	106	71	−118	177	−59	53
$(8 1^4)$	8	−50	262	−375	123	−185	379	−73	−81	118	−148	41	−34
$(8 2 1^2)$	−28	120	−663	1017	−365	471	−1057	227	240	−300	429	−131	86
$(8 2^2)$	12	−30	180	−318	156	−114	288	−100	−52	84	−144	40	−23
$(8 3 1)$	24	−60	341	−524	166	−281	657	−115	−193	170	−257	107	−47
$(8 4)$	−20	20	−120	200	−80	100	−240	52	68	−80	136	−56	18
$(9 1^3)$	−4	42	−222	319	−107	161	−323	65	65	−110	132	−33	34
$(9 2 1)$	8	−63	358	−569	212	−257	605	−139	−137	167	−251	75	−52
$(9 3)$	−4	21	−126	210	−75	105	−279	63	90	−57	99	−51	18
$(10, 1^2)$	4	−33	177	−256	89	−134	260	−56	−47	101	−114	24	−34
$(10, 2)$	−4	22	−132	230	−108	90	−224	76	36	−68	112	−24	23
$(11, 1)$	−4	23	−127	186	−59	104	−210	36	47	−81	94	−24	29
(12)	4	−12	72	−120	48	−60	144	−36	−36	48	−72	24	−18

Table 1·4·12 (cont.)

$w=12$ (v)	$h_5^2 h_2$	$h_6 h_1^6$	$h_6 h_2 h_1^4$	$h_6 h_2^2 h_1^2$	$h_6 h_2^3$	$h_6 h_3 h_1^3$	$h_6 h_3 h_2 h_1$	$h_6 h_3^2$	$h_6 h_4 h_1^2$	$h_6 h_4 h_2$	$h_6 h_5 h_1$	h_6^2	$h_7 h_1^5$	$h_7 h_2 h_1^3$
(1^{12})	16632	665280	332640	166320	83160	110880	55440	18480	27720	13860	5544	924	95040	47520
(21^{10})	10962	408240	206640	104580	52920	70560	35700	12180	18270	9240	3822	672	62640	31680
$(2^2 1^8)$	7224	246960	126840	65128	33432	44520	22848	8008	11984	6146	2632	490	40560	20808
$(2^3 1^6)$	4758	147510	77040	40227	21000	27870	14547	5257	7821	4080	1809	358	25830	13482
$(2^4 1^4)$	3132	87120	46356	24672	13134	17316	9220	3448	5076	2704	1240	262	16200	8628
$(2^5 1^2)$	2061	50940	27660	15040	8190	10680	5820	2260	3275	1790	847	192	10020	5460
(2^6)	1356	29520	16380	9120	5100	6540	3660	1480	2100	1185	576	141	6120	3420
(31^9)	5922	206640	105840	54180	27720	37128	18984	6636	10038	5124	2226	420	36000	18360
(321^7)	3906	121800	63560	33138	17262	23079	12012	4340	6538	3395	1533	308	22620	11753
$(32^2 1^5)$	2572	71070	37860	20151	10716	14252	7574	2839	4235	2246	1052	226	14000	7436
$(32^3 1^3)$	1692	41100	22386	12192	6639	8741	4760	1858	2726	1484	719	166	8550	4656
$(32^4 1)$	1113	23580	13148	7344	4110	5324	2982	1216	1743	980	489	122	5160	2888
$(3^2 1^6)$	2120	57200	30500	16288	8662	11646	6180	2326	3530	1863	896	196	11960	6338
$(3^2 2 1^4)$	1392	32750	17928	9793	5338	7112	3871	1521	2268	1229	612	144	7210	3928
$(3^2 2^2 1^2)$	914	18620	10454	5866	3289	4312	2418	996	1446	810	416	106	4300	2414
$(3^2 2^3)$	601	10520	6062	3502	2029	2596	1506	652	915	534	281	78	2540	1472
$(3^3 1^3)$	750	14550	8244	4653	2616	3480	1953	813	1200	669	354	92	3540	1998
$(3^3 2 1)$	492	8160	4752	2766	1608	2085	1214	534	756	440	239	68	2070	1209
(3^4)	264	3480	2112	1284	780	988	604	288	388	238	136	44	960	588
(41^8)	2814	93240	48300	25004	12936	17388	8988	3220	4914	2534	1162	238	19080	9804
(421^6)	1860	53265	28250	14966	7920	10614	5610	2093	3174	1672	801	176	11565	6086
$(42^2 1^4)$	1226	30210	16431	8929	4848	6443	3495	1363	2037	1102	549	130	6920	3743
$(42^3 1^2)$	807	17025	9507	5312	2970	3887	2173	889	1298	726	374	96	4095	2283
(42^4)	531	9540	5474	3152	1824	2330	1348	580	821	479	253	71	2400	1382
(431^5)	1016	23525	12900	7048	3838	5180	2811	1107	1693	911	471	114	5720	3098
$(432 1^3)$	666	13160	7424	4177	2344	3116	1745	723	1076	599	320	84	3345	1874
$(432^2 1)$	437	7325	4253	2470	1435	1861	1081	473	678	394	216	62	1940	1126
$(43^2 1^2)$	358	5580	3286	1927	1125	1484	865	385	560	324	184	54	1550	910
$(43^2 2)$	235	3075	1866	1134	690	876	534	253	348	213	123	40	885	540
$(4^2 1^4)$	492	8930	5092	2886	1627	2206	1236	518	800	442	250	68	2510	1406
$(4^2 2 1^2)$	321	4910	2890	1695	991	1310	763	339	502	290	168	50	1430	834
$(4^2 2^2)$	210	2690	1634	994	607	772	470	222	312	191	112	37	810	492
$(4^3 1)$	171	2010	1244	767	471	610	374	181	256	156	95	32	630	390
(4^3)	81	690	453	297	195	243	159	85	114	75	48	19	240	159
(51^7)	1218	37632	19796	10402	5460	7371	3864	1428	2198	1148	560	126	9276	4817
(521^5)	806	20841	11286	6102	3294	4421	2383	925	1407	755	386	94	5411	2897
$(52^2 1^3)$	532	11490	6411	3575	1992	2636	1468	601	893	496	264	70	3126	1731
$(52^3 1)$	351	6309	3629	2092	1209	1562	903	391	562	326	179	52	1791	1028
(531^4)	442	8645	4896	2758	1546	2089	1166	485	741	409	228	62	2486	1387
$(532 1^2)$	290	4724	2762	1611	937	1234	717	317	464	268	154	46	1411	819
(532^2)	191	2573	1553	940	571	724	440	207	288	176	103	34	796	481
$(53^2 1)$	156	1908	1174	721	441	569	349	169	236	144	88	30	616	379
(541^3)	216	3014	1802	1068	628	850	496	224	344	197	122	38	991	581
(5421)	141	1628	1007	621	382	496	304	147	212	129	81	28	551	338
(543)	75	642	420	275	180	224	147	79	105	69	45	18	231	152
$(5^2 1^2)$	96	926	590	372	232	310	192	96	144	87	60	22	352	220
$(5^2 2)$	**63**/77	494	326	215	142	178	117	63	87	57	39	16	192	126
(61^6)	24	**13327**/64	7186	3865	2074	2820	1509	583	909	483	255	64	4051	2140
(621^4)	-120	-276	**4011**/1246	2233	1240	1666	923	378	576	317	175	48	2276	1250
$(62^2 1^2)$	142	279	-1338	**1290**/1584	745	977	564	246	361	208	119	36	1271	727
(62^3)	-22	-38	198	-276	**451**/84	569	344	160	224	137	80	27	706	421
(631^3)	93	186	-858	909	-120	**766**/675	444	198	299	171	103	32	971	565
(6321)	-127	-228	1137	-1381	200	-875	**271**/1460	130	184	112	69	24	221	326
(63^2)	12	30	-159	207	-24	138	-279	**70**/84	91	60	39	16	221	144
(641^2)	-76	-114	543	-593	98	-451	545	-72	**136**/383	82	55	20	351	219
(642)	34	60	-318	430	-108	244	-408	60	-210	**54**/220	36	15	191	125
(651)	59	60	-301	350	-48	266	-375	60	-231	120	**27**/196	12	112	76
(6^2)	-6	-12	66	-90	18	-60	108	-24	54	-48	-48	**7**/21	32	24
(71^5)	-24	-57	240	-234	30	-156	180	-21	90	-42	-42	6	**1546**/57	841
(721^3)	103	190	-842	873	-120	561	-695	84	-335	168	164	-24	-190	**480**/680
$(72^2 1)$	-91	-114	543	-633	100	-348	510	-63	215	-128	-118	18	114	-450
(731^2)	-76	-114	516	-518	62	-406	473	-63	257	-114	-129	18	114	-423
(732)	69	57	-285	345	-44	208	-342	51	-117	68	83	-12	-57	242
(741)	59	57	-267	274	-40	224	-233	24	-189	80	94	-12	-57	224
(75)	-47	-19	95	-107	12	-83	107	-12	71	-24	-59	6	19	-83
(81^4)	17	50	-212	213	-30	135	-166	21	-76	42	35	-6	-50	162
(821^2)	-51	-120	543	-594	98	-351	475	-63	204	-134	-94	18	120	-423
(82^2)	22	30	-150	198	-56	84	-128	-62	60	24	-6	-30	120	
(811)	24	60	-281	303	-40	221	-303	51	-126	80	59	-12	-60	221
(84)	-12	-20	100	-120	28	-80	104	-12	68	-56	-24	6	20	-80
(91^3)	-17	-42	180	-181	22	-119	150	-21	68	-34	-35	6	42	-138
(921)	34	63	-295	337	-44	194	-305	51	-104	68	59	-12	-63	232
(93)	-12	-21	105	-126	12	-84	153	-39	36	-24	-24	6	21	-84
$(10,1^2)$	17	33	-144	145	-22	101	-114	12	-68	34	35	-6	-33	111
$(10,2)$	-22	-22	110	-142	32	-68	112	-12	46	-44	-24	6	22	-88
$(11,1)$	-12	-23	104	-105	12	-81	94	-12	58	-24	-35	6	23	-81
(12)	12	12	-60	72	-12	48	-72	12	-36	24	24	-6	-12	48

Table 1·4·12 (*cont.*)

$w=12$ (vi)	$h_7h_2^2h_1$	$h_7h_3h_1^2$	$h_7h_3h_2$	$h_7h_4h_1$	h_7h_5	$h_8h_1^4$	$h_8h_2h_1^2$	$h_8h_2^2$	$h_8h_3h_1$	h_8h_4	$h_9h_1^3$	$h_9h_2h_1$	h_9h_3	$h_{10}h_1^2$	$h_{10}h_2$	$h_{11}h_1$	h_{12}
(1^{12})	23760	15840	7920	3960	792	11880	5940	2970	1980	495	1320	660	220	132	66	12	1
(21^{10})	16020	10800	5460	2790	582	8460	4275	2160	1455	375	1020	515	175	111	56	11	1
(2^21^8)	10672	7288	3736	1956	428	5910	3028	1551	1058	283	774	396	138	92	47	10	1
(2^31^6)	7035	4872	2541	1365	315	4050	2112	1101	762	213	576	300	108	75	39	9	1
(2^41^4)	4596	3228	1720	948	232	2724	1452	774	544	160	420	224	84	60	32	8	1
(2^51^2)	2980	2120	1160	655	171	1800	985	540	385	120	300	165	65	47	26	7	1
(2^6)	1920	1380	780	450	126	1170	660	375	270	90	210	120	50	36	21	6	1
(31^9)	9360	6384	3252	1704	372	5508	2799	1422	966	255	748	379	130	91	46	10	1
(321^7)	6102	4232	2194	1185	274	3734	1932	999	690	191	553	285	101	74	38	9	1
(32^21^5)	3946	2784	1475	821	202	2482	1314	695	489	143	400	211	78	59	31	8	1
(32^31^3)	2535	1817	989	566	149	1620	882	480	344	107	283	154	60	46	25	7	1
(32^41)	1620	1176	662	388	110	1040	585	330	240	80	196	111	46	35	20	6	1
(3^21^6)	3352	2380	1254	708	176	2246	1180	619	436	127	380	198	72	58	30	8	1
(3^221^4)	2135	1546	837	488	130	1446	783	423	305	95	266	143	55	45	24	7	1
$(3^22^21^2)$	1354	996	558	334	96	916	514	288	212	71	182	102	42	34	19	6	1
(3^22^3)	856	636	372	227	71	572	334	196	146	53	122	72	32	25	15	5	1
(3^31^3)	1122	840	468	288	84	798	447	249	186	63	168	93	38	33	18	6	1
(3^321)	705	534	311	195	62	492	288	168	128	47	111	65	29	24	14	5	1
(3^4)	360	280	172	112	40	252	156	96	76	31	64	40	20	16	10	4	1
(41^8)	5036	3468	1780	954	218	3393	1733	885	605	163	529	269	93	73	37	9	1
(421^6)	3200	2250	1181	657	161	2219	1160	606	423	121	379	197	71	58	30	8	1
(42^21^4)	2023	1451	783	451	119	1423	766	412	294	90	265	142	54	45	24	7	1
(42^31^2)	1274	929	519	308	88	897	500	279	203	67	181	101	41	34	19	6	1
(42^4)	800	590	344	209	65	557	323	189	139	50	121	71	31	25	15	5	1
(431^5)	1673	1216	653	386	104	1255	671	358	257	79	248	131	49	44	23	7	1
(4321^3)	1047	777	432	264	77	779	433	240	177	59	167	92	37	33	18	6	1
(432^21)	654	492	286	179	57	477	277	161	121	44	110	64	28	24	14	5	1
(43^21^2)	531	410	237	154	50	403	235	136	105	39	99	57	25	23	13	5	1
(43^22)	330	256	157	103	37	241	148	91	71	29	63	39	19	16	10	4	1
(4^21^4)	784	590	326	206	62	648	356	195	144	48	152	82	32	32	17	6	1
(4^221^2)	485	372	215	140	46	388	224	129	98	36	98	56	24	23	13	5	1
(4^22^2)	300	232	142	94	34	230	140	86	66	27	62	38	18	16	10	4	1
(4^231)	240	192	117	81	30	188	116	71	57	24	54	33	16	15	9	4	1
(4^3)	105	87	57	42	18	81	54	36	30	15	27	18	10	9	6	3	1
(51^7)	2500	1744	904	498	120	1961	1000	519	358	99	358	183	64	57	29	8	1
(521^5)	1549	1108	591	340	89	1231	653	346	245	73	247	130	48	44	23	7	1
(52^21^3)	958	700	387	231	66	759	418	230	167	54	166	91	36	33	18	6	1
(52^31)	592	439	254	156	49	461	265	153	113	40	109	63	27	24	14	5	1
(531^4)	770	576	317	107	58	647	355	194	143	47	152	82	32	32	17	6	1
(5321^2)	474	361	208	133	43	387	223	128	97	35	98	56	24	23	13	5	1
(532^2)	292	224	137	89	32	229	139	85	65	26	62	38	18	16	10	4	1
(53^21)	232	184	112	76	28	187	115	70	56	23	54	33	16	15	9	4	1
(541^3)	338	265	152	103	35	304	174	99	76	28	86	48	20	22	12	5	1
(5421)	207	164	100	69	26	176	107	65	51	21	53	32	15	15	9	4	1
(543)	100	82	54	39	17	80	53	35	29	14	27	18	10	9	6	3	1
(5^21^2)	136	112	68	50	20	128	78	47	38	16	44	26	12	14	8	4	1
(5^22)	83	68	45	33	15	72	47	31	25	12	26	17	9	9	6	3	1
(61^6)	1129	804	423	243	63	1045	544	283	198	57	229	118	42	43	22	7	1
(621^4)	685	502	274	165	47	627	340	184	133	42	151	81	31	32	17	6	1
(62^21^2)	416	311	178	111	35	371	211	120	89	31	97	55	23	23	13	5	1
(62^3)	253	191	116	74	26	217	130	79	59	23	61	37	17	16	10	4	1
(631^3)	326	253	144	95	31	303	173	98	75	27	86	48	20	22	12	5	1
(6321)	198	155	94	63	23	175	106	64	50	20	53	32	15	15	9	4	1
(63^2)	94	76	50	35	15	79	52	34	28	13	27	18	10	9	6	3	1
(641^2)	135	111	67	49	19	128	78	47	38	16	44	26	12	14	8	4	1
(642)	82	67	44	32	14	72	47	31	25	12	26	17	9	9	6	3	1
(651)	51	44	29	23	11	48	32	21	18	9	20	13	7	8	5	3	1
(6^2)	18	16	12	10	6	16	12	9	8	5	8	6	4	4	3	2	1
(71^5)	456	336	181	111	32	501	266	141	101	31	136	71	26	31	16	6	1
(721^3)	273	207	117	75	24	287	161	90	67	23	85	47	19	22	12	5	1
(72^21)	$\dfrac{164}{366}$	126	76	50	18	163	97	58	44	17	52	31	14	15	9	4	1
(731^2)	263	$\dfrac{102}{329}$	61	43	16	127	77	46	37	15	44	26	12	14	8	4	1
(732)	−216	−165	$\dfrac{40}{172}$	28	12	71	46	30	24	11	26	17	9	9	6	3	1
(741)	−148	−189	83	$\dfrac{22}{154}$	10	48	32	21	18	9	20	13	7	8	5	3	1
(75)	71	71	−59	−59	$\dfrac{6}{47}$	16	12	9	8	5	8	6	4	4	3	2	1
(81^4)	−93	−93	43	43	−12	$\dfrac{209}{50}$	115	63	47	16	73	39	15	21	11	5	1
(821^2)	267	255	−129	−126	36	−120	$\dfrac{68}{327}$	40	31	12	43	25	11	14	8	4	1
(82^2)	−100	−62	44	40	−12	30	−98	$\dfrac{26}{56}$	20	9	25	16	8	9	6	3	1
(831)	−126	−177	75	91	−24	60	−177	40	$\dfrac{17}{142}$	8	20	13	7	8	5	3	1
(84)	52	68	−24	−56	12	−20	68	−28	−56	$\dfrac{5}{44}$	8	6	4	4	3	2	1
(91^3)	85	77	−43	−35	12	−42	96	−22	−44	12	$\dfrac{34}{42}$	19	8	13	7	4	1
(921)	−168	−131	95	59	−24	63	−169	44	86	−24	−63	$\dfrac{12}{124}$	6	8	5	3	1
(93)	63	63	−51	−24	12	−21	63	−12	−51	12	21	−51	$\dfrac{4}{39}$	4	3		1
$(10,1^2)$	−67	−68	34	35	−12	33	−78	22	35	−12	−33	45	−12	$\dfrac{7}{33}$	4	3	1
$(10,2)$	76	46	−44	−24	12	−22	66	−32	−24	12	22	−44	12	−22	$\dfrac{3}{32}$	2	1
$(11,1)$	47	58	−24	−35	12	−23	58	−12	−35	12	23	−35	12	−23	12	$\dfrac{2}{23}$	1
(12)	−36	−36	24	24	−12	12	−36	12	24	−12	−12	24	−12	12	−12	−12	$\dfrac{1}{12}$

Note that the final digit (or digits) in the subheadings indicates the weight, w.

To express a product of a-functions in terms of s-functions, the appropriate coefficients are found by reading vertically downwards up to and including the diagonal figure in bold type, and dividing the coefficient by w! E.g. from Table 1·5·5

$$a_3 a_2 \, 5! = 10(1)^5 - 40(2)(1)^3 + 30(2)^2(1) + 20(3)(1)^2 - 20(3)(2).$$

To express a product of s-functions in terms of a-functions, the appropriate coefficients are found by reading horizontally up to and including the diagonal figure in bold type, thus

$$(3)(2) = a_1^5 - 5a_2 a_1^3 + 6a_2^2 a_1 + 3a_3 a_1^2 - 6a_3 a_2.$$

The two diagonal figures are separated by a bold horizontal rule.

Table 1·5·2

$w=2$	a_1^2	a_2
$(1)^2$	$\frac{2}{1}$	1
(2)	1	$\frac{-1}{-2}$

Table 1·5·3

$w=3$	a_1^3	a_2a_1	a_3
$(1)^3$	$\frac{6}{1}$	3	1
$(2)(1)$	1	$\frac{-3}{-2}$	-3
(3)	1	-3	$\frac{2}{3}$

Table 1·5·4

$w=4$	a_1^4	$a_2a_1^2$	a_2^2	a_3a_1	a_4
$(1)^4$	$\frac{24}{1}$	12	6	4	1
$(2)(1)^2$	1	$\frac{-12}{-2}$	-12	-12	-6
$(2)^2$	1	-4	$\frac{6}{4}$	$\cdot$	3
$(3)(1)$	1	-3	$\cdot$	$\frac{8}{3}$	8
(4)	1	-4	2	4	$\frac{-6}{-4}$

Table 1·5·5

$w=5$	a_1^5	$a_2a_1^3$	$a_2^2a_1$	$a_3a_1^2$	a_3a_2	a_4a_1	a_5
$(1)^5$	$\frac{120}{1}$	60	30	20	10	5	1
$(2)(1)^3$	1	$\frac{-60}{-2}$	-60	-60	-40	-30	-10
$(2)^2(1)$	1	-4	$\frac{30}{4}$	$\cdot$	30	15	15
$(3)(1)^2$	1	-3	$\cdot$	$\frac{40}{3}$	20	40	20
$(3)(2)$	1	-5	6	3	$\frac{-20}{-6}$	$\cdot$	-20
$(4)(1)$	1	-4	2	4	$\cdot$	$\frac{-30}{-4}$	-30
(5)	1	-5	5	5	-5	-5	$\frac{24}{5}$

Table 1·5·6

$w=6$	a_1^6	$a_2a_1^4$	$a_2^2a_1^2$	a_2^3	$a_3a_1^3$	$a_3a_2a_1$	a_3^2	$a_4a_1^2$	a_4a_2	a_5a_1	a_6
$(1)^6$	$\frac{720}{1}$	360	180	90	120	60	20	30	15	6	1
$(2)(1)^4$	1	$\frac{-360}{-2}$	-360	-270	-360	-240	-120	-180	-105	-60	-15
$(2)^2(1)^2$	1	-4	$\frac{180}{4}$	270	$\cdot$	180	180	90	135	90	45
$(2)^3$	1	-6	12	$\frac{-90}{-8}$	$\cdot$	$\cdot$	$\cdot$	$\cdot$	-45	$\cdot$	-15
$(3)(1)^3$	1	-3	$\cdot$	$\cdot$	$\frac{240}{3}$	120	80	240	120	120	40
$(3)(2)(1)$	1	-5	6	$\cdot$	3	$\frac{-120}{-6}$	-240	$\cdot$	-120	-120	-120
$(3)^2$	1	-6	9	$\cdot$	6	-18	$\frac{80}{9}$	$\cdot$	$\cdot$	$\cdot$	40
$(4)(1)^2$	1	-4	2	$\cdot$	4	$\cdot$	$\cdot$	$\frac{-180}{-4}$	-90	-180	-90
$(4)(2)$	1	-6	10	-4	4	-8	$\cdot$	-4	$\frac{90}{8}$	$\cdot$	90
$(5)(1)$	1	-5	5	$\cdot$	5	-5	$\cdot$	-5	$\cdot$	$\frac{144}{5}$	144
(6)	1	-6	9	-2	6	-12	3	-6	6	6	$\frac{-120}{-6}$

Table 1·5·7

$w=7$	a_1^7	$a_2a_1^5$	$a_2^2a_1^3$	$a_2^3a_1$	$a_3a_1^4$	$a_3a_2a_1^2$	$a_3a_2^2$	$a_3^2a_1$	$a_4a_1^3$	$a_4a_2a_1$	a_4a_3	$a_5a_1^2$	a_5a_2	a_6a_1	a_7
$(1)^7$	$\frac{5040}{1}$	2520	1260	630	840	420	210	140	210	105	35	42	21	7	1
$(2)(1)^5$	1	$\frac{-2520}{-2}$	-2520	-1890	-2520	-1680	-1050	-840	-1260	-735	-315	-420	-231	-105	-21
$(2)^2(1)^3$	1	-4	$\frac{1260}{4}$	1890	$\cdot$	1260	1470	1260	630	945	735	630	525	315	105
$(2)^3(1)$	1	-6	12	$\frac{-630}{-8}$	$\cdot$	$\cdot$	-630	$\cdot$	$\cdot$	-315	-315	$\cdot$	-315	-105	-105
$(3)(1)^4$	1	-3	$\cdot$	$\cdot$	$\frac{1680}{3}$	840	420	560	1680	840	350	840	420	280	70
$(3)(2)(1)^2$	1	-5	6	$\cdot$	3	$\frac{-840}{-6}$	-840	-1680	$\cdot$	-840	-1260	-840	-840	-840	-420
$(3)(2)^2$	1	-7	16	-12	3	-12	$\frac{420}{12}$	$\cdot$	$\cdot$	$\cdot$	210	$\cdot$	420	$\cdot$	210
$(3)^2(1)$	1	-6	9	$\cdot$	6	-18	$\cdot$	$\frac{560}{9}$	$\cdot$	$\cdot$	560	$\cdot$	$\cdot$	280	280
$(4)(1)^3$	1	-4	2	$\cdot$	4	$\cdot$	$\cdot$	$\cdot$	$\frac{-1260}{-4}$	-630	-210	-1260	-630	-630	-210
$(4)(2)(1)$	1	-6	10	-4	4	-8	$\cdot$	$\cdot$	-4	$\frac{630}{8}$	630	$\cdot$	630	630	630
$(4)(3)$	1	-7	14	-6	7	-24	6	12	-4	12	$\frac{-420}{-12}$	$\cdot$	$\cdot$	$\cdot$	-420
$(5)(1)^2$	1	-5	5	$\cdot$	5	-5	$\cdot$	$\cdot$	-5	$\cdot$	$\cdot$	$\frac{1008}{5}$	504	1008	504
$(5)(2)$	1	-7	15	-10	5	-15	10	$\cdot$	-5	10	$\cdot$	5	$\frac{-504}{-10}$	$\cdot$	-504
$(6)(1)$	1	-6	9	-2	6	-12	$\cdot$	3	-6	6	$\cdot$	6	$\cdot$	$\frac{-840}{-6}$	-840
(7)	1	-7	14	-7	7	-21	7	7	-7	14	-7	7	-7	-7	$\frac{720}{7}$

Table 1·5·8

$w = 8$ (i)

$w=8$ (i)	a_1^8	$a_2 a_1^6$	$a_2^2 a_1^4$	$a_2^3 a_1^2$	a_2^4	$a_3 a_1^5$	$a_3 a_2 a_1^3$	$a_3 a_2^2 a_1$	$a_3^2 a_1^2$	$a_3^2 a_2$	$a_4 a_1^4$
$(1)^8$	$\frac{40320}{I}$	20160	10080	5040	2520	6720	3360	1680	1120	560	1680
$(2)(1)^6$	I	$\frac{-20160}{-2}$	−20160	−15120	−10080	−20160	−13440	−8400	−6720	−3920	−10080
$(2)^2(1)^4$	I	−4	$\frac{10080}{4}$	15120	15120	·	10080	11760	10080	8400	5040
$(2)^3(1)^2$	I	−6	12	$\frac{-5040}{-8}$	−10080	·	·	−5040	·	−5040	·
$(2)^4$	I	−8	24	−32	$\frac{2520}{16}$	·	·	·	·	·	·
$(3)(1)^5$	I	−3	·	·	·	$\frac{13440}{3}$	6720	3360	4480	2240	13440
$(3)(2)(1)^3$	I	−5	6	·	·	3	$\frac{-6720}{-6}$	−6720	−13440	−8960	·
$(3)(2)^2(1)$	I	−7	16	−12	·	3	−12	$\frac{3360}{12}$	·	6720	·
$(3)^2(1)^2$	I	−6	9	·	·	6	−18	·	$\frac{4480}{9}$	2240	·
$(3)^2(2)$	I	−8	21	−18	·	6	−30	36	9	$\frac{-2240}{-18}$	·
$(4)(1)^4$	I	−4	2	·	·	4	·	·	·	·	$\frac{-10080}{-4}$
$(4)(2)(1)^2$	I	−6	10	−4	·	4	−8	·	·	·	−4
$(4)(2)^2$	I	−8	22	−24	8	4	−16	16	·	·	−4
$(4)(3)(1)$	I	−7	14	−6	·	7	−24	6	12	·	−4
$(4)^2$	I	−8	20	−16	4	8	−32	16	16	·	−8
$(5)(1)^3$	I	−5	5	−10	·	5	−5	·	·	·	−5
$(5)(2)(1)$	I	−7	15	−10	·	5	−15	10	·	·	−5
$(5)(3)$	I	−8	20	−15	·	8	−35	30	15	−15	−5
$(6)(1)^2$	I	−6	9	−2	·	6	−12	·	3	·	−6
$(6)(2)$	I	−8	21	−20	4	6	−24	24	3	−6	−6
$(7)(1)$	I	−7	14	−7	·	7	−21	7	7	·	−7
(8)	I	−8	20	−16	2	8	−32	24	12	−8	−8

$w = 8$ (ii)

$w=8$ (ii)	$a_4 a_2 a_1^2$	$a_4 a_2^2$	$a_4 a_3 a_1$	a_4^2	$a_5 a_1^3$	$a_5 a_2 a_1$	$a_5 a_3$	$a_6 a_1^2$	$a_6 a_2$	$a_7 a_1$	a_8
$(1)^8$	840	420	280	70	336	168	56	56	28	8	I
$(2)(1)^6$	−5880	−3360	−2520	−840	−3360	−1848	−728	−840	−448	−168	−28
$(2)^2(1)^4$	7560	6720	5880	2940	5040	4200	2520	2520	1680	840	210
$(2)^3(1)^2$	−2520	−5040	−2520	−2520	·	−2520	−2520	−840	−1680	−840	−420
$(2)^4$	·	1260	·	630	·	·	·	·	420	·	105
$(3)(1)^5$	6720	3360	2800	1120	6720	3360	1232	2240	1120	560	112
$(3)(2)(1)^3$	−6720	−6720	−10080	−6720	−6720	−6720	−5600	−6720	−4480	−3360	−1120
$(3)(2)^2(1)$	·	3360	1680	3360	·	3360	5040	·	3360	1680	1680
$(3)^2(1)^2$	·	·	4480	4480	·	·	2240	2240	1120	2240	1120
$(3)^2(2)$	·	·	·	·	·	·	−2240	·	−1120	·	−1120
$(4)(1)^4$	−5040	−2520	−1680	−840	−10080	−5040	−1680	−5040	−2520	−1680	−420
$(4)(2)(1)^2$	$\frac{5040}{8}$	5040	5040	5040	·	5040	5040	5040	5040	5040	2520
$(4)(2)^2$	16	$\frac{-2520}{-16}$	·	−2520	·	·	·	·	−2520	·	−1260
$(4)(3)(1)$	12	·	$\frac{-3360}{-12}$	−6720	·	·	−3360	·	·	−3360	−3360
$(4)^2$	32	−16	−32	$\frac{2520}{16}$	·	·	·	·	·	·	1260
$(5)(1)^3$	·	·	·	·	$\frac{8064}{5}$	4032	1344	8064	4032	4032	1344
$(5)(2)(1)$	10	·	·	·	5	$\frac{-4032}{-10}$	−4032	·	−4032	−4032	−4032
$(5)(3)$	15	·	−15	·	5	−15	$\frac{2688}{15}$	·	·	·	2688
$(6)(1)^2$	6	·	·	·	6	·	·	$\frac{-6720}{-6}$	−3360	−6720	−3360
$(6)(2)$	18	−12	·	·	6	−12	·	−6	$\frac{3360}{12}$	·	3360
$(7)(1)$	14	·	−7	·	7	−7	·	−7	·	$\frac{5760}{7}$	5760
(8)	24	−8	−16	4	8	−16	8	−3	8	8	$\frac{-5040}{-8}$

Table 1·5·9

$w=9$ (i)	a_1^9	$a_2a_1^7$	$a_2^2a_1^5$	$a_2^3a_1^3$	$a_2^4a_1$	$a_3a_1^6$	$a_3a_2a_1^4$	$a_3a_2^2a_1^2$	$a_3a_2^3$	$a_3^2a_1^3$	$a_3^2a_2a_1$	a_3^3	$a_4a_1^5$	$a_4a_2a_1^3$	$a_4a_2^2a_1$
$(1)^9$	362880 / 1	181440	90720	45360	22680	60480	30240	15120	7560	10080	5040	1680	15120	7560	3780
$(2)(1)^7$	1	−181440 / −2	−181440	−136080	−90720	−181440	−120960	−75600	−45360	−60480	−35280	−15120	−90720	−52920	−30240
$(2)^2(1)^5$	1	−4	90720 / 4	136080	136080	.	90720	105840	90720	90720	75600	45360	45360	68040	60480
$(2)^3(1)^3$	1	−6	12	−45360 / −8	−90720	.	.	−45360	−75600	.	−45360	−45360	.	−22680	−45360
$(2)^4(1)$	1	−8	24	−32	22680 / 16	.	.	.	22680	.	.	.	.	.	11340
$(3)(1)^6$	1	−3	.	.	.	120960 / 3	60480	30240	15120	40320	20160	10080	120960	60480	30240
$(3)(2)(1)^4$	1	−5	6	.	.	3	−60480 / −6	−60480	−45360	−120960	−80640	−60480	.	−60480	−60480
$(3)(2)^2(1)^2$	1	−7	16	−12	.	3	−12	30240 / 12	45360	.	60480	90720	.	.	30240
$(3)(2)^3$	1	−9	30	−44	24	3	−18	36	−15120 / −24	.	.	.	.	.	.
$(3)^2(1)^3$	1	−6	9	.	.	6	−18	.	.	40320 / 9	20160	20160	.	.	.
$(3)^2(2)(1)$	1	−8	21	−18	.	6	−30	36	.	9	−20160 / −18	−60480	.	.	.
$(3)^3$	1	−9	27	−27	.	9	−54	81	.	27	−81	13440 / 27	.	.	.
$(4)(1)^5$	1	−4	2	.	.	4	.	.	.	.	.	.	−90720 / −4	−45360	−22680
$(4)(2)(1)^3$	1	−6	10	−4	.	4	−8	.	.	.	.	.	−4	45360 / 8	45360
$(4)(2)^2(1)$	1	−8	22	−24	8	4	−16	16	.	.	.	.	−4	16	−22680 / −16
$(4)(3)(1)^2$	1	−7	14	−6	.	7	−24	6	.	12	.	.	−4	12	.
$(4)(3)(2)$	1	−9	28	−34	12	7	−38	54	−12	12	−24	.	−4	20	−24
$(4)^2(1)$	1	−8	20	−16	4	8	−32	16	.	16	.	.	−8	32	−16
$(5)(1)^4$	1	−5	5	.	.	5	−5	.	.	.	.	.	−5	.	.
$(5)(2)(1)^2$	1	−7	15	−10	.	5	−15	10	.	.	.	.	−5	10	.
$(5)(2)^2$	1	−9	29	−40	20	5	−25	40	−20	.	.	.	−5	20	−20
$(5)(3)(1)$	1	−8	20	−15	.	8	−35	30	.	15	−15	.	−5	15	.
$(5)(4)$	1	−9	27	−30	10	9	−45	50	−10	20	−20	.	−9	40	−30
$(6)(1)^3$	1	−6	9	−2	.	6	−12	.	.	3	.	.	−6	6	.
$(6)(2)(1)$	1	−8	21	−20	4	6	−24	24	.	3	−6	.	−6	18	−12
$(6)(3)$	1	−9	27	−29	6	9	−48	63	−6	21	−45	9	−6	24	−18
$(7)(1)^2$	1	−7	14	−7	.	7	−21	7	.	7	.	.	−7	14	.
$(7)(2)$	1	−9	28	−35	14	7	−35	49	−14	7	−14	.	−7	28	−28
$(8)(1)$	1	−8	20	−16	2	8	−32	24	.	12	−8	.	−8	24	−8
(9)	1	−9	27	−30	9	9	−45	54	−9	18	−27	3	−9	36	−27

$w=9$ (ii)	$a_4a_3a_1^2$	$a_4a_3a_2$	$a_4^2a_1$	$a_5a_1^4$	$a_5a_2a_1^2$	$a_5a_2^2$	$a_5a_3a_1$	a_5a_4	$a_6a_1^3$	$a_6a_2a_1$	a_6a_3	$a_7a_1^2$	a_7a_2	a_8a_1	a_9
$(1)^9$	2520	1260	630	3024	1512	756	504	126	504	252	84	72	36	9	1
$(2)(1)^7$	−22680	−12600	−7560	−30240	−16632	−9072	−6552	−2016	−7560	−4032	−1512	−1512	−792	−252	−36
$(2)^2(1)^5$	52920	37800	26460	45360	37800	27216	22680	9828	22680	15120	7560	7560	4536	1890	378
$(2)^3(1)^3$	−22680	−37800	−22680	.	−22680	−30240	−22680	−15120	−7560	−15120	−12600	−7560	−7560	−3780	−1260
$(2)^4(1)$	.	11340	5670	.	.	11340	.	5670	.	3780	3780	.	3780	945	945
$(3)(1)^6$	25200	12600	10080	60480	30240	15120	11088	3528	20160	10080	3528	5040	2520	1008	168
$(3)(2)(1)^4$	−90720	−57960	−60480	−60480	−60480	−45360	−50400	−27720	−60480	−40320	−22680	−30240	−17640	−10080	−2520
$(3)(2)^2(1)^2$	15120	52920	30240	.	30240	45360	45360	37800	.	30240	37800	15120	22680	15120	7560
$(3)(2)^3$	.	−7560	.	.	.	−15120	.	−7560	.	.	−2520	.	−7560	.	−2520
$(3)^2(1)^3$	40320	20160	40320	.	.	.	20160	20160	20160	10080	10080	20160	10080	10080	3360
$(3)^2(2)(1)$	.	−20160	.	.	.	.	−20160	−20160	.	−10080	−30240	.	−10080	−10080	−10080
$(3)^3$	.	.	.	.	.	.	.	.	.	.	6720	.	.	.	2240
$(4)(1)^5$	−15120	−7560	−7560	−90720	−45360	−22680	−15120	−4536	−45360	−22680	−7560	−15120	−7560	−3780	−756
$(4)(2)(1)^3$	45360	30240	45360	.	45360	45360	45360	30240	45360	45360	30240	45360	30240	22680	7560
$(4)(2)^2(1)$	.	−22680	−22680	.	.	−22680	.	−22680	.	−22680	−22680	.	−22680	−11340	−11340
$(4)(3)(1)^2$	−30240 / −12	−15120	−60480	.	.	.	−30240	−45360	.	.	−15120	−30240	−15120	−30240	−15120
$(4)(3)(2)$	−12	15120 / 24	.	.	.	.	.	15120	.	.	15120	.	15120	.	15120
$(4)^2(1)$	−32	.	22680 / 16	.	.	.	.	22680	.	.	.	.	.	11340	11340
$(5)(1)^4$	.	.	.	72576 / 5	36288	18144	12096	3024	72576	36288	12096	36288	18144	12096	3024
$(5)(2)(1)^2$	.	.	.	5	−36288 / −10	−36288	−36288	−18144	.	−36288	−36288	−36288	−36288	−36288	−18144
$(5)(2)^2$	.	.	.	5	−20	18144 / 20	.	9072	.	.	.	.	18144	.	9072
$(5)(3)(1)$	−15	.	.	5	−15	.	24192 / 15	24192	.	.	24192	.	.	24192	24192
$(5)(4)$	−40	20	20	5	−20	10	20	−18144 / −20	.	.	.	.	.	.	−18144
$(6)(1)^3$	.	.	.	6	.	.	.	.	−60480 / −6	−30240	−10080	−60480	−30240	−30240	−10080
$(6)(2)(1)$	.	.	.	6	−12	.	.	.	−6	30240 / 12	30240	.	30240	30240	30240
$(6)(3)$	−18	18	.	6	−18	.	18	.	−6	18	−20160 / −18	.	.	.	−20160
$(7)(1)^2$	−7	.	.	7	−7	.	.	.	−7	.	.	51840 / 7	25920	51840	25920
$(7)(2)$	−7	14	.	7	−21	14	.	.	−7	14	.	7	−25920 / −14	.	−25920
$(8)(1)$	−16	.	4	8	−16	.	8	.	−8	8	.	8	.	−45360 / −8	−45360
(9)	−27	18	9	9	−27	9	18	−9	−9	18	−9	9	−9	−9	40320 / 9

Table 1·5·10

$w=10$ (i)	a_1^{10}	$a_2 a_1^8$	$a_2^2 a_1^6$	$a_2^3 a_1^4$	$a_2^4 a_1^2$	a_2^5	$a_3 a_1^7$	$a_3 a_2 a_1^5$	$a_3 a_2^2 a_1^3$	$a_3 a_2^3 a_1$
$(1)^{10}$	3628800	1814400	907200	453600	226800	113400	604800	302400	151200	75600
	1									
$(2)(1)^8$		−1814400	−1814400	−1360800	−907200	−567000	−1814400	−1209600	−756000	−453600
	1	−2								
$(2)^2(1)^6$			907200	1360800	1360800	1134000	·	907200	1058400	907200
	1	−4	4							
$(2)^3(1)^4$				−453600	−907200	−1134000	·	·	−453600	−756000
	1	−6	12	−8						
$(2)^4(1)^2$					226800	567000	·	·	·	226800
	1	−8	24	−32	16					
$(2)^5$						−113400	·	·	·	·
	1	−10	40	−80	80	−32				
$(3)(1)^7$							1209600	604800	302400	151200
	1						3			
$(3)(2)(1)^5$								−604800	−604800	−453600
	1	−5	6	·	·	·	3	−6		
$(3)(2)^2(1)^3$									302400	453600
	1	−7	16	−12	·	·	3	−12	12	
$(3)(2)^3(1)$										−151200
	1	−9	30	−44	24	·	3	−18	36	−24
$(3)^2(1)^4$	1	−6	9	·	·	·	6	−18		
$(3)^2(2)(1)^2$	1	−8	21	−18	·	·	6	−30	36	
$(3)^2(2)^2$	1	−10	37	−60	36	·	6	−42	96	−72
$(3)^3(1)$	1	−9	27	−27	·	·	9	−54	81	·
$(4)(1)^6$	1	−4	2	·	·	·	4	·	·	·
$(4)(2)(1)^4$	1	−6	10	−4	·	·	4	−8	·	·
$(4)(2)^2(1)^2$	1	−8	22	−24	8	·	4	−16	16	·
$(4)(2)^3$	1	−10	38	−68	56	−16	4	−24	48	−32
$(4)(3)(1)^3$	1	−7	14	−6	·	·	7	−24	6	·
$(4)(3)(2)(1)$	1	−9	28	−34	12	·	7	−38	54	−12
$(4)(3)^2$	1	−10	35	−48	18	·	10	−66	120	−36
$(4)^2(1)^2$	1	−8	20	−16	4	·	8	−32	16	·
$(4)^2(2)$	1	−10	36	−56	36	−8	8	−48	80	−32
$(5)(1)^5$	1	−5	5	·	·	·	5	−5	·	·
$(5)(2)(1)^3$	1	−7	15	−10	·	·	5	−15	10	·
$(5)(2)^2(1)$	1	−9	29	−40	20	·	5	−25	40	−20
$(5)(3)(1)^2$	1	−8	20	−15	·	·	8	−35	30	·
$(5)(3)(2)$	1	−10	36	−55	30	·	8	−51	100	−60
$(5)(4)(1)$	1	−9	27	−30	10	·	9	−45	50	−10
$(5)^2$	1	−10	35	−50	25	·	10	−60	100	−50
$(6)(1)^4$	1	−6	9	−2	·	·	6	−12	·	·
$(6)(2)(1)^2$	1	−8	21	−20	4	·	6	−24	24	·
$(6)(2)^2$	1	−10	37	−62	44	−8	6	−36	72	−48
$(6)(3)(1)$	1	−9	27	−29	6	·	9	−48	63	−6
$(6)(4)$	1	−10	35	−50	26	−4	10	−60	96	−32
$(7)(1)^3$	1	−7	14	−7	·	·	7	−21	7	·
$(7)(2)(1)$	1	−9	28	−35	14	·	7	−35	49	−14
$(7)(3)$	1	−10	35	−49	21	·	10	−63	112	−42
$(8)(1)^2$	1	−8	20	−16	2	·	8	−32	24	·
$(8)(2)$	1	−10	36	−56	34	−4	8	−48	88	−48
$(9)(1)$	1	−9	27	−30	9	·	9	−45	54	−9
(10)	1	−10	35	−50	25	−2	10	−60	100	−40

Table 1·5·10 (cont.)

$w=10$ (ii)	$a_3^2 a_1^4$	$a_3^2 a_2 a_1^2$	$a_3^2 a_2^2$	$a_3^3 a_1$	$a_4 a_1^6$	$a_4 a_2 a_1^4$	$a_4 a_2^2 a_1^2$	$a_4 a_2^3$	$a_4 a_3 a_1^3$	$a_4 a_3 a_2 a_1$
$(1)^{10}$	100800	50400	25200	16800	151200	75600	37800	18900	25200	12600
$(2)(1)^8$	-604800	-352800	-201600	-151200	-907200	-529200	-302400	-170100	-226800	-126000
$(2)^2(1)^6$	907200	756000	554400	453600	453600	680400	604800	453600	529200	378000
$(2)^3(1)^4$	.	-453600	-604800	-453600	.	-226800	-453600	-529200	-226800	-378000
$(2)^4(1)^2$	.	.	226800	.	.	.	113400	283500	.	113400
$(2)^5$	.	.	.	.	.	.	.	-56700	.	.
$(3)(1)^7$	403200	201600	100800	100800	1209600	604800	302400	151200	252000	126000
$(3)(2)(1)^5$	-1209600	-806400	-504000	-604800	.	-604800	-604800	-453600	-907200	-579600
$(3)(2)^2(1)^3$	.	604800	705600	907200	.	.	302400	453600	151200	529200
$(3)(2)^3(1)$	.	.	-302400	.	.	.	.	-151200	.	-75600
$(3)^2(1)^4$	403200 9	201600	100800	201600	.	.	.	.	403200	201600
$(3)^2(2)(1)^2$	9	-201600 -18	-201600	-604800	.	.	.	.	.	-201600
$(3)^2(2)^2$	9	-36	100800 36	.	.	.	.	.	.	.
$(3)^3(1)$	27	-81	.	134400 27	.	.	.	.	.	.
$(4)(1)^6$	.	.	.	.	-907200 -4	-453600	-226800	-113400	-151200	-75600
$(4)(2)(1)^4$	.	.	.	.	-4	453600 8	453600	340200	453600	302400
$(4)(2)^2(1)^2$	.	.	.	.	-4	16	-226800 -16	-340200	.	-226800
$(4)(2)^3$	.	.	.	.	-4	16	-16	113400 32	.	.
$(4)(3)(1)^3$	12	.	.	.	-4	24	-48	32	-302400 -12	-151200
$(4)(3)(2)(1)$	12	-24	.	.	-4	20	-24	.	-12	151200 24
$(4)(3)^2$	33	-108	18	36	-4	24	-36	.	-24	72
$(4)^2(1)^2$	16	.	.	.	-8	32	-16	.	-32	.
$(4)^2(2)$	16	-32	.	.	-8	48	-80	32	-32	64
$(5)(1)^5$	.	.	.	.	-5	.	.	.	.	.
$(5)(2)(1)^3$	.	.	.	.	-5	10	.	.	.	.
$(5)(2)^2(1)$	.	.	.	.	-5	20	-20	.	.	.
$(5)(3)(1)^2$	15	-15	.	.	-5	15	.	.	-15	.
$(5)(3)(2)$	15	-45	30	.	-5	25	-30	.	-15	30
$(5)(4)(1)$	20	-20	.	.	-9	40	-30	.	-40	20
$(5)^2$	25	-50	25	.	-10	50	-50	.	-50	50
$(6)(1)^4$	3	.	.	.	-6	6	.	.	.	.
$(6)(2)(1)^2$	3	-6	.	.	-6	18	-12	.	.	.
$(6)(2)^2$	3	-12	12	.	-6	30	-48	24	-18	18
$(6)(3)(1)$	21	-45	.	9	-6	24	-18	.	-18	18
$(6)(4)$	27	-60	6	12	-10	54	-72	20	-48	72
$(7)(1)^3$	7	.	.	.	-7	14	.	.	-7	.
$(7)(2)(1)$	7	-14	.	.	-7	28	-28	.	-7	14
$(7)(3)$	28	-84	21	21	-7	35	-42	.	-28	63
$(8)(1)^2$	12	-8	.	.	-8	24	-8	.	-16	.
$(8)(2)$	12	-32	16	.	-8	40	-56	16	-16	32
$(9)(1)$	18	-27	.	3	-9	36	-27	.	-27	18
(10)	25	-60	15	10	-10	50	-60	10	-40	60

Table 1·5·10 (*cont.*)

$w=10$ (iii)	$a_4 a_3^2$	$a_4^2 a_1^2$	$a_4^2 a_2$	$a_5 a_1^5$	$a_5 a_2 a_1^3$	$a_5 a_2^2 a_1$	$a_5 a_3 a_1^2$	$a_5 a_3 a_2$	$a_5 a_4 a_1$	a_5^2	$a_6 a_1^4$
$(1)^{10}$	4200	6300	3150	30240	15120	7560	5040	2520	1260	252	5040
$(2)(1)^8$	−50400	−75600	−40950	−302400	−166320	−90720	−65520	−35280	−20160	−5040	−75600
$(2)^2(1)^6$	201600	264600	170100	453600	378000	272160	226800	146160	98280	32760	226800
$(2)^3(1)^4$	−302400	−226800	−245700	·	−226800	·	−302400	−226800	−226800	−151200	−75600
$(2)^4(1)^2$	113400	56700	141750	·	·	113400	·	113400	56700	56700	·
$(2)^5$	·	·	−28350	·	·	·	·	·	·	·	·
$(3)(1)^7$	50400	100800	50400	604800	302400	151200	110880	55440	35280	10080	201600
$(3)(2)(1)^5$	−352800	−604800	−352800	−604800	−604800	−453600	−504000	−307440	−277200	−110880	−604800
$(3)(2)^2(1)^3$	655200	302400	453600	·	302400	453600	453600	478800	378000	252000	·
$(3)(2)^3(1)$	−151200	·	−151200	·	·	−151200	·	−226800	−75600	−151200	·
$(3)^2(1)^4$	151200	403200	201600	·	·	·	201600	100800	201600	100800	201600
$(3)^2(2)(1)^2$	−504000	·	−201600	·	·	·	−201600	−201600	−201600	−201600	·
$(3)^2(2)^2$	50400	·	·	·	·	·	·	100800	·	100800	·
$(3)^3(1)$	134400	·	·	·	·	·	·	·	·	·	·
$(4)(1)^6$	−25200	−75600	−37800	−907200	−453600	−226800	−151200	−75600	−45360	−15120	−453600
$(4)(2)(1)^4$	151200	453600	264600	·	453600	453600	453600	302400	302400	151200	453600
$(4)(2)^2(1)^2$	−226800	−226800	−340200	·	·	−226800	−226800	−226800	−226800	−226800	−226800
$(4)(2)^3$	·	·	113400	·	·	·	·	·	·	·	·
$(4)(3)(1)^3$	−100800	−604800	−302400	·	·	·	−302400	−151200	−453600	−302400	·
$(4)(3)(2)(1)$	302400	·	302400	·	·	·	·	151200	151200	302400	·
$(4)(3)^2$	$\dfrac{-100800}{-36}$	·	·	·	·	·	·	·	·	·	·
$(4)^2(1)^2$	·	$\dfrac{226800}{16}$	113400	·	·	·	·	·	226800	226800	·
$(4)^2(2)$	·	·	$\dfrac{-113400}{-32}$	·	·	·	·	·	·	·	·
$(5)(1)^5$	·	·	·	$\dfrac{725760}{5}$	362880	181440	120960	60480	30240	12096	725760
$(5)(2)(1)^3$	·	·	·	5	$\dfrac{-362880}{-10}$	−362880	−362880	−241920	−181440	−120960	·
$(5)(2)^2(1)$	·	·	·	5	−20	$\dfrac{181440}{20}$	·	181440	90720	181440	·
$(5)(3)(1)^2$	·	·	·	5	−15	·	$\dfrac{241920}{15}$	120960	241920	241920	·
$(5)(3)(2)$	·	·	·	5	−25	30	15	$\dfrac{-120960}{-30}$	·	−241920	·
$(5)(4)(1)$	·	20	·	5	−20	10	20	·	$\dfrac{-181440}{-20}$	−362880	·
$(5)^2$	·	25	·	10	−50	50	50	−50	−50	$\dfrac{145152}{25}$	·
$(6)(1)^4$	·	·	·	6	·	·	·	·	·	·	$\dfrac{-604800}{-6}$
$(6)(2)(1)^2$	·	·	·	6	−12	·	·	·	·	·	−6
$(6)(2)^2$	·	·	·	6	−24	24	·	·	·	·	−6
$(6)(3)(1)$	·	·	·	6	−18	·	18	·	·	·	−6
$(6)(4)$	−12	24	−24	6	−24	12	24	·	−24	·	−6
$(7)(1)^3$	·	·	·	7	−7	·	·	·	·	·	−7
$(7)(2)(1)$	·	·	·	7	−21	14	·	·	·	·	−7
$(7)(3)$	−21	·	·	7	−28	21	21	−21	·	·	−7
$(8)(1)^2$	·	4	·	8	−16	·	8	·	·	·	−8
$(8)(2)$	·	4	−8	8	−32	32	8	−16	·	·	−8
$(9)(1)$	·	9	·	9	−27	9	18	·	−9	·	−9
(10)	−10	15	−10	10	−40	30	30	−20	−20	5	−10

Table 1·5·10 (*cont.*)

$w=10$ (iv)	$a_6 a_2 a_1^2$	$a_6 a_2^2$	$a_6 a_3 a_1$	$a_6 a_4$	$a_7 a_1^3$	$a_7 a_2 a_1$	$a_7 a_3$	$a_8 a_1^2$	$a_8 a_2$	$a_9 a_1$	a_{10}
$(1)^{10}$	2520	1260	840	210	720	360	120	90	45	10	1
$(2)(1)^8$	−40320	−21420	−15120	−4410	−15120	−7920	−2880	−2520	−1305	−360	−45
$(2)^2(1)^6$	151200	95760	75600	28980	75600	45360	20160	18900	10710	3780	630
$(2)^3(1)^4$	−151200	−151200	−126000	−69300	−75600	.	−50400	−37800	−28350	−12600	−3150
$(2)^4(1)^2$	37800	94500	37800	47250	.	37800	37800	9450	23625	9450	4725
$(2)^5$	.	−18900	.	−9450	.	.	.	.	−4725	.	−945
$(3)(1)^7$	100800	50400	35280	10080	50400	25200	8640	10080	5040	1680	240
$(3)(2)(1)^5$	−403200	−252000	−226800	−100800	−302400	−176400	−80640	−100800	−55440	−25200	−5040
$(3)(2)^2(1)^3$	302400	352800	378000	252000	151200	226800	201600	151200	126000	75600	25200
$(3)(2)^3(1)$	.	−151200	−25200	−100800	.	−75600	.	.	−75600	−25200	−25200
$(3)^2(1)^4$	100800	50400	100800	75600	201600	100800	50400	100800	50400	33600	8400
$(3)^2(2)(1)^2$	−100800	−100800	−302400	−252000	.	−100800	−201600	−100800	−100800	−100800	−50400
$(3)^2(2)^2$	.	50400	.	25200	.	.	.	.	50400	.	25200
$(3)^3(1)$	.	.	67200	67200	.	.	67200	.	.	22400	22400
$(4)(1)^6$	−226800	−113400	−75600	−20160	−151200	−75600	−25200	−37800	−18900	−7560	−1260
$(4)(2)(1)^4$	453600	340200	302400	151200	453600	302400	151200	226800	132300	75600	18900
$(4)(2)^2(1)^2$	−226800	−340200	−226800	−226800	.	−226800	−226800	−113400	−170100	−113400	−56700
$(4)(2)^3$	.	113400	.	75600	.	.	.	.	56700	.	18900
$(4)(3)(1)^3$	.	.	−151200	−201600	−302400	−151200	−100800	−302400	−151200	−151200	−50400
$(4)(3)(2)(1)$	.	.	151200	302400	.	151200	302400	.	151200	151200	151200
$(4)(3)^2$	.	.	.	−50400	.	.	−100800	.	.	.	−50400
$(4)^2(1)^2$	.	.	.	113400	.	.	.	113400	56700	113400	56700
$(4)^2(2)$	.	.	.	−113400	.	.	.	.	−56700	.	−56700
$(5)(1)^5$	362880	181440	120960	30240	362880	181440	60480	120960	60480	30240	6048
$(5)(2)(1)^3$	−362880	−362880	−362880	−181440	−362880	−362880	−241920	−362880	−241920	−181440	−60480
$(5)(2)^2(1)$	.	181440	.	90720	.	181440	181440	.	181440	90720	90720
$(5)(3)(1)^2$	.	.	241920	241920	.	.	120960	241920	120960	241920	120960
$(5)(3)(2)$	.	.	.	.	.	.	−120960	.	−120960	.	−120960
$(5)(4)(1)$	.	.	.	−181440	.	.	.	.	.	−181440	−181440
$(5)^2$	.	.	.	.	.	.	.	.	.	.	72576
$(6)(1)^4$	−302400	−151200	−100800	−25200	−604800	−302400	−100800	−302400	−151200	−100800	−25200
$(6)(2)(1)^2$	$\frac{302400}{12}$	302400	302400	151200	.	302400	302400	302400	302400	302400	151200
$(6)(2)^2$	24	$\frac{-151200}{-24}$	.	−75600	.	.	.	.	−151200	.	−75600
$(6)(3)(1)$	18	.	$\frac{-201600}{-18}$	−201600	.	.	−201600	.	.	−201600	−201600
$(6)(4)$	24	−12	−24	$\frac{151200}{24}$	.	.	.	.	.	.	151200
$(7)(1)^3$	.	.	.	.	$\frac{518400}{7}$	259200	86400	518400	259200	259200	86400
$(7)(2)(1)$	14	.	.	.	7	$\frac{-259200}{-14}$	−259200	.	−259200	−259200	−259200
$(7)(3)$	21	.	−21	.	7	−21	$\frac{172800}{21}$	.	.	.	172800
$(8)(1)^2$	8	.	.	.	8	.	.	$\frac{-453600}{-8}$	−226800	−453600	−226800
$(8)(2)$	24	−16	.	.	8	−16	.	−8	$\frac{226800}{16}$	.	226800
$(9)(1)$	18	.	−9	.	9	−9	.	−9	.	$\frac{403200}{9}$	403200
(10)	30	−10	−20	10	10	−20	10	−10	10	10	$\frac{-362880}{-10}$

Table 1.5.11

Principal (leading) entries — upper-triangle values:

$w = 11$ (i)	a_1^{11}	$a_2 a_1^9$	$a_2^2 a_1^7$	$a_2^3 a_1^5$	$a_2^4 a_1^3$	$a_2^5 a_1$	$a_3 a_1^8$	$a_3 a_2 a_1^6$	$a_3 a_2^2 a_1^4$	$a_3 a_2^3 a_1^2$	$a_3 a_2^4$
$(1)^{11}$	39916800	19958400	9979200	4989600	2494800	1247400	6652800	3326400	1663200	831600	415800
$(2)(1)^9$		−19958400	−19958400	−14668800	−9979200	−6237000	−19958400	−13305600	−8316000	−4989600	−2910600
$(2)^2(1)^7$			9979200	14668800	14668800	12474000	•	9979200	11642400	9979200	7484400
$(2)^3(1)^5$				−4989600	−9979200	−12474000	•	•	−4989600	−8316000	−9147600
$(2)^4(1)^3$					2494800	6237000	•	•	•	2494800	5405400
$(2)^5(1)$						−1247400	•	•	•	•	−1247400
$(3)(1)^8$							13305600	6652800	3326400	1663200	831600
$(3)(2)(1)^6$								−6652800	−6652800	−4989600	−3326400
$(3)(2)^2(1)^4$									3326400	4989600	4989600
$(3)(2)^3(1)^2$										−1663200	−3326400
$(3)(2)^4$											831600

Coefficient table (smaller entries):

$w = 11$ (i)	a_1^{11}	$a_2 a_1^9$	$a_2^2 a_1^7$	$a_2^3 a_1^5$	$a_2^4 a_1^3$	$a_2^5 a_1$	$a_3 a_1^8$	$a_3 a_2 a_1^6$	$a_3 a_2^2 a_1^4$	$a_3 a_2^3 a_1^2$	$a_3 a_2^4$
$(1)^{11}$	1										
$(2)(1)^9$	1	−2									
$(2)^2(1)^7$	1	−4	4								
$(2)^3(1)^5$	1	−6	12	−8							
$(2)^4(1)^3$	1	−8	24	−32	16						
$(2)^5(1)$	1	−10	40	−80	80	−32					
$(3)(1)^8$	1	−3	•	•	•	•	3				
$(3)(2)(1)^6$	1	−5	6	•	•	•	3	−6			
$(3)(2)^2(1)^4$	1	−7	16	−12	•	•	3	−12	12		
$(3)(2)^3(1)^2$	1	−9	30	−44	24	•	3	−18	36	−24	
$(3)(2)^4$	1	−11	48	−104	112	−48	3	−24	72	−96	48
$(3)^2(1)^5$	1	−6	9	•	•	•	6	−18	•	•	•
$(3)^2(2)(1)^3$	1	−8	21	−18	•	•	6	−30	36	•	•
$(3)^2(2)^2(1)$	1	−10	37	−60	36	•	6	−42	96	−72	•
$(3)^3(1)^2$	1	−9	27	−27	•	•	9	−54	81	•	•
$(3)^3(2)$	1	−11	45	−81	54	•	9	−72	189	−162	•
$(4)(1)^7$	1	−4	2	•	•	•	4	•	•	•	•
$(4)(2)(1)^5$	1	−6	10	−4	•	•	4	−8	•	•	•
$(4)(2)^2(1)^3$	1	−8	22	−24	8	•	4	−16	16	•	•
$(4)(2)^3(1)$	1	−10	38	−68	56	−16	4	−24	48	−32	•
$(4)(3)(1)^4$	1	−7	14	−6	•	•	7	−24	6	•	•
$(4)(3)(2)(1)^2$	1	−9	28	−34	12	•	7	−38	54	−12	•
$(4)(3)(2)^2$	1	−11	46	−90	80	−24	7	−52	130	−120	24
$(4)(3)^2(1)$	1	−10	35	−48	18	•	10	−66	120	−36	•
$(4)^2(1)^3$	1	−8	20	−16	4	•	8	−32	16	•	•
$(4)^2(2)(1)$	1	−10	36	−56	36	−8	8	−48	80	−32	•
$(4)^2(3)$	1	−11	44	−76	52	−12	11	−80	172	−96	12
$(5)(1)^6$	1	−5	5	•	•	•	5	−5	•	•	•
$(5)(2)(1)^4$	1	−7	15	−10	•	•	5	−15	10	•	•
$(5)(2)^2(1)^2$	1	−9	29	−40	20	•	5	−25	40	−20	•
$(5)(2)^3$	1	−11	47	−98	100	−40	5	−35	90	−100	40
$(5)(3)(1)^3$	1	−8	20	−15	•	•	8	−35	30	•	•
$(5)(3)(2)(1)$	1	−10	36	−55	30	•	8	−51	100	−60	•
$(5)(3)^2$	1	−11	44	−75	45	•	11	−83	195	−135	•
$(5)(4)(1)^2$	1	−9	27	−30	10	•	9	−45	50	−10	•
$(5)(4)(2)$	1	−11	45	−84	70	−20	9	−63	140	−110	20
$(5)^2(1)$	1	−10	35	−50	25	•	10	−60	100	−50	•
$(6)(1)^5$	1	−6	9	−2	•	•	6	−12	•	•	•
$(6)(2)(1)^3$	1	−8	21	−20	4	•	6	−24	24	•	•
$(6)(2)^2(1)$	1	−10	37	−62	44	−8	6	−36	72	−48	•
$(6)(3)(1)^2$	1	−9	27	−29	6	•	9	−48	63	−6	•
$(6)(3)(2)$	1	−11	45	−83	64	−12	9	−66	159	−132	12
$(6)(4)(1)$	1	−10	35	−50	26	−4	10	−60	96	−32	•
$(6)(5)$	1	−11	44	−77	55	−10	11	−77	165	−115	10
$(7)(1)^4$	1	−7	14	−7	•	•	7	−21	7	•	•
$(7)(2)(1)^2$	1	−9	28	−35	14	•	7	−35	49	−14	•
$(7)(2)^2$	1	−11	46	−91	84	−28	7	−49	119	−112	28
$(7)(3)(1)$	1	−10	35	−49	21	•	10	−63	112	−42	•
$(7)(4)$	1	−11	44	−77	56	−14	11	−77	161	−98	14
$(8)(1)^3$	1	−8	20	−16	2	•	8	−32	24	•	•
$(8)(2)(1)$	1	−10	36	−56	34	−4	8	−48	88	−48	•
$(8)(3)$	1	−11	44	−76	50	−6	11	−80	180	−120	6
$(9)(1)^2$	1	−9	27	−30	9	•	9	−45	54	−9	•
$(9)(2)$	1	−11	45	−84	69	−18	9	−63	144	−117	18
$(10)(1)$	1	−10	35	−50	25	−2	10	−60	100	−40	•
(11)	1	−11	44	−77	55	−11	11	−77	165	−110	11

Table 1·5·11 (cont.)

$w=11$ (ii)	$a_3^2 a_1^5$	$a_3^2 a_2 a_1^3$	$a_3^2 a_2^2 a_1$	$a_3^3 a_1^2$	$a_3^3 a_2$	$a_4 a_1^7$	$a_4 a_2 a_1^5$	$a_4 a_2^2 a_1^3$	$a_4 a_2^3 a_1$	$a_4 a_3 a_1^4$	$a_4 a_3 a_2 a_1^2$
$(1)^{11}$	1108800	554400	277200	184800	92400	1663200	831600	415800	207900	277200	138600
$(2)(1)^9$	−6652800	−3880800	−2217600	−1663200	−924000	−9979200	−5821200	−3326400	−1871100	−2494800	−1386000
$(2)^2(1)^7$	9979200	8316000	6098400	4989600	3326400	4989600	7484400	6652800	4989600	5821200	4158000
$(2)^3(1)^5$	·	−4989600	−6652800	−4989600	−4989600	·	−2494800	−4989600	−5821200	−2494800	−4158000
$(2)^4(1)^3$	·	·	2494800	·	2494800	·	·	1247400	3118500	·	1247400
$(2)^5(1)$	·	·	·	·	·	·	·	·	−623700	·	·
$(3)(1)^8$	4435200	2217600	1108800	1108800	554400	13305600	6652800	3326400	1663200	2772000	1386000
$(3)(2)(1)^6$	−13305600	−8870400	−5544000	−6652800	−3880800	·	−6652800	−6652800	−4989600	−9979200	−6375600
$(3)(2)^2(1)^4$	·	6652800	7761600	9979200	8316000	·	·	3326400	4989600	1663200	5821200
$(3)(2)^3(1)^2$	·	·	−3326400	·	−4989600	·	·	·	−1663200	·	−831600
$(3)(2)^4$	·	·	·	·	·	·	·	·	·	·	·
$(3)^2(1)^5$	$\frac{4435200}{9}$	2217600	1108800	2217600	1108800	·	·	·	·	4435200	2217600
$(3)^2(2)(1)^3$	9	$\frac{-2217600}{-18}$	−2217600	−6652800	−4435200	·	·	·	·	·	−2217600
$(3)^2(2)^2(1)$	9	−36	$\frac{1108800}{36}$	·	3326400	·	·	·	·	·	·
$(3)^3(1)^2$	27	−81	·	$\frac{1478400}{27}$	739200	·	·	·	·	·	·
$(3)^3(2)$	27	−135	162	27	$\frac{-739200}{-54}$	·	·	·	·	·	·
$(4)(1)^7$	·	·	·	·	·	$\frac{-9979200}{-4}$	−4989600	−2494800	−1247400	−1663200	−831600
$(4)(2)(1)^5$	·	·	·	·	·	−4	$\frac{4989600}{8}$	4989600	3742200	4989600	3326400
$(4)(2)^2(1)^3$	·	·	·	·	·	−4	16	$\frac{-2494800}{-16}$	−3742200	·	−2494800
$(4)(2)^3(1)$	·	·	·	·	·	−4	24	−48	$\frac{1247400}{32}$	·	·
$(4)(3)(1)^4$	12	·	·	·	·	−4	12	·	·	$\frac{-3326400}{-12}$	−1663200
$(4)(3)(2)(1)^2$	12	−24	·	·	·	−4	20	−24	·	−12	$\frac{1663200}{24}$
$(4)(3)(2)^2$	12	−48	48	·	·	−4	28	−64	48	−12	48
$(4)(3)^2(1)$	33	−108	18	36	·	−4	24	−36	·	−24	72
$(4)^2(1)^3$	16	·	·	·	·	−8	32	−16	·	−32	·
$(4)^2(2)(1)$	16	−32	·	·	·	−8	48	−80	32	−32	64
$(4)^2(3)$	40	−144	48	48	·	−8	56	−112	48	−56	192
$(5)(1)^6$	·	·	·	·	·	−5	·	·	·	·	·
$(5)(2)(1)^4$	·	·	·	·	·	−5	10	·	·	·	·
$(5)(2)^2(1)^2$	·	·	·	·	·	−5	20	−20	·	·	·
$(5)(2)^3$	·	·	·	·	·	−5	30	−60	40	·	·
$(5)(3)(1)^3$	15	−15	·	·	·	−5	15	·	·	−15	·
$(5)(3)(2)(1)$	15	−45	30	·	·	−5	25	−30	·	−15	30
$(5)(3)^2$	39	−165	135	45	−45	−5	30	−45	·	−30	90
$(5)(4)(1)^2$	20	−20	·	·	·	−9	40	−30	·	−40	20
$(5)(4)(2)$	20	−60	40	·	·	−9	58	−110	60	−40	100
$(5)^2(1)$	25	−50	25	·	·	−10	50	−50	·	−50	50
$(6)(1)^5$	3	·	·	·	·	−6	6	·	·	·	·
$(6)(2)(1)^3$	3	−6	·	·	·	−6	18	−12	·	·	·
$(6)(2)^2(1)$	3	−12	12	·	·	−6	30	−48	24	·	·
$(6)(3)(1)^2$	21	−45	90	9	·	−6	24	−18	·	−18	18
$(6)(3)(2)$	21	−87	90	9	−18	−6	36	−66	36	−18	54
$(6)(4)(1)$	27	−60	6	12	·	−10	54	−72	20	−48	72
$(6)(5)$	33	−105	75	15	−15	−11	66	−105	40	−60	120
$(7)(1)^4$	7	·	·	·	·	−7	14	·	·	·	·
$(7)(2)(1)^2$	7	−14	·	·	·	−7	28	−28	·	−7	14
$(7)(2)^2$	7	−28	28	·	·	−7	42	−84	56	−7	28
$(7)(3)(1)$	28	−84	21	21	·	−7	35	−42	·	−28	63
$(7)(4)$	35	−112	42	28	·	−11	70	−126	56	−63	168
$(8)(1)^3$	12	−8	·	·	·	−8	24	−8	·	−16	·
$(8)(2)(1)$	12	−32	16	·	·	−8	40	−56	16	−16	32
$(8)(3)$	36	−140	96	36	−24	−8	48	−80	24	−40	120
$(9)(1)^2$	18	−27	·	3	·	−9	36	−27	·	−27	18
$(9)(2)$	18	−63	54	3	−6	−9	54	−99	54	−27	72
$(10)(1)$	25	−60	15	10	·	−10	50	−60	10	−40	60
(11)	33	−110	66	22	−11	−11	66	−110	44	−55	132

Table 1·5·11 (cont.)

$w=11$ (iii)	$a_4a_3a_2^2$	$a_4a_3a_2a_1$	$a_4^2a_1^3$	$a_4^2a_2a_1$	$a_4^2a_3$	$a_5a_1^6$	$a_5a_2a_1^4$	$a_5a_2^2a_1^2$	$a_5a_2^3$	$a_5a_3a_1^3$	$a_5a_3a_2a_1$
$(1)^{11}$	69300	46200	69300	34650	11550	332640	166320	83160	41580	55440	27720
$(2)(1)^9$	−762300	−554400	−831600	−450450	−173250	−3326400	−1829520	−997920	−540540	−720720	−388080
$(2)^2(1)^7$	2772000	2217600	2910600	1871100	900900	4989600	4158000	2993760	1995840	2494800	1607760
$(2)^3(1)^5$	−4158000	−3326400	−2494800	−2702700	−1871100	.	−2494800	−3326400	−3160080	−2494800	−2494800
$(2)^4(1)^3$	2702700	1247400	623700	1559250	1351350	.	.	1247400	2286900	.	1247400
$(2)^5(1)$	−623700	.	.	−311850	−311850	.	.	.	−623700	.	.
$(3)(1)^8$	693000	554400	1108800	554400	207900	6652800	3326400	1663200	831600	1219680	609840
$(3)(2)(1)^6$	−3880800	−3880800	−6652800	−3880800	−1940400	−6652800	−6652800	−4989600	−3326400	−5544000	−3381840
$(3)(2)^2(1)^4$	6098400	7207200	3326400	4989600	4851000	.	3326400	4989600	4989600	4989600	5266800
$(3)(2)^3(1)^2$	−3326400	−1663200	.	−1663200	−2494800	.	.	−1663200	−3326400	.	−2494800
$(3)(2)^4$	415800	.	.	.	207900	.	.	.	831600	.	.
$(3)^2(1)^5$	1108800	1663200	4435200	2217600	1108800	.	.	.	.	2217600	1108800
$(3)^2(2)(1)^3$	−2217600	−5544000	.	−2217600	−4435200	.	.	.	.	−2217600	−2217600
$(3)^2(2)^2(1)$	1108800	554400	.	.	1108800	.	.	.	.	.	1108800
$(3)^3(1)^2$	.	1478400	.	.	1478400	.	.	.	.	.	.
$(3)^3(2)$	.	.	.	.	.	.	.	.	.	.	.
$(4)(1)^7$	−415800	−277200	−831600	−415800	−138600	−9979200	−4989600	−2494800	−1247400	−1663200	−831600
$(4)(2)(1)^5$	2079000	1663200	4989600	2910600	1247400	.	4989600	4989600	3742200	4989600	3326400
$(4)(2)^2(1)^3$	−2910600	−2494800	−2494800	−3742200	−2910600	.	.	−2494800	−3742200	.	−2494800
$(4)(2)^3(1)$	1247400	.	.	1247400	1247400	.	.	.	1247400	.	.
$(4)(3)(1)^4$	−831600	−1108800	−6652800	−3326400	−1386000	.	.	.	.	−3326400	−1663200
$(4)(3)(2)(1)^2$	1663200	3326400	.	3326400	4989600	.	.	.	.	.	1663200
$(4)(3)(2)^2$	−831600	.	.	.	−831600	.	.	.	.	.	.
	−48	.	.	.	.	.	.	.	.	.	.
$(4)(3)^2(1)$	.	−1108800	.	−2217600	.	.	.	.	.	.	.
	.	−36	.	.	.	.	.	.	.	.	.
$(4)^2(1)^3$	.	.	2494800	1247400	415800	.	.	.	.	.	.
	.	.	16	.	.	.	.	.	.	.	.
$(4)^2(2)(1)$	.	.	.	−1247400	−1247400	.	.	.	.	.	.
	.	.	16	−32	.	.	.	.	.	.	.
$(4)^2(3)$	.	.	.	.	831600	.	.	.	.	.	.
	−48	−96	16	−48	48	.	.	.	.	.	.
$(5)(1)^6$	.	.	.	.	.	7983360	3991680	1995840	997920	1330560	665280
	.	.	.	.	.	5	.	.	.	.	.
$(5)(2)(1)^4$	.	.	.	.	.	5	−3991680	−3991680	−2993760	−3991680	−2661120
	.	.	.	.	.	.	−10	.	.	.	.
$(5)(2)^2(1)^2$	.	.	.	.	.	5	−20	1995840	2993760	.	1995840
	.	.	.	.	.	.	.	20	.	.	.
$(5)(2)^3$	.	.	.	.	.	5	−30	60	−997920	−20	.
	.	.	.	.	.	.	.	.	−40	.	.
$(5)(3)(1)^3$	.	.	.	.	.	5	−15	.	.	2661120	1330560
	.	.	.	.	.	.	.	.	.	15	.
$(5)(3)(2)(1)$	.	.	.	.	.	5	−25	.	.	.	−1330560
	.	.	.	.	.	.	.	.	.	.	−30
$(5)(3)^2$	.	−45	.	.	.	5	−30	30	.	15	−30
$(5)(4)(1)^2$	.	.	20	.	.	5	−30	45	.	30	−90
$(5)(4)(2)$	−40	.	20	−40	.	5	−20	10	.	20	−40
$(5)^2(1)$	.	.	25	.	.	10	−50	50	.	50	−50
$(6)(1)^5$	.	.	.	.	.	6	.	.	.	.	.
$(6)(2)(1)^3$	.	.	.	.	.	6	−12	.	.	.	.
$(6)(2)^2(1)$	.	.	.	.	.	6	−24	24	.	.	.
$(6)(3)(1)^2$	.	.	.	.	.	6	−18	.	.	.	.
$(6)(3)(2)$	−36	.	.	.	.	6	−30	36	.	18	−36
$(6)(4)(1)$	.	−12	24	−24	.	6	−24	12	.	24	.
$(6)(5)$	−30	−15	30	−30	.	11	−60	75	.	60	−90
$(7)(1)^4$	.	.	.	.	.	7	−7	.	.	.	.
$(7)(2)(1)^2$	.	.	.	.	.	7	−21	14	.	.	.
$(7)(2)^2$	−28	.	.	.	.	7	−35	56	.	.	−28
$(7)(3)(1)$	.	−21	28	.	.	7	−28	21	.	21	−21
$(7)(4)$	−42	−56	28	−56	28	7	−35	42	.	28	−28
$(8)(1)^3$	.	.	4	.	.	8	−16	.	.	8	.
$(8)(2)(1)$	.	.	4	−8	.	8	−32	32	.	8	−16
$(8)(3)$	−24	−48	4	−12	12	8	−40	48	.	32	−72
$(9)(1)^2$	.	.	9	.	.	9	−27	9	.	18	.
$(9)(2)$	−36	.	9	−18	.	9	−45	63	.	18	−36
$(10)(1)$	.	−10	15	−10	.	10	−40	30	.	30	−20
(11)	−33	−33	22	−33	11	11	−55	66	.	44	−66

Table 1·5·11 (*cont.*)

$w = 11$ (iv)	$a_5a_3^2$	$a_5a_4a_1^2$	$a_5a_4a_2$	$a_5^2a_1$	$a_6a_1^5$	$a_6a_2a_1^3$	$a_6a_2^2a_1$	$a_6a_3a_1^2$	$a_6a_3a_2$	$a_6a_4a_1$	a_6a_5
$(1)^{11}$	9240	13860	6930	2772	55440	27720	13860	9240	4620	2310	462
$(2)(1)^9$	-147840	-221760	-117810	-55440	-831600	-443520	-235620	-166320	-87780	-48510	-11550
$(2)^2(1)^7$	776160	1081080	651420	360360	2494800	1663200	1053360	831600	498960	318780	97020
$(2)^3(1)^5$	-1663200	-1663200	-1372140	-831600	-831600	-1663200	-1663200	-1386000	-1108800	-762300	-318780
$(2)^4(1)^3$	1247400	623700	1143450	623700	.	415800	1039500	415800	900900	519750	381150
$(2)^5(1)$	.	.	-311850	.	.	.	-207900	.	-207900	-103950	-103950
$(3)(1)^8$	221760	388080	194040	110880	2217600	1108800	554400	388080	194040	110880	27720
$(3)(2)(1)^6$	-1774080	-3049200	-1718640	-1219680	-6652800	-4435200	-2772000	-2494800	-1441440	-1108800	-388080
$(3)(2)^2(1)^4$	4435200	4158000	3603600	2772000	.	3326400	.	3880800	4158000	2772000	1386000
$(3)(2)^3(1)^2$	-3326400	-831600	-2494800	-1663200	.	.	-1663200	-277200	-2217600	-1108800	-1386000
$(3)(2)^4$	.	.	415800	1108800	2217600	1108800	554400	1108800	138600	831600	138600
$(3)^2(1)^5$	776160	2217600	1108800	1108800	.	.	554400	1108800	554400	831600	388080
$(3)^2(2)(1)^3$	-3326400	-2217600	-2217600	-2217600	.	-1108800	-1108800	-3326400	-2217600	-2772000	-1663200
$(3)^2(2)^2(1)$	2772000	.	1108800	1108800	.	.	554400	.	277200	277200	1386000
$(3)^3(1)^2$	739200	.	.	.	.	.	.	739200	369600	739200	369600
$(3)^3(2)$	-739200	.	.	.	.	.	.	-831600	-369600	-369600	-369600
$(4)(1)^7$	-277200	-498960	-249480	-166320	-4989600	-2494800	-1247400	-831600	-415800	-221760	-55440
$(4)(2)(1)^5$	1663200	3326400	1912680	1663200	4989600	4989600	3742200	3326400	2079000	1663200	665280
$(4)(2)^2(1)^3$	-2494800	-2494800	-2910600	-2494800	.	-2494800	-3742200	-2494800	-2910600	-2494800	-1663200
$(4)(2)^3(1)$	.	.	1247400	.	.	.	1247400	.	1247400	831600	831600
$(4)(3)(1)^4$	-1108800	-4989600	-2494800	-3326400	.	.	.	-1663200	-831600	-2217600	-1386000
$(4)(3)(2)(1)^2$	3326400	1663200	3326400	3326400	.	.	.	1663200	1663200	3326400	3326400
$(4)(3)(2)^2$	.	.	-831600	.	.	.	.	.	-831600	-554400	-831600
$(4)(3)^2(1)$	-1108800	.	.	.	.	.	.	.	.	.	-554400
$(4)^2(1)^3$	.	2494800	1247400	2494800	.	.	.	.	.	1247400	1247400
$(4)^2(2)(1)$	.	.	-1247400	.	.	.	.	.	.	-1247400	-1247400
$(4)^2(3)$	.	.	.	.	.	.	.	.	.	.	.
$(5)(1)^6$	221760	332640	166320	133056	7983360	3991680	1995840	1330560	665280	332640	77616
$(5)(2)(1)^4$	-1330560	-1995840	-1164240	-1330560	.	-3991680	-3991680	-3991680	-2661120	-1995840	-831600
$(5)(2)^2(1)^2$	1995840	997920	1496880	1995840	.	.	1995840	.	.	997920	1496880
$(5)(2)^3$	.	.	-498960	.	.	.	.	.	.	.	-166320
$(5)(3)(1)^3$	887040	2661120	1330560	2661120	.	.	.	2661120	1330560	2661120	1774080
$(5)(3)(2)(1)$	-2661120	.	-1330560	-2661120	.	.	.	.	-1330560	.	-2661120
$(5)(3)^2$	887040 (45)	.	.	.	.	.	.	.	.	.	443520
$(5)(4)(1)^2$	.	-1995840 (-20)	-997920	-3991680	.	.	.	.	.	-1995840	-2993760
$(5)(4)(2)$	.	-20	997920 (40)	.	.	.	.	.	.	.	997920
$(5)^2(1)$	.	-50	.	1596672 (25)	.	.	.	.	.	.	1596672
$(6)(1)^5$	.	.	.	.	-6652800 (-6)	-3326400	-1663200	-1108800	-554400	-277200	-55440
$(6)(2)(1)^3$	.	.	.	.	-6	3326400 (12)	3326400	3326400	2217600	1663200	554400
$(6)(2)^2(1)$	.	.	.	.	-6	-24	-1663200 (-24)	.	-1663200	-831600	-831600
$(6)(3)(1)^2$	.	.	.	.	-6	18	.	-2217600 (-18)	-1108800	-2217600	-1108800
$(6)(3)(2)$	.	.	.	.	-6	30	-36	-18	1108800 (36)	1663200	1663200
$(6)(4)(1)$	.	-24	.	.	-6	24	-12	-24	.	1663200 (24)	1108800
$(6)(5)$	15	-60	30	30	-6	30	-30	-30	30	30	-1330560 (-30)
$(7)(1)^4$	.	.	.	.	-7	.	.	.	.	.	.
$(7)(2)(1)^2$	.	.	.	.	-7	14	.	.	.	.	.
$(7)(2)^2$	.	.	.	.	-7	28	-28	.	.	.	.
$(7)(3)(1)$	.	.	.	.	-7	21	.	-21	.	.	.
$(7)(4)$	.	-28	28	.	-7	28	-14	-28	.	28	.
$(8)(1)^3$	.	.	.	.	-8	8	.	.	.	.	.
$(8)(2)(1)$	.	.	.	.	-8	24	-16	.	.	.	.
$(8)(3)$	24	.	.	.	-8	32	-24	-24	24	.	.
$(9)(1)^2$	.	-9	.	.	-9	18	.	-9	.	.	.
$(9)(2)$	.	.	18	.	-9	36	-36	-9	18	.	.
$(10)(1)$	.	-20	.	5	-10	30	-10	-20	.	10	.
(11)	11	-33	22	11	-11	44	-33	-33	22	22	-11

Table 1.5.11 (cont.)

$w=11$ (v)	$a_7a_1^4$	$a_7a_2a_1^2$	$a_7a_2^2$	$a_7a_3a_1$	a_7a_4	$a_8a_1^3$	$a_8a_2a_1$	a_8a_3	$a_9a_1^2$	a_9a_2	$a_{10}a_1$	a_{11}
$(1)^{11}$	7920	3960	1980	1320	330	990	495	165	110	55	11	1
$(2)(1)^9$	−166320	−87120	−45540	−31680	−8910	−27720	−14355	−5115	−3960	−2035	−495	−55
$(2)^2(1)^7$	831600	498960	293040	221760	77220	207900	117810	48510	41580	22770	6930	990
$(2)^3(1)^5$	−831600	−831600	−665280	−554400	−263340	−415800	−311850	−173250	−138600	−90090	−34650	−6930
$(2)^4(1)^3$	.	415800	623700	415800	311850	103950	259875	225225	103950	121275	51975	17325
$(2)^5(1)$	.	.	−207900	.	−103950	.	−51975	−51975	.	−51975	−10395	−10395
$(3)(1)^8$	554400	277200	138600	95040	25740	110880	55440	18810	18480	9240	2640	330
$(3)(2)(1)^6$	−3326400	−1940400	−1108800	−887040	−332640	−1108800	−609840	−249480	−277200	−147840	−55440	−9240
$(3)(2)^2(1)^4$	1663200	2494800	2217600	2217600	1247400	1663200	1386000	900900	831600	554400	277200	69300
$(3)(2)^3(1)^2$	.	−831600	−1663200	−1108800	−1108800	.	−831600	−970200	−277200	−554400	−277200	−138600
$(3)(2)^4$	.	.	415800	.	207900	.	.	34650	.	138600	.	34650
$(3)^2(1)^5$	2217600	1108800	554400	554400	277200	1108800	554400	221760	369600	184800	92400	18480
$(3)^2(2)(1)^3$	.	−1108800	−1108800	−2217600	−1663200	−1108800	−1108800	−1108800	−1108800	−739200	−554400	−184800
$(3)^2(2)^2(1)$	.	.	.	554400	831600	.	554400	1108800	.	554400	277200	277200
$(3)^3(1)^2$	.	.	.	739200	739200	.	.	369600	246400	123200	246400	123200
$(3)^3(2)$	.	.	.	.	.	.	.	−369600	.	−123200	.	−123200
$(4)(1)^7$	−1663200	−831600	−415800	−277200	−71280	−415800	−207900	−69300	−83160	−41580	−13860	−1980
$(4)(2)(1)^5$	4989600	3326400	2079000	1663200	665280	2494800	1455300	623700	831600	455380	207900	41580
$(4)(2)^2(1)^3$	.	−2494800	−2910600	−2494800	−1663200	−1247400	−1871100	−1455300	−1247400	−1039500	−623700	−207900
$(4)(2)^3(1)$	.	.	1247400	.	831600	.	.	623700	.	623700	207900	207900
$(4)(3)(1)^4$	−3326400	−1663200	−831600	−1108800	−831600	−3326400	−1663200	−693000	−1663200	−831600	−554400	−138600
$(4)(3)(2)(1)^2$	.	1663200	1663200	3326400	3326400	.	1663200	2494800	1663200	1663200	1663200	831600
$(4)(3)(2)^2$	.	.	−831600	.	−831600	.	.	−415800	.	−831600	.	−415800
$(4)(3)^2(1)$	.	.	.	−1108800	−1663200	.	.	−1108800	.	.	.	−554400
$(4)^2(1)^3$	.	.	.	.	415800	1247400	623700	207900	1247400	623700	623700	207900
$(4)^2(2)(1)$	.	.	.	.	−1247400	.	−623700	−623700	.	−623700	−623700	−623700
$(4)^2(3)$	.	.	.	.	831600	.	.	415800	.	.	.	415800
$(5)(1)^6$	3991680	1995840	997920	665280	166320	1330560	665280	221760	332640	166320	66528	11088
$(5)(2)(1)^4$	−3991680	−3991680	−2993760	−2661120	−1164240	−3991680	−2661120	−1330560	−1995840	−1164240	−66528	−166320
$(5)(2)^2(1)^2$	.	1995840	2993760	1995840	1496880	.	1995840	1995840	997920	1496880	997920	498960
$(5)(2)^3$	.	.	−997920	.	−498960	.	.	.	.	−498960	.	−166320
$(5)(3)(1)^3$	.	.	.	1330560	1330560	2661120	1330560	887040	2661120	1330560	1330560	443520
$(5)(3)(2)(1)$	.	.	.	−1330560	−1330560	.	−1330560	−2661120	.	−1330560	−1330560	−1330560
$(5)(3)^2$	.	.	.	.	.	.	.	887040	.	.	.	443520
$(5)(4)(1)^2$	.	.	.	.	−997920	.	.	.	−1995840	−997920	−1995840	−997920
$(5)(4)(2)$	.	.	.	.	997920	.	.	.	.	997920	.	997920
$(5)^2(1)$	.	.	.	.	.	.	.	.	.	.	798336	798336
$(6)(1)^5$	−6652800	−3326400	−1663200	−1108800	−277200	−3326400	−1663200	−554400	−1108800	−554400	−277200	−55440
$(6)(2)(1)^3$	.	3326400	3326400	3326400	1663200	3326400	3326400	2217600	3326400	2217600	1663200	554400
$(6)(2)^2(1)$	.	.	−1663200	.	−831600	.	.	−1663200	.	−1663200	−831600	−831600
$(6)(3)(1)^2$	.	.	.	2217600	−2217600	.	.	−1108800	−2217600	−1108800	−2217600	−1108800
$(6)(3)(2)$	.	.	.	.	.	.	.	1108800	.	1108800	.	1108800
$(6)(4)(1)$	.	.	.	.	1663200	.	.	.	.	.	1663200	1663200
$(6)(5)$	.	.	.	.	.	.	.	.	.	.	.	−1330560
$(7)(1)^4$	5702400 / 7	2851200	1425600	950400	237600	5702400	2851200	950400	2851200	1425600	950400	237600
$(7)(2)(1)^2$	7	−2851200 / −14	−2851200	−2851200	−1425600	.	−2851200	−2851200	−2851200	−2851200	−2851200	−1425600
$(7)(2)^2$	7	−28	1425600 / 28	.	712800	.	.	.	.	1425600	.	712800
$(7)(3)(1)$	7	−21	.	1900800 / 21	1900800	.	.	1900800	.	.	1900800	1900800
$(7)(4)$	7	−28	14	28	−1425600 / −28	.	.	.	.	.	.	−1425600
$(8)(1)^3$	8	.	.	.	.	−4989600 / −8	−2494800	−831600	−4989600	−2494800	−2494800	−831600
$(8)(2)(1)$	8	−16	.	.	.	.	2494800 / 16	2494800	.	2494800	2494800	2494800
$(8)(3)$	8	−24	.	24	.	−8	24	−1663200 / −24	.	.	.	−1663200
$(9)(1)^2$	9	−9	.	.	.	.	.	.	4435200 / 9	2217600	4435200	2217600
$(9)(2)$	9	−27	18	.	.	.	.	.	.	−2217600 / −18	.	−2217600
$(10)(1)$	10	−20	.	10	.	−10	10	.	10	.	−3991680 / −10	−3991680
(11)	11	−33	11	22	−11	−11	22	−11	11	−11	−11	3628800 / 11

Table 1·5·12

$w=12$ (i)	a_1^{12}	$a_2a_1^{10}$	$a_2^2a_1^8$	$a_2^3a_1^6$	$a_2^4a_1^4$	$a_2^5a_1^2$	a_2^6	$a_3a_1^9$	$a_3a_2a_1^7$	$a_3a_2^2a_1^5$	$a_3a_2^3a_1^3$
$(1)^{12}$	479001600	239500800	119750400	59875200	29937600	14968800	7484400	79833600	39916800	19958400	9979200
	1										
$(2)(1)^{10}$		−239500800	−239500800	−179625600	−119750400	−74844000	−44906400	−239500800	−159667200	−99792000	−59875200
	1	−2									
$(2)^2(1)^8$			119750400	179625600	179625600	149688000	112266000	·	119750400	139708800	119750400
	1	−4	4								
$(2)^3(1)^6$				−59875200	−119750400	−149688000	−149688000		·	−59875200	−99792000
	1	−6	12	−8							
$(2)^4(1)^4$					29937600	74844000	112266000			·	29937600
	1	−8	24	−32	16						
$(2)^5(1)^2$						−14968800	−44906400				·
	1	−10	40	−80	80	−32					
$(2)^6$							7484400				
	1	−12	60	−160	240	−192	64				
$(3)(1)^9$								159667200	79833600	39916800	19958400
	1	−3	3					3			
$(3)(2)(1)^7$									−79833600	−79833600	−59875200
	1	−5	6					3	−6		
$(3)(2)^2(1)^5$										39916800	59875200
	1	−7	16	−12				3	−12	12	
$(3)(2)^3(1)^3$											−19958400
	1	−9	30	−44	24	·	·	3	−18	36	−24
$(3)(2)^4(1)$	1	−11	48	−104	112	−48	·	3	−24	72	−96
$(3)^2(1)^6$	1	−6	9	·	·	·	·	6	−18	·	·
$(3)^2(2)(1)^4$	1	−8	21	−18	·	·	·	6	−30	36	·
$(3)^2(2)^2(1)^2$	1	−10	37	−60	36	·	·	6	−42	96	−72
$(3)^2(2)^3$	1	−12	57	−134	156	−72	·	6	−54	180	−264
$(3)^3(1)^3$	1	−9	27	−27	·	·	·	9	−54	81	·
$(3)^3(2)(1)$	1	−11	45	−81	54	·	·	9	−72	189	−162
$(3)^4$	1	−12	54	−108	81	·	·	12	−108	324	−324
$(4)(1)^8$	1	−4	2	·	·	·	·	4	·	·	·
$(4)(2)(1)^6$	1	−6	10	−4	·	·	·	4	−8	·	·
$(4)(2)^2(1)^4$	1	−8	22	−24	8	·	·	4	−16	16	·
$(4)(2)^3(1)^2$	1	−10	38	−68	56	−16	·	4	−24	48	−32
$(4)(2)^4$	1	−12	58	−144	192	−128	32	4	−32	96	−128
$(4)(3)(1)^5$	1	−7	14	−6	·	·	·	7	−24	6	·
$(4)(3)(2)(1)^3$	1	−9	28	−34	12	·	·	7	−38	54	−12
$(4)(3)(2)^2(1)$	1	−11	46	−90	80	−24	·	7	−52	130	−120
$(4)(3)^2(1)^2$	1	−10	35	−48	18	·	·	10	−66	120	−36
$(4)(3)^2(2)$	1	−12	55	−118	114	−36	·	10	−86	252	−276
$(4)^2(1)^4$	1	−8	20	−16	4	·	·	8	−32	16	·
$(4)^2(2)(1)^2$	1	−10	36	−56	36	−8	·	8	−48	80	−32
$(4)^2(2)^2$	1	−12	56	−128	148	−80	16	8	−64	176	−192
$(4)^2(3)(1)$	1	−11	44	−76	52	−12	·	11	−80	172	−96
$(4)^3$	1	−12	54	−112	108	−48	8	12	−96	240	−192
$(5)(1)^7$	1	−5	5	·	·	·	·	5	−5	·	·
$(5)(2)(1)^5$	1	−7	15	−10	·	·	·	5	−15	10	·
$(5)(2)^2(1)^3$	1	−9	29	−40	20	·	·	5	−25	40	−20
$(5)(2)^3(1)$	1	−11	47	−98	100	−40	·	5	−35	90	−100
$(5)(3)(1)^4$	1	−8	20	−15	·	·	·	8	−35	30	·
$(5)(3)(2)(1)^2$	1	−10	36	−55	30	·	·	8	−51	100	−60
$(5)(3)(2)^2$	1	−12	56	−127	140	−60	·	8	−67	202	−260
$(5)(3)^2(1)$	1	−11	44	−75	45	·	·	11	−83	195	−135
$(5)(4)(1)^3$	1	−9	27	−30	10	·	·	9	−45	50	−10
$(5)(4)(2)(1)$	1	−11	45	−84	70	−20	·	9	−63	140	−110
$(5)(4)(3)$	1	−12	54	−111	100	−30	·	12	−99	266	−250
$(5)^2(1)^2$	1	−10	35	−50	25	·	·	10	−60	100	−50
$(5)^2(2)$	1	−12	55	−120	125	−50	·	10	−80	220	−250
$(6)(1)^6$	1	−6	9	−2	·	·	·	6	−12	·	·
$(6)(2)(1)^4$	1	−8	21	−20	4	·	·	6	−24	24	·
$(6)(2)^2(1)^2$	1	−10	37	−62	44	−8	·	6	−36	72	−48
$(6)(2)^3$	1	−12	57	−136	168	−96	16	6	−48	144	−192
$(6)(3)(1)^3$	1	−9	27	−29	6	·	·	9	−48	63	−6
$(6)(3)(2)(1)$	1	−11	45	−83	64	−12	·	9	−66	159	−132
$(6)(3)^2$	1	−12	54	−110	93	−18	·	12	−102	288	−282
$(6)(4)(1)^2$	1	−10	35	−50	26	−4	·	10	−60	96	−32
$(6)(4)(2)$	1	−12	55	−120	126	−56	8	10	−80	216	−224
$(6)(5)(1)$	1	−11	44	−77	55	−10	·	11	−77	165	−115
$(6)^2$	1	−12	54	−112	105	−36	4	12	−96	252	−240
$(7)(1)^5$	1	−7	14	−7	·	·	·	7	−21	7	·
$(7)(2)(1)^3$	1	−9	28	−35	14	·	·	7	−35	49	−14
$(7)(2)^2(1)$	1	−11	46	−91	84	−28	·	7	−49	119	−112
$(7)(3)(1)^2$	1	−10	35	−49	21	·	·	10	−63	112	−42
$(7)(3)(2)$	1	−12	55	−119	119	−42	·	10	−83	238	−266
$(7)(4)(1)$	1	−11	44	−77	56	−14	·	11	−77	161	−98
$(7)(5)$	1	−12	54	−112	105	−35	·	12	−96	252	−245
$(8)(1)^4$	1	−8	20	−16	2	·	·	8	−32	24	·
$(8)(2)(1)^2$	1	−10	36	−56	34	−4	·	8	−48	88	−48
$(8)(2)^2$	1	−12	56	−128	146	−72	8	8	−64	184	−224
$(8)(3)(1)$	1	−11	44	−76	50	−6	·	11	−80	180	−120
$(8)(4)$	1	−12	54	−112	106	−40	4	12	−96	248	−224
$(9)(1)^3$	1	−9	27	−30	9	·	·	9	−45	54	−9
$(9)(2)(1)$	1	−11	45	−84	69	−18	·	9	−63	144	−117
$(9)(3)$	1	−12	54	−111	99	−27	·	12	−99	270	−261
$(10)(1)^2$	1	−10	35	−50	25	−2	·	10	−60	100	−40
$(10)(2)$	1	−12	55	−120	125	−52	4	10	−80	220	−240
$(11)(1)$	1	−11	44	−77	55	−11	·	11	−77	165	−110
(12)	1	−12	54	−112	105	−36	2	12	−96	252	−240

Table 1·5·12 (cont.)

$w = 12$ (ii)	$a_3 a_2^4 a_1$	$a_3^2 a_1^6$	$a_3^2 a_2 a_1^4$	$a_3^2 a_2^2 a_1^2$	$a_3^2 a_2^3$	$a_3^3 a_1^3$	$a_3^3 a_2 a_1$	a_3^4	$a_4 a_1^8$	$a_4 a_2 a_1^6$	$a_4 a_2^2 a_1^4$
$(1)^{12}$	4989600	13305600	6652800	3326400	1663200	2217600	1108800	369600	19958400	9979200	4989600
$(2)(1)^{10}$	−34927200	−79833600	−46569600	−26611200	−14968800	−19958400	−11088000	−4435200	−119750400	−69854400	−39916800
$(2)^2(1)^8$	89812800	119750400	99792000	71808000	49896000	59875200	39916800	19958400	59875200	89812800	79833600
$(2)^3(1)^6$	−109771200	·	−59875200	−79833600	−76507200	−59875200	−59875200	−39916800	·	−59875200	−59875200
$(2)^4(1)^4$	64864800	·	·	29937600	54885600	·	·	29937600	·	−29937600	14968800
$(2)^5(1)^2$	−14968800	·	·	·	−14968800	·	·	·	·	·	·
$(2)^6$	·	·	·	·	·	·	·	·	·	·	·
$(3)(1)^9$	9979200	53222400	26611200	13305600	6652800	13305600	6652800	2956800	159667200	79833600	39916800
$(3)(2)(1)^7$	−39916800	−159667200	−106444800	−66528000	−39916800	−79833600	−46569600	−26611200	·	−79833600	−79833600
$(3)(2)^2(1)^5$	59875200	·	79833600	93139200	79833600	119750400	99792000	79833600	·	·	39916800
$(3)(2)^3(1)^3$	−39916800	·	·	·	−39916800	−66528000	·	−59875200	−79833600	·	·
$(3)(2)^4(1)$	9979200 ÷48	·	·	·	19958400	·	·	·	·	·	·
$(3)^2(1)^6$	·	53222400 ÷9	26611200	13305600	6652800	26611200	13305600	8870400	·	·	·
$(3)^2(2)(1)^4$	·	9	−26611200 ÷−18	−26611200	−19958400	−79833600	−53222400	−53222400	·	·	·
$(3)^2(2)^2(1)^2$	·	9	−36	13305600 ÷−36	13305600	19958400	·	39916800	79833600	·	·
$(3)^2(2)^3$	144	9	−54	108	−6652800 ÷−72	·	·	·	·	·	·
$(3)^3(1)^3$	·	27	−81	·	·	17740800 ÷27	8870400	11827200	·	·	·
$(3)^3(2)(1)$	·	27	−135	162	·	27	−8870400 ÷−54	−35481600	·	·	·
$(3)^4$	·	54	−324	486	·	108	−324	5913600 ÷81	·	·	·
$(4)(1)^8$	·	·	·	·	·	·	·	·	−119750400 ÷−4	−59875200	−29937600
$(4)(2)(1)^6$	·	·	·	·	·	·	·	·	−4	59875200 ÷8	59875200
$(4)(2)^2(1)^4$	·	·	·	·	·	·	·	·	−4	16	−29937600 ÷−16
$(4)(2)^3(1)^2$	·	·	·	·	·	·	·	·	−4	24	−48
$(4)(2)^4$	64	·	·	·	·	·	·	·	−4	32	−96
$(4)(3)(1)^5$	·	12	·	·	·	·	·	·	−4	12	·
$(4)(3)(2)(1)^3$	·	12	−24	·	·	·	·	·	−4	20	−24
$(4)(3)(2)^2(1)$	24	12	−48	48	·	·	·	·	−4	28	−64
$(4)(3)^2(1)^2$	·	33	−108	18	·	36	·	·	−4	24	−36
$(4)(3)^2(2)$	72	33	−174	234	−36	36	−72	·	−4	32	−84
$(4)^2(1)^4$	·	16	·	·	·	·	·	·	−8	32	−16
$(4)^2(2)(1)^2$	·	16	−32	64	·	·	·	·	−8	48	−80
$(4)^2(2)^2$	64	16	−64	64	·	·	·	·	−8	64	−176
$(4)^2(3)(1)$	12	40	−144	48	·	48	·	·	−8	56	−112
$(4)^3$	48	48	−192	96	·	64	·	·	−12	96	−240
$(5)(1)^7$	·	·	·	·	·	·	·	·	−5	·	·
$(5)(2)(1)^5$	·	·	·	·	·	·	·	·	−5	10	·
$(5)(2)^2(1)^3$	·	·	·	·	·	·	·	·	−5	20	−20
$(5)(2)^3(1)$	40	·	·	·	·	·	·	·	−5	30	−60
$(5)(3)(1)^4$	·	15	−15	·	·	·	·	·	−5	15	·
$(5)(3)(2)(1)^2$	·	15	−45	30	·	·	·	·	−5	25	−30
$(5)(3)(2)^2$	120	15	−75	120	−60	·	·	·	−5	35	−80
$(5)(3)^2(1)$	·	39	−165	135	·	45	−45	·	−5	30	−45
$(5)(4)(1)^3$	·	20	−20	·	·	·	·	·	−9	40	−30
$(5)(4)(2)(1)$	20	20	−60	40	·	60	−60	·	−9	58	−110
$(5)(4)(3)$	60	47	−215	210	−30	60	−60	·	−9	67	−150
$(5)^2(1)^2$	·	25	−50	25	·	·	·	·	−10	50	−50
$(5)^2(2)$	100	25	−100	125	−50	·	·	·	−10	70	−150
$(6)(1)^6$	·	3	·	·	·	·	·	·	−6	6	·
$(6)(2)(1)^4$	·	3	−6	·	·	·	·	·	−6	18	−12
$(6)(2)^2(1)^2$	·	3	−12	12	·	·	·	·	−6	30	−48
$(6)(2)^3$	96	3	−18	36	−24	·	·	·	−6	42	−108
$(6)(3)(1)^3$	·	21	−45	·	·	9	·	·	−6	24	−18
$(6)(3)(2)(1)$	12	21	−87	90	·	9	−18	·	−6	36	−66
$(6)(3)^2$	36	48	−252	324	−18	72	−162	27	−6	42	−90
$(6)(4)(1)^2$	·	27	−60	6	·	12	·	·	−10	54	−72
$(6)(4)(2)$	64	27	−114	126	−12	12	−24	·	−10	74	−180
$(6)(5)(1)$	10	33	−105	75	·	15	−15	·	−11	66	−105
$(6)^2$	48	42	−180	198	−12	36	−72	9	−12	84	−180
$(7)(1)^5$	·	7	·	·	·	·	·	·	−7	14	·
$(7)(2)(1)^3$	·	7	−14	·	·	·	·	·	−7	28	−28
$(7)(2)^2(1)$	28	7	−28	28	·	·	·	·	−7	42	−84
$(7)(3)(1)^2$	·	28	−84	21	·	21	·	·	−7	35	−42
$(7)(3)(2)$	84	28	−140	189	−42	21	−42	·	−7	49	−112
$(7)(4)(1)$	14	35	−112	42	·	28	·	·	−11	70	−126
$(7)(5)$	70	42	−175	175	−35	35	−35	·	−12	84	−175
$(8)(1)^4$	·	12	−8	·	·	·	·	·	−8	24	−8
$(8)(2)(1)^2$	·	12	−32	16	·	·	·	·	−8	40	−56
$(8)(2)^2$	96	12	−56	80	−32	·	·	·	−8	56	−136
$(8)(3)(1)$	6	36	−140	96	·	36	−24	·	−8	48	−80
$(8)(4)$	56	44	−184	152	−16	48	−32	·	−12	88	−200
$(9)(1)^3$	·	18	−27	·	·	3	·	·	−9	36	−27
$(9)(2)(1)$	18	18	−63	54	·	3	−6	·	−9	54	−99
$(9)(3)$	54	45	−216	243	−27	57	−90	9	−9	63	−135
$(10)(1)^2$	·	25	−60	15	·	10	·	·	−10	50	−60
$(10)(2)$	80	25	−110	135	−30	10	−20	·	−10	70	−160
$(11)(1)$	11	33	−110	66	·	22	−11	·	−11	66	−110
(12)	60	42	−180	180	−24	40	−48	3	−12	84	−180

Table 1·5·12 (cont.)

$w=12$ (iii)	$a_4 a_3^3 a_1^2$	$a_4 a_2^4$	$a_4 a_3 a_1^5$	$a_4 a_3 a_2 a_1^3$	$a_4 a_3 a_2^2 a_1$	$a_4 a_3^2 a_1^2$	$a_4 a_3^2 a_2$	$a_4^2 a_1^4$	$a_4^2 a_2 a_1^2$	$a_4^2 a_2^2$	$a_4^2 a_3 a_1$
$(1)^{12}$	2494800	1247400	3326400	1663200	831600	554400	277200	831600	415800	207900	138600
$(2)(1)^{10}$	−22453200	−12474000	−29937600	−16632000	−9147600	−6652800	−3603600	−9979200	−5405400	−2910600	−2079000
$(2)^2(1)^8$	59875200	41164200	69854400	49896000	33264000	26611200	16632000	34927200	22453200	13929300	10810800
$(2)^3(1)^6$	−69854400	−64864800	−29937600	−49896000	−49896000	−39916800	−33264000	−29937600	−32432400	−27442800	−22453200
$(2)^4(1)^4$	37422000	53638200	·	14968800	32432400	14968800	27442800	7484400	18711000	25571700	16216200
$(2)^5(1)^2$	−7484400	−22453200	·	·	−7484400	·	−7484400	·	−3742200	−11226600	−3742200
$(2)^6$	·	3742200	·	·	·	·	·	·	·	1871100	·
$(3)(1)^9$	19958400	9979200	33264000	16632000	8316000	6652800	3326400	13305600	6652800	3326400	2494800
$(3)(2)(1)^7$	−59875200	−39916800	−119750400	−76507200	−46569600	−46569600	−26611200	−79833600	−46569600	−26611200	−23284800
$(3)(2)^2(1)^5$	59875200	59875200	19958400	69854400	73180800	86486400	66528000	39916800	59875200	53222400	58212000
$(3)(2)^3(1)^3$	−19958400	−39916800	·	−9979200	−39916800	−19958400	−53222400	·	−19958400	−39916800	−29937600
$(3)(2)^4(1)$	·	9979200	·	·	4989600	·	9979200	·	·	9979200	2494800
$(3)^2(1)^6$	·	·	53222400	26611200	13305600	19958400	9979200	53222400	26611200	13305600	13305600
$(3)^2(2)(1)^4$	·	·	·	−26611200	−26611200	−66528000	−43243200	·	−26611200	−26611200	−53222400
$(3)^2(2)^2(1)^2$	·	·	·	·	13305600	6652800	36590400	·	·	13305600	13305600
$(3)^2(2)^3$	·	·	·	·	·	·	−3326400	·	·	·	·
$(3)^3(1)^3$	·	·	·	·	·	17740800	8870400	·	·	·	17740800
$(3)^3(2)(1)$	·	·	·	·	·	·	−8870400	·	·	·	·
$(3)^4$	·	·	·	·	·	·	·	·	·	·	·
$(4)(1)^8$	−14968800	−7484400	−19958400	−9979200	−4989600	−3326400	−1663200	−9979200	−4989600	−2494800	−1663200
$(4)(2)(1)^6$	44906400	29937600	59875200	39916800	24948000	19958400	11642400	59875200	34927200	19958400	14968800
$(4)(2)^2(1)^4$	−44906400	−44906400	·	−29937600	−34927200	−29937600	−24948000	−29937600	−44906400	−39916800	−34927200
$(4)(2)^3(1)^2$	14968800 (÷32)	29937600	·	·	14968800	·	14968800	·	14968800	29937600	14968800
$(4)(2)^4$	128	−7484400 (÷−64)	·	·	·	·	·	·	·	−7484400	·
$(4)(3)(1)^5$	·	·	−39916800 (÷−12)	−19958400	−9979200	−13305600	−6652800	−79833600	−39916800	−19958400	−16632000
$(4)(3)(2)(1)^3$	·	·	−12	19958400 (÷24)	19958400	39916800	26611200	·	39916800	39916800	59875200
$(4)(3)(2)^2(1)$	48	·	−12	48	−9979200 (÷−48)	·	−19958400	·	·	−19958400	−9979200
$(4)(3)^2(1)^2$	·	·	−24	72	·	−13305600 (÷−36)	−6652800	·	·	·	−26611200
$(4)(3)^2(2)$	72	·	−24	120	−144	−36	6652800 (÷72)	·	·	·	4989600
$(4)^2(1)^4$	·	·	−32	·	·	·	·	29937600 (÷16)	14968800	7484400	4989600
$(4)^2(2)(1)^2$	32	·	−32	64	·	·	·	16	−14968800 (÷−32)	−14968800	−14968800
$(4)^2(2)^2$	192	−64	−32	128	−128	·	·	16	−64	7484400 (÷64)	·
$(4)^2(3)(1)$	48	·	−56	192	−48	−96	·	16	−48	·	9979200 (÷48)
$(4)^3$	192	−48	−96	384	−192	−192	·	48	−192	96	192
$(5)(1)^7$	·	·	·	·	·	·	·	·	·	·	·
$(5)(2)(1)^5$	·	·	·	·	·	·	·	·	·	·	·
$(5)(2)^2(1)^3$	·	·	·	·	·	·	·	·	·	·	·
$(5)(2)^3(1)$	40	·	·	·	·	·	·	·	·	·	·
$(5)(3)(1)^4$	·	·	−15	·	·	·	·	·	·	·	·
$(5)(3)(2)(1)^2$	·	·	−15	30	·	·	·	·	·	·	·
$(5)(3)(2)^2$	60	·	−15	60	−60	·	·	·	·	·	·
$(5)(3)^2(1)$	·	·	−30	90	·	·	−45	·	·	·	·
$(5)(4)(1)^3$	·	·	−40	20	·	·	·	20	·	·	·
$(5)(4)(2)(1)$	60	·	−40	100	−40	·	·	20	−40	·	·
$(5)(4)(3)$	90	·	−67	260	−150	−120	60	20	−60	·	60
$(5)^2(1)^2$	·	·	−50	50	·	·	·	25	·	·	·
$(5)^2(2)$	100	·	−50	150	−100	·	·	25	−50	·	·
$(6)(1)^6$	·	·	·	·	·	·	·	·	·	·	·
$(6)(2)(1)^4$	·	·	·	·	·	·	·	·	·	·	·
$(6)(2)^2(1)^2$	24	·	·	·	·	·	·	·	·	·	·
$(6)(2)^3$	120	−48	·	·	·	·	·	·	·	·	·
$(6)(3)(1)^3$	·	·	−18	18	·	·	·	·	·	·	·
$(6)(3)(2)(1)$	36	·	−18	54	−36	·	·	·	·	·	·
$(6)(3)^2$	54	·	−36	144	−108	−54	54	·	·	·	·
$(6)(4)(1)^2$	20	·	−48	72	·	−12	·	24	−24	·	·
$(6)(4)(2)$	164	−40	−48	168	−144	−12	24	24	−72	48	·
$(6)(5)(1)$	40	·	−60	120	−30	−15	·	30	−30	·	·
$(6)^2$	132	−24	−72	216	−144	−36	36	36	−72	36	·
$(7)(1)^5$	·	·	−7	·	·	·	·	·	·	·	·
$(7)(2)(1)^3$	·	·	−7	14	·	·	·	·	·	·	·
$(7)(2)^2(1)$	56	·	−7	28	−28	·	·	·	·	·	·
$(7)(3)(1)^2$	·	·	−28	63	·	−21	·	·	·	·	·
$(7)(3)(2)$	84	·	−28	119	−126	−21	42	28	−56	·	·
$(7)(4)(1)$	56	·	−63	168	−42	−56	·	35	−70	·	·
$(7)(5)$	105	·	−77	245	−140	−70	35	35	−70	·	35
$(8)(1)^4$	·	·	−16	·	·	·	·	4	·	·	·
$(8)(2)(1)^2$	16	·	−16	32	·	·	·	4	−8	·	·
$(8)(2)^2$	128	−32	−16	64	−64	·	·	4	−16	16	·
$(8)(3)(1)$	24	·	−40	120	−24	−48	·	4	−12	·	12
$(8)(4)$	144	−24	−80	288	−160	−112	32	36	−112	40	80
$(9)(1)^3$	·	·	−27	18	·	·	·	9	·	·	·
$(9)(2)(1)$	54	·	−27	72	−36	·	·	9	−18	·	·
$(9)(3)$	81	·	−54	207	−135	−81	54	9	−27	·	27
$(10)(1)^2$	10	·	−40	60	·	−10	·	15	−10	·	·
$(10)(2)$	130	−20	−40	140	−120	−10	20	15	−40	20	·
$(11)(1)$	44	−12	−55	132	−33	−33	·	22	−33	·	11
(12)	120	−12	−72	240	−144	−72	36	30	−72	18	36

Table 1·5·12 (*cont.*)

$w=12$ (iv)	a_4^3	$a_5a_1^7$	$a_5a_2a_1^5$	$a_5a_2^2a_1^3$	$a_5a_2^3a_1$	$a_5a_3a_1^4$	$a_5a_3a_2a_1^2$	$a_5a_3a_2^2$	$a_5a_3^2a_1$	$a_5a_4a_1^3$	$a_5a_4a_2a_1$
$(1)^{12}$	34650	3991680	1995840	997920	498960	665280	332640	166320	110880	166320	83160
$(2)(1)^{10}$	−623700	−39916800	−21954240	−11975040	−6486480	−8648640	−4656960	−2494800	−1774080	−2661120	−1413720
$(2)^2(1)^8$	4054050	59875200	49896000	35925120	23950080	29937600	19293120	11975040	9313920	12972060	7817040
$(2)^3(1)^6$	−11226600	·	−29937600	−39916800	−37920960	−29937600	−29937600	−24615360	−19958400	−19958400	−16465680
$(2)^4(1)^4$	12162150	·	·	14968800	27442800	·	14968800	·	22453200	14968800	13721400
$(2)^5(1)^2$	−5613300	·	·	·	−7484400	·	·	−7484400	·	·	−3742200
$(2)^6$	935550	·	·	·	·	·	·	·	·	·	·
$(3)(1)^9$	831600	79833600	39916800	19958400	9979200	14636160	7318080	3659040	2661120	4656960	2328480
$(3)(2)(1)^7$	−9979200	−79833600	−79833600	−59875200	−39916800	−66528000	−40582080	−23950080	−21288960	−36590400	−20623680
$(3)(2)^2(1)^5$	34927200	·	39916800	59875200	59875200	59875200	63201600	51891840	53222400	49896000	43243200
$(3)(2)^3(1)^3$	−29937600	·	·	−19958400	−39916800	·	−29937600	−46569600	−39916800	−9979200	−29937600
$(3)(2)^4(1)$	7484400	·	·	·	9979200	·	·	14968800	·	·	4989600
$(3)^2(1)^6$	6652800	·	·	·	·	26611200	13305600	6652800	9313920	26611200	13305600
$(3)^2(2)(1)^4$	−39916800	·	·	·	·	−26611200	−26611200	−19958400	−39916800	−26611200	−26611200
$(3)^2(2)^2(1)^2$	19958400	·	·	·	·	·	13305600	19958400	33264000	·	13305600
$(3)^2(2)^3$	·	·	·	·	·	·	·	−6652800	·	·	·
$(3)^3(1)^3$	17740800	·	·	·	·	·	·	·	8870400	·	·
$(3)^3(2)(1)$	·	·	·	·	·	·	·	·	−8870400	·	·
$(3)^4$	·	·	·	·	·	·	·	·	·	·	·
$(4)(1)^8$	−623700	−119750400	−59875200	−29937600	−14968800	−19958400	−9979200	−4989600	−3326400	−5987520	−2993760
$(4)(2)(1)^6$	7484400	·	59875200	59875200	44906400	59875200	39916800	24948000	19958400	39916800	22952160
$(4)(2)^2(1)^4$	−26195400	·	·	−29937600	−44906400	·	−29937600	−34927200	−29937600	−29937600	−34927200
$(4)(2)^3(1)^2$	22453200	·	·	·	14968800	·	·	14968800	·	·	14968800
$(4)(2)^4$	−5613300	·	·	·	·	·	·	·	·	·	·
$(4)(3)(1)^5$	−9979200	·	·	·	·	−39916800	−19958400	−9979200	−13305600	−59875200	−29937600
$(4)(3)(2)(1)^3$	59875200	·	·	·	·	·	19958400	19958400	39916800	19958400	39916800
$(4)(3)(2)^2(1)$	−29937600	·	·	·	·	·	·	−9979200	·	·	−9979200
$(4)(3)^2(1)^2$	−39916800	·	·	·	·	·	·	·	−13305600	·	·
$(4)(3)^2(2)$	·	·	·	·	·	·	·	·	·	·	·
$(4)^2(1)^4$	3742200	·	·	·	·	·	·	·	·	29937600	14968800
$(4)^2(2)(1)^2$	−22453200	·	·	·	·	·	·	·	·	·	−14968800
$(4)^2(2)^2$	11226600	·	·	·	·	·	·	·	·	·	·
$(4)^2(3)(1)$	29937600	·	·	·	·	·	·	·	·	·	·
$(4)^3$	−7484400 / −64	·	·	·	·	·	·	·	·	·	·
$(5)(1)^7$	·	95800320 / 5	47900160	23950080	11975040	15966720	7983360	3991680	2661120	3991680	1995840
$(5)(2)(1)^5$	·	5	−47900160 / −10	−47900160	−35925120	−47900160	−31933440	−19958400	−15966720	−23950080	−13970880
$(5)(2)^2(1)^3$	·	5	−20	23950080 / 20	35925120	·	23950080	27941760	23950080	11975040	17962560
$(5)(2)^3(1)$	·	5	−30	60	−11975040 / −40	·	·	−11975040	·	·	−5987520
$(5)(3)(1)^4$	·	5	−15	·	·	31933440 / 15	15966720	7983360	10644480	31933440	15966720
$(5)(3)(2)(1)^2$	·	5	−25	30	·	15	−15966720 / −30	−15966720	−31933440	·	−15966720
$(5)(3)(2)^2$	·	5	−35	80	−60	15	−60	7983360 / 60	·	·	·
$(5)(3)^2(1)$	·	5	−30	45	·	30	−90	·	10644480 / 45	·	·
$(5)(4)(1)^3$	·	5	−20	10	·	20	·	·	·	−23950080 / −20	−11975040
$(5)(4)(2)(1)$	·	5	−30	50	−20	20	−40	·	·	−20	11975040 / 40
$(5)(4)(3)$	·	5	−35	70	−30	35	−120	30	60	−20	60
$(5)^2(1)^2$	·	10	−50	50	·	50	−50	·	·	−50	·
$(5)^2(2)$	·	10	−70	150	−100	50	−150	100	·	−50	100
$(6)(1)^6$	·	6	·	·	·	·	·	·	·	·	·
$(6)(2)(1)^4$	·	6	−12	·	·	·	·	·	·	·	·
$(6)(2)^2(1)^2$	·	6	−24	24	·	·	·	·	·	·	·
$(6)(2)^3$	·	6	−36	72	−48	·	·	·	·	·	·
$(6)(3)(1)^3$	·	6	−18	·	·	18	·	·	·	·	·
$(6)(3)(2)(1)$	·	6	−30	36	·	18	−36	·	·	·	·
$(6)(3)^2$	·	6	−36	54	·	36	−108	·	54	·	·
$(6)(4)(1)^2$	·	6	−24	12	·	24	·	·	·	−24	·
$(6)(4)(2)$	·	6	−36	60	−24	24	−48	·	·	−24	48
$(6)(5)(1)$	·	11	−60	75	−10	60	−90	·	15	−60	30
$(6)^2$	·	12	−72	108	−24	72	−144	·	36	−72	72
$(7)(1)^5$	·	7	−7	·	·	·	·	·	·	·	·
$(7)(2)(1)^3$	·	7	−21	14	·	·	·	·	·	·	·
$(7)(2)^2(1)$	·	7	−35	56	−28	·	·	·	·	·	·
$(7)(3)(1)^2$	·	7	−28	21	·	21	−21	·	·	·	·
$(7)(3)(2)$	·	7	−42	77	−42	21	−63	42	·	·	·
$(7)(4)(1)$	·	7	−35	42	−14	28	−28	·	·	−28	28
$(7)(5)$	·	12	−77	140	−70	70	−175	70	35	−70	105
$(8)(1)^4$	·	8	−16	·	·	·	·	·	·	·	·
$(8)(2)(1)^2$	·	8	−32	32	·	8	−16	·	·	·	·
$(8)(2)^2$	·	8	−48	96	−64	8	−32	32	·	·	·
$(8)(3)(1)$	·	8	−40	48	·	32	−72	·	24	·	·
$(8)(4)$	−16	8	−48	80	−32	40	−96	16	32	−32	64
$(9)(1)^3$	·	9	−27	9	·	18	·	·	·	−9	·
$(9)(2)(1)$	·	9	−45	63	−18	18	−36	·	·	−9	18
$(9)(3)$	·	9	−54	90	−27	45	−135	27	54	−9	27
$(10)(1)^2$	·	10	−40	30	·	30	−20	·	·	−20	·
$(10)(2)$	·	10	−60	110	−60	30	−80	40	·	−20	40
$(11)(1)$	·	11	−55	66	−11	44	−66	·	11	−33	22
(12)	−4	12	−72	120	−48	60	−144	36	36	−48	72

Table 1·5·12 (*cont.*)

$w=12$ (v)	$a_5a_4a_3$	$a_5^2a_1^2$	$a_5^2a_2$	$a_6a_1^6$	$a_6a_2a_1^4$	$a_6a_2^2a_1^2$	$a_6a_2^3$	$a_6a_3a_1^3$	$a_6a_3a_2a_1$	$a_6a_3^2$	$a_6a_4a_1^2$
$(1)^{12}$	27720	33264	16632	665280	332640	166320	83160	110880	55440	18480	27720
$(2)(1)^{10}$	-526680	-665280	-349272	-9979200	-5322240	-2827440	-1496880	-1995840	-1053360	-388080	-582120
$(2)^2(1)^8$	3492720	4324320	2494800	29937600	19958400	12640320	7733880	9979200	5987520	2661120	3825360
$(2)^3(1)^6$	-9812880	-9979200	-7151760	-9979200	-19958400	-16299360	-16209360	-16632000	-13305600	-7761600	-9147600
$(2)^4(1)^4$	11226600	7484400	8731800	.	4989600	12474000	16216200	4989600	10810800	9147600	6237000
$(2)^5(1)^2$	-3742200	.	-3742200	.	.	-2494800	-7484400	.	-2494800	-2494800	-1247400
$(2)^6$	.	.	.	.	.	.	1247400	.	.	.	.
$(3)(1)^9$	831600	1330560	665280	26611200	13305600	6652800	3326400	4656960	2328480	813120	1330560
$(3)(2)(1)^7$	-9313920	-14636160	-7983360	-79833600	-53222400	-33264000	-19958400	-29937600	-17297280	-7983360	-13305600
$(3)(2)^2(1)^5$	30935520	33264000	23950080	.	39916800	46569600	39916800	49896000	39916800	26611200	33264000
$(3)(2)^3(1)^3$	-33264000	-19958400	-26611200	.	.	-19958400	-33264000	-3326400	-26611200	-31046400	-13305600
$(3)(2)^4(1)$	7484400	.	9979200	.	.	.	9979200	.	1663200	3326400	.
$(3)^2(1)^6$	5987520	13305600	6652800	26611200	13305600	6652800	3326400	13305600	6652800	3769920	9979200
$(3)^2(2)(1)^4$	-29937600	-26611200	-19958400	.	-13305600	-13305600	-13305600	-39916800	.	-23284800	-33264000
$(3)^2(2)^2(1)^2$	29937600	13305600	19958400	.	.	6652800	6652800	9979200	19958400	36590400	3326400
$(3)^2(2)^3$	-3326400	.	-6652800	.	.	.	.	-3326400	.	-1108800	.
$(3)^3(1)^3$	8870400	.	.	.	.	.	.	8870400	4435200	5913600	8870400
$(3)^3(2)(1)$	-8870400	.	.	.	.	.	.	.	-4435200	-17740800	.
$(3)^4$	.	.	.	.	.	.	.	.	.	2956800	.
$(4)(1)^8$	-997920	-1995840	-997920	-59875200	-29937600	-14968800	-7484400	-9979200	-4989600	-1663200	-2661120
$(4)(2)(1)^6$	9646560	19958400	10077120	59875200	59875200	44906400	29937600	39916800	24948000	11642400	19958400
$(4)(2)^2(1)^4$	-24948000	-29937600	-24948000	.	-29937600	-44906400	-44906400	-29937600	-34927200	-24948000	-29937600
$(4)(2)^3(1)^2$	14968800	.	14968800	.	.	14968800	29937600	.	14968800	14968800	9979200
$(4)(2)^4$	.	.	.	.	.	.	-7484400	.	.	.	.
$(4)(3)(1)^5$	-11975040	-39916800	-19958400	.	.	.	.	-19958400	-9979200	-6652800	-26611200
$(4)(3)(2)(1)^3$	46569600	39916800	39916800	.	.	.	.	19958400	19958400	26611200	39916800
$(4)(3)(2)^2(1)$	-19958400	.	-19958400	.	.	.	.	.	-9979200	-19958400	-6652800
$(4)(3)^2(1)^2$	-19958400	.	.	.	.	.	.	.	.	-6652800	-6652800
$(4)(3)^2(2)$	6652800	.	.	.	.	.	.	.	.	6652800	.
$(4)^2(1)^4$	4989600	29937600	14968800	.	.	.	.	.	.	.	14968800
$(4)^2(2)(1)^2$	-14968800	.	-14968800	.	.	.	.	.	.	.	-14968800
$(4)^2(2)^2$	.	.	.	.	.	.	.	.	.	.	.
$(4)^2(3)(1)$	9979200	.	.	.	.	.	.	.	.	.	.
$(4)^3$	.	.	.	.	.	.	.	.	.	.	.
$(5)(1)^7$	665280	1596672	798336	95800320	47900160	23950080	11975040	15966720	798336	2661120	3991680
$(5)(2)(1)^5$	-5987520	-15966720	-8781696	.	-47900160	-47900160	-35925120	-47900160	-31933440	-15966720	-23950080
$(5)(2)^2(1)^3$	13070880	23950080	19958400	.	.	.	23950080	35925120	23950080	23950080	11975040
$(5)(2)^3(1)$	-5987520	.	-11975040	.	.	.	-11975040	.	.	.	.
$(5)(3)(1)^4$	6652800	31933440	15966720	.	.	.	.	31933440	15966720	10644480	31933440
$(5)(3)(2)(1)^2$	-23950080	-31933440	-31933440	.	.	.	.	.	-15966720	-31933440	.
$(5)(3)(2)^2$	3991680	.	15966720	.	.	.	.	.	.	10644480	.
$(5)(3)^2(1)$	10644480	.	.	.	.	.	.	.	.	10644480	.
$(5)(4)(1)^3$	-3991680	-47900160	-23950080	.	.	.	.	.	.	.	-23950080
$(5)(4)(2)(1)$	11975040	.	23950080	.	.	.	.	.	.	.	.
$(5)(4)(3)$	-7983360 / -60	.	.	.	.	.	.	.	.	.	.
$(5)^2(1)^2$	.	19160064 / 25	9580032	.	.	.	.	.	.	.	.
$(5)^2(2)$	.	25	-9580032 / -50	.	.	.	.	.	.	.	.
$(6)(1)^6$	.	.	.	-79833600 / -6	-39916800	-19958400	-9979200	-13305600	-6652800	-2217600	-3326400
$(6)(2)(1)^4$	.	.	.	-6	39916800 / 12	39916800	29937600	39916800	26611200	13305600	19958400
$(6)(2)^2(1)^2$	.	.	.	-6	24	-19958400 / -24	-29937600	.	-19958400	-19958400	-9979200
$(6)(2)^3$	.	.	.	-6	36	-72	9979200 / 48	.	.	.	.
$(6)(3)(1)^3$	.	.	.	-6	18	.	.	-26611200 / -18	-13305600	-8870400	-26611200
$(6)(3)(2)(1)$	.	.	.	-6	30	-36	.	-18	13305600 / 36	26611200	.
$(6)(3)^2$	.	.	.	-6	36	-54	.	-36	108	-8870400 / -54	19958400
$(6)(4)(1)^2$	.	.	.	-6	24	-12	.	-24	.	.	24
$(6)(4)(2)$	.	.	.	-6	36	-60	24	-24	48	.	24
$(6)(5)(1)$	.	30	.	-6	30	-30	24	-30	30	-36	30
$(6)^2$	.	36	.	-12	72	-108	24	-72	144	.	72
$(7)(1)^5$	.	.	.	-7	.	.	.	.	.	.	.
$(7)(2)(1)^3$	.	.	.	-7	14	.	.	.	.	.	.
$(7)(2)^2(1)$	.	.	.	-7	28	-28	.	-21	.	.	.
$(7)(3)(1)^2$	.	.	.	-7	21	.	.	-21	42	.	.
$(7)(3)(2)$	.	.	.	-7	35	-42	.	-28	.	.	28
$(7)(4)(1)$	.	.	.	-7	28	-14	.	-28	.	.	35
$(7)(5)$	-35	35	-35	-7	35	-35	.	.	35	.	35
$(8)(1)^4$	.	.	.	-8	.	.	.	.	.	.	.
$(8)(2)(1)^2$	.	.	.	-8	24	-16	.	.	.	.	.
$(8)(2)^2$	.	.	.	-8	40	-64	32	.	.	.	.
$(8)(3)(1)$	.	.	.	-8	32	-24	.	-24	24	.	.
$(8)(4)$	-32	.	.	-8	40	-48	16	-32	32	.	32
$(9)(1)^3$	.	.	.	-9	18	-36	.	-9	.	.	.
$(9)(2)(1)$	.	.	.	-9	36	-54	.	-36	18	.	.
$(9)(3)$	-27	.	.	-9	45	.	.	-36	81	-27	.
$(10)(1)^2$	.	5	.	-10	30	-10	.	-20	.	.	10
$(10)(2)$	.	5	-10	-10	50	-70	20	-20	40	.	10
$(11)(1)$	.	11	-12	-11	44	-33	12	-33	22	-12	22
(12)	-24	18	-12	-12	60	-72	12	-48	72	-12	36

Table 1·5·12 (cont.)

Main table (large entries). Columns are headed by the monomial products shown; the first column gives the partition of $w = 12$.

$w=12$ (vi)	$a_6a_4a_2$	$a_6a_5a_1$	a_6^2	$a_7a_1^5$	$a_7a_2a_1^3$	$a_7a_2^2a_1$	$a_7a_3a_1^2$	$a_7a_3a_2$	$a_7a_4a_1$	a_7a_5	$a_8a_1^4$
$(1)^{12}$	13860	5544	924	95040	47520	23760	15840	7920	3960	792	11880
$(2)(1)^{10}$	−304920	−138600	−27720	−1995840	−1045440	−546480	−380160	−198000	−106920	−24552	−332640
$(2)^2(1)^8$	2203740	1164240	291060	9979200	5987520	3516480	2661120	1520640	926640	261360	2494800
$(2)^3(1)^6$	−6486480	−3825360	−1275120	−9979200	−9979200	−7983360	−6652800	−4656960	−3160080	−1164240	−4989600
$(2)^4(1)^4$	7692300	4573800	2286900	.	4989600	7484400	4989600	5821200	3742200	2079000	1247400
$(2)^5(1)^2$	−3742200	−1247400	−1247400	.	.	−2494800	.	−2494800	−1247400	−1247400	.
$(2)^6$	623700	.	207900	.	.	.	.	.	.	.	.
$(3)(1)^9$	665280	332640	73920	6652800	3326400	1663200	1140480	570240	308880	71280	1330560
$(3)(2)(1)^7$	−7318080	−4656960	−1330560	−39916800	−23284800	−13305600	−10644480	−5892480	−3991680	−1235520	−13305600
$(3)(2)^2(1)^5$	23284800	16632000	6652800	19958400	29937600	26611200	26611200	18627840	14968800	6320160	19958400
$(3)(2)^3(1)^3$	−23284800	−16632000	−11088000	.	−9979200	−19958400	−13305600	−19958400	−13305600	−9979200	.
$(3)(2)^4(1)$	6652800	1663200	3326400	.	.	4989600	.	.	2494800	4158000	.
$(3)^2(1)^6$	4989600	4656960	1552320	26611200	13305600	6652800	6652800	3326400	3326400	1330560	13305600
$(3)^2(2)(1)^4$	−21621600	−19958400	−9979200	.	−13305600	−13305600	−26611200	6652800	−16632000	−19958400	−13305600
$(3)^2(2)^2(1)^2$	18295200	16632000	16632000	.	.	6652800	.	6652800	16632000	9979200	13305600
$(3)^2(2)^3$	−1663200	.	−1108800	.	.	.	.	−3326400	.	.	.
$(3)^3(1)^3$	4435200	4435200	2956800	.	.	.	.	8870400	4435200	8870400	4435200
$(3)^3(2)(1)$	−4435200	−4435200	−8870400	.	.	.	.	.	−4435200	.	−4435200
$(3)^4$	.	.	1478400	.	.	.	.	.	.	.	.
$(4)(1)^8$	−1330560	−665280	−166320	−19958400	−9979200	−4989600	−3326400	−1663200	−855360	−190080	−4989600
$(4)(2)(1)^6$	11309760	7983360	2661120	59875200	39916800	24948000	19958400	11642400	7983360	2661120	29937600
$(4)(2)^2(1)^4$	−24948000	−19958400	−9979200	.	−29937600	−34927200	−29937600	−24948000	−19958400	−9979200	−14968800
$(4)(2)^3(1)^2$	19958400	9979200	9979200	.	.	14968800	.	14968800	9979200	9979200	.
$(4)(2)^4$	−4989600	.	−2494800	.	.	.	.	.	.	.	.
$(4)(3)(1)^5$	−13305600	−16632000	−6652800	−39916800	−19958400	−9979200	−13305600	−6652800	−9979200	−5322240	−39916800
$(4)(3)(2)(1)^3$	33264000	39916800	26611200	.	19958400	19958400	39916800	26611200	39916800	26611200	.
$(4)(3)(2)^2(1)$	−19958400	−9979200	−19958400	.	.	−9979200	.	−19958400	−9979200	−19958400	.
$(4)(3)^2(1)^2$	−3326400	−6652800	−6652800	.	.	.	−13305600	.	−6652800	−13305600	.
$(4)(3)^2(2)$	3326400	.	6652800	.	.	.	.	6652800	.	6652800	.
$(4)^2(1)^4$	7484400	14968800	7484400	.	.	.	.	.	4989600	4989600	14968800
$(4)^2(2)(1)^2$	−14968800	−14968800	−14968800	.	.	.	.	.	−14968800	−14968800	14968800
$(4)^2(2)^2$	7484400	.	7484400	.	.	.	.	.	.	.	.
$(4)^2(3)(1)$	.	.	.	.	.	.	.	.	9979200	9979200	.
$(4)^3$	.	.	.	.	.	.	.	.	.	.	.
$(5)(1)^7$	1995840	931392	266112	47900160	23950080	11975040	7983360	3991680	1995840	418176	15966720
$(5)(2)(1)^5$	−13970880	−9979200	−3991680	−47900160	−47900160	−35925120	−31933440	−19958400	−13970880	−4790016	−47900160
$(5)(2)^2(1)^3$	17962560	17962560	11975040	.	23950080	35925120	23950080	27941760	17962560	11975040	.
$(5)(2)^3(1)$	−5987520	−1995840	−3991680	.	.	−11975040	.	−11975040	−5987520	−7983360	.
$(5)(3)(1)^4$	15966720	21288960	10644480	.	.	.	15966720	7983360	15966720	9313920	31933440
$(5)(3)(2)(1)^2$	−15966720	−31933440	−31933440	.	.	.	−15966720	−15966720	−15966720	−23950080	−23950080
$(5)(3)^2$	.	.	.	.	.	.	.	7983360	.	.	.
$(5)(3)^2(1)$	.	5322240	1064480	.	.	.	.	.	.	11975040	.
$(5)(4)(1)^3$	−11975040	−35925120	−23950080	.	.	.	.	.	−11975040	−15966720	5322240
$(5)(4)(2)(1)$	11975040	11975040	23950080	.	.	.	.	.	11975040	23950080	.
$(5)(4)(3)$	.	.	.	.	.	.	.	.	.	−7983360	.
$(5)^2(1)^2$	.	19160064	19160064	.	.	.	.	.	.	9580032	.
$(5)^2(2)$	.	.	.	.	.	.	.	.	.	−9580032	.
$(6)(1)^6$	−1663200	−665280	−221760	−79833600	−39916800	−19958400	−13305600	−6652800	−3326400	−665280	−39916800
$(6)(2)(1)^4$	11642400	6652800	3326400	.	39916800	39916800	39916800	26611200	19958400	6652800	39916800
$(6)(2)^2(1)^2$	−14968800	−9979200	−9979200	.	.	−19958400	.	−19958400	−9979200	−9979200	.
$(6)(2)^3$	4989600	.	3326400	.	.	.	.	.	.	.	.
$(6)(3)(1)^3$	−13305600	−13305600	−8870400	.	.	.	−26611200	−13305600	−26611200	−13305600	.
$(6)(3)(2)(1)$	13305600	13305600	26611200	.	.	.	.	13305600	.	13305600	.
$(6)(3)^2$	.	.	−8870400	.	.	.	.	.	.	.	.
$(6)(4)(1)^2$	9979200	19958400	19958400	.	.	.	.	.	19958400	19958400	.
$(6)(4)(2)$	−9979200	.	−19958400	.	.	.	.	.	.	.	.
$(6)(5)(1)$	.	−15966720	−31933440	.	.	.	.	.	.	−15966720	.
$(6)^2$	.	.	1330560	.	.	.	.	.	.	.	.
$(7)(1)^5$	.	.	.	68428800	34214400	17107200	11404800	5702400	2851200	570240	68428800
$(7)(2)(1)^3$	.	.	.	.	−34214400	−34214400	−34214400	−22809600	−17107200	−5702400	.
$(7)(2)^2(1)$	.	.	.	.	.	17107200	.	17107200	8553600	8553600	.
$(7)(3)(1)^2$	.	.	.	.	.	.	22809600	11404800	22809600	11404800	.
$(7)(3)(2)$	.	.	.	.	.	.	.	−11404800	.	−11404800	.
$(7)(4)(1)$	.	.	.	.	.	.	.	.	−17107200	−17107200	.
$(7)(5)$	.	.	.	.	.	.	.	.	.	13685760	.
$(8)(1)^4$	.	.	.	.	.	.	.	.	.	.	−59875200

Companion integer entries (printed as small figures within the lower rows):

$w=12$ (vi)	$a_6a_4a_2$	$a_6a_5a_1$	a_6^2	$a_7a_1^5$	$a_7a_2a_1^3$	$a_7a_2^2a_1$	$a_7a_3a_1^2$	$a_7a_3a_2$	$a_7a_4a_1$	a_7a_5	$a_8a_1^4$
$(6)(4)(2)$	−48	.	.	.	.	.	.	.	.	.	.
$(6)(5)(1)$	.	−30	.	.	.	.	.	.	.	.	.
$(6)^2$	−72	−72	36	.	.	.	.	.	.	.	.
$(7)(1)^5$	.	.	.	7	.	.	.	.	.	.	.
$(7)(2)(1)^3$	.	.	.	7	−14	.	.	.	.	.	.
$(7)(2)^2(1)$	.	.	.	7	−28	28	.	.	.	.	.
$(7)(3)(1)^2$	.	.	.	7	−21	.	21	.	.	.	.
$(7)(3)(2)$	.	.	.	7	−35	42	21	−42	.	.	.
$(7)(4)(1)$	.	.	.	7	−28	14	28	.	−28	.	.
$(7)(5)$	.	−35	.	7	−35	35	35	−35	−35	35	.
$(8)(1)^4$	.	.	.	8	.	.	.	.	.	.	−8
$(8)(2)(1)^2$	.	.	.	8	−16	.	.	.	.	.	−8
$(8)(2)^2$	.	.	.	8	−32	32	.	.	.	.	−8
$(8)(3)(1)$	.	.	.	8	−24	.	24	.	.	.	−8
$(8)(4)$	−32	.	.	8	−32	16	32	.	−32	.	−8
$(9)(1)^3$	.	.	.	9	−9	.	.	.	.	.	−9
$(9)(2)(1)$	.	.	.	9	−27	18	.	.	.	.	−9
$(9)(3)$	.	.	.	9	−36	27	27	.	−27	.	−9
$(10)(1)^2$	.	.	.	10	−20	.	10	.	.	.	−10
$(10)(2)$	−20	.	.	10	−40	40	10	.	−20	.	−10
$(11)(1)$	.	−11	.	11	−33	11	22	.	−11	.	−11
(12)	−24	−24	6	12	−48	36	36	−24	−24	12	−12

Table 1·5·12 (*cont.*)

$w = 12$ (vii)	$a_8 a_2 a_1^2$	$a_8 a_2^2$	$a_8 a_3 a_1$	$a_8 a_4$	$a_9 a_1^3$	$a_9 a_2 a_1$	$a_9 a_3$	$a_{10} a_1^2$	$a_{10} a_2$	$a_{11} a_1$	a_{12}
$(1)^{12}$	5940	2970	1980	495	1320	660	220	132	66	12	1
$(2)(1)^{10}$	−172260	−89100	−61380	−16830	−47520	−24420	−8580	−5940	−3036	−660	−66
$(2)^2(1)^8$	1413720	792990	582120	188595	498960	273240	106920	83160	44550	11880	1485
$(2)^3(1)^6$	−3742200	−2577960	−2079000	−873180	−1663200	−1081080	−526680	−415800	−249480	−83160	−13860
$(2)^4(1)^4$	3118500	3430350	2702700	1611225	1247400	.	1455300	1039500	623700	207900	51975
$(2)^5(1)^2$	−623700	−1871100	−623700	−935550	.	−623700	−623700	−124740	−374220	−124740	−62370
$(2)^6$	.	311850	.	155925	.	.	.	.	62370	.	10395
$(3)(1)^9$	665280	332640	225720	59400	221760	110880	37400	31680	15840	3960	440
$(3)(2)(1)^7$	−7318080	−3991680	−2993760	−997920	−3326400	−1774080	−681120	−665280	−348480	−110880	−15840
$(3)(2)^2(1)^5$	16632000	11975040	10810800	5155920	9979200	6652800	3492720	3326400	1995840	831600	166320
$(3)(2)^3(1)^3$	−9979200	−13305600	−11642400	−8316000	−3326400	−6652800	−6652800	−3326400	−3326400	−1663200	−554400
$(3)(2)^4(1)$	.	4989600	415800	2910600	.	1663200	2079000	.	1663200	415800	415800
$(3)^2(1)^6$	6652800	3326400	2661120	997920	4435200	2217600	813120	1108800	554400	221760	36960
$(3)^2(2)(1)^4$	−13305600	−9979200	−13305600	−8316000	−13305600	.	−8870400	−6652800	−3880800	−2217600	−554400
$(3)^2(2)^2(1)^2$	6652800	9979200	13305600	11642400	.	6652800	6652800	3326400	4989600	3326400	1663200
$(3)^2(2)^3$	.	−3326400	.	−1663200	.	.	−1108800	.	−1663200	.	−554400
$(3)^3(1)^3$	.	.	4435200	4435200	2956800	1478400	1971200	2956800	1478400	1478400	492800
$(3)^3(2)(1)$	.	.	−4435200	−4435200	.	−1478400	−5913600	.	−1478400	−1478400	−1478400
$(3)^4$	.	.	.	.	.	.	985600	.	.	.	246400
$(4)(1)^8$	−2494800	−1247400	−831600	−210870	−997920	−498960	−166320	−166320	−83160	−23760	−2970
$(4)(2)(1)^6$	17463600	9979200	7484400	2577960	9979200	5488560	2162160	2494800	1330560	498960	83160
$(4)(2)^2(1)^4$	−22453200	−19958400	−17463600	−9355200	−14968800	−12474000	−7484400	−7484400	−4989600	−2494800	−623700
$(4)(2)^3(1)^2$	7484400	14968800	7484400	8731800	.	7484400	7484400	2494800	4989600	2494800	1247400
$(4)(2)^4$	.	−3742200	.	−2182950	.	.	.	.	−1247400	.	−311850
$(4)(3)(1)^5$	−19958400	−9979200	−8316000	−3659040	−19958400	−9979200	−3659040	−6652800	−3326400	−1663200	−332640
$(4)(3)(2)(1)^3$	19958400	1958400	29937600	23284800	19958400	19958400	16632000	16632000	13305600	9979200	3326400
$(4)(3)(2)^2(1)$	.	−9979200	−4989600	−14968800	.	.	−9979200	−14968800	−9979200	−4989600	−4989600
$(4)(3)^2(1)^2$	.	.	−13305600	−16632000	.	.	.	−6652800	−6652800	−6652800	−3326400
$(4)(3)^2(2)$	.	.	.	3326400	.	.	.	6652800	−3326400	.	3326400
$(4)^2(1)^4$	7484400	3742200	2494800	1871100	14968800	7484400	2494800	7484400	3742200	2494800	623700
$(4)^2(2)(1)^2$	−7484400	−7484400	−7484400	−11226600	.	−7484400	−7484400	−7484400	−7484400	−7484400	−3742200
$(4)^2(2)^2$	.	3742200	.	5613300	.	.	.	.	3742200	.	1871100
$(4)^2(3)(1)$	.	.	4989600	14968800	.	.	4989600	.	.	4989600	4989600
$(4)^3$	.	.	.	−3742200	.	.	.	.	.	.	−1247400
$(5)(1)^7$	7983360	3991680	2661120	665280	3991680	1995840	665280	798336	399168	133056	19008
$(5)(2)(1)^5$	−31933440	−19958400	−15966720	−5987520	−23950080	−13970880	−5987520	−7983360	−4390848	−1995840	−399168
$(5)(2)^2(1)^3$	23950080	27941760	23950080	13970880	11975040	17962560	13970880	13970880	9979200	5987520	1995840
$(5)(2)^3(1)$	.	−11975040	.	−5987520	.	−5987520	−5987520	.	−5987520	−1995840	−1995840
$(5)(3)(1)^4$	15966720	7983360	10644480	6652800	31933440	15966720	6652800	15966720	7983360	5322240	1330560
$(5)(3)(2)(1)^2$	−15966720	−15966720	−31933440	−23950080	.	−15966720	−23950080	−15966720	−15966720	−15966720	−7983360
$(5)(3)(2)^2$	.	7983360	.	3991680	.	.	3991680	.	7983360	.	3991680
$(5)(3)^2(1)$	.	.	10644480	10644480	.	.	10644480	.	.	5322240	5322240
$(5)(4)(1)^3$	.	.	−3991680	−3991680	−23950080	−11975040	−3991680	−23950080	−11975040	−11975040	−3991680
$(5)(4)(2)(1)$	.	.	.	11975040	.	11975040	11975040	.	11975040	11975040	11975040
$(5)(4)(3)$	.	.	.	−7983360	.	.	−7983360	.	.	.	−7983360
$(5)^2(1)^2$	.	.	.	.	.	.	.	9580032	4790016	9580032	4790016
$(5)^2(2)$	.	.	.	.	.	.	.	.	−4790016	.	−4790016
$(6)(1)^6$	−19958400	−9979200	−6652800	−1663200	−13305600	−6652800	−2217600	−3326400	−1663200	−6652800	−110880
$(6)(2)(1)^4$	39916800	29937600	26611200	11642400	39916800	26611200	13305600	19958400	11642400	6652800	1663200
$(6)(2)^2(1)^2$	−19958400	−29937600	−19958400	−14968800	.	−19958400	−19958400	−9979200	−14968800	−9979200	−4989600
$(6)(2)^3$	.	.	.	4989600	.	.	.	.	4989600	.	1663200
$(6)(3)(1)^3$	.	9979200	−13305600	−13305600	−26611200	−13305600	−8870400	−26611200	−13305600	−13305600	−4435200
$(6)(3)(2)(1)$	.	.	13305600	13305600	.	13305600	26611200	.	13305600	13305600	13305600
$(6)(3)^2$	.	.	.	.	.	.	−8870400	.	.	.	−4435200
$(6)(4)(1)^2$	.	.	.	9979200	.	.	.	19958400	9979200	19958400	9979200
$(6)(4)(2)$	.	.	.	−9979200	.	.	.	.	−9979200	.	−9979200
$(6)(5)(1)$	.	.	.	.	.	.	.	.	.	−15966720	−15966720
$(6)^2$	.	.	.	.	.	.	.	.	.	.	6652800
$(7)(1)^5$	34214400	17107200	11404800	2851200	34214400	17107200	5702400	11404800	5702400	2851200	570240
$(7)(2)(1)^3$	−34214400	−34214400	−34214400	−17107200	−34214400	−34214400	−22809600	−34214400	−22809600	−17107200	−5702400
$(7)(2)^2(1)$	.	17107200	.	8553600	.	17107200	17107200	.	17107200	8553600	8553600
$(7)(3)(1)^2$	.	.	22809600	22809600	.	.	11404800	22809600	11404800	22809600	11404800
$(7)(3)(2)$	.	.	.	.	.	.	−11404800	.	−11404800	.	−11404800
$(7)(4)(1)$	.	.	.	−17107200	.	.	.	.	.	−17107200	−17107200
$(7)(5)$	.	.	.	.	.	.	.	.	.	.	13685760
$(8)(1)^4$	−29937600	−14968800	−9979200	−2494800	−59875200	−29937600	−9979200	−29937600	−14968800	−9979200	−2494800
$(8)(2)(1)^2$	29937600	29937600	29937600	14968800	.	29937600	29937600	29937600	29937600	29937600	14968800
$(8)(2)^2$	16	−14968800	−32	−7484400	.	.	.	.	−14968800	.	−7484400
$(8)(3)(1)$	24	.	−19958400	−19958400	.	.	−19958400	.	.	.	−19958400
$(8)(4)$	32	−16	−32	14968800	.	.	.	.	.	.	14968800
$(9)(1)^3$	.	.	.	.	53222400	26611200	8870400	53222400	26611200	26611200	8870400
$(9)(2)(1)$	18	.	.	.	9	−26611200	−26611200	.	−26611200	−26611200	−26611200
$(9)(3)$	27	.	−27	.	9	−27	17740800	27	.	.	17740800
$(10)(1)^2$	10	.	.	.	10	.	.	−47900160	−23950080	−47900160	−23950080
$(10)(2)$	30	−20	.	.	10	−20	.	−10	23950080	20	23950080
$(11)(1)$	22	.	−11	.	11	−11	.	−11	.	43545600	43545600
(12)	36	−12	−24	12	12	−24	12	−12	12	12	−39916800

Table 1·6. *Bipartite augmented symmetric functions in terms of the bipartite s-functions and conversely (up to and including weight 8) (Tables 1·6·4, 1·6·5, 1·6·6, 1·6·7, 1·6·8)*

The weight of the symmetric functions tabulated and the appropriate partition of this weight are indicated at the top of each table.

To express a product of the bipartite power sums in terms of binomial symmetric function it is necessary to read horizontally up to and including the bold figured diagonal. Thus we denote

$$s_{pq} = \sum_i x_i^p y_i^q$$

and, for example, from 1·6·5, (32) partition,

$$s_{30}s_{01}^2 = [32] + 2[31, 01] + [30, 02] + [30, (01)^2].$$

To express binomial symmetric functions in terms of the bipartite power sums the table is read vertically downwards as far as and including the bold figured diagonal. Thus, for example, from the same 1·6·5, (32) partition we have

$$[30, (01)^2] = 2s_{32} - 2s_{31}s_{01} - s_{30}s_{02} + s_{30}s_{01}^2.$$

Table 1·6·4. *Partitions (31), (22)*

$w = 4$ (31)	[31]	[30, 01]	[21, 10]	[20, 11]	[20, 10, 01]	[11, (10)²]	[(10)³, 01]
s_{31}	**1**	−1	−1	−1	2	2	−6
$s_{30}s_{01}$	1	**1**	.	.	−1	.	2
$s_{21}s_{10}$	1	.	**1**	.	−1	−2	6
$s_{20}s_{11}$	1	.	.	**1**	−1	−1	3
$s_{20}s_{10}s_{01}$	1	1	1	1	**1**	.	−3
$s_{11}s_{10}^2$	1	.	2	1	.	**1**	−3
$s_{10}^3 s_{01}$	1	1	3	3	3	3	**1**

$w = 4$ (22)	[22]	[21, 01]	[20, 02]	[12, 10]	[(11)²]	[20, (01)²]	[11, 10, 01]	[(10)², 02]	[(10)²(01)²]
s_{22}	**1**	−1	−1	−1	−1	2	2	2	−6
$s_{21}s_{01}$	1	**1**	.	.	.	−2	−1	.	4
$s_{20}s_{02}$	1	.	**1**	.	.	−1	.	−1	1
$s_{12}s_{10}$	1	.	.	**1**	.	.	−1	−2	4
s_{11}^2	1	.	.	.	**1**	.	−1	.	2
$s_{20}s_{01}^2$	1	2	1	.	.	**1**	.	.	−1
$s_{11}s_{10}s_{01}$	1	1	.	1	1	.	**1**	.	−4
$s_{10}^2 s_{02}$	1	.	1	2	.	.	.	**1**	−1
$s_{10}^2 s_{01}^2$	1	2	1	2	2	1	4	1	**1**

Table 1·6·5. *Partitions* (41), (32)

$w=5$ (41)	[41]	[40, 01]	[31, 10]	[30, 11]	[21, 20]	[30, 10, 01]	[21 (10)²]	[(20)² 01]	[20, 11, 10]	[20, (10)³ 01]	[11, (10)³]	[(10)⁴, 01]
S_{41}	1	-1	-1	-1	-1	2	2	2	2	-6	-6	24
$S_{40}S_{01}$	1	1	.	.	.	-1	.	-1	.	2	-6	-6
$S_{31}S_{10}$	1	.	1	.	.	-1	-2	.	-1	4	6	-24
$S_{30}S_{11}$	1	.	.	1	.	-1	.	.	-1	2	2	-8
$S_{21}S_{20}$	1	.	.	.	1	.	-1	-2	-1	3	3	-12
$S_{30}S_{10}S_{01}$	1	1	1	1	.	1	.	.	.	-2	.	8
$S_{21}S_{10}^2$	1	.	2	.	1	.	1	.	.	-1	-3	12
$S_{20}^2S_{01}$	1	1	.	.	2	.	.	1	.	-1	.	3
$S_{20}S_{11}S_{10}$	1	.	1	1	1	.	.	.	1	-2	-3	12
$S_{20}S_{10}^2S_{01}$	1	1	2	2	2	2	1	1	2	1	.	-6
$S_{11}S_{10}^3$	1	.	3	1	3	.	3	.	3	.	1	-4
$S_{10}^4S_{01}$	1	1	4	4	6	4	6	3	12	6	4	1

$w=5$ (32)	[32]	[31, 01]	[30, 02]	[22, 10]	[21, 11]	[20, 12]	[30, (01)²]	[21, 10, 01]	[20, 11, 01]	[20, 10, 02]	[12, (10)²]	[(11)², 10]	[20, 10, (01)²]	[11, (10)², 01]	[(10)³, 02]	[(10)³, (01)²]
S_{32}	1	-1	-1	-1	-1	-1	2	2	2	2	2	2	-6	-6	-6	24
$S_{31}S_{01}$	1	1	.	.	.	.	-2	-1	-1	.	.	.	4	2	.	-12
$S_{30}S_{02}$	1	.	1	.	.	.	-1	.	.	-1	.	.	1	2	2	-2
$S_{22}S_{10}$	1	.	.	1	.	.	.	-1	.	-1	-2	-1	2	4	6	-18
$S_{21}S_{11}$	1	.	.	.	1	.	.	-1	-1	.	.	-2	2	4	.	-12
$S_{20}S_{12}$	1	.	.	.	.	1	.	.	-1	-1	-1	-1	2	1	3	-6
$S_{30}S_{01}^2$	1	2	1	.	.	.	1	.	.	.	.	.	-1	.	.	2
$S_{21}S_{10}S_{01}$	1	1	.	1	1	.	.	1	.	.	.	.	-2	-2	.	12
$S_{20}S_{11}S_{01}$	1	1	.	.	1	.	.	.	1	.	.	.	-2	-1	.	6
$S_{20}S_{10}S_{02}$	1	.	1	1	.	1	.	.	.	1	.	.	-1	.	-3	3
$S_{12}S_{10}^2$	1	.	.	2	.	1	.	.	.	.	1	.	.	-1	-3	6
$S_{11}^2S_{10}$	1	.	.	1	2	.	.	.	.	.	.	1	.	-2	.	6
$S_{20}S_{10}S_{01}^2$	1	2	1	1	2	1	1	2	2	1	.	.	1	.	.	-3
$S_{11}S_{10}^2S_{01}$	1	1	.	2	3	1	.	2	1	.	1	2	.	1	.	-6
$S_{10}^3S_{02}$	1	.	1	3	.	3	.	.	.	3	3	.	.	.	1	-1
$S_{10}^3S_{01}^2$	1	2	1	3	6	3	1	6	6	3	3	6	3	6	1	1

Table 1·6·6. *Partitions* (51), (42), (33)

w = 6 (51)

	$[51]$	$[50, 01]$	$[41, 10]$	$[40, 11]$	$[31, 20]$	$[30, 21]$	$[40, 10, 01]$	$[31, (10)^2]$	$[30, 20, 01]$	$[30, 11, 10]$	$[21, 20, 10]$	$[(20)^2, 11]$	$[30, (10)^2, 01]$	$[21, (10)^3]$	$[(20)^2, 10, 01]$	$[20, 11, (10)^2]$	$[20, (10)^3, 01]$	$[11, (10)^4]$	$[(10)^5, 01]$
s_{51}	I	−1	−1	−1	−1	−1	2	2	2	2	2	2	−6	−6	−6	−6	24	24	−120
$s_{50}s_{01}$	I	I	.	.	.	.	−1	.	−1	.	.	.	2	.	2	.	−6	.	24
$s_{41}s_{10}$	I	.	I	.	.	.	−1	−2	.	−1	−1	.	4	6	2	4	−18	−24	120
$s_{40}s_{11}$	I	.	.	I	.	.	−1	.	.	−1	.	−1	2	.	1	2	−6	−6	30
$s_{31}s_{20}$	I	.	.	.	I	.	.	−1	−1	.	−1	−2	1	3	4	3	−12	−12	60
$s_{30}s_{21}$	I	.	.	.	.	I	.	.	−1	−1	−1	.	2	2	2	2	−8	−8	40
$s_{40}s_{10}s_{01}$	I	I	I	I	.	.	I	.	.	.	.	.	−2	.	−1	.	6	.	−30
$s_{31}s_{10}^2$	I	.	2	.	I	.	.	I	.	.	.	.	−1	−3	.	−1	6	12	−60
$s_{30}s_{20}s_{01}$	I	I	.	.	I	.	.	.	I	.	.	.	−1	.	−2	.	5	.	−20
$s_{30}s_{11}s_{10}$	I	.	I	I	.	I	.	.	.	I	.	.	−2	.	.	−2	6	8	−40
$s_{21}s_{20}s_{10}$	I	.	I	.	I	I	.	.	.	.	I	.	.	−3	−2	−2	9	12	−60
$s_{20}^2 s_{11}$	I	.	.	I	2	.	.	.	.	.	.	I	.	−1	−1	3	3	.	−15
$s_{30}s_{10}^2 s_{01}$	I	I	2	2	I	.	2	I	.	I	2	.	I	.	.	.	−3	.	20
$s_{21}s_{10}^3$	I	.	3	.	3	I	.	3	.	.	3	.	.	I	.	.	−1	−4	20
$s_{20}^2 s_{10}s_{01}$	I	I	I	2	2	I	.	2	.	2	2	I	.	.	I	.	−3	.	15
$s_{20}s_{11}s_{10}^2$	I	.	2	I	2	2	.	I	.	2	2	I	.	.	.	I	−3	−6	30
$s_{20}s_{10}^3 s_{01}$	I	I	3	3	4	4	3	3	4	6	6	3	3	I	3	3	I	.	−10
$s_{11}s_{10}^4$	I	.	4	I	6	4	.	6	.	4	12	3	.	4	.	6	.	I	−5
$s_{10}^5 s_{01}$	I	I	5	5	10	10	5	10	10	20	30	15	10	10	15	30	10	5	I

w = 6 (42)

	$[42]$	$[41, 01]$	$[40, 02]$	$[32, 10]$	$[31, 11]$	$[30, 12]$	$[22, 20]$	$[(21)^2]$	$[40, (01)^2]$	$[31, 10, 01]$	$[30, 11, 01]$	$[30, 10, 02]$	$[22, (10)^2]$	$[21, 20, 01]$	$[21, 11, 10]$	$[(20)^2, 02]$	$[20, 12, 10]$	$[20, (11)^2]$	$[30, 10, (01)^2]$	$[21, (10)^2, 01]$	$[(20)^2, (01)^2]$	$[20, 11, 10, 01]$	$[20, (10)^2, 02]$	$[12, (10)^3]$	$[(11)^2, (10)^2]$	$[20, (10)^2, (01)^2]$	$[11, (10)^3, 01]$	$[(10)^4, 02]$	$[(10)^4, (01)^2]$
s_{42}	I	−1	−1	−1	−1	−1	−1	−1	2	2	2	2	2	2	2	2	2	2	−6	−6	−6	−6	−6	−6	−6	24	24	24	−120
$s_{41}s_{01}$	I	I	.	.	.	.	.	.	−2	−1	−1	.	.	−1	.	.	.	.	4	2	4	2	.	.	.	−12	−6	.	48
$s_{40}s_{02}$	I	.	I	.	.	.	.	.	−1	.	.	−1	.	.	.	−1	.	.	1	.	1	.	2	.	.	−2	.	−6	6
$s_{32}s_{10}$	I	.	.	I	.	.	.	.	.	−1	.	.	−1	−2	.	.	−1	.	2	4	.	2	4	6	.	−12	−18	−24	96
$s_{31}s_{11}$	I	.	.	.	I	.	.	.	.	−1	−1	.	.	.	−1	.	.	−2	2	2	.	3	.	.	4	−8	−12	.	48
$s_{30}s_{12}$	I	.	.	.	.	I	.	.	.	.	−1	−1	.	.	.	−1	.	.	2	.	.	1	2	2	.	−4	−2	−8	16
$s_{22}s_{20}$	I	.	.	.	.	.	I	.	.	.	.	.	−1	−1	.	−2	−1	−1	.	1	4	2	3	3	1	−8	−6	−12	36
s_{21}^2	I	.	.	.	.	.	.	I	.	.	.	.	.	−1	−1	.	.	.	.	.	.	2	2	1	.	2	−4	−6	24
$s_{40}s_{01}^2$	I	2	I	.	.	.	.	.	I	.	.	.	.	.	.	.	.	.	−1	.	−1	.	.	.	.	2	.	.	−6
$s_{31}s_{10}s_{01}$	I	I	.	I	.	.	.	.	.	I	.	.	.	.	.	.	.	.	−2	−2	.	−1	.	.	.	8	1	.	−10
$s_{30}s_{11}s_{01}$	I	I	.	.	I	.	.	.	.	.	I	.	.	.	.	.	.	.	−2	.	.	−1	.	.	.	4	2	8	−16
$s_{30}s_{10}s_{02}$	I	.	I	I	.	.	.	.	.	.	.	I	.	.	.	.	.	.	−1	.	.	.	−2	.	.	2	.	8	−8
$s_{22}s_{10}^2$	I	.	.	2	.	.	I	.	.	.	.	.	I	.	.	.	.	.	.	.	.	−1	.	−3	−1	2	6	12	−36
$s_{21}s_{20}s_{01}$	I	I	.	.	.	.	I	.	.	.	.	.	.	I	.	.	.	.	−1	−4	−1	.	.	.	.	6	3	.	−24
$s_{21}s_{11}s_{10}$	I	.	.	I	I	.	.	.	.	.	.	.	.	.	I	.	.	.	−2	−1	.	.	.	.	−4	4	.	12	−48
$s_{20}^2 s_{02}$	I	.	I	.	.	.	2	.	.	.	.	.	.	.	.	I	.	.	.	.	−1	.	−1	.	.	1	.	3	−3
$s_{20}s_{12}s_{10}$	I	.	.	.	I	.	.	.	.	.	.	.	.	.	I	.	I	.	.	.	.	−1	−2	−3	.	4	3	12	−24
$s_{20}s_{11}^2$	I	.	.	.	2	.	.	.	.	.	.	.	.	.	I	.	.	I	.	.	.	−1	.	.	.	2	3	.	−12
$s_{30}s_{10}s_{01}^2$	I	2	I	I	2	I	.	.	I	2	2	I	.	.	.	2	.	.	I	.	.	.	.	.	.	−2	.	.	8
$s_{21}s_{10}^2 s_{01}$	I	2	.	2	2	.	I	.	.	2	.	.	I	2	2	.	.	.	.	I	.	.	.	.	.	−2	−3	.	24
$s_{20}^2 s_{01}^2$	I	2	I	.	.	.	2	2	I	.	.	.	.	.	.	4	.	.	.	.	I	.	.	.	.	−1	.	.	3
$s_{20}s_{11}s_{10}s_{01}$	I	I	.	I	2	.	I	.	.	I	.	.	.	I	I	.	.	.	I	.	.	.	.	.	.	−4	−3	.	24
$s_{20}s_{10}^2 s_{02}$	I	.	I	2	.	.	.	.	.	2	.	I	.	.	.	2	.	.	.	.	.	.	I	.	.	I	.	6	6
$s_{12}s_{10}^3$	I	.	.	3	.	I	3	.	.	.	.	.	.	.	3	.	.	.	.	3	.	.	.	I	.	.	−1	−4	8
$s_{11}^2 s_{10}^2$	I	.	.	2	2	.	I	2	.	.	.	.	I	.	.	4	.	I	.	.	.	.	.	.	I	.	−3	.	12
$s_{20}s_{10}^3 s_{01}$	I	2	.	4	2	.	2	2	I	4	4	I	2	2	2	2	I	4	I	.	.	.	.	3	3	I	.	.	−6
$s_{11}s_{10}^3 s_{01}$	I	I	.	3	4	I	.	3	.	3	3	.	3	3	9	.	3	3	.	3	.	3	.	3	3	.	I	.	−8
$s_{10}^4 s_{02}$	I	.	I	4	.	4	6	.	.	.	.	4	6	.	.	3	12	.	.	.	.	.	.	6	4	.	.	I	−1
$s_{10}^4 s_{01}^2$	I	2	I	4	8	4	6	6	I	8	8	4	6	12	24	3	12	12	4	12	3	24	6	4	12	6	8	I	I

Table 1·6·6 (*cont.*)

$w = 6$ (33)	[33]	[32, 01]	[31, 02]	[30, 03]	[23, 10]	[22, 11]	[21, 12]	[20, 13]	[31, $(01)^2$]	[30, 02, 01]	[22, 10, 01]	[21, 11, 01]	[21, 10, 02]	[20, 12, 01]	[20, 11, 02]	[20, 10, 03]	[13, $(10)^2$]	[12, 11, 10]	[$(11)^3$]	[30, $(01)^3$]	[21, 10, $(01)^2$]	[20, 11, $(01)^2$]	[20, 10, 02, 01]	[12, $(10)^2$, 01]	[$(11)^2$, 10, 01]	[11, $(10)^2$, 02]	[$(10)^3$, 03]	[20, 10, $(01)^3$]	[11, $(10)^2$, $(01)^2$]	[$(10)^3$, 02, 01]	[$(10)^3$, $(01)^3$]
s_{33}	1	-1	-1	-1	-1	-1	-1	-1	2	2	2	2	2	2	2	2	2	2	·	-6	-6	-6	-6	-6	-6	-6	-6	24	24	24	-120
$s_{32}s_{01}$	1	1	·	·	·	·	·	·	-2	-1	-1	-1	·	-1	·	·	·	·	·	6	4	4	2	2	2	·	·	-18	-1	-6	72
$s_{31}s_{02}$	1	·	1	·	·	·	·	·	-1	-1	·	·	-1	·	-1	·	·	·	·	3	1	1	2	·	·	2	·	-6	-1	-6	18
$s_{30}s_{03}$	1	·	·	1	·	·	·	·	·	-1	·	·	·	·	·	-1	·	·	·	2	·	·	1	·	·	2	-2	-2	·	-2	4
$s_{23}s_{10}$	1	·	·	·	1	·	·	·	·	·	-1	·	-1	·	-1	-2	-1	·	·	·	2	·	2	4	2	4	6	-6	-2	-18	72
$s_{22}s_{11}$	1	·	·	·	·	1	·	·	·	·	-1	-1	·	·	-1	·	-1	-3	·	·	2	2	2	2	·	2	·	-6	·	-6	54
$s_{21}s_{12}$	1	·	·	·	·	·	1	·	·	·	·	-1	-1	-1	·	·	-1	·	·	·	2	2	1	2	2	2	·	-6	-4	-6	36
$s_{20}s_{13}$	1	·	·	·	·	·	·	1	·	·	·	·	·	-1	-1	-1	-1	·	·	·	2	2	1	5	1	3	·	-6	-2	-6	18
$s_{31}s_{01}^2$	1	2	1	·	·	·	·	·	1	·	·	·	·	·	·	·	·	·	·	-3	-1	-1	·	·	·	·	·	6	2	·	-18
$s_{30}s_{02}s_{01}$	1	1	1	·	·	·	·	·	·	1	·	·	·	·	·	·	·	·	·	-3	·	-1	·	·	·	·	·	3	·	2	-6
$s_{22}s_{10}s_{01}$	1	1	·	·	1	1	·	·	·	·	1	·	·	·	·	·	·	·	·	·	-2	·	-1	-2	-1	·	·	6	8	6	-54
$s_{21}s_{11}s_{01}$	1	1	·	·	·	1	1	·	·	·	·	1	·	·	·	·	·	·	·	·	-2	-2	·	·	-2	·	·	6	8	·	-36
$s_{21}s_{10}s_{02}$	1	1	·	·	1	·	·	·	·	·	·	·	1	·	·	·	·	·	·	·	-1	·	·	-1	·	-2	·	3	2	6	-18
$s_{20}s_{12}s_{01}$	1	1	·	·	·	·	1	1	·	·	·	·	·	1	·	·	·	·	·	·	-2	-1	-1	·	·	·	·	6	2	3	-18
$s_{20}s_{11}s_{02}$	1	·	1	·	·	1	·	1	·	·	·	·	·	·	1	·	·	·	·	·	-1	-1	·	·	·	-1	·	3	1	3	-9
$s_{20}s_{10}s_{03}$	1	·	·	1	1	·	·	1	·	·	·	·	·	·	·	1	·	·	·	·	·	-1	·	·	·	-1	-3	2	·	3	-6
$s_{13}s_{10}^2$	1	·	·	·	2	·	·	1	·	·	·	·	·	·	·	·	1	·	·	·	·	·	·	-1	·	-1	-3	·	2	6	-18
$s_{12}s_{11}s_{10}$	1	·	·	·	1	1	1	·	·	·	·	·	·	·	·	·	·	1	·	·	·	·	·	-2	-2	-2	·	·	8	6	-36
s_{11}^3	1	·	·	·	·	3	·	·	·	·	·	·	·	·	·	·	·	·	1	·	·	·	·	·	-1	·	·	·	2	·	-6
$s_{30}s_{01}^3$	1	3	3	1	·	·	2	·	3	3	·	·	·	·	·	·	·	·	·	1	·	·	·	·	·	·	·	-1	·	·	2
$s_{21}s_{10}s_{01}^2$	1	2	1	·	1	2	1	1	·	·	2	2	1	·	·	·	·	·	·	·	1	·	·	·	·	·	·	-3	-2	·	18
$s_{20}s_{11}s_{01}^2$	1	2	1	·	·	1	2	1	1	·	·	2	·	2	1	·	·	·	·	·	·	1	·	·	·	·	·	-3	-1	·	9
$s_{20}s_{10}s_{02}s_{01}$	1	1	1	1	1	1	1	1	·	1	1	·	1	1	1	1	·	·	·	·	·	·	1	·	·	·	·	-3	·	-3	9
$s_{12}s_{10}^2s_{01}$	1	1	·	·	2	2	1	1	·	·	·	2	·	·	1	·	·	1	2	·	·	·	·	1	·	·	·	·	-2	-3	18
$s_{11}^2s_{10}s_{01}$	1	1	·	·	1	3	2	·	·	·	·	1	2	·	·	·	·	2	1	·	·	·	·	1	·	·	·	·	-4	·	18
$s_{11}s_{10}^2s_{02}$	1	·	1	·	2	1	2	1	·	·	·	·	·	·	2	·	·	1	·	·	1	·	·	·	·	·	·	·	-1	-3	9
$s_{10}^3s_{03}$	1	·	·	1	3	·	·	3	·	·	·	·	·	·	·	3	3	·	·	·	·	·	·	·	1	·	·	·	·	-1	2
$s_{20}s_{10}s_{01}^3$	1	3	3	1	1	3	3	1	3	3	3	6	3	3	3	1	·	6	·	1	3	3	3	·	2	4	·	1	·	·	-3
$s_{11}s_{10}^2s_{01}^2$	1	2	1	·	2	5	4	1	1	·	4	6	2	2	2	1	·	6	2	·	2	1	·	3	·	2	4	1	·	1	-9
$s_{10}^3s_{02}s_{01}$	1	1	1	1	3	3	3	3	·	1	3	·	3	3	3	3	3	6	·	·	·	3	3	·	3	1	·	·	1	-3
$s_{10}^3s_{01}^3$	1	3	3	1	3	9	9	3	3	3	9	18	9	9	9	3	3	18	6	1	9	9	9	9	18	9	1	3	9	3	1

Table 1·6·7. *Partitions* (61), (52), (43)

$w = 7$ (61) (i)

	[61]	[60, 01]	[51, 10]	[50, 11]	[41, 20]	[40, 21]	[31, 30]	[50, 10, 01]	[41, (10)²]	[40, 20, 01]	[40, 11, 10]	[31, 20, 10]	[(30)², 01]	[30, 21, 10]	[30, 20, 11]	[21, (20)²]
s_{61}	1	−1	−1	−1	−1	−1	−1	2	2	2	2	2	2	2	2	2
$s_{60}s_{01}$	1	1	·	·	·	·	·	−1	·	−1	·	·	−1	·	·	·
$s_{51}s_{10}$	1	·	1	·	·	·	·	−1	−2	·	−1	−1	·	−1	·	·
$s_{50}s_{11}$	1	·	·	1	·	·	·	−1	·	−1	−1	−1	·	·	−1	·
$s_{41}s_{20}$	1	·	·	·	1	·	·	·	−1	−1	·	−1	·	·	−1	−2
$s_{40}s_{21}$	1	·	·	·	·	1	·	·	·	·	−1	−1	·	−1	·	−1
$s_{31}s_{30}$	1	·	·	·	·	·	1	·	·	·	·	·	−1	−2	−1	−1
$s_{50}s_{10}s_{01}$	1	1	1	1	·	·	·	1	·	·	·	·	·	·	·	·
$s_{41}s_{10}^2$	1	·	2	·	1	·	·	·	1	·	·	·	·	·	·	·
$s_{40}s_{20}s_{01}$	1	1	·	·	1	1	·	·	·	1	·	·	·	·	·	·
$s_{40}s_{11}s_{10}$	1	·	1	1	·	1	·	·	·	·	1	·	·	·	·	·
$s_{31}s_{20}s_{10}$	1	·	1	·	1	·	1	·	·	·	·	1	·	·	·	·
$s_{30}^2 s_{01}$	1	1	·	·	·	·	2	·	·	·	·	·	1	·	·	·
$s_{30}s_{21}s_{10}$	1	·	1	·	·	·	1	1	·	·	·	·	·	1	·	·
$s_{30}s_{20}s_{11}$	1	·	·	1	1	1	1	·	·	·	·	·	·	·	1	·
$s_{21}s_{20}^2$	1	·	·	·	2	1	·	·	·	·	·	·	·	·	·	1
$s_{40}s_{10}^2 s_{01}$	1	1	2	2	1	1	·	2	1	1	2	·	·	·	·	·
$s_{31}s_{10}^3$	1	·	3	·	3	·	1	·	3	·	·	3	·	·	1	·
$s_{30}s_{20}s_{10}s_{01}$	1	1	1	1	1	1	2	1	·	1	·	1	1	1	·	·
$s_{30}s_{11}s_{10}^2$	1	·	2	1	1	2	1	·	1	·	2	·	·	2	1	·
$s_{21}s_{20}s_{10}^2$	1	·	2	·	2	1	2	·	1	·	·	2	·	2	·	1
$s_{20}^3 s_{01}$	1	1	·	·	3	3	·	·	·	3	·	·	·	·	·	3
$s_{20}^2 s_{11}s_{10}$	1	·	1	1	2	1	2	·	·	·	1	2	·	·	2	1
$s_{30}s_{10}^3 s_{01}$	1	1	3	3	3	3	2	3	3	3	6	3	1	3	3	·
$s_{21}s_{10}^4$	1	·	4	·	6	1	4	·	6	·	·	12	·	4	·	3
$s_{20}^2 s_{10}^2 s_{01}$	1	1	2	2	3	3	4	2	1	3	2	4	2	4	4	3
$s_{20}s_{11}s_{10}^3$	1	·	3	1	4	3	4	·	3	·	3	6	·	6	4	3
$s_{20}s_{10}^4 s_{01}$	1	1	4	4	7	7	8	4	6	7	12	16	4	16	16	9
$s_{11}s_{10}^5$	1	·	5	1	10	5	10	·	10	·	5	30	·	20	10	15
$s_{10}^6 s_{01}$	1	1	6	6	15	15	20	6	15	15	30	60	10	60	60	45

$w = 7$ (61) (ii)

	[40, (10)², 01]	[31, (10)³]	[30, 20, 10, 01]	[30, 11, (10)²]	[21, 20, (10)²]	[(20)³, 01]	[(20)², 11, 10]	[30, (10)³, 01]	[21, (10)⁴]	[(20)², (10)², 01]	[20, 11, (10)³]	[20, (10)⁴, 01]	[11, (10)⁵]	[[(10)⁶, 01]
s_{61}	−6	−6	−6	−6	−6	−6	−6	24	24	24	24	−120	−120	720
$s_{60}s_{01}$	2	·	2	·	·	2	·	−6	·	−6	−18	24	·	−120
$s_{51}s_{10}$	4	6	2	4	4	·	2	−18	−24	−12	−12	96	120	−720
$s_{50}s_{11}$	2	·	1	2	·	·	2	−6	·	−4	−6	24	24	−144
$s_{41}s_{20}$	1	3	2	1	3	6	4	−6	−12	−14	−12	60	60	−360
$s_{40}s_{21}$	2	·	1	2	2	3	1	−6	−6	−6	−6	30	30	−180
$s_{31}s_{30}$	·	2	3	2	2	·	2	−8	−8	−8	−8	40	40	−240
$s_{50}s_{10}s_{01}$	−2	·	−1	·	·	·	·	6	·	4	6	−24	·	144
$s_{41}s_{10}^2$	−1	−3	·	−1	−1	·	·	6	12	2	6	−36	−60	360
$s_{40}s_{20}s_{01}$	−1	·	−1	·	·	−3	·	3	·	5	·	−18	·	90
$s_{40}s_{11}s_{10}$	−2	·	·	−2	·	·	−1	6	·	2	6	−24	−30	180
$s_{31}s_{20}s_{10}$	·	−3	−1	·	−2	·	−2	3	12	8	9	−48	−60	360
$s_{30}^2 s_{01}$	·	·	−1	·	·	·	·	2	·	2	·	−8	·	40
$s_{30}s_{21}s_{10}$	·	·	−1	−2	−2	·	·	6	8	4	6	−32	−40	240
$s_{30}s_{20}s_{11}$	·	·	−1	−1	·	·	−2	3	·	4	5	−20	−20	120
$s_{21}s_{20}^2$	·	·	·	·	−1	−3	−1	·	3	4	3	−15	−15	90
$s_{40}s_{10}^2 s_{01}$	1	·	·	·	·	·	·	−3	·	−1	·	12	·	−90
$s_{31}s_{10}^3$	·	1	·	·	·	·	·	−1	−4	−4	−1	8	20	−120
$s_{30}s_{20}s_{10}s_{01}$	·	·	1	·	·	·	·	−3	·	·	·	20	20	−120
$s_{30}s_{11}s_{10}^2$	·	·	·	1	·	·	·	−3	·	·	−3	12	20	−120
$s_{21}s_{20}s_{10}^2$	·	·	·	·	1	·	·	·	−6	−2	−3	18	30	−180
$s_{20}^3 s_{01}$	·	·	·	·	·	1	·	·	·	−1	·	3	·	−15
$s_{20}^2 s_{11}s_{10}$	·	·	·	·	·	·	1	·	·	−2	−3	12	15	−90
$s_{30}s_{10}^3 s_{01}$	3	1	3	3	6	·	·	1	·	·	·	−4	·	40
$s_{21}s_{10}^4$	·	4	·	·	6	·	·	·	1	·	·	−1	−5	30
$s_{20}^2 s_{10}^2 s_{01}$	1	·	4	·	2	1	2	·	·	1	·	−6	·	45
$s_{20}s_{11}s_{10}^3$	·	1	·	3	3	·	3	·	·	·	1	−4	−10	60
$s_{20}s_{10}^4 s_{01}$	6	4	16	12	12	3	12	4	1	6	4	1	·	−15
$s_{11}s_{10}^5$	·	10	·	10	30	·	15	·	5	·	10	·	1	−6
$s_{10}^6 s_{01}$	15	20	60	60	90	15	90	20	15	45	60	15	6	1

Table 1·6·7 (*cont.*)

$w=7$ (52) (i)	[52]	[51, 01]	[50, 02]	[42, 10]	[41, 11]	[40, 12]	[32, 20]	[31, 21]	[30, 22]	[50, $(01)^2$]	[41, 10, 01]	[40, 11, 01]	[40, 10, 02]	[32 $(10)^2$]	[31, 20, 01]	[31, 11, 10]	[30, 21, 01]	[30, 20, 02]	[30, 12, 10]	[30 $(11)^2$]	[22, 20, 10]	[$(21)^2$, 10]	[21, 20, 11]	[$(20)^2$, 12]
s_{52}	1	−1	−1	−1	−1	−1	−1	−1	−1	2	2	2	2	2	2	2	2	2	2	2	2	2	2	2
$s_{51}s_{01}$	1	1	.	.	.	.	.	.	.	−2	−1	−1	.	.	−1	.	−1	.	.	.	.	.	.	.
$s_{50}s_{02}$	1	.	1	.	.	.	.	.	.	−1	.	.	−1	.	.	.	.	−1	.	.	.	.	.	.
$s_{42}s_{10}$	1	.	.	1	.	.	.	.	.	.	−1	.	−1	−2	.	−1	.	.	−1	.	.	.	−1	.
$s_{41}s_{11}$	1	.	.	.	1	.	.	.	.	.	−1	−1	−1	.	.	−1	.	.	−1	−2	.	.	.	−1
$s_{40}s_{12}$	1	.	.	.	.	1	.	.	.	.	.	−1	−1	.	.	.	−1	−1	.	.	.	.	−1	−1
$s_{32}s_{20}$	1	.	.	.	.	.	1	.	.	.	.	.	.	.	.	−1	−1	.	.	−1	.	−1	.	−2
$s_{31}s_{21}$	1	.	.	.	.	.	.	1	.	.	.	.	.	.	.	−1	−1	−1	.	−1	.	−2	−1	.
$s_{30}s_{22}$	1	.	.	.	.	.	.	.	1	.	.	.	.	.	.	.	−1	−1	−1	−1	−1	.	.	.
$s_{50}s_{01}^2$	1	2	1	.	.	.	.	.	.	1	.	.	.	.	.	.	.	.	.	.	.	.	.	.
$s_{41}s_{10}s_{01}$	1	1	.	1	1	.	.	.	.	.	1	.	.	.	.	.	.	.	.	.	.	.	.	.
$s_{40}s_{11}s_{01}$	1	1	.	.	1	1	.	.	.	.	.	1	.	.	.	.	.	.	.	.	.	.	.	.
$s_{40}s_{10}s_{02}$	1	.	1	1	.	.	.	.	.	.	.	.	1	.	.	.	.	.	.	.	.	.	.	.
$s_{32}s_{10}^2$	1	.	.	2	.	.	1	.	.	.	.	.	.	1	.	.	.	.	.	.	.	.	.	.
$s_{31}s_{20}s_{01}$	1	1	.	.	.	.	1	1	.	.	.	.	.	.	1	.	.	.	.	.	.	.	.	.
$s_{31}s_{11}s_{10}$	1	.	.	1	1	.	1	1	.	.	.	.	.	.	.	1	.	.	.	.	.	.	.	.
$s_{30}s_{21}s_{01}$	1	1	.	.	.	.	1	1	1	.	.	.	.	.	.	.	1	.	.	.	.	.	.	.
$s_{30}s_{20}s_{02}$	1	.	1	.	.	.	1	.	1	.	.	.	.	.	.	.	.	1	.	.	.	.	.	.
$s_{30}s_{12}s_{10}$	1	.	.	1	.	1	.	.	1	.	.	.	.	.	.	.	.	.	1	.	.	.	.	.
$s_{30}s_{11}^2$	1	.	.	.	2	.	.	.	1	.	.	.	.	.	.	.	.	.	.	1	.	.	.	.
$s_{22}s_{20}s_{10}$	1	.	.	1	.	.	1	.	1	.	.	.	.	.	.	.	.	.	.	.	1	.	.	.
$s_{21}^2s_{10}$	1	.	.	1	.	.	.	2	.	.	.	.	.	.	.	.	.	.	.	.	.	1	.	.
$s_{21}s_{20}s_{11}$	1	.	.	.	1	.	1	1	.	.	.	.	.	.	.	.	.	.	.	.	.	.	1	.
$s_{20}^2s_{12}$	1	.	.	.	.	1	2	.	.	.	.	.	.	.	.	.	.	.	.	.	.	.	.	1
$s_{40}s_{10}s_{01}^2$	1	2	1	1	2	1	.	.	.	1	2	2	1	.	.	.	.	.	.	.	.	.	.	.
$s_{31}s_{10}^2s_{01}$	1	1	.	2	2	.	1	1	.	.	2	.	.	.	1	1	2	.	.	.	.	.	.	.
$s_{30}s_{20}s_{01}^2$	1	2	1	.	.	.	1	2	1	1	.	.	.	.	2	.	2	1	.	.	.	.	.	.
$s_{30}s_{11}s_{10}s_{01}$	1	1	.	1	2	1	1	.	1	.	1	1	.	.	1	1	.	1	1	1	.	.	.	.
$s_{30}s_{10}^2s_{02}$	1	.	1	2	.	2	1	.	1	.	.	.	2	1	.	.	.	1	2	.	.	.	.	.
$s_{22}s_{10}^3$	1	.	.	3	.	.	3	.	1	.	.	.	.	3	.	.	.	.	.	.	3	1	1	.
$s_{21}s_{20}s_{10}s_{01}$	1	1	.	2	1	.	1	2	.	.	2	.	.	.	1	.	2	.	.	.	.	.	.	1
$s_{21}s_{11}s_{10}^2$	1	.	.	2	1	.	1	3	.	.	.	.	.	.	1	.	2	.	.	.	.	.	2	1
$s_{20}^2s_{11}s_{01}$	1	1	.	.	1	1	2	2	.	.	.	1	.	.	.	2	.	.	.	.	.	.	2	1
$s_{20}^2s_{10}s_{02}$	1	.	1	1	.	1	2	.	2	.	.	.	1	.	.	.	.	2	.	.	2	.	.	1
$s_{20}s_{12}s_{10}^2$	1	.	.	2	1	1	2	.	2	.	.	.	.	.	.	2	.	.	2	.	.	2	.	1
$s_{20}s_{11}^2s_{10}$	1	.	.	1	2	.	1	2	1	.	.	.	.	.	.	2	.	.	.	.	1	1	2	.
$s_{30}s_{10}^2s_{01}^2$	1	2	1	2	4	2	1	2	1	1	4	4	2	1	2	4	2	1	2	2	.	.	.	.
$s_{21}s_{10}^3s_{01}$	1	1	.	3	3	.	3	4	1	.	3	.	1	3	3	6	1	.	.	.	3	3	3	1
$s_{20}^2s_{10}s_{01}^2$	1	2	1	1	2	.	2	4	2	1	2	2	1	.	4	4	4	2	.	2	2	2	4	1
$s_{20}s_{11}s_{10}^2s_{01}$	1	1	.	2	3	1	2	4	2	.	2	1	.	1	2	4	2	.	2	2	2	2	4	1
$s_{20}s_{10}^2s_{02}$	1	.	1	3	.	3	4	.	4	.	.	.	3	3	.	.	.	4	6	.	6	.	.	3
$s_{12}s_{10}^4$	1	.	.	4	.	1	6	.	4	.	.	.	.	6	.	6	.	.	.	.	4	.	12	3
$s_{11}^2s_{10}^3$	1	.	.	3	2	.	3	6	1	.	.	.	.	3	.	6	.	.	.	.	1	3	6	6
$s_{20}s_{10}^3s_{01}^2$	1	2	1	3	6	3	4	8	4	1	6	6	3	3	8	12	8	4	6	6	6	6	12	3
$s_{11}s_{10}^4s_{01}$	1	1	.	4	5	1	6	10	4	.	4	1	.	6	6	16	4	.	10	20	4	12	18	3
$s_{10}^5s_{02}$	1	.	1	5	.	5	10	.	10	.	.	.	5	10	.	.	.	10	20	.	30	.	.	15
$s_{10}^5s_{01}^2$	1	2	1	5	10	5	10	20	10	1	10	10	5	10	20	40	20	10	20	20	30	30	60	15

Table 1·6·7 (cont.)

$w = 7$ (52) (ii)	$[40, 10, (01)^2]$	$[31, (10)^2, 01]$	$[30, 20, (01)^2]$	$[30, 11, 10, 01]$	$[30, (10)^2, 02]$	$[22, (10)^3]$	$[21, 20, 10, 01]$	$[21, 11, (10)^2]$	$[(20)^2, 11, 01]$	$[(20)^2, 10, 02]$	$[20, 12, (10)^2]$	$[20, (11)^2, 10]$	$[30, (10)^2, (01)^2]$	$[21, (10)^3, 01]$	$[(20)^2, 10, (01)^2]$	$[20, 11, (10)^2, 01]$	$[20, (10)^3, 02]$	$[12, (10)^4]$	$[(11)^2, (10)^3]$	$[20, (10)^3, (01)^2]$	$[11, (10)^4, 01]$	$[(10)^5, 02]$	$[(10)^5, (01)^2]$
s_{52}	-6	-6	-6	-6	-6	-6	-6	-6	-6	-6	-6	-6	24	24	24	24	24	24	24	-120	-120	-120	720
$s_{51}s_{01}$	4	2	4	2	.	.	2	.	2	.	.	.	-12	-6	-12	-6	.	.	.	48	24	.	-240
$s_{50}s_{02}$	1	.	1	.	2	.	.	4	.	2	.	2	-2	.	-2	.	-6	.	.	6	.	24	-24
$s_{42}s_{10}$	2	4	.	2	4	6	2	4	.	2	4	.	-12	-6	-6	-12	-18	-24	-18	72	96	120	-600
$s_{41}s_{11}$	2	2	.	3	.	.	1	2	2	.	.	4	-8	-6	-4	-10	.	.	-12	36	48	.	-240
$s_{40}s_{12}$	2	.	.	1	2	.	.	1	1	2	.	.	-4	.	-2	-2	-6	-6	.	12	6	30	-60
$s_{32}s_{20}$	.	1	2	.	1	3	2	1	4	4	3	2	-2	-6	-12	-8	-12	-12	-6	42	36	60	-240
$s_{31}s_{21}$	.	2	2	1	.	.	3	4	2	.	.	2	-4	-12	-8	-8	.	.	-12	36	48	.	-240
$s_{30}s_{22}$	.	.	2	2	2	2	1	.	.	2	2	1	-6	-2	-4	-4	-8	-8	-2	22	16	40	-120
$s_{50}s_{01}^2$	-1	.	-1	.	.	.	.	.	.	.	.	.	2	.	2	.	.	.	.	-6	.	.	24
$s_{41}s_{10}s_{01}$	-2	-2	.	-1	.	.	-1	.	.	.	.	.	8	6	4	4	.	.	.	-36	-24	.	240
$s_{40}s_{11}s_{01}$	-2	.	.	-1	.	.	.	.	-1	.	.	.	4	.	2	2	.	.	.	-12	-6	.	60
$s_{40}s_{10}s_{02}$	-1	.	.	.	-2	.	.	.	.	-1	.	.	2	6	1	.	6	.	.	-6	.	-30	30
$s_{32}s_{10}^2$	.	-1	.	.	-1	-3	.	-1	.	.	.	-1	2	6	.	2	6	12	6	-18	-36	-60	240
$s_{31}s_{20}s_{01}$	.	-1	-2	.	.	.	-1	.	-2	.	.	.	2	3	8	3	.	.	.	-24	-12	.	120
$s_{31}s_{11}s_{10}$	.	-2	.	-1	.	.	.	-2	.	.	.	-2	4	6	.	6	.	.	12	-24	-48	.	240
$s_{30}s_{21}s_{01}$	.	.	-2	-1	.	.	-1	.	.	.	.	.	4	2	4	2	.	.	.	-16	-8	.	80
$s_{30}s_{20}s_{02}$	.	.	-1	.	-1	.	.	.	.	-2	.	.	1	.	2	.	5	.	.	-5	.	-20	20
$s_{30}s_{12}s_{10}$	.	.	-1	-2	.	.	.	.	.	.	-2	.	4	.	2	6	8	.	.	-12	-8	-40	80
$s_{30}s_{11}^2$	.	.	.	-1	.	.	.	.	.	.	.	-1	.	.	2	.	.	.	2	-6	-8	.	40
$s_{22}s_{20}s_{10}$	.	.	.	.	.	-3	-1	.	.	-2	-2	-1	.	6	4	4	9	12	3	-24	-24	-60	180
$s_{21}^2s_{10}$	.	.	.	.	.	.	-1	-2	.	.	.	.	.	6	2	2	.	.	6	-12	-24	.	120
$s_{21}s_{20}s_{11}$	.	.	.	.	.	.	-1	-1	-2	.	.	-2	.	3	4	5	.	.	6	-18	-24	.	120
$s_{20}^2s_{12}$	.	.	.	.	.	.	.	-1	-1	-1	.	.	.	2	1	3	3	.	.	-6	-3	-15	30
$s_{40}s_{10}s_{01}^2$	1	.	.	.	.	.	.	.	.	.	.	.	-2	.	-1	.	.	.	.	6	.	.	-30
$s_{31}s_{10}^2s_{01}$	.	1	.	.	.	.	.	.	.	.	.	.	-2	-3	.	-1	.	.	.	12	12	.	-120
$s_{30}s_{20}s_{01}^2$	.	.	1	.	.	.	.	.	.	.	.	.	-1	.	-2	.	.	.	.	5	.	.	-20
$s_{30}s_{11}s_{10}s_{01}$	.	.	.	1	.	.	.	.	.	.	.	.	-4	.	.	-2	.	.	.	12	8	.	-80
$s_{30}s_{10}^2s_{02}$	.	.	.	.	1	.	.	.	.	.	.	.	-1	.	.	.	-3	.	.	3	8	20	-20
$s_{22}s_{10}^3$	.	.	.	.	.	1	.	.	.	.	.	.	.	-1	.	.	-1	-4	-1	2	8	20	-60
$s_{21}s_{20}s_{10}s_{01}$	.	.	.	.	.	.	1	.	.	.	.	.	.	-3	-4	-2	.	.	.	18	12	.	-120
$s_{21}s_{11}s_{10}^2$	.	.	.	.	.	.	.	1	.	.	.	.	.	-3	.	-1	.	.	-6	6	24	.	-120
$s_{20}^2s_{11}s_{01}$	.	.	.	.	.	.	.	.	1	.	.	.	.	.	-2	-1	.	.	.	6	3	.	-30
$s_{20}^2s_{10}s_{02}$	.	.	.	.	.	.	.	.	.	1	.	.	.	.	-1	.	-3	.	.	3	.	15	-15
$s_{20}s_{12}s_{10}^2$	.	.	.	.	.	.	.	.	.	.	1	.	.	.	.	-1	-3	-6	.	6	6	30	-60
$s_{20}s_{11}^2s_{10}$	.	.	.	.	.	.	.	.	.	.	.	1	.	.	.	.	.	-2	-3	6	12	.	-60
$s_{30}s_{10}^2s_{01}^2$	2	2	1	4	1	.	.	.	.	.	.	.	1	.	.	.	.	.	.	-3	.	.	20
$s_{21}s_{10}^3s_{01}$	.	3	.	.	.	1	3	3	.	.	.	.	.	1	.	.	.	.	.	-2	-4	.	40
$s_{20}^2s_{10}s_{01}^2$	1	.	2	.	.	.	4	.	2	1	.	2	.	.	1	.	.	.	.	-3	.	.	15
$s_{20}s_{11}s_{10}^2s_{01}$	.	1	.	2	.	.	2	1	1	.	1	2	.	.	.	1	.	.	.	-6	-6	.	60
$s_{20}s_{10}^3s_{02}$	.	.	.	.	3	1	.	.	.	3	3	.	.	.	.	.	1	.	.	-1	.	-10	10
$s_{12}s_{10}^4$	.	.	.	.	.	4	.	6	.	.	.	3	.	.	.	.	.	1	.	.	-1	-5	10
$s_{11}^2s_{10}^3$	.	.	.	.	.	.	1	.	6	.	.	3	.	.	.	.	.	.	1	.	-4	.	20
$s_{20}s_{10}^3s_{01}^2$	3	6	4	12	3	1	12	6	6	3	3	6	3	2	3	6	1	.	.	1	.	.	-10
$s_{11}s_{10}^4s_{01}$	.	6	.	4	.	4	12	18	3	.	6	12	.	4	.	6	.	1	4	.	1	.	-10
$s_{10}^5s_{02}$	.	.	.	.	10	10	.	.	.	15	30	.	.	.	.	.	10	5	.	.	.	1	-1
$s_{10}^5s_{01}^2$	5	20	10	40	10	10	60	60	30	15	30	60	10	20	15	60	10	5	20	10	10	1	1

Table 1·6·7 (cont.)

$w=7$ (43)(i)	$[43]$	$[42,01]$	$[41,02]$	$[40,03]$	$[33,10]$	$[32,11]$	$[31,12]$	$[30,13]$	$[23,20]$	$[22,21]$	$[41,(01)^2]$	$[40,02,01]$	$[32,10,01]$	$[31,11,01]$	$[31,10,02]$	$[30,12,01]$	$[30,11,02]$	$[30,10,03]$	$[23,(10)^2]$	$[22,20,01]$	$[22,11,10]$	$[(21)^2,01]$	$[21,20,02]$	$[21,12,10]$	$[21,(11)^2]$	$[(20)^2,03]$	$[20,13,10]$	$[20,12,11]$
S_{43}	1	-1	-1	-1	-1	-1	-1	-1	-1	-1	2	2	2	2	2	2	2	2	2	2	2	2	2	2	2	2	2	2
$S_{42}S_{01}$	1	1	.	.	.	.	.	.	.	.	-2	-1	-1	-1	.	-1	.	.	.	.	-1	.	-1	.	.	.	.	.
$S_{41}S_{02}$	1	.	1	.	.	.	.	.	.	.	-1	-1	.	.	-1	.	-1	.	.	.	.	.	-1	.	.	.	.	.
$S_{40}S_{03}$	1	.	.	1	.	.	.	.	.	.	.	-1	.	.	.	.	.	-1	.	.	.	.	.	.	.	-1	.	.
$S_{33}S_{10}$	1	.	.	.	1	.	.	.	.	.	.	.	-1	.	.	-1	.	.	-1	-2	.	-1	.	.	-1	.	-1	.
$S_{32}S_{11}$	1	.	.	.	.	1	.	.	.	.	.	.	-1	-1	.	.	-1	.	.	.	-1	.	.	.	-2	.	.	-1
$S_{31}S_{12}$	1	.	.	.	.	.	1	.	.	.	.	.	.	-1	-1	-1	.	.	.	.	.	.	-1	.	.	.	.	-1
$S_{30}S_{13}$	1	.	.	.	.	.	.	1	.	.	.	.	.	.	.	-1	-1	-1	.	.	.	.	.	.	.	.	-1	-1
$S_{23}S_{20}$	1	.	.	.	.	.	.	.	1	.	.	.	.	.	.	.	.	.	-1	-1	.	-2	.	-1	.	-2	-1	-1
$S_{22}S_{21}$	1	.	.	.	.	.	.	.	.	1	.	.	.	.	.	.	.	.	.	-1	-1	-2	-1	-1	-1	.	-1	-1
$S_{41}S_{01}^2$	1	2	1	.	.	.	.	.	.	.	1	.	.	.	.	.	.	.	.	.	.	.	.	.	.	.	.	.
$S_{40}S_{02}S_{01}$	1	1	1	1	.	.	.	.	.	.	.	1	.	.	.	.	.	.	.	.	.	.	.	.	.	.	.	.
$S_{32}S_{10}S_{01}$	1	1	.	.	1	1	.	.	.	.	.	.	1	.	.	.	.	.	.	.	.	.	.	.	.	.	.	.
$S_{31}S_{11}S_{01}$	1	1	.	.	.	1	1	.	.	.	.	.	.	1	.	.	.	.	.	.	.	.	.	.	.	.	.	.
$S_{31}S_{10}S_{02}$	1	.	1	.	1	.	1	.	.	.	.	.	.	.	1	.	.	.	.	.	.	.	.	.	.	.	.	.
$S_{30}S_{12}S_{01}$	1	1	.	.	.	.	1	1	.	.	.	.	.	.	.	1	.	.	.	.	.	.	.	.	.	.	.	.
$S_{30}S_{11}S_{02}$	1	.	1	.	.	1	.	1	.	.	.	.	.	.	.	.	1	.	.	.	.	.	.	.	.	.	.	.
$S_{30}S_{10}S_{03}$	1	.	.	1	1	.	.	1	.	.	.	.	.	.	.	.	.	1	.	.	.	.	.	.	.	.	.	.
$S_{23}S_{10}^2$	1	.	.	.	2	.	.	.	1	.	.	.	.	.	.	.	.	.	1	.	.	.	.	.	.	.	.	.
$S_{22}S_{20}S_{01}$	1	1	.	.	.	.	.	.	1	1	.	.	.	.	.	.	.	.	.	1	.	.	.	.	.	.	.	.
$S_{22}S_{11}S_{10}$	1	.	.	.	1	1	.	.	.	1	.	.	.	.	.	.	.	.	.	.	1	.	.	.	.	.	.	.
$S_{21}^2S_{01}$	1	1	.	.	.	.	.	.	.	2	.	.	.	.	.	.	.	.	.	.	.	1	.	.	.	.	.	.
$S_{21}S_{20}S_{02}$	1	.	1	.	.	.	.	.	1	1	.	.	.	.	.	.	.	.	.	.	.	.	1	.	.	.	.	.
$S_{21}S_{12}S_{10}$	1	.	.	.	1	.	1	.	.	1	.	.	.	.	.	.	.	.	.	.	.	.	.	1	.	.	.	.
$S_{21}S_{11}^2$	1	.	.	.	.	2	.	.	.	1	.	.	.	.	.	.	.	.	.	.	.	.	.	.	1	.	.	.
$S_{20}^2S_{03}$	1	.	.	1	.	.	.	.	2	.	.	.	.	.	.	.	.	.	.	.	.	.	.	.	.	1	.	.
$S_{20}S_{13}S_{10}$	1	.	.	.	1	.	.	1	1	.	.	.	.	.	.	.	.	.	.	.	.	.	.	.	.	.	1	.
$S_{20}S_{12}S_{11}$	1	.	.	.	.	1	1	.	1	.	.	.	.	.	.	.	.	.	.	.	.	.	.	.	.	.	.	1
$S_{40}S_{01}^3$	1	3	3	1	.	.	.	.	.	.	3	3	.	.	.	.	.	.	.	.	.	.	.	.	.	.	.	.
$S_{31}S_{10}S_{01}^2$	1	2	1	.	1	2	1	.	.	.	1	.	2	2	1	.	.	.	.	.	.	.	.	.	.	.	.	.
$S_{30}S_{11}S_{01}^2$	1	2	1	.	.	1	2	1	.	.	1	.	.	2	.	2	1	.	.	.	.	.	.	.	.	.	.	.
$S_{30}S_{10}S_{02}S_{01}$	1	1	1	1	.	1	1	1	.	.	.	1	.	1	1	1	1	1	.	.	.	.	.	.	.	.	.	.
$S_{22}S_{10}^2S_{01}$	1	1	.	.	2	2	.	.	1	1	.	.	2	.	.	.	.	.	.	1	1	2	.	.	.	.	.	.
$S_{21}S_{20}S_{01}^2$	1	2	1	.	.	.	.	.	1	3	1	.	.	.	.	.	.	.	.	.	.	2	.	2	1	.	.	.
$S_{21}S_{11}S_{10}S_{01}$	1	1	.	.	1	2	1	.	.	2	.	.	1	1	.	2	.	.	.	1	.	.	1	2	.	.	.	.
$S_{21}S_{10}^2S_{02}$	1	.	1	.	2	.	2	.	1	1	.	.	.	.	2	.	.	.	.	1	.	.	1	2	.	.	.	.
$S_{20}^2S_{02}S_{01}$	1	1	1	1	.	.	.	.	2	2	.	1	.	.	.	.	.	.	.	2	.	.	2	.	.	1	.	.
$S_{20}S_{12}S_{10}S_{01}$	1	1	.	.	1	1	1	1	1	1	.	.	1	.	.	1	.	.	.	1	.	.	1	.	.	.	1	1
$S_{20}S_{11}^2S_{01}$	1	1	.	.	.	2	2	.	1	1	.	.	.	2	.	.	.	.	.	1	.	.	.	1	.	.	.	2
$S_{20}S_{11}S_{10}S_{02}$	1	.	1	.	1	1	1	1	1	1	.	.	.	1	.	1	.	1	.	1	.	1	.	1	.	.	1	1
$S_{20}S_{10}^2S_{03}$	1	.	.	1	2	.	.	2	2	.	.	.	.	.	.	.	.	2	.	.	2	.	.	.	.	1	2	.
S_{13}^3	1	.	.	.	3	.	.	3	3	.	.	.	.	.	.	.	.	.	.	3	.	3	.	.	3	.	3	.
$S_{12}S_{11}S_{10}^2$	1	.	.	.	2	1	1	.	1	2	.	.	.	.	.	.	.	.	.	1	.	2	.	.	2	.	.	1
$S_{11}^3S_{10}$	1	.	.	.	1	3	.	.	.	3	.	.	.	.	.	.	.	.	.	.	3	.	.	.	3	.	.	.
$S_{30}S_{10}S_{01}^3$	1	3	3	1	2	6	6	2	.	.	3	3	6	6	3	3	3	1	.	2	4	2	1	2	2	.	.	.
$S_{21}S_{10}^2S_{01}^2$	1	2	1	.	2	4	2	.	1	3	1	.	4	4	2	.	.	.	1	2	4	2	1	2	2	.	1	.
$S_{20}^2S_{01}^3$	1	3	3	1	.	.	.	.	2	6	3	3	.	.	.	.	.	.	2	6	6	6	6	6	.	1	.	.
$S_{20}S_{11}S_{10}S_{01}^2$	1	2	1	.	1	3	3	1	1	3	1	.	2	4	1	2	1	.	.	2	2	.	2	2	2	.	1	3
$S_{20}S_{10}^2S_{02}S_{01}$	1	1	1	1	2	2	2	2	2	2	.	1	2	.	2	2	2	2	1	2	2	.	2	2	.	1	2	2
$S_{12}S_{10}^3S_{01}$	1	1	.	.	3	3	1	.	3	3	.	.	3	.	.	1	.	.	3	3	6	.	.	3	.	.	.	3
$S_{11}^2S_{10}^3S_{01}$	1	1	.	.	2	4	2	.	1	5	.	.	2	2	.	.	.	.	1	1	6	2	.	.	4	5	.	3
$S_{11}S_{10}^3S_{02}$	1	.	1	.	3	3	1	.	3	3	.	.	.	3	.	.	1	.	3	3	.	3	.	.	3	6	.	3
$S_{10}^4S_{03}$	1	.	.	1	4	.	.	4	6	.	.	.	.	.	.	.	.	4	6	6	.	.	.	.	.	3	12	.
$S_{20}S_{10}^2S_{01}^3$	1	3	3	1	2	6	6	2	2	6	3	3	6	12	6	6	6	2	1	6	6	6	6	6	6	1	2	6
$S_{11}S_{10}^3S_{01}^2$	1	2	1	.	3	7	5	1	3	9	1	.	6	8	3	2	1	.	3	6	15	6	3	12	12	.	3	9
$S_{10}^4S_{02}S_{01}$	1	1	1	1	4	4	4	4	6	6	.	1	4	.	4	4	4	4	6	6	12	.	6	12	.	3	12	12
$S_{10}^4S_{01}^3$	1	3	3	1	4	12	12	4	6	18	3	3	12	24	12	12	12	4	6	18	36	18	18	36	36	3	12	36

Table 1·6·7 (cont.)

$w=7$ (43)(ii)	$[[40,(01)^3]]$	$[31,10,(01)^2]$	$[30,11,(01)^2]$	$[30,10,02,01]$	$[22,(10)^2,01]$	$[21,20,(01)_2]$	$[21,11,10,01]$	$[21,(10)^2,02]$	$[[20]^2,02,01]$	$[20,12,10,01]$	$[20,(11)^2,01]$	$[20,11,10,02]$	$[20,(10)^2,03]$	$[13,(10)^3]$	$[12,11,(10)^2]$	$[[11]^3,10]$	$[30,10,(01)^3]$	$[21,(10)^2,(01)^2]$	$[[20]^2,(01)^3]$	$[20,11,10,(01)^2]$	$[20,(10)^2,02,01]$	$[12,(10)^3,01]$	$[[11]^2,(10)^2,01]$	$[11,(10)^3,02]$	$[[10]^4,03]$	$[20,(10)^2,(01)^3]$	$[11,(10)^3,(01)^2]$	$[[10]^4,02,01]$	$[[10]^4,(01)^3]$
s_{43}	-6	-6	-6	-6	-6	-6	-6	-6	-6	-6	-6	-6	-6	-6	-6	-6	24	24	24	24	24	24	24	24	24	-120	-120	-120	720
$s_{42}s_{01}$	6	4	4	2	2	4	2	.	2	2	2	.	.	.	.	.	-18	-12	-18	-12	-6	-6	-6	.	.	72	48	24	-360
$s_{41}s_{02}$	3	1	1	2	.	1	.	2	2	.	.	2	.	.	.	.	-6	-2	-6	-2	-6	.	.	-6	.	18	6	24	-72
$s_{40}s_{03}$	2	.	.	1	.	.	.	1	.	.	2	.	.	.	.	.	-2	.	-2	.	-2	.	.	.	-6	4	.	6	-12
$s_{33}s_{10}$	.	2	.	2	4	.	2	4	.	2	.	2	4	6	4	2	-6	-12	.	-6	-12	-18	-12	-18	-24	48	72	96	-480
$s_{32}s_{11}$	.	2	2	1	2	.	3	.	.	1	4	2	.	.	2	6	-6	-8	.	-10	-4	-6	-16	-6	.	36	60	24	-288
$s_{31}s_{12}$	.	2	2	1	.	.	1	2	.	2	2	1	.	.	2	.	-6	-4	.	-6	-4	-6	-4	-6	.	24	24	24	-144
$s_{30}s_{13}$	.	.	2	2	.	.	.	.	.	1	.	1	2	2	.	.	-6	.	-2	-4	-2	.	.	-2	-8	12	4	16	-48
$s_{23}s_{20}$	.	.	.	1	2	.	1	4	2	2	.	2	3	3	1	.	.	-2	-12	-6	-8	-6	-2	-6	-12	30	18	36	-144
$s_{22}s_{21}$	.	.	.	.	2	4	3	2	2	1	1	.	.	.	2	3	.	-10	-12	-6	-4	-6	-10	-6	.	30	42	24	-216
$s_{41}s_{01}{}^2$	-3	-1	-1	.	.	-1	.	.	.	.	.	.	.	.	.	.	6	2	6	2	.	.	.	.	.	-18	-6	.	72
$s_{40}s_{02}s_{01}$	-3	.	.	-1	.	.	.	-1	.	.	.	.	.	.	.	.	3	.	3	.	2	.	.	.	.	-6	.	-6	18
$s_{32}s_{10}s_{01}$	.	-2	.	-1	-2	.	-1	.	.	-1	.	.	.	.	.	.	6	8	.	4	4	6	4	.	.	-36	-36	-24	288
$s_{31}s_{11}s_{01}$	.	-2	-2	.	.	.	-1	.	.	.	-2	.	.	.	.	.	6	4	.	6	.	.	4	.	.	-24	-24	.	144
$s_{31}s_{10}s_{02}$	.	-1	.	-1	.	.	.	-2	.	.	-1	.	.	.	.	.	3	2	.	1	4	.	.	6	.	-12	-6	-24	72
$s_{30}s_{12}s_{01}$	.	.	-2	-1	.	.	.	.	.	-1	.	.	.	.	.	.	6	.	.	2	2	2	.	.	.	-12	-4	-8	48
$s_{30}s_{11}s_{02}$	.	.	-1	-1	.	.	.	.	.	-1	.	.	.	.	.	.	3	.	.	1	2	.	.	2	.	-6	-2	-8	24
$s_{30}s_{10}s_{03}$	.	.	.	-1	.	.	.	.	.	.	-2	.	.	.	.	.	2	.	.	.	2	.	.	.	8	-4	.	-8	16
$s_{23}s_{10}{}^2$	.	.	.	-1	.	.	-1	.	.	.	-1	-3	-1	.	.	.	.	2	.	.	6	2	6	12	.	-6	-18	-36	144
$s_{22}s_{20}s_{01}$	.	.	.	-1	-2	.	.	-2	-1	-1	.	.	.	.	.	.	.	2	12	4	3	3	1	.	.	-24	-12	-12	108
$s_{22}s_{11}s_{10}$	.	.	.	-2	.	-1	.	.	.	-1	.	.	.	.	-2	-3	.	4	.	2	2	6	10	6	.	-12	-42	-24	216
$s_{21}{}^2s_{10}$	.	.	.	.	-2	-1	.	.	.	.	.	.	.	.	.	.	.	4	6	2	.	.	2	.	.	-12	-12	.	72
$s_{21}s_{20}s_{02}$	.	.	.	.	-1	.	-1	-2	.	.	.	-1	.	.	.	.	.	1	6	1	3	.	.	3	.	-9	-3	-12	36
$s_{21}s_{12}s_{10}$	.	.	.	.	-1	-2	.	.	-1	.	.	.	-2	.	.	.	.	4	.	2	2	6	4	6	.	-12	-24	-24	144
$s_{21}s_{11}{}^2$	.	.	.	.	-1	.	.	.	-1	.	.	.	.	.	-3	.	.	2	.	2	.	.	6	.	.	-6	-18	.	72
$s_{20}{}^2s_{03}$	.	.	.	.	.	.	-1	.	.	.	-1	.	.	.	.	.	.	.	2	.	1	.	.	.	3	-2	.	-3	6
$s_{20}s_{13}s_{10}$	.	.	.	.	.	.	.	.	.	-1	.	-1	-2	-3	.	.	.	2	.	4	3	.	3	12	.	-12	-6	-24	72
$s_{20}s_{12}s_{11}$	.	.	.	.	.	.	.	.	-1	-2	-1	.	.	-1	.	.	.	4	.	2	3	2	3	.	.	-12	-12	-12	72
$s_{40}s_{01}{}^3$	1	.	.	.	.	.	.	.	.	.	.	.	.	.	.	.	-1	.	-1	.	.	.	.	.	.	2	6	.	-6
$s_{31}s_{10}s_{01}{}^2$	.	1	.	.	.	.	.	.	.	.	.	.	.	.	.	.	-3	-2	.	-1	.	.	.	.	.	12	6	.	-72
$s_{30}s_{11}s_{01}{}^2$	.	.	1	.	.	.	.	.	.	.	.	.	.	.	.	.	-3	.	.	.	.	-2	.	.	.	6	2	8	-24
$s_{30}s_{10}s_{02}s_{01}$	.	.	.	1	.	.	.	.	.	.	.	.	.	.	.	.	-3	.	.	.	-2	.	.	.	.	6	.	8	-24
$s_{22}s_{10}{}^2s_{01}$	.	.	.	.	1	.	.	.	.	.	.	.	.	.	.	.	.	-2	.	-1	.	-3	-1	.	.	6	12	12	-108
$s_{21}s_{20}s_{01}{}^2$	.	.	.	.	.	1	.	.	.	.	.	.	.	.	.	.	.	-1	-6	-1	.	.	.	.	.	9	3	.	-36
$s_{21}s_{11}s_{10}s_{01}$	.	.	.	.	.	.	1	.	.	.	.	.	.	.	.	.	.	-4	.	-2	.	.	-4	.	.	12	24	.	-144
$s_{21}s_{10}{}^2s_{02}$	.	.	.	.	.	.	.	1	.	.	.	.	.	.	.	.	.	-1	.	.	-1	.	.	-3	.	3	3	12	-36
$s_{20}{}^2s_{02}s_{01}$	.	.	.	.	.	.	.	.	1	.	.	.	.	.	.	.	.	.	.	-3	.	-1	.	.	.	3	3	.	-9
$s_{20}s_{12}s_{10}s_{01}$	.	.	.	.	.	.	.	.	.	1	.	.	.	.	.	.	.	.	.	-2	-2	.	-3	.	.	12	6	12	-72
$s_{20}s_{11}{}^2s_{01}$	.	.	.	.	.	.	.	.	.	.	1	.	.	.	.	.	.	.	.	-2	.	.	-1	.	.	6	6	.	-36
$s_{20}s_{11}s_{10}s_{02}$	.	.	.	.	.	.	.	.	.	.	.	1	.	.	.	.	.	.	.	.	-1	-2	.	-3	.	6	3	12	-36
$s_{20}s_{10}{}^2s_{03}$	.	.	.	.	.	.	.	.	.	.	.	.	1	.	.	.	.	.	.	.	-1	.	.	.	-6	2	.	6	-12
$s_{13}s_{10}{}^3$	.	.	.	.	.	.	.	.	.	.	.	.	.	1	.	.	.	.	.	.	.	.	-1	-1	-4	.	2	8	-24
$s_{12}s_{11}s_{10}{}^2$	.	.	.	.	.	.	.	.	.	.	.	.	.	.	1	.	.	.	.	.	.	.	-3	-2	-3	.	12	12	-72
$s_{11}{}^3s_{10}$	.	.	.	.	.	.	.	.	.	.	.	.	.	.	.	1	.	.	.	.	.	.	.	-2	.	.	6	.	-24
$s_{30}s_{10}s_{01}{}^3$	1	3	3	3	.	2	.	.	.	.	.	.	.	.	.	.	1	.	.	.	.	.	.	.	.	-2	.	.	8
$s_{21}s_{10}{}^2s_{01}{}^2$	.	2	.	.	2	1	4	1	.	.	.	.	.	.	.	.	.	1	.	.	.	.	.	.	.	-3	-3	.	36
$s_{20}{}^2s_{01}{}^3$	1	.	.	.	6	.	.	.	3	.	.	.	.	.	.	.	.	.	1	.	.	.	.	.	.	-1	.	.	3
$s_{20}s_{11}s_{10}s_{01}{}^2$	.	1	1	.	1	2	.	2	2	1	.	.	.	.	.	.	.	.	.	1	.	.	.	.	.	-6	-3	.	36
$s_{20}s_{10}{}^2s_{02}s_{01}$	.	.	.	2	1	.	1	1	2	.	2	1	.	.	.	.	.	.	.	.	1	.	.	.	.	-3	.	-6	18
$s_{12}s_{10}{}^3s_{01}$	.	.	.	3	.	.	.	.	3	.	.	.	1	3	.	.	.	.	.	.	.	1	.	.	.	.	-2	-4	24
$s_{11}{}^2s_{10}{}^2s_{01}$	.	.	.	.	1	.	4	.	.	.	3	.	.	2	2	.	.	.	.	.	.	.	1	.	.	.	-6	.	36
$s_{11}s_{10}{}^3s_{02}$	.	.	.	.	.	.	3	.	.	3	1	3	.	.	.	.	.	.	.	.	.	.	.	1	.	.	-1	-4	12
$s_{10}{}^4s_{03}$	.	.	.	.	.	.	.	.	.	.	6	4	.	.	.	.	.	.	.	.	.	.	.	.	1	.	.	-1	.
$s_{20}s_{10}{}^2s_{01}{}^3$	1	6	6	6	3	6	12	3	3	6	6	6	1	.	.	.	2	3	1	6	3	.	.	.	.	1	.	.	-6
$s_{11}s_{10}{}^3s_{01}{}^2$	.	3	1	.	6	3	18	3	.	6	6	3	.	1	9	6	.	3	.	3	.	2	6	1	.	.	1	.	-12
$s_{10}{}^4s_{02}s_{01}$	.	.	.	4	6	.	.	6	3	12	.	12	6	4	12	.	.	.	.	.	6	4	.	4	1	.	.	1	-3
$s_{10}{}^4s_{01}{}^3$	1	12	12	12	18	18	72	18	9	36	36	36	6	4	36	24	4	18	3	36	18	12	36	12	1	6	12	3	1

Table 1·6·8. *Partitions* (71), (62), (53), (44)

$w = 8$ (71)(i)	[71]	[70, 01]	[61, 10]	[60, 11]	[51, 20]	[50, 21]	[41, 30]	[40, 31]	[60, 10, 01]	[51, (10)²]	[50, 20, 01]	[50, 11, 10]	[41, 20, 10]	[40, 30, 01]	[40, 21, 10]	[40, 20, 11]	[31, 30, 10]	[31, (20)²]	[(30)², 11]	[30, 21, 20]
s_{71}	1	-1	-1	-1	-1	-1	-1	-1	2	2	2	2	2	2	2	2	2	2	2	2
$s_{70}s_{01}$	1	1	·	·	·	·	·	·	-1	·	-1	·	·	-1	·	·	·	·	·	·
$s_{61}s_{10}$	1	·	1	·	·	·	·	·	-1	-2	·	-1	-1	·	-1	·	-1	·	·	·
$s_{60}s_{11}$	1	·	·	1	·	·	·	·	-1	·	·	-1	·	-1	·	-1	·	-1	·	·
$s_{51}s_{20}$	1	·	·	·	1	·	·	·	·	-1	·	-1	-1	·	-1	·	·	-2	·	-1
$s_{50}s_{21}$	1	·	·	·	·	1	·	·	·	·	·	-1	·	-1	·	-1	·	-1	-2	-1
$s_{41}s_{30}$	1	·	·	·	·	·	1	·	·	·	·	·	·	-1	-1	-1	-1	-1	·	·
$s_{40}s_{31}$	1	·	·	·	·	·	·	1	·	·	·	·	·	·	·	·	·	·	·	·
$s_{60}s_{10}s_{01}$	1	1	1	1	·	·	·	·	1	·	·	·	·	·	·	·	·	·	·	·
$s_{51}s_{10}^{2}$	1	·	2	·	1	·	·	·	·	1	·	·	·	·	·	·	·	·	·	·
$s_{50}s_{20}s_{01}$	1	1	·	1	1	1	·	·	·	·	1	·	·	·	·	·	·	·	·	·
$s_{50}s_{11}s_{10}$	1	·	1	1	·	1	·	·	·	·	·	1	·	·	·	·	·	·	·	·
$s_{41}s_{20}s_{10}$	1	·	1	·	1	·	1	·	·	·	·	·	1	·	·	·	·	·	·	·
$s_{40}s_{30}s_{01}$	1	1	·	·	·	1	·	1	·	·	·	·	·	1	·	·	·	·	·	·
$s_{40}s_{21}s_{10}$	1	·	1	·	1	·	1	·	·	·	·	·	·	·	1	·	·	·	·	·
$s_{40}s_{20}s_{11}$	1	·	·	1	1	·	·	1	·	·	·	·	·	·	·	1	·	·	·	·
$s_{31}s_{30}s_{10}$	1	·	1	·	·	·	1	1	·	·	·	·	·	·	·	·	1	·	·	·
$s_{31}s_{20}^{2}$	1	·	·	·	2	·	·	1	·	·	·	·	·	·	·	·	·	1	·	·
$s_{30}^{2}s_{11}$	1	·	·	1	·	·	2	·	·	·	·	·	·	·	·	·	·	·	1	·
$s_{30}s_{21}s_{20}$	1	·	·	·	1	1	1	·	·	·	·	·	·	·	·	·	·	·	·	1
$s_{50}s_{10}^{2}s_{01}$	1	1	2	2	1	1	·	·	2	1	1	2	·	·	·	·	·	·	·	·
$s_{41}s_{10}^{3}$	1	·	3	·	3	·	1	·	·	3	·	·	3	·	·	·	·	·	·	·
$s_{40}s_{20}s_{10}s_{01}$	1	1	1	1	1	1	1	1	1	·	1	·	1	1	1	1	·	·	·	·
$s_{40}s_{11}s_{10}^{2}$	1	·	2	1	1	2	·	1	·	1	·	2	·	2	1	·	2	1	·	·
$s_{31}s_{20}s_{10}^{2}$	1	·	2	·	2	·	2	1	·	1	·	·	2	·	2	2	·	1	·	1
$s_{30}s_{10}^{2}s_{01}$	1	1	1	·	·	·	2	2	1	·	1	·	·	2	·	2	·	·	·	2
$s_{30}s_{21}s_{10}^{2}$	1	·	2	·	1	1	1	2	·	1	·	·	·	1	·	1	1	·	1	2
$s_{30}s_{20}^{2}s_{01}$	1	1	·	·	2	2	1	1	·	·	2	·	1	·	1	·	1	1	1	2
$s_{30}s_{20}s_{11}s_{10}$	1	·	1	1	1	1	2	1	·	·	·	1	1	·	2	·	1	1	1	2
$s_{21}s_{20}^{2}s_{10}$	1	·	1	·	2	1	2	1	·	·	·	·	2	·	·	3	·	3	·	·
$s_{20}^{3}s_{11}$	1	·	·	1	3	·	·	3	·	·	·	·	·	·	·	·	3	·	3	·
$s_{40}s_{10}^{3}s_{01}$	1	1	3	3	3	3	1	1	3	3	3	6	3	1	3	3	·	4	3	·
$s_{31}s_{10}^{4}$	1	·	4	·	6	·	4	1	·	6	·	2	2	·	2	2	4	1	2	2
$s_{30}s_{20}s_{10}^{2}s_{01}$	1	1	2	2	2	2	3	3	2	1	2	2	3	·	6	3	4	3	1	3
$s_{30}s_{11}s_{10}^{3}$	1	·	3	·	3	2	3	3	·	3	·	3	·	·	6	·	3	3	·	4
$s_{21}s_{20}s_{10}^{3}$	1	·	3	·	4	1	4	3	·	3	·	·	6	3	3	·	6	3	·	6
$s_{20}^{2}s_{10}^{2}s_{01}$	1	1	1	1	3	3	4	3	1	1	3	·	3	3	2	3	4	3	2	4
$s_{20}^{2}s_{11}s_{10}^{2}$	1	·	2	1	3	2	4	3	·	1	·	3	2	·	2	3	4	3	·	4
$s_{30}s_{10}^{4}s_{01}$	1	1	4	4	6	6	5	5	4	6	6	12	12	5	12	12	8	3	4	6
$s_{21}s_{10}^{5}$	1	·	5	·	10	1	10	5	·	10	·	·	30	·	5	·	20	15	6	10
$s_{20}^{2}s_{10}^{3}s_{01}$	1	·	3	1	5	5	7	7	3	3	5	6	9	7	9	12	12	7	6	14
$s_{20}s_{11}s_{10}^{4}$	1	·	4	1	7	4	8	7	·	6	·	4	16	·	12	7	16	9	4	16
$s_{20}s_{10}^{5}s_{01}$	1	1	5	5	11	11	15	15	5	10	11	20	35	15	35	35	40	25	20	50
$s_{11}s_{10}^{6}$	1	·	6	1	15	6	20	15	·	15	·	6	60	·	30	15	60	45	10	60
$s_{10}^{7}s_{01}$	1	1	7	7	21	21	35	35	7	21	21	42	105	35	105	105	140	105	70	210

Table 1·6·8 (*cont.*)

$w = 8$ (71) (ii)	$[50, (10)^2, 01]$	$[41, (10)^3]$	$[40, 20, 10, 01]$	$[40, 11, (10)^2]$	$[31, 20, (10)^2]$	$[(30)^2, 10, 01]$	$[30, 21, (10)^2]$	$[30, (20)^2, 01]$	$[30, 20, 11, 10]$	$[21, (20)^2, 10]$	$[(20)^3, 11]$	$[40, (10)^3, 01]$	$[31, (10)^4]$	$[30, 20, (10)^2, 01]$	$[30, 11, (10)^3]$	$[21, 20, (10)^3]$	$[20^3, 10, 01]$	$[20^2, 11, 10^2]$
s_{71}	−6	−6	−6	−6	−6	−6	−6	−6	−6	−6	−6	24	24	24	24	24	24	24
$s_{70}s_{01}$	2	·	2	·	·	2	·	2	·	·	·	−6	·	−6	·	·	−6	·
$s_{61}s_{10}$	4	6	2	4	4	2	4	·	2	2	·	−18	−24	−12	−18	−18	−6	−12
$s_{60}s_{11}$	2	·	1	2	·	1	·	·	2	2	2	−6	·	−4	−6	·	−2	−6
$s_{51}s_{20}$	1	3	2	1	3	·	·	1	4	2	4	6	−6	−12	−8	−6	−12	−18	−14
$s_{50}s_{21}$	2	·	1	2	·	·	·	2	2	1	2	·	−6	·	−4	−6	−6	−6	−4
$s_{41}s_{30}$	·	2	1	·	2	4	2	2	3	2	·	−2	−8	−10	−8	−8	−6	−8
$s_{40}s_{31}$	·	·	2	2	2	2	2	1	1	1	3	−6	−6	−6	−6	−6	−6	−6
$s_{60}s_{10}{}^2$	−2	·	−1	·	·	−1	·	·	·	·	·	6	·	4	·	·	2	·
$s_{51}s_{10}{}^2$	−1	−3	·	−1	−1	·	−1	·	·	·	·	6	12	2	6	6	·	2
$s_{50}s_{20}s_{01}$	−1	·	−1	·	·	·	·	−2	·	·	·	3	·	3	·	·	6	·
$s_{50}s_{11}s_{10}$	−2	·	·	−2	·	·	·	·	−1	·	·	6	·	2	6	·	·	4
$s_{41}s_{20}s_{10}$	·	−3	−1	·	−2	·	·	·	−1	−2	·	3	12	4	3	9	6	8
$s_{40}s_{30}s_{01}$	·	·	−1	·	·	−2	·	−1	·	·	·	2	·	4	·	·	3	·
$s_{40}s_{21}s_{10}$	·	·	−1	−2	·	·	−2	·	−1	·	·	6	·	2	6	6	·	2
$s_{40}s_{20}s_{11}$	·	·	−1	−1	·	·	·	·	−1	·	−3	3	·	2	3	3	3	5
$s_{31}s_{30}s_{10}$	·	·	·	·	−2	−2	−2	·	−1	·	·	·	8	6	6	6	3	4
$s_{31}s_{20}{}^2$	·	·	·	·	−1	·	·	−1	·	−1	−3	·	3	1	·	3	6	4
$s_{30}{}^2 s_{11}$	·	·	·	·	·	−1	·	·	−1	·	·	·	·	2	2	·	6	2
$s_{30}s_{21}s_{20}$	·	·	·	·	·	·	−1	−2	−1	−2	·	·	·	3	3	5	6	4
$s_{50}s_{10}{}^2 s_{01}$	1	·	·	·	·	·	·	·	·	·	·	−3	·	−1	·	·	·	·
$s_{41}s_{10}{}^3$	·	1	·	·	·	·	·	·	·	·	·	−1	−4	·	−1	−1	·	·
$s_{40}s_{20}s_{10}s_{01}$	·	·	1	·	·	·	·	·	·	·	·	−3	·	−2	·	·	−3	·
$s_{40}s_{11}s_{10}{}^2$	·	·	·	1	·	·	·	·	·	·	·	−3	·	·	−3	·	·	−1
$s_{31}s_{20}s_{10}{}^2$	·	·	·	·	1	·	·	·	·	·	·	·	−6	−1	·	−3	·	−2
$s_{30}{}^2 s_{10}s_{01}$	·	·	·	·	·	1	·	·	·	·	·	·	·	−2	·	·	·	·
$s_{30}s_{21}s_{10}{}^2$	·	·	·	·	·	·	1	·	·	·	·	·	·	−1	−3	−3	·	·
$s_{30}s_{20}{}^2 s_{01}$	·	·	·	·	·	·	·	1	·	·	·	·	·	−1	·	·	−3	·
$s_{30}s_{20}s_{11}s_{10}$	·	·	·	·	·	·	·	·	1	·	·	·	·	−2	−3	·	·	−4
$s_{21}s_{20}{}^2 s_{10}$	·	·	·	·	·	·	·	·	·	1	·	·	·	·	·	−3	−3	−2
$s_{20}{}^3 s_{11}$	·	·	·	·	·	·	·	·	·	·	1	·	·	·	·	·	−1	−1
$s_{40}s_{10}{}^3 s_{01}$	3	1	3	3	·	·	·	·	·	·	·	1	·	·	·	·	·	·
$s_{31}s_{10}{}^4$	·	4	·	·	6	·	·	·	·	·	·	·	1	·	·	·	·	·
$s_{30}s_{20}s_{10}{}^2 s_{01}$	1	·	2	·	1	2	1	1	2	·	·	·	·	1	·	·	·	·
$s_{30}s_{11}s_{10}{}^3$	·	1	·	3	·	3	3	·	3	·	·	·	·	·	1	·	·	·
$s_{21}s_{20}s_{10}{}^3$	·	1	·	·	3	·	3	·	3	·	·	·	·	·	·	1	·	·
$s_{20}{}^3 s_{10}s_{01}$	·	·	3	·	·	·	·	3	·	3	1	·	·	·	·	·	1	·
$s_{20}{}^2 s_{11}s_{10}{}^2$	·	·	·	1	2	·	·	·	4	2	1	·	·	·	·	·	·	1
$s_{30}s_{10}{}^4 s_{01}$	6	4	12	12	6	4	6	3	12	·	·	4	1	6	4	·	·	·
$s_{21}s_{10}{}^5$	·	10	·	·	30	·	10	·	·	15	·	·	5	·	·	10	·	·
$s_{20}{}^2 s_{10}{}^3 s_{01}$	3	1	9	3	6	6	6	7	12	9	3	1	·	6	·	2	3	3
$s_{20}s_{11}s_{10}{}^4$	·	4	·	6	9	·	15	·	16	12	3	·	1	·	4	4	·	6
$s_{20}s_{10}{}^5 s_{01}$	10	10	35	30	40	20	40	25	80	45	15	10	5	40	20	20	15	30
$s_{11}s_{10}{}^6$	·	20	·	15	90	·	60	·	60	90	15	·	15	·	20	60	·	45
$s_{10}{}^7 s_{01}$	21	35	105	105	210	70	210	105	420	315	105	35	35	210	140	210	105	315

Table 1·6·8 (*cont.*)

$w = 8$ (71) (iii)	$[30, (10)^4, 01]$	$[21, (10)^5]$	$[(20)^2, (10)^3, 01]$	$[20, 11, (10)^4]$	$[20, (10)^5, 01]$	$[11, (10)^6]$	$[(10)^7, 01]$
s_{71}	−120	−120	−120	−120	720	720	−5040
$s_{70}s_{01}$	24	.	24	.	−120	.	720
$s_{61}s_{10}$	96	120	72	96	−600	−720	5040
$s_{60}s_{11}$	24	.	18	24	−120	−120	840
$s_{51}s_{20}$	36	60	66	66	−390	−450	3150
$s_{50}s_{21}$	24	24	24	18	−114	−54	378
$s_{41}s_{30}$	40	40	40	40	−240	−240	1680
$s_{40}s_{31}$	30	30	30	30	−180	−180	1260
$s_{60}s_{10}s_{01}$	−24	.	−18	.	120	.	−840
$s_{51}s_{10}^2$	−36	−60	−18	−36	240	360	−2520
$s_{50}s_{20}s_{01}$	−12	.	−18	.	84	.	−504
$s_{50}s_{11}s_{10}$	−24	.	−12	−24	120	144	−1008
$s_{41}s_{20}s_{10}$	−24	−60	−42	−54	330	450	−3150
$s_{40}s_{30}s_{01}$	−14	.	−14	.	70	.	−420
$s_{40}s_{21}s_{10}$	−24	−30	−18	−18	120	90	−630
$s_{40}s_{20}s_{11}$	−12	.	−15	−18	90	90	−630
$s_{31}s_{30}s_{10}$	−32	−40	−24	−32	200	240	−1680
$s_{31}s_{20}^2$	−3	−15	−18	−18	105	135	−945
$s_{30}^2s_{11}$	−8	.	−6	−8	40	40	−280
$s_{30}s_{21}s_{20}$	−12	−20	−22	−17	105	75	525
$s_{50}s_{10}^2s_{01}$	12	.	6	.	−60	.	504
$s_{41}s_{10}^3$	8	20	2	8	−60	−120	840
$s_{40}s_{20}s_{10}s_{01}$	12	.	15	.	−90	.	630
$s_{40}s_{11}s_{10}^2$	12	.	3	12	−60	−90	630
$s_{31}s_{20}s_{10}^2$	6	30	12	21	−135	−225	1575
$s_{30}^2s_{10}s_{01}$	8	.	6	.	−40	.	280
$s_{30}s_{21}s_{10}^2$	12	20	6	9	−65	−75	525
$s_{30}s_{20}^2s_{01}$	3	.	8	.	−35	.	210
$s_{30}s_{20}s_{11}s_{10}$	12	.	12	20	−100	−120	840
$s_{21}s_{20}^2s_{10}$	.	15	12	12	−75	−90	630
$s_{20}^3s_{11}$	.	.	3	3	−15	−15	105
$s_{40}s_{10}^3s_{01}$	−4	.	−1	.	20	.	−210
$s_{31}s_{10}^4$	−1	−5	.	−1	10	30	−210
$s_{30}s_{20}s_{10}^2s_{01}$	−6	.	−6	.	50	.	−420
$s_{30}s_{11}s_{10}^3$	−4	.	.	−4	20	40	−280
$s_{21}s_{20}s_{10}^3$	.	−10	−2	−4	30	60	−420
$s_{20}^3s_{10}s_{01}$	.	.	−3	.	15	.	−105
$s_{20}^2s_{11}s_{10}$	.	.	−3	−6	30	45	−315
$s_{30}s_{10}^4s_{01}$	1	.	.	.	−5	.	70
$s_{21}s_{10}^5$	.	1	.	.	−1	−6	42
$s_{20}^2s_{10}^3s_{01}$	.	.	1	.	−10	.	105
$s_{20}s_{11}s_{10}^4$	.	.	.	1	−5	−15	105
$s_{20}s_{10}^5s_{01}$	5	1	10	5	1	.	−21
$s_{11}s_{10}^6$	.	6	.	15	.	1	−7
$s_{10}^7s_{01}$	35	21	105	105	21	7	1

Table 1·6·8 (cont.)

$w=8$ (62)(i)	[62]	[61, 01]	[60, 02]	[52, 10]	[51, 11]	[50, 12]	[42, 20]	[41, 21]	[40, 22]	[32, 30]	$[(31)^2]$	$[60, (01)^2]$	[51, 10, 01]	[50, 11, 01]	[50, 10, 02]	$[42, (10)^2]$	[41, 20, 01]	[41, 11, 10]	[40, 21, 01]	[40, 20, 02]	[40, 12, 10]	$[40, (11)^2]$	[32, 20, 10]	[31, 30, 01]
s_{62}	1	-1	-1	-1	-1	-1	-1	-1	-1	-1	-1	2	2	2	2	2	2	2	2	2	2	2	2	2
$s_{61}s_{01}$	1	1	.	.	.	.	.	.	.	.	.	-2	-1	-1	.	.	-1	.	-1	.	.	.	.	-1
$s_{60}s_{02}$	1	.	1	.	.	.	.	.	.	.	.	-1	.	.	-1	.	.	.	.	-1	.	.	.	.
$s_{52}s_{10}$	1	.	.	1	.	.	.	.	.	.	.	.	-1	.	-1	-2	.	-1	.	.	-1	.	-1	.
$s_{51}s_{11}$	1	.	.	.	1	.	.	.	.	.	.	.	-1	-1	.	.	-1	.	.	.	.	-2	.	.
$s_{50}s_{12}$	1	.	.	.	.	1	.	.	.	.	.	.	.	-1	-1	.	.	.	.	-1	.	.	.	.
$s_{42}s_{20}$	1	.	.	.	.	.	1	.	.	.	.	.	.	.	.	-1	-1	.	.	-1	.	.	-1	.
$s_{41}s_{21}$	1	.	.	.	.	.	.	1	.	.	.	.	.	.	.	.	-1	-1	-1	.	.	.	.	.
$s_{40}s_{22}$	1	.	.	.	.	.	.	.	1	.	.	.	.	.	.	.	.	.	-1	-1	-1	-1	.	.
$s_{32}s_{30}$	1	.	.	.	.	.	.	.	.	1	.	.	.	.	.	.	.	.	.	.	.	.	-1	-1
$s_{31}{}^2$	1	.	.	.	.	.	.	.	.	.	1	.	.	.	.	.	.	.	.	.	.	.	.	-1
$s_{60}s_{01}{}^2$	1	2	1	.	.	.	.	.	.	.	.	1	.	.	.	.	.	.	.	.	.	.	.	.
$s_{51}s_{10}s_{01}$	1	1	.	1	1	.	.	.	.	.	.	.	1	.	.	.	.	.	.	.	.	.	.	.
$s_{50}s_{11}s_{01}$	1	1	.	.	1	1	.	.	.	.	.	.	.	1	.	.	.	.	.	.	.	.	.	.
$s_{50}s_{10}s_{02}$	1	.	1	1	.	1	.	.	.	.	.	.	.	.	1	.	.	.	.	.	.	.	.	.
$s_{42}s_{10}{}^2$	1	.	.	2	.	.	1	.	.	.	.	.	.	.	.	1	.	.	.	.	.	.	.	.
$s_{41}s_{20}s_{01}$	1	1	.	1	1	.	1	1	.	.	.	.	.	.	.	.	1	.	.	.	.	.	.	.
$s_{41}s_{11}s_{10}$	1	.	.	1	1	.	.	1	.	.	.	.	.	.	.	.	.	1	.	.	.	.	.	.
$s_{40}s_{21}s_{01}$	1	1	.	.	.	.	.	1	1	.	.	.	.	.	.	.	.	.	1	.	.	.	.	.
$s_{40}s_{20}s_{02}$	1	.	1	.	.	.	1	.	1	.	.	.	.	.	.	.	.	.	.	1	.	.	.	.
$s_{40}s_{12}s_{10}$	1	.	.	1	.	1	.	.	1	.	.	.	.	.	.	.	.	.	.	.	1	.	.	.
$s_{40}s_{11}{}^2$	1	.	.	.	2	.	.	.	1	.	.	.	.	.	.	.	.	.	.	.	.	1	.	.
$s_{32}s_{20}s_{10}$	1	.	.	1	.	.	1	.	.	1	.	.	.	.	.	.	.	.	.	.	.	.	1	.
$s_{31}s_{30}s_{01}$	1	1	.	.	.	.	.	.	.	1	1	.	.	.	.	.	.	.	.	.	.	.	.	1
$s_{31}s_{21}s_{10}$	1	.	.	1	.	.	.	1	.	.	1	.	.	.	.	.	.	.	.	.	.	.	.	.
$s_{31}s_{20}s_{11}$	1	.	.	.	1	.	1	.	.	.	1	.	.	.	.	.	.	.	.	.	.	.	.	.
$s_{30}{}^2s_{02}$	1	.	1	.	.	.	.	.	.	.	2	.	.	.	.	.	.	.	.	.	.	.	.	.
$s_{30}s_{21}s_{11}$	1	.	.	1	.	.	.	.	1	1	.	.	.	.	.	.	.	.	.	.	.	.	.	.
$s_{30}s_{20}s_{12}$	1	.	.	.	1	.	1	.	.	.	1	.	.	.	.	.	.	.	.	.	.	.	.	.
$s_{22}s_{20}{}^2$	1	.	.	.	.	.	2	.	1	.	.	.	.	.	.	.	.	.	.	.	.	.	.	.
$s_{21}{}^2s_{20}$	1	.	.	.	.	.	1	2	.	.	.	.	.	.	.	.	.	.	.	.	.	.	.	.
$s_{50}s_{10}s_{01}{}^2$	1	2	1	1	2	1	.	.	.	.	.	1	2	2	1	.	.	.	.	.	.	.	.	.
$s_{41}s_{10}{}^2s_{01}$	1	1	.	2	2	.	1	1	.	.	.	.	2	.	.	1	1	2	.	.	.	.	.	.
$s_{40}s_{20}s_{01}{}^2$	1	2	1	.	.	.	1	2	1	.	.	1	.	.	.	.	2	.	2	1	.	.	.	.
$s_{40}s_{11}s_{10}s_{01}$	1	1	.	1	2	1	1	.	1	1	.	.	1	1	.	.	1	1	.	1	.	.	.	.
$s_{40}s_{10}{}^2s_{02}$	1	.	1	2	.	2	1	.	1	.	.	.	.	.	2	1	.	.	.	1	2	.	.	.
$s_{32}s_{10}{}^3$	1	.	.	1	.	.	3	.	.	1	.	.	.	.	.	3	.	.	.	.	.	.	3	.
$s_{31}s_{20}s_{10}s_{01}$	1	1	.	1	1	.	1	1	.	1	1	.	1	.	.	.	1	.	1	.	.	.	1	1
$s_{31}s_{11}s_{10}{}^2$	1	.	.	2	1	.	1	2	.	1	2	.	.	.	.	1	1	2	.	.	.	.	1	.
$s_{30}{}^2s_{01}{}^2$	1	2	1	.	.	.	.	.	.	2	2	1	.	.	.	.	.	.	.	.	.	.	.	4
$s_{30}s_{21}s_{10}s_{01}$	1	1	.	1	1	.	1	1	1	1	1	.	1	.	.	.	1	.	1	.	.	.	1	1
$s_{30}s_{20}s_{11}s_{01}$	1	1	.	.	1	1	1	1	.	1	1	.	.	1	.	.	1	.	1	.	.	.	1	1
$s_{30}s_{20}s_{10}s_{02}$	1	.	1	1	.	1	1	.	1	2	1	.	.	.	1	1	.	.	.	1	.	.	1	.
$s_{30}s_{12}s_{10}{}^2$	1	.	.	2	2	.	.	2	1	1	.	.	.	.	.	.	2	.	.	.	1	.	2	.
$s_{30}s_{11}{}^2s_{10}$	1	.	.	1	2	.	.	2	1	1	.	.	.	.	.	.	1	.	2	.	.	1	.	.
$s_{22}s_{20}s_{10}{}^2$	1	.	.	2	.	.	2	.	1	2	.	.	.	.	.	1	.	.	.	.	.	.	2	.
$s_{21}{}^2s_{10}{}^2$	1	.	.	2	.	.	1	2	.	2	.	.	.	.	.	1	.	.	.	.	.	.	.	.
$s_{21}s_{20}{}^2s_{01}$	1	1	.	.	.	.	2	3	1	.	.	.	.	.	.	.	2	.	2	1	.	.	1	.
$s_{21}s_{20}s_{11}s_{10}$	1	1	.	1	1	.	1	2	.	1	1	.	.	.	.	.	1	.	1	.	.	.	1	.
$s_{20}{}^3s_{02}$	1	.	1	.	.	.	3	.	3	.	.	.	.	.	.	.	.	.	.	.	3	.	.	.
$s_{20}{}^2s_{12}s_{10}$	1	.	.	1	.	1	2	.	1	2	.	.	.	.	.	.	.	.	.	.	.	1	.	2
$s_{20}{}^2s_{11}{}^2$	1	.	.	.	2	.	2	.	1	.	2	.	.	.	.	.	.	.	.	.	.	1	.	.
$s_{40}s_{10}{}^2s_{01}{}^2$	1	2	1	2	4	2	1	2	1	.	.	1	4	4	2	1	2	4	2	1	2	2	.	.
$s_{31}s_{10}{}^3s_{01}$	1	1	.	3	3	.	3	3	.	1	1	.	3	.	.	3	3	6	.	.	.	.	3	1
$s_{30}s_{20}s_{10}s_{01}{}^2$	1	2	1	1	2	1	1	2	1	2	2	1	2	2	1	.	2	.	2	1	.	.	1	4
$s_{30}s_{11}s_{10}{}^2s_{01}$	1	1	.	2	3	1	1	3	2	1	1	.	2	1	.	1	1	4	2	.	2	2	.	1
$s_{30}s_{10}{}^3s_{02}$	1	.	1	3	.	3	3	.	3	2	.	.	.	.	3	3	.	.	.	3	6	.	3	.
$s_{22}s_{10}{}^4$	1	.	.	4	.	.	6	.	1	4	.	.	.	.	.	6	.	.	.	.	.	.	12	.
$s_{21}s_{20}s_{10}{}^2s_{01}$	1	1	.	2	2	.	2	3	1	2	2	.	2	.	.	1	2	2	1	.	.	.	2	2
$s_{21}s_{11}s_{10}{}^3$	1	.	.	3	1	.	3	4	.	1	3	.	.	.	.	3	3	.	.	.	.	.	3	.
$s_{20}{}^3s_{01}{}^2$	1	2	1	.	.	.	3	6	3	.	.	1	.	.	.	.	6	.	6	3	.	.	2	.
$s_{20}{}^2s_{11}s_{10}s_{01}$	1	1	.	1	2	1	2	3	1	2	2	.	1	1	.	.	2	1	1	.	1	1	2	2
$s_{20}{}^2s_{10}{}^2s_{02}$	1	.	1	2	.	2	3	.	3	4	.	.	.	.	2	1	.	.	.	3	2	.	4	.
$s_{20}s_{12}s_{10}{}^3$	1	.	.	3	.	1	4	.	3	4	.	.	.	.	.	3	.	.	.	.	3	.	6	.
$s_{20}s_{11}{}^2s_{10}{}^2$	1	.	.	2	2	.	2	4	1	2	2	.	.	.	.	1	4	.	.	.	.	1	2	.
$s_{30}s_{10}{}^3s_{01}{}^2$	1	2	1	3	6	3	3	6	3	2	2	1	6	6	3	3	6	12	6	3	6	6	3	4
$s_{21}s_{10}{}^4s_{01}$	1	1	.	4	4	.	6	7	1	4	4	.	4	.	.	6	6	12	1	.	.	.	12	4
$s_{20}{}^2s_{10}{}^2s_{01}{}^2$	1	2	1	2	4	2	3	6	3	4	4	1	4	4	2	1	6	4	2	3	2	2	4	8
$s_{20}s_{11}s_{10}{}^3s_{01}$	1	1	.	3	4	1	4	7	3	4	4	.	3	1	.	3	4	9	3	.	3	3	6	4
$s_{20}s_{10}{}^4s_{02}$	1	.	1	4	.	4	7	.	7	8	.	.	.	.	4	6	.	.	.	7	12	.	16	.
$s_{12}s_{10}{}^5$	1	.	.	5	.	1	10	.	5	10	.	.	.	.	.	10	.	.	.	.	5	.	30	.
$s_{11}{}^2s_{10}{}^4$	1	.	.	4	2	.	6	8	1	4	6	.	.	.	.	6	8	.	.	.	.	1	12	.
$s_{20}s_{10}{}^4s_{01}{}^2$	1	2	1	4	8	4	7	14	7	8	8	1	8	8	4	6	14	24	14	7	12	12	16	16
$s_{11}s_{10}{}^5s_{01}$	1	1	.	5	6	1	10	15	5	10	10	.	5	1	.	10	10	25	5	.	5	5	30	10
$s_{10}{}^6s_{02}$	1	.	1	6	.	6	15	.	15	20	.	.	.	.	6	15	.	.	.	15	30	.	60	.
$s_{10}{}^6s_{01}{}^2$	1	2	1	6	12	6	15	30	15	20	20	1	12	12	6	15	30	60	30	15	30	30	60	40

Table 1·6·8 (*cont.*)

$w=8$ (62) (ii)	[31, 21, 10]	[31, 20, 11]	[(30)², 02]	[30, 22, 10]	[30, 21, 11]	[30, 20, 12]	[22, (20)²]	[(21)², 20]	[50, 10, (01)²]	[41, (10)², 01]	[40, 20, (01)²]	[40, 11, 10, 01]	[40, (10)², 02]	[32, (10)²]	[31, 20, 10, 01]	[31, 11, (10)²]	[(30)², (01)²]	[30, 21, 10, 01]	[30, 20, 11, 01]
s_{62}	2	2	2	2	2	2	2	2	−6	−6	−6	−6	−6	−6	−6	−6	−6	−6	−6
$s_{61}s_{01}$	·	·	·	·	·	·	·	·	4	2	4	2	·	·	2	·	4	2	2
$s_{60}s_{02}$	·	·	−1	·	·	·	·	·	1	·	1	·	2	·	2	4	1	2	2
$s_{52}s_{10}$	−1	·	·	−1	·	·	·	·	2	4	·	2	4	6	2	4	·	2	2
$s_{51}s_{11}$	·	−1	·	·	−1	·	·	·	2	2	·	3	2	·	1	2	·	1	2
$s_{50}s_{12}$	·	·	·	·	·	−1	·	−1	2	·	·	1	·	·	·	·	·	·	2
$s_{42}s_{20}$	·	−1	·	·	·	−1	−2	·	·	1	2	1	·	3	1	2	·	·	2
$s_{41}s_{21}$	−1	·	·	·	−1	·	·	−2	·	2	2	1	2	·	1	2	·	2	1
$s_{40}s_{22}$	·	·	·	−1	·	−1	·	−1	·	·	·	·	·	2	1	·	4	2	2
$s_{32}s_{30}$	·	·	−2	−1	−1	−1	·	·	·	·	·	·	·	·	2	2	2	1	1
s_{31}^2	−1	−1	·	·	·	·	·	·	·	·	·	·	·	·	2	2	·	1	1
$s_{60}s_{01}^2$	·	·	·	·	·	·	·	·	−1	·	−1	−1	·	·	−1	·	−1	−1	·
$s_{51}s_{10}s_{01}$	·	·	·	·	·	·	·	·	−2	−2	·	−1	·	·	−1	·	·	·	−1
$s_{50}s_{11}s_{01}$	·	·	·	·	·	·	·	·	−2	·	−2	·	·	·	·	·	·	·	·
$s_{50}s_{10}s_{02}$	·	·	·	·	·	·	·	·	−1	·	·	·	−1	−3	·	−1	·	·	−1
$s_{42}s_{10}^2$	·	·	·	·	·	·	·	·	·	−1	·	·	·	·	−1	·	·	·	−1
$s_{41}s_{20}s_{01}$	·	·	·	·	·	·	·	·	·	−1	−2	·	·	·	−1	−2	·	·	·
$s_{41}s_{11}s_{10}$	·	·	·	·	·	·	·	·	·	−2	·	−1	·	·	·	·	·	−1	·
$s_{40}s_{21}s_{01}$	·	·	·	·	·	·	·	·	·	·	−1	−2	−1	·	·	·	·	·	·
$s_{40}s_{20}s_{02}$	·	·	·	·	·	·	·	·	·	·	·	−1	−2	·	·	·	·	·	·
$s_{40}s_{12}s_{10}$	·	·	·	·	·	·	·	·	·	·	·	·	−1	·	·	·	·	·	·
$s_{40}s_{11}^2$	·	·	·	·	·	·	·	·	·	·	·	·	·	−3	−1	·	·	·	·
$s_{32}s_{20}s_{10}$	·	·	·	·	·	·	·	·	·	·	·	·	·	·	−1	−1	−4	−1	−1
$s_{31}s_{30}s_{01}$	·	·	·	·	·	·	·	·	·	·	·	·	·	·	−1	−2	·	−1	−1
$s_{31}s_{21}s_{10}$	1	·	·	·	·	·	·	·	·	·	·	·	·	·	−1	·	·	−1	−1
$s_{31}s_{20}s_{11}$	·	1	·	·	·	·	·	·	·	·	·	·	·	·	−1	−1	·	·	−1
$s_{30}^2s_{02}$	·	·	1	·	·	·	·	·	·	·	·	·	·	·	·	·	−1	−1	·
$s_{30}s_{22}s_{10}$	·	·	·	1	·	·	·	·	·	·	·	·	·	·	·	·	·	−1	−1
$s_{30}s_{20}s_{11}$	·	·	·	·	1	·	·	·	·	·	·	·	·	·	·	·	·	−1	−1
$s_{30}s_{20}s_{12}$	·	·	·	·	·	1	·	·	·	·	·	·	·	·	·	·	·	·	·
$s_{22}s_{20}^2$	·	·	·	·	·	·	1	·	·	·	·	·	·	·	·	·	·	·	·
$s_{21}^2s_{20}$	·	·	·	·	·	·	·	1	·	·	·	·	·	·	·	·	·	·	·
$s_{50}s_{10}s_{01}^2$	·	·	·	·	·	·	·	·	1	·	·	·	·	·	·	·	·	·	·
$s_{41}s_{10}^2s_{01}$	·	·	·	·	·	·	·	·	·	1	·	·	·	·	·	·	·	·	·
$s_{40}s_{20}s_{01}^2$	·	·	·	·	·	·	·	·	·	·	1	·	·	·	·	·	·	·	·
$s_{40}s_{11}s_{10}s_{01}$	·	·	·	·	·	·	·	·	·	·	·	1	·	·	·	·	·	·	·
$s_{40}s_{10}^2s_{02}$	·	·	·	·	·	·	·	·	·	·	·	·	1	·	·	·	·	·	·
$s_{32}s_{10}^3$	·	·	·	·	·	·	·	·	·	·	·	·	·	1	·	·	·	·	·
$s_{31}s_{20}s_{10}s_{01}$	1	1	·	·	·	·	·	·	·	·	·	·	·	·	1	·	·	·	·
$s_{31}s_{11}s_{10}^2$	2	1	·	·	·	·	·	·	·	·	·	·	·	·	·	1	·	·	·
$s_{30}^2s_{01}^2$	·	·	·	1	1	·	·	·	·	·	·	·	·	·	·	·	1	·	·
$s_{30}s_{21}s_{10}s_{01}$	1	·	·	1	1	·	·	·	·	·	·	·	·	·	·	·	·	1	·
$s_{30}s_{20}s_{11}s_{01}$	·	1	·	1	·	1	1	·	·	·	·	·	·	·	·	·	·	·	1
$s_{30}s_{20}s_{10}s_{02}$	·	·	1	1	·	1	1	·	·	·	·	·	·	·	·	·	·	·	·
$s_{30}s_{12}s_{10}^2$	·	·	·	2	·	1	·	·	·	·	·	·	·	·	·	·	·	·	·
$s_{30}s_{11}^2s_{10}$	·	·	·	2	2	·	·	·	·	·	·	·	·	·	·	·	·	·	·
$s_{22}s_{20}s_{10}^2$	·	·	·	2	·	·	1	·	·	·	·	·	·	·	·	·	·	·	·
$s_{21}^2s_{10}^2$	4	·	·	·	·	·	·	1	·	·	·	·	·	·	·	·	·	·	·
$s_{21}s_{20}^2s_{01}$	·	·	·	·	·	·	1	2	·	·	·	·	·	·	·	·	·	·	·
$s_{21}s_{20}s_{11}s_{10}$	1	1	·	·	·	1	·	1	·	·	·	·	·	·	·	·	·	·	·
$s_{20}^3s_{02}$	·	·	·	·	·	·	3	·	·	·	·	·	·	·	·	·	·	·	·
$s_{20}^2s_{12}s_{10}$	·	·	·	·	·	2	1	1	·	·	·	·	·	·	·	·	·	·	·
$s_{20}^2s_{11}^2$	·	4	·	·	·	·	1	·	·	·	·	·	·	·	·	·	·	·	·
$s_{40}s_{10}^2s_{01}^2$	·	·	·	·	·	·	·	·	2	2	1	4	1	·	1	3	3	·	·
$s_{31}s_{10}^3s_{01}$	3	3	·	·	·	·	·	·	·	3	·	·	·	·	·	2	3	·	·
$s_{30}s_{20}s_{10}^2s_{01}$	2	2	1	1	2	1	·	·	1	·	1	·	·	·	1	·	1	2	2
$s_{30}s_{11}s_{10}^2s_{01}$	2	1	·	2	3	1	·	·	·	1	·	2	·	3	1	·	1	2	1
$s_{30}s_{10}^3s_{02}$	·	·	1	3	·	3	·	·	·	·	·	·	·	·	3	·	·	·	·
$s_{22}s_{10}^4$	·	·	·	4	·	·	3	·	·	·	·	·	·	·	4	·	·	·	·
$s_{21}s_{20}s_{10}^2s_{01}$	4	2	·	2	2	·	1	2	·	1	·	·	·	·	1	2	·	2	·
$s_{21}s_{11}s_{10}^3$	9	3	·	·	1	·	3	3	·	·	3	·	·	·	1	·	3	·	·
$s_{20}^3s_{01}^2$	·	·	·	·	·	·	3	6	·	·	·	·	·	·	·	·	·	·	2
$s_{20}^2s_{11}s_{10}s_{01}$	2	4	·	·	2	·	2	2	·	·	·	1	·	·	2	·	·	·	·
$s_{20}^2s_{10}^2s_{02}$	·	·	2	4	·	4	3	·	·	·	·	·	1	·	1	·	·	·	·
$s_{20}s_{12}s_{10}^3$	·	·	·	6	·	4	3	·	·	·	·	·	·	·	·	2	·	·	·
$s_{20}s_{11}^2s_{10}^2$	4	4	·	2	4	·	1	2	·	·	·	·	·	·	·	·	·	·	·
$s_{30}s_{10}^3s_{01}^2$	6	6	1	3	6	3	·	6	3	6	3	12	3	1	6	6	1	6	6
$s_{21}s_{10}^4s_{01}$	16	12	·	4	4	·	3	6	·	6	6	4	1	4	12	12	·	4	8
$s_{20}^2s_{10}^3s_{01}$	8	8	2	4	8	4	3	6	2	2	3	4	1	1	8	6	2	8	4
$s_{20}s_{11}s_{10}^3s_{01}$	12	10	·	6	10	4	3	6	·	3	·	3	6	1	6	6	·	6	4
$s_{20}s_{10}^4s_{02}$	·	·	4	16	·	16	9	·	·	·	·	·	·	10	·	·	·	·	·
$s_{12}s_{10}^5$	·	·	·	20	·	10	15	·	·	·	·	·	·	4	·	12	·	·	·
$s_{11}^2s_{10}^4$	24	12	·	4	8	8	3	12	·	·	·	·	·	4	·	·	·	·	·
$s_{20}s_{10}^4s_{01}^2$	32	32	4	·	20	16	9	18	4	12	7	24	6	4	32	24	4	32	32
$s_{11}s_{10}^5s_{01}$	50	40	·	20	30	10	15	30	·	10	·	5	·	10	30	40	·	20	10
$s_{10}^6s_{02}$	·	·	10	60	·	60	45	·	·	·	·	·	15	20	·	·	·	·	·
$s_{10}^6s_{01}^2$	120	120	10	60	120	60	45	90	6	30	15	60	15	20	120	120	10	120	120

Table 1·6·8 (*cont.*)

$w = 8$ (62) (iii)	[30, 20, 10, 02]	[30, 12, (10)²]	[30, (11)², 10]	[22, 20, (10)²]	[(21)², (10)²]	[21, (20)², 01]	[21, 20, 11, 10]	[(20)³, 02]	[(20)², 12, 10]	[(20)², (11)²]	[40, (10)², (01)²]	[31, (10)³, 01]	[30, 20, 10, (01)²]	[30, 11, (10)², 01]	[30, (10)³, 02]	[22, (10)⁴]	[21, 20, (10)², 01]	[21, 11, (10)³]	[(20)³, (01)²]
s_{62}	−6	−6	−6	−6	−6	−6	−6	−6	−6	−6	24	24	24	24	24	24	24	24	24
$s_{61}s_{01}$	·	·	·	·	·	2	·	·	·	·	−12	−6	−12	−6	·	·	−6	·	−12
$s_{60}s_{02}$	2	·	·	·	·	·	·	2	·	·	−2	·	−2	·	−6	·	·	·	−2
$s_{52}s_{10}$	2	4	2	4	4	·	2	·	·	2	−12	−18	−6	−12	−18	−24	−12	−18	·
$s_{51}s_{11}$	·	·	4	·	·	·	2	·	·	4	−8	−6	−4	−10	·	·	−4	−6	·
$s_{50}s_{12}$	1	2	·	·	·	·	·	·	·	·	−4	·	−2	−2	−6	·	·	·	·
$s_{42}s_{20}$	2	1	·	3	1	4	2	6	4	4	−2	−6	−6	−2	−6	−12	−8	−6	−18
$s_{41}s_{21}$	·	·	2	·	4	4	3	·	·	·	−4	−6	−4	−6	·	·	−10	−12	−12
$s_{40}s_{22}$	1	2	1	2	·	1	·	3	1	1	−6	·	−2	−4	−6	−6	−2	−2	−6
$s_{32}s_{30}$	3	2	2	2	·	·	1	·	2	·	·	−2	−8	−6	−8	−8	−2	−2	·
$s_{31}{}^{2}$	·	·	·	·	2	·	1	·	·	2	−6	−4	−2	·	·	·	−4	−6	·
$s_{60}s_{01}{}^{2}$	·	·	·	·	·	·	·	·	·	·	2	·	2	·	·	·	·	·	2
$s_{51}s_{10}s_{01}$	·	·	·	·	·	·	·	·	·	·	8	6	4	4	·	·	4	·	·
$s_{50}s_{11}s_{01}$	·	·	·	·	·	·	·	·	·	·	4	·	2	2	·	·	·	·	·
$s_{50}s_{10}s_{02}$	−1	·	·	·	·	·	·	·	·	·	2	·	·	·	6	·	·	·	·
$s_{42}s_{10}{}^{2}$	·	−1	·	−1	−1	·	·	·	·	·	2	6	1	2	6	12	2	6	·
$s_{41}s_{20}s_{01}$	·	·	·	·	−2	·	·	·	·	·	2	3	4	1	·	·	3	2	12
$s_{41}s_{11}s_{10}$	·	·	−2	·	·	·	−1	·	·	·	4	6	·	6	·	·	2	6	·
$s_{40}s_{21}s_{01}$	·	·	·	·	−1	·	·	·	·	·	4	·	2	2	·	·	2	·	6
$s_{40}s_{20}s_{02}$	−1	·	·	·	·	·	·	−3	·	·	1	·	1	·	3	·	·	·	3
$s_{40}s_{12}s_{10}$	·	−2	·	·	·	·	·	·	−1	·	4	·	·	2	6	·	·	·	·
$s_{40}s_{11}{}^{2}$	·	·	−1	·	·	·	·	·	·	−1	2	·	·	2	·	·	·	·	·
$s_{32}s_{20}s_{10}$	−1	·	·	−2	·	·	−1	·	−2	·	·	3	2	·	3	12	4	3	·
$s_{31}s_{30}s_{01}$	·	·	·	·	·	·	·	·	·	·	·	2	6	2	·	·	2	·	·
$s_{31}s_{21}s_{10}$	·	·	·	−4	·	·	−1	·	·	·	·	6	2	2	·	·	6	12	·
$s_{31}s_{20}s_{11}$	·	·	·	·	·	·	−1	·	·	−4	·	3	2	1	·	·	2	3	·
$s_{30}{}^{2}s_{02}$	−1	·	·	·	·	·	·	·	·	·	·	·	1	·	2	·	·	·	·
$s_{30}s_{22}s_{10}$	−1	−2	−1	−2	·	·	·	·	·	·	·	·	2	4	6	8	2	·	·
$s_{30}s_{21}s_{11}$	·	−2	·	·	·	·	−1	·	·	·	·	·	2	4	3	·	2	·	·
$s_{30}s_{20}s_{12}$	−1	−1	·	·	·	·	·	·	−2	·	·	·	2	1	3	·	2	2	·
$s_{22}s_{20}{}^{2}$	·	·	·	−1	−1	·	−3	−1	−1	·	·	·	·	·	·	3	1	·	6
$s_{21}{}^{2}s_{20}$	·	·	·	·	−1	−2	−1	−3	−1	−1	·	·	·	·	·	·	3	3	6
$s_{50}s_{10}s_{01}{}^{2}$	·	·	·	·	·	·	·	·	·	·	−2	·	−1	·	·	·	·	·	·
$s_{41}s_{10}{}^{2}s_{01}$	·	·	·	·	·	·	·	·	·	·	−2	−3	·	·	·	·	−1	·	·
$s_{40}s_{20}s_{01}{}^{2}$	·	·	·	·	·	·	·	·	·	·	−1	·	−1	·	·	·	−1	·	−3
$s_{40}s_{11}s_{10}s_{01}$	·	·	·	·	·	·	·	·	·	·	−4	·	·	−2	·	·	·	·	·
$s_{40}s_{10}{}^{2}s_{02}$	·	·	·	·	·	·	·	·	·	·	−1	·	·	·	·	−3	·	·	·
$s_{32}s_{10}{}^{3}$	·	·	·	·	·	·	·	·	·	·	−1	·	·	·	−1	−4	·	−1	·
$s_{31}s_{20}s_{10}s_{01}$	·	·	·	·	·	·	·	·	·	·	·	−3	−2	·	·	·	−2	·	·
$s_{31}s_{11}s_{10}{}^{2}$	·	·	·	·	·	·	·	·	·	·	·	−3	·	−1	·	·	·	−3	·
$s_{30}{}^{2}s_{01}{}^{2}$	·	·	·	·	·	·	·	·	·	·	·	−1	·	·	·	·	·	·	·
$s_{30}s_{21}s_{10}s_{01}$	·	·	·	·	·	·	·	·	·	·	·	−2	−2	·	·	·	−2	·	·
$s_{30}s_{20}s_{11}s_{01}$	·	·	·	·	·	·	·	·	·	·	·	−2	−1	·	·	·	·	·	·
$s_{30}s_{20}s_{10}s_{02}$	*1*	·	·	·	·	·	·	·	·	·	·	−1	·	·	−3	·	·	·	·
$s_{30}s_{12}s_{10}{}^{2}$	·	*1*	·	·	·	·	·	·	·	·	·	·	−1	−3	·	·	·	·	·
$s_{30}s_{11}{}^{2}s_{10}$	·	·	*1*	·	·	·	·	·	·	·	·	·	−2	·	·	·	·	·	·
$s_{22}s_{20}s_{10}{}^{2}$	·	·	·	*1*	·	·	·	·	·	·	·	·	·	·	·	−6	−1	·	·
$s_{21}{}^{2}s_{10}{}^{2}$	·	·	·	·	*1*	·	·	·	·	·	·	·	·	·	·	·	−1	−3	·
$s_{21}s_{20}{}^{2}s_{01}$	·	·	·	·	·	*1*	·	·	·	·	·	·	·	·	·	·	−1	·	−6
$s_{21}s_{20}s_{11}s_{10}$	·	·	·	·	·	·	*1*	·	·	·	·	·	·	·	·	·	−2	−3	·
$s_{20}{}^{3}s_{02}$	·	·	·	·	·	·	·	*1*	·	·	·	·	·	·	·	·	·	·	−1
$s_{20}{}^{2}s_{12}s_{10}$	·	·	·	·	·	·	·	·	*1*	·	·	·	·	·	·	·	·	·	·
$s_{20}{}^{2}s_{11}{}^{2}$	·	·	·	·	·	·	·	·	·	*1*	·	·	·	·	·	·	·	·	·
$s_{40}s_{10}{}^{2}s_{01}{}^{2}$	·	·	·	·	·	·	·	·	·	·	*1*	·	·	·	·	·	·	·	·
$s_{31}s_{10}{}^{3}s_{01}$	·	·	·	·	·	·	·	·	·	·	·	*1*	·	·	·	·	·	·	·
$s_{30}s_{20}s_{10}s_{01}{}^{2}$	1	·	·	·	·	·	·	·	·	·	·	·	*1*	·	·	·	·	·	·
$s_{30}s_{11}s_{10}{}^{2}s_{01}$	·	1	2	·	·	·	·	·	·	·	·	·	·	*1*	·	·	·	·	·
$s_{30}s_{10}{}^{3}s_{02}$	3	3	·	·	·	·	·	·	·	·	·	·	·	·	*1*	·	·	·	·
$s_{22}s_{10}{}^{4}$	·	·	·	6	·	·	·	·	·	·	·	·	·	·	·	*1*	·	·	·
$s_{21}s_{20}s_{10}{}^{2}s_{01}$	·	·	·	1	1	1	2	·	·	·	·	·	·	·	·	·	*1*	·	·
$s_{21}s_{11}s_{10}{}^{3}$	·	·	·	·	3	3	·	·	·	·	·	·	·	·	·	·	·	*1*	·
$s_{20}{}^{3}s_{01}{}^{2}$	·	·	·	·	·	6	·	1	·	·	·	·	·	·	·	·	·	·	*1*
$s_{20}{}^{2}s_{11}s_{10}s_{01}$	·	·	·	·	·	1	2	1	1	·	·	·	·	·	·	·	·	·	·
$s_{20}{}^{2}s_{10}{}^{2}s_{02}$	4	·	·	2	·	·	·	1	2	·	·	·	·	·	·	·	·	·	·
$s_{20}s_{12}s_{10}{}^{3}$	·	3	·	3	·	·	·	·	3	·	·	·	·	·	·	·	·	·	·
$s_{20}s_{11}{}^{2}s_{10}{}^{2}$	·	·	2	1	·	·	4	·	·	1	·	·	·	·	·	·	·	·	·
$s_{30}s_{10}{}^{3}s_{01}{}^{2}$	3	3	6	·	·	·	·	·	·	·	3	2	3	6	1	·	·	·	·
$s_{21}s_{10}{}^{4}s_{01}$	·	·	·	6	6	3	12	·	·	·	·	4	·	·	·	1	·	6	4
$s_{20}{}^{2}s_{10}{}^{2}s_{01}{}^{2}$	4	·	·	2	2	6	8	1	2	2	1	·	4	·	·	·	4	1	·
$s_{20}s_{11}s_{10}{}^{3}s_{01}$	·	3	6	3	3	·	12	·	·	·	·	1	·	·	3	·	3	1	1
$s_{20}s_{10}{}^{4}s_{02}$	16	12	·	12	·	·	·	3	12	·	·	·	·	3	·	4	1	·	·
$s_{12}s_{10}{}^{5}$	·	10	·	30	·	·	·	·	15	·	·	·	·	·	·	5	·	·	·
$s_{11}{}^{2}s_{10}{}^{4}$	·	·	4	6	12	·	24	·	·	3	·	·	·	·	·	·	·	8	·
$s_{20}s_{10}{}^{4}s_{01}{}^{2}$	16	12	24	12	12	18	48	3	12	12	6	8	16	24	4	1	24	8	3
$s_{11}s_{10}{}^{5}s_{01}$	·	10	20	30	30	15	90	·	15	15	·	10	·	10	·	5	30	30	·
$s_{10}{}^{6}s_{02}$	60	60	·	90	·	·	·	15	90	·	·	·	·	·	·	20	15	·	·
$s_{10}{}^{6}s_{01}{}^{2}$	60	60	120	90	90	90	360	15	90	90	15	40	60	120	20	15	180	120	15

Table 1·6·8 (*cont.*)

$w = 8$ (62) (iv)	[(20)², 11, 10, 01]	[(20)², (10)², 02]	[20, 12, (10)³]	[20, (11)², (10)²]	[30, (10)³, (01)²]	[21, (10)⁴, (01)]	[(20)², (10)², (01)²]	[20, 11, (10)³, 01]	[20, (10)⁴, 02]	[12, (10)⁶]	[(11)², (10)⁴]	[20, (10)⁴, (01)²]	[11, (10)⁵, 01]	[(10)⁶, 02]	[(10)⁶, (01)²]
s_{62}	24	24	24	24	−120	−120	−120	−120	−120	−120	−120	720	720	720	−5040
$s_{61}s_{01}$	−6	·	·	·	48	24	48	24	·	·	·	−240	−120	·	1440
$s_{60}s_{02}$	·	−6	·	·	6	·	6	·	24	·	·	−24	·	−120	120
$s_{52}s_{10}$	−6	−12	−18	−12	72	96	48	72	96	120	96	−480	−600	−720	4320
$s_{51}s_{11}$	−8	·	·	−12	36	24	24	42	·	·	48	−192	−240	·	1440
$s_{50}s_{12}$	−2	−4	−6	·	12	·	8	6	24	24	·	−48	−24	−144	288
$s_{42}s_{20}$	−12	−14	−12	−8	18	36	54	42	60	60	36	−264	−240	−360	1800
$s_{41}s_{21}$	−6	·	·	−8	24	48	32	36	·	·	48	−192	−240	·	1440
$s_{40}s_{22}$	−2	−6	−6	−2	18	6	14	12	30	30	6	−84	−60	−180	540
$s_{32}s_{30}$	−4	−8	−8	−4	28	16	24	22	40	40	16	−144	−120	−240	960
$s_{31}{}^{2}$	−4	·	·	−4	12	24	16	18	·	·	24	−96	−120	·	720
$s_{60}s_{01}{}^{2}$	·	·	·	·	−6	·	−6	·	·	·	·	24	·	·	−120
$s_{51}s_{10}s_{01}$	2	·	·	·	−36	−24	−24	−18	·	·	·	192	120	·	−1440
$s_{50}s_{11}s_{01}$	2	·	·	·	−12	·	−8	−6	·	·	·	48	24	·	−288
$s_{50}s_{10}s_{02}$	·	4	·	·	−6	·	−4	·	−24	·	·	24	·	144	−144
$s_{42}s_{10}{}^{2}$	·	2	6	2	−18	−36	−6	−18	−36	−60	−36	144	240	360	−1800
$s_{41}s_{20}s_{01}$	4	·	·	·	−12	−12	−28	−12	·	·	·	120	60	·	−720
$s_{41}s_{11}s_{10}$	2	·	·	8	−24	−24	−8	−30	·	·	−48	144	240	·	−1440
$s_{40}s_{21}s_{01}$	1	·	·	·	−12	−6	−12	−6	·	·	·	60	30	·	−360
$s_{40}s_{20}s_{02}$	·	5	·	·	−3	·	−5	·	−18	·	·	18	·	90	−90
$s_{40}s_{12}s_{10}$	1	2	6	·	−12	·	−4	−6	−24	−30	·	48	30	180	−360
$s_{40}s_{11}{}^{2}$	1	·	·	2	−6	·	−2	−6	·	·	−6	24	30	·	−180
$s_{32}s_{20}s_{10}$	4	8	9	4	−6	−24	−24	−24	−48	−60	−24	168	180	360	−1440
$s_{31}s_{30}s_{01}$	2	·	·	·	−16	−8	−16	−8	·	·	·	80	40	·	−480
$s_{31}s_{21}s_{10}$	2	·	·	4	−12	−48	−16	−24	·	·	−48	144	240	·	−1440
$s_{31}s_{20}s_{11}$	6	·	·	6	−6	−12	−16	−21	·	·	−24	96	120	·	−720
$s_{30}{}^{2}s_{02}$	·	2	·	·	−2	·	−2	·	−8	·	·	8	·	40	−40
$s_{30}s_{22}s_{10}$	·	4	6	2	−18	−8	−8	−12	−32	−40	−8	88	80	240	−720
$s_{30}s_{21}s_{11}$	2	·	·	4	−12	−8	−8	−14	·	·	−16	64	80	·	−480
$s_{30}s_{20}s_{12}$	2	4	5	·	−6	·	−8	−5	−20	−20	·	40	20	120	−240
$s_{22}s_{20}{}^{2}$	2	4	3	1	·	−3	−10	−6	−15	−15	−3	42	30	90	−270
$s_{21}{}^{2}s_{20}$	2	·	·	2	·	−12	−10	−9	·	·	−12	48	60	·	−360
$s_{50}s_{10}s_{01}{}^{2}$	·	·	·	·	6	·	4	·	·	·	·	−24	·	·	144
$s_{41}s_{10}{}^{2}s_{01}$	·	·	·	·	12	12	4	6	·	·	·	−72	−60	·	720
$s_{40}s_{20}s_{01}{}^{2}$	·	·	·	·	3	·	5	·	·	·	·	−18	·	·	90
$s_{40}s_{11}s_{10}s_{01}$	−1	·	·	·	12	·	4	6	·	·	·	−48	−30	·	360
$s_{40}s_{10}{}^{2}s_{02}$	·	−1	·	·	3	·	1	·	12	·	·	−12	·	−90	90
$s_{32}s_{10}{}^{3}$	·	·	−1	·	2	8	·	2	8	20	8	−24	−60	−120	480
$s_{31}s_{20}s_{10}s_{01}$	−2	·	·	·	6	12	16	9	·	·	·	−96	−60	·	720
$s_{31}s_{11}s_{10}{}^{2}$	·	·	·	−2	6	12	·	9	·	·	24	−48	−120	·	720
$s_{30}{}^{2}s_{01}{}^{2}$	·	·	·	·	2	·	2	·	·	·	·	−8	·	·	40
$s_{30}s_{21}s_{10}s_{01}$	·	·	·	·	12	8	8	6	·	·	·	−64	−40	·	480
$s_{30}s_{20}s_{11}s_{01}$	−2	·	·	·	6	·	8	5	·	·	·	−40	−20	·	240
$s_{30}s_{20}s_{10}s_{02}$	·	−4	·	·	3	·	4	·	20	·	·	−20	·	−120	120
$s_{30}s_{12}s_{10}{}^{2}$	·	·	−3	·	6	·	·	3	12	20	·	−24	−20	−120	240
$s_{30}s_{11}{}^{2}s_{10}$	·	·	·	−2	6	·	·	6	·	·	8	−24	−40	·	240
$s_{22}s_{20}s_{10}{}^{2}$	·	−2	−3	−1	·	6	4	6	18	30	6	−48	−60	−180	540
$s_{21}{}^{2}s_{10}{}^{2}$	·	·	·	·	·	12	2	3	·	·	12	−24	−60	·	360
$s_{21}s_{20}{}^{2}s_{01}$	−1	·	·	·	·	3	8	3	·	·	·	−30	−15	·	180
$s_{21}s_{20}s_{11}s_{10}$	−2	·	·	−4	·	12	8	15	·	·	24	−72	−120	·	720
$s_{20}{}^{3}s_{02}$	·	−1	·	·	·	·	1	·	3	·	·	−3	·	−15	15
$s_{20}{}^{2}s_{12}s_{10}$	−1	−2	−3	·	·	·	4	3	12	15	·	−24	−15	−90	180
$s_{20}{}^{2}s_{11}{}^{2}$	−1	·	·	−1	·	·	2	3	·	·	3	−12	−15	·	90
$s_{40}s_{10}{}^{2}s_{01}{}^{2}$	·	·	·	·	−3	·	−1	·	·	·	·	12	·	·	−90
$s_{31}s_{10}{}^{3}s_{01}$	·	·	·	·	−2	−4	·	−1	·	·	·	16	20	·	−240
$s_{30}s_{20}s_{10}s_{01}{}^{2}$	·	·	·	·	−3	·	−4	·	·	·	·	20	·	·	−120
$s_{30}s_{11}s_{10}{}^{2}s_{01}$	·	·	·	·	−6	·	·	−3	·	·	·	24	20	·	−240
$s_{30}s_{10}{}^{3}s_{02}$	·	·	·	·	−1	·	·	·	−4	·	·	4	·	40	−40
$s_{22}s_{10}{}^{4}$	·	·	·	·	·	−1	·	·	−1	−5	−1	2	10	30	−90
$s_{21}s_{20}s_{10}{}^{2}s_{01}$	·	·	·	·	·	−6	−4	−3	·	·	·	36	30	·	−360
$s_{21}s_{11}s_{10}{}^{3}$	·	·	·	·	·	−4	·	−1	·	·	−8	8	40	·	−240
$s_{20}{}^{3}s_{01}{}^{2}$	·	·	·	·	·	·	−1	·	·	·	·	3	·	·	−15
$s_{20}{}^{2}s_{11}s_{10}s_{01}$	1	·	·	·	·	·	−4	−3	·	·	·	24	15	·	−180
$s_{20}{}^{2}s_{10}{}^{2}s_{02}$	·	1	·	·	·	·	−1	·	−6	·	·	6	·	45	−45
$s_{20}s_{12}s_{10}{}^{3}$	·	·	1	·	·	·	·	−1	−4	−10	·	8	10	60	−120
$s_{20}s_{11}{}^{2}s_{10}{}^{2}$	·	·	·	1	·	·	·	−3	·	·	−6	12	30	·	−180
$s_{30}s_{10}{}^{3}s_{01}{}^{2}$	·	·	·	·	1	·	·	·	·	·	·	−4	·	·	40
$s_{21}s_{10}{}^{4}s_{01}$	·	·	·	·	·	1	·	·	·	·	·	−2	−5	·	60
$s_{20}{}^{2}s_{10}{}^{2}s_{01}{}^{2}$	4	1	·	·	·	·	1	·	·	·	·	−6	·	·	45
$s_{20}s_{11}s_{10}{}^{3}s_{01}$	3	·	1	3	·	·	·	1	·	·	·	−8	−10	·	120
$s_{20}s_{10}{}^{4}s_{02}$	·	6	4	·	·	·	·	·	1	·	·	−1	·	−15	15
$s_{12}s_{10}{}^{5}$	·	·	10	·	·	·	·	·	·	1	·	·	−1	−6	12
$s_{11}{}^{2}s_{10}{}^{4}$	·	·	·	6	·	·	·	·	·	·	1	·	−5	·	30
$s_{20}s_{10}{}^{4}s_{01}{}^{2}$	24	6	4	12	4	2	6	8	1	·	·	1	·	·	−15
$s_{11}s_{10}{}^{5}s_{01}$	15	·	10	30	·	5	·	10	·	1	5	·	1	·	−12
$s_{10}{}^{6}s_{02}$	·	45	60	·	·	·	·	·	15	6	·	·	·	1	−1
$s_{10}{}^{6}s_{01}{}^{2}$	180	45	60	180	20	30	45	120	15	6	30	15	12	1	1

Table 1·6·8 (*cont.*)

$w = 8$ (53) (i)	[53]	[52, 01]	[51, 02]	[50, 03]	[43, 10]	[42, 11]	[41, 12]	[40, 13]	[33, 20]	[32, 21]	[31, 22]	[30, 23]	[51, (01)²]	[50, 02, 01]	[42, 10, 01]	[41, 11, 01]	[41, 10, 02]	[40, 12, 01]	[40, 11, 02]	[40, 10, 03]	[33, (10)²]	[32, 20, 01]	[32, 11, 10]
s_{53}	1	−1	−1	−1	−1	−1	−1	−1	−1	−1	−1	−1	2	2	2	2	2	2	2	2	2	2	2
$s_{52}s_{01}$	1	*1*	.	.	.	.	.	.	.	.	.	.	−2	−1	−1	−1	.	−1	.	.	.	−1	.
$s_{51}s_{02}$	1	.	*1*	.	.	.	.	.	.	.	.	.	−1	−1	.	.	−1	.	−1	.	.	.	.
$s_{50}s_{03}$	1	.	.	*1*	.	.	.	.	.	.	.	.	.	−1	.	.	.	.	.	−1	.	.	.
$s_{43}s_{10}$	1	.	.	.	*1*	.	.	.	.	.	.	.	.	.	−1	.	−1	.	.	−1	−2	.	−1
$s_{42}s_{11}$	1	.	.	.	.	*1*	.	.	.	.	.	.	.	.	−1	−1	.	.	−1	.	.	.	−1
$s_{41}s_{12}$	1	.	.	.	.	.	*1*	.	.	.	.	.	.	.	.	−1	−1	−1	.	.	.	.	.
$s_{40}s_{13}$	1	.	.	.	.	.	.	*1*	.	.	.	.	.	.	.	.	.	−1	−1	−1	.	.	.
$s_{33}s_{20}$	1	.	.	.	.	.	.	.	*1*	.	.	.	.	.	.	.	.	.	.	.	−1	−1	.
$s_{32}s_{21}$	1	.	.	.	.	.	.	.	.	*1*	.	.	.	.	.	.	.	.	.	.	.	−1	−1
$s_{31}s_{22}$	1	.	.	.	.	.	.	.	.	.	*1*	.	.	.	.	.	.	.	.	.	.	.	.
$s_{30}s_{23}$	1	.	.	.	.	.	.	.	.	.	.	*1*	.	.	.	.	.	.	.	.	.	.	.
$s_{51}s_{01}^2$	1	2	1	.	.	.	.	.	.	.	.	.	*1*	.	.	.	.	.	.	.	.	.	.
$s_{50}s_{02}s_{01}$	1	1	1	1	.	.	.	.	.	.	.	.	.	*1*	.	.	.	.	.	.	.	.	.
$s_{42}s_{10}s_{01}$	1	1	.	.	1	1	.	.	.	.	.	.	.	.	*1*	.	.	.	.	.	.	.	.
$s_{41}s_{11}s_{01}$	1	1	.	.	.	1	1	.	.	.	.	.	.	.	.	*1*	.	.	.	.	.	.	.
$s_{41}s_{10}s_{02}$	1	.	1	.	1	.	1	.	.	.	.	.	.	.	.	.	*1*	.	.	.	.	.	.
$s_{40}s_{12}s_{01}$	1	1	.	.	.	.	1	1	.	.	.	.	.	.	.	.	.	*1*	.	.	.	.	.
$s_{40}s_{11}s_{02}$	1	.	1	.	.	1	.	1	.	.	.	.	.	.	.	.	.	.	*1*	.	.	.	.
$s_{40}s_{10}s_{03}$	1	.	.	1	1	.	.	1	.	.	.	.	.	.	.	.	.	.	.	*1*	.	.	.
$s_{33}s_{10}^2$	1	.	.	.	2	.	.	.	1	.	.	.	.	.	.	.	.	.	.	.	*1*	.	.
$s_{32}s_{20}s_{01}$	1	1	.	.	.	.	.	.	1	1	.	.	.	.	.	.	.	.	.	.	.	*1*	.
$s_{32}s_{11}s_{10}$	1	.	.	.	1	1	.	.	.	1	.	.	.	.	.	.	.	.	.	.	.	.	*1*
$s_{31}s_{21}s_{01}$	1	1	.	.	.	.	.	.	.	1	1	.	.	.	.	.	.	.	.	.	.	.	.
$s_{31}s_{20}s_{02}$	1	.	1	.	.	.	.	.	1	.	1	.	.	.	.	.	.	.	.	.	.	.	.
$s_{31}s_{12}s_{10}$	1	.	.	.	1	.	1	.	.	.	1	.	.	.	.	.	.	.	.	.	.	.	.
$s_{31}s_{11}^2$	1	.	.	.	.	2	.	.	.	.	1	.	.	.	.	.	.	.	.	.	.	.	.
$s_{30}s_{22}s_{01}$	1	1	.	.	.	.	.	.	.	.	1	1	.	.	.	.	.	.	.	.	.	.	.
$s_{30}s_{21}s_{02}$	1	.	1	.	.	.	.	.	.	1	.	1	.	.	.	.	.	.	.	.	.	.	.
$s_{30}s_{20}s_{03}$	1	.	.	1	.	.	.	.	1	.	.	1	.	.	.	.	.	.	.	.	.	.	.
$s_{30}s_{13}s_{10}$	1	.	.	.	1	.	.	1	.	.	.	1	.	.	.	.	.	.	.	.	.	.	.
$s_{30}s_{12}s_{11}$	1	.	.	.	.	1	1	.	.	.	.	1	.	.	.	.	.	.	.	.	.	.	.
$s_{23}s_{20}s_{10}$	1	.	.	.	1	.	.	.	1	.	.	1	.	.	.	.	.	.	.	.	.	.	.
$s_{22}s_{21}s_{10}$	1	.	.	.	1	.	.	.	.	1	1	.	.	.	.	.	.	.	.	.	.	.	.
$s_{22}s_{20}s_{11}$	1	.	.	.	.	1	.	.	1	.	1	.	.	.	.	.	.	.	.	.	.	.	.
$s_{21}^2s_{11}$	1	.	.	.	.	1	.	.	.	2	.	.	.	.	.	.	.	.	.	.	.	.	.
$s_{21}s_{20}s_{12}$	1	.	.	.	.	.	1	.	1	1	.	.	.	.	.	.	.	.	.	.	.	.	.
$s_{20}^2s_{13}$	1	.	.	.	.	.	.	1	2	.	.	.	.	.	.	.	.	.	.	.	.	.	.
$s_{50}s_{01}^3$	1	3	3	1	.	.	.	.	.	.	.	.	3	3	.	.	.	.	.	.	.	.	.
$s_{41}s_{10}s_{01}^2$	1	2	1	.	1	2	1	.	.	.	.	.	1	.	2	2	1	.	.	.	.	.	.
$s_{40}s_{11}s_{01}^2$	1	2	1	.	.	1	2	1	.	.	.	.	1	.	.	2	.	2	1	.	.	.	.
$s_{40}s_{10}s_{02}s_{01}$	1	1	1	1	1	1	1	1	.	.	.	.	.	1	1	.	1	1	1	1	.	.	.
$s_{32}s_{10}^2s_{01}$	1	1	.	.	2	2	.	.	1	1	.	.	.	.	2	.	.	.	.	.	1	1	2
$s_{31}s_{20}s_{01}^2$	1	2	1	.	.	.	.	.	1	2	1	.	1	.	.	.	.	.	.	.	.	2	.
$s_{31}s_{11}s_{10}s_{01}$	1	1	.	.	1	2	1	.	.	1	1	.	.	.	1	1	.	.	.	.	.	.	1
$s_{31}s_{10}^2s_{02}$	1	.	1	.	2	.	2	.	1	.	1	.	.	.	.	.	2	.	.	.	1	.	.
$s_{30}s_{21}s_{01}^2$	1	2	1	.	.	1	.	.	.	1	2	1	1	.	.	.	.	.	.	.	.	.	.
$s_{30}s_{20}s_{02}s_{01}$	1	1	1	1	.	.	.	.	1	1	1	1	.	1	.	.	.	.	.	.	.	1	.
$s_{30}s_{12}s_{10}s_{01}$	1	1	.	.	1	1	1	1	.	.	1	1	.	.	1	.	.	1	.	.	.	.	.
$s_{30}s_{11}^2s_{01}$	1	1	.	.	.	2	2	.	.	.	1	1	.	.	.	2	.	.	.	.	.	.	.
$s_{30}s_{11}s_{10}s_{02}$	1	.	1	.	1	1	1	1	.	1	.	1	.	.	.	.	1	.	1	.	.	.	1
$s_{30}s_{10}^2s_{03}$	1	.	.	1	2	.	.	2	1	.	.	1	.	.	.	.	.	.	.	2	1	.	.
$s_{23}s_{10}^3$	1	.	.	.	3	.	.	.	3	.	.	1	.	.	.	.	.	.	.	.	3	.	.
$s_{22}s_{20}s_{10}s_{01}$	1	1	.	.	1	1	.	.	1	1	1	1	.	.	1	.	.	.	.	.	.	1	.
$s_{22}s_{11}s_{10}^2$	1	.	.	.	2	1	.	.	1	2	1	.	.	.	.	.	.	.	.	.	1	.	2
$s_{21}^2s_{10}s_{01}$	1	1	.	.	1	1	.	.	.	2	2	.	.	.	1	.	.	.	.	.	.	.	.
$s_{21}s_{20}s_{11}s_{01}$	1	1	.	.	.	1	1	.	1	2	1	.	.	.	.	1	.	.	.	.	.	1	.
$s_{21}s_{20}s_{10}s_{02}$	1	.	1	.	1	.	1	.	1	1	1	1	.	.	.	.	1	.	.	.	.	.	.
$s_{21}s_{12}s_{10}^2$	1	.	.	.	2	.	1	.	1	1	2	.	.	.	.	.	.	.	.	.	1	.	.
$s_{21}s_{11}^2s_{10}$	1	.	.	.	1	2	.	.	.	3	1	.	.	.	.	.	.	.	.	.	.	.	2
$s_{20}^2s_{12}s_{01}$	1	1	.	.	.	.	1	1	2	2	.	.	.	.	.	.	.	1	.	.	.	2	.
$s_{20}^2s_{11}s_{02}$	1	.	1	.	.	1	.	1	2	.	2	.	.	.	.	.	.	.	1	.	.	.	.
$s_{20}^2s_{10}s_{03}$	1	.	.	1	1	.	.	1	2	.	.	2	.	.	.	.	.	.	.	1	.	.	.
$s_{20}s_{13}s_{10}^2$	1	.	.	.	2	.	.	1	2	.	.	2	.	.	.	.	.	.	.	.	1	.	.
$s_{20}s_{12}s_{11}s_{10}$	1	.	.	.	1	1	1	.	1	1	1	1	.	.	.	.	.	.	.	.	.	.	1
$s_{20}s_{11}^3$	1	.	.	.	.	3	.	.	1	.	3	.	.	.	.	.	.	.	.	.	.	.	.
$s_{40}s_{10}s_{01}^3$	1	3	3	1	1	3	3	1	.	.	.	.	3	3	3	6	3	3	3	1	.	.	.
$s_{31}s_{10}^2s_{01}^2$	1	2	1	.	2	4	2	.	1	2	1	.	1	.	4	4	2	.	.	.	1	2	4
$s_{30}s_{20}s_{01}^3$	1	3	3	1	.	.	.	.	1	3	3	1	3	3	.	.	.	.	.	.	.	3	.
$s_{30}s_{11}s_{10}s_{01}^2$	1	2	1	.	1	3	3	1	.	1	2	1	1	.	2	4	1	2	1	.	.	.	1
$s_{30}s_{10}^2s_{02}s_{01}$	1	1	1	1	2	2	2	2	1	1	1	1	.	1	2	.	2	2	2	2	1	1	2
$s_{22}s_{10}^3s_{01}$	1	1	.	.	3	3	.	.	3	3	1	1	.	.	3	.	.	.	.	.	3	3	6
$s_{21}s_{20}s_{10}s_{01}^2$	1	2	1	.	1	2	1	.	1	3	3	1	1	.	2	2	1	.	.	.	.	2	.
$s_{21}s_{11}s_{10}^2s_{01}$	1	1	.	.	2	3	1	.	1	4	3	.	.	.	2	1	.	.	.	.	1	1	4
$s_{21}s_{10}^3s_{02}$	1	.	1	.	3	.	3	.	3	1	3	1	.	.	.	.	3	.	.	.	3	.	.
$s_{20}^2s_{11}s_{01}^2$	1	2	1	.	.	1	2	1	2	4	2	.	1	.	.	2	.	2	1	.	.	4	.
$s_{20}^2s_{10}s_{02}s_{01}$	1	1	1	1	1	1	1	1	2	2	2	2	.	1	1	.	1	1	1	1	.	2	.
$s_{20}s_{12}s_{10}^2s_{01}$	1	1	.	.	2	2	1	1	2	2	2	2	.	.	2	.	.	1	.	.	1	2	2
$s_{20}s_{11}^2s_{10}s_{01}$	1	1	.	.	1	3	2	.	1	3	3	1	.	.	1	2	.	.	.	.	.	1	2
$s_{20}s_{11}s_{10}^2s_{02}$	1	.	1	.	2	1	2	1	2	2	2	2	.	.	.	.	2	.	1	.	1	.	2
$s_{20}s_{10}^3s_{03}$	1	.	.	1	3	.	.	3	4	.	.	4	.	.	.	.	.	.	.	3	3	.	.
$s_{13}s_{10}^4$	1	.	.	.	4	.	.	1	6	.	.	4	.	.	.	.	.	.	.	.	6	.	.
$s_{12}s_{11}s_{10}^3$	1	.	.	.	3	1	1	.	3	3	3	1	.	.	.	.	.	.	.	.	3	.	3
$s_{11}^3s_{10}^2$	1	.	.	.	2	6	.	.	1	3	3	.	.	.	.	.	.	.	.	.	1	.	6
$s_{30}s_{10}^2s_{01}^3$	1	3	3	1	2	6	6	2	1	3	3	1	3	3	6	12	6	6	6	2	1	3	6
$s_{21}s_{10}^3s_{01}^2$	1	2	1	.	3	6	3	.	3	7	5	1	1	.	6	6	3	.	.	.	3	6	12
$s_{20}^2s_{10}s_{01}^3$	1	3	3	1	1	3	3	1	2	6	6	2	3	3	3	6	3	3	3	1	.	6	.
$s_{20}s_{11}s_{10}^2s_{01}^2$	1	2	1	.	2	5	4	1	2	6	6	2	1	.	4	6	2	2	1	.	1	4	6
$s_{20}s_{10}^3s_{02}s_{01}$	1	1	1	1	3	3	3	3	4	4	4	4	.	1	3	.	3	3	3	3	3	4	6
$s_{12}s_{10}^4s_{01}$	1	1	.	.	4	4	1	1	6	6	4	4	.	.	4	.	.	1	.	.	6	6	12
$s_{11}^2s_{10}^3s_{01}$	1	1	.	.	3	5	2	.	3	9	7	1	.	.	3	2	.	.	.	.	3	3	12
$s_{11}s_{10}^4s_{02}$	1	.	1	.	4	1	4	1	6	4	6	4	.	.	.	.	4	.	1	.	6	.	4
$s_{10}^5s_{03}$	1	.	.	1	5	.	.	5	10	.	.	10	.	.	.	.	.	.	.	5	10	.	.
$s_{20}s_{10}^3s_{01}^3$	1	3	3	1	3	9	9	3	4	12	12	4	3	3	9	18	9	9	9	3	3	12	18
$s_{11}s_{10}^4s_{01}^2$	1	2	1	.	4	9	6	1	6	16	14	4	1	.	8	10	4	2	1	.	6	12	28
$s_{10}^5s_{02}s_{01}$	1	1	1	1	5	5	5	5	10	10	10	10	.	1	5	.	5	5	5	5	10	10	20
$s_{10}^5s_{01}^3$	1	3	3	1	5	15	15	5	10	30	30	10	3	3	15	30	15	15	15	5	10	30	60

Table 1·6·8 (*cont.*)

$w = 8$ (53) (ii)	[31, 21, 01]	[31, 20, 02]	[31, 12, 10]	[31, $(11)^2$]	[30, 22, 01]	[30, 21, 02]	[30, 20, 03]	[30, 13, 10]	[30, 12, 11]	[23, 20, 10]	[22, 21, 10]	[22, 20, 11]	[$(21)^2$, 11]	[21, 20, 12]	[$(20)^2$, 13]	[50, $(01)^3$]	[41, 10, $(01)^2$]	[40, 11, $(01)^2$]	[40, 10, 02, 01]	[32, $(10)^2$, (01)]	[31, 20, $(01)^2$]	[31, 11, 10, 01]	[31, $(10)^2$, 02]	[30, 21, $(01)^2$]
s_{53}	2	2	2	2	2	2	2	2	2	2	2	2	2	2	2	−6	−6	−6	−6	−6	−6	−6	−6	−6
$s_{52}s_{01}$	−1	.	.	.	−1	.	.	.	.	.	.	.	.	.	.	6	4	4	2	2	4	2	.	4
$s_{51}s_{02}$	.	−1	.	.	.	−1	.	.	.	.	.	.	.	.	.	3	1	1	2	.	1	.	2	1
$s_{50}s_{03}$	.	.	.	.	.	.	−1	.	.	.	.	.	.	.	.	2	.	.	1	.	.	.	.	.
$s_{43}s_{10}$	.	.	−1	.	.	.	.	−1	.	−1	−1	.	.	.	.	.	2	.	2	4	.	2	4	.
$s_{42}s_{11}$	.	.	.	−2	.	.	.	.	−1	.	.	−1	−1	.	.	.	2	2	1	2	.	3	.	.
$s_{41}s_{12}$	.	.	−1	.	.	.	.	.	−1	.	.	.	.	−1	.	.	2	2	1	.	.	1	2	.
$s_{40}s_{13}$	.	.	.	.	.	.	.	−1	.	.	.	.	.	.	−1	.	.	2	2	.	.	.	.	.
$s_{33}s_{20}$	.	−1	.	.	.	.	−1	.	.	−1	.	−1	.	−1	−2	.	.	.	.	1	2	.	1	.
$s_{32}s_{21}$	−1	.	.	.	.	−1	.	.	.	.	−1	.	−2	−1	.	.	.	.	.	2	2	1	.	2
$s_{31}s_{22}$	−1	−1	−1	−1	−1	.	.	.	.	.	−1	−1	.	.	.	.	.	.	.	.	2	2	2	2
$s_{30}s_{23}$	.	.	.	.	−1	−1	−1	−1	−1	−1	.	.	.	.	.	.	.	.	.	.	.	.	.	2
$s_{51}s_{01}^2$	.	.	.	.	.	.	.	.	.	.	.	.	.	.	.	−3	−1	−1	.	.	−1	.	.	−1
$s_{50}s_{02}s_{01}$	.	.	.	.	.	.	.	.	.	.	.	.	.	.	.	−3	.	.	−1	.	.	.	.	.
$s_{42}s_{10}s_{01}$	.	.	.	.	.	.	.	.	.	.	.	.	.	.	.	.	−2	.	−1	−2	.	−1	.	.
$s_{41}s_{11}s_{01}$	.	.	.	.	.	.	.	.	.	.	.	.	.	.	.	.	−2	−2	.	.	.	−1	.	.
$s_{41}s_{10}s_{02}$	.	.	.	.	.	.	.	.	.	.	.	.	.	.	.	.	−1	.	−1	.	.	.	−2	.
$s_{40}s_{12}s_{01}$	.	.	.	.	.	.	.	.	.	.	.	.	.	.	.	.	.	−2	−1	.	.	.	.	.
$s_{40}s_{11}s_{02}$	.	.	.	.	.	.	.	.	.	.	.	.	.	.	.	.	.	−1	−1	.	.	.	.	.
$s_{40}s_{10}s_{03}$	.	.	.	.	.	.	.	.	.	.	.	.	.	.	.	.	.	.	−1	.	.	.	.	.
$s_{33}s_{10}^2$	.	.	.	.	.	.	.	.	.	.	.	.	.	.	.	.	.	.	.	−1	.	.	−1	.
$s_{32}s_{20}s_{01}$	.	.	.	.	.	.	.	.	.	.	.	.	.	.	.	.	.	.	.	−1	−2	.	.	.
$s_{32}s_{11}s_{10}$	.	.	.	.	.	.	.	.	.	.	.	.	.	.	.	.	.	.	.	−2	.	−1	.	.
$s_{31}s_{21}s_{01}$	*1*	.	.	.	.	.	.	.	.	.	.	.	.	.	.	.	.	.	.	.	−2	−1	.	−2
$s_{31}s_{20}s_{02}$	.	*1*	.	.	.	.	.	.	.	.	.	.	.	.	.	.	.	.	.	.	−1	.	−1	.
$s_{31}s_{12}s_{10}$	.	.	*1*	.	.	.	.	.	.	.	.	.	.	.	.	.	.	.	.	.	.	−1	−2	.
$s_{31}s_{11}^2$	.	.	.	*1*	.	.	.	.	.	.	.	.	.	.	.	.	.	.	.	.	.	−1	.	.
$s_{30}s_{22}s_{01}$	.	.	.	.	*1*	.	.	.	.	.	.	.	.	.	.	.	.	.	.	.	.	.	.	−2
$s_{30}s_{21}s_{02}$	.	.	.	.	.	*1*	.	.	.	.	.	.	.	.	.	.	.	.	.	.	.	.	.	−1
$s_{30}s_{20}s_{03}$	.	.	.	.	.	.	*1*	.	.	.	.	.	.	.	.	.	.	.	.	.	.	.	.	.
$s_{30}s_{13}s_{10}$	.	.	.	.	.	.	.	*1*	.	.	.	.	.	.	.	.	.	.	.	.	.	.	.	.
$s_{30}s_{12}s_{11}$	.	.	.	.	.	.	.	.	*1*	.	.	.	.	.	.	.	.	.	.	.	.	.	.	.
$s_{23}s_{20}s_{10}$	.	.	.	.	.	.	.	.	.	*1*	.	.	.	.	.	.	.	.	.	.	.	.	.	.
$s_{22}s_{21}s_{10}$	.	.	.	.	.	.	.	.	.	.	*1*	.	.	.	.	.	.	.	.	.	.	.	.	.
$s_{22}s_{20}s_{11}$	.	.	.	.	.	.	.	.	.	.	.	*1*	.	.	.	.	.	.	.	.	.	.	.	.
$s_{21}^2s_{11}$	.	.	.	.	.	.	.	.	.	.	.	.	*1*	.	.	.	.	.	.	.	.	.	.	.
$s_{21}s_{20}s_{12}$	.	.	.	.	.	.	.	.	.	.	.	.	.	*1*	.	.	.	.	.	.	.	.	.	.
$s_{20}^2s_{13}$	.	.	.	.	.	.	.	.	.	.	.	.	.	.	*1*	.	.	.	.	.	.	.	.	.
$s_{50}s_{01}^3$	.	.	.	.	.	.	.	.	.	.	.	.	.	.	.	*1*	.	.	.	.	.	.	.	.
$s_{41}s_{10}s_{01}^2$	.	.	.	.	.	.	.	.	.	.	.	.	.	.	.	.	*1*	.	.	.	.	.	.	.
$s_{40}s_{11}s_{01}^2$	.	.	.	.	.	.	.	.	.	.	.	.	.	.	.	.	.	*1*	.	.	.	.	.	.
$s_{40}s_{10}s_{02}s_{01}$	.	.	.	.	.	.	.	.	.	.	.	.	.	.	.	.	.	.	*1*	.	.	.	.	.
$s_{32}s_{10}^2s_{01}$	.	.	.	.	.	.	.	.	.	.	.	.	.	.	.	.	.	.	.	*1*	.	.	.	.
$s_{31}s_{20}s_{01}^2$	2	1	.	.	.	.	.	.	.	.	.	.	.	.	.	.	.	.	.	.	*1*	.	.	.
$s_{31}s_{11}s_{10}s_{01}$	1	.	1	1	.	.	.	.	.	.	.	.	.	.	.	.	.	.	.	.	.	*1*	.	.
$s_{31}s_{10}^2s_{02}$	.	1	2	.	.	.	.	.	.	.	.	.	.	.	.	.	.	.	.	.	.	.	*1*	.
$s_{30}s_{21}s_{01}^2$	2	.	.	.	2	1	.	.	.	.	.	.	.	.	.	.	.	.	.	.	.	.	.	*1*
$s_{30}s_{20}s_{02}s_{01}$	.	1	.	.	1	1	1	.	.	.	.	.	.	.	.	.	.	.	.	.	.	.	.	.
$s_{30}s_{12}s_{10}s_{01}$	.	.	1	.	1	.	.	1	1	.	.	.	.	.	.	.	.	.	.	.	.	.	.	.
$s_{30}s_{11}^2s_{01}$	.	.	.	1	1	.	.	.	2	.	.	.	.	.	.	.	.	.	.	.	.	.	.	.
$s_{30}s_{11}s_{10}s_{02}$	.	.	.	.	.	1	.	1	1	.	.	.	.	.	.	.	.	.	.	.	.	.	.	.
$s_{30}s_{10}^2s_{03}$	.	.	.	.	.	.	1	2	.	.	.	.	.	.	.	.	.	.	.	.	.	.	.	.
$s_{23}s_{10}^3$	.	.	.	.	.	.	.	.	.	3	.	.	.	.	.	.	.	.	.	.	.	.	.	.
$s_{22}s_{20}s_{10}s_{01}$	.	.	.	.	1	.	.	.	.	1	1	1	.	.	.	.	.	.	.	.	.	.	.	.
$s_{22}s_{11}s_{10}^2$	.	.	.	.	.	.	.	.	.	.	2	1	.	.	.	.	.	.	.	.	.	.	.	.
$s_{21}^2s_{10}s_{01}$	2	.	.	.	.	.	.	.	.	.	2	.	1	.	.	.	.	.	.	.	.	.	.	.
$s_{21}s_{20}s_{11}s_{01}$	1	.	.	.	.	.	.	.	.	.	.	1	1	1	.	.	.	.	.	.	.	.	.	.
$s_{21}s_{20}s_{10}s_{02}$	.	1	.	.	.	1	.	.	.	1	1	.	.	1	.	.	.	.	.	.	.	.	.	.
$s_{21}s_{12}s_{10}^2$	.	.	2	.	.	.	.	.	.	.	2	.	.	1	.	.	.	.	.	.	.	.	.	.
$s_{21}s_{11}^2s_{10}$	.	.	.	1	.	.	.	.	.	.	1	.	2	.	.	.	.	.	.	.	.	.	.	.
$s_{20}^2s_{12}s_{01}$	.	.	.	.	.	.	.	.	.	.	.	.	.	2	1	.	.	.	.	.	.	.	.	.
$s_{20}^2s_{11}s_{02}$	.	2	.	.	.	.	.	.	.	.	.	2	.	.	1	.	.	.	.	.	.	.	.	.
$s_{20}^2s_{10}s_{03}$	.	.	.	.	.	.	2	.	.	2	.	.	.	.	1	.	.	.	.	.	.	.	.	.
$s_{20}s_{13}s_{10}^2$	.	.	.	.	.	.	.	2	.	2	.	.	.	.	1	.	.	.	.	.	.	.	.	.
$s_{20}s_{12}s_{11}s_{10}$	.	.	1	.	.	.	.	.	1	1	.	1	.	1	.	.	.	.	.	.	.	.	.	.
$s_{20}s_{11}^3$	.	.	.	3	.	.	.	.	.	.	.	3	.	.	.	.	.	.	.	.	.	.	.	.
$s_{40}s_{10}s_{01}^3$	.	.	.	.	.	.	.	.	.	.	.	.	.	.	.	1	3	3	3	.	.	.	.	.
$s_{31}s_{10}^2s_{01}^2$	2	1	2	2	.	.	.	.	.	.	.	.	.	.	.	.	2	.	.	2	1	4	1	.
$s_{30}s_{20}s_{01}^3$	6	3	.	.	3	3	1	.	.	.	.	.	.	.	.	1	.	.	.	.	3	.	.	3
$s_{30}s_{11}s_{10}s_{01}^2$	2	.	2	2	2	1	.	1	3	.	.	.	.	.	.	.	1	1	.	.	.	2	.	1
$s_{30}s_{10}^2s_{02}s_{01}$	.	1	2	.	1	1	1	2	2	.	.	.	.	.	.	.	.	.	2	1	.	.	1	.
$s_{22}s_{10}^3s_{01}$	.	.	.	.	1	.	.	.	.	3	3	3	.	.	.	.	.	.	.	3	.	.	.	.
$s_{21}s_{20}s_{10}s_{01}^2$	4	1	.	.	2	1	.	.	.	1	3	2	2	1	.	.	1	.	.	.	1	.	.	1
$s_{21}s_{11}s_{10}^2s_{01}$	3	.	2	2	.	.	.	.	.	.	4	1	3	1	.	.	.	.	.	1	.	2	.	.
$s_{21}s_{10}^3s_{02}$	.	3	6	.	.	1	.	.	.	3	3	.	.	3	.	.	.	.	.	.	.	.	3	.
$s_{20}^2s_{11}s_{01}^2$	4	2	.	.	.	.	.	.	.	.	.	2	2	4	1	.	.	1	.	.	2	.	.	.
$s_{20}^2s_{10}s_{02}s_{01}$	.	2	.	.	2	2	2	.	.	2	2	2	.	2	1	.	.	.	1	.	.	.	.	.
$s_{20}s_{12}s_{10}^2s_{01}$	.	.	2	.	2	.	.	2	2	2	2	2	.	2	1	.	.	.	.	1	.	.	.	.
$s_{20}s_{11}^2s_{10}s_{01}$	2	.	2	3	1	.	.	.	2	1	1	3	2	2	.	.	.	.	.	.	.	2	.	.
$s_{20}s_{11}s_{10}^2s_{02}$	.	2	2	.	.	2	.	2	2	2	2	2	.	2	1	.	.	.	.	.	.	.	1	.
$s_{20}s_{10}^3s_{03}$	.	.	.	.	.	.	4	6	.	6	.	.	.	.	3	.	.	.	.	.	.	.	.	.
$s_{13}s_{10}^4$	.	.	.	.	.	.	.	4	.	12	.	.	.	.	3	.	.	.	.	.	.	.	.	.
$s_{12}s_{11}s_{10}^3$	.	.	3	.	.	.	.	.	1	3	6	3	.	3	.	.	.	.	.	.	.	.	.	.
$s_{11}^3s_{10}^2$	.	.	.	3	.	.	.	.	.	.	6	3	6	.	.	.	.	.	.	.	.	.	.	.
$s_{30}s_{10}^2s_{01}^3$	6	3	6	6	3	3	1	2	6	.	.	.	.	.	.	1	6	6	6	3	3	12	3	3
$s_{21}s_{10}^3s_{01}^2$	8	3	6	6	2	1	.	.	.	3	9	6	6	3	.	.	3	.	.	6	3	12	3	1
$s_{20}^2s_{10}s_{01}^3$	12	6	.	.	6	6	2	.	.	2	6	6	6	6	1	1	3	3	3	.	6	.	.	6
$s_{20}s_{11}s_{10}^2s_{01}^2$	8	2	6	6	4	2	.	2	6	2	6	6	6	6	1	.	2	1	.	2	2	8	1	2
$s_{20}s_{10}^3s_{02}s_{01}$	.	4	6	.	4	4	4	6	6	6	6	6	.	6	3	.	.	.	3	3	.	.	3	.
$s_{12}s_{10}^4s_{01}$	.	.	4	.	4	.	.	4	4	12	12	12	.	6	3	.	.	.	.	6	.	.	.	.
$s_{11}^2s_{10}^3s_{01}$	6	.	6	7	1	.	.	.	2	3	15	9	12	6	.	.	.	.	.	3	.	6	.	.
$s_{11}s_{10}^4s_{02}$	.	6	12	.	.	4	.	4	4	12	12	6	.	12	3	.	.	.	.	.	.	.	6	.
$s_{10}^5s_{03}$	.	.	.	.	.	.	10	20	.	30	.	.	.	.	15	.	.	.	.	.	.	.	.	.
$s_{20}s_{10}^3s_{01}^3$	24	12	18	18	12	12	4	6	18	6	18	18	18	18	3	1	9	9	9	9	12	36	9	12
$s_{11}s_{10}^4s_{01}^2$	20	6	20	20	8	4	.	4	12	12	36	30	30	24	3	.	4	1	.	12	6	32	6	4
$s_{10}^5s_{02}s_{01}$	.	10	20	.	10	10	10	20	20	30	30	30	.	30	15	.	.	.	5	10	.	.	10	.
$s_{10}^5s_{01}^3$	60	30	60	60	30	30	10	20	60	30	90	90	90	90	15	1	15	15	15	30	30	120	30	30

Table 1·6·8 (*cont.*)

$w = 8$ (53) (iii)	[30, 20, 02, 01]	[30, 12, 10, 01]	[30, (11)², 01]	[30, 11, 10, 02]	[30, (10)², 03]	[23, (10)³]	[22, 20, 10, 01]	[22, 11, (10)²]	[(21)², 10, 01]	[21, 20, 11, 01]	[21, 20, 10, 02]	[21, 12, (10)²]	[21, (11)², 10]	[(20)², 12, 01]	[(20)², 11, 02]	[(20)², 10, 03]	[20, 13, (10)²]	[20, 12, 11, 10]	[20, (11)³]
s_{53}	−6	−6	−6	−6	−6	−6	−6	−6	−6	−6	−6	−6	−6	−6	−6	−6	−6	−6	−6
$s_{52}s_{01}$	2	2	2	·	·	·	2	·	2	2	·	·	·	2	·	·	·	·	·
$s_{51}s_{02}$	1	·	·	2	·	·	·	·	·	·	2	·	·	·	·	2	·	·	·
$s_{50}s_{03}$	1	·	·	·	2	·	·	·	·	·	·	·	·	·	·	·	2	·	·
$s_{43}s_{10}$	·	2	·	2	4	6	2	4	2	·	2	4	2	·	·	2	4	2	6
$s_{42}s_{11}$	·	1	4	2	·	·	1	2	1	2	·	4	·	2	·	·	·	2	6
$s_{41}s_{12}$	·	2	2	1	·	·	·	·	·	1	1	2	·	2	·	·	·	2	·
$s_{40}s_{13}$	·	1	·	1	2	·	·	·	·	·	·	·	·	1	1	1	2	2	·
$s_{33}s_{20}$	2	·	·	·	1	3	2	1	·	2	2	1	·	4	4	4	3	2	2
$s_{32}s_{21}$	2	·	·	1	·	·	1	2	4	3	2	2	4	2	·	·	·	1	·
$s_{31}s_{22}$	1	1	1	1	·	·	2	2	2	1	2	2	1	·	2	·	·	1	3
$s_{30}s_{23}$	2	2	2	2	2	2	1	·	·	·	1	·	·	·	·	2	2	1	·
$s_{51}s_{01}^2$	−1	·	·	·	·	·	·	·	·	·	·	·	·	·	·	·	·	·	·
$s_{50}s_{02}s_{01}$	·	−1	·	·	·	·	−1	·	−1	·	·	·	·	·	·	·	·	·	·
$s_{42}s_{10}s_{01}$	·	−1	−2	·	·	·	·	·	·	−1	·	·	·	·	·	·	·	·	·
$s_{41}s_{11}s_{01}$	·	·	−2	·	·	·	·	·	·	·	−1	·	·	·	·	·	·	·	·
$s_{41}s_{10}s_{02}$	·	·	·	−1	·	·	·	·	·	−1	·	·	·	·	·	·	·	·	·
$s_{40}s_{12}s_{01}$	·	−1	·	·	·	·	·	·	·	·	·	·	·	−1	·	·	·	·	·
$s_{40}s_{11}s_{02}$	·	·	·	−1	·	·	·	·	·	·	·	·	·	·	−1	·	·	·	·
$s_{40}s_{10}s_{03}$	·	·	·	·	−2	·	·	·	·	·	·	·	·	·	·	−1	·	·	·
$s_{33}s_{10}^2$	·	·	·	·	−1	−3	·	−1	·	·	·	·	−1	·	·	·	−1	·	·
$s_{32}s_{20}s_{01}$	−1	·	·	·	·	·	−1	·	−1	·	·	·	−2	·	·	·	·	·	·
$s_{32}s_{11}s_{10}$	·	·	·	−1	·	·	·	−2	·	·	·	·	−2	·	·	·	·	−1	·
$s_{31}s_{21}s_{01}$	·	·	·	·	·	·	·	−2	−1	·	·	·	·	·	·	·	·	·	·
$s_{31}s_{20}s_{02}$	−1	·	·	·	·	·	·	·	·	·	−1	·	·	·	−2	·	·	·	·
$s_{31}s_{12}s_{10}$	·	−1	·	·	·	·	·	·	·	·	·	·	−2	·	·	·	·	−1	·
$s_{31}s_{11}^2$	·	·	−1	·	·	·	·	·	·	·	·	·	−1	·	·	·	·	·	−3
$s_{30}s_{22}s_{01}$	−1	−1	−1	·	·	·	−1	·	·	·	·	·	·	·	·	·	·	·	·
$s_{30}s_{21}s_{02}$	−1	·	·	−1	·	·	·	·	·	·	−1	·	·	·	·	·	·	·	·
$s_{30}s_{20}s_{03}$	−1	·	·	·	−1	·	·	·	·	·	·	·	·	·	·	·	−2	·	·
$s_{30}s_{13}s_{10}$	·	−1	·	−1	−2	·	·	·	·	·	·	·	·	·	·	·	−2	·	·
$s_{30}s_{12}s_{11}$	·	−1	−2	−1	·	·	·	·	·	·	·	·	·	·	·	·	·	−1	·
$s_{23}s_{20}s_{10}$	·	·	·	·	·	−3	−1	·	·	·	−1	·	·	·	·	−2	−2	−1	·
$s_{22}s_{21}s_{10}$	·	·	·	·	·	·	−1	−2	−2	·	−1	−2	−1	·	·	·	·	−1	−3
$s_{22}s_{20}s_{11}$	·	·	·	·	·	·	−1	−1	·	−1	·	·	−2	·	−2	·	·	−1	−3
$s_{21}^2s_{11}$	·	·	·	·	·	·	·	−1	−1	·	·	·	−2	·	·	·	·	·	·
$s_{21}s_{20}s_{12}$	·	·	·	·	·	·	·	·	−1	−1	−1	·	·	−2	·	·	·	−1	·
$s_{20}^2s_{13}$	·	·	·	·	·	·	·	·	·	·	·	·	·	−1	−1	−1	−1	·	·
$s_{50}s_{01}^3$	·	·	·	·	·	·	·	·	·	·	·	·	·	·	·	·	·	·	·
$s_{41}s_{10}s_{01}^2$	·	·	·	·	·	·	·	·	·	·	·	·	·	·	·	·	·	·	·
$s_{40}s_{11}s_{01}^2$	·	·	·	·	·	·	·	·	·	·	·	·	·	·	·	·	·	·	·
$s_{40}s_{10}s_{02}s_{01}$	·	·	·	·	·	·	·	·	·	·	·	·	·	·	·	·	·	·	·
$s_{32}s_{10}^2s_{01}$	·	·	·	·	·	·	·	·	·	·	·	·	·	·	·	·	·	·	·
$s_{31}s_{20}s_{01}^2$	·	·	·	·	·	·	·	·	·	·	·	·	·	·	·	·	·	·	·
$s_{31}s_{11}s_{10}s_{01}$	·	·	·	·	·	·	·	·	·	·	·	·	·	·	·	·	·	·	·
$s_{31}s_{10}^2s_{02}$	·	·	·	·	·	·	·	·	·	·	·	·	·	·	·	·	·	·	·
$s_{30}s_{21}s_{01}^2$	·	·	·	·	·	·	·	·	·	·	·	·	·	·	·	·	·	·	·
$s_{30}s_{20}s_{02}s_{01}$	*1*	·	·	·	·	·	·	·	·	·	·	·	·	·	·	·	·	·	·
$s_{30}s_{12}s_{10}s_{01}$	·	*1*	·	·	·	·	·	·	·	·	·	·	·	·	·	·	·	·	·
$s_{30}s_{11}^2s_{01}$	·	·	*1*	·	·	·	·	·	·	·	·	·	·	·	·	·	·	·	·
$s_{30}s_{11}s_{10}s_{02}$	·	·	·	*1*	·	·	·	·	·	·	·	·	·	·	·	·	·	·	·
$s_{30}s_{10}^2s_{03}$	·	·	·	·	*1*	·	·	·	·	·	·	·	·	·	·	·	·	·	·
$s_{23}s_{10}^3$	·	·	·	·	·	*1*	·	·	·	·	·	·	·	·	·	·	·	·	·
$s_{22}s_{20}s_{10}s_{01}$	·	·	·	·	·	·	*1*	·	·	·	·	·	·	·	·	·	·	·	·
$s_{22}s_{11}s_{10}^2$	·	·	·	·	·	·	·	*1*	·	·	·	·	·	·	·	·	·	·	·
$s_{21}^2s_{10}s_{01}$	·	·	·	·	·	·	·	·	*1*	·	·	·	·	·	·	·	·	·	·
$s_{21}s_{20}s_{11}s_{01}$	·	·	·	·	·	·	·	·	·	*1*	·	·	·	·	·	·	·	·	·
$s_{21}s_{20}s_{10}s_{02}$	·	·	·	·	·	·	·	·	·	·	*1*	·	·	·	·	·	·	·	·
$s_{21}s_{12}s_{10}^2$	·	·	·	·	·	·	·	·	·	·	·	*1*	·	·	·	·	·	·	·
$s_{21}s_{11}^2s_{10}$	·	·	·	·	·	·	·	·	·	·	·	·	*1*	·	·	·	·	·	·
$s_{20}^2s_{12}s_{01}$	·	·	·	·	·	·	·	·	·	·	·	·	·	*1*	·	·	·	·	·
$s_{20}^2s_{11}s_{02}$	·	·	·	·	·	·	·	·	·	·	·	·	·	·	*1*	·	·	·	·
$s_{20}^2s_{10}s_{03}$	·	·	·	·	·	·	·	·	·	·	·	·	·	·	·	*1*	·	·	·
$s_{20}s_{13}s_{10}^2$	·	·	·	·	·	·	·	·	·	·	·	·	·	·	·	·	*1*	·	·
$s_{20}s_{12}s_{11}s_{10}$	·	·	·	·	·	·	·	·	·	·	·	·	·	·	·	·	·	*1*	·
$s_{20}s_{11}^3$	·	·	·	·	·	·	·	·	·	·	·	·	·	·	·	·	·	·	*1*
$s_{40}s_{10}s_{01}^3$	·	·	·	·	·	·	·	·	·	·	·	·	·	·	·	·	·	·	·
$s_{31}s_{10}^2s_{01}^2$	3	·	·	·	·	·	·	·	·	·	·	·	·	·	·	·	·	·	·
$s_{30}s_{20}s_{01}^3$	·	2	2	1	·	·	·	·	·	·	·	·	·	·	·	·	·	·	·
$s_{30}s_{11}s_{10}s_{01}^2$	1	2	·	2	1	·	·	·	·	·	·	·	·	·	·	·	·	·	·
$s_{30}s_{10}^2s_{02}s_{01}$	·	·	·	·	·	·	1	3	3	·	·	·	·	·	·	·	·	·	·
$s_{22}s_{10}^3s_{01}$	·	·	·	·	·	·	2	·	·	2	2	1	·	·	·	·	·	·	·
$s_{21}s_{20}s_{10}s_{01}^2$	·	·	·	·	·	·	·	1	2	2	1	·	1	2	·	·	·	·	·
$s_{21}s_{11}s_{10}^2s_{01}$	·	·	·	·	·	1	·	·	·	·	·	3	3	·	2	1	·	·	·
$s_{21}s_{10}^3s_{02}$	2	·	·	·	·	·	2	·	·	·	4	·	2	·	1	1	1	·	·
$s_{20}^2s_{11}s_{01}^2$	·	2	·	·	·	·	2	·	·	2	·	1	·	1	1	·	1	2	·
$s_{20}s_{12}s_{10}^2s_{01}$	·	·	1	·	·	·	1	·	·	·	2	·	1	·	1	·	2	2	1
$s_{20}s_{11}^2s_{10}s_{01}$	·	·	·	2	·	·	·	1	·	·	2	·	·	1	1	2	·	2	2
$s_{20}s_{11}s_{10}^2s_{02}$	·	·	·	·	3	1	·	·	·	·	·	·	·	·	·	3	3	·	·
$s_{13}s_{10}^4$	·	·	·	·	·	4	·	3	·	·	·	·	·	·	·	·	6	·	·
$s_{12}s_{11}s_{10}^3$	·	·	·	·	·	1	·	3	·	·	·	3	·	·	·	·	·	3	·
$s_{11}^3s_{10}^2$	·	·	·	·	·	·	·	3	·	·	·	·	6	·	·	·	·	·	1
$s_{30}s_{10}^2s_{01}^3$	3	6	6	6	1	·	6	6	6	6	3	3	6	·	·	·	·	·	·
$s_{21}s_{10}^3s_{01}^2$	·	·	·	·	·	1	6	6	6	12	6	3	3	·	3	3	1	1	6
$s_{20}^2s_{10}s_{01}^3$	6	4	4	2	·	·	6	·	6	12	6	·	·	3	3	1	·	2	2
$s_{20}s_{11}s_{10}^2s_{01}^2$	·	4	4	2	·	4	4	1	4	8	2	2	4	2	1	1	1	6	2
$s_{20}s_{10}^3s_{02}s_{01}$	4	6	·	6	3	1	6	3	·	·	6	·	3	3	3	3	·	6	3
$s_{12}s_{10}^4s_{01}$	·	4	·	·	·	4	12	12	·	·	·	6	6	3	·	·	6	12	·
$s_{11}^2s_{10}^3s_{01}$	·	·	1	·	·	1	3	9	6	6	·	6	15	·	·	·	6	6	3
$s_{11}s_{10}^4s_{02}$	·	·	·	4	·	4	·	6	·	·	12	12	·	3	·	6	12	·	·
$s_{10}^5s_{03}$	·	·	·	·	10	10	·	·	·	·	·	·	·	·	15	30	·	·	·
$s_{20}s_{10}^3s_{01}^3$	12	18	18	18	3	1	18	9	18	36	18	9	18	9	9	3	3	18	6
$s_{11}s_{10}^4s_{01}^2$	·	8	8	4	·	4	24	30	24	36	12	24	48	6	3	3	3	36	12
$s_{10}^5s_{02}s_{01}$	10	20	·	20	10	10	30	30	·	·	42	30	·	15	15	15	30	48	·
$s_{10}^5s_{01}^3$	30	60	60	60	10	10	90	90	90	180	90	90	180	45	45	15	30	180	60

Table 1·6·8 (*cont.*)

$w = 8$
(53) (iv)

	$[40, 10, (01)^3]$	$[31, (10)^2, (01)^2]$	$[30, 20, (01)^3]$	$[30, 11, 10, (01)^2]$	$[30, (10)^2, 02, 01]$	$[22, (10)^3, 01]$	$[21, 20, 10, (01)^2]$	$[21, 11, (10)^2, 01]$	$[21, (10)^3, 02]$	$[(20)^2, 11, (01)^2]$	$[(20)^2, 10, 02, 01]$	$[20, 12, (10)^2, 01]$	$[20, (11)^2, 10, 01]$	$[20, 11, (10)^2, 02]$	$[20, (10)^3, 03]$	$[13, (10)^4]$	$[12, 11, (10)^3]$	$[(11)^3, (10)^2]$
s_{53}	24	24	24	24	24	24	24	24	24	24	24	24	24	24	24	24	24	24
$s_{52}s_{01}$	−18	−12	−18	−12	−6	−6	−12	−6	·	−12	−6	−6	−6	·	·	·	·	·
$s_{51}s_{02}$	−6	−2	−3	−2	−5	·	−2	·	−6	−2	−4	·	·	−6	·	·	·	·
$s_{50}s_{03}$	−2	·	−2	·	−2	·	·	·	·	·	−2	·	·	·	−6	·	·	·
$s_{43}s_{10}$	−6	−12	·	−6	−12	−18	−6	−12	−18	·	−6	−12	−6	−12	−18	−24	−18	−12
$s_{42}s_{11}$	−6	−8	·	−10	−4	−6	−4	−10	·	−6	−2	−4	−14	−6	·	·	−6	−21
$s_{41}s_{12}$	−6	−4	·	−6	−4	·	−2	−2	−6	·	−4	−2	−6	−4	·	·	−6	·
$s_{40}s_{13}$	−6	·	·	−2	−4	·	·	·	·	−2	−2	−2	·	−2	−6	−6	·	·
$s_{33}s_{20}$	·	−2	−6	·	−2	−6	−6	−2	−6	−12	−12	−8	−6	−8	−12	−12	−6	−2
$s_{32}s_{21}$	·	−4	−9	−2	−3	−6	−10	−12	−6	−8	−6	−4	−6	−4	·	·	−6	−9
$s_{31}s_{22}$	·	−6	−6	−4	−2	−6	−6	−6	−6	−4	−4	−4	−6	−4	·	·	−6	−6
$s_{30}s_{23}$	·	·	−6	−6	−6	−2	−2	·	−2	·	−4	−4	−2	−4	−8	−8	−2	·
$s_{51}s_{01}^2$	6	2	6	2	·	·	2	·	·	2	·	·	·	·	·	·	·	·
$s_{50}s_{02}s_{01}$	3	·	3	·	·	·	·	·	·	·	2	·	·	·	·	·	·	·
$s_{42}s_{10}s_{01}$	6	8	·	4	4	6	4	4	·	·	2	4	2	·	·	·	·	·
$s_{41}s_{11}s_{01}$	6	4	·	6	·	·	2	2	·	4	·	·	4	·	·	·	·	·
$s_{41}s_{10}s_{02}$	3	2	·	1	4	·	1	·	6	·	2	·	·	4	·	·	·	·
$s_{40}s_{12}s_{01}$	6	·	·	2	2	·	·	·	·	2	1	2	·	·	·	·	·	·
$s_{40}s_{11}s_{02}$	3	·	·	1	2	·	·	·	·	1	1	·	·	2	·	·	·	·
$s_{40}s_{10}s_{03}$	2	·	·	·	2	·	·	·	·	1	1	·	·	·	6	·	·	·
$s_{33}s_{10}^2$	·	2	·	·	2	6	·	2	6	·	·	2	·	2	6	12	6	2
$s_{32}s_{20}s_{01}$	·	2	6	·	1	3	4	1	·	8	4	3	2	·	2	·	·	·
$s_{32}s_{11}s_{10}$	·	4	6	2	2	6	·	6	·	·	·	·	2	4	4	·	6	12
$s_{31}s_{21}s_{01}$	·	4	6	2	·	·	6	4	·	4	·	·	2	·	·	·	·	·
$s_{31}s_{20}s_{02}$	·	1	3	·	1	·	1	·	3	2	4	·	3	·	·	·	·	·
$s_{31}s_{12}s_{10}$	·	4	·	2	2	·	·	2	6	·	·	4	2	2	·	·	6	·
$s_{31}s_{11}^2$	·	2	·	2	·	·	·	2	·	·	·	·	4	·	·	·	·	6
$s_{30}s_{22}s_{01}$	·	·	6	4	2	2	2	·	·	2	2	1	·	·	·	·	·	·
$s_{30}s_{21}s_{02}$	·	·	3	1	2	·	1	·	2	2	·	·	2	·	·	·	·	·
$s_{30}s_{20}s_{03}$	·	·	2	·	1	·	·	·	·	2	·	·	·	·	5	·	·	·
$s_{30}s_{13}s_{10}$	·	·	·	2	4	·	·	·	·	·	·	2	·	2	6	8	·	·
$s_{30}s_{12}s_{11}$	·	·	·	4	2	·	·	·	3	·	·	2	2	2	·	·	2	·
$s_{23}s_{20}s_{10}$	·	·	·	·	·	3	2	·	3	·	4	4	2	4	9	12	3	·
$s_{22}s_{21}s_{10}$	·	·	·	·	·	6	4	6	6	·	2	2	1	2	·	·	6	6
$s_{22}s_{20}s_{11}$	·	·	·	·	·	3	2	1	·	4	2	2	5	3	·	·	3	3
$s_{21}^2s_{11}$	·	·	·	·	·	·	2	4	·	2	·	·	2	·	·	·	·	6
$s_{21}s_{20}s_{12}$	·	·	·	·	·	·	2	1	3	4	2	3	2	2	·	·	3	·
$s_{20}^2s_{13}$	·	·	·	·	·	·	·	·	·	2	2	1	·	1	3	3	·	·
$s_{50}s_{01}^3$	−1	·	−1	·	·	·	·	·	·	·	·	·	·	·	·	·	·	·
$s_{41}s_{10}s_{01}^2$	−3	−2	·	−1	·	·	−1	·	·	·	·	·	·	·	·	·	·	·
$s_{40}s_{11}s_{01}^2$	−3	·	·	−1	·	·	·	·	·	−1	·	·	·	·	·	·	·	·
$s_{40}s_{10}s_{02}s_{01}$	−3	·	·	·	−2	·	·	·	·	·	−1	·	−1	·	·	·	·	·
$s_{32}s_{10}^2s_{01}$	·	−2	−3	·	−1	−3	−1	−1	·	−2	·	−1	·	·	·	·	·	·
$s_{31}s_{20}s_{01}^2$	·	−1	−3	·	·	·	−1	·	·	−2	·	·	·	·	·	·	·	·
$s_{31}s_{11}s_{10}s_{01}$	·	−4	·	−2	·	·	·	−2	·	·	·	·	−2	·	·	·	·	·
$s_{31}s_{10}^2s_{02}$	·	−1	·	·	−1	·	·	·	−3	·	·	·	·	−1	·	·	·	·
$s_{30}s_{21}s_{01}^2$	·	·	−3	−1	·	·	−1	·	·	·	−2	·	·	·	·	·	·	·
$s_{30}s_{20}s_{02}s_{01}$	·	·	−3	−1	·	·	·	·	·	·	·	·	−2	·	·	·	·	·
$s_{30}s_{12}s_{10}s_{01}$	·	·	·	−2	−2	·	·	·	·	·	·	−2	·	·	·	·	·	·
$s_{30}s_{11}^2s_{01}$	·	·	·	−2	·	·	·	·	·	·	·	−1	·	·	·	·	·	·
$s_{30}s_{11}s_{10}s_{02}$	·	·	·	−1	−2	·	·	·	·	·	·	·	−2	·	·	·	·	·
$s_{30}s_{10}^2s_{03}$	·	·	·	·	−1	·	·	·	·	·	·	·	·	−3	·	·	·	·
$s_{23}s_{10}^3$	·	·	·	·	·	−1	·	·	−1	·	·	·	·	−1	−4	−1	·	·
$s_{22}s_{20}s_{10}s_{01}$	·	·	·	·	·	−3	−2	·	·	·	−2	−2	−1	·	·	·	·	·
$s_{22}s_{11}s_{10}^2$	·	·	·	·	·	−3	·	−1	·	·	·	·	−1	·	·	−3	−3	·
$s_{21}^2s_{10}s_{01}$	·	·	·	·	·	·	−2	−2	·	·	·	·	·	·	·	·	·	·
$s_{21}s_{20}s_{11}s_{01}$	·	·	·	·	·	·	−2	−1	·	−4	·	·	−2	·	·	·	·	·
$s_{21}s_{20}s_{10}s_{02}$	·	·	·	·	·	·	−1	·	−3	·	−2	·	−2	·	·	·	·	·
$s_{21}s_{12}s_{10}^2$	·	·	·	·	·	·	·	−1	−3	·	·	−1	·	·	·	·	−3	·
$s_{21}s_{11}^2s_{10}$	·	·	·	·	·	·	·	−2	·	·	·	·	−1	·	·	·	·	−6
$s_{20}^2s_{12}s_{01}$	·	·	·	·	·	·	·	·	·	−2	−1	−1	·	·	·	·	·	·
$s_{20}^2s_{11}s_{02}$	·	·	·	·	·	·	·	·	·	−1	−1	·	−1	·	·	·	·	·
$s_{20}^2s_{10}s_{03}$	·	·	·	·	·	·	·	·	·	−1	·	·	·	−3	·	·	·	·
$s_{20}s_{13}s_{10}^2$	·	·	·	·	·	·	·	·	·	·	−1	·	−1	−3	−6	·	·	·
$s_{20}s_{12}s_{11}s_{10}$	·	·	·	·	·	·	·	·	·	·	−2	·	−2	·	·	·	−3	·
$s_{20}s_{11}^3$	·	·	·	·	·	·	·	·	·	·	·	·	−1	·	·	·	·	−1
$s_{40}s_{10}s_{01}^3$	1	·	·	·	·	·	·	·	·	·	·	·	·	·	·	·	·	·
$s_{31}s_{10}^2s_{01}^2$	·	1	·	·	·	·	·	·	·	·	·	·	·	·	·	·	·	·
$s_{30}s_{20}s_{01}^3$	·	·	1	·	·	·	·	·	·	·	·	·	·	·	·	·	·	·
$s_{30}s_{11}s_{10}s_{01}^2$	·	·	·	1	·	·	·	·	·	·	·	·	·	·	·	·	·	·
$s_{30}s_{10}^2s_{02}s_{01}$	·	·	·	·	1	·	·	·	·	·	·	·	·	·	·	·	·	·
$s_{22}s_{10}^3s_{01}$	·	·	·	·	·	1	·	·	·	·	·	·	·	·	·	·	·	·
$s_{21}s_{20}s_{10}s_{01}^2$	·	·	·	·	·	·	1	·	·	·	·	·	·	·	·	·	·	·
$s_{21}s_{11}s_{10}^2s_{01}$	·	·	·	·	·	·	·	1	·	·	·	·	·	·	·	·	·	·
$s_{21}s_{10}^3s_{02}$	·	·	·	·	·	·	·	·	1	·	·	·	·	·	·	·	·	·
$s_{20}^2s_{11}s_{01}^2$	·	·	·	·	·	·	·	·	·	1	·	·	·	·	·	·	·	·
$s_{20}^2s_{10}s_{02}s_{01}$	·	·	·	·	·	·	·	·	·	·	1	·	·	·	·	·	·	·
$s_{20}s_{12}s_{10}^2s_{01}$	·	·	·	·	·	·	·	·	·	·	·	1	·	·	·	·	·	·
$s_{20}s_{11}^2s_{10}s_{01}$	·	·	·	·	·	·	·	·	·	·	·	·	1	·	·	·	·	·
$s_{20}s_{11}s_{10}^2s_{02}$	·	·	·	·	·	·	·	·	·	·	·	·	·	1	·	·	·	·
$s_{20}s_{10}^3s_{03}$	·	·	·	·	·	·	·	·	·	·	·	·	·	·	1	·	·	·
$s_{13}s_{10}^4$	·	·	·	·	·	·	·	·	·	·	·	·	·	·	·	1	·	·
$s_{12}s_{11}s_{10}^3$	·	·	·	·	·	·	·	·	·	·	·	·	·	·	·	·	1	·
$s_{11}^3s_{10}^2$	·	·	·	·	·	·	·	·	·	·	·	·	·	·	·	·	·	1
$s_{20}s_{10}^2s_{01}^3$	2	3	1	6	3	·	2	3	6	1	·	·	·	·	·	·	·	·
$s_{21}s_{10}^3s_{01}$	·	3	·	·	·	2	3	6	·	·	·	·	·	·	·	·	·	·
$s_{20}^2s_{10}s_{01}^3$	1	·	2	·	·	·	6	·	·	3	3	·	·	·	·	·	·	·
$s_{20}s_{11}s_{10}^2s_{01}^2$	·	1	·	2	·	·	2	2	·	1	·	3	2	4	1	·	·	·
$s_{20}s_{10}^3s_{02}s_{01}$	·	·	·	·	3	1	·	·	·	1	·	3	3	3	1	·	·	·
$s_{12}s_{10}^4s_{01}$	·	·	·	·	·	4	·	·	·	·	·	·	6	·	·	1	4	·
$s_{11}^2s_{10}^3s_{01}$	·	·	·	·	·	1	·	6	·	·	·	·	·	3	·	·	2	3
$s_{11}s_{10}^4s_{02}$	·	·	·	·	·	·	·	·	4	·	·	·	·	·	6	1	4	·
$s_{10}^5s_{03}$	·	·	·	·	·	·	·	·	·	·	·	·	·	·	·	10	5	·
$s_{20}s_{10}^3s_{01}^3$	3	9	4	18	9	3	18	18	3	9	9	9	18	9	1	·	·	·
$s_{11}s_{10}^4s_{01}^2$	·	6	·	4	·	8	12	36	4	3	·	12	24	6	·	1	12	12
$s_{10}^5s_{02}s_{01}$	·	·	·	·	10	10	·	·	10	·	15	30	·	30	10	5	20	·
$s_{10}^5s_{01}^3$	5	30	10	60	30	30	90	180	30	45	45	90	180	90	10	5	60	60

Table 1·6·8 (cont.)

$w=8$ (53)(v)	$[30,(10)^2,(01)^3]$	$[21,(10)^3,(01)^2]$	$[(20)^2,10(01)^3]$	$[20,11,(10)^2,(01)^2]$	$[20,(10)^3,02,01]$	$[12,(10)^4,01]$	$[(11)^2,(10)^3,01]$	$[11,(10)^4,02]$	$[(10)^5,03]$	$[20,(10)^3,(01)^3]$	$[11,(10)^4,(01)^2]$	$[(10)^5,02,01]$	$[(10)^6,(01)^3]$
s_{53}	−120	−120	−120	−120	−120	−120	−120	−120	−120	720	720	720	−5040
$s_{52}s_{01}$	72	48	72	48	24	24	24	.	.	−360	−240	−120	2160
$s_{51}s_{02}$	15	6	12	6	19	.	.	24	.	−57	−24	−124	372
$s_{50}s_{03}$	4	.	4	.	6	.	.	.	24	−12	.	−24	48
$s_{43}s_{10}$	48	72	24	48	72	96	72	96	120	−360	−480	−600	3600
$s_{42}s_{11}$	36	36	18	48	18	24	75	24	.	−216	−348	−96	1908
$s_{41}s_{12}$	24	12	12	20	18	24	12	24	.	−108	−96	−108	684
$s_{40}s_{13}$	12	.	6	4	12	6	.	6	30	−36	−12	−60	180
$s_{33}s_{20}$	6	18	48	30	42	36	18	36	60	−192	−144	−240	1200
$s_{32}s_{21}$	15	48	42	32	23	24	39	24	.	−195	−204	−152	1356
$s_{31}s_{22}$	18	30	24	26	18	24	30	24	.	−144	−168	−120	1080
$s_{30}s_{23}$	24	4	12	12	22	16	4	16	40	−84	−48	−120	480
$s_{51}s_{01}^{2}$	−18	−6	−18	−6	.	.	.	.	.	72	24	.	−360
$s_{50}s_{02}s_{01}$	−6	.	−6	.	−6	.	.	.	.	18	.	24	−72
$s_{42}s_{10}s_{01}$	−36	−36	−18	−24	−18	−24	−18	.	.	216	192	120	−1800
$s_{41}s_{11}s_{01}$	−24	−12	−12	−20	.	.	−12	.	.	108	96	.	−720
$s_{41}s_{10}s_{02}$	−12	−6	−6	−4	−18	.	.	−24	.	54	24	132	−396
$s_{40}s_{12}s_{01}$	−12	.	−6	−4	−6	−6	.	.	.	36	12	30	−180
$s_{40}s_{11}s_{02}$	−6	.	−3	−2	−6	.	.	−6	.	18	6	30	−90
$s_{40}s_{10}s_{03}$	−4	.	−2	.	−6	.	.	.	−30	12	.	30	−60
$s_{33}s_{10}^{2}$	−6	−18	.	−6	−18	−36	−18	−36	−60	72	144	240	−1200
$s_{32}s_{20}s_{01}$	−6	−12	−36	−16	−12	−12	−6	.	.	126	72	60	−720
$s_{32}s_{11}s_{10}$	−12	−24	.	−20	−12	−24	−48	−24	.	108	240	108	−1404
$s_{31}s_{21}s_{01}$	−12	−24	−24	−16	.	.	−12	.	.	108	96	.	−720
$s_{31}s_{20}s_{02}$	−3	−3	−12	−3	−3	.	.	−12	.	36	12	72	−216
$s_{31}s_{12}s_{10}$	−12	−12	.	−12	−12	−24	−12	−24	.	72	96	108	−684
$s_{31}s_{11}^{2}$	−6	−6	.	−10	.	.	−18	.	.	36	72	.	−360
$s_{30}s_{22}s_{01}$	−18	−4	−12	−8	−8	−8	−2	.	.	66	32	40	−360
$s_{30}s_{21}s_{02}$	−6	−2	−6	−2	−8	.	.	−8	.	24	8	52	−156
$s_{30}s_{20}s_{03}$	−2	.	−4	.	−5	.	.	.	−20	10	.	20	−40
$s_{30}s_{13}s_{10}$	−12	.	.	−4	−12	−8	.	−8	−40	36	16	80	−240
$s_{30}s_{12}s_{11}$	−12	.	.	−8	−6	−8	−4	−8	.	36	32	28	−204
$s_{22}s_{20}s_{10}$	.	−6	−12	−12	−24	−24	−6	−24	−60	90	72	180	−720
$s_{22}s_{21}s_{10}$	.	−30	−12	−12	−12	−24	−30	−24	.	90	168	132	−1116
$s_{22}s_{20}s_{11}$	.	−6	−12	−16	−9	−12	−15	−12	.	72	84	48	−604
$s_{21}^{2}s_{11}$	.	−12	−6	−8	.	.	−18	.	.	36	72	.	−360
$s_{21}s_{20}s_{12}$	.	−6	−12	−10	−9	−12	−6	−12	.	54	48	60	−360
$s_{20}^{2}s_{13}$	.	.	−6	−2	−6	−3	.	−3	−15	18	6	30	−90
$s_{50}s_{01}^{3}$	2	.	2	.	.	.	.	.	.	−6	.	.	24
$s_{41}s_{10}s_{01}^{2}$	12	6	6	4	.	.	.	.	.	−54	−24	.	360
$s_{40}s_{11}s_{01}^{2}$	6	.	3	2	.	.	.	.	.	−18	−6	.	90
$s_{40}s_{10}s_{02}s_{01}$	6	.	3	.	6	.	.	.	.	−18	.	−30	90
$s_{32}s_{10}^{2}s_{01}$	6	12	.	4	6	12	6	.	.	−54	−72	−60	720
$s_{31}s_{20}s_{01}^{2}$	3	3	12	3	.	.	.	.	.	−36	−12	.	180
$s_{31}s_{11}s_{10}s_{01}$	12	12	.	12	.	.	12	.	.	−72	−96	.	120
$s_{31}s_{10}^{2}s_{02}$	3	3	.	1	6	.	.	12	.	−18	−12	−60	180
$s_{30}s_{21}s_{01}^{2}$	6	2	6	2	.	.	.	.	.	−18	−12	.	720
$s_{30}s_{20}s_{02}s_{01}$	3	.	6	.	5	.	.	.	.	−15	.	−20	60
$s_{30}s_{12}s_{10}s_{01}$	12	.	.	4	6	8	.	.	.	−36	−16	−40	240
$s_{30}s_{11}^{2}s_{01}$	6	.	.	4	6	.	2	.	.	−18	−16	.	120
$s_{30}s_{11}s_{10}s_{02}$	6	.	.	2	6	.	.	8	.	−18	−8	−40	120
$s_{30}s_{10}^{2}s_{03}$	2	.	.	.	3	.	.	.	20	−6	.	−20	40
$s_{22}s_{10}^{3}$	.	2	.	.	2	8	2	8	20	−6	−24	−60	240
$s_{22}s_{20}s_{10}s_{01}$	.	6	12	8	9	12	3	.	.	−72	−48	−60	540
$s_{22}s_{11}s_{10}^{2}$	.	6	.	2	3	12	15	12	.	−18	−84	−60	540
$s_{21}^{2}s_{10}s_{01}$	.	12	6	4	.	.	6	.	.	−36	−48	.	360
$s_{21}s_{20}s_{11}s_{01}$	.	6	12	10	.	.	6	.	.	−54	−48	.	360
$s_{21}s_{20}s_{10}s_{02}$	.	3	6	2	9	.	.	12	.	−27	−12	−72	216
$s_{21}s_{12}s_{10}^{2}$	.	6	.	2	3	12	6	12	.	−18	−48	−60	360
$s_{21}s_{11}^{2}s_{10}$	.	6	.	.	4	.	18	.	.	−18	−72	.	360
$s_{20}^{2}s_{12}s_{01}$	.	.	6	2	3	3	.	.	.	−18	−6	−15	90
$s_{20}^{2}s_{11}s_{02}$	.	.	3	1	3	.	.	3	.	−9	−3	−15	45
$s_{20}^{2}s_{10}s_{03}$	.	.	2	.	3	.	.	.	15	−6	.	−15	30
$s_{20}s_{13}s_{10}^{2}$	.	.	.	2	6	6	.	6	30	−18	−12	−60	180
$s_{20}s_{12}s_{11}s_{10}$	.	.	.	8	6	12	6	12	.	−36	−48	−48	324
$s_{20}s_{11}^{3}$	.	.	.	2	.	.	3	.	.	−6	−12	.	60
$s_{40}s_{10}s_{01}^{3}$	−2	.	−1	.	.	.	.	.	.	6	.	.	−30
$s_{31}s_{10}^{2}s_{01}^{2}$	−3	−3	.	−1	.	.	.	.	.	18	12	.	−180
$s_{30}s_{20}s_{01}^{3}$	−1	.	−2	.	.	.	.	.	.	5	.	.	−20
$s_{30}s_{11}s_{10}s_{01}^{2}$	−6	.	.	−2	.	.	.	.	.	18	8	.	−120
$s_{30}s_{10}^{2}s_{02}s_{01}$	−3	.	.	.	−3	.	.	.	.	9	.	20	−60
$s_{22}s_{10}^{3}s_{01}$	.	−2	.	.	−1	−4	−1	.	.	6	16	20	−180
$s_{21}s_{20}s_{10}s_{01}^{2}$	.	−3	−6	−2	.	.	.	.	.	27	12	.	−180
$s_{21}s_{11}s_{10}^{2}s_{01}$	.	−6	.	−2	.	.	−6	.	.	18	48	.	−360
$s_{21}s_{10}^{3}s_{02}$	.	−1	.	.	−1	.	.	−4	.	3	4	20	−60
$s_{20}^{2}s_{11}s_{01}^{2}$	.	.	−3	−1	.	.	.	.	.	9	3	.	−45
$s_{20}^{2}s_{10}s_{02}s_{01}$	.	.	−3	.	−3	.	.	.	.	9	.	15	−45
$s_{20}s_{12}s_{10}^{2}s_{01}$	.	.	.	−2	−3	−6	.	.	.	18	12	30	−180
$s_{20}s_{11}^{2}s_{10}s_{01}$	.	.	.	−4	.	.	−3	.	.	18	24	.	−180
$s_{20}s_{11}s_{10}^{2}s_{02}$	.	.	.	−1	−3	.	.	−6	.	9	6	30	−90
$s_{20}s_{10}^{3}s_{03}$	.	.	.	.	−1	.	.	.	−10	2	.	10	−20
$s_{13}s_{10}^{4}$	.	.	.	.	.	−1	.	−1	−5	.	2	10	−30
$s_{12}s_{11}s_{10}^{3}$	.	.	.	.	.	−4	−2	−4	.	.	16	20	−120
$s_{11}^{3}s_{10}^{2}$	.	.	.	.	.	.	−3	.	.	.	12	.	−60
$s_{30}s_{10}^{2}s_{01}^{3}$	1	.	.	.	.	.	.	.	.	−3	.	.	20
$s_{21}s_{10}^{3}s_{01}^{2}$	.	1	.	.	.	.	.	.	.	−3	−4	.	60
$s_{20}^{2}s_{10}s_{01}^{3}$	.	.	1	.	.	.	.	.	.	−3	.	.	15
$s_{20}s_{11}s_{10}^{2}s_{01}^{2}$	.	.	.	1	.	.	.	.	.	−9	−6	.	90
$s_{20}s_{10}^{3}s_{02}s_{01}$	.	.	.	.	1	.	.	.	.	−3	.	−10	30
$s_{12}s_{10}^{4}s_{01}$	.	.	.	.	.	1	.	.	.	.	−2	−5	30
$s_{11}^{2}s_{10}^{3}s_{01}$	.	.	.	.	.	.	1	.	.	.	−8	.	60
$s_{11}s_{10}^{4}s_{02}$	.	.	.	.	.	.	.	1	.	.	−1	−5	15
$s_{10}^{5}s_{03}$	.	.	.	.	.	.	.	.	1	.	.	−1	2
$s_{20}s_{10}^{2}s_{01}^{3}$	3	3	3	9	3	.	.	.	.	1	.	.	−10
$s_{11}s_{10}^{4}s_{01}^{2}$	.	4	.	6	.	2	(8)	1	.	.	1	.	−15
$s_{10}^{5}s_{02}s_{01}$	.	.	.	.	10	5	.	5	1	.	.	1	−3
$s_{10}^{5}s_{01}^{3}$	10	30	15	90	30	15	60	15	1	10	15	3	1

Table 1·6·8 (*cont.*)

$w = 8$ (44) (i)	[44]	[43, 01]	[42, 02]	[41, 03]	[40, 04]	[34, 10]	[33, 11]	[32, 12]	[31, 13]	[30, 14]	[24, 20]	[23, 21]	[(22)²]	[42, (01)²]	[41, 02, 01]	[40, 03, 01]	[40, (02)²]	[33, 10, 01]	[32, 11, 01]	[32, 10, 02]	[31, 12, 01]
s_{44}	1	−1	−1	−1	−1	−1	−1	−1	−1	−1	−1	−1	−1	2	2	2	2	2	2	2	2
$s_{43}s_{01}$	1	1	.	.	.	.	.	.	.	.	.	.	.	−2	−1	−1	.	−1	−1	.	−1
$s_{42}s_{02}$	1	.	1	.	.	.	.	.	.	.	.	.	.	−1	−1	.	−2	.	.	−1	.
$s_{41}s_{03}$	1	.	.	1	.	.	.	.	.	.	.	.	.	.	−1	−1	.	.	.	.	.
$s_{40}s_{04}$	1	.	.	.	1	.	.	.	.	.	.	.	.	.	.	−1	−1	.	.	.	.
$s_{34}s_{10}$	1	.	.	.	.	1	.	.	.	.	.	.	.	.	.	.	.	−1	.	−1	.
$s_{33}s_{11}$	1	.	.	.	.	.	1	.	.	.	.	.	.	.	.	.	.	−1	−1	.	.
$s_{32}s_{12}$	1	.	.	.	.	.	.	1	.	.	.	.	.	.	.	.	.	.	−1	−1	−1
$s_{31}s_{13}$	1	.	.	.	.	.	.	.	1	.	.	.	.	.	.	.	.	.	.	.	−1
$s_{30}s_{14}$	1	.	.	.	.	.	.	.	.	1	.	.	.	.	.	.	.	.	.	.	.
$s_{24}s_{20}$	1	.	.	.	.	.	.	.	.	.	1	.	.	.	.	.	.	.	.	.	.
$s_{23}s_{21}$	1	.	.	.	.	.	.	.	.	.	.	1	.	.	.	.	.	.	.	.	.
s_{22}^2	1	.	.	.	.	.	.	.	.	.	.	.	1	.	.	.	.	.	.	.	.
$s_{42}s_{01}^2$	1	2	1	.	.	.	.	.	.	.	.	.	.	1	.	.	.	.	.	.	.
$s_{41}s_{02}s_{01}$	1	1	1	1	.	.	.	.	.	.	.	.	.	.	1	.	.	.	.	.	.
$s_{40}s_{03}s_{01}$	1	1	.	1	1	.	.	.	.	.	.	.	.	.	.	1	.	.	.	.	.
$s_{40}s_{02}^2$	1	.	2	.	1	.	.	.	.	.	.	.	.	.	.	.	1	.	.	.	.
$s_{33}s_{10}s_{01}$	1	1	.	.	.	1	1	.	.	.	.	.	.	.	.	.	.	1	.	.	.
$s_{32}s_{11}s_{01}$	1	1	.	.	.	.	1	1	.	.	.	.	.	.	.	.	.	.	1	.	.
$s_{32}s_{10}s_{02}$	1	.	1	.	.	1	.	1	.	.	.	.	.	.	.	.	.	.	.	1	.
$s_{31}s_{12}s_{01}$	1	1	.	.	.	.	.	1	1	.	.	.	.	.	.	.	.	.	.	.	1
$s_{31}s_{11}s_{02}$	1	.	1	.	.	.	1	.	1	.	.	.	.	.	.	.	.	.	.	.	.
$s_{31}s_{10}s_{03}$	1	.	.	1	.	1	.	.	1	.	.	.	.	.	.	.	.	.	.	.	.
$s_{30}s_{13}s_{01}$	1	1	.	.	.	.	.	.	1	1	.	.	.	.	.	.	.	.	.	.	.
$s_{30}s_{12}s_{02}$	1	.	1	.	.	.	.	1	.	1	.	.	.	.	.	.	.	.	.	.	.
$s_{30}s_{11}s_{03}$	1	.	.	1	.	.	1	.	.	1	.	.	.	.	.	.	.	.	.	.	.
$s_{30}s_{10}s_{04}$	1	.	.	.	1	1	.	.	.	1	.	.	.	.	.	.	.	.	.	.	.
$s_{24}s_{10}^2$	1	.	.	.	.	2	.	.	.	.	1	.	.	.	.	.	.	.	.	.	.
$s_{23}s_{20}s_{01}$	1	1	.	.	.	.	.	.	.	.	1	1	.	.	.	.	.	.	.	.	.
$s_{23}s_{11}s_{10}$	1	.	.	.	.	1	1	.	.	.	.	1	.	.	.	.	.	.	.	.	.
$s_{22}s_{21}s_{01}$	1	1	.	.	.	.	.	.	.	.	.	1	1	.	.	.	.	.	.	.	.
$s_{22}s_{20}s_{02}$	1	.	1	.	.	.	.	.	.	.	1	.	1	.	.	.	.	.	.	.	.
$s_{22}s_{12}s_{10}$	1	.	.	.	.	1	.	1	.	.	.	.	1	.	.	.	.	.	.	.	.
$s_{22}s_{11}^2$	1	.	.	.	.	.	2	.	.	.	.	.	1	.	.	.	.	.	.	.	.
$s_{21}^2s_{02}$	1	.	1	.	.	.	.	.	.	.	.	2	.	.	.	.	.	.	.	.	.
$s_{21}s_{20}s_{03}$	1	.	.	1	.	.	.	.	.	.	1	1	.	.	.	.	.	.	.	.	.
$s_{21}s_{13}s_{10}$	1	.	.	.	.	1	.	.	1	.	.	1	.	.	.	.	.	.	.	.	.
$s_{21}s_{12}s_{11}$	1	.	.	.	.	.	1	1	.	.	.	1	.	.	.	.	.	.	.	.	.
$s_{20}^2s_{04}$	1	.	.	.	1	.	.	.	.	.	2	.	.	.	.	.	.	.	.	.	.
$s_{20}s_{14}s_{10}$	1	.	.	.	.	1	.	.	.	1	1	.	.	.	.	.	.	.	.	.	.
$s_{20}s_{13}s_{11}$	1	.	.	.	.	.	1	.	1	.	1	.	.	.	.	.	.	.	.	.	.
$s_{20}s_{12}^2$	1	.	.	.	.	.	.	2	.	.	1	.	.	.	.	.	.	.	.	.	.
$s_{41}s_{01}^3$	1	3	3	1	.	.	.	.	.	.	.	.	.	3	3	.	.	.	.	.	.
$s_{40}s_{02}s_{01}^2$	1	2	2	2	1	.	.	.	.	.	.	.	.	1	2	2	1	.	.	.	.
$s_{32}s_{10}s_{01}^2$	1	2	1	.	.	1	2	1	.	.	.	.	.	1	.	.	.	2	2	1	.
$s_{31}s_{11}s_{01}^2$	1	2	1	.	.	.	1	2	1	.	.	.	.	1	.	.	.	.	2	.	2
$s_{31}s_{10}s_{02}s_{01}$	1	1	1	1	.	1	1	1	1	.	.	.	.	.	1	.	.	1	.	1	1
$s_{30}s_{12}s_{01}^2$	1	2	1	.	.	.	.	1	2	1	.	.	.	1	.	.	.	.	.	.	2
$s_{30}s_{11}s_{02}s_{01}$	1	1	1	1	.	.	1	1	1	1	.	.	.	.	1	.	.	.	1	.	.
$s_{30}s_{10}s_{03}s_{01}$	1	1	.	1	1	1	1	.	1	1	.	.	.	.	.	1	.	1	.	.	.
$s_{30}s_{10}s_{02}^2$	1	.	2	.	1	1	.	2	.	1	.	.	.	.	.	.	1	.	.	2	.
$s_{23}s_{10}^2s_{01}$	1	1	.	.	.	2	2	.	.	.	1	1	.	.	.	.	.	2	.	.	.
$s_{22}s_{20}s_{01}^2$	1	2	1	.	.	.	.	.	.	.	1	2	1	1	.	.	.	.	.	.	.
$s_{22}s_{11}s_{10}s_{01}$	1	1	.	.	.	1	2	1	.	.	.	1	1	.	.	.	.	1	1	.	.
$s_{22}s_{10}^2s_{02}$	1	.	1	.	.	2	.	2	.	.	1	.	1	.	.	.	.	.	.	2	.
$s_{21}^2s_{01}^2$	1	2	1	.	.	.	.	.	.	.	.	2	2	1	.	.	.	.	.	.	.
$s_{21}s_{20}s_{02}s_{01}$	1	1	1	1	.	.	.	.	.	.	1	2	1	.	1	.	.	.	.	.	.
$s_{21}s_{12}s_{10}s_{01}$	1	1	.	.	.	1	1	1	1	.	.	1	1	.	.	.	.	1	.	.	1
$s_{21}s_{11}^2s_{01}$	1	1	.	.	.	.	2	2	.	.	.	1	1	.	.	.	.	.	2	.	.
$s_{21}s_{11}s_{10}s_{02}$	1	.	1	.	.	1	1	1	1	.	.	2	.	.	.	.	.	.	.	1	.
$s_{21}s_{10}^2s_{03}$	1	.	.	1	.	2	.	.	2	.	1	1	.	.	.	.	.	.	.	.	.
$s_{20}^2s_{03}s_{01}$	1	1	.	1	1	.	.	.	.	.	2	2	.	.	.	1	.	.	.	.	.
$s_{20}^2s_{02}^2$	1	.	2	.	1	.	.	.	.	.	2	.	2	.	.	.	1	.	.	.	.
$s_{20}s_{13}s_{10}s_{01}$	1	1	.	.	.	1	1	.	1	1	1	1	.	.	.	.	.	1	.	.	.
$s_{20}s_{12}s_{11}s_{01}$	1	1	.	.	.	.	1	2	1	.	1	1	.	.	.	.	.	.	1	.	1
$s_{20}s_{12}s_{10}s_{02}$	1	.	1	.	.	1	.	2	.	1	1	.	1	.	.	.	.	.	.	1	.
$s_{20}s_{11}^2s_{02}$	1	.	1	.	.	.	2	.	2	.	1	.	1	.	.	.	.	.	.	.	.
$s_{20}s_{11}s_{10}s_{03}$	1	.	.	1	.	1	1	.	1	1	1	1	.	.	.	.	.	.	.	.	.
$s_{20}s_{10}^2s_{04}$	1	.	.	.	1	2	.	.	.	2	2	.	.	.	.	.	.	.	.	.	.
$s_{14}s_{10}^3$	1	.	.	.	.	3	.	.	.	1	3	.	.	.	.	.	.	.	.	.	.
$s_{13}s_{11}s_{10}^2$	1	.	.	.	.	2	1	.	1	.	1	2	.	.	.	.	.	.	.	.	.
$s_{12}^2s_{10}^2$	1	.	.	.	.	2	.	2	.	.	1	.	2	.	.	.	.	.	.	.	.
$s_{12}s_{11}^2s_{10}$	1	.	.	.	.	1	2	1	.	.	.	2	1	.	.	.	.	.	.	.	.
s_{11}^4	1	.	.	.	.	.	4	.	.	.	.	.	3	.	.	.	.	.	.	.	.
$s_{40}s_{01}^4$	1	4	6	4	1	.	.	.	.	.	.	.	.	6	12	4	3	.	.	.	.
$s_{31}s_{10}s_{01}^3$	1	3	3	1	.	1	3	3	1	.	.	.	.	3	3	.	.	3	6	3	3
$s_{30}s_{11}s_{01}^3$	1	3	3	1	.	.	1	3	3	1	.	.	.	3	3	.	.	.	3	.	6
$s_{30}s_{10}s_{02}s_{01}^2$	1	2	2	2	1	1	2	2	2	1	.	.	.	1	2	2	1	2	2	2	2
$s_{22}s_{10}^2s_{01}^2$	1	2	1	.	.	2	4	2	.	.	1	2	1	1	.	.	.	4	4	2	.
$s_{21}s_{20}s_{01}^3$	1	3	3	1	.	.	.	.	.	.	1	4	3	3	3	.	.	.	.	.	.
$s_{21}s_{11}s_{10}s_{01}^2$	1	2	1	.	.	1	3	3	1	.	.	2	2	1	.	.	.	2	4	1	2
$s_{21}s_{10}^2s_{02}s_{01}$	1	1	1	1	.	2	2	2	2	.	1	2	1	.	1	.	.	2	.	2	2
$s_{20}^2s_{02}s_{01}^2$	1	2	2	2	1	.	.	.	.	.	2	4	2	1	2	2	1	.	.	.	.
$s_{20}s_{12}s_{10}s_{01}^2$	1	2	1	.	.	1	2	2	2	1	1	2	1	1	.	.	.	2	2	1	2
$s_{20}s_{11}^2s_{01}^2$	1	2	1	.	.	.	2	4	2	.	1	2	1	1	.	.	.	.	4	.	4
$s_{20}s_{11}s_{10}s_{02}s_{01}$	1	1	1	1	.	1	2	2	2	1	1	2	1	.	1	.	.	1	1	1	1
$s_{20}s_{10}^2s_{03}s_{01}$	1	1	.	1	1	2	2	.	2	2	2	2	.	.	.	1	.	2	.	.	.
$s_{20}s_{10}^2s_{02}^2$	1	.	2	.	1	2	.	4	.	2	2	.	2	.	.	.	1	.	.	4	.
$s_{13}s_{10}^3s_{01}$	1	1	.	.	.	3	3	.	1	1	3	3	.	.	.	.	.	3	.	.	.
$s_{12}s_{11}s_{10}^2s_{01}$	1	1	.	.	.	2	3	2	1	.	1	3	2	.	.	.	.	2	1	.	1
$s_{12}s_{10}^3s_{02}$	1	.	1	.	.	3	.	4	.	1	3	.	3	.	.	.	.	.	.	3	.
$s_{11}^3s_{10}s_{01}$	1	1	.	.	.	1	4	3	.	.	.	3	3	.	.	.	.	1	3	.	.
$s_{11}^2s_{10}^2s_{02}$	1	.	1	.	.	2	2	2	2	.	1	4	1	.	.	.	.	.	.	2	.
$s_{11}s_{10}^3s_{03}$	1	.	.	1	.	3	1	.	3	1	3	3	.	.	.	.	.	.	.	.	.
$s_{10}^4s_{04}$	1	.	.	.	1	4	.	.	.	4	6	.	.	.	.	.	.	.	.	.	.
$s_{30}s_{10}s_{01}^4$	1	4	6	4	1	1	4	6	4	1	.	.	.	6	12	4	3	4	12	6	12
$s_{21}s_{10}^2s_{01}^3$	1	3	3	1	.	2	6	6	2	.	1	4	3	3	3	.	.	6	12	6	6
$s_{20}^2s_{01}^4$	1	4	6	4	1	.	.	.	.	.	2	8	6	6	12	4	3	.	.	.	.
$s_{20}s_{11}s_{10}s_{01}^3$	1	3	3	1	.	1	4	6	4	1	1	4	3	3	3	.	.	3	9	3	9
$s_{20}s_{10}^2s_{02}s_{01}^2$	1	2	2	2	1	2	4	4	4	2	2	4	2	1	2	2	1	4	4	4	4
$s_{12}s_{10}^3s_{01}^2$	1	2	1	.	.	3	6	4	2	1	3	6	3	1	.	.	.	6	6	3	2
$s_{11}^2s_{10}^2s_{01}^2$	1	2	1	.	.	2	6	6	2	.	1	6	5	1	.	.	.	4	8	2	4
$s_{11}s_{10}^3s_{02}s_{01}$	1	1	1	1	.	3	4	4	4	1	3	6	3	.	1	.	.	3	1	3	3
$s_{10}^4s_{03}s_{01}$	1	1	.	1	1	4	4	.	4	4	6	6	.	.	.	1	.	4	.	.	.
$s_{10}^4s_{02}^2$	1	.	2	.	1	4	.	8	.	4	6	.	6	.	.	.	1	.	.	8	.
$s_{20}s_{10}^2s_{01}^4$	1	4	6	4	1	2	8	12	8	2	2	8	6	6	12	4	3	8	24	12	24
$s_{11}s_{10}^3s_{01}^3$	1	3	3	1	.	3	10	12	6	1	3	12	9	3	3	.	.	9	21	9	15
$s_{10}^4s_{02}s_{01}^2$	1	2	2	2	1	4	8	8	8	4	6	12	6	1	2	2	1	8	8	8	8
$s_{10}^4s_{01}^4$	1	4	6	4	1	4	16	24	16	4	6	24	18	6	12	4	3	16	48	24	48

Table 1·6·8 (*cont.*)

$w = 8$ (44) (ii)	[31, 11, 02]	[31, 10, 03]	[30, 13, 01]	[30, 12, 02]	[30, 11, 03]	[30, 10, 04]	[24, $(10)^2$]	[23, 20, 01]	[23, 11, 10]	[22, 21, 01]	[22, 20, 02]	[22, 12, 10]	[22, $(11)^2$]	[$(21)^2$, 02]	[21, 20, 03]	[21, 13, 10]	[21, 12, 11]	[$(20)^2$, 04]	[20, 14, 10]	[20, 13, 11]	[20, $(12)^2$]
s_{44}	2	2	2	2	2	2	2	2	2	2	2	2	2	2	2	2	2	2	2	2	2
$s_{43}s_{01}$	.	.	−1	.	.	.	.	−1	.	−1	.	.	.	.	.	.	.	.	.	.	.
$s_{42}s_{02}$	−1	.	.	−1	.	.	.	.	.	.	−1	.	.	−1	.	.	.	.	.	.	.
$s_{41}s_{03}$	.	−1	.	.	−1	.	.	.	.	.	.	.	.	.	−1	.	.	.	.	.	.
$s_{40}s_{04}$	.	.	.	.	.	−1	.	.	.	.	.	.	.	.	.	.	.	−1	.	.	.
$s_{34}s_{10}$	.	−1	.	.	.	−1	−2	.	−1	.	.	−1	.	.	.	−1	.	.	−1	.	.
$s_{33}s_{11}$	−1	.	.	.	−1	.	.	.	−1	.	.	.	−2	.	.	.	−1	.	.	−1	.
$s_{32}s_{12}$	.	.	.	−1	.	.	.	.	.	.	.	−1	.	.	.	.	−1	.	.	.	−2
$s_{31}s_{13}$	−1	−1	−1	.	.	.	.	.	.	.	.	.	.	.	.	−1	.	.	.	−1	.
$s_{30}s_{14}$	.	.	−1	−1	−1	−1	.	.	.	.	.	.	.	.	.	.	.	.	−1	.	.
$s_{24}s_{20}$	.	.	.	.	.	.	−1	−1	.	.	−1	.	.	.	−1	.	.	−2	−1	−1	−1
$s_{23}s_{21}$	.	.	.	.	.	.	.	−1	−1	−1	.	.	.	−2	−1	−1	−1	.	.	.	.
s_{22}^2	.	.	.	.	.	.	.	.	.	−1	−1	−1	−1	.	.	.	.	.	.	.	.
$s_{42}s_{01}^2$	.	.	.	.	.	.	.	.	.	.	.	.	.	.	.	.	.	.	.	.	.
$s_{41}s_{02}s_{01}$	.	.	.	.	.	.	.	.	.	.	.	.	.	.	.	.	.	.	.	.	.
$s_{40}s_{03}s_{01}$	.	.	.	.	.	.	.	.	.	.	.	.	.	.	.	.	.	.	.	.	.
$s_{40}s_{02}^2$	.	.	.	.	.	.	.	.	.	.	.	.	.	.	.	.	.	.	.	.	.
$s_{33}s_{10}s_{01}$	.	.	.	.	.	.	.	.	.	.	.	.	.	.	.	.	.	.	.	.	.
$s_{32}s_{11}s_{01}$	.	.	.	.	.	.	.	.	.	.	.	.	.	.	.	.	.	.	.	.	.
$s_{32}s_{10}s_{02}$	.	.	.	.	.	.	.	.	.	.	.	.	.	.	.	.	.	.	.	.	.
$s_{31}s_{12}s_{01}$	.	.	.	.	.	.	.	.	.	.	.	.	.	.	.	.	.	.	.	.	.
$s_{31}s_{11}s_{02}$	*1*	.	.	.	.	.	.	.	.	.	.	.	.	.	.	.	.	.	.	.	.
$s_{31}s_{10}s_{03}$	.	*1*	.	.	.	.	.	.	.	.	.	.	.	.	.	.	.	.	.	.	.
$s_{30}s_{13}s_{01}$	.	.	*1*	.	.	.	.	.	.	.	.	.	.	.	.	.	.	.	.	.	.
$s_{30}s_{12}s_{02}$	.	.	.	*1*	.	.	.	.	.	.	.	.	.	.	.	.	.	.	.	.	.
$s_{30}s_{11}s_{03}$	.	.	.	.	*1*	.	.	.	.	.	.	.	.	.	.	.	.	.	.	.	.
$s_{30}s_{10}s_{04}$	.	.	.	.	.	*1*	.	.	.	.	.	.	.	.	.	.	.	.	.	.	.
$s_{24}s_{10}^2$	.	.	.	.	.	.	*1*	.	.	.	.	.	.	.	.	.	.	.	.	.	.
$s_{23}s_{20}s_{01}$	.	.	.	.	.	.	.	*1*	.	.	.	.	.	.	.	.	.	.	.	.	.
$s_{23}s_{11}s_{10}$	.	.	.	.	.	.	.	.	*1*	.	.	.	.	.	.	.	.	.	.	.	.
$s_{22}s_{21}s_{01}$	.	.	.	.	.	.	.	.	.	*1*	.	.	.	.	.	.	.	.	.	.	.
$s_{22}s_{20}s_{02}$	.	.	.	.	.	.	.	.	.	.	*1*	.	.	.	.	.	.	.	.	.	.
$s_{22}s_{12}s_{10}$	.	.	.	.	.	.	.	.	.	.	.	*1*	.	.	.	.	.	.	.	.	.
$s_{22}s_{11}^2$	.	.	.	.	.	.	.	.	.	.	.	.	*1*	.	.	.	.	.	.	.	.
$s_{21}^2s_{02}$	.	.	.	.	.	.	.	.	.	.	.	.	.	*1*	.	.	.	.	.	.	.
$s_{21}s_{20}s_{03}$	.	.	.	.	.	.	.	.	.	.	.	.	.	.	*1*	.	.	.	.	.	.
$s_{21}s_{13}s_{10}$	.	.	.	.	.	.	.	.	.	.	.	.	.	.	.	*1*	.	.	.	.	.
$s_{21}s_{12}s_{11}$	.	.	.	.	.	.	.	.	.	.	.	.	.	.	.	.	*1*	.	.	.	.
$s_{20}^2s_{04}$	.	.	.	.	.	.	.	.	.	.	.	.	.	.	.	.	.	*1*	.	.	.
$s_{20}s_{14}s_{10}$	.	.	.	.	.	.	.	.	.	.	.	.	.	.	.	.	.	.	*1*	.	.
$s_{20}s_{13}s_{11}$	.	.	.	.	.	.	.	.	.	.	.	.	.	.	.	.	.	.	.	*1*	.
$s_{20}s_{12}^2$	.	.	.	.	.	.	.	.	.	.	.	.	.	.	.	.	.	.	.	.	*1*
$s_{41}s_{01}^3$	.	.	.	.	.	.	.	.	.	.	.	.	.	.	.	.	.	.	.	.	.
$s_{40}s_{02}s_{01}^2$	.	.	.	.	.	.	.	.	.	.	.	.	.	.	.	.	.	.	.	.	.
$s_{32}s_{10}s_{01}^2$	.	.	.	.	.	.	.	.	.	.	.	.	.	.	.	.	.	.	.	.	.
$s_{31}s_{11}s_{01}^2$	1	.	.	.	.	.	.	.	.	.	.	.	.	.	.	.	.	.	.	.	.
$s_{31}s_{10}s_{02}s_{01}$	1	1	.	.	.	.	.	.	.	.	.	.	.	.	.	.	.	.	.	.	.
$s_{30}s_{12}s_{01}^2$	.	.	2	1	.	.	.	.	.	.	.	.	.	.	.	.	.	.	.	.	.
$s_{30}s_{11}s_{02}s_{01}$	1	.	1	1	1	.	.	.	.	.	.	.	.	.	.	.	.	.	.	.	.
$s_{30}s_{10}s_{03}s_{01}$	.	1	1	.	1	1	.	.	.	.	.	.	.	.	.	.	.	.	.	.	.
$s_{30}s_{10}s_{02}^2$	.	.	.	2	.	1	.	.	.	.	.	.	.	.	.	.	.	.	.	.	.
$s_{23}s_{10}^2s_{01}$	.	.	.	.	.	.	1	1	2	.	.	.	.	.	.	.	.	.	.	.	.
$s_{22}s_{20}s_{01}^2$	.	.	.	.	.	.	.	2	.	2	1	.	.	.	.	.	.	.	.	.	.
$s_{22}s_{11}s_{10}s_{01}$	.	.	.	.	.	.	.	.	1	1	.	1	1	.	.	.	.	.	.	.	.
$s_{22}s_{10}^2s_{02}$	.	.	.	.	.	.	1	.	.	.	1	2	.	.	.	.	.	.	.	.	.
$s_{21}^2s_{01}^2$	.	.	.	.	.	.	.	.	.	4	.	.	.	1	.	.	.	.	.	.	.
$s_{21}s_{20}s_{02}s_{01}$	.	.	.	.	.	.	.	1	.	1	1	.	.	1	1	.	.	.	.	.	.
$s_{21}s_{12}s_{10}s_{01}$	.	.	.	.	.	.	.	.	.	1	.	1	.	.	.	1	1	.	.	.	.
$s_{21}s_{11}^2s_{01}$	.	.	.	.	.	.	.	.	.	1	.	.	1	.	.	.	2	.	.	.	.
$s_{21}s_{11}s_{10}s_{02}$	1	.	.	.	.	.	.	.	1	.	.	.	.	1	.	1	1	.	.	.	.
$s_{21}s_{10}^2s_{03}$	.	2	.	.	.	.	1	.	.	.	.	.	.	.	1	2	.	.	.	.	.
$s_{20}^2s_{03}s_{01}$	.	.	.	.	.	.	.	2	.	.	.	.	.	.	2	.	.	1	.	.	.
$s_{20}^2s_{02}^2$	.	.	.	.	.	.	.	.	.	.	4	.	.	.	.	.	.	1	.	.	.
$s_{20}s_{13}s_{10}s_{01}$	.	.	1	.	.	.	.	1	.	.	.	.	.	.	.	1	.	.	1	1	.
$s_{20}s_{12}s_{11}s_{01}$	.	.	.	.	.	.	.	1	.	.	.	.	.	.	.	.	1	.	.	1	1
$s_{20}s_{12}s_{10}s_{02}$	.	.	.	1	.	.	.	.	.	.	1	1	.	.	.	.	.	.	1	.	1
$s_{20}s_{11}^2s_{02}$	2	.	.	.	.	.	.	.	.	.	1	.	1	.	.	.	.	.	.	2	.
$s_{20}s_{11}s_{10}s_{03}$	.	1	.	.	1	.	.	.	1	.	.	.	.	.	1	.	.	.	1	1	.
$s_{20}s_{10}^2s_{04}$	.	.	.	.	.	2	1	.	.	.	.	.	.	.	.	.	.	1	2	.	.
$s_{14}s_{10}^3$	.	.	.	.	.	.	3	.	.	.	.	.	.	.	.	.	.	.	3	.	.
$s_{13}s_{11}s_{10}^2$	.	.	.	.	.	.	1	.	2	.	.	.	.	.	.	2	.	.	.	1	.
$s_{12}^2s_{10}^2$	.	.	.	.	.	.	1	.	.	.	.	4	.	.	.	.	.	.	.	.	1
$s_{12}s_{11}^2s_{10}$	.	.	.	.	.	.	.	.	2	.	.	1	1	.	.	.	2	.	.	.	.
s_{11}^4	.	.	.	.	.	.	.	.	.	.	.	.	6	.	.	.	.	.	.	.	.
$s_{40}s_{01}^4$	.	.	.	.	.	.	.	.	.	.	.	.	.	.	.	.	.	.	.	.	.
$s_{31}s_{10}s_{01}^3$	3	1	.	.	.	.	.	.	.	.	.	.	.	.	.	.	.	.	.	.	.
$s_{30}s_{11}s_{01}^3$	3	.	3	3	1	.	.	.	.	.	.	.	.	.	.	.	.	.	.	.	.
$s_{30}s_{10}s_{02}s_{01}^2$	2	2	2	2	2	1	.	.	.	.	.	.	.	.	.	.	.	.	.	.	.
$s_{22}s_{10}^2s_{01}^2$	.	.	.	.	.	.	1	2	4	2	1	2	2	.	.	.	.	.	.	.	.
$s_{21}s_{20}s_{01}^3$	.	.	.	.	.	.	.	3	.	9	3	.	.	3	1	.	.	.	.	.	.
$s_{21}s_{11}s_{10}s_{01}^2$	1	.	.	.	.	.	.	.	1	4	.	2	2	1	.	1	3	.	.	.	.
$s_{21}s_{10}^2s_{02}s_{01}$	2	2	.	.	.	.	1	1	2	1	1	2	.	1	1	2	2	.	.	.	.
$s_{20}^2s_{02}s_{01}^2$	.	.	.	.	.	.	.	4	.	4	4	.	.	2	4	.	.	1	.	.	.
$s_{20}s_{12}s_{10}s_{01}^2$	.	.	2	1	.	.	.	2	.	2	1	1	.	.	.	2	2	.	1	2	1
$s_{20}s_{11}^2s_{01}^2$	2	.	.	.	.	.	.	2	.	2	1	.	1	.	.	.	4	.	.	2	2
$s_{20}s_{11}s_{10}s_{02}s_{01}$	2	1	1	1	1	.	.	1	1	1	1	1	1	1	1	1	1	.	1	2	1
$s_{20}s_{10}^2s_{03}s_{01}$	.	2	2	.	2	2	1	2	2	.	.	.	.	.	2	2	.	1	2	2	.
$s_{20}s_{10}^2s_{02}^2$	.	.	.	4	.	2	1	.	.	.	4	4	.	.	.	.	.	1	2	.	2
$s_{13}s_{10}^3s_{01}$	.	.	1	.	.	.	3	3	6	.	.	.	.	.	.	3	.	.	3	3	.
$s_{12}s_{11}s_{10}^2s_{01}$	.	.	.	.	.	.	1	1	4	2	.	4	2	.	.	2	3	.	.	1	1
$s_{12}s_{10}^3s_{02}$	.	.	.	1	.	.	3	.	.	.	3	9	.	.	.	.	.	.	3	.	3
$s_{11}^3s_{10}s_{01}$	.	.	.	.	.	.	.	.	3	3	.	3	6	.	.	.	6	.	.	.	.
$s_{11}^2s_{10}^2s_{02}$	2	.	.	.	.	.	1	.	4	.	1	2	1	2	.	4	4	.	.	2	.
$s_{11}s_{10}^3s_{03}$	.	3	.	.	1	.	3	.	3	.	.	.	.	.	3	6	.	.	3	3	.
$s_{10}^4s_{04}$	.	.	.	.	.	4	6	.	.	.	.	.	.	.	.	.	.	3	12	.	.
$s_{30}s_{10}s_{01}^4$	12	4	4	6	4	1	.	.	.	.	.	.	.	.	.	.	.	.	.	.	.
$s_{21}s_{10}^2s_{01}^3$	6	2	.	.	.	.	1	3	6	9	3	6	6	3	1	2	6	.	.	.	.
$s_{20}^2s_{01}^4$	.	.	.	.	.	.	.	8	.	24	12	.	.	12	8	.	.	1	.	.	.
$s_{20}s_{11}s_{10}s_{01}^3$	6	1	3	3	1	.	.	3	1	9	3	3	3	3	1	3	9	.	1	4	3
$s_{20}s_{10}^2s_{02}s_{01}^2$	4	4	4	4	4	2	1	4	4	4	4	4	2	2	4	4	4	1	2	4	2
$s_{12}s_{10}^3s_{01}^2$	.	.	2	1	.	.	3	6	12	6	3	9	6	.	.	6	6	.	3	6	3
$s_{11}^2s_{10}^2s_{01}^2$	2	.	.	.	.	.	1	2	8	10	1	10	11	2	.	4	16	.	.	2	2
$s_{11}s_{10}^3s_{02}s_{01}$	4	3	1	1	1	.	3	3	9	3	3	9	3	3	3	9	9	.	3	6	3
$s_{10}^4s_{03}s_{01}$	.	4	4	.	4	4	6	6	12	.	.	.	.	.	6	12	.	3	12	12	.
$s_{10}^4s_{02}^2$	.	.	.	8	.	4	6	.	.	.	12	24	.	.	.	.	.	3	12	.	12
$s_{20}s_{10}^2s_{01}^4$	24	8	8	12	8	2	1	8	8	24	12	12	12	12	8	8	24	1	2	8	6
$s_{11}s_{10}^3s_{01}^3$	12	3	3	3	1	.	3	9	21	27	9	27	27	9	3	15	45	.	3	12	9
$s_{10}^4s_{02}s_{01}^2$	8	8	8	8	8	4	6	12	24	12	12	24	12	6	12	24	24	3	12	24	12
$s_{10}^4s_{01}^4$	48	16	16	24	16	4	6	24	48	72	36	72	72	36	24	48	144	3	12	48	36

Table 1·6·8 (cont.)

$w = 8$ (44) (iii)	$[41, (01)^3]$	$[40, 02, (01)^2]$	$[32, 10, (01)^2]$	$[31, 11, (01)^2]$	$[31, 10, 02, 01]$	$[30, 12, (01)^2]$	$[30, 11, 02, 01]$	$[30, 10, 03, 01]$	$[30, 10, (02)^2]$	$[23, (10)^2, 01]$	$[22, 20, (01)^2]$	$[22, 11, 10, 01]$	$[22, (10)^2, 02]$	$[(21)^2, (01)^2]$	$[21, 20, 02, 01]$	$[21, 12, 10, 01]$
s_{44}	-6	-6	-6	-6	-6	-6	-6	-6	-6	-6	-6	-6	-6	-6	-6	-6
$s_{43}s_{01}$	6	4	4	4	2	4	2	2	.	2	4	2	.	4	2	2
$s_{42}s_{02}$	3	3	1	1	2	1	2	.	4	.	1	.	2	1	2	.
$s_{41}s_{03}$	2	2	.	.	1	.	1	2	.	.	.	.	.	.	1	.
$s_{40}s_{04}$	.	2	.	.	.	.	.	1	1	.	.	.	.	.	.	.
$s_{24}s_{10}$	.	.	2	.	2	.	.	2	2	4	.	2	4	.	.	2
$s_{33}s_{11}$	.	.	2	2	1	.	2	1	.	2	.	3	.	.	.	1
$s_{32}s_{12}$	.	.	2	2	1	2	1	.	2	.	.	1	2	.	.	2
$s_{31}s_{13}$	.	.	.	2	2	2	1	1	.	.	.	.	.	.	.	1
$s_{30}s_{14}$	.	.	.	.	2	2	2	2	.	.	.	.	.	.	.	.
$s_{24}s_{20}$	.	.	.	.	.	.	.	.	.	1	2	.	1	.	2	2
$s_{23}s_{21}$	.	.	.	.	.	.	.	.	.	2	2	1	.	4	3	2
s_{22}^2	.	.	.	.	.	.	.	.	.	.	2	2	2	2	1	1
$s_{42}s_{01}^2$	-3	-1	-1	-1	.	-1	.	.	.	.	.	-1	.	.	-1	.
$s_{41}s_{02}s_{01}$	-3	-2	.	.	-1	.	-1	.	.	.	.	.	.	.	-1	.
$s_{40}s_{03}s_{01}$	.	-2	.	.	.	.	.	-1	.	.	.	.	.	.	.	.
$s_{40}s_{02}^2$	.	-1	.	.	.	.	.	.	-1	.	.	.	.	.	.	.
$s_{33}s_{10}s_{01}$	.	.	-2	.	-1	.	.	-1	.	-2	.	-1	.	.	.	-1
$s_{32}s_{11}s_{01}$	.	.	-2	-2	.	.	-1	.	.	.	.	-1	.	.	.	.
$s_{32}s_{10}s_{02}$	.	.	-1	-1	.	.	.	-2	.	.	.	.	-2	.	.	.
$s_{31}s_{12}s_{01}$	.	.	.	-2	-1	-2	.	.	.	.	.	.	.	.	.	-1
$s_{31}s_{11}s_{02}$	.	.	.	-1	-1	.	-1	.	.	.	.	.	.	.	.	.
$s_{31}s_{10}s_{03}$	.	.	.	.	-1	.	.	-1	.	.	.	.	.	.	.	.
$s_{30}s_{13}s_{01}$	.	.	.	.	.	-2	-1	-1	.	.	.	.	.	.	.	.
$s_{30}s_{12}s_{02}$	.	.	.	.	.	-1	-1	.	-2	.	.	.	.	.	.	.
$s_{30}s_{11}s_{03}$	.	.	.	.	.	.	-1	-1	.	.	.	.	.	.	.	.
$s_{30}s_{10}s_{04}$	.	.	.	.	.	.	-1	-1	.	.	.	.	.	.	.	.
$s_{24}s_{10}^2$	.	.	.	.	.	.	.	.	.	-1	-2	.	-1	.	.	.
$s_{23}s_{20}s_{01}$	.	.	.	.	.	.	.	.	.	-1	-2	.	.	.	-1	.
$s_{23}s_{11}s_{10}$	.	.	.	.	.	.	.	.	.	-2	.	-1	.	.	.	.
$s_{22}s_{21}s_{01}$	.	.	.	.	.	.	.	.	.	.	-2	-1	.	-4	-1	-1
$s_{22}s_{20}s_{02}$	.	.	.	.	.	.	.	.	.	.	-1	.	-1	.	-1	-1
$s_{22}s_{12}s_{10}$	.	.	.	.	.	.	.	.	.	.	.	-1	-2	.	.	-1
$s_{22}s_{11}^2$	.	.	.	.	.	.	.	.	.	.	.	-1	.	.	.	.
$s_{21}^2s_{02}$	.	.	.	.	.	.	.	.	.	.	.	.	.	-1	-1	.
$s_{21}s_{20}s_{03}$	.	.	.	.	.	.	.	.	.	.	.	.	.	.	-1	.
$s_{21}s_{13}s_{10}$	.	.	.	.	.	.	.	.	.	.	.	.	.	.	.	-1
$s_{21}s_{12}s_{11}$	.	.	.	.	.	.	.	.	.	.	.	.	.	.	.	-1
$s_{20}^2s_{04}$	.	.	.	.	.	.	.	.	.	.	.	.	.	.	.	.
$s_{20}s_{14}s_{10}$	.	.	.	.	.	.	.	.	.	.	.	.	.	.	.	.
$s_{20}s_{13}s_{11}$	.	.	.	.	.	.	.	.	.	.	.	.	.	.	.	.
$s_{20}s_{12}^2$	.	.	.	.	.	.	.	.	.	.	.	.	.	.	.	.
$s_{41}s_{01}^3$	1	.	.	.	.	.	.	.	.	.	.	.	.	.	.	.
$s_{40}s_{02}s_{01}^2$	.	1	.	.	.	.	.	.	.	.	.	.	.	.	.	.
$s_{32}s_{10}s_{01}^2$	.	.	1	.	.	.	.	.	.	.	.	.	.	.	.	.
$s_{31}s_{11}s_{01}^2$	.	.	.	1	.	.	.	.	.	.	.	.	.	.	.	.
$s_{31}s_{10}s_{02}s_{01}$	.	.	.	.	1	.	.	.	.	.	.	.	.	.	.	.
$s_{30}s_{12}s_{01}^2$	.	.	.	.	.	1	.	.	.	.	.	.	.	.	.	.
$s_{30}s_{11}s_{02}s_{01}$	.	.	.	.	.	.	1	.	.	.	.	.	.	.	.	.
$s_{30}s_{10}s_{03}s_{01}$	.	.	.	.	.	.	.	1	.	.	.	.	.	.	.	.
$s_{30}s_{10}s_{02}^2$	.	.	.	.	.	.	.	.	1	.	.	.	.	.	.	.
$s_{23}s_{10}^2s_{01}$	.	.	.	.	.	.	.	.	.	1	.	.	.	.	.	.
$s_{22}s_{20}s_{01}^2$	.	.	.	.	.	.	.	.	.	.	1	.	.	.	.	.
$s_{22}s_{11}s_{10}s_{01}$	.	.	.	.	.	.	.	.	.	.	.	1	.	.	.	.
$s_{22}s_{10}^2s_{02}$	.	.	.	.	.	.	.	.	.	.	.	.	1	.	.	.
$s_{21}^2s_{01}^2$	.	.	.	.	.	.	.	.	.	.	.	.	.	1	.	.
$s_{21}s_{20}s_{02}s_{01}$	.	.	.	.	.	.	.	.	.	.	.	.	.	.	1	.
$s_{21}s_{12}s_{10}s_{01}$	.	.	.	.	.	.	.	.	.	.	.	.	.	.	.	1
$s_{21}s_{11}^2s_{01}$	.	.	.	.	.	.	.	.	.	.	.	.	.	.	.	.
$s_{21}s_{11}s_{10}s_{02}$	.	.	.	.	.	.	.	.	.	.	.	.	.	.	.	.
$s_{21}s_{10}^2s_{03}$	.	.	.	.	.	.	.	.	.	.	.	.	.	.	.	.
$s_{20}^2s_{03}s_{01}$	.	.	.	.	.	.	.	.	.	.	.	.	.	.	.	.
$s_{20}^2s_{02}^2$	.	.	.	.	.	.	.	.	.	.	.	.	.	.	.	.
$s_{20}s_{13}s_{10}s_{01}$	.	.	.	.	.	.	.	.	.	.	.	.	.	.	.	.
$s_{20}s_{12}s_{11}s_{01}$	.	.	.	.	.	.	.	.	.	.	.	.	.	.	.	.
$s_{20}s_{12}s_{10}s_{02}$	.	.	.	.	.	.	.	.	.	.	.	.	.	.	.	.
$s_{20}s_{11}^2s_{02}$	.	.	.	.	.	.	.	.	.	.	.	.	.	.	.	.
$s_{20}s_{11}s_{10}s_{03}$	.	.	.	.	.	.	.	.	.	.	.	.	.	.	.	.
$s_{20}s_{10}^2s_{04}$	.	.	.	.	.	.	.	.	.	.	.	.	.	.	.	.
$s_{14}s_{10}^3$	.	.	.	.	.	.	.	.	.	.	.	.	.	.	.	.
$s_{13}s_{11}s_{10}^2$	.	.	.	.	.	.	.	.	.	.	.	.	.	.	.	.
$s_{12}^2s_{10}^2$	.	.	.	.	.	.	.	.	.	.	.	.	.	.	.	.
$s_{12}s_{11}^2s_{10}$	.	.	.	.	.	.	.	.	.	.	.	.	.	.	.	.
s_{11}^4	.	.	.	.	.	.	.	.	.	.	.	.	.	.	.	.
$s_{40}s_{01}^4$	4	6	.	.	.	.	.	.	.	.	.	.	.	.	.	.
$s_{31}s_{10}s_{01}^3$	1	.	3	3	3	.	.	.	.	.	.	.	.	.	.	.
$s_{30}s_{11}s_{01}^3$	1	.	3	3	.	3	3	.	.	.	.	.	.	.	.	.
$s_{30}s_{10}s_{02}s_{01}^2$	.	1	1	.	2	1	2	2	1	.	.	.	.	.	.	.
$s_{22}s_{10}^2s_{01}^2$	.	.	2	.	.	.	.	.	.	2	1	4	1	.	.	.
$s_{21}s_{20}s_{01}^3$	1	.	.	.	.	.	.	.	.	.	3	.	.	3	3	.
$s_{21}s_{11}s_{10}s_{01}^2$	.	.	1	1	.	.	.	.	.	.	.	2	.	1	1	2
$s_{21}s_{10}^2s_{02}s_{01}$	.	.	.	.	2	.	.	.	.	1	.	.	1	.	1	2
$s_{20}^2s_{02}s_{01}^2$	.	1	.	.	.	.	.	.	.	.	2	.	.	.	4	.
$s_{20}s_{12}s_{10}s_{01}^2$	.	.	1	.	.	1	.	.	.	.	1	.	.	.	.	2
$s_{20}s_{11}^2s_{01}^2$	.	.	.	2	.	.	.	.	.	.	1	.	.	.	.	.
$s_{20}s_{11}s_{10}s_{02}s_{01}$	.	.	.	.	1	.	1	.	.	1	.	1	.	.	1	.
$s_{20}s_{10}^2s_{03}s_{01}$	.	.	.	.	.	.	.	2	.	1	.	.	.	.	.	.
$s_{20}s_{10}^2s_{02}^2$	.	.	.	.	.	.	.	.	2	.	.	.	2	.	.	.
$s_{13}s_{10}^3s_{01}$	.	.	.	.	.	.	.	.	.	3	.	.	.	.	.	.
$s_{12}s_{11}s_{10}^2s_{01}$	.	.	.	.	.	.	.	.	.	1	.	2	.	.	.	2
$s_{12}s_{10}^3s_{02}$	.	.	.	.	.	.	.	.	.	.	.	.	3	.	.	.
$s_{11}^3s_{10}s_{01}$	.	.	.	.	.	.	.	.	.	.	.	3	.	.	.	.
$s_{11}^2s_{10}^2s_{02}$	.	.	.	.	.	.	.	.	.	.	.	.	1	.	.	.
$s_{11}s_{10}^3s_{03}$	.	.	.	.	.	.	.	.	.	.	.	.	.	.	.	.
$s_{10}^4s_{04}$	.	.	.	.	.	.	.	.	.	.	.	.	.	.	.	.
$s_{30}s_{10}s_{01}^4$	4	6	6	12	12	6	12	4	3	.	.	.	.	.	.	.
$s_{21}s_{10}^2s_{01}^3$	1	.	6	6	6	.	.	.	.	3	3	12	3	3	3	6
$s_{20}^2s_{01}^4$	4	6	.	.	.	.	.	.	.	.	12	.	.	12	24	.
$s_{20}s_{11}s_{10}s_{01}^3$	1	.	3	6	3	3	3	.	.	.	3	3	.	3	3	6
$s_{20}s_{10}^2s_{02}s_{01}^2$	.	1	2	.	4	2	4	4	2	2	2	4	2	.	4	4
$s_{12}s_{10}^3s_{01}^2$	.	.	3	.	.	1	.	.	.	6	3	12	3	.	.	6
$s_{11}^2s_{10}^2s_{01}^2$	.	.	2	2	.	.	.	.	.	2	1	12	1	2	.	8
$s_{11}s_{10}^3s_{02}s_{01}$	.	.	.	.	3	.	1	.	.	3	.	3	3	.	3	6
$s_{10}^4s_{03}s_{01}$	.	.	.	.	.	.	.	4	.	6	.	.	.	.	.	.
$s_{10}^4s_{02}^2$	.	.	.	.	.	.	.	.	4	.	.	.	12	.	.	.
$s_{20}s_{10}^2s_{01}^4$	4	6	12	24	24	12	24	8	6	4	12	24	6	12	24	24
$s_{11}s_{10}^3s_{01}^3$	1	.	9	12	9	3	3	8	.	9	9	45	9	9	9	36
$s_{10}^4s_{02}s_{01}^2$	.	1	4	.	8	4	8	8	4	12	6	24	12	.	12	24
$s_{10}^4s_{01}^4$	4	6	24	48	48	24	48	16	12	24	36	144	36	36	72	144

Table 1·6·8 (cont.)

$w=8$ (44)(iv)	$[21,(11)^2,01]$	$[21,11,10,02]$	$[21,(10)^2,03]$	$[(20)^2,03,01]$	$[(20)^2,(02)^2]$	$[20,13,10,01]$	$[20,12,11,01]$	$[20,12,10,02]$	$[20,(11)^2,02]$	$[20,11,10,03]$	$[20,(10)^2,04]$	$[14,(10)^3]$	$[13,11,(10)^2]$	$[(12)^2,(10)^2]$	$[12,(11)^2,10]$	$[11^4]$
s_{44}	-6	-6	-6	-6	-6	-6	-6	-6	-6	-6	-6	-6	-6	-6	-6	-6
$s_{43}s_{01}$	2	.	.	2	.	2	2	.	.	.	.	.	.	.	.	.
$s_{42}s_{02}$	.	2	.	.	4	.	.	2	2	.	.	.	.	.	.	.
$s_{41}s_{03}$	.	.	2	2	.	.	.	.	.	2	.	.	.	.	.	.
$s_{40}s_{04}$	.	.	.	1	1	.	.	.	.	.	2	.	.	.	.	.
$s_{34}s_{10}$	.	2	4	.	.	2	2	.	2	4	6	4	4	2	.	.
$s_{33}s_{11}$	4	2	.	.	.	1	2	.	4	2	.	.	2	.	4	8
$s_{32}s_{12}$	2	1	.	.	.	.	3	3	.	.	.	.	.	4	2	.
$s_{31}s_{13}$	.	1	2	.	.	2	1	.	2	1	.	.	2	.	.	.
$s_{30}s_{14}$	.	.	.	.	.	1	.	1	.	1	2	2	.	.	.	.
$s_{24}s_{20}$	.	.	1	4	4	2	2	2	2	2	3	3	1	1	.	.
$s_{23}s_{21}$	2	3	2	2	.	1	1	.	1	.	.	.	2	.	2	.
$s_{22}{}^2$	1	.	.	.	2	.	.	1	1	.	.	.	.	2	1	3
$s_{42}s_{01}{}^2$	1	.	.	.	.	.	.	.	.	.	.	.	.	.	.	.
$s_{41}s_{02}s_{01}$	.	.	.	.	.	.	.	.	.	.	.	.	.	.	.	.
$s_{40}s_{03}s_{01}$	.	.	.	-1	.	.	.	.	.	.	.	.	.	.	.	.
$s_{40}s_{02}{}^2$	.	.	.	.	-1	.	.	.	.	.	.	.	.	.	.	.
$s_{33}s_{10}s_{01}$	.	.	.	.	.	-1	.	.	.	.	.	.	.	.	.	.
$s_{32}s_{11}s_{01}$	-2	.	.	.	.	.	-1	.	.	.	.	.	.	.	.	.
$s_{32}s_{10}s_{02}$	.	-1	.	.	.	.	.	-1	.	.	.	.	.	.	.	.
$s_{31}s_{12}s_{01}$	.	.	.	.	.	.	-1	.	.	.	.	.	.	.	.	.
$s_{31}s_{11}s_{02}$	.	-1	.	.	.	.	.	.	-2	.	.	.	.	.	.	.
$s_{31}s_{10}s_{03}$	.	.	-2	.	.	.	.	.	-1	.	.	.	.	.	.	.
$s_{30}s_{13}s_{01}$	.	.	.	.	.	-1	.	.	.	.	.	.	.	.	.	.
$s_{30}s_{12}s_{02}$	.	.	.	.	.	.	.	-1	.	.	.	.	.	.	.	.
$s_{30}s_{11}s_{03}$	.	.	.	.	.	.	.	.	-1	.	.	.	.	.	.	.
$s_{30}s_{10}s_{04}$	.	.	.	.	.	.	.	.	.	.	-2	.	.	.	.	.
$s_{24}s_{10}{}^2$	.	.	-1	.	.	.	.	.	.	.	-1	-3	-1	-1	.	.
$s_{23}s_{20}s_{01}$	.	.	.	-2	.	-1	-1	.	.	.	.	.	.	.	.	.
$s_{23}s_{11}s_{10}$	.	-1	.	.	.	.	.	.	.	-1	.	.	-2	.	-2	.
$s_{22}s_{21}s_{01}$	-1	.	.	.	.	.	.	.	.	.	.	.	.	.	.	.
$s_{22}s_{20}s_{02}$	.	.	.	.	-4	.	.	-1	-1	.	.	.	.	.	.	.
$s_{22}s_{12}s_{10}$	.	.	.	.	.	.	.	-1	.	.	.	.	.	-4	-1	.
$s_{22}{}^2 s_{11}$	-1	.	.	.	.	.	.	.	-1	.	.	.	.	.	-1	-6
$s_{21}{}^2 s_{02}$	.	-1	.	.	.	.	.	.	.	.	.	.	.	.	.	.
$s_{21}s_{20}s_{03}$	.	.	-1	-2	.	.	.	.	.	-1	.	.	.	.	.	.
$s_{21}s_{13}s_{10}$	.	-1	-2	.	.	-1	.	.	.	.	.	.	-2	.	.	.
$s_{21}s_{12}s_{11}$	-2	-1	.	.	.	.	-1	.	.	.	.	.	.	.	-2	.
$s_{20}{}^2 s_{04}$	.	.	.	-1	-1	.	.	.	.	.	-1	.	.	.	.	.
$s_{20}s_{14}s_{10}$	.	.	.	.	.	-1	.	-1	.	-1	-2	-3	.	.	.	.
$s_{20}s_{13}s_{11}$	.	.	.	.	.	-1	-1	.	-2	-1	.	.	-1	.	.	.
$s_{20}s_{12}{}^2$	.	.	.	.	.	-1	-1	.	.	.	.	.	.	-1	.	.
$s_{41}s_{01}{}^3$	.	.	.	.	.	.	.	.	.	.	.	.	.	.	.	.
$s_{40}s_{02}s_{01}{}^2$	.	.	.	.	.	.	.	.	.	.	.	.	.	.	.	.
$s_{32}s_{10}s_{01}{}^2$	.	.	.	.	.	.	.	.	.	.	.	.	.	.	.	.
$s_{31}s_{11}s_{01}{}^2$	.	.	.	.	.	.	.	.	.	.	.	.	.	.	.	.
$s_{31}s_{10}s_{02}s_{01}$	.	.	.	.	.	.	.	.	.	.	.	.	.	.	.	.
$s_{30}s_{12}s_{01}{}^2$	.	.	.	.	.	.	.	.	.	.	.	.	.	.	.	.
$s_{30}s_{11}s_{02}s_{01}$	.	.	.	.	.	.	.	.	.	.	.	.	.	.	.	.
$s_{30}s_{10}s_{03}s_{01}$	.	.	.	.	.	.	.	.	.	.	.	.	.	.	.	.
$s_{30}s_{10}s_{02}{}^2$	.	.	.	.	.	.	.	.	.	.	.	.	.	.	.	.
$s_{23}s_{10}{}^2 s_{01}$	.	.	.	.	.	.	.	.	.	.	.	.	.	.	.	.
$s_{22}s_{20}s_{01}{}^2$	.	.	.	.	.	.	.	.	.	.	.	.	.	.	.	.
$s_{22}s_{11}s_{10}s_{01}$	.	.	.	.	.	.	.	.	.	.	.	.	.	.	.	.
$s_{22}s_{10}{}^2 s_{02}$	.	.	.	.	.	.	.	.	.	.	.	.	.	.	.	.
$s_{21}{}^2 s_{01}{}^2$	.	.	.	.	.	.	.	.	.	.	.	.	.	.	.	.
$s_{21}s_{20}s_{02}s_{01}$	.	.	.	.	.	.	.	.	.	.	.	.	.	.	.	.
$s_{21}s_{12}s_{10}s_{01}$	.	.	.	.	.	.	.	.	.	.	.	.	.	.	.	.
$s_{21}s_{11}{}^2 s_{01}$	1	.	.	.	.	.	.	.	.	.	.	.	.	.	.	.
$s_{21}s_{11}s_{10}s_{02}$	.	1	.	.	.	.	.	.	.	.	.	.	.	.	.	.
$s_{21}s_{10}{}^2 s_{03}$	.	.	1	.	.	.	.	.	.	.	.	.	.	.	.	.
$s_{20}{}^2 s_{03}s_{01}$	.	.	.	1	.	.	.	.	.	.	.	.	.	.	.	.
$s_{20}{}^2 s_{02}{}^2$	.	.	.	.	1	.	.	.	.	.	.	.	.	.	.	.
$s_{20}s_{13}s_{10}s_{01}$	.	.	.	.	.	1	.	.	.	.	.	.	.	.	.	.
$s_{20}s_{12}s_{11}s_{01}$	.	.	.	.	.	.	1	.	.	.	.	.	.	.	.	.
$s_{20}s_{12}s_{10}s_{02}$	.	.	.	.	.	.	.	1	.	.	.	.	.	.	.	.
$s_{20}s_{11}{}^2 s_{02}$	.	.	.	.	.	.	.	.	1	.	.	.	.	.	.	.
$s_{20}s_{11}s_{10}s_{03}$	.	.	.	.	.	.	.	.	.	1	.	.	.	.	.	.
$s_{20}s_{10}{}^2 s_{04}$	.	.	.	.	.	.	.	.	.	.	1	.	.	.	.	.
$s_{14}s_{10}{}^3$	.	.	.	.	.	.	.	.	.	.	.	1	.	.	.	.
$s_{13}s_{11}s_{10}{}^2$	.	.	.	.	.	.	.	.	.	.	.	.	1	.	.	.
$s_{12}{}^2 s_{10}{}^2$	.	.	.	.	.	.	.	.	.	.	.	.	.	1	.	.
$s_{12}s_{11}{}^2 s_{10}$	.	.	.	.	.	.	.	.	.	.	.	.	.	.	1	.
$s_{11}{}^4$	.	.	.	.	.	.	.	.	.	.	.	.	.	.	.	1
$s_{40}s_{01}{}^4$	.	.	.	.	.	.	.	.	.	.	.	.	.	.	.	.
$s_{31}s_{10}s_{01}{}^3$	.	.	.	.	.	.	.	.	.	.	.	.	.	.	.	.
$s_{30}s_{11}s_{01}{}^3$	.	.	.	.	.	.	.	.	.	.	.	.	.	.	.	.
$s_{30}s_{10}s_{02}s_{01}{}^2$	.	.	.	.	.	.	.	.	.	.	.	.	.	.	.	.
$s_{22}s_{10}{}^2 s_{01}{}^2$	.	.	.	.	.	.	.	.	.	.	.	.	.	.	.	.
$s_{21}s_{20}s_{01}{}^3$	.	.	.	.	.	.	.	.	.	.	.	.	.	.	.	.
$s_{21}s_{11}s_{10}s_{01}{}^2$	2	1	.	.	.	.	.	.	.	.	.	.	.	.	.	.
$s_{21}s_{10}{}^2 s_{02}s_{01}$	.	2	1	.	.	.	.	.	.	.	.	.	.	.	.	.
$s_{20}{}^2 s_{02}s_{01}{}^2$	.	.	.	2	1	.	.	.	.	.	.	.	.	.	.	.
$s_{20}s_{12}s_{10}s_{01}{}^2$	.	.	.	.	.	2	2	1	.	.	.	.	.	.	.	.
$s_{20}s_{11}{}^2 s_{01}{}^2$	2	.	.	.	.	.	4	.	1	.	.	.	.	.	.	.
$s_{20}s_{11}s_{10}s_{02}s_{01}$	.	1	.	.	.	1	1	1	1	1	.	.	.	.	.	.
$s_{20}s_{10}{}^2 s_{03}s_{01}$	.	.	1	1	.	2	.	.	.	2	1	.	.	.	.	.
$s_{20}s_{10}{}^2 s_{02}{}^2$	.	.	.	.	1	.	.	4	.	.	1	.	.	.	.	.
$s_{13}s_{10}{}^3 s_{01}$	.	.	.	.	.	3	.	.	.	.	.	1	3	.	2	.
$s_{12}s_{11}s_{10}{}^2 s_{01}$	.	.	.	.	.	.	1	.	.	.	.	1	1	1	2	.
$s_{12}s_{10}{}^3 s_{02}$	.	.	.	.	.	.	.	3	.	.	.	1	3	.	.	.
$s_{11}{}^3 s_{10}s_{01}$	3	.	.	.	.	.	.	.	.	.	.	.	.	.	3	1
$s_{11}{}^2 s_{10}{}^2 s_{02}$	.	4	.	.	.	.	.	.	.	1	.	.	2	.	2	.
$s_{11}s_{10}{}^3 s_{03}$	.	.	3	.	.	.	.	.	.	.	3	1	3	.	.	.
$s_{10}{}^4 s_{04}$	.	.	.	.	.	.	.	.	.	.	6	4	.	.	.	.
$s_{21}s_{10}{}^2 s_{01}{}^3$	6	6	1	.	.	.	.	.	.	.	.	.	.	.	.	.
$s_{20}{}^2 s_{01}{}^4$	.	.	.	4	3	.	.	.	.	.	.	.	.	.	.	.
$s_{20}s_{11}s_{10}s_{01}{}^3$	6	3	.	.	.	3	9	3	3	1	.	.	.	.	.	.
$s_{20}s_{10}{}^2 s_{02}s_{01}{}^2$	.	4	2	2	1	4	4	4	2	4	1	.	.	.	.	.
$s_{12}s_{10}{}^3 s_{01}{}^2$	.	.	.	.	.	.	6	.	.	.	.	.	1	6	3	6
$s_{11}{}^2 s_{10}{}^2 s_{01}{}^2$	10	4	.	.	.	.	4	.	.	.	1	.	2	2	10	2
$s_{11}s_{10}{}^3 s_{02}s_{01}$	.	9	3	.	.	3	3	3	3	.	.	1	6	3	6	.
$s_{10}{}^4 s_{03}s_{01}$	.	.	6	3	.	12	.	.	.	.	12	6	4	12	.	.
$s_{10}{}^4 s_{02}{}^2$	.	.	.	.	3	.	.	.	24	.	.	6	4	.	12	.
$s_{20}s_{10}{}^2 s_{01}{}^4$	24	24	4	4	3	8	24	12	12	8	1	.	.	.	.	.
$s_{11}s_{10}{}^3 s_{01}{}^3$	36	27	3	.	3	9	27	9	9	3	.	1	12	9	36	6
$s_{10}{}^4 s_{02}s_{01}{}^2$	.	24	12	6	3	24	24	12	24	6	4	24	12	24	.	.
$s_{10}{}^4 s_{01}{}^4$	144	144	24	12	9	48	144	72	72	48	6	4	48	36	144	24

Table 1·6·8 (*cont.*)

$w = 8$ (44) (v)	[40, (01)⁴]	[31, 10, (01)³]	[30, 11, (01)³]	[30, 10, 02, (01)²]	[22, (10)², (01)²]	[21, 20, (01)³]	[21, 11, 10, (01)²]	[21, (10)², 02, 01]	[(20)², 02, (01)²]	[20, 12, 10, (01)²]	[20, (11)², (01)²]	[20, 11, 10, 02, 01]	[20, (10)², 03, 01]	[20, (10)², (02)²]	[13, (10)³, 01]	[12, 11, (10)², 01]	[12, (10)³, 02]	[(11)³, 10, 01]	[(11)², (10)², 02]	[11, (10)³, 03]	[(10)⁴, 04]
s_{44}	24	24	24	24	24	24	24	24	24	24	24	24	24	24	24	24	24	24	24	24	24
$s_{43}s_{01}$	−24	−18	−18	−12	−12	−18	−12	−6	−12	−12	−12	−6	−6	.	−6	−6	.	−6	.	.	.
$s_{42}s_{02}$	−12	−6	−6	−8	−2	−6	−2	−6	−8	−2	−2	−6	.	−12	.	.	−6	.	−6	.	.
$s_{41}s_{03}$	−8	−2	−2	−4	.	−2	.	−2	−4	.	.	−2	−6	.	.	.	.	.	.	−6	.
$s_{40}s_{04}$	−6	.	.	−2	.	.	.	.	−2	.	.	.	−2	−2	.	.	.	.	.	.	−6
$s_{34}s_{10}$	.	−6	.	−6	−12	.	−6	−12	.	−6	.	−6	−12	−12	−18	−12	−18	−6	−12	−18	−24
$s_{33}s_{11}$	.	−6	−6	−4	−8	.	−10	−4	.	−4	−12	−8	−4	.	−6	−10	.	−20	−12	−6	.
$s_{32}s_{12}$	.	−6	−6	−4	−4	.	−6	−4	.	−8	−8	−4	.	−8	.	−8	−12	−6	−4	.	.
$s_{31}s_{13}$	.	−6	−6	−4	.	.	−2	−4	.	−4	−4	−4	−4	.	−6	−2	.	.	−4	−6	.
$s_{30}s_{14}$	.	.	−6	−6	.	.	.	.	.	−2	.	−2	−4	−4	−2	.	−2	.	.	−2	−8
$s_{24}s_{20}$	.	.	.	.	−2	−6	.	−2	−12	−6	−6	−6	−8	−8	−6	−2	−6	.	−2	−6	−12
$s_{23}s_{21}$	.	.	.	.	−4	−12	−8	−8	−8	−4	−4	−4	−4	.	−6	−6	.	−6	−8	−6	.
s_{22}^2	.	.	.	.	−6	−6	−4	−2	−4	−2	−2	−2	.	−4	.	−4	−6	−6	−2	.	.
$s_{42}s_{01}^2$	12	6	6	2	2	6	2	.	2	2	2	.	.	.	.	.	.	.	.	.	.
$s_{41}s_{02}s_{01}$	12	3	3	4	.	3	.	2	4	.	.	2	.	.	.	.	.	.	.	.	.
$s_{40}s_{03}s_{01}$	8	.	.	2	.	.	.	.	2	.	.	.	2	.	.	.	.	.	.	.	.
$s_{40}s_{02}^2$	3	.	.	1	.	.	.	.	1	.	.	.	.	2	.	.	.	.	.	.	.
$s_{33}s_{10}s_{01}$	.	6	.	4	8	.	4	4	.	4	.	2	4	.	6	4	.	2	.	.	.
$s_{32}s_{11}s_{01}$	.	6	6	2	4	.	6	.	.	2	8	2	.	.	.	2	.	6	.	.	.
$s_{32}s_{10}s_{02}$	.	3	.	3	2	.	1	4	.	1	.	2	.	8	.	.	6	.	4	.	.
$s_{31}s_{12}s_{01}$	.	6	6	2	.	.	2	2	.	4	4	1	.	.	.	2	.	.	.	.	.
$s_{31}s_{11}s_{02}$	.	3	3	2	.	.	1	2	.	.	2	3	.	.	.	.	.	.	4	.	.
$s_{31}s_{10}s_{03}$	.	2	.	2	.	.	.	2	.	.	.	1	4	.	.	.	.	.	.	6	.
$s_{30}s_{13}s_{01}$	.	.	6	4	.	.	.	.	.	2	.	1	2	.	2	.	.	.	.	.	.
$s_{30}s_{12}s_{02}$	.	.	3	3	.	.	.	.	.	1	.	1	.	4	.	.	2	.	.	.	.
$s_{30}s_{11}s_{03}$	.	.	2	2	.	.	.	.	.	.	.	1	2	.	.	.	.	.	.	2	.
$s_{30}s_{10}s_{04}$	.	.	.	2	.	.	.	.	.	.	.	.	2	2	.	.	.	.	.	.	8
$s_{24}s_{10}^2$	.	.	.	.	2	.	.	2	.	.	.	.	2	2	6	2	6	.	2	6	12
$s_{23}s_{20}s_{01}$	.	.	.	.	2	6	.	1	8	4	4	2	3	.	3	1	.	.	.	.	.
$s_{23}s_{11}s_{10}$	.	.	.	.	4	.	2	2	.	.	.	2	2	.	6	6	.	6	8	6	.
$s_{22}s_{21}s_{01}$	.	.	.	.	4	12	6	2	4	2	2	1	.	.	.	2	.	3	.	.	.
$s_{22}s_{20}s_{02}$	.	.	.	.	1	3	.	1	6	1	1	2	.	6	.	.	3	.	1	.	.
$s_{22}s_{12}s_{10}$	.	.	.	.	4	.	2	2	.	2	.	1	.	4	.	6	12	3	2	.	.
$s_{22}s_{11}^2$	.	.	.	.	2	.	2	.	.	.	2	1	.	.	.	2	.	9	2	.	.
$s_{21}^2s_{02}$	.	.	.	.	.	3	1	2	2	.	.	1	.	.	.	.	.	.	2	.	.
$s_{21}s_{20}s_{03}$	.	.	.	.	.	2	.	1	4	.	.	1	3	.	.	.	.	.	.	3	.
$s_{21}s_{13}s_{10}$	.	.	.	.	.	.	2	4	.	2	.	1	2	.	6	2	.	.	4	6	.
$s_{21}s_{12}s_{11}$	.	.	.	.	.	.	4	2	.	2	4	1	.	.	.	4	.	6	4	.	.
$s_{20}^2s_{04}$	.	.	.	.	.	.	.	.	2	.	.	.	1	1	.	.	.	.	.	.	3
$s_{20}s_{14}s_{10}$	.	.	.	.	.	.	.	.	.	2	.	2	4	4	3	.	3	.	.	3	12
$s_{20}s_{13}s_{11}$	.	.	.	.	.	.	.	.	.	2	4	3	2	.	3	1	.	.	2	3	.
$s_{20}s_{12}^2$	.	.	.	.	.	.	.	.	.	2	2	1	.	2	.	1	3	.	.	.	.
$s_{41}s_{01}^3$	−4	−1	−1	.	.	−1	.	.	.	.	.	.	.	.	.	.	.	.	.	.	.
$s_{40}s_{02}s_{01}^2$	−6	.	.	−1	.	.	.	.	−1	.	.	.	.	.	.	.	.	.	.	.	.
$s_{32}s_{10}s_{01}^2$	.	−3	.	−1	−2	.	−1	.	.	−1	.	.	.	.	.	.	.	.	.	.	.
$s_{31}s_{11}s_{01}^2$	.	−3	−3	.	.	.	−1	.	.	.	−2	.	.	.	.	.	.	.	.	.	.
$s_{31}s_{10}s_{02}s_{01}$	.	−3	.	−2	.	.	.	−2	.	.	.	−1	.	.	.	.	.	.	.	.	.
$s_{30}s_{12}s_{01}^2$	.	.	−3	−1	.	.	.	.	.	−1	.	.	.	.	.	.	.	.	.	.	.
$s_{30}s_{11}s_{02}s_{01}$	.	.	−3	−2	.	.	.	.	.	.	.	−1	.	.	.	.	.	.	.	.	.
$s_{30}s_{10}s_{03}s_{01}$	.	.	.	−2	.	.	.	.	.	.	.	.	−2	.	.	.	.	.	.	.	.
$s_{30}s_{10}s_{02}^2$	.	.	.	−1	.	.	.	.	.	.	.	.	.	−2	.	.	.	.	.	.	.
$s_{23}s_{10}^2s_{01}$	.	.	.	.	−2	.	.	−1	.	.	.	.	−1	.	−3	−1	.	.	.	.	.
$s_{22}s_{20}s_{01}^2$	.	.	.	.	−1	−3	.	.	−2	−1	−1	.	.	.	.	.	.	.	.	.	.
$s_{22}s_{11}s_{10}s_{01}$	.	.	.	.	−4	.	−2	.	.	.	.	−1	.	.	.	−2	.	−3	.	.	.
$s_{22}s_{10}^2s_{02}$	.	.	.	.	−1	.	.	−1	.	.	.	.	.	−2	.	.	−3	.	−1	.	.
$s_{21}^2s_{01}^2$	.	.	.	.	.	−3	−1	.	.	.	.	.	.	.	.	.	.	.	.	.	.
$s_{21}s_{20}s_{02}s_{01}$	.	.	.	.	.	−3	.	−1	−4	.	.	−1	.	.	.	.	.	.	.	.	.
$s_{21}s_{12}s_{10}s_{01}$	.	.	.	.	.	.	−2	−2	.	−2	.	.	.	.	.	−2	.	.	.	.	.
$s_{21}s_{11}^2s_{01}$	.	.	.	.	.	.	−2	.	.	.	−2	.	.	.	.	.	.	−3	.	.	.
$s_{21}s_{11}s_{10}s_{02}$	.	.	.	.	.	.	−1	−2	.	.	.	−1	.	.	.	.	.	.	−4	.	.
$s_{21}s_{10}^2s_{03}$	.	.	.	.	.	.	.	−1	.	.	.	.	−1	.	.	.	.	.	.	−3	.
$s_{20}^2s_{03}s_{01}$	.	.	.	.	.	.	.	.	−2	.	.	.	−1	.	.	.	.	.	.	.	.
$s_{20}^2s_{02}^2$	.	.	.	.	.	.	.	.	−1	.	.	.	.	−1	.	.	.	.	.	.	.
$s_{20}s_{13}s_{10}s_{01}$	.	.	.	.	.	.	.	.	.	−2	.	−1	−2	.	−3	.	.	.	.	.	.
$s_{20}s_{12}s_{11}s_{01}$	.	.	.	.	.	.	.	.	.	−2	−4	−1	.	.	.	−1	.	.	.	.	.
$s_{20}s_{12}s_{10}s_{02}$	.	.	.	.	.	.	.	.	.	−1	.	−1	.	−4	.	.	−3	.	.	.	.
$s_{20}s_{11}^2s_{02}$	.	.	.	.	.	.	.	.	.	.	−1	−1	.	.	.	.	.	.	−1	.	.
$s_{20}s_{11}s_{10}s_{03}$	.	.	.	.	.	.	.	.	.	.	.	−1	−2	.	.	.	.	.	.	−3	.
$s_{20}s_{10}^2s_{04}$	.	.	.	.	.	.	.	.	.	.	.	.	−1	−1	.	.	.	.	.	.	−6
$s_{14}s_{10}^3$	.	.	.	.	.	.	.	.	.	.	.	.	.	.	−1	.	−1	.	.	−1	−4
$s_{13}s_{11}s_{10}^2$	.	.	.	.	.	.	.	.	.	.	.	.	.	.	−3	−1	.	.	−2	−3	.
$s_{12}^2s_{10}^2$	.	.	.	.	.	.	.	.	.	.	.	.	.	.	.	−1	−3	.	.	.	.
$s_{12}s_{11}^2s_{10}$	.	.	.	.	.	.	.	.	.	.	.	.	.	.	.	−2	.	−3	−2	.	.
s_{11}^4	.	.	.	.	.	.	.	.	.	.	.	.	.	.	.	.	.	−1	.	.	.
$s_{40}s_{01}^4$	*1*	.	.	.	.	.	.	.	.	.	.	.	.	.	.	.	.	.	.	.	.
$s_{31}s_{10}s_{01}^3$	.	*1*	.	.	.	.	.	.	.	.	.	.	.	.	.	.	.	.	.	.	.
$s_{30}s_{11}s_{01}^3$	.	.	*1*	.	.	.	.	.	.	.	.	.	.	.	.	.	.	.	.	.	.
$s_{30}s_{10}s_{02}s_{01}^2$	.	.	.	*1*	.	.	.	.	.	.	.	.	.	.	.	.	.	.	.	.	.
$s_{22}s_{10}^2s_{01}^2$	.	.	.	.	*1*	.	.	.	.	.	.	.	.	.	.	.	.	.	.	.	.
$s_{21}s_{20}s_{01}^3$	.	.	.	.	.	*1*	.	.	.	.	.	.	.	.	.	.	.	.	.	.	.
$s_{21}s_{11}s_{10}s_{01}^2$	.	.	.	.	.	.	*1*	.	.	.	.	.	.	.	.	.	.	.	.	.	.
$s_{21}s_{10}^2s_{02}s_{01}$	.	.	.	.	.	.	.	*1*	.	.	.	.	.	.	.	.	.	.	.	.	.
$s_{20}^2s_{02}s_{01}^2$	.	.	.	.	.	.	.	.	*1*	.	.	.	.	.	.	.	.	.	.	.	.
$s_{20}s_{12}s_{10}s_{01}^2$	.	.	.	.	.	.	.	.	.	*1*	.	.	.	.	.	.	.	.	.	.	.
$s_{20}s_{11}^2s_{01}^2$	.	.	.	.	.	.	.	.	.	.	*1*	.	.	.	.	.	.	.	.	.	.
$s_{20}s_{11}s_{10}s_{02}s_{01}$	.	.	.	.	.	.	.	.	.	.	.	*1*	.	.	.	.	.	.	.	.	.
$s_{20}s_{10}^2s_{03}s_{01}$	.	.	.	.	.	.	.	.	.	.	.	.	*1*	.	.	.	.	.	.	.	.
$s_{20}s_{10}^2s_{02}^2$	.	.	.	.	.	.	.	.	.	.	.	.	.	*1*	.	.	.	.	.	.	.
$s_{13}s_{10}^3s_{01}$	.	.	.	.	.	.	.	.	.	.	.	.	.	.	*1*	.	.	.	.	.	.
$s_{12}s_{11}s_{10}^2s_{01}$	.	.	.	.	.	.	.	.	.	.	.	.	.	.	.	*1*	.	.	.	.	.
$s_{12}s_{10}^3s_{02}$	.	.	.	.	.	.	.	.	.	.	.	.	.	.	.	.	*1*	.	.	.	.
$s_{11}^3s_{10}s_{01}$	.	.	.	.	.	.	.	.	.	.	.	.	.	.	.	.	.	*1*	.	.	.
$s_{11}^2s_{10}^2s_{02}$	.	.	.	.	.	.	.	.	.	.	.	.	.	.	.	.	.	.	*1*	.	.
$s_{11}s_{10}^3s_{03}$	.	.	.	.	.	.	.	.	.	.	.	.	.	.	.	.	.	.	.	*1*	.
$s_{10}^4s_{04}$	.	.	.	.	.	.	.	.	.	.	.	.	.	.	.	.	.	.	.	.	*1*
$s_{30}s_{10}s_{01}^4$	1	4	4	6	.	.	.	.	.	.	.	.	.	.	.	.	.	.	.	.	.
$s_{21}s_{10}^2s_{01}^3$	.	2	.	.	3	1	6	3	.	.	.	.	.	.	.	.	.	.	.	.	.
$s_{20}^2s_{01}^4$	1	.	.	.	.	8	.	.	6	.	.	.	.	.	.	.	.	.	.	.	.
$s_{20}s_{11}s_{10}s_{01}^3$	.	1	1	.	.	1	3	.	.	3	3	3	.	.	.	.	.	.	.	.	.
$s_{20}s_{10}^2s_{02}s_{01}^2$	.	.	.	2	1	.	.	2	1	2	.	4	2	1	.	.	.	.	.	.	.
$s_{12}s_{10}^3s_{01}^2$	.	.	.	.	3	.	.	.	.	3	.	.	.	.	2	6	1	.	.	.	.
$s_{11}^2s_{10}^2s_{01}^2$	.	.	.	.	1	.	4	.	.	.	1	.	.	.	.	4	.	4	1	.	.
$s_{11}s_{10}^3s_{02}s_{01}$	.	.	.	.	.	.	.	3	.	.	.	3	.	.	1	3	1	.	3	1	.
$s_{10}^4s_{03}s_{01}$	.	.	.	.	.	.	.	.	.	.	.	.	6	.	4	.	.	.	.	4	1
$s_{10}^4s_{02}^2$	.	.	.	.	.	.	.	.	.	.	.	.	.	6	.	.	8	.	.	.	1
$s_{20}s_{10}^2s_{01}^4$	1	8	8	12	6	8	24	12	6	12	12	24	4	3	.	.	.	.	.	.	.
$s_{11}s_{10}^3s_{01}^3$	.	3	1	.	9	3	27	9	.	9	9	9	.	.	3	27	3	18	9	1	.
$s_{10}^4s_{02}s_{01}^2$	.	.	.	4	6	.	.	12	3	12	.	24	12	6	8	24	8	.	12	8	1
$s_{10}^4s_{01}^4$	1	16	16	24	36	24	144	72	18	72	72	144	24	18	16	144	24	96	72	16	1

Table 1·6·8 (cont.)

$w=8$ (44) (vi)	$[30, 10, (01)^4]$	$[21, (10)^2, (01)^3]$	$[(20)^2, (01)^4]$	$[20, 11, 10, (01)^3]$	$[20, (10)^2, 02, (01)^2]$	$[12, (10)^3, (01)^2]$	$[(11)^2, (10)^2, (01)^2]$	$[11, (10)^3, 02, 01]$	$[(10)^4, 03, 01]$	$[(10)^4, (02)^2]$	$[20, (10)^2, (01)^2]$	$[11, (10)^3, (01)^2]$	$[(10^4), 02, (01)^2]$	$[(10)^4, (01)^4]$
s_{44}	−120	−120	−120	−120	−120	−120	−120	−120	−120	−120	720	720	720	−5040
$s_{43}s_{01}$	96	72	96	72	48	48	48	24	24	.	−480	−360	−240	2880
$s_{42}s_{02}$	36	18	36	18	30	6	6	24	.	48	−144	−72	−144	720
$s_{41}s_{03}$	16	4	16	4	12	.	.	6	24	.	−48	−12	−48	192
$s_{40}s_{04}$	6	.	6	.	4	.	.	.	.	6	−12	.	−12	36
$s_{34}s_{10}$	24	48	.	24	48	72	48	72	96	96	−240	−360	−480	2880
$s_{33}s_{11}$	24	36	.	42	24	36	72	42	24	.	−192	−336	−192	1920
$s_{32}s_{12}$	24	24	.	30	20	36	32	24	.	48	−144	−180	−144	1152
$s_{31}s_{13}$	24	12	.	18	16	12	8	18	24	.	−96	−72	−96	576
$s_{30}s_{14}$	24	.	.	6	12	4	.	4	16	16	−48	−12	−48	192
$s_{24}s_{20}$	.	6	48	24	30	18	6	18	36	36	−144	−72	−144	720
$s_{23}s_{21}$	.	36	48	24	20	24	32	30	24	.	−144	−180	−144	1152
$s_{22}{}^2$	.	18	24	12	10	18	22	12	.	24	−72	−108	−72	648
$s_{42}s_{01}{}^2$	−36	−18	−36	−18	−6	−6	−6	.	.	.	144	72	24	−720
$s_{41}s_{02}s_{01}$	−24	−6	−24	−6	−12	.	.	−6	.	.	72	18	48	−288
$s_{40}s_{03}s_{01}$	−8	.	−8	.	−4	.	.	.	−6	.	16	.	12	−48
$s_{40}s_{02}{}^2$	−3	.	−3	.	−2	.	.	.	.	−6	6	.	6	−18
$s_{33}s_{10}s_{01}$	−24	−36	.	−18	−24	−36	−24	−18	−24	.	192	216	192	−1920
$s_{32}s_{11}s_{01}$	−24	−24	.	−30	−8	−12	−32	−6	.	.	144	180	48	−1152
$s_{32}s_{10}s_{02}$	−12	−12	.	−6	−16	−6	−4	−18	.	−48	72	54	120	−576
$s_{31}s_{12}s_{01}$	−24	−12	.	−18	−8	−12	−8	−6	.	.	96	72	48	−576
$s_{31}s_{11}s_{02}$	−12	−6	.	−9	−8	.	−4	−12	.	.	48	36	48	−288
$s_{31}s_{10}s_{03}$	−8	−4	.	−2	−8	.	.	−6	−24	.	32	12	48	−192
$s_{30}s_{13}s_{01}$	−24	.	.	−6	−8	−4	.	−2	−8	.	48	12	32	−192
$s_{30}s_{12}s_{02}$	−12	.	.	−3	−6	−2	.	−2	.	−16	24	6	24	−96
$s_{30}s_{11}s_{03}$	−8	.	.	−2	−4	.	.	−2	−8	.	16	4	16	−64
$s_{30}s_{10}s_{04}$	−6	.	.	.	−4	.	.	.	−8	−8	12	.	16	−48
$s_{24}s_{10}{}^2$	.	−6	.	−6	−18	−6	−18	−36	−36	.	24	72	144	−720
$s_{23}s_{20}s_{01}$	.	−6	−48	−18	−16	−12	−4	−6	−12	.	120	54	72	−576
$s_{23}s_{11}s_{10}$	.	−12	.	−6	−8	−24	−32	−30	−24	.	48	180	144	−1152
$s_{22}s_{21}s_{10}$	.	−30	−48	−18	−8	−12	−20	−6	.	.	120	126	48	−864
$s_{22}s_{20}s_{02}$	.	−3	−24	−6	−11	−3	−1	−6	.	−24	48	18	48	−216
$s_{22}s_{12}s_{10}$	.	−12	.	−6	−8	−30	−20	−18	.	−48	48	126	120	−864
$s_{22}s_{11}{}^2$	.	−6	.	−6	−2	−6	−24	−6	.	.	24	90	24	−432
$s_{21}{}^2s_{02}$	.	−6	−12	−3	−4	.	−2	−6	.	.	24	18	24	−144
$s_{21}s_{20}s_{03}$	.	−2	−16	−2	−6	.	.	−3	−12	.	24	6	24	−96
$s_{21}s_{13}s_{10}$	.	−12	.	−6	−8	−12	−8	−18	−24	.	48	72	96	−576
$s_{21}s_{12}s_{11}$	.	−12	.	−12	−4	−12	−24	−12	.	.	48	108	48	−576
$s_{20}{}^2s_{04}$	.	.	−6	.	−2	.	.	.	−3	−3	6	.	6	−18
$s_{20}s_{14}s_{10}$	.	.	.	−6	−12	−6	.	−6	−24	−24	48	18	72	−288
$s_{20}s_{13}s_{11}$	.	.	.	−12	−8	−6	−4	−9	−12	.	48	36	48	−288
$s_{20}s_{12}{}^2$	.	.	.	−6	−4	−6	−2	−3	.	−12	24	18	24	−144
$s_{41}s_{01}{}^3$	8	2	8	2	.	.	.	.	.	.	−24	−6	.	96
$s_{40}s_{02}s_{01}{}^2$	6	.	6	.	2	.	.	.	.	.	−12	.	−6	36
$s_{32}s_{10}s_{01}{}^2$	12	12	.	6	4	6	4	.	.	.	−72	−54	−24	576
$s_{31}s_{11}s_{01}{}^2$	12	6	.	9	.	4	.	.	.	.	−48	−36	.	288
$s_{31}s_{10}s_{02}s_{01}$	12	6	.	3	8	.	6	.	.	.	−48	−18	−48	288
$s_{30}s_{12}s_{01}{}^2$	12	.	.	3	2	2	.	.	.	.	−24	−6	−8	96
$s_{30}s_{11}s_{02}s_{01}$	12	.	.	3	4	.	2	.	.	.	−24	−6	−16	96
$s_{30}s_{10}s_{03}s_{01}$	8	.	.	.	4	.	.	.	8	.	−16	.	−16	64
$s_{30}s_{10}s_{02}{}^2$	3	.	.	.	2	.	.	.	.	8	−6	.	−8	24
$s_{23}s_{10}{}^2s_{01}$	.	6	.	.	4	12	4	6	12	.	−24	−54	−72	576
$s_{22}s_{20}s_{01}{}^2$	.	3	24	6	3	3	1	.	.	.	−48	−18	−12	216
$s_{22}s_{11}s_{10}s_{01}$	.	12	.	6	4	12	20	6	.	.	−48	−126	−48	864
$s_{22}s_{10}{}^2s_{02}$	.	3	.	.	3	3	1	6	.	24	−12	−18	−48	216
$s_{21}{}^2s_{01}{}^2$	.	6	12	3	.	.	2	.	.	.	−24	−18	.	144
$s_{21}s_{20}s_{03}s_{01}$	.	3	24	3	6	.	.	3	.	.	−36	−9	−24	144
$s_{21}s_{12}s_{10}s_{01}$	.	12	.	6	4	12	8	6	.	.	−48	−72	−48	576
$s_{21}s_{11}{}^2s_{01}$	.	6	.	6	.	.	12	.	.	.	−24	−54	.	288
$s_{21}s_{11}s_{10}s_{02}$	.	6	.	3	4	.	4	12	.	.	−24	−36	−48	288
$s_{21}s_{10}{}^2s_{03}$	.	2	.	.	2	.	.	3	12	.	−8	−6	−24	96
$s_{20}{}^2s_{03}s_{01}$	.	.	8	.	2	.	.	.	3	.	−8	.	−6	24
$s_{20}{}^2s_{02}{}^2$	.	.	3	.	1	.	.	.	.	3	−3	.	−3	9
$s_{20}s_{13}s_{10}s_{01}$	.	.	.	6	8	6	.	3	12	.	−48	−18	−48	288
$s_{20}s_{12}s_{11}s_{01}$	.	.	.	12	4	6	4	3	.	.	−48	−36	−24	288
$s_{20}s_{12}s_{10}s_{02}$	.	.	.	3	6	3	.	3	.	24	−24	−9	−36	144
$s_{20}s_{11}{}^2s_{02}$	.	.	.	3	2	.	1	3	.	.	−12	−9	−12	72
$s_{20}s_{11}s_{10}s_{03}$	.	.	.	2	4	.	.	3	12	.	−16	−6	−24	96
$s_{20}s_{10}{}^2s_{04}$	.	.	.	.	2	.	.	.	6	6	−6	.	−12	36
$s_{14}s_{10}{}^2$	.	.	.	.	.	2	.	2	8	8	.	−6	−24	96
$s_{13}s_{11}s_{10}{}^2$	.	.	.	.	.	6	4	9	12	.	.	−36	−48	288
$s_{12}{}^2s_{10}{}^2$	.	.	.	.	.	6	2	3	.	12	.	−18	−24	144
$s_{12}s_{11}{}^2s_{10}$	.	.	.	.	.	6	12	6	.	.	.	−54	−24	288
$s_{11}{}^4$	.	.	.	.	.	2	.	.	.	.	.	−6	.	24
$s_{40}s_{01}{}^4$	−1	.	−1	.	.	.	.	.	.	.	2	.	.	−6
$s_{31}s_{10}s_{01}{}^3$	−4	−2	.	−1	.	.	.	.	.	.	16	6	.	−96
$s_{30}s_{11}s_{01}{}^3$	−4	.	.	−1	.	.	.	.	.	.	8	2	.	−32
$s_{30}s_{10}s_{02}s_{01}{}^2$	−6	.	.	.	−2	.	.	.	.	.	12	.	8	−48
$s_{22}s_{10}{}^2s_{01}{}^2$	.	−3	.	.	−1	−3	−1	.	.	.	12	18	12	−216
$s_{21}s_{20}s_{01}{}^3$	.	−1	−8	−1	.	.	.	.	.	.	12	3	.	−48
$s_{21}s_{11}s_{10}s_{01}{}^2$	.	−6	.	−3	.	.	−4	.	.	.	24	36	.	−288
$s_{21}s_{10}{}^2s_{02}s_{01}$	.	−3	.	.	−2	.	.	−3	.	.	12	9	24	−144
$s_{20}{}^2s_{02}s_{01}{}^2$	.	.	−6	.	−1	.	.	.	.	.	6	.	3	−18
$s_{20}s_{12}s_{10}s_{01}{}^2$	.	.	.	−3	−2	−3	.	.	.	.	24	9	12	−144
$s_{20}s_{11}{}^2s_{01}{}^2$	.	.	.	−3	.	.	−1	.	.	.	12	9	.	−72
$s_{20}s_{11}s_{10}s_{02}s_{01}$	.	.	.	−3	−4	.	.	−3	.	.	24	9	24	−144
$s_{20}s_{10}{}^2s_{03}s_{01}$	.	.	.	.	−2	.	.	.	−6	.	8	.	12	−48
$s_{20}s_{10}{}^2s_{02}{}^2$	.	.	.	.	−1	.	.	.	.	−6	3	.	6	−18
$s_{13}s_{10}{}^3s_{01}$	.	.	.	.	.	−2	.	−1	−4	.	.	6	16	−96
$s_{12}s_{11}s_{10}{}^2s_{01}$	.	.	.	.	.	−6	−4	−3	.	.	.	36	24	−288
$s_{12}s_{10}{}^3s_{02}$	.	.	.	.	.	−1	.	−1	.	−8	.	3	12	−48
$s_{11}{}^3s_{10}s_{01}$	.	.	.	.	.	.	−4	.	.	.	.	18	.	−96
$s_{11}{}^2s_{10}{}^2s_{02}$	.	.	.	.	.	.	−1	−3	.	.	.	9	12	−72
$s_{11}s_{10}{}^3s_{03}$	.	.	.	.	.	.	.	−1	−4	.	.	2	8	−32
$s_{10}{}^4s_{04}$	.	.	.	.	.	.	.	.	−1	−1	.	.	2	−6
$s_{30}s_{10}s_{01}{}^4$	1	.	.	.	.	.	.	.	.	.	−2	.	.	8
$s_{21}s_{10}{}^2s_{01}{}^3$	.	1	.	.	.	.	.	.	.	.	−4	−3	.	48
$s_{20}{}^2s_{01}{}^4$	.	.	1	.	.	.	.	.	.	.	−1	.	.	3
$s_{20}s_{11}s_{10}s_{01}{}^3$	.	.	.	1	.	.	.	.	.	.	−8	−3	.	48
$s_{20}s_{10}{}^2s_{02}s_{01}{}^2$	.	.	.	.	1	.	.	.	.	.	−6	.	−6	36
$s_{12}s_{10}{}^3s_{01}{}^2$	.	.	.	.	.	1	.	.	.	.	.	−3	−4	48
$s_{11}{}^2s_{10}{}^2s_{01}{}^2$	.	.	.	.	.	.	1	.	.	.	.	−9	.	72
$s_{11}s_{10}{}^3s_{02}s_{01}$	.	.	.	.	.	.	.	1	.	.	.	−3	−8	48
$s_{10}{}^4s_{03}s_{01}$	.	.	.	.	.	.	.	.	1	.	.	.	−2	8
$s_{10}{}^4s_{02}{}^2$	.	.	.	.	.	.	.	.	.	1	.	.	−1	3
$s_{20}s_{10}{}^2s_{01}{}^4$	2	4	1	8	6	.	.	3	.	.	1	.	.	−6
$s_{11}s_{10}{}^3s_{01}{}^3$	.	3	.	3	.	3	9	3	.	.	.	1	.	−16
$s_{10}{}^4s_{02}s_{01}{}^2$	.	.	.	.	6	.	4	.	8	2	.	.	1	−6
$s_{10}{}^4s_{01}{}^4$	4	24	3	48	36	24	72	48	4	3	6	16	6	1

Table 1·7. *Monomial symmetric functions in term of powers and products of monomial symmetric functions of lower weight (up to and including weight 8)*

Note that the final digit in the subheadings indicates the weight, w.

The monomial symmetric functions are *not* augmented in the sense of David and Kendall. The tables should be read horizontally only, from the left border of the table to the right, and in 1·7·8 the desired expansion may extend over two pages.

Table 1·7·2

$w = 2$	(2)	(1^2)
$(1)(1)$	1	2

Table 1·7·3

$w = 3$	(3)	(21)	(1^3)
$(2)(1)$	1	1	.
$(1^2)(1)$	.	1	3
$(1)^3$	1	3	6

Table 1·7·4

$w = 4$	(4)	(31)	(2^2)	(21^2)	(1^4)
$(3)(1)$	1	1	.	.	.
$(21)(1)$	.	1	2	2	.
$(2)^2$	1	.	2	.	.
$(2)(1^2)$	.	1	.	1	.
$(1^3)(1)$	.	.	.	1	4
$(1^2)^2$	.	.	1	2	6
$(2)(1)^2$	1	2	2	2	.
$(1^2)(1)^2$	.	1	2	5	12
$(1)^4$	1	4	6	12	24

Table 1·7·5

$w = 5$	(5)	(41)	(32)	(31^2)	(2^21)	(21^3)	(1^5)
$(4)(1)$	1	1	.	.	.	.	.
$(31)(1)$	.	1	1	2	.	.	.
$(3)(2)$	1	.	1	.	.	.	.
$(3)(1^2)$	.	1	.	1	.	.	.
$(2^2)(1)$	.	.	1	.	1	.	.
$(21^2)(1)$	.	.	.	1	2	3	.
$(21)(2)$	.	1	1	.	2	.	.
$(21)(1^2)$	.	.	1	2	2	3	.
$(2)(1^3)$	.	.	.	1	.	1	.
$(1^4)(1)$	.	.	.	.	.	1	5
$(1^3)(1^2)$	.	.	.	.	1	3	10
$(3)(1)^2$	1	2	1	2	.	.	.
$(21)(1)^2$	.	1	3	4	6	6	.
$(2)^2(1)$	1	1	2	.	2	.	.
$(2)(1^2)(1)$	.	1	1	3	2	3	.
$(1^3)(1)^2$	.	.	.	1	2	7	20
$(1^2)^2(1)$	.	.	1	2	5	12	30
$(2)(1)^3$	1	3	4	6	6	6	.
$(1^2)(1)^3$	.	1	3	7	12	27	60
$(1)^5$	1	5	10	20	30	60	120

Table 1·7·6

$w = 6$	(6)	(51)	(42)	(41^2)	(3^2)	(321)	(31^3)	(2^3)	(2^21^2)	(21^4)	(1^6)
$(5)(1)$	1	1	.	.	.	.	.	.	.	.	.
$(41)(1)$	.	1	1	2	.	.	.	.	.	.	.
$(4)(2)$	1	1	1	.	.	.	.	.	.	.	.
$(4)(1^2)$	.	1	.	1	.	.	.	.	.	.	.
$(32)(1)$	.	.	1	.	2	1	.	.	.	.	.
$(31^2)(1)$	.	.	.	1	.	1	3	.	.	.	.
$(31)(2)$	.	1	.	.	2	1	.	.	.	.	.
$(31)(1^2)$	.	.	1	2	.	1	3	.	.	.	.
$(3)^2$	1	.	.	.	2	.	.	.	.	.	.
$(3)(21)$	.	1	1	.	.	1	.	.	.	.	.
$(3)(1^3)$	.	.	.	1	.	.	1	.	.	.	.
$(2^21)(1)$	.	.	.	.	.	1	.	3	2	.	.
$(2^2)(2)$	.	.	1	.	.	.	.	3	.	.	.
$(2^2)(1^2)$	.	.	.	.	1	1	.	.	1	.	.
$(21^3)(1)$	.	.	.	.	.	.	1	.	2	4	.
$(21^2)(2)$	.	.	.	1	.	.	1	.	2	.	.
$(21^2)(1^2)$	.	.	.	.	.	1	3	3	4	6	.
$(21)^2$	.	.	1	2	2	2	.	6	4	.	.
$(21)(1^3)$	.	.	.	.	.	1	3	.	2	4	.
$(2)(1^4)$	.	.	.	.	.	.	1	.	.	1	.
$(1^5)(1)$	.	.	.	.	.	.	.	.	.	1	6
$(1^4)(1^2)$	.	.	.	.	.	.	.	.	1	4	15
$(1^3)^2$	.	.	.	.	.	.	.	1	2	6	20
$(4)(1)^2$	1	2	1	2	.	.	.	.	.	.	.
$(31)(1)^2$	.	1	2	4	2	3	6	.	.	.	.
$(3)(2)(1)$	1	1	1	.	2	1	.	.	.	.	.
$(3)(1^2)(1)$	.	1	1	3	.	1	3	.	.	.	.
$(2^2)(1)^2$	.	.	1	.	2	2	.	3	2	.	.
$(21^2)(1)^2$	.	.	.	1	.	3	6	.	10	12	.
$(21)(2)(1)$	.	1	2	2	2	3	.	6	4	.	.
$(21)(1^2)(1)$	.	.	1	2	2	5	9	6	10	12	.
$(2)^3$	1	.	3	.	.	.	.	6	.	.	.
$(2)^2(1^2)$	.	.	1	.	2	2	.	.	2	.	.
$(2)(1^3)(1)$	.	.	.	1	.	1	4	.	2	4	.
$(2)(1^2)^2$	.	.	1	2	.	2	6	3	4	6	.
$(1^4)(1)^2$	.	.	.	.	.	.	1	.	2	9	30
$(1^3)(1^2)(1)$	.	.	.	.	.	1	3	3	8	22	60
$(1^2)^3$	.	.	.	.	1	3	6	6	15	36	90
$(3)(1)^3$	1	3	3	6	2	3	.	.	.	.	.
$(21)(1)^3$	.	1	4	6	6	13	18	18	24	24	.
$(2)^2(1)^2$	1	2	3	2	4	4	.	6	4	.	.
$(2)(1^2)(1)^2$	.	1	2	5	2	6	12	6	10	12	.
$(1^3)(1)^3$	.	.	.	1	.	3	10	6	18	48	120
$(1^2)^2(1)^2$	.	.	1	2	2	8	18	15	34	78	180
$(2)(1^4)$	1	4	7	12	8	16	24	18	24	24	.
$(1^2)(1)^4$	.	1	4	9	6	22	48	36	78	168	360
$(1)^6$	1	6	15	30	20	60	120	90	180	360	720

Table 1·7·7

$w = 7$ (i)

	(7)	(61)	(52)	(51^2)	(43)	(421)	(41^3)	(3^21)	(32^2)	(321^2)	(31^4)	(2^31)	(2^21^3)	(21^5)	(1^7)
$(6)(1)$	1	1	·	·	·	·	·	·	·	·	·	·	·	·	·
$(51)(1)$	·	1	1	2	·	·	·	·	·	·	·	·	·	·	·
$(5)(2)$	1	·	1	·	·	·	·	·	·	·	·	·	·	·	·
$(5)(1^2)$	·	1	·	1	·	·	·	·	·	·	·	·	·	·	·
$(42)(1)$	·	·	1	·	1	1	·	·	·	·	·	·	·	·	·
$(41^2)(1)$	·	·	·	1	·	1	3	·	·	·	·	·	·	·	·
$(41)(2)$	·	1	·	·	1	1	·	·	·	·	·	·	·	·	·
$(41)(1^2)$	·	·	1	2	·	1	3	·	·	·	·	·	·	·	·
$(4)(3)$	1	·	·	·	1	·	·	·	·	·	·	·	·	·	·
$(4)(21)$	·	1	1	·	·	1	·	·	·	·	·	·	·	·	·
$(4)(1^3)$	·	·	·	1	·	·	1	1	·	·	·	·	·	·	·
$(3^2)(1)$	·	·	·	·	1	·	·	1	·	·	·	·	·	·	·
$(321)(1)$	·	·	·	·	·	1	·	·	2	2	2	·	·	·	·
$(32)(2)$	·	·	1	·	1	·	·	·	2	2	·	·	·	·	·
$(32)(1^2)$	·	·	·	·	1	1	·	·	2	·	1	·	·	·	·
$(31^3)(1)$	·	·	·	·	·	·	1	·	·	·	1	·	·	·	·
$(31^2)(2)$	·	·	·	1	·	·	·	·	2	·	1	4	·	·	·
$(31^2)(1^2)$	·	·	·	·	·	1	3	·	1	2	6	·	·	·	·
$(31)(3)$	·	1	·	·	1	·	·	2	·	2	·	·	·	·	·
$(31)(21)$	·	·	1	2	1	1	·	2	2	2	·	·	·	·	·
$(31)(1^3)$	·	·	·	·	·	1	3	·	·	1	4	·	·	·	·
$(3)(2^2)$	·	·	1	·	1	·	·	1	·	·	1	·	·	·	·
$(3)(21^2)$	·	·	·	1	·	1	·	·	·	1	·	·	·	·	·
$(3)(1^4)$	·	·	·	·	·	·	1	·	·	·	1	·	·	·	·
$(2^3)(1)$	·	·	·	·	·	·	·	1	·	·	·	1	·	·	·
$(2^21^2)(1)$	·	·	·	·	·	·	·	·	·	1	·	·	3	3	·
$(2^21)(2)$	·	·	·	·	·	1	·	·	1	·	·	3	3	·	·
$(2^21)(1^2)$	·	·	·	·	·	·	·	1	2	2	·	3	3	·	·
$(2^2)(21)$	·	·	·	·	1	1	·	·	2	2	2	3	·	·	·
$(2^2)(1^3)$	·	·	·	·	·	·	·	1	·	1	·	·	1	·	·
$(21^4)(1)$	·	·	·	·	·	·	·	·	·	·	1	1	2	5	·
$(21^3)(2)$	·	·	·	·	·	·	1	·	·	·	1	·	2	·	·
$(21^3)(1^2)$	·	·	·	·	·	·	·	·	·	1	4	3	6	10	·
$(21^2)(21)$	·	·	·	·	·	1	3	2	2	2	·	6	6	·	·
$(21^2)(1^3)$	·	·	·	·	·	·	·	·	1	2	3	6	6	10	·
$(21)(1^4)$	·	·	·	·	·	·	·	·	·	·	1	4	6	5	·
$(2)(1^5)$	·	·	·	·	·	·	·	·	·	·	1	·	2	5	·
$(1^6)(1)$	·	·	·	·	·	·	·	·	·	·	·	·	·	1	7
$(1^5)(1^2)$	·	·	·	·	·	·	·	·	·	·	·	·	1	5	21
$(1^4)(1^3)$	·	·	·	·	·	·	·	·	·	·	·	1	3	10	35
$(5)(1)^2$	1	2	1	2	·	·	·	·	·	·	·	·	·	·	·
$(41)(1)^2$	1	1	2	4	1	3	6	·	·	·	·	·	·	·	·
$(4)(2)(1)$	1	1	1	·	1	1	·	·	·	·	·	·	·	·	·
$(4)(1^2)(1)$	·	1	1	3	1	3	·	·	·	·	·	·	·	·	·
$(32)(1)^2$	·	·	1	·	3	2	·	4	2	2	·	·	·	·	·
$(31^2)(1)^2$	·	·	·	·	·	2	6	2	2	5	12	·	·	·	·
$(31)(2)(1)$	·	1	1	2	2	1	·	4	2	2	·	·	·	·	·
$(31)(1^2)(1)$	·	·	1	2	1	4	9	2	2	5	12	·	·	·	·
$(3)^2(1)$	1	1	·	·	2	·	·	2	·	·	·	·	·	·	·
$(3)(21)(1)$	·	1	2	2	1	2	·	2	2	2	·	·	·	·	·

$w = 7$ (ii)

	(7)	(61)	(52)	(51^2)	(43)	(421)	(41^3)	(3^21)	(32^2)	(321^2)	(31^4)	(2^31)	(2^21^3)	(21^5)	(1^7)
$(3)(2)^2$	1	·	2	·	1	·	·	·	2	·	·	·	·	·	·
$(3)(2)(1^2)$	·	1	·	1	1	1	·	2	·	1	·	·	·	·	·
$(3)(1^3)(1)$	·	·	·	1	·	1	4	·	·	·	1	4	·	·	·
$(3)(1^2)^2$	·	·	1	2	·	2	6	·	1	2	6	·	·	·	·
$(2^21)(1)^2$	·	·	·	·	·	1	·	2	5	4	·	9	6	·	·
$(2^2)(2)(1)$	·	·	1	·	1	1	·	·	3	·	·	3	·	·	·
$(2^2)(1^2)(1)$	·	·	·	·	1	1	·	3	2	3	·	3	3	·	·
$(21^3)(1^2)$	·	·	·	·	·	·	1	·	·	3	8	6	14	20	·
$(21^2)(2)(1)$	·	·	·	1	·	2	3	2	2	4	·	6	6	·	·
$(21^2)(1^2)(1)$	·	·	·	·	·	1	3	2	5	9	18	15	24	30	·
$(21)^2(1)$	·	·	1	2	3	5	6	6	10	8	·	18	12	·	·
$(21)(2)^2$	·	1	1	·	2	3	·	·	4	·	·	6	·	·	·
$(21)(2)(1^2)$	·	·	1	2	1	2	3	4	4	5	·	6	6	·	·
$(21)(1^3)(1)$	·	·	·	·	·	1	3	2	2	7	16	6	14	20	·
$(21)(1^2)^2$	·	·	·	·	1	3	6	4	6	12	24	15	24	30	·
$(2^2)(1^3)$	·	·	·	1	·	·	1	2	·	2	·	·	2	·	·
$(2)(1^4)(1)$	·	·	·	·	·	·	1	·	·	1	5	2	2	5	·
$(2)(1^3)(1^2)$	·	·	·	·	·	1	3	·	1	3	10	3	6	10	·
$(1^5)(1^2)$	·	·	·	·	·	·	·	·	·	·	1	·	2	11	42
$(1^4)(1^2)(1)$	·	·	·	·	·	·	·	·	·	·	1	4	11	35	105
$(1^3)^2(1)$	·	·	·	·	·	·	·	·	1	2	6	7	18	50	140
$(1^3)(1^2)^2$	·	·	·	·	·	·	·	1	2	5	12	12	31	80	210
$(4)(1)^3$	1	3	3	6	·	3	6	·	·	·	·	·	·	·	·
$(31)(1)^3$	·	1	3	6	4	9	18	8	6	12	24	·	·	·	·
$(3)(2)(1)^2$	1	2	2	3	3	2	·	4	2	2	·	·	·	·	·
$(3)(1^2)(1)^2$	·	1	2	5	1	5	12	2	2	5	12	·	·	·	·
$(2^2)(1)^3$	·	·	1	·	3	3	·	6	7	6	6	·	9	6	·
$(21^2)(1)^3$	·	·	·	1	·	4	9	6	12	22	36	36	54	60	·
$(21)(2)(1)^2$	·	1	3	4	4	7	6	8	12	10	·	18	12	·	·
$(21)(1^2)(1)^2$	·	·	1	2	3	8	15	12	16	29	48	36	54	60	·
$(2)^3(1)$	1	1	3	·	3	3	·	·	6	·	·	6	·	·	·
$(2)^2(1^2)(1)$	·	1	1	3	2	3	3	6	4	6	·	6	6	20	·
$(2)(1^3)(1^2)$	·	·	·	1	·	2	7	2	2	8	20	6	14	20	·
$(2)(1^2)^2(1)$	·	·	1	2	1	5	12	4	7	14	30	15	24	30	·
$(1^4)(1^3)$	·	·	·	·	·	·	1	·	·	3	13	6	24	75	210
$(1^3)(1^2)(1)^2$	·	·	·	·	·	1	3	2	5	13	34	27	68	170	420
$(1^2)^3(1)$	·	·	·	·	1	3	6	7	12	27	60	51	117	270	630
$(3)(1)^4$	1	4	6	12	5	12	24	8	6	12	24	·	·	·	·
$(2)^2(1)^3$	1	3	5	6	7	9	6	12	14	12	·	18	12	·	·
$(21)(1)^4$	·	1	5	8	10	23	36	32	44	68	96	90	120	120	·
$(2)(1^2)(1)^3$	·	1	3	7	4	13	27	14	18	34	60	36	54	60	·
$(1^3)(1)^4$	·	·	·	1	·	4	13	6	12	34	88	60	150	360	840
$(1^2)^2(1)^3$	·	·	1	2	3	11	24	18	31	68	150	117	258	570	1260
$(2)(1)^5$	1	5	11	20	15	35	60	40	50	80	120	90	120	120	·
$(1^2)(1)^5$	·	1	5	11	10	35	75	50	80	170	360	270	570	1200	2520
$(1)^7$	1	7	21	42	35	105	210	140	210	420	840	630	1260	2520	5040

Table 1·7·8

$w = 8$ (i)	(8)	(71)	(62)	(61^2)	(53)	(521)	(51^3)	(4^2)	(431)	(42^2)	(421^2)
(7)(1)	1	1	.	.	.	.	.	.	.	.	.
(61)(1)	.	1	1	2	.	.	.	.	.	.	.
(6)(2)	1	.	1	.	.	.	.	.	.	.	.
$(6)(1^2)$	.	1	.	1	.	.	.	.	.	.	.
(52)(1)	.	.	1	.	1	1	.	.	.	.	.
$(51^2)(1)$	.	.	.	1	.	1	3	.	.	.	.
(51)(2)	.	1	.	.	1	1	.	.	.	.	.
$(51)(1^2)$	.	.	1	2	.	1	3	.	.	.	.
(5)(3)	1	.	.	.	1	.	1	.	.	.	.
(5)(21)	.	1	1	.	.	1	.	.	.	.	.
$(5)(1^3)$	.	.	.	1	.	.	1	.	.	.	.
(43)(1)	.	.	.	.	1	.	.	2	1	2	2
(421)(1)	.	.	.	.	.	1	.	.	1	2	2
(42)(2)	.	.	1	.	.	.	.	2	.	2	.
$(42)(1^2)$	.	.	.	.	1	1	.	.	1	.	1
$(41^3)(1)$	.	.	.	.	.	.	1	.	.	.	1
$(41^2)(2)$	.	.	.	1	.	.	.	.	1	.	1
$(41^2)(1^2)$	.	.	.	.	.	1	3	.	.	1	2
(41)(3)	.	1	.	.	.	.	.	2	1	2	.
(41)(21)	.	.	1	2	1	1	.	.	1	2	2
$(41)(1^3)$	.	.	.	.	.	1	3	.	.	.	1
$(4)^2$	1	.	.	.	.	.	.	2	.	.	.
(4)(31)	.	1	.	.	1	.	.	.	1	.	.
$(4)(2^2)$	.	.	1	.	.	.	.	.	.	1	.
$(4)(21^2)$	.	.	.	1	.	1	.	.	.	.	1
$(4)(1^4)$	.	.	.	.	.	.	1	.	.	.	.
$(3^21)(1)$	.	.	.	.	.	.	.	.	1	.	.
$(3^2)(2)$	.	.	.	.	1	.	.	.	.	.	.
$(3^2)(1^2)$	.	.	.	.	.	.	.	1	1	.	.
$(32^2)(1)$	.	.	.	.	.	.	.	.	.	1	.
$(321^2)(1)$	.	.	.	.	.	.	.	.	.	.	1
(321)(2)	.	.	.	.	.	1	.	.	1	.	.
$(321)(1^2)$	.	.	.	.	.	.	.	.	1	2	2
(32)(3)	.	.	1	.	1	.	.	.	.	.	.
(32)(21)	.	.	.	.	1	1	.	2	1	2	.
$(32)(1^3)$	.	.	.	.	.	.	.	.	1	.	1
$(31^4)(1)$	.	.	.	.	.	.	.	.	.	.	.
$(31^3)(2)$	.	.	.	.	.	.	1	.	.	.	.
$(31^3)(1^2)$	.	.	.	.	.	.	.	.	.	.	1
$(31^2)(3)$	.	.	.	1	.	.	.	.	1	.	.
$(31^2)(21)$	.	.	.	.	.	1	3	.	1	.	1
$(31^2)(1^3)$	.	.	.	.	.	.	.	.	.	1	2
$(31)^2$	.	.	1	2	.	.	.	2	2	.	.
$(31)(2^2)$	.	.	.	.	1	1	.	.	.	1	.
$(31)(21^2)$	.	.	.	.	.	1	3	.	1	2	2
$(31)(1^4)$	.	.	.	.	.	1	.	.	.	.	1
$(3)(2^21)$	.	.	.	.	.	1	.	.	.	1	.
$(3)(21^3)$	.	.	.	.	.	.	1	.	.	.	1
$(3)(1^5)$	.	.	.	.	.	.	.	.	.	.	.
$(2^21)(1)$	.	.	.	.	.	.	.	.	.	.	.
$(2^3)(2)$	.	.	.	.	.	.	.	.	.	1	.
$(2^3)(1^2)$	.	.	.	.	.	.	.	.	.	.	.
$(2^21^3)(1)$	.	.	.	.	.	.	.	.	.	.	.
$(2^21^2)(2)$	.	.	.	.	.	.	.	.	.	.	1
$(2^21^2)(1^2)$	.	.	.	.	.	.	.	.	.	.	.
$(2^21)(21)$	.	.	.	.	.	.	.	.	1	2	2
$(2^21)(1^3)$	.	.	.	.	.	.	.	.	.	.	.
$(2^2)^2$	.	.	.	.	.	.	.	1	.	2	.
$(2^2)(21^2)$	.	.	.	.	.	.	.	.	1	.	1
$(2^2)(1^4)$	.	.	.	.	.	.	.	.	.	.	.
$(21^5)(1)$	.	.	.	.	.	.	.	.	.	.	.
$(21^4)(2)$	.	.	.	.	.	.	.	.	.	.	.
$(21^4)(1^3)$	.	.	.	.	.	.	.	.	.	.	.
$(21^3)(21)$	.	.	.	.	.	.	.	.	.	.	1
$(21^3)(1^3)$	.	.	.	.	.	.	.	.	.	.	.
$(21^2)^2$	.	.	.	.	.	.	.	.	.	1	2
$(21^2)(1^4)$	.	.	.	.	.	.	.	.	.	.	.
$(21)(1^5)$	.	.	.	.	.	.	.	.	.	.	.
$(2)(1^6)$	.	.	.	.	.	.	.	.	.	.	.
$(1^7)(1)$	.	.	.	.	.	.	.	.	.	.	.
$(1^6)(1^2)$	.	.	.	.	.	.	.	.	.	.	.
$(1^5)(1^3)$	.	.	.	.	.	.	.	.	.	.	.
$(1^4)^2$	.	.	.	.	.	.	.	.	.	.	.

Table 1·7·8 (*cont.*)

$w = 8$ (ii)	(41^4)	$(3^2 2)$	$(3^2 1^2)$	$(32^2 1)$	(321^3)	(31^5)	(2^4)	$(2^3 1^2)$	$(2^2 1^4)$	(21^6)	(1^8)
$(7)(1)$	.	.	.	.	.	.	.	.	.	.	.
$(61)(1)$	.	.	.	.	.	.	.	.	.	.	.
$(6)(2)$	.	.	.	.	.	.	.	.	.	.	.
$(6)(1^2)$	.	.	.	.	.	.	.	.	.	.	.
$(52)(1)$	.	.	.	.	.	.	.	.	.	.	.
$(51^2)(1)$	.	.	.	.	.	.	.	.	.	.	.
$(51)(2)$	.	.	.	.	.	.	.	.	.	.	.
$(51)(1^2)$	.	.	.	.	.	.	.	.	.	.	.
$(5)(3)$	.	.	.	.	.	.	.	.	.	.	.
$(5)(21)$	.	.	.	.	.	.	.	.	.	.	.
$(5)(1^3)$	.	.	.	.	.	.	.	.	.	.	.
$(43)(1)$	.	.	.	.	.	.	.	.	.	.	.
$(421)(1)$	.	.	.	.	.	.	.	.	.	.	.
$(42)(2)$	.	.	.	.	.	.	.	.	.	.	.
$(42)(1^2)$	.	.	.	.	.	.	.	.	.	.	.
$(41^3)(1)$	4	.	.	.	.	.	.	.	.	.	.
$(41^2)(2)$	.	.	.	.	.	.	.	.	.	.	.
$(41^2)(1^2)$	6	.	.	.	.	.	.	.	.	.	.
$(41)(3)$	.	.	.	.	.	.	.	.	.	.	.
$(41)(21)$	.	.	.	.	.	.	.	.	.	.	.
$(41)(1^3)$	4	.	.	.	.	.	.	.	.	.	.
$(4)^2$	.	.	.	.	.	.	.	.	.	.	.
$(4)(31)$	.	.	.	.	.	.	.	.	.	.	.
$(4)(2^2)$	.	.	.	.	.	.	.	.	.	.	.
$(4)(21^2)$	.	.	.	.	.	.	.	.	.	.	.
$(4)(1^4)$	1	.	.	.	.	.	.	.	.	.	.
$(3^2 1)(1)$	.	1	2	.	.	.	.	.	.	.	.
$(3^2)(2)$	.	1	1	.	.	.	.	.	.	.	.
$(3^2)(1^2)$	.	.	1	.	.	.	.	.	.	.	.
$(32^2)(1)$	.	2	.	1	.	.	.	.	.	.	.
$(321^2)(1)$	.	.	2	2	3	.	.	.	.	.	.
$(321)(2)$	.	2	.	2	2	.	.	.	.	.	.
$(321)(1^2)$	.	2	4	2	3	.	.	.	.	.	.
$(32)(3)$	.	2	.	.	.	.	.	.	.	.	.
$(32)(21)$	.	2	.	2	.	.	.	.	.	.	.
$(32)(1^3)$	.	.	2	.	1	.	.	.	.	.	.
$(31^4)(1)$	1	.	.	.	1	5	.	.	.	.	.
$(31^3)(2)$	.	.	2	.	1	.	.	.	.	.	.
$(31^3)(1^2)$	4	.	.	1	3	10	.	.	.	.	.
$(31^2)(3)$	.	.	.	2	.	.	.	.	.	.	.
$(31^2)(21)$	.	2	4	2	3	.	.	.	.	.	.
$(31^2)(1^3)$	6	.	.	1	3	10	.	.	.	.	.
$(31)^2$	.	2	4	.	.	.	.	.	.	.	.
$(31)(2^2)$	.	2	.	1	.	.	.	.	.	.	.
$(31)(21^2)$	.	.	2	2	3	.	.	.	.	.	.
$(31)(1^4)$	4	.	.	.	1	5	.	.	.	.	.
$(3)(2^2 1)$	.	.	.	1	1	.	.	.	.	.	.
$(3)(21^3)$	.	.	.	.	1	.	.	.	.	.	.
$(3)(1^5)$	1	.	.	.	.	1	.	.	.	.	.
$(2^3 1)(1)$	.	.	.	1	.	.	4	2	.	.	.
$(2^3)(2)$	.	.	.	.	.	.	4	.	.	.	.
$(2^3)(1^2)$	.	1	.	1	.	.	.	1	.	.	.
$(2^2 1^3)(1)$	.	.	.	.	1	.	.	3	4	.	.
$(2^2 1^2)(2)$	.	.	.	1	.	.	.	3	.	.	.
$(2^2 1^2)(1^2)$	.	.	1	2	3	.	6	6	6	.	.
$(2^2 1)(21)$	.	2	.	3	3	.	12	6	.	.	.
$(2^2 1)(1^3)$	.	1	2	2	3	.	6	3	4	.	.
$(2^2)^2$	.	.	.	.	.	.	6	.	.	.	.
$(2^2)(21^2)$	.	1	.	2	.	.	.	3	.	.	.
$(2^2)(1^4)$	.	.	1	1	.	.	.	.	1	.	.
$(21^5)(1)$	.	.	.	.	.	1	.	.	2	6	.
$(21^4)(2)$	1	.	.	.	1	.	.	.	2	.	.
$(21^4)(1^2)$	.	.	.	.	1	5	.	3	8	15	.
$(21^3)(21)$	4	.	2	2	4	.	.	6	8	.	.
$(21^3)(1^3)$	.	.	.	.	3	10	4	6	12	20	.
$(21^2)^2$	6	2	4	4	6	.	12	12	12	.	.
$(21^2)(1^4)$	.	.	.	1	3	10	.	3	8	15	.
$(21)(1^5)$	.	.	.	.	1	5	.	.	2	6	.
$(2)(1^6)$	.	.	.	.	.	1	.	.	.	1	.
$(1^7)(1)$	.	.	.	.	.	.	.	.	.	1	8
$(1^6)(1^2)$	.	.	.	.	.	.	.	.	1	6	28
$(1^5)(1^3)$	.	.	.	.	.	.	.	1	4	15	56
$(1^4)^2$	.	.	.	.	.	.	1	2	6	20	70

Table 1.7.8 (*cont.*)

$w = 8$ (iii)	(8)	(71)	(62)	(61^2)	(53)	(521)	(51^3)	(4^2)	(431)	(42^2)	(421^2)
$(6)(1)^2$	1	2	1	2	·	·	·	·	·	·	·
$(51)(1)^2$	·	1	2	4	1	3	6	·	·	·	·
$(5)(2)(1)$	1	1	1	·	1	1	·	·	·	·	·
$(5)(1^3)(1)$	·	1	1	1	·	1	3	·	·	·	·
$(42)(1)^2$	·	·	1	·	2	2	·	2	2	2	2
$(41^2)(1)^2$	·	·	·	1	·	2	6	·	1	2	5
$(41)(2)(1)$	·	1	1	2	1	1	·	2	2	2	2
$(41)(1^2)(1)$	·	1	1	2	1	4	9	·	1	2	5
$(4)(3)(1)$	1	1	2	·	1	2	·	2	1	2	·
$(4)(21)(1)$	·	1	2	2	1	2	·	·	1	2	2
$(4)(2)^2$	1	·	2	·	·	·	·	2	·	2	·
$(4)(2)(1^2)$	·	1	·	1	1	1	·	·	1	·	1
$(4)(1^3)(1)$	·	·	·	1	·	1	4	·	·	1	1
$(4)(1^2)^2$	·	·	1	2	·	2	6	·	·	1	2
$(3^2)(1)^2$	·	·	·	·	1	·	·	2	2	·	·
$(321)(1)^2$	·	·	·	·	·	1	·	·	3	4	4
$(32)(2)(1)$	·	·	1	·	2	1	·	2	1	2	·
$(32)(1^2)(1)$	·	·	·	1	1	1	·	2	4	2	3
$(31^3)(1)^2$	·	·	·	·	·	·	1	·	·	·	2
$(31^2)(2)(1)$	·	·	·	1	·	1	3	·	2	·	1
$(31^2)(1^2)(1)$	·	·	·	·	·	1	3	·	1	3	7
$(31)(3)(1)$	·	1	1	2	1	·	·	2	3	·	·
$(31)(21)(1)$	·	·	1	2	2	4	6	2	4	4	4
$(31)(2)^2$	·	1	·	·	3	2	·	1	·	·	·
$(31)(2)(1^2)$	·	·	1	2	·	1	3	2	3	2	2
$(31)(1^3)(1)$	·	·	·	·	·	1	3	·	1	2	6
$(31)(1^2)^2$	·	·	·	·	1	3	6	·	2	4	10
$(3)^2(2)$	1	·	1	·	2	·	·	·	·	·	·
$(3)^2(1^2)$	·	1	1	1	·	·	·	·	2	2	·
$(3)(2^2)(1)$	·	·	1	·	1	1	·	·	·	1	·
$(3)(21^2)(1)$	·	·	·	1	·	2	3	·	1	2	3
$(3)(21)(2)$	·	1	1	·	1	2	2	2	2	2	3
$(3)(21)(1^3)$	·	·	1	2	1	2	3	·	2	2	3
$(3)(2)(1^3)$	·	·	·	1	·	·	1	·	1	·	1
$(3)(1^4)(1)$	·	·	·	·	·	·	1	·	·	·	·
$(3)(1^3)(1^2)$	·	·	·	·	·	1	3	·	·	1	3
$(2^3)(1^2)$	·	·	·	·	·	·	·	·	·	1	1
$(2^21^2)(1)^2$	·	·	·	·	·	1	·	·	·	·	2
$(2^21)(2)(1)$	·	·	·	·	·	·	·	·	1	3	2
$(2^21)(1^2)(1)$	·	·	·	·	·	·	·	·	1	2	2
$(2^2)(21)(1)$	·	·	·	·	1	1	·	2	2	4	2
$(2^2)(2)^2$	·	·	1	·	·	·	·	2	·	5	·
$(2^2)(2)(1^2)$	·	·	·	·	1	1	·	·	1	·	1
$(2^2)(1^3)(1)$	·	·	·	·	·	·	·	·	1	1	1
$(2^2)(1^2)^2$	·	·	·	·	·	·	·	1	2	2	2
$(21^4)(1)^2$	·	·	·	·	·	·	·	·	·	·	·
$(21^3)(2)(1)$	·	·	·	·	·	·	1	·	·	·	2
$(21^3)(1^2)(1)$	·	·	·	·	·	·	·	·	·	·	1
$(21^2)(21)(1)$	·	·	·	·	·	1	3	·	3	4	8
$(21^2)(2)^2$	·	·	·	1	·	1	·	·	2	·	3
$(21^2)(2)(1^2)$	·	·	·	·	·	1	3	·	1	3	4
$(21^2)(1^3)(1)$	·	·	·	·	·	·	·	·	·	1	2
$(21^2)(1^2)^2$	·	·	·	·	·	·	·	·	1	2	5
$(21)^2(2)$	·	·	1	2	2	2	·	2	4	8	6
$(21)^2(1^2)$	·	·	·	·	1	3	6	2	5	6	9
$(21)(2)(1^3)$	·	·	·	·	·	1	3	·	1	·	2
$(21)(1^4)(1)$	·	·	·	·	·	·	·	·	·	·	1
$(21)(1^3)(1^2)$	·	·	·	·	·	·	·	·	1	2	5
$(2)^2(1^4)$	·	·	·	·	·	·	1	·	·	·	·
$(2)(1^5)(1)$	·	·	·	·	·	·	·	·	·	·	·
$(2)(1^4)(1^2)$	·	·	·	·	·	·	·	·	·	·	1
$(2)(1^3)^2$	·	·	·	·	·	·	·	·	·	1	2
$(1^6)(1)^2$	·	·	·	·	·	·	·	·	·	·	·
$(1^5)(1^2)(1)$	·	·	·	·	·	·	·	·	·	·	·
$(1^4)(1^3)(1)$	·	·	·	·	·	·	·	·	·	·	·
$(1^4)(1^2)^2$	·	·	·	·	·	·	·	·	·	·	·
$(1^3)^2(1^2)$	·	·	·	·	·	·	·	·	·	·	·

Table 1·7·8 (*cont.*)

$w = 8$ (iv)	(41^4)	(3^22)	(3^21^2)	(32^21)	(321^3)	(31^5)	(2^4)	(2^31^2)	(2^21^4)	(21^6)	(1^8)
(6) (1)²	·	·	·	·	·	·	·	·	·	·	·
(51) (1)²	·	·	·	·	·	·	·	·	·	·	·
(5) (2) (1)	·	·	·	·	·	·	·	·	·	·	·
(5) (1²) (1)	·	·	·	·	·	·	·	·	·	·	·
(42) (1)²	·	·	·	·	·	·	·	·	·	·	·
(41²) (1)²	12	·	·	·	·	·	·	·	·	·	·
(41) (2) (1)	·	·	·	·	·	·	·	·	·	·	·
(41) (1²) (1)	12	·	·	·	·	·	·	·	·	·	·
(4) (3) (1)	·	·	·	·	·	·	·	·	·	·	·
(4) (21) (1)	·	·	·	·	·	·	·	·	·	·	·
(4) (2)²	·	·	·	·	·	·	·	·	·	·	·
(4) (2) (1²)	·	·	·	·	·	·	·	·	·	·	·
(4) (1³) (1)	4	·	·	·	·	·	·	·	·	·	·
(4) (1²)²	6	·	·	·	·	·	·	·	·	·	·
(3²) (1)²	·	1	2	·	·	·	·	·	·	·	·
(321) (1)²	·	6	8	6	6	·	·	·	·	·	·
(32) (2) (1)	·	4	·	2	·	·	·	·	·	·	·
(32) (1²) (1)	·	2	6	2	3	·	·	·	·	·	·
(31²) (1)²	8	·	2	2	7	20	·	·	·	·	·
(31²) (2) (1)	·	2	6	2	3	·	·	·	·	·	·
(31²) (1²) (1)	18	2	4	5	12	30	·	·	·	·	·
(31) (3) (1)	·	2	4	·	·	·	·	·	·	·	·
(31) (21) (1)	·	6	8	6	6	·	·	·	·	·	·
(31) (2)²	·	4	·	2	·	·	·	·	·	·	·
(31) (2) (1²)	·	2	6	2	3	·	·	·	·	·	·
(31) (1³) (1)	16	·	2	2	7	20	·	·	·	·	·
(31) (1²)²	24	2	4	5	12	30	·	·	·	·	·
(3²) (2)	·	2	·	·	·	·	·	·	·	·	·
(3²) (1²)	·	·	2	·	·	·	·	·	·	·	·
(3) (2²) (1)	·	2	·	1	·	·	·	·	·	·	·
(3) (21²) (1)	·	·	2	2	3	·	·	·	·	·	·
(3) (21) (2)	·	2	·	2	·	·	·	·	·	·	·
(3) (21) (1²)	·	2	4	2	3	·	·	·	·	·	·
(3) (2) (1³)	·	·	2	·	1	·	·	·	·	·	·
(3) (1⁴) (1)	5	·	·	·	1	5	·	·	·	·	·
(3) (1³) (1²)	10	·	·	1	3	10	·	·	·	·	·
(2³) (1)²	·	2	·	2	·	·	4	2	·	·	·
(2²1²) (1)²	·	·	2	5	6	·	12	15	12	·	·
(2²1) (2) (1)	·	2	·	4	·	·	12	6	·	·	·
(2²1) (1²) (1)	·	5	6	9	9	·	12	15	12	·	·
(2²) (21) (1)	·	4	·	5	·	·	12	6	·	·	·
(2²) (2)²	·	·	·	·	·	·	12	·	·	·	·
(2²) (2) (1²)	·	3	·	3	·	·	·	3	·	·	·
(2²) (1³) (1)	·	1	4	2	4	·	·	3	4	·	·
(2²) (1²)²	·	2	6	4	6	·	6	6	6	·	·
(21⁴) (1)²	1	·	·	·	3	10	·	6	18	30	·
(21³) (2) (1)	4	·	2	2	5	·	·	6	8	·	·
(21³) (1²) (1)	4	·	2	5	13	30	12	24	44	60	·
(21²) (21) (1)	12	6	10	14	15	·	24	30	24	·	·
(21²) (2)²	·	2	·	4	·	·	·	6	·	·	·
(21²) (2) (1²)	6	2	6	6	9	·	12	12	12	·	·
(21²) (1³) (1)	6	2	4	8	18	40	12	24	44	60	·
(21²) (1²)²	12	5	8	16	30	60	24	45	72	90	·
(21)² (2)	·	6	·	8	·	·	24	12	·	·	·
(21)² (1²)	12	10	14	18	18	·	24	30	24	·	·
(21) (2) (1³)	4	2	6	4	7	·	·	6	8	·	·
(21) (1⁴) (1)	4	·	2	2	9	25	·	6	18	30	·
(21) (1³) (1²)	12	2	6	9	22	50	12	24	44	60	·
(2)² (1⁴)	1	·	2	·	2	·	·	·	2	·	·
(2) (1⁵) (1)	1	·	·	1	1	6	·	·	2	6	·
(2) (1⁴) (1²)	4	·	·	1	4	15	·	3	8	15	·
(2) (1³)²	6	·	·	2	6	20	4	6	12	20	·
(1⁶) (1)²	·	·	·	·	·	1	·	·	2	13	56
(1⁵) (1²) (1)	·	·	·	·	1	5	·	3	14	51	168
(1⁴) (1³) (1)	·	·	·	1	3	10	4	11	32	95	280
(1⁴) (1²)²	·	·	1	2	7	20	6	18	53	150	420
(1³)² (1²)	·	1	2	5	12	30	12	31	80	210	560

Table 1·7·8 (*cont.*)

$w = 8$ (v)	(8)	(71)	(62)	(61²)	(53)	(521)	(51³)	(4²)	(431)	(42²)	(421²)
(5)(1)³	1	3	3	6	1	3	6	·	·	6	12
(41)(1)³	·	1	3	6	3	9	18	2	4	3	2
(4)(2)(1)²	1	2	2	2	2	2	·	2	2	3	5
(4)(1²)(1)²	·	1	2	5	1	5	12	·	1	6	6
(32)(1)³	·	·	1	·	4	3	·	6	9	6	15
(31²)(1)³	·	·	·	1	·	3	9	·	4	6	15
(31)(2)(1)²	·	1	2	4	3	4	6	4	7	4	4
(31)(1²)(1)²	·	1	1	2	2	7	15	2	7	10	22
(3)²(1)²	1	2	1	2	2	2	·	4	4	·	·
(3)(21)(1)³	·	1	3	4	3	6	6	2	5	6	6
(3)(2)²(1)	1	1	2	·	3	2	·	2	1	2	·
(3)(2)(1²)(1)	·	1	1	3	1	2	3	2	4	2	3
(3)(1³)(1)²	·	·	·	1	·	2	7	·	1	2	7
(3)(1²)²(1)	·	·	1	2	1	5	12	·	2	5	12
(2²1)(1)³	·	·	·	·	·	1	1	·	3	7	6
(2²)(2)(1)²	·	·	1	·	2	2	·	2	2	5	2
(2²)(1²)(1)²	·	·	·	·	1	1	·	2	5	4	5
(21³)(1)³	·	·	·	·	·	·	1	·	·	·	4
(21²)(2)(1)²	·	·	·	1	·	3	6	·	4	6	11
(21²)(1²)(1)²	·	·	·	·	·	1	3	·	3	7	14
(21)³(1)²	·	·	1	2	4	8	12	6	14	20	24
(21)(2)²(1)	·	1	2	2	3	4	·	4	5	10	6
(21)(2)(1²)(1)	·	·	1	2	2	5	9	2	7	8	12
(21)(1³)(1)²	·	·	·	·	·	1	3	·	3	4	12
(21)(1²)²(1)	·	·	·	·	1	3	6	2	8	12	24
(2)⁴	1	·	4	·	3	3	·	6	·	12	·
(2)³(1²)	·	1	·	1	3	3	·	·	3	·	3
(2)²(1³)(1)	·	·	·	1	·	1	4	·	2	·	3
(2)²(1²)²	·	·	1	2	·	2	6	2	4	5	6
(2)(1⁴)(1)²	·	·	·	·	·	·	1	·	·	·	2
(2)(1³)(1²)(1)	·	·	·	·	·	1	3	·	1	3	8
(2)(1²)³	·	·	·	·	1	3	6	·	3	6	15
(1⁵)(1)³	·	·	·	·	·	·	·	·	·	·	·
(1⁴)(1²)(1)²	·	·	·	·	·	·	·	·	·	·	1
(1³)²(1)²	·	·	·	·	·	·	·	·	·	1	2
(1³)(1²)²(1)	·	·	·	·	·	·	·	·	1	2	5
(1²)⁴	·	·	·	·	·	·	·	1	4	6	12
(4)(1)⁴	1	4	6	12	4	12	24	2	4	6	12
(31)(1)⁴	·	1	4	8	7	18	36	8	21	24	48
(3)(2)(1)³	1	3	4	6	5	6	6	6	9	6	6
(3)(1²)(1)³	·	1	3	7	3	12	27	2	8	12	27
(2²)(1)⁴	·	·	1	·	4	4	·	6	12	13	12
(21²)(1)⁴	·	·	·	1	·	5	12	·	10	20	39
(21)(2)(1)³	·	1	4	6	7	14	18	8	19	26	30
(21)(1²)(1)³	·	·	1	2	4	11	21	6	23	32	60
(2)³(1²)	1	2	4	2	6	6	·	6	6	12	6
(2)²(1²)(1)²	·	1	2	5	3	7	12	4	11	10	15
(2)(1³)(1)³	·	·	·	1	·	3	10	·	4	6	19
(2)(1²)²(1)²	·	·	1	2	2	8	18	2	10	17	36
(1⁴)(1)⁴	·	·	·	·	·	·	1	·	·	·	4
(1³)(1²)(1)³	·	·	·	·	·	1	3	·	3	7	18
(1²)²(1²)	·	·	·	·	1	3	6	2	11	18	39
(3)(1)⁵	1	5	10	20	11	30	60	10	25	30	60
(21)(1)⁵	·	1	6	10	15	36	60	20	65	90	150
(2)²(1)⁴	1	4	8	12	12	20	24	14	28	32	36
(2)(1²)(1)⁴	·	1	4	9	7	23	48	8	31	44	87
(1³)(1)⁵	·	·	·	1	·	5	16	·	10	20	55
(1²)²(1)⁴	·	·	1	2	4	14	30	6	32	53	114
(2)(1)⁶	1	6	16	30	26	66	120	30	90	120	210
(1²)(1)⁶	·	1	6	13	15	51	108	20	95	150	315
(1)⁸	1	8	28	56	56	168	336	70	280	420	840

Table 1·7·8 (*cont.*)

$w=8$ (vi)	(41^4)	$(3^2 2)$	$(3^2 1^2)$	$(32^2 1)$	(321^3)	(31^5)	(2^4)	$(2^3 1^2)$	$(2^2 1^4)$	(21^6)	(1^8)
$(5)(1)^3$	·	·	·	·	·	·	·	·	·	·	·
$(41)(1)^3$	24	·	·	·	·	·	·	·	·	·	·
$(4)(2)(1)^2$	·	·	·	·	·	·	·	·	·	·	·
$(4)(1^2)(1)^2$	12	·	·	·	·	·	·	·	·	·	·
$(32)(1)^3$	·	8	12	6	6	·	·	·	·	·	·
$(31^2)(1)^3$	36	6	14	12	27	60	·	·	·	·	·
$(31)(2)(1)^2$	·	8	12	6	6	·	·	·	·	·	·
$(31)(1^2)(1)^2$	48	6	14	12	27	60	·	·	·	·	·
$(3)^2(1)^2$	·	2	4	·	·	·	·	·	·	·	·
$(3)(21)(1)^2$	·	6	8	6	6	·	·	·	·	·	·
$(3)(2)^2(1)$	·	4	·	2	·	·	·	·	·	·	·
$(3)(2)(1^2)(1)$	·	2	6	2	3	·	·	·	·	·	·
$(3)(1^3)(1)^2$	20	·	2	2	7	20	·	·	·	·	·
$(3)(1^2)^2(1)$	30	2	4	5	12	30	·	·	·	·	·
$(2^2 1)(1)^3$	·	12	12	22	18	·	36	36	24	·	·
$(2^2)(2)(1)^2$	·	6	·	6	·	·	12	6	·	·	·
$(2^2)(1^2)(1)^2$	·	7	12	11	12	·	12	15	12	·	·
$(21^3)(1)^3$	12	·	6	12	31	60	24	54	96	120	·
$(21^2)(2)(1)^2$	12	6	12	16	18	·	24	30	24	·	·
$(21^2)(1^2)(1)^2$	30	12	22	38	69	120	60	102	156	180	·
$(21)^2(1)^2$	24	26	28	44	36	·	72	72	48	·	·
$(21)(2)^2(1)$	·	8	·	10	·	·	24	12	·	·	·
$(21)(2)(1^2)(1)$	12	12	18	20	21	·	24	30	24	·	·
$(21)(1^3)(1)^2$	28	6	18	22	51	100	24	54	96	120	·
$(21)(1^2)^2(1)$	48	16	32	45	84	150	60	102	156	180	·
$(2)^4$	·	·	·	·	·	·	24	·	·	·	·
$(2)^3(1^2)$	·	6	·	6	·	·	·	6	·	·	·
$(2)^2(1^3)(1)$	4	2	8	4	8	·	·	6	8	·	·
$(2)^2(1^2)^2$	6	4	12	8	12	·	12	12	12	·	·
$(2)(1^4)(1)^2$	9	·	2	2	10	30	·	6	18	30	·
$(2)(1^3)(1^2)(1)$	22	2	6	10	25	60	12	24	44	60	·
$(2)(1^2)^3$	36	7	12	21	42	90	24	45	72	90	·
$(1^5)(1)^3$	1	·	·	·	3	16	·	6	30	108	336
$(1^4)(1^2)(1)^2$	4	·	2	5	18	55	12	39	114	315	840
$(1^3)^2(1^2)^2$	6	2	4	12	30	80	28	68	172	440	1120
$(1^3)(1^2)^2(1)$	12	5	12	24	58	140	48	117	284	690	1680
$(1^2)^4$	24	12	28	48	108	240	90	204	468	1080	2520
$(4)(1)^4$	24	·	·	·	·	·	·	·	·	·	·
$(31)(1)^4$	96	20	40	30	60	120	·	·	·	·	·
$(3)(2)(1)^3$	·	8	12	6	6	·	·	·	·	·	·
$(3)(1^2)(1)^3$	60	6	14	12	27	60	·	·	·	·	·
$(2^2)(1)^4$	·	20	24	28	24	·	36	36	24	·	·
$(21^2)(1)^4$	72	30	56	92	156	240	144	234	336	360	·
$(21)(2)(1)^3$	24	32	36	50	42	·	72	72	48	·	·
$(21)(1^2)(1)^3$	108	44	82	110	189	300	144	234	336	360	·
$(2)^3(1)^2$	·	12	·	12	·	·	24	12	·	·	·
$(2)^2(1^2)(1)^2$	12	14	24	22	24	·	24	30	24	·	·
$(2)(1^3)(1)^3$	48	6	20	24	58	120	24	54	96	120	·
$(2)(1^2)^2(1)^2$	78	18	36	50	96	180	60	102	156	180	·
$(1^4)(1)^4$	17	·	6	12	46	140	24	84	246	660	1680
$(1^3)(1^2)(1)^3$	46	12	30	58	141	340	108	258	612	1440	3360
$(1^2)^3(1)^2$	84	31	68	117	258	570	204	453	1008	2250	5040
$(3)(1)^5$	120	20	40	30	60	120	·	·	·	·	·
$(21)(1)^5$	240	120	200	270	420	600	360	540	720	720	·
$(2^2)(1)^4$	24	40	48	56	48	·	72	72	48	·	·
$(2)(1^2)(1)^4$	168	50	96	122	216	360	144	234	336	360	·
$(1^3)(1)^5$	140	30	80	140	340	800	240	570	1320	3000	6720
$(1^2)^2(1)^4$	246	80	172	284	612	1320	468	1008	2172	4680	10080
$(2)(1)^6$	360	140	240	300	480	720	360	540	720	720	·
$(1^2)(1)^6$	660	210	440	690	1440	3000	1080	2250	4680	9720	20160
$(1)^8$	1680	560	1120	1680	3360	6720	2520	5040	10080	20160	40320

For brevity in this table $\{pq\}$ indicates κ_{pq}. For moments about the means, κ_{10} and κ_{01}, all terms containing suffices which have unity and zero as parts and no other are to be ignored. Thus, for example

$$\mu'_{21} = \{21\} + \{20\}\{01\} + 2\{11\}\{10\} + \{10\}^2\{01\}$$

$$= \kappa_{21} + \kappa_{20}\kappa_{01} + 2\kappa_{11}\kappa_{10} + \kappa_{10}^2\kappa_{01}$$

and

$$\mu_{21} = \kappa_{21}.$$

$\mu'_{10} = \{10\}.$

$\mu'_{20} = \{20\} + \{10\}^2.$

$\mu'_{11} = \{11\} + \{10\}\{01\}.$

$\mu'_{30} = \{30\} + 3\{20\}\{10\} + \{10\}^3.$

$\mu'_{21} = \{21\} + \{20\}\{01\} + 2\{11\}\{10\} + \{10\}^2\{01\}.$

$\mu'_{40} = \{40\} + 4\{30\}\{10\} + 3\{20\}^2 + 6\{20\}\{10\}^2 + \{10\}^4.$

$\mu'_{31} = \{31\} + \{30\}\{01\} + 3\{21\}\{10\} + 3\{20\}\{11\} + 3\{20\}\{10\}\{01\} + 3\{11\}\{10\}^2 + \{10\}^3\{01\}.$

$\mu'_{22} = \{22\} + 2\{21\}\{01\} + \{20\}\{02\} + 2\{12\}\{10\} + 2\{11\}^2 + \{20\}\{01\}^2 + 4\{11\}\{10\}\{01\} + \{10\}^2\{02\} + \{10\}^2\{01\}^2.$

$\mu'_{50} = \{50\} + 5\{40\}\{10\} + 10\{30\}\{20\} + 10\{30\}\{10\}^2 + 15\{20\}^2\{10\} + 10\{20\}\{10\}^3 + \{10\}^5.$

$\mu'_{41} = \{41\} + \{40\}\{01\} + 4\{31\}\{10\} + 4\{30\}\{11\} + 6\{21\}\{20\} + 4\{30\}\{10\}\{01\} + 6\{21\}\{10\}^2 + 3\{20\}^2\{01\}$
$\qquad + 12\{20\}\{11\}\{10\} + 6\{20\}\{10\}^2\{01\} + 4\{11\}\{10\}^3 + \{10\}^4\{01\}.$

$\mu'_{32} = \{32\} + 2\{31\}\{01\} + \{30\}\{02\} + 3\{22\}\{10\} + 6\{21\}\{11\} + 3\{20\}\{12\} + \{30\}\{01\}^2 + 6\{21\}\{10\}\{01\}$
$\qquad + 6\{20\}\{11\}\{01\} + 3\{20\}\{10\}\{02\} + 3\{12\}\{10\}^2 + 6\{11\}^2\{10\} + 3\{20\}\{10\}\{01\}^2 + 6\{11\}\{10\}^2\{01\}$
$\qquad + \{10\}^3\{02\} + \{10\}^3\{01\}^2.$

$\mu'_{60} = \{60\} + 6\{50\}\{10\} + 15\{40\}\{20\} + 10\{30\}^2 + 15\{40\}\{10\}^2 + 60\{30\}\{20\}\{10\} + 15\{20\}^3 + 20\{30\}\{10\}^3$
$\qquad + 45\{20\}^2\{10\}^2 + 15\{20\}\{10\}^4 + \{10\}^6.$

$\mu'_{51} = \{51\} + \{50\}\{01\} + 5\{41\}\{10\} + 5\{40\}\{11\} + 10\{31\}\{20\} + 10\{30\}\{21\} + 5\{40\}\{10\}\{01\} + 10\{31\}\{10\}^2$
$\qquad + 10\{30\}\{20\}\{01\} + 20\{30\}\{11\}\{10\} + 30\{21\}\{20\}\{10\} + 15\{20\}^2\{11\} + 10\{30\}\{10\}^2\{01\} + 10\{21\}\{10\}^3$
$\qquad + 15\{20\}^2\{10\}\{01\} + 30\{20\}\{11\}\{10\}^2 + 10\{20\}\{10\}^3\{01\} + 5\{11\}\{10\}^4 + \{10\}^5\{01\}.$

$\mu'_{42} = \{42\} + 2\{41\}\{01\} + \{40\}\{02\} + 4\{32\}\{10\} + 8\{31\}\{11\} + 4\{30\}\{12\} + 6\{22\}\{20\} + 6\{21\}^2 + \{40\}\{01\}^2$
$\qquad + 8\{31\}\{10\}\{01\} + 8\{30\}\{11\}\{01\} + 4\{30\}\{10\}\{02\} + 6\{22\}\{10\}^2 + 12\{21\}\{20\}\{01\} + 24\{21\}\{11\}\{10\}$
$\qquad + 3\{20\}^2\{02\} + 12\{20\}\{12\}\{10\} + 12\{20\}\{11\}^2 + 4\{30\}\{10\}\{01\}^2 + 12\{21\}\{10\}^2\{01\} + 3\{20\}^2\{01\}^2$
$\qquad + 24\{20\}\{11\}\{10\}\{01\} + 6\{20\}\{10\}^2\{02\} + 4\{12\}\{10\}^3 + 12\{11\}^2\{10\}^2 + 6\{20\}\{10\}^2\{01\}^2$
$\qquad + 8\{11\}\{10\}^3\{01\} + \{10\}^4\{02\} + \{10\}^4\{01\}^2.$

$\mu'_{33} = \{33\} + 3\{32\}\{01\} + 3\{31\}\{02\} + \{30\}\{03\} + 3\{23\}\{10\} + 9\{22\}\{11\} + 9\{21\}\{12\} + 3\{20\}\{13\} + 3\{31\}\{01\}^2$
$\qquad + 3\{30\}\{02\}\{01\} + 9\{22\}\{10\}\{01\} + 18\{21\}\{11\}\{01\} + 9\{21\}\{10\}\{02\} + 9\{20\}\{12\}\{01\}$
$\qquad + 9\{20\}\{11\}\{02\} + 3\{20\}\{10\}\{03\} + 3\{13\}\{10\}^2 + 18\{12\}\{11\}\{10\} + 6\{11\}^3 + \{30\}\{01\}^3$
$\qquad + 9\{21\}\{10\}\{01\}^2 + 9\{20\}\{11\}\{01\}^2 + 9\{20\}\{10\}\{02\}\{01\} + 9\{12\}\{10\}^2\{01\} + 18\{11\}^2\{10\}\{01\}$
$\qquad + 9\{11\}\{10\}^2\{02\} + \{10\}^3\{03\} + 3\{20\}\{10\}\{01\}^3 + 9\{11\}\{10\}^2\{01\}^2 + 3\{10\}^3\{02\}\{01\} + \{10\}^3\{01\}^3.$

$\mu'_{70} = \{70\} + 7\{60\}\{10\} + 21\{50\}\{20\} + 35\{40\}\{30\} + 21\{50\}\{10\}^2 + 105\{40\}\{20\}\{10\} + 70\{30\}^2\{10\}$
$\qquad + 105\{30\}\{20\}^2 + 35\{40\}\{10\}^3 + 210\{30\}\{20\}\{10\}^2 + 105\{20\}^3\{10\} + 35\{30\}\{10\}^4 + 105\{20\}^2\{10\}^3$
$\qquad + 21\{20\}\{10\}^5 + \{10\}^7.$

$$\mu'_{61} = \{61\} + \{60\}\{01\} + 6\{51\}\{10\} + 6\{50\}\{11\} + 15\{41\}\{20\} + 15\{40\}\{21\} + 20\{31\}\{30\} + 6\{50\}\{10\}\{01\}$$
$$+ 15\{41\}\{10\}^2 + 15\{40\}\{20\}\{01\} + 30\{40\}\{11\}\{10\} + 60\{31\}\{20\}\{10\} + 10\{30\}^2\{01\} + 60\{30\}\{21\}\{10\}$$
$$+ 60\{30\}\{20\}\{11\} + 45\{21\}\{20\}^2 + 15\{40\}\{10\}^2\{01\} + 20\{31\}\{10\}^3 + 60\{30\}\{20\}\{10\}\{01\}$$
$$+ 60\{30\}\{11\}\{10\}^2 + 90\{21\}\{20\}\{10\}^2 + 15\{20\}^3\{01\} + 90\{20\}^2\{11\}\{10\} + 20\{30\}\{10\}^3\{01\}$$
$$+ 15\{21\}\{10\}^4 + 45\{20\}^2\{10\}^2\{01\} + 60\{20\}\{11\}\{10\}^3 + 15\{20\}\{10\}^4\{01\} + 6\{11\}\{10\}^5 + \{10\}^6\{01\}.$$

$$\mu'_{52} = \{52\} + 2\{51\}\{01\} + \{50\}\{02\} + 5\{42\}\{10\} + 10\{41\}\{11\} + 5\{40\}\{12\} + 10\{32\}\{20\} + 20\{31\}\{21\}$$
$$+ 10\{30\}\{22\} + \{50\}\{01\}^2 + 10\{41\}\{10\}\{01\} + 10\{40\}\{11\}\{01\} + 5\{40\}\{10\}\{02\} + 10\{32\}\{10\}^2$$
$$+ 20\{31\}\{20\}\{01\} + 40\{31\}\{11\}\{10\} + 20\{30\}\{21\}\{01\} + 10\{30\}\{20\}\{02\} + 20\{30\}\{12\}\{10\}$$
$$+ 20\{30\}\{11\}^2 + 30\{22\}\{20\}\{10\} + 30\{21\}^2\{10\} + 60\{21\}\{20\}\{11\} + 15\{20\}^2\{12\} + 5\{40\}\{10\}\{01\}^2$$
$$+ 20\{31\}\{10\}^2\{01\} + 10\{30\}\{20\}\{01\}^2 + 40\{30\}\{11\}\{10\}\{01\} + 10\{30\}\{10\}^2\{02\} + 10\{22\}\{10\}^3$$
$$+ 60\{21\}\{20\}\{10\}\{01\} + 60\{21\}\{11\}\{10\}^2 + 30\{20\}^2\{11\}\{01\} + 15\{20\}^2\{10\}\{02\} + 30\{20\}\{12\}\{10\}^2$$
$$+ 60\{20\}\{11\}^2\{10\} + 10\{30\}\{10\}^2\{01\}^2 + 20\{21\}\{10\}^3\{01\} + 15\{20\}^2\{10\}\{01\}^2 + 60\{20\}\{11\}\{10\}^2\{01\}$$
$$+ 10\{20\}\{10\}^3\{02\} + 5\{12\}\{10\}^4 + 20\{11\}^2\{10\}^3 + 10\{20\}\{10\}^3\{01\}^2 + 10\{11\}\{10\}^4\{01\} + \{10\}^5\{02\}$$
$$+ \{10\}^5\{01\}^2.$$

$$\mu'_{43} = \{43\} + 3\{42\}\{01\} + 3\{41\}\{02\} + \{40\}\{03\} + 4\{33\}\{10\} + 12\{32\}\{11\} + 12\{31\}\{12\} + 4\{30\}\{13\} + 6\{23\}\{20\}$$
$$+ 18\{22\}\{21\} + 3\{41\}\{01\}^2 + 3\{40\}\{02\}\{01\} + 12\{32\}\{10\}\{01\} + 24\{31\}\{11\}\{01\} + 12\{31\}\{10\}\{02\}$$
$$+ 12\{30\}\{12\}\{01\} + 12\{30\}\{11\}\{02\} + 4\{30\}\{10\}\{03\} + 6\{23\}\{10\}^2 + 18\{22\}\{20\}\{01\}$$
$$+ 36\{22\}\{11\}\{10\} + 18\{21\}^2\{01\} + 18\{21\}\{20\}\{02\} + 36\{21\}\{12\}\{10\} + 36\{21\}\{11\}^2 + 3\{20\}^2\{03\}$$
$$+ 12\{20\}\{13\}\{10\} + 36\{20\}\{12\}\{11\} + 4\{40\}\{01\}^3 + 12\{31\}\{10\}\{01\}^2 + 12\{30\}\{11\}\{01\}^2$$
$$+ 12\{30\}\{10\}\{02\}\{01\} + 18\{22\}\{10\}^2\{01\} + 18\{21\}\{20\}\{01\}^2 + 72\{21\}\{11\}\{10\}\{01\} + 18\{21\}\{10\}^2\{02\}$$
$$+ 9\{20\}^2\{02\}\{01\} + 36\{20\}\{12\}\{10\}\{01\} + 36\{20\}\{11\}^2\{01\} + 36\{20\}\{11\}\{10\}\{02\} + 6\{20\}\{10\}^2\{03\}$$
$$+ 4\{13\}\{10\}^3 + 36\{12\}\{11\}\{10\}^2 + 24\{11\}^3\{10\} + 4\{30\}\{10\}\{01\}^3 + 18\{21\}\{10\}^2\{01\}^2 + 3\{20\}^2\{01\}^3$$
$$+ 36\{20\}\{11\}\{10\}\{01\}^2 + 18\{20\}\{10\}^2\{02\}\{01\} + 12\{12\}\{10\}^3\{01\} + 36\{11\}^2\{10\}^2\{01\}$$
$$+ 12\{11\}\{10\}^3\{02\} + \{10\}^4\{03\} + 6\{20\}\{10\}^2\{01\}^3 + 12\{11\}\{10\}^3\{01\}^2 + 3\{10\}^4\{02\}\{01\} + \{10\}^4\{01\}^3.$$

$$\mu'_{80} = \{80\} + 8\{70\}\{10\} + 28\{60\}\{20\} + 56\{50\}\{30\} + 35\{40\}^2 + 28\{60\}\{10\}^2 + 168\{50\}\{20\}\{10\}$$
$$+ 280\{40\}\{30\}\{10\} + 210\{40\}\{20\}^2 + 280\{30\}^2\{20\} + 56\{50\}\{10\}^3 + 420\{40\}\{20\}\{10\}^2$$
$$+ 280\{30\}^2\{10\}^2 + 840\{30\}\{20\}^2\{10\} + 105\{20\}^4 + 70\{40\}\{10\}^4 + 560\{30\}\{20\}\{10\}^3 + 420\{20\}^3\{10\}^2$$
$$+ 56\{30\}\{10\}^5 + 210\{20\}^2\{10\}^4 + 28\{20\}\{10\}^6 + \{10\}^8.$$

$$\mu'_{71} = \{71\} + \{70\}\{01\} + 7\{61\}\{10\} + 7\{60\}\{11\} + 21\{51\}\{20\} + 21\{50\}\{21\} + 35\{41\}\{30\} + 35\{40\}\{31\}$$
$$+ 7\{60\}\{10\}\{01\} + 21\{51\}\{10\}^2 + 21\{50\}\{20\}\{01\} + 42\{50\}\{11\}\{10\} + 105\{41\}\{20\}\{10\}$$
$$+ 35\{40\}\{30\}\{01\} + 105\{40\}\{21\}\{10\} + 105\{40\}\{20\}\{11\} + 140\{31\}\{30\}\{10\} + 105\{31\}\{20\}^2$$
$$+ 70\{30\}^2\{11\} + 210\{30\}\{21\}\{20\} + 21\{50\}\{10\}^2\{01\} + 35\{41\}\{10\}^3 + 105\{40\}\{20\}\{10\}\{01\}$$
$$+ 105\{40\}\{11\}\{10\}^2 + 210\{31\}\{20\}\{10\}^2 + 70\{30\}^2\{10\}\{01\} + 210\{30\}\{21\}\{10\}^2 + 105\{30\}\{20\}^2\{01\}$$
$$+ 420\{30\}\{20\}\{11\}\{10\} + 315\{21\}\{20\}^2\{10\} + 105\{20\}^3\{11\} + 35\{40\}\{10\}^3\{01\} + 35\{31\}\{10\}^4$$
$$+ 210\{30\}\{20\}\{10\}^2\{01\} + 140\{30\}\{11\}\{10\}^3 + 210\{21\}\{20\}\{10\}^3 + 105\{20\}^3\{10\}\{01\}$$
$$+ 315\{20\}^2\{11\}\{10\}^2 + 35\{30\}\{10\}^4\{01\} + 21\{21\}\{10\}^5 + 105\{20\}^2\{10\}^3\{01\} + 105\{20\}\{11\}\{10\}^4$$
$$+ 21\{20\}\{10\}^5\{01\} + 7\{11\}\{10\}^6 + \{10\}^7\{01\}.$$

$$\mu'_{62} = \{62\} + 2\{61\}\{01\} + \{60\}\{02\} + 6\{52\}\{10\} + 12\{51\}\{11\} + 6\{50\}\{12\} + 15\{42\}\{20\} + 30\{41\}\{21\}$$
$$+ 15\{40\}\{22\} + 20\{32\}\{30\} + 20\{31\}^2 + \{60\}\{01\}^2 + 12\{51\}\{10\}\{01\} + 12\{50\}\{11\}\{01\}$$
$$+ 6\{50\}\{10\}\{02\} + 15\{42\}\{10\}^2 + 30\{41\}\{20\}\{01\} + 60\{41\}\{11\}\{10\} + 30\{40\}\{21\}\{01\}$$
$$+ 15\{40\}\{20\}\{02\} + 30\{40\}\{12\}\{10\} + 30\{40\}\{11\}^2 + 60\{32\}\{20\}\{10\} + 40\{31\}\{30\}\{01\}$$
$$+ 120\{31\}\{21\}\{10\} + 120\{31\}\{20\}\{11\} + 10\{30\}^2\{02\} + 60\{30\}\{22\}\{10\} + 120\{30\}\{21\}\{11\}$$
$$+ 60\{30\}\{20\}\{12\} + 45\{22\}\{20\}^2 + 90\{21\}^2\{20\} + 6\{50\}\{10\}\{01\}^2 + 30\{41\}\{10\}^2\{01\}$$
$$+ 15\{40\}\{20\}\{01\}^2 + 60\{40\}\{11\}\{10\}\{01\} + 15\{40\}\{10\}^2\{02\} + 20\{32\}\{10\}^3 + 120\{31\}\{20\}\{10\}\{01\}$$
$$+ 120\{31\}\{11\}\{10\}^2 + 10\{30\}^2\{01\}^2 + 120\{30\}\{21\}\{10\}\{01\} + 120\{30\}\{20\}\{11\}\{01\}$$
$$+ 60\{30\}\{20\}\{10\}\{02\} + 60\{30\}\{12\}\{10\}^2 + 120\{30\}\{11\}^2\{10\} + 90\{22\}\{20\}\{10\}^2 + 90\{21\}^2\{10\}^2$$
$$+ 90\{21\}\{20\}^2\{01\} + 360\{21\}\{20\}\{11\}\{10\} + 15\{20\}^3\{02\} + 90\{20\}^2\{12\}\{10\} + 90\{20\}^2\{11\}^2$$

$$+ 15\{40\}\{10\}^2\{01\}^2 + 40\{31\}\{10\}^3\{01\} + 60\{30\}\{20\}\{10\}\{01\}^2 + 120\{30\}\{11\}\{10\}^2\{01\}$$
$$+ 20\{30\}\{10\}^3\{02\} + 15\{22\}\{10\}^4 + 180\{21\}\{20\}\{10\}^2\{01\} + 120\{21\}\{11\}\{10\}^3 + 15\{20\}^3\{01\}^2$$
$$+ 180\{20\}^2\{11\}\{10\}\{01\} + 45\{20\}^2\{10\}^2\{02\} + 60\{20\}\{12\}\{10\}^3 + 180\{20\}\{11\}^2\{10\}^2$$
$$+ 20\{30\}\{10\}^3\{01\}^2 + 30\{21\}\{10\}^4\{01\} + 45\{20\}^2\{10\}^2\{01\}^2 + 120\{20\}\{11\}\{10\}^3\{01\}$$
$$+ 15\{20\}\{10\}^4\{02\} + 6\{12\}\{10\}^5 + 30\{11\}^2\{10\}^4 + 15\{20\}\{10\}^4\{01\}^2 + 12\{11\}\{10\}^5\{01\}$$
$$+ \{10\}^6\{02\} + \{10\}^6\{01\}^2.$$

$\mu'_{53} = \{53\} + 3\{52\}\{01\} + 3\{51\}\{02\} + \{50\}\{03\} + 5\{43\}\{10\} + 15\{42\}\{11\} + 15\{41\}\{12\} + 5\{40\}\{13\} + 10\{33\}\{20\}$
$$+ 30\{32\}\{21\} + 30\{31\}\{22\} + 10\{30\}\{23\} + 3\{51\}\{01\}^2 + 3\{50\}\{02\}\{01\} + 15\{42\}\{10\}\{01\}$$
$$+ 30\{41\}\{11\}\{01\} + 15\{41\}\{10\}\{02\} + 15\{40\}\{12\}\{01\} + 15\{40\}\{11\}\{02\} + 5\{40\}\{10\}\{03\}$$
$$+ 10\{33\}\{10\}^2 + 30\{32\}\{20\}\{01\} + 60\{32\}\{11\}\{10\} + 60\{31\}\{21\}\{01\} + 30\{31\}\{20\}\{02\}$$
$$+ 60\{31\}\{12\}\{10\} + 60\{31\}\{11\}^2 + 30\{30\}\{22\}\{01\} + 30\{30\}\{21\}\{02\} + 10\{30\}\{20\}\{03\}$$
$$+ 20\{30\}\{13\}\{10\} + 60\{30\}\{12\}\{11\} + 30\{23\}\{20\}\{10\} + 90\{22\}\{21\}\{10\} + 90\{22\}\{20\}\{11\}$$
$$+ 90\{21\}^2\{11\} + 90\{21\}\{20\}\{12\} + 15\{20\}^2\{13\} + \{50\}\{01\}^3 + 15\{41\}\{10\}\{01\}^2 + 15\{40\}\{11\}\{01\}^2$$
$$+ 15\{40\}\{10\}\{02\}\{01\} + 30\{32\}\{10\}^2\{01\} + 30\{31\}\{20\}\{01\}^2 + 120\{31\}\{11\}\{10\}\{01\}$$
$$+ 30\{31\}\{10\}^2\{02\} + 30\{30\}\{21\}\{01\}^2 + 30\{30\}\{20\}\{02\}\{01\} + 60\{30\}\{12\}\{10\}\{01\} + 60\{30\}\{11\}^2\{01\}$$
$$+ 60\{30\}\{11\}\{10\}\{02\} + 10\{30\}\{10\}^2\{03\} + 10\{23\}\{10\}^3 + 90\{22\}\{20\}\{10\}\{01\} + 90\{22\}\{11\}\{10\}^2$$
$$+ 90\{21\}^2\{10\}\{01\} + 180\{21\}\{20\}\{11\}\{01\} + 90\{21\}\{20\}\{10\}\{02\} + 90\{21\}\{12\}\{10\}^2$$
$$+ 180\{21\}\{11\}^2\{10\} + 45\{20\}^2\{12\}\{01\} + 45\{20\}^2\{11\}\{02\} + 15\{20\}^2\{10\}\{03\} + 30\{20\}\{13\}\{10\}^2$$
$$+ 180\{20\}\{12\}\{11\}\{10\} + 60\{20\}\{11\}^3 + 5\{40\}\{10\}\{01\}^3 + 30\{31\}\{10\}^2\{01\}^2 + 10\{30\}\{20\}\{01\}^3$$
$$+ 60\{30\}\{11\}\{10\}\{01\}^2 + 30\{30\}\{10\}^2\{02\}\{01\} + 30\{22\}\{10\}^3\{01\} + 90\{21\}\{20\}\{10\}\{01\}^2$$
$$+ 180\{21\}\{11\}\{10\}^2\{01\} + 30\{21\}\{10\}^3\{02\} + 45\{20\}^2\{11\}\{01\}^2 + 45\{20\}^2\{10\}\{02\}\{01\}$$
$$+ 90\{20\}\{12\}\{10\}^2\{01\} + 180\{20\}\{11\}^2\{10\}\{01\} + 90\{20\}\{11\}\{10\}^2\{02\} + 10\{20\}\{10\}^3\{03\}$$
$$+ 5\{13\}\{10\}^4 + 60\{12\}\{11\}\{10\}^3 + 60\{11\}^3\{10\}^2 + 10\{30\}\{10\}^2\{01\}^3 + 30\{21\}\{10\}^3\{01\}^2$$
$$+ 15\{20\}^2\{10\}\{01\}^3 + 90\{20\}\{11\}\{10\}^2\{01\}^2 + 30\{20\}\{10\}^3\{02\}\{01\} + 15\{12\}\{10\}^4\{01\}$$
$$+ 60\{11\}^2\{10\}^3\{01\} + 15\{11\}\{10\}^4\{02\} + \{10\}^5\{03\} + 10\{20\}\{10\}^3\{01\}^3 + 15\{11\}\{10\}^4\{01\}^2$$
$$+ 3\{10\}^5\{02\}\{01\} + \{10\}^5\{01\}^3.$$

$\mu'_{44} = \{44\} + 4\{43\}\{01\} + 6\{42\}\{02\} + 4\{41\}\{03\} + \{40\}\{04\} + 4\{34\}\{10\} + 16\{33\}\{11\} + 24\{32\}\{12\}$
$$+ 16\{31\}\{13\} + 4\{30\}\{14\} + 6\{24\}\{20\} + 24\{23\}\{21\} + 18\{22\}^2 + 6\{42\}\{01\}^2 + 12\{41\}\{02\}\{01\}$$
$$+ 4\{40\}\{03\}\{01\} + 3\{40\}\{02\}^2 + 16\{33\}\{10\}\{01\} + 48\{32\}\{11\}\{01\} + 24\{32\}\{10\}\{02\}$$
$$+ 48\{31\}\{12\}\{01\} + 48\{31\}\{11\}\{02\} + 16\{31\}\{10\}\{03\} + 16\{30\}\{13\}\{01\} + 24\{30\}\{12\}\{02\}$$
$$+ 16\{30\}\{11\}\{03\} + 4\{30\}\{10\}\{04\} + 6\{24\}\{10\}^2 + 24\{23\}\{20\}\{01\} + 48\{23\}\{11\}\{10\}$$
$$+ 72\{22\}\{21\}\{01\} + 36\{22\}\{20\}\{02\} + 72\{22\}\{12\}\{10\} + 72\{22\}\{11\}^2 + 36\{21\}^2\{02\} + 24\{21\}\{20\}\{03\}$$
$$+ 48\{21\}\{13\}\{10\} + 144\{21\}\{12\}\{11\} + 3\{20\}^2\{04\} + 12\{20\}\{14\}\{10\} + 48\{20\}\{13\}\{11\} + 36\{20\}\{12\}^2$$
$$+ 4\{41\}\{01\}^3 + 6\{40\}\{02\}\{01\}^2 + 24\{32\}\{10\}\{01\}^2 + 48\{31\}\{11\}\{01\}^2 + 48\{31\}\{10\}\{02\}\{01\}$$
$$+ 24\{30\}\{12\}\{01\}^2 + 48\{30\}\{11\}\{02\}\{01\} + 16\{30\}\{10\}\{03\}\{01\} + 12\{30\}\{10\}\{02\}^2$$
$$+ 24\{23\}\{10\}^2\{01\} + 36\{22\}\{20\}\{01\}^2 + 144\{22\}\{11\}\{10\}\{01\} + 36\{22\}\{10\}^2\{02\} + 36\{21\}^2\{01\}^2$$
$$+ 72\{21\}\{20\}\{02\}\{01\} + 144\{21\}\{12\}\{10\}\{01\} + 144\{21\}\{11\}^2\{01\} + 144\{21\}\{11\}\{10\}\{02\}$$
$$+ 24\{21\}\{10\}^2\{03\} + 12\{20\}^2\{03\}\{01\} + 9\{20\}^2\{02\}^2 + 48\{20\}\{13\}\{10\}\{01\} + 144\{20\}\{12\}\{11\}\{01\}$$
$$+ 72\{20\}\{12\}\{10\}\{02\} + 72\{20\}\{11\}^2\{02\} + 48\{20\}\{11\}\{10\}\{03\} + 6\{20\}\{10\}^2\{04\} + 4\{14\}\{10\}^3$$
$$+ 48\{13\}\{11\}\{10\}^2 + 36\{12\}^2\{10\}^2 + 144\{12\}\{11\}^2\{10\} + 24\{11\}^4 + \{40\}\{01\}^4 + 16\{31\}\{10\}\{01\}^3$$
$$+ 16\{30\}\{11\}\{01\}^3 + 24\{30\}\{10\}\{02\}\{01\}^2 + 36\{22\}\{10\}^2\{01\}^2 + 24\{21\}\{20\}\{01\}^3$$
$$+ 144\{21\}\{11\}\{10\}\{01\}^2 + 72\{21\}\{10\}^2\{02\}\{01\} + 18\{20\}^2\{02\}\{01\}^2 + 72\{20\}\{12\}\{10\}\{01\}^2$$
$$+ 72\{20\}\{11\}^2\{01\}^2 + 144\{20\}\{11\}\{10\}\{02\}\{01\} + 24\{20\}\{10\}^2\{03\}\{01\} + 18\{20\}\{10\}^2\{02\}^2$$
$$+ 16\{13\}\{10\}^3\{01\} + 144\{12\}\{11\}\{10\}^2\{01\} + 24\{12\}\{10\}^3\{02\} + 96\{11\}^3\{10\}\{01\} + 72\{11\}^2\{10\}^2\{02\}$$
$$+ 16\{11\}\{10\}^3\{03\} + \{10\}^4\{04\} + 4\{30\}\{10\}\{01\}^4 + 24\{21\}\{10\}^2\{01\}^3 + 3\{20\}^2\{01\}^4$$
$$+ 48\{20\}\{11\}\{10\}\{01\}^3 + 36\{20\}\{10\}^2\{02\}\{01\}^2 + 24\{12\}\{10\}^3\{01\}^2 + 72\{11\}^2\{10\}^2\{01\}^2$$
$$+ 48\{11\}\{10\}^3\{02\}\{01\} + 4\{10\}^4\{03\}\{01\} + 3\{10\}^4\{02\}^2 + 6\{20\}\{10\}^2\{01\}^4 + 16\{11\}\{10\}^3\{01\}^3$$
$$+ 6\{10\}^4\{02\}\{01\}^2 + \{10\}^4\{01\}^4.$$

Table 2·1·2. *Bivariate cumulants in terms of bivariate crude moments*

For brevity in this table $\{pq\}$ denotes μ'_{pq}. For moments about the means, μ'_{10} and μ'_{01}, all terms containing suffices which have unity or zero as parts and no other are to be ignored. Thus for example

$$\kappa_{21} = \{21\} - \{20\}\{01\} - 2\{11\}\{10\} + 2\{10\}^2\{01\}$$

$$= \mu'_{21} - \mu'_{20}\mu'_{01} - 2\mu'_{11}\mu'_{10} + 2\mu'^2_{10}\mu'_{01}$$

and

$$\kappa_{21} = \mu_{21}.$$

$\kappa_{10} = \{10\}.$ $\kappa_{20} = \{20\} - \{10\}^2.$

$\kappa_{11} = \{11\} - \{10\}\{01\}.$

$\kappa_{30} = \{30\} - 3\{20\}\{10\} + 2\{10\}^3.$

$\kappa_{21} = \{21\} - \{20\}\{01\} - 2\{11\}\{10\} + 2\{10\}^2\{01\}.$

$\kappa_{40} = \{40\} - 4\{30\}\{10\} - 3\{20\}^2 + 12\{20\}\{10\}^2 - 6\{10\}^4.$

$\kappa_{31} = \{31\} - \{30\}\{01\} - 3\{21\}\{10\} - 3\{20\}\{11\} + 6\{20\}\{10\}\{01\} + 6\{11\}\{10\}^2 - 6\{10\}^3\{01\}.$

$\kappa_{22} = \{22\} - 2\{21\}\{01\} - \{20\}\{02\} - 2\{12\}\{10\} - 2\{11\}^2 + 2\{20\}\{01\}^2 + 8\{11\}\{10\}\{01\} + 2\{10\}^2\{02\}$
$\qquad - 6\{10\}^2\{01\}^2.$

$\kappa_{50} = \{50\} - 5\{40\}\{10\} - 10\{30\}\{20\} + 20\{30\}\{10\}^2 + 30\{20\}^2\{10\} - 60\{20\}\{10\}^3 + 24\{10\}^5.$

$\kappa_{41} = \{41\} - \{40\}\{01\} - 4\{31\}\{10\} - 4\{30\}\{11\} - 6\{21\}\{20\} + 8\{30\}\{10\}\{01\} + 12\{21\}\{10\}^2 + 6\{20\}^2\{01\}$
$\qquad + 24\{20\}\{11\}\{10\} - 36\{20\}\{10\}^2\{01\} - 24\{11\}\{10\}^3 + 24\{10\}^4\{01\}.$

$\kappa_{32} = \{32\} - 2\{31\}\{01\} - \{30\}\{02\} - 3\{22\}\{10\} - 6\{21\}\{11\} - 3\{20\}\{12\} + 2\{30\}\{01\}^2 + 12\{21\}\{10\}\{01\}$
$\qquad + 12\{20\}\{11\}\{01\} + 6\{20\}\{10\}\{02\} + 6\{12\}\{10\}^2 + 12\{11\}^2\{10\} - 18\{20\}\{10\}\{01\}^2 - 36\{11\}\{10\}^2\{01\}$
$\qquad - 6\{10\}^3\{02\} + 24\{10\}^3\{01\}^2.$

$\kappa_{60} = \{60\} - 6\{50\}\{10\} - 15\{40\}\{20\} - 10\{30\}^2 + 30\{40\}\{10\}^2 + 120\{30\}\{20\}\{10\} + 30\{20\}^3 - 120\{30\}\{10\}^3$
$\qquad - 270\{20\}^2\{10\}^2 + 360\{20\}\{10\}^4 - 120\{10\}^6.$

$\kappa_{51} = \{51\} - \{50\}\{01\} - 5\{41\}\{10\} - 5\{40\}\{11\} - 10\{31\}\{20\} - 10\{30\}\{21\} + 10\{40\}\{10\}\{01\} + 20\{31\}\{10^2\}$
$\qquad + 20\{30\}\{20\}\{01\} + 40\{30\}\{11\}\{10\} + 60\{21\}\{20\}\{10\} + 30\{20\}^2\{11\} - 60\{30\}\{10\}^2\{01\}$
$\qquad - 60\{21\}\{10\}^3 - 90\{20\}^2\{10\}\{01\} - 180\{20\}\{11\}\{10\}^2 + 240\{20\}\{10\}^3\{01\} + 120\{11\}\{10\}^4$
$\qquad - 120\{10\}^5\{01\}.$

$\kappa_{42} = \{42\} - 2\{41\}\{01\} - \{40\}\{02\} - 4\{32\}\{10\} - 8\{31\}\{11\} - 4\{30\}\{12\} - 6\{22\}\{20\} - 6\{21\}^2 + 2\{40\}\{01\}^2$
$\qquad + 16\{31\}\{10\}\{01\} + 16\{30\}\{11\}\{01\} + 8\{30\}\{10\}\{02\} + 12\{22\}\{10\}^2 + 24\{21\}\{20\}\{01\}$
$\qquad + 48\{21\}\{11\}\{10\} + 6\{20\}^2\{02\} + 24\{20\}\{12\}\{10\} + 24\{20\}\{11\}^2 - 24\{30\}\{10\}\{01\}^2 - 72\{21\}\{10\}^2\{01\}$
$\qquad - 18\{20\}^2\{01\}^2 - 144\{20\}\{11\}\{10\}\{01\} - 36\{20\}\{10\}^2\{02\} - 24\{12\}\{10\}^3 - 72\{11\}^2\{10\}^2$
$\qquad + 144\{20\}\{10\}^2\{01\}^2 + 192\{11\}\{10\}^3\{01\} + 24\{10\}^4\{02\} - 120\{10\}^4\{01\}^2.$

$\kappa_{33} = \{33\} - 3\{32\}\{01\} - 3\{31\}\{02\} - \{30\}\{03\} - 3\{23\}\{10\} - 9\{22\}\{11\} - 9\{21\}\{12\} - 3\{20\}\{13\} + 6\{31\}\{01\}^2$
$\qquad + 6\{30\}\{02\}\{01\} + 18\{22\}\{10\}\{01\} + 36\{21\}\{11\}\{01\} + 18\{21\}\{10\}\{02\} + 18\{20\}\{12\}\{01\}$
$\qquad + 18\{20\}\{11\}\{02\} + 6\{20\}\{10\}\{03\} + 6\{13\}\{10\}^2 + 36\{12\}\{11\}\{10\} + 12\{11\}^3 - 6\{30\}\{01\}^3$
$\qquad - 54\{21\}\{10\}\{01\}^2 - 54\{20\}\{11\}\{01\}^2 - 54\{20\}\{10\}\{02\}\{01\} - 54\{12\}\{10\}^2\{01\} - 108\{11\}^2\{10\}\{01\}$
$\qquad - 54\{11\}\{10\}^2\{02\} - 6\{10\}^3\{03\} + 72\{20\}\{10\}\{01\}^3 + 216\{11\}\{10\}^2\{01\}^2 + 72\{10\}^3\{02\}\{01\}$
$\qquad - 120\{10\}^3\{01\}^3.$

$\kappa_{70} = \{70\} - 7\{60\}\{10\} - 21\{50\}\{20\} - 35\{40\}\{30\} + 42\{50\}\{10\}^2 + 210\{40\}\{20\}\{10\} + 140\{30\}^2\{10\}$
$\qquad + 210\{30\}\{20\}^2 - 210\{40\}\{10\}^3 - 1260\{30\}\{20\}\{10\}^2 - 630\{20\}^3\{10\} + 840\{30\}\{10\}^4$
$\qquad + 2520\{20\}^2\{10\}^3 - 2520\{20\}\{10\}^5 + 720\{10\}^7.$

$\kappa_{61} = \{61\} - \{60\}\{01\} - 6\{51\}\{10\} - 6\{50\}\{11\} - 15\{41\}\{20\} - 15\{40\}\{21\} - 20\{31\}\{30\} + 12\{50\}\{10\}\{01\}$
$\qquad + 30\{41\}\{10\}^2 + 30\{40\}\{20\}\{01\} + 60\{40\}\{11\}\{10\} + 120\{31\}\{20\}\{10\} + 20\{30\}^2\{01\}$
$\qquad + 120\{30\}\{21\}\{10\} + 120\{30\}\{20\}\{11\} + 90\{21\}\{20\}^2 - 90\{40\}\{10\}^2\{01\} - 120\{31\}\{10\}^3$

$$-360\{30\}\{20\}\{10\}\{01\} - 360\{30\}\{11\}\{10\}^2 - 540\{21\}\{20\}\{10\}^2 - 540\{20\}^2\{11\}\{10\} - 90\{20\}^3\{01\}$$
$$+480\{30\}\{10\}^3\{01\} + 360\{21\}\{10\}^4 + 1080\{20\}^2\{10\}^2\{01\} + 1440\{20\}\{11\}\{10\}^3$$
$$-1800\{20\}\{10\}^4\{01\} - 720\{11\}\{10\}^5 + 720\{10\}^6\{01\}.$$

$$\kappa_{52} = \{52\} - 2\{51\}\{01\} - \{50\}\{02\} - 5\{42\}\{10\} - 10\{41\}\{11\} - 5\{40\}\{12\} - 10\{32\}\{20\} - 20\{31\}\{21\} - 10\{30\}\{22\}$$
$$+2\{50\}\{01\}^2 + 20\{41\}\{10\}\{01\} + 20\{40\}\{11\}\{01\} + 10\{40\}\{10\}\{02\} + 20\{32\}\{10\}^2 + 40\{31\}\{20\}\{01\}$$
$$+80\{31\}\{11\}\{10\} + 40\{30\}\{21\}\{01\} + 20\{30\}\{20\}\{02\} + 40\{30\}\{12\}\{10\} + 40\{30\}\{11\}^2$$
$$+60\{22\}\{20\}\{10\} + 60\{21\}^2\{10\} + 120\{21\}\{20\}\{11\} + 30\{20\}^2\{12\} - 30\{40\}\{10\}\{01\}^2$$
$$-120\{31\}\{10\}^2\{01\} - 60\{30\}\{20\}\{01\}^2 - 240\{30\}\{11\}\{10\}\{01\} - 60\{30\}\{10\}^2\{02\} - 60\{22\}\{10\}^3$$
$$-360\{21\}\{20\}\{10\}\{01\} - 360\{21\}\{11\}\{10\}^2 - 180\{20\}^2\{11\}\{01\} - 90\{20\}^2\{10\}\{02\}$$
$$-180\{20\}\{12\}\{10\}^2 - 360\{20\}\{11\}^2\{10\} + 240\{30\}\{10\}^2\{01\}^2 + 480\{21\}\{10\}^3\{01\}$$
$$+360\{20\}^2\{10\}\{01\}^2 + 1440\{20\}\{11\}\{10\}^2\{01\} + 240\{20\}\{10\}^3\{02\} + 120\{12\}\{10\}^4 + 480\{11\}^2\{10\}^3$$
$$-1200\{20\}\{10\}^3\{01\}^2 - 1200\{11\}\{10\}^4\{01\} - 120\{10\}^5\{02\} + 720\{10\}^5\{01\}^2.$$

$$\kappa_{43} = \{43\} - 3\{42\}\{01\} - 3\{41\}\{02\} - \{40\}\{03\} - 4\{33\}\{10\} - 12\{32\}\{11\} - 12\{31\}\{12\} - 4\{30\}\{13\} - 6\{23\}\{20\}$$
$$-18\{22\}\{21\} + 6\{41\}\{01\}^2 + 6\{40\}\{02\}\{01\} + 24\{32\}\{10\}\{01\} + 48\{31\}\{11\}\{01\} + 24\{31\}\{10\}\{02\}$$
$$+24\{30\}\{12\}\{01\} + 24\{30\}\{11\}\{02\} + 8\{30\}\{10\}\{03\} + 12\{23\}\{10\}^2 + 36\{22\}\{20\}\{01\}$$
$$+72\{22\}\{11\}\{10\} + 36\{21\}^2\{01\} + 36\{21\}\{20\}\{02\} + 72\{21\}\{12\}\{10\} + 72\{21\}\{11\}^2 + 6\{20\}^2\{03\}$$
$$+24\{20\}\{13\}\{10\} + 72\{20\}\{12\}\{11\} - 6\{40\}\{01\}^3 - 72\{31\}\{10\}\{01\}^2 - 72\{30\}\{11\}\{01\}^2$$
$$-72\{30\}\{10\}\{02\}\{01\} - 108\{22\}\{10\}^2\{01\} - 108\{21\}\{20\}\{01\}^2 - 432\{21\}\{11\}\{10\}\{01\}$$
$$-108\{21\}\{10\}^2\{02\} - 54\{20\}^2\{02\}\{01\} - 216\{20\}\{12\}\{10\}\{01\} - 216\{20\}\{11\}^2\{01\}$$
$$-216\{20\}\{11\}\{10\}\{02\} - 36\{20\}\{10\}^2\{03\} - 24\{13\}\{10\}^3 - 216\{12\}\{11\}\{10\}^2 - 144\{11\}^3\{10\}$$
$$+96\{30\}\{10\}\{01\}^3 + 432\{21\}\{10\}^2\{01\}^2 + 72\{20\}^2\{01\}^3 + 864\{20\}\{11\}\{10\}\{01\}^2$$
$$+432\{20\}\{10\}^2\{02\}\{01\} + 288\{12\}\{10\}^3\{01\} + 864\{11\}^2\{10\}^2\{01\} + 288\{11\}\{10\}^3\{02\} + 24\{10\}^4\{03\}$$
$$-720\{20\}\{10\}^2\{01\}^3 - 1440\{11\}\{10\}^3\{01\}^2 - 360\{10\}^4\{02\}\{01\} + 720\{10\}^4\{01\}^3.$$

$$\kappa_{80} = \{80\} - 8\{70\}\{10\} - 28\{60\}\{20\} - 56\{50\}\{30\} - 35\{40\}^2 + 56\{60\}\{10\}^2 + 336\{50\}\{20\}\{10\}$$
$$+560\{40\}\{30\}\{10\} + 420\{40\}\{20\}^2 + 560\{30\}^2\{20\} - 336\{50\}\{10\}^3 - 2520\{40\}\{20\}\{10\}^2$$
$$-1680\{30\}^2\{10\}^2 - 5040\{30\}\{20\}^2\{10\} - 630\{20\}^4 + 1680\{40\}\{10\}^4 + 13{,}440\{30\}\{20\}\{10\}^3$$
$$+10{,}080\{20\}^3\{10\}^2 - 6720\{30\}\{10\}^5 - 25{,}200\{20\}^2\{10\}^4 + 20{,}160\{20\}\{10\}^6 - 5040\{10\}^8.$$

$$\kappa_{71} = \{71\} - \{70\}\{01\} - 7\{61\}\{10\} - 7\{60\}\{11\} - 21\{51\}\{20\} - 21\{50\}\{21\} - 35\{41\}\{30\} - 35\{40\}\{31\}$$
$$+14\{60\}\{10\}\{01\} + 42\{51\}\{10\}^2 + 42\{50\}\{20\}\{01\} + 84\{50\}\{11\}\{10\} + 210\{41\}\{20\}\{10\}$$
$$+70\{40\}\{30\}\{01\} + 210\{40\}\{21\}\{10\} + 210\{40\}\{20\}\{11\} + 280\{31\}\{30\}\{10\} + 210\{31\}\{20\}^2$$
$$+140\{30\}^2\{11\} + 420\{30\}\{21\}\{20\} - 126\{50\}\{10\}^2\{01\} - 210\{41\}\{10\}^3 - 630\{40\}\{20\}\{10\}\{01\}$$
$$-630\{40\}\{11\}\{10\}^2 - 1260\{31\}\{20\}\{10\}^2 - 420\{30\}^2\{10\}\{01\} - 1260\{30\}\{21\}\{10\}^2$$
$$-630\{30\}\{20\}^2\{01\} - 2520\{30\}\{20\}\{11\}\{10\} - 1890\{21\}\{20\}^2\{10\} - 630\{20\}^3\{11\}$$
$$+840\{40\}\{10\}^3\{01\} + 840\{31\}\{10\}^4 + 5040\{30\}\{20\}\{10\}^2\{01\} + 3360\{30\}\{11\}\{10\}^3$$
$$+5040\{21\}\{20\}\{10\}^3 + 2520\{20\}^3\{10\}\{01\} + 7560\{20\}^2\{11\}\{10\}^2 - 4200\{30\}\{10\}^4\{01\}$$
$$-2520\{21\}\{10\}^5 - 12{,}600\{20\}^2\{10\}^3\{01\} - 12{,}600\{20\}\{11\}\{10\}^4 + 15{,}120\{20\}\{10\}^5\{01\}$$
$$+5040\{11\}\{10\}^6 - 5040\{10\}^7\{01\}.$$

$$\kappa_{62} = \{62\} - 2\{61\}\{01\} - \{60\}\{02\} - 6\{52\}\{10\} - 12\{51\}\{11\} - 6\{50\}\{12\} - 15\{42\}\{20\} - 30\{41\}\{21\}$$
$$-15\{40\}\{22\} - 20\{32\}\{30\} - 20\{31\}^2 + 2\{60\}\{01\}^2 + 24\{51\}\{10\}\{01\} + 24\{50\}\{11\}\{01\}$$
$$+12\{50\}\{10\}\{02\} + 30\{42\}\{10\}^2 + 60\{41\}\{20\}\{01\} + 120\{41\}\{11\}\{10\} + 60\{40\}\{21\}\{01\}$$
$$+30\{40\}\{20\}\{02\} + 60\{40\}\{12\}\{10\} + 60\{40\}\{11\}^2 + 120\{32\}\{20\}\{10\} + 80\{31\}\{30\}\{01\}$$
$$+240\{31\}\{21\}\{10\} + 240\{31\}\{20\}\{11\} + 120\{30\}\{22\}\{10\} + 240\{30\}\{21\}\{11\} + 20\{30\}^2\{02\}$$
$$+120\{30\}\{20\}\{12\} + 90\{22\}\{20\}^2 + 180\{21\}^2\{20\} - 36\{50\}\{10\}\{01\}^2 - 180\{41\}\{10\}^2\{01\}$$
$$-90\{40\}\{20\}\{01\}^2 - 360\{40\}\{11\}\{10\}\{01\} - 90\{40\}\{10\}^2\{02\} - 120\{32\}\{10\}^3 - 720\{31\}\{20\}\{10\}\{01\}$$
$$-720\{31\}\{11\}\{10\}^2 - 60\{30\}^2\{01\}^2 - 720\{30\}\{21\}\{10\}\{01\} - 720\{30\}\{20\}\{11\}\{01\}$$
$$-360\{30\}\{20\}\{10\}\{02\} - 360\{30\}\{12\}\{10\}^2 - 720\{30\}\{11\}^2\{10\} - 540\{22\}\{20\}\{10\}^2$$

$$-540\{21\}^2\{10\}^2 - 540\{21\}\{20\}^2\{01\} - 2160\{21\}\{20\}\{11\}\{10\} - 90\{20\}^3\{02\} - 540\{20\}^2\{12\}\{10\}$$
$$-540\{20\}^2\{11\}^2 + 360\{40\}\{10\}^2\{01\}^2 + 960\{31\}\{10\}^3\{01\} + 1440\{30\}\{20\}\{10\}\{01\}^2$$
$$+2880\{30\}\{11\}\{10\}^2\{01\} + 480\{30\}\{10\}^3\{02\} + 360\{22\}\{10\}^4 + 4320\{21\}\{20\}\{10\}^2\{01\}$$
$$+2880\{21\}\{11\}\{10\}^3 + 360\{20\}^3\{01\}^2 + 4320\{20\}^2\{11\}\{10\}\{01\} + 1080\{20\}^2\{10\}^2\{02\}$$
$$+1440\{20\}\{12\}\{10\}^3 + 4320\{20\}\{11\}^2\{10\}^2 - 2400\{30\}\{10\}^3\{01\}^2 - 3600\{21\}\{10\}^4\{01\}$$
$$-5400\{20\}^2\{10\}^2\{01\}^2 - 14{,}400\{20\}\{11\}\{10\}^3\{01\} - 1800\{20\}\{10\}^4\{02\} - 720\{12\}\{10\}^5$$
$$-3600\{11\}^2\{10\}^4 + 10{,}800\{20\}\{10\}^4\{01\}^2 + 8640\{11\}\{10\}^5\{01\} + 720\{10\}^6\{02\} - 5040\{10\}^6\{01\}^2.$$

$$\kappa_{53} = \{53\} - 3\{52\}\{01\} - 3\{51\}\{02\} - \{50\}\{03\} - 5\{43\}\{10\} - 15\{42\}\{11\} - 15\{41\}\{12\} - 5\{40\}\{13\} - 10\{33\}\{20\}$$
$$-30\{32\}\{21\} - 30\{31\}\{22\} - 10\{30\}\{23\} + 6\{51\}\{01\}^2 + 6\{50\}\{02\}\{01\} + 30\{42\}\{10\}\{01\}$$
$$+60\{41\}\{11\}\{01\} + 30\{41\}\{10\}\{02\} + 30\{40\}\{12\}\{01\} + 30\{40\}\{11\}\{02\} + 10\{40\}\{10\}\{03\}$$
$$+20\{33\}\{10\}^2 + 60\{32\}\{20\}\{01\} + 120\{32\}\{11\}\{10\} + 120\{31\}\{21\}\{01\} + 60\{31\}\{20\}\{02\}$$
$$+120\{31\}\{12\}\{10\} + 120\{31\}\{11\}^2 + 60\{30\}\{22\}\{01\} + 60\{30\}\{21\}\{02\} + 20\{30\}\{20\}\{03\}$$
$$+40\{30\}\{13\}\{10\} + 120\{30\}\{12\}\{11\} + 60\{23\}\{20\}\{10\} + 180\{22\}\{21\}\{10\} + 180\{22\}\{20\}\{11\}$$
$$+180\{21\}^2\{11\} + 180\{21\}\{20\}\{12\} + 30\{20\}^2\{13\} - 6\{50\}\{01\}^3 - 90\{41\}\{10\}\{01\}^2 - 90\{40\}\{11\}\{01\}^2$$
$$-90\{40\}\{10\}\{02\}\{01\} - 180\{32\}\{10\}^2\{01\} - 180\{31\}\{20\}\{01\}^2 - 720\{31\}\{11\}\{10\}\{01\}$$
$$-180\{31\}\{10\}^2\{02\} - 180\{30\}\{21\}\{01\}^2 - 180\{30\}\{20\}\{02\}\{01\} - 360\{30\}\{12\}\{10\}\{01\}$$
$$-360\{30\}\{11\}^2\{01\} - 360\{30\}\{11\}\{10\}\{02\} - 60\{30\}\{10\}^2\{03\} - 60\{23\}\{10\}^3 - 540\{22\}\{20\}\{10\}\{01\}$$
$$-540\{22\}\{11\}\{10\}^2 - 540\{21\}^2\{10\}\{01\} - 1080\{21\}\{20\}\{11\}\{01\} - 540\{21\}\{20\}\{10\}\{02\}$$
$$-540\{21\}\{12\}\{10\}^2 - 1080\{21\}\{11\}^2\{10\} - 270\{20\}^2\{12\}\{01\} - 270\{20\}^2\{11\}\{02\} - 90\{20\}^2\{10\}\{03\}$$
$$-180\{20\}\{13\}\{10\}^2 - 1080\{20\}\{12\}\{11\}\{10\} - 360\{20\}\{11\}^3 + 120\{40\}\{10\}\{01\}^3 + 720\{31\}\{10\}^2\{01\}^2$$
$$+240\{30\}\{20\}\{01\}^3 + 1440\{30\}\{11\}\{10\}\{01\}^2 + 720\{30\}\{10\}^2\{02\}\{01\} + 720\{22\}\{10\}^3\{01\}$$
$$+2160\{21\}\{20\}\{10\}\{01\}^2 + 4320\{21\}\{11\}\{10\}^2\{01\} + 720\{21\}\{10\}^3\{02\} + 1080\{20\}^2\{11\}\{01\}^2$$
$$+1080\{20\}^2\{10\}\{02\}\{01\} + 2160\{20\}\{12\}\{10\}^2\{01\} + 4320\{20\}\{11\}^2\{10\}\{01\}$$
$$+2160\{20\}\{11\}\{10\}^2\{02\} + 240\{20\}\{10\}^3\{03\} + 120\{13\}\{10\}^4 + 1440\{12\}\{11\}\{10\}^3$$
$$+1440\{11\}^3\{10\}^2 - 1200\{30\}\{10\}^2\{01\}^3 - 3600\{21\}\{10\}^3\{01\}^2 - 1800\{20\}^2\{10\}\{01\}^3$$
$$-10{,}800\{20\}\{11\}\{10\}^2\{01\}^2 - 3600\{20\}\{10\}^3\{02\}\{01\} - 1800\{12\}\{10\}^4\{01\} - 7200\{11\}^2\{10\}^3\{01\}$$
$$-1800\{11\}\{10\}^4\{02\} - 120\{10\}^5\{03\} + 7200\{20\}\{10\}^3\{01\}^3 + 10{,}800\{11\}\{10\}^4\{01\}^2$$
$$+2160\{10\}^5\{02\}\{01\} - 5040\{10\}^5\{01\}^3.$$

$$\kappa_{44} = \{44\} - 4\{43\}\{01\} - 6\{42\}\{02\} - 4\{41\}\{03\} - \{40\}\{04\} - 4\{34\}\{10\} - 16\{33\}\{11\} - 24\{32\}\{12\}$$
$$-16\{31\}\{13\} - 4\{30\}\{14\} - 6\{24\}\{20\} - 24\{23\}\{21\} - 18\{22\}^2 + 12\{42\}\{01\}^2 + 24\{41\}\{02\}\{01\}$$
$$+8\{40\}\{03\}\{01\} + 6\{40\}\{02\}^2 + 32\{33\}\{10\}\{01\} + 96\{32\}\{11\}\{01\} + 48\{32\}\{10\}\{02\}$$
$$+96\{31\}\{12\}\{01\} + 96\{31\}\{11\}\{02\} + 32\{31\}\{10\}\{03\} + 32\{30\}\{13\}\{01\} + 48\{30\}\{12\}\{02\}$$
$$+32\{30\}\{11\}\{03\} + 8\{30\}\{10\}\{04\} + 12\{24\}\{10\}^2 + 48\{23\}\{20\}\{01\} + 96\{23\}\{11\}\{10\}$$
$$+144\{22\}\{21\}\{01\} + 72\{22\}\{20\}\{02\} + 144\{22\}\{12\}\{10\} + 144\{22\}\{11\}^2 + 72\{21\}^2\{02\}$$
$$+48\{21\}\{20\}\{03\} + 96\{21\}\{13\}\{10\} + 288\{21\}\{12\}\{11\} + 6\{20\}^2\{04\} + 24\{20\}\{14\}\{10\}$$
$$+96\{20\}\{13\}\{11\} + 72\{20\}\{12\}^2 - 24\{41\}\{01\}^3 - 36\{40\}\{02\}\{01\}^2 - 144\{32\}\{10\}\{01\}^2$$
$$-288\{31\}\{11\}\{01\}^2 - 288\{31\}\{10\}\{02\}\{01\} - 144\{30\}\{12\}\{01\}^2 - 288\{30\}\{11\}\{02\}\{01\}$$
$$-96\{30\}\{10\}\{03\}\{01\} - 72\{30\}\{10\}\{02\}^2 - 144\{23\}\{10\}^2\{01\} - 216\{22\}\{20\}\{01\}^2$$
$$-864\{22\}\{11\}\{10\}\{01\} - 216\{22\}\{10\}^2\{02\} - 216\{21\}^2\{01\}^2 - 432\{21\}\{20\}\{02\}\{01\}$$
$$-864\{21\}\{12\}\{10\}\{01\} - 864\{21\}\{11\}^2\{01\} - 864\{21\}\{11\}\{10\}\{02\} - 144\{21\}\{10\}^2\{03\}$$
$$-72\{20\}^2\{03\}\{01\} - 54\{20\}^2\{02\}^2 - 288\{20\}\{13\}\{10\}\{01\} - 864\{20\}\{12\}\{11\}\{01\}$$
$$-432\{20\}\{12\}\{10\}\{02\} - 432\{20\}\{11\}^2\{02\} - 288\{20\}\{11\}\{10\}\{03\} - 36\{20\}\{10\}^2\{04\}$$
$$-24\{14\}\{10\}^3 - 288\{13\}\{11\}\{10\}^2 - 216\{12\}^2\{10\}^2 - 864\{12\}\{11\}^2\{10\} - 144\{11\}^4 + 24\{40\}\{01\}^4$$
$$+384\{31\}\{10\}\{01\}^3 + 384\{30\}\{11\}\{01\}^3 + 576\{30\}\{10\}\{02\}\{01\}^2 + 864\{22\}\{10\}^2\{01\}^2$$
$$+576\{21\}\{20\}\{01\}^3 + 3456\{21\}\{11\}\{10\}\{01\}^2 + 1728\{21\}\{10\}^2\{02\}\{01\} + 432\{20\}^2\{02\}\{01\}^2$$
$$+1728\{20\}\{12\}\{10\}\{01\}^2 + 1728\{20\}\{11\}^2\{01\}^2 + 3456\{20\}\{11\}\{10\}\{02\}\{01\} + 576\{20\}\{10\}^2\{03\}\{01\}$$
$$+432\{20\}\{10\}^2\{02\}^2 + 384\{13\}\{10\}^3\{01\} + 3456\{12\}\{11\}\{10\}^2\{01\} + 576\{12\}\{10\}^3\{02\}$$
$$+2304\{11\}^3\{10\}\{01\} + 1728\{11\}^2\{10\}^2\{02\} + 384\{11\}\{10\}^3\{03\} + 24\{10\}^4\{04\} - 480\{30\}\{10\}\{01\}^4$$
$$-2880\{21\}\{10\}^2\{01\}^3 - 360\{20\}^2\{01\}^4 - 5760\{20\}\{11\}\{10\}\{01\}^3 - 4320\{20\}\{10\}^2\{02\}\{01\}^2$$
$$-2880\{12\}\{10\}^3\{01\}^2 - 8640\{11\}^2\{10\}^2\{01\}^2 - 5760\{11\}\{10\}^3\{02\}\{01\} - 480\{10\}^4\{03\}\{01\}$$
$$-360\{10\}^4\{02\}^2 + 4320\{20\}\{10\}^2\{01\}^4 + 11{,}520\{11\}\{10\}^3\{01\}^3 + 4320\{10\}^4\{02\}\{01\}^2$$
$$-5040\{10\}^4\{01\}^4.$$

Table 2·1·3. *k-statistics in terms of products of s-functions*
(*up to and including weight 11*)

The arrangement of Table 2·1·3 has been made to allow the expression of the *k*-statistics in terms of products of the *s*-functions to be made succinctly. The layout is, to a large extent, self-explanatory. Thus, for example

$$k_3 = \frac{1}{n^{(3)}}\{n^2 s_3 - 3n s_2 s_1 + 2s_1^3\}$$

and similarly for the others.

k statistic	Dividing factor	Product of s functions	Coefficients of						
			n^0	n^1	n^2	n^3	n^4	n^5	n^6
k_1	n	s_1	1	.	.	.	.	.	.
k_2	$n^{(2)}$	s_2	.	1	.	.	.	.	.
		s_1^2	-1	.	.	.	.	.	.
k_3	$n^{(3)}$	s_3	.	.	1	.	.	.	.
		$s_2 s_1$	.	-3	.	.	.	.	.
		s_1^3	2	.	.	.	.	.	.
k_4	$n^{(4)}$	s_4	.	.	1	1	.	.	.
		$s_3 s_1$	.	-4	-4	.	.	.	.
		s_2^2	.	3	-3	.	.	.	.
		$s_2 s_1^2$	.	12	.	.	.	.	.
		s_1^4	-6	.	.	.	.	.	.
k_5	$n^{(5)}$	s_5	.	.	.	5	1	.	.
		$s_4 s_1$	.	.	-25	-5	.	.	.
		$s_3 s_2$	.	.	10	-10	.	.	.
		$s_3 s_1^2$	.	40	20	.	.	.	.
		$s_2^2 s_1$	.	-30	30	.	.	.	.
		$s_2 s_1^3$	.	-60	.	.	.	.	.
		s_1^5	24	.	.	.	.	.	.
k_6	$n^{(6)}$	s_6	.	.	-4	11	16	1	.
		$s_5 s_1$	.	24	-66	-96	-6	.	.
		$s_4 s_2$	.	-60	105	-30	-15	.	.
		s_3^2	.	40	-50	20	-10	.	.
		$s_4 s_1^2$	.	60	270	30	.	.	.
		$s_3 s_2 s_1$	.	-120	.	120	.	.	.
		s_2^3	.	60	-90	30	.	.	.
		$s_3 s_1^3$	.	-360	-120	.	.	.	.
		$s_2^2 s_1^2$	.	270	-270	.	.	.	.
		$s_2 s_1^4$	.	360	.	.	.	.	.
		s_1^6	-120	.	.	.	.	.	.
k_7	$n^{(7)}$	s_7	.	.	.	-42	119	42	1
		$s_6 s_1$	.	.	294	-833	-294	-7	.
		$s_5 s_2$	.	.	-378	651	-252	-21	.
		$s_4 s_3$	.	.	210	-175	.	-35	.
		$s_5 s_1^2$	.	-504	1848	1134	42	.	.
		$s_4 s_2 s_1$	.	1260	-2730	1260	210	.	.
		$s_3^2 s_1$	.	-840	700	.	140	.	.
		$s_3 s_2^2$	.	.	420	-630	210	.	.
		$s_4 s_1^3$	.	-1260	-2730	-210	.	.	.
		$s_3 s_2 s_1^2$	.	2520	-1260	-1260	.	.	.
		$s_2^3 s_1$	.	-1260	1890	-630	.	.	.
		$s_3 s_1^4$	.	3360	840	.	.	.	.
		$s_2^2 s_1^3$	.	-2520	2520	.	.	.	.
		$s_2 s_1^5$	.	-2520	.	.	.	.	.
		s_1^7	720	.	.	.	.	.	.

Table 2·1·3 (cont.)

k statistic	Dividing factor	Product of s functions	Coefficients of								
			n^0	n^1	n^2	n^3	n^4	n^5	n^6	n^7	n^8
k_8	$n^{(8)}$	s_8	·	·	120	-398	141	757	99	1	·
		$s_7 s_1$		-960	3184	-1128	-6056	-792	-8	·	·
		$s_6 s_2$		3360	-7784	4396	1092	-1036	-28	·	·
		$s_5 s_3$		-6720	12208	-6216	1288	-504	-56	·	·
		$s_4{}^2$		4200	-7210	4235	-1155	-35	-35	·	·
		$s_6 s_1{}^2$		-3360	-448	20104	3808	56	-35	·	·
		$s_5 s_2 s_1$		10080	-7728	-10416	7728	336	·	·	·
		$s_4 s_3 s_1$		-3360	-2800	2800	2800	560	·	·	·
		$s_4 s_2{}^2$		-10080	19320	-10500	840	420	·	·	·
		$s_3{}^2 s_2$		6720	-11200	6160	-2240	560	·	·	·
		$s_5 s_1{}^3$		6048	-33264	-12768	-336	·	·	·	·
		$s_4 s_2 s_1{}^2$		-15120	42840	-25200	-2520	·	·	·	·
		$s_3{}^2 s_1{}^2$		16800	-11760	-3360	-1680	·	·	·	·
		$s_3 s_2{}^2 s_1$		-10080	5040	10080	-5040	·	·	·	·
		$s_2{}^4$		3780	-6930	3780	-630	·	·	·	·
		$s_4 s_1{}^4$		20160	28560	1680	·	·	·	·	·
		$s_3 s_2 s_1{}^3$		-40320	26880	13440	·	·	·	·	·
		$s_2{}^3 s_1{}^2$		20160	-30240	10080	·	·	·	·	·
		$s_3 s_1{}^5$		-33600	-6720	·	·	·	·	·	·
		$s_2{}^2 s_1{}^4$		25200	-25200	·	·	·	·	·	·
		$s_2 s_1{}^6$		20160	·	·	·	·	·	·	·
		$s_1{}^8$	-5040	·	·	·	·	·	·	·	·
k_9	$n^{(9)}$	s_9	·	·	·	2160	-7250	6189	3721	219	1
		$s_8 s_1$			-19440	65250	-55701	-33489	-1971	-9	·
		$s_7 s_2$			43200	-98856	69012	-9972	-3348	-36	·
		$s_7 s_1{}^2$		34560	-162144	153792	143928	11232	72	·	·
		$s_6 s_3$			-60480	101976	-45612	6972	-2772	-84	·
		$s_6 s_2 s_1$		-120960	386664	-346248	48888	31752	504	·	·
		$s_6 s_1{}^3$		120960	-128016	-368424	-47376	-504	·	·	·
		$s_5 s_4$			30240	-46620	25326	-7686	-1134	-126	·
		$s_5 s_3 s_1$		241920	-425376	172368	-11088	21168	1008	·	·
		$s_5 s_2{}^2$			-78624	149688	-85428	13608	756	·	·
		$s_5 s_2 s_1{}^2$		-362880	480816	40824	-154224	-4536	·	·	·
		$s_5 s_1{}^4$		-48384	532224	148176	3024	·	·	·	·
		$s_4{}^2 s_1$		-151200	233100	-126630	38430	5670	630	·	·
		$s_4 s_3 s_2$			10080	·	-12600	·	2520	·	·
		$s_4 s_3 s_1{}^2$		120960	75600	-113400	-75600	-7560	·	·	·
		$s_4 s_2{}^2 s_1$		362880	-748440	464940	-68040	-11340	·	·	·
		$s_4 s_2 s_1{}^3$		120960	-574560	423360	30240	·	·	·	·
		$s_4 s_1{}^5$		-302400	-317520	-15120	·	·	·	·	·
		$s_3{}^3$			22400	-3696	1736	-3360	560	·	·
		$s_3{}^2 s_2 s_1$		-241920	332640	-105840	30240	-15120	·	·	·
		$s_3{}^2 s_1{}^3$		-322560	221760	80640	20160	·	·	·	·
		$s_3 s_2{}^3$			45360	-83160	45360	-7560	·	·	·
		$s_3 s_2{}^2 s_1{}^2$		362880	-362880	-90702	90720	·	·	·	·
		$s_3 s_2 s_1{}^4$		604800	-453600	-151200	·	·	·	·	·
		$s_3 s_1{}^6$		362880	60480	·	·	·	·	·	·
		$s_2{}^4 s_1$		-136080	249480	-136080	22680	·	·	·	·
		$s_2{}^3 s_1{}^3$		-302400	453600	-151200	·	·	·	·	·
		$s_2{}^2 s_1{}^5$		-272160	272160	·	·	·	·	·	·
		$s_2 s_1{}^7$		-181440	·	·	·	·	·	·	·
		$s_1{}^9$	40320	·	·	·	·	·	·	·	·

Table 2·1·3 (*cont.*)

k statistic	Dividing factor	Product of s functions	n^0	n^1	n^2	n^3	n^4	n^5	n^6	n^7	n^8	n^9
k_{10}	$n^{(10)}$	s_{10}	·	·	−12096	45624	−41186	−41171	72976	15706	466	1
		$s_9 s_1$	·	120960	−456240	411860	411710	−729760	−157060	−4660	−10	·
		$s_8 s_2$	·	−544320	1508760	−1219410	−38655	412470	−109260	−9540	−45	·
		$s_8 s_1^2$	·	544320	−633960	−1814040	2871450	816030	30510	90	·	·
		$s_7 s_3$	·	1451520	−3297600	2250960	−326280	−63120	−4800	−10560	−120	·
		$s_7 s_2 s_1$	·	−2177280	3002400	1288080	−3110400	888480	108000	720	·	·
		$s_7 s_1^3$	·	−311040	3978720	−5583600	−2768400	−153360	−720	·	·	·
		$s_6 s_4$	·	−2540160	5226480	−3649380	1142610	−154140	−18480	−6720	−210	·
		$s_6 s_3 s_1$	·	2177280	−1159200	−2286480	1058400	107520	100800	1680	·	·
		$s_6 s_2^2$	·	2177280	−4929120	3520440	−585900	−255780	71820	1260	·	·
		$s_6 s_2 s_1^2$	·	1088640	−8119440	10470600	−2759400	−672840	−7560	·	·	·
		$s_6 s_1^4$	·	−2903040	4536000	6224400	604800	5040	·	−2016	−126	·
		s_5^2	·	1524096	−3027024	2127636	−750834	160524	60480	2520	·	·
		$s_5 s_4 s_1$	·	−1088640	619920	52680	−680400	433440	75600	5040	·	·
		$s_5 s_3 s_2$	·	−4354560	8766600	−6027840	2041200	−500040	−32256	·	·	·
		$s_5 s_3 s_1^2$	·	−6531840	11581920	−3855600	−680400	−498960	−15120	·	·	·
		$s_5 s_2^2 s_1$	·	3265920	−3039120	−2608200	3061800	−657720	−22680	·	·	·
		$s_5 s_2 s_1^3$	·	8709120	−13608000	2116800	2721600	60480	·	·	·	·
		$s_5 s_1^5$	·	·	−8316000	−1814400	−30240	·	·	·	·	·
		$s_4^2 s_2$	·	2721600	−5632200	3981600	−1181250	97650	9450	3150	·	·
		$s_4^2 s_1^2$	·	4082400	−5613300	2882250	−1181250	−160650	−9450	·	·	·
		$s_4 s_3^2$	·	201600	−109200	−189000	105000	−12600	−75600	4200	·	·
		$s_4 s_3 s_2 s_1$	·	·	−1058400	378000	1134000	−378000	·	·	·	·
		$s_4 s_3 s_1^3$	·	−3628800	−1512000	3528000	1512000	100800	−18900	·	·	·
		$s_4 s_2^3$	·	−2721600	5783400	−4063500	1039500	−18900	·	·	·	·
		$s_4 s_2^2 s_1^2$	·	−8164800	18144000	−12474000	2268000	226800	·	·	·	·
		$s_4 s_2 s_1^4$	·	·	7182000	−6804000	−378000	·	·	·	·	·
		$s_4 s_1^6$	·	4536000	3780000	151200	·	·	·	·	·	·
		$s_3^3 s_1$	·	−806400	436800	756000	−420000	50400	−16800	·	·	·
		$s_3^2 s_2^2$	·	1814400	−3553200	2457000	−945000	264600	−37800	·	·	·
		$s_3^2 s_2 s_1^2$	·	7257600	−9072000	1512000	·	302400	·	·	·	·
		$s_3^2 s_1^4$	·	6048000	−4284000	−1512000	−252000	·	·	·	·	·
		$s_3 s_2^3 s_1$	·	−1814400	1512000	1512000	−1512000	302400	·	·	·	·
		$s_3 s_2^2 s_1^3$	·	−9072000	10584000	·	−1512000	·	·	·	·	·
		$s_3 s_2 s_1^5$	·	−9072000	7257600	1814400	·	·	·	·	·	·
		$s_3 s_1^7$	·	−4233600	−604800	·	·	·	·	·	·	·
		s_2^5	·	544320	−1134000	793800	−226800	22680	·	·	·	·
		$s_2^4 s_1^2$	·	3402000	−6237000	3402000	−567000	·	·	·	·	·
		$s_2^3 s_1^4$	·	4536000	−6804000	2268000	·	·	·	·	·	·
		$s_2^2 s_1^6$	·	3175200	−3175200	·	·	·	·	·	·	·
		$s_2 s_1^8$	·	1814400	·	·	·	·	·	·	·	·
		s_1^{10}	−362880	·	·	·	·	·	·	·	·	·

Table 2·1·3 (cont.)

k statistic	Dividing factor	Product of s functions	n^0	n^1	n^2	n^3	n^4	n^5	n^6	n^7	n^8	n^9	n^{10}
k_{11}	$n^{(11)}$	s_{11}	.	.	.	-332640	1261788	-1594648	371569	595760	60082	968	1
		$s_{10}s_1$	.	.	3659040	-13879668	17541128	-4087259	-6553360	-660902	-10648	-11	.
		s_9s_2	.	.	-11642400	31997460	-30546560	10268555	634150	-685960	-25190	-55	.
		$s_9s_1^2$	.	-6652800	3740880	-57159080	10167740	32132650	3990470	78430	110	.	.
		s_8s_3	.	.	24948000	-54859860	39087840	-9734835	862950	-269940	-33990	-165	.
		$s_8s_2s_1$	.	29937600	-123397560	157655520	-63212490	-8296200	6983460	328680	990	.	.
		$s_8s_1^3$	.	-29937600	66373560	11638440	-90867150	-16627050	-454410	-990	.	.	.
		s_7s_4	.	.	-29937600	58342680	-38089920	11616330	-1798500	-104280	-28380	-330	.
		$s_7s_3s_1$	.	-79833600	205508160	-160343040	31413360	290400	2576640	385440	2640	.	.
		$s_7s_2^2$	.	.	32788800	-73711440	56319120	-16384500	702900	283140	1980	.	.
		$s_7s_2s_1^2$	.	119750400	-242684640	93091680	65518200	-33204600	-2459160	-11880	.	.	.
		$s_7s_1^4$	.	-11404800	-69822720	148975200	49856400	2138400	7920	.	.	.	.
		s_6s_5	.	.	13970880	-25385976	15286656	-4249938	697620	-301224	-17556	-462	.
		$s_6s_4s_1$	.	139708800	-281468880	190196160	-60064620	9101400	2236080	286440	4620	.	.
		$s_6s_3s_2$	.	.	-36590400	68486880	-40711440	10857000	-2541000	489720	9240	.	.
		$s_6s_3s_1^2$	.	-119750400	112321440	50893920	-30076200	-10949400	-2411640	-27720	.	.	.
		$s_6s_2^2s_1$	.	-119750400	310519440	-272099520	82120500	2702700	-3451140	-41580	.	.	.
		$s_6s_2s_1^3$	.	39916800	94691520	-232293600	84823200	12751200	110880	.	.	.	.
		$s_6s_1^5$	.	59875200	-115647840	-103728240	-8094240	-55440	.	.	.	.	.
		$s_5^2s_1$	.	-83825280	152315856	-91719936	25499628	-4185720	1807344	105336	2772	.	.
		$s_5s_4s_2$	.	.	22952160	-48953520	36230040	-11157300	679140	235620	13860	.	.
		$s_5s_4s_1^2$	.	59875200	-63035280	16465680	4781700	-16424100	-1621620	-41580	.	.	.
		$s_5s_3^2$	.	.	-18627840	39214560	-27128640	7253400	-822360	101640	9240	.	.
		$s_5s_3s_2s_1$	.	239500800	-450394560	262120320	-64033200	17463600	-4490640	-166320	.	.	.
		$s_5s_3s_1^3$	.	159667200	-298488960	96465600	32155200	9979200	221760	.	.	.	.
		$s_5s_2^3$	.	.	-25446960	53887680	-38461500	11018700	-956340	-41580	.	.	.
		$s_5s_2^2s_1^2$	.	-179625600	261455040	-34927200	-67359600	19958400	498960	.	.	.	.
		$s_5s_2s_1^4$	.	-179625600	311018400	-83991600	-46569600	-831600	.	.	.	.	.
		$s_5s_1^6$	.	11975040	131725440	23617440	332640	.	.	.	.	.	.
		$s_4^2s_3$	.	.	11642400	-20836200	12520200	-4100250	773850	-11550	11550	.	.
		$s_4^2s_2s_1$	.	-149688000	307276200	-218710800	68087250	-5717250	-1143450	-103950	.	.	.
		$s_4^2s_1^3$	.	-99792000	118364400	-53361000	31185000	3465000	138600	.	.	.	.
		$s_4s_3^2s_1$	.	.	-29383200	35481600	-3465000	-2079000	-415800	-138600	.	.	.
		$s_4s_3s_2^2$	.	.	-7484400	19958400	-17671500	5197500	207900	-207900	.	.	.
		$s_4s_3s_2s_1^2$	.	.	73180800	-66528000	-24948000	16632000	1663200	.	.	.	.
		$s_4s_3s_1^4$	.	99792000	19404000	-90090000	-27720000	-1386000	.	.	.	.	.
		$s_4s_2^3s_1$	.	149688000	-329313600	245322000	-70686000	4158000	831600	.	.	.	.
		$s_4s_2^2s_1^3$	.	149688000	-365904000	278586000	-58212000	-4158000	.	.	.	.	.
		$s_4s_2s_1^5$	.	-29937600	-84823200	109771200	4989600	.	.	.	.	.	.
		$s_4s_1^7$	.	-69854400	-48232800	-1663200	.	.	.	.	.	.	.
		$s_3^3s_2$	.	.	14414400	-28089600	18018000	-5082000	831600	-92400	.	.	.
		$s_3^3s_1^2$	.	44352000	-42873600	-11088000	9240000	.	369600	.	.	.	.
		$s_3^2s_2^2s_1$	.	-99792000	172927800	-91476000	24948000	-8316000	1663200	.	.	.	.
		$s_3^2s_2s_1^3$	.	-199584000	243936000	-27720000	-11088000	-5544000	.	.	.	.	.
		$s_3^2s_1^5$	.	-113097600	83160000	26611200	3326400	.	.	.	.	.	.
		$s_3s_2^4$	.	.	9979200	-20790000	14553000	-4158000	415800	.	.	.	.
		$s_3s_2^3s_1^2$	.	99792000	-133056000	8316000	33264000	-8316000	.	.	.	.	.
		$s_3s_2^2s_1^4$	.	199584000	-249480000	24948000	24948000	.	.	.	.	.	.
		$s_3s_2s_1^6$	.	139708800	-116424000	-23284800	.	.	.	.	.	.	.
		$s_3s_1^8$	.	53222400	6652800	.	.	.	.	.	.	.	.
		$s_2^5s_1$	.	-29937600	62370000	-43659000	12474000	-1247400	.	.	.	.	.
		$s_2^4s_1^3$	.	-74844000	137214000	-74844000	12474000	.	.	.	.	.	.
		$s_2^3s_1^5$	.	-69854400	104781600	-34927200	.	.	.	.	.	.	.
		$s_2^2s_1^7$	.	-39916800	39916800	.	.	.	.	.	.	.	.
		$s_2s_1^9$	.	-19958400	.	.	.	.	.	.	.	.	.
		s_1^{11}	3628800	.	.	.	.	.	.	.	.	.	.

For any given augmented monomial symmetric function of weight 12 an expression in terms of
the polykay functions may be obtained by reading horizontally across the table up to and including
the unity in bold type. Thus

$$[321^7]/n^{(9)} = k_{1^{12}} + 4k_{21^{10}} + 3k_{2^21^8} + k_{31^9} + k_{321^7}.$$

All tables of lesser weight can be deduced from this basic table. Thus, for example

$$[321]/n^{(3)} = k_{1^6} + 4k_{21^4} + 3k_{2^21^2} + k_{31^3} + k_{321},$$

or again $$[32]/n^{(2)} = k_{1^5} + 4k_{21^3} + 3k_{2^21} + k_{31^2} + k_{321}.$$

Conversely, any polykay function may be expressed in terms of the augmented monomial symmetric functions by reading vertically downwards from the top of the table up to and including the
unity in bold type. Thus

$$k_{321^7} = -\frac{2[1^{12}]}{n^{(12)}} + \frac{5[21^{10}]}{n^{(11)}} - \frac{3[2^21^8]}{n^{(10)}} - \frac{[31^9]}{n^{(10)}} + \frac{[321^7]}{n^{(9)}},$$

from which, for the sake of example, we deduce

$$k_{321} = -\frac{2[1^6]}{n^{(6)}} + \frac{5[21^4]}{n^{(5)}} - \frac{3[2^21^2]}{n^{(4)}} - \frac{[31^3]}{n^{(4)}} + \frac{[321]}{n^{(3)}},$$

or $$k_{32} = -\frac{2[1^5]}{n^{(5)}} + \frac{5[21^3]}{n^{(4)}} - \frac{3[2^21]}{n^{(3)}} - \frac{[31^2]}{n^{(3)}} + \frac{[32]}{n^{(2)}}.$$

A complete set of tables was computed by Abdel-Aty but because of the ease with which those of
lower weight can be deduced from those of weight 12, only those of weight 12 were published.
A manuscript set does, however, exist.

Table 2·2

Weight 12 (i)	k_1^{12}	$k_2 1^{10}$	$k_2^2 1^8$	$k_2^3 1^6$	$k_2^4 1^4$	$k_2^5 1^2$	k_2^6	$k_3 1^9$	$k_3 2 1^7$	$k_3 2^2 1^5$	$k_3 2^3 1^3$	$k_3 2^4 1$	$k_3^2 1^6$	$k_3^2 2 1^4$	$k_3^2 2^2 1^2$
$[1^{12}]/n^{(12)}$	1	−1	1	−1	1	−1	1	2	−2	2	−2	2	4	−4	4
$[21^{10}]/n^{(11)}$	1	1	−2	3	−4	5	−6	−3	5	−7	9	−11	−12	16	−20
$[2^2 1^8]/n^{(10)}$	1	2	1	−3	6	−10	15	·	−3	8	−15	24	9	−21	37
$[2^3 1^6]/n^{(9)}$	1	3	3	1	−4	10	−20	·	·	−3	11	−26	·	9	−30
$[2^4 1^4]/n^{(8)}$	1	4	6	4	1	−5	15	·	·	·	−3	14	·	·	9
$[2^5 1^2]/n^{(7)}$	1	5	10	10	5	1	−6	·	·	·	·	−3	·	·	·
$[2^6]/n^{(6)}$	1	6	15	20	15	6	1	·	·	·	·	·	·	·	·
$[31^9]/n^{(10)}$	1	3	·	·	·	·	·	1	−1	1	−1	1	4	−4	4
$[321^7]/n^{(9)}$	1	4	3	·	·	·	·	1	1	−2	3	−4	−6	10	−14
$[32^2 1^5]/n^{(8)}$	1	5	7	3	·	·	·	1	2	1	−3	6	·	−6	16
$[32^3 1^3]/n^{(7)}$	1	6	12	10	3	·	·	1	3	3	1	−4	·	·	−6
$[32^4 1]/n^{(6)}$	1	7	18	22	13	3	·	1	4	6	4	1	·	·	·
$[3^2 1^6]/n^{(8)}$	1	6	9	·	·	·	·	2	6	·	·	·	1	−1	1
$[3^2 21^4]/n^{(7)}$	1	7	15	9	·	·	·	2	8	6	·	·	1	1	−2
$[3^2 2^2 1^2]/n^{(6)}$	1	8	22	24	9	·	·	2	10	14	6	·	1	2	1
$[3^2 2^3]/n^{(5)}$	1	9	30	46	33	9	·	2	12	24	20	6	1	3	3
$[3^3 1^3]/n^{(6)}$	1	9	27	27	·	·	·	3	18	27	·	·	3	9	·
$[3^3 21]/n^{(5)}$	1	10	36	54	27	·	·	3	21	45	27	·	3	12	9
$[3^4]/n^{(4)}$	1	12	54	108	81	·	·	4	36	108	108	·	6	36	54
$[41^8]/n^{(9)}$	1	6	3	·	·	·	·	4	·	·	·	·	·	·	·
$[421^6]/n^{(8)}$	1	7	9	3	·	·	·	4	4	·	·	·	·	·	·
$[42^2 1^4]/n^{(7)}$	1	8	16	12	3	·	·	4	8	4	·	·	·	·	·
$[42^3 1^2]/n^{(6)}$	1	9	24	28	15	3	·	4	12	12	4	·	·	·	·
$[42^4]/n^{(5)}$	1	10	33	52	43	18	3	4	16	24	16	4	·	·	·
$[431^5]/n^{(7)}$	1	9	21	9	·	·	·	5	18	3	·	·	4	·	·
$[4321^3]/n^{(6)}$	1	10	30	30	9	·	·	5	23	21	3	·	4	4	·
$[432^2 1]/n^{(5)}$	1	11	40	60	39	9	·	5	28	44	24	3	4	8	4
$[43^2 1^2]/n^{(5)}$	1	12	48	72	27	·	·	6	42	78	18	·	9	30	3
$[43^2 2]/n^{(4)}$	1	13	60	120	99	27	·	6	48	120	96	18	9	39	33
$[4^2 1^4]/n^{(6)}$	1	12	42	36	9	·	·	8	48	24	·	·	16	·	·
$[4^2 21^2]/n^{(5)}$	1	13	54	78	45	9	·	8	56	72	24	·	16	16	16
$[4^2 2^2]/n^{(4)}$	1	14	67	132	123	54	9	8	64	128	96	24	16	32	16
$[4^2 31]/n^{(4)}$	1	15	78	162	117	27	·	9	84	210	108	9	24	96	24
$[4^3]/n^{(3)}$	1	18	117	324	351	162	27	12	144	504	432	108	48	288	144
$[51^7]/n^{(8)}$	1	10	15	·	·	·	·	10	10	·	·	·	·	·	·
$[521^5]/n^{(7)}$	1	11	25	15	·	·	·	10	20	10	·	·	·	·	·
$[52^2 1^3]/n^{(6)}$	1	12	36	40	15	·	·	10	30	30	10	·	·	·	·
$[52^3 1]/n^{(5)}$	1	13	48	76	55	15	·	10	40	60	40	10	·	·	·
$[531^4]/n^{(6)}$	1	13	45	45	·	·	·	11	50	45	·	·	10	10	·
$[5321^2]/n^{(5)}$	1	14	58	90	45	·	·	11	61	95	45	·	10	20	10
$[532^2]/n^{(4)}$	1	15	72	148	135	45	·	11	72	156	140	45	10	30	30
$[53^2 1]/n^{(4)}$	1	16	84	180	135	·	·	12	96	240	180	·	21	90	75
$[541^3]/n^{(5)}$	1	16	78	120	45	·	·	14	110	150	30	·	40	40	·
$[5421]/n^{(4)}$	1	17	94	198	165	45	·	14	124	260	180	30	40	80	40
$[543]/n^{(3)}$	1	19	126	354	405	135	·	15	168	558	600	135	54	270	270
$[5^2 1^2]/n^{(4)}$	1	20	130	300	225	·	·	20	220	500	300	·	100	200	100
$[5^2 2]/n^{(3)}$	1	21	150	430	525	225	·	20	240	720	800	300	100	300	300
$[61^6]/n^{(7)}$	1	15	45	15	·	·	·	20	60	·	·	·	10	·	·
$[621^4]/n^{(6)}$	1	16	60	60	15	·	·	20	80	60	·	·	10	10	·
$[62^2 1^2]/n^{(5)}$	1	17	76	120	75	15	·	20	100	140	60	·	10	20	10
$[62^3]/n^{(4)}$	1	18	93	196	195	90	15	20	120	240	200	60	10	30	30
$[631^3]/n^{(5)}$	1	18	90	150	45	·	·	21	135	225	15	·	30	90	·
$[6321]/n^{(4)}$	1	19	108	240	195	45	·	21	156	360	240	15	30	120	90
$[63^2]/n^{(3)}$	1	21	144	420	495	135	·	22	216	720	840	90	51	315	495
$[641^2]/n^{(4)}$	1	21	138	330	225	45	·	24	240	600	240	·	90	300	30
$[642]/n^{(3)}$	1	22	159	468	555	270	45	24	264	840	840	240	90	390	330
$[651]/n^{(3)}$	1	25	210	690	825	225	·	30	420	1500	1500	150	210	900	750
$[6^2]/n^{(2)}$	1	30	315	1380	2475	1350	225	40	720	3600	6000	1800	420	2700	4500
$[71^5]/n^{(6)}$	1	21	105	105	·	·	·	35	210	105	·	·	70	·	·
$[721^3]/n^{(5)}$	1	22	126	210	105	·	·	35	245	315	105	·	70	70	·
$[72^2 1]/n^{(4)}$	1	23	148	336	315	105	·	35	280	560	420	105	70	140	70
$[731^2]/n^{(4)}$	1	24	168	420	315	·	·	36	336	840	420	·	105	420	105
$[732]/n^{(3)}$	1	25	192	588	735	315	·	36	372	1176	1260	420	105	525	525
$[741]/n^{(3)}$	1	27	234	798	945	315	·	39	504	1890	1680	315	210	1260	630
$[75]/n^{(2)}$	1	31	330	1470	2625	1575	·	45	780	3990	6300	2625	420	3150	4200
$[81^4]/n^{(5)}$	1	28	210	420	105	·	·	56	560	840	·	·	280	280	·
$[821^2]/n^{(4)}$	1	29	238	630	525	105	·	56	616	1400	840	·	280	560	280
$[82^2]/n^{(3)}$	1	30	267	868	1155	630	105	56	672	2016	2240	840	280	840	840
$[831]/n^{(3)}$	1	31	294	1050	1365	315	·	57	756	2730	2940	105	336	1680	1680
$[84]/n^{(2)}$	1	34	381	1764	3255	1890	315	60	1008	5208	8400	2940	504	4200	5880
$[91^3]/n^{(4)}$	1	36	378	1260	945	·	·	84	1260	3780	1260	·	840	2520	·
$[921]/n^{(3)}$	1	37	414	1638	2205	945	·	84	1344	5040	5040	1260	840	3360	2520
$[93]/n^{(2)}$	1	39	486	2394	4725	2835	·	85	1548	7938	13860	4725	924	6300	11340
$[10,1^2]/n^{(3)}$	1	45	630	3150	4725	945	·	120	2520	12600	12600	·	2100	12600	6300
$[10,2]/n^{(2)}$	1	46	675	3780	7875	5670	945	120	2640	15120	25200	12600	2100	14700	18900
$[11,1]/n^{(2)}$	1	55	990	6930	17325	10395	·	165	4620	34650	69300	17325	4620	46200	69300
$[12]/n$	1	66	1485	13860	51975	62370	10395	220	7920	83160	277200	207900	9240	138600	415800

Table 2·2 (cont.)

Weight 12 (ii)	$k_{3^2 2^3}$	$k_{3^3 1^3}$	$k_{3^3 21}$	k_{3^4}	k_{41^8}	k_{421^6}	$k_{42^2 1^4}$	$k_{42^3 1^2}$	k_{42^4}	k_{431^5}	k_{4321^3}	$k_{432^2 1}$	$k_{43^2 1^2}$
$[1^{12}]/n^{(12)}$	-4	8	-8	16	-6	6	-6	6	-6	-12	12	-12	-24
$[21^{10}]/n^{(11)}$	24	-36	44	-96	12	-18	24	-30	36	42	-54	66	120
$[2^2 1^8]/n^{(10)}$	-57	54	-90	216	-3	15	-33	57	-87	-42	84	-138	-210
$[2^3 1^6]/n^{(9)}$	67	-27	81	-216	·	-3	18	-51	108	9	-51	135	144
$[2^4 1^4]/n^{(8)}$	-39	·	-27	81	·	·	-3	21	-72	·	9	-60	-27
$[2^5 1^2]/n^{(7)}$	9	·	·	·	·	·	·	-3	24	·	·	9	·
$[2^6]/n^{(6)}$	·	·	·	·	·	·	·	·	-3	·	·	·	·
$[31^9]/n^{(10)}$	-4	12	-12	32	-4	4	-4	4	-4	-14	14	-14	-40
$[321^7]/n^{(9)}$	18	-36	48	-144	·	-4	8	-12	16	24	-38	52	132
$[32^2 1^5]/n^{(8)}$	-30	27	-63	216	·	·	-4	12	-24	-3	27	-65	-120
$[32^3 1^3]/n^{(7)}$	22	·	27	-108	·	·	·	-4	16	·	-3	30	18
$[32^4 1]/n^{(6)}$	-6	·	·	·	·	·	·	·	-4	·	·	-3	·
$[3^2 1^6]/n^{(8)}$	-1	6	-6	24	·	·	·	·	·	-4	4	-4	-22
$[3^2 21^4]/n^{(7)}$	3	-9	15	-72	·	·	·	·	·	·	-4	8	36
$[3^2 2^2 1^2]/n^{(6)}$	-3	·	-9	54	·	·	·	·	·	·	·	-4	-3
$[3^2 2^3]/n^{(5)}$	1	·	·	·	·	·	·	·	·	·	·	·	·
$[3^3 1^3]/n^{(6)}$	·	1	-1	8	·	·	·	·	·	·	·	·	-4
$[3^3 21]/n^{(5)}$	·	1	1	-12	·	·	·	·	·	·	·	·	·
$[3^4]/n^{(4)}$	·	4	12	1	·	·	·	·	·	·	·	·	·
$[41^8]/n^{(9)}$	·	·	·	·	1	-1	1	-1	1	2	-2	2	4
$[421^6]/n^{(8)}$	·	·	·	·	1	1	-2	3	-4	-3	5	-7	-12
$[42^2 1^4]/n^{(7)}$	·	·	·	·	1	2	1	-3	6	-3	-3	8	9
$[42^3 1^2]/n^{(6)}$	·	·	·	·	1	3	3	1	-4	-4	·	-3	·
$[42^4]/n^{(5)}$	·	·	·	·	1	4	6	4	1	1	·	·	·
$[431^5]/n^{(7)}$	·	·	·	·	1	3	·	·	·	1	-1	1	4
$[4321^3]/n^{(6)}$	·	·	·	·	1	4	3	·	·	1	1	-2	-6
$[432^2 1]/n^{(5)}$	·	·	·	·	1	5	7	3	·	1	2	1	1
$[43^2 1^2]/n^{(5)}$	·	·	·	·	1	6	9	·	·	2	6	·	1
$[43^2 2]/n^{(4)}$	3	4	4	·	1	7	15	9	·	2	8	6	1
$[4^2 1^4]/n^{(6)}$	·	·	·	·	2	12	6	·	·	8	·	·	·
$[4^2 21^2]/n^{(5)}$	·	·	·	·	2	14	18	6	·	8	8	8	·
$[4^2 2^2]/n^{(4)}$	·	·	·	·	2	16	32	24	6	8	16	8	·
$[4^2 31]/n^{(4)}$	·	16	·	·	2	18	42	18	·	10	36	6	8
$[4^3]/n^{(3)}$	·	64	·	·	3	36	126	108	27	24	144	72	48
$[51^7]/n^{(8)}$	·	·	·	·	5	·	·	·	·	·	·	·	·
$[521^5]/n^{(7)}$	·	·	·	·	5	5	·	·	·	·	·	·	·
$[52^2 1^3]/n^{(6)}$	·	·	·	·	5	10	5	·	·	·	·	·	·
$[52^3 1]/n^{(5)}$	·	·	·	·	5	15	15	5	·	·	·	·	·
$[531^4]/n^{(6)}$	·	·	·	·	5	15	15	·	·	5	·	·	·
$[5321^2]/n^{(5)}$	·	·	·	·	5	20	15	·	·	5	5	·	·
$[532^2]/n^{(4)}$	10	·	·	·	5	25	35	15	·	5	10	5	·
$[53^2 1]/n^{(4)}$	·	10	10	·	5	30	45	·	·	10	30	·	5
$[541^3]/n^{(5)}$	·	·	·	·	6	40	30	·	·	30	10	·	·
$[5421]/n^{(4)}$	·	·	·	·	6	46	70	30	·	30	40	10	5
$[543]/n^{(3)}$	30	40	40	·	6	58	150	90	·	36	140	60	30
$[5^2 1^2]/n^{(4)}$	·	·	·	·	10	100	150	·	·	100	100	·	·
$[5^2 2]/n^{(3)}$	100	·	·	·	10	110	250	150	·	100	200	100	·
$[61^6]/n^{(7)}$	·	·	·	·	15	15	·	·	·	·	·	·	·
$[621^4]/n^{(6)}$	·	·	·	·	15	30	15	·	·	·	·	·	·
$[62^2 1^2]/n^{(5)}$	·	·	·	·	15	45	45	15	·	·	·	·	·
$[62^3]/n^{(4)}$	10	·	·	·	15	60	90	60	15	·	·	·	·
$[631^3]/n^{(5)}$	·	10	·	·	15	60	45	·	·	15	15	·	·
$[6321]/n^{(4)}$	·	10	10	·	15	75	105	45	·	15	30	15	·
$[63^2]/n^{(3)}$	15	40	120	10	15	105	225	135	·	30	120	90	15
$[641^2]/n^{(4)}$	·	40	·	·	16	120	180	60	·	80	120	10	10
$[642]/n^{(3)}$	30	40	40	·	16	136	300	240	60	80	200	120	10
$[651]/n^{(3)}$	·	100	100	·	20	240	600	300	·	250	600	150	50
$[6^2]/n^{(2)}$	300	400	1200	100	30	480	1800	1800	450	600	2400	1800	300
$[71^5]/n^{(6)}$	·	·	·	·	35	105	·	·	·	35	·	·	·
$[721^3]/n^{(5)}$	·	·	·	·	35	140	105	·	·	35	35	·	·
$[72^2 1]/n^{(4)}$	·	·	·	·	35	175	245	105	·	35	70	35	·
$[731^2]/n^{(4)}$	·	70	·	·	35	210	315	·	·	70	210	·	35
$[732]/n^{(3)}$	105	70	70	·	35	245	525	315	·	70	280	210	35
$[741]/n^{(3)}$	·	280	·	·	36	336	840	420	·	210	840	210	210
$[75]/n^{(2)}$	1050	700	700	·	40	560	2100	2100	·	560	2800	2100	700
$[81^4]/n^{(5)}$	·	·	·	·	70	420	210	·	·	280	·	·	·
$[821^2]/n^{(4)}$	·	·	·	·	70	490	630	210	·	280	280	·	·
$[82^2]/n^{(3)}$	280	·	·	·	70	560	1120	840	210	280	560	280	280
$[831]/n^{(3)}$	·	280	280	·	70	630	1470	630	·	350	1260	210	280
$[84]/n^{(2)}$	840	1120	1120	·	71	868	3150	2940	735	616	3920	2520	1400
$[91^3]/n^{(4)}$	·	280	280	·	126	1260	1890	·	·	1260	1260	·	·
$[921]/n^{(3)}$	·	280	280	·	126	1386	3150	1890	·	1260	2520	1260	·
$[93]/n^{(2)}$	1260	1120	3360	280	126	1638	5670	5670	·	1386	6300	5670	1260
$[10,1^2]/n^{(3)}$	·	2800	·	·	210	3150	9450	3150	·	4200	12600	·	2100
$[10,2]/n^{(2)}$	6300	2800	2800	·	210	3360	12600	12600	3150	4200	16800	12600	2100
$[11,1]/n^{(2)}$	·	15400	15400	·	330	6930	34650	34650	·	11550	69300	34650	23100
$[12/n]$	138600	61600	184800	15400	495	13860	103950	207900	51975	27720	277200	415800	138600

Table 2·2 (cont.)

Weight 12 (iii)	$k_{43^2 2}$	$k_{4^2 1^4}$	$k_{4^2 21^2}$	$k_{4^2 2^2}$	$k_{4^2 31}$	k_{4^3}	k_{51^7}	k_{521^5}	$k_{52^2 1^3}$	$k_{52^3 1}$	k_{531^4}	k_{5321^2}	k_{532^2}
$[1^{12}]/n^{(12)}$	24	36	−36	36	72	−216	24	−24	24	−24	48	−48	48
$[21^{10}]/n^{(11)}$	−144	−144	180	−216	−396	1296	−60	84	−108	132	−192	240	−288
$[2^2 1^8]/n^{(10)}$	330	180	−324	504	792	−2916	30	−90	174	−282	240	−432	672
$[2^3 1^6]/n^{(9)}$	−354	−72	252	−576	−684	3024	.	30	−120	294	−90	330	−762
$[2^4 1^4]/n^{(8)}$	171	9	−81	333	234	−1458	.	.	30	−150	.	−90	420
$[2^5 1^2]/n^{(7)}$	−27	.	9	−90	−27	324	.	.	.	30	.	.	−90
$[2^6]/n^{(6)}$	.	.	.	9	9	−27	.	.	.	.	.	.	.
$[31^9]/n^{(10)}$	40	48	−48	48	132	−432	20	−20	20	−20	64	−64	64
$[321^7]/n^{(9)}$	−172	−96	144	−192	−480	1728	−10	30	−50	70	−140	204	−268
$[32^2 1^5]/n^{(8)}$	252	24	−120	264	516	−2160	.	−10	40	−90	60	−200	404
$[32^3 1^3]/n^{(7)}$	−138	.	24	−144	−144	864	.	.	−10	50	.	60	−260
$[32^4 1]/n^{(6)}$	18	.	.	24	9	−108	.	.	.	−10	.	.	60
$[3^2 1^6]/n^{(8)}$	22	16	−16	16	80	−288	.	.	.	.	20	−20	20
$[3^2 21^4]/n^{(7)}$	−58	.	16	−32	−144	576	.	.	.	.	−10	30	−50
$[3^2 2^2 1^2]/n^{(6)}$	39	.	.	16	24	−144	.	.	.	.	.	−10	40
$[3^2 2^3]/n^{(5)}$	−3	.	.	.	.	.	.	.	.	.	.	.	−10
$[3^3 1^3]/n^{(6)}$	4	.	.	.	16	−64	.	.	.	.	.	.	.
$[3^3 21]/n^{(5)}$	−4	.	.	.	.	.	.	.	.	.	.	.	.
$[3^4]/n^{(4)}$	.	.	.	.	.	.	.	.	.	.	.	.	.
$[41^8]/n^{(9)}$	−4	−12	12	−12	−24	108	−5	5	−5	5	−10	10	−10
$[421^6]/n^{(8)}$	16	24	−36	48	84	−432	.	−5	10	−15	15	−25	35
$[42^2 1^4]/n^{(7)}$	−21	−6	30	−66	−84	540	.	.	−5	15	.	15	−40
$[42^3 1^2]/n^{(6)}$	9	.	−6	36	18	−216	.	.	.	−5	.	.	15
$[42^4]/n^{(5)}$	.	.	.	−6	.	27	.	.	.	.	.	.	.
$[431^5]/n^{(7)}$	−4	−8	8	−8	−28	144	.	.	.	.	−5	5	−5
$[4321^3]/n^{(6)}$	10	.	−8	16	48	−288	.	.	.	.	.	−5	10
$[432^2 1]/n^{(5)}$	−6	.	.	−8	−6	72	.	.	.	.	.	.	−5
$[43^2 1^2]/n^{(5)}$	−1	.	.	.	−8	48	.	.	.	.	.	.	.
$[43^2 2]/n^{(4)}$	1	.	.	.	.	.	.	.	.	.	.	.	.
$[4^2 1^4]/n^{(6)}$	.	1	−1	1	2	−18	.	.	.	.	.	.	.
$[4^2 21^2]/n^{(5)}$	.	1	1	−2	−3	36	.	.	.	.	.	.	.
$[4^2 2^2]/n^{(4)}$	.	1	2	1	.	−9	.	.	.	.	.	.	.
$[4^2 31]/n^{(4)}$	.	1	3	.	1	−12	.	.	.	.	.	.	.
$[4^3]/n^{(3)}$	.	3	18	9	12	1	.	.	.	.	.	.	.
$[51^7]/n^{(8)}$	.	.	.	.	.	.	1	−1	1	−1	2	−2	2
$[521^5]/n^{(7)}$	.	.	.	.	.	.	1	1	−2	3	−3	5	−7
$[52^2 1^3]/n^{(6)}$	.	.	.	.	.	.	1	2	1	−3	.	−3	8
$[52^3 1]/n^{(5)}$	.	.	.	.	.	.	1	3	3	1	.	.	−3
$[531^4]/n^{(6)}$	.	.	.	.	.	.	1	3	.	.	1	−1	1
$[5321^2]/n^{(5)}$	.	.	.	.	.	.	1	4	3	.	1	1	−2
$[532^2]/n^{(4)}$	.	.	.	.	.	.	1	5	7	3	1	2	1
$[53^2 1]/n^{(5)}$	.	.	.	.	.	.	1	6	9	.	2	6	.
$[541^3]/n^{(5)}$	.	5	.	.	.	.	1	6	3	.	4	.	.
$[5421]/n^{(4)}$	.	5	5	.	.	.	1	7	9	3	4	4	.
$[543]/n^{(3)}$	10	5	15	.	5	.	1	9	21	9	5	18	3
$[5^2 1^2]/n^{(4)}$	.	25	.	.	.	.	2	20	30	.	20	20	.
$[5^2 2]/n^{(3)}$	.	25	25	.	.	.	2	22	50	30	20	40	20
$[61^6]/n^{(7)}$	.	.	.	.	.	.	6	.	.	.	.	.	.
$[621^4]/n^{(6)}$	.	.	.	.	.	.	6	6	.	.	.	.	.
$[62^2 1^2]/n^{(5)}$	.	.	.	.	.	.	6	12	6	.	.	.	.
$[62^3]/n^{(4)}$	.	.	.	.	.	.	6	18	18	6	.	.	.
$[631^3]/n^{(5)}$	.	.	.	.	.	.	6	18	.	.	6	.	.
$[6321]/n^{(4)}$	.	.	.	.	.	.	6	24	18	.	6	6	.
$[63^2]/n^{(3)}$	15	.	.	.	.	.	6	36	54	.	12	36	.
$[641^2]/n^{(4)}$	.	15	15	.	.	.	6	36	18	.	24	.	.
$[642]/n^{(3)}$	10	15	30	15	.	.	6	42	54	18	24	24	.
$[651]/n^{(3)}$	.	75	75	.	.	.	7	75	135	15	80	120	.
$[6^2]/n^{(2)}$	300	225	450	225	.	.	12	180	540	180	240	720	.
$[71^5]/n^{(6)}$	.	.	.	.	.	.	21	21	.	.	.	.	.
$[721^3]/n^{(5)}$	.	.	.	.	.	.	21	42	21	.	.	.	.
$[72^2 1]/n^{(4)}$	.	.	.	.	.	.	21	63	63	21	.	.	.
$[731^2]/n^{(4)}$	.	.	.	.	.	.	21	84	63	.	21	21	.
$[732]/n^{(3)}$	35	.	.	.	.	.	21	105	147	63	21	42	21
$[741]/n^{(3)}$	.	35	105	.	35	.	21	147	189	63	84	84	.
$[75]/n^{(2)}$	350	175	525	.	175	.	22	252	630	420	245	630	315
$[81^4]/n^{(5)}$	.	35	.	.	.	.	56	168	.	.	56	.	.
$[821^2]/n^{(4)}$	.	35	35	.	.	.	56	224	168	.	56	56	.
$[82^2]/n^{(3)}$	.	35	70	35	.	.	56	280	392	168	56	112	56
$[831]/n^{(3)}$	.	35	105	.	35	.	56	336	504	.	112	336	.
$[84]/n^{(2)}$	280	105	630	315	420	35	56	504	1176	504	280	1008	168
$[91^3]/n^{(4)}$	.	315	.	.	.	.	126	756	378	.	504	.	.
$[921]/n^{(3)}$	.	315	315	.	.	.	126	882	1134	378	504	504	.
$[93]/n^{(2)}$	1260	315	945	.	315	.	126	1134	2646	1134	630	2268	378
$[10,1^2]/n^{(3)}$	.	1575	1575	.	.	.	252	2520	3780	.	2520	2520	.
$[10,2]/n^{(2)}$	2100	1575	3150	1575	.	.	252	2772	6300	3780	2520	5040	2520
$[11,1]/n^{(2)}$	.	5775	17325	.	5775	.	462	6930	20790	6930	9240	27720	.
$[12]/n$	138600	17325	103950	51975	69300	5775	792	16632	83160	83160	27720	166320	83160

Table 2·2 (cont.)

Weight 12 (iv)	$k_{53^2 1}$	k_{541^3}	k_{5421}	k_{543}	$k_{5^2 1^2}$	$k_{5^2 2}$	k_{61^6}	k_{621^4}	$k_{62^2 1^2}$	k_{62^3}	k_{631^3}	k_{6321}	k_{63^2}
$[1^{12}]/n^{(12)}$	96	−144	144	−288	576	−576	−120	120	−120	120	−240	240	−480
$[21^{10}]/n^{(11)}$	−528	648	−792	1728	−2880	3456	360	−480	600	−720	1080	−1320	2880
$[2^2 1^8]/n^{(10)}$	1056	−972	1620	−3888	5040	−7920	−270	630	−1110	1710	−1620	2700	−6480
$[2^3 1^6]/n^{(9)}$	−900	540	−1512	3996	−3600	8640	30	−300	930	−2040	870	−2490	6600
$[2^4 1^4]/n^{(8)}$	270	−90	630	−1800	900	−4500	·	30	−330	1260	−90	960	−2790
$[2^5 1^2]/n^{(7)}$	·	·	−90	270	·	900	·	·	30	−360	·	·	270
$[2^6]/n^{(6)}$	·	·	·	·	·	·	·	·	·	30	·	·	·
$[31^9]/n^{(10)}$	176	−216	216	−576	960	−960	−120	120	−120	120	−360	360	−960
$[321^7]/n^{(9)}$	−664	540	−756	2376	−2880	3840	120	−240	360	−480	960	−1320	4080
$[32^2 1^5]/n^{(8)}$	780	−300	840	−3192	2400	−5280	·	120	−360	720	−630	1590	−5760
$[32^3 1^3]/n^{(7)}$	−270	30	−330	1500	−600	3000	·	·	120	−480	30	−660	2820
$[32^4 1]/n^{(6)}$	·	·	·	30	·	−180	·	·	·	120	·	30	−180
$[3^2 1^6]/n^{(8)}$	104	−80	80	−376	400	−400	−10	10	−10	10	−140	140	−640
$[3^2 21^4]/n^{(7)}$	−220	40	−120	860	−400	800	·	−10	20	−30	150	−290	1680
$[3^2 2^2 1^2]/n^{(6)}$	90	·	40	−420	100	−500	·	·	−10	30	·	150	−1080
$[3^2 2^3]/n^{(5)}$	·	·	·	30	·	100	·	·	·	−10	·	·	30
$[3^3 1^3]/n^{(6)}$	20	·	·	−80	·	·	·	·	·	·	−10	10	−160
$[3^3 21]/n^{(5)}$	−10	·	·	40	·	·	·	·	·	·	·	−10	180
$[3^4]/n^{(4)}$	·	·	·	·	·	·	·	·	·	·	·	·	−10
$[41^8]/n^{(9)}$	−20	54	−54	108	−240	240	30	−30	30	−30	60	−60	120
$[421^6]/n^{(8)}$	60	−120	174	−402	600	−840	−15	45	−75	105	−120	180	−420
$[42^2 1^4]/n^{(7)}$	−45	45	−165	450	−300	900	·	−15	60	−135	45	−165	450
$[42^3 1^2]/n^{(6)}$	·	·	45	−135	·	−300	·	·	−15	75	·	45	−135
$[42^4]/n^{(5)}$	·	·	·	·	·	·	·	·	·	−15	·	·	·
$[431^5]/n^{(7)}$	−20	40	−40	134	−200	200	·	·	·	·	30	−30	120
$[4321^3]/n^{(6)}$	30	−10	50	−260	100	−300	·	·	·	·	−15	45	−240
$[432^2 1]/n^{(5)}$	·	·	−10	75	·	100	·	·	·	·	·	−15	90
$[43^2 1^2]/n^{(5)}$	−5	·	·	40	·	·	·	·	·	·	·	·	30
$[43^2 2]/n^{(4)}$	·	·	·	−10	·	·	·	·	·	·	·	·	−15
$[4^2 1^4]/n^{(6)}$	·	−5	5	−10	25	−25	·	·	·	·	·	·	·
$[4^2 21^2]/n^{(5)}$	·	·	−5	15	·	25	·	·	·	·	·	·	·
$[4^2 2^2]/n^{(4)}$	·	·	·	−5	·	·	·	·	·	·	·	·	·
$[4^2 31]/n^{(4)}$	·	·	·	·	−5	·	·	·	·	·	·	·	·
$[4^3]/n^{(3)}$	·	·	·	·	·	·	·	·	·	·	·	·	·
$[51^7]/n^{(8)}$	4	−6	6	−12	48	−48	−6	6	−6	6	−12	12	−24
$[521^5]/n^{(7)}$	−12	12	−18	42	−120	168	·	−6	12	−18	18	−30	72
$[52^2 1^3]/n^{(6)}$	9	−3	15	−42	60	−180	·	·	−6	18	·	18	−54
$[52^3 1]/n^{(5)}$	·	·	−3	9	·	60	·	·	·	−6	·	·	−24
$[531^4]/n^{(6)}$	4	−4	4	−14	40	−40	·	·	·	·	−6	6	−24
$[5321^2]/n^{(5)}$	−6	·	−4	24	−20	60	·	·	·	·	·	−6	36
$[532^2]/n^{(4)}$	·	·	·	−3	·	−20	·	·	·	·	·	·	−6
$[53^2 1]/n^{(4)}$	1	·	−1	−4	·	·	·	·	·	·	·	·	·
$[541^3]/n^{(5)}$	·	1	−1	2	−10	10	·	·	·	·	·	·	·
$[5421]/n^{(4)}$	·	1	1	−3	1	·	·	·	·	·	·	·	·
$[543]/n^{(3)}$	4	1	3	1	·	·	·	·	·	·	·	·	·
$[5^2 1^2]/n^{(4)}$	·	10	·	·	1	−1	·	·	·	·	·	·	·
$[5^2 2]/n^{(3)}$	·	10	10	·	1	1	·	·	·	·	·	·	·
$[61^6]/n^{(7)}$	·	·	·	·	·	·	1	−1	1	−1	2	−2	4
$[621^4]/n^{(6)}$	·	·	·	·	·	·	1	1	−2	3	−3	5	−12
$[62^2 1^2]/n^{(5)}$	·	·	·	·	·	·	1	2	1	−3	·	−3	9
$[62^3]/n^{(4)}$	·	·	·	·	·	·	1	3	3	1	·	·	4
$[631^3]/n^{(5)}$	·	·	·	·	·	·	1	3	3	·	1	−1	−6
$[6321]/n^{(4)}$	·	·	·	·	·	·	1	4	3	·	1	1	·
$[63^2]/n^{(3)}$	6	·	·	·	·	·	1	6	9	·	2	6	1
$[641^2]/n^{(4)}$	·	6	6	·	·	·	1	6	3	·	4	·	·
$[642]/n^{(3)}$	·	6	·	·	·	·	1	7	9	3	4	4	·
$[651]/n^{(3)}$	10	45	15	·	6	·	1	10	90	30	40	120	20
$[6^2]/n^{(2)}$	120	180	180	·	36	·	2	30	90	30	40	120	20
$[71^5]/n^{(6)}$	·	·	·	·	·	·	7	7	·	·	·	·	·
$[721^3]/n^{(5)}$	·	·	·	·	·	·	7	7	·	·	·	·	·
$[72^2 1]/n^{(4)}$	·	·	·	·	·	·	7	14	7	·	7	·	·
$[731^2]/n^{(4)}$	·	·	·	·	·	·	7	21	21	·	7	7	·
$[732]/n^{(3)}$	·	·	·	·	·	·	7	28	21	·	·	7	·
$[741]/n^{(3)}$	·	21	21	·	·	·	7	42	21	·	28	70	·
$[75]/n^{(2)}$	70	140	210	35	21	21	7	70	105	·	70	70	·
$[81^4]/n^{(5)}$	·	·	·	·	·	·	28	28	·	·	·	·	·
$[821^2]/n^{(4)}$	·	·	·	·	·	·	28	56	28	·	·	·	·
$[82^2]/n^{(3)}$	·	·	·	·	·	·	28	84	84	28	·	·	·
$[831]/n^{(3)}$	56	·	·	·	·	·	28	112	84	84	28	28	·
$[84]/n^{(2)}$	224	56	168	56	·	·	28	196	252	84	112	112	·
$[91^3]/n^{(4)}$	·	126	·	·	·	·	84	252	·	·	84	·	·
$[921]/n^{(3)}$	·	126	126	·	·	·	84	336	252	·	84	84	·
$[93]/n^{(2)}$	504	126	378	126	·	·	84	504	756	·	168	504	84
$[10,1^2]/n^{(3)}$	·	1260	·	·	126	·	210	1260	630	630	840	840	·
$[10,2]/n^{(2)}$	·	1260	1260	·	126	126	210	1470	1890	630	840	840	·
$[11,1]/n^{(2)}$	4620	6930	6930	·	1386	·	462	4620	6930	·	4620	4620	·
$[12]/n$	55440	27720	83160	27720	8316	8316	924	13860	41580	13860	18480	55440	9240

Table 2·2 (*cont.*)

Weight 12 (ν)	k_{641^2}	k_{642}	k_{651}	k_{6^2}	k_{71^5}	k_{721^3}	k_{72^21}	k_{731^2}	k_{732}	k_{741}	k_{75}	k_{81^4}	k_{821^2}
$[1^{12}]/n^{(12)}$	720	−720	−2880	14400	720	−720	720	1440	−1440	−4320	17280	−5040	5040
$[21^{10}]/n^{(11)}$	−3600	4320	15840	−86400	−2520	3240	−3960	−7200	8640	23760	−103680	20160	−25200
$[2^21^8]/n^{(10)}$	6300	−9900	−31680	194400	2520	−5040	8280	12600	−19800	−47520	233280	−25200	45360
$[2^31^6]/n^{(9)}$	−4500	10800	27720	−201600	−630	3150	−8190	−8820	21420	41580	−241920	·	−35280
$[2^41^4]/n^{(8)}$	1170	−5670	−9900	94500	·	−630	3780	1890	−10710	−15120	113400	−630	10710
$[2^51^2]/n^{(7)}$	−90	1260	900	−16200	·	·	−630	·	1890	1890	−18900	·	−630
$[2^6]/n^{(6)}$	·	−90	·	900	·	·	·	·	−630	·	·	·	·
$[31^9]/n^{(10)}$	1200	−1200	−5280	28800	840	−840	840	2400	−2400	−7920	34560	−6720	6720
$[321^7]/n^{(9)}$	−3600	4800	18480	−115200	−1260	2100	−2940	−7560	9960	27720	−138240	13440	−20160
$[32^21^5]/n^{(8)}$	2880	−6480	−19800	151200	210	−1470	3570	6720	−14280	−28980	181440	−5040	18480
$[32^31^3]/n^{(7)}$	−480	3360	6900	−72000	·	210	−1680	−1260	7980	8820	−88200	·	−5040
$[32^41]/n^{(6)}$	·	−480	−300	7200	·	·	210	·	−1260	−630	12600	·	·
$[3^21^6]/n^{(8)}$	540	−540	−2640	16800	140	−140	140	1120	−1120	−4200	20160	−1680	1680
$[3^221^4]/n^{(7)}$	−600	1140	4200	−36000	·	140	−280	−1680	2800	6720	−42000	560	−2240
$[3^22^21^2]/n^{(6)}$	30	−630	−1500	19800	·	·	140	210	−1890	−1260	21000	·	560
$[3^22^3]/n^{(5)}$	·	30	·	−600	·	·	·	·	210	·	−2100	·	·
$[3^31^3]/n^{(6)}$	40	−40	−200	2400	·	·	·	·	·	140	−560	·	·
$[3^321]/n^{(5)}$	·	40	100	−2400	·	·	·	·	140	·	2800	·	·
$[3^4]/n^{(4)}$	·	·	·	100	·	·	·	·	·	140	−1400	·	·
$[41^8]/n^{(9)}$	−300	300	1320	−7200	−210	210	−210	−420	420	1980	−8640	1680	−1680
$[421^6]/n^{(8)}$	810	−1110	−3960	25200	210	−420	630	1050	−1470	−6300	30240	−2520	4200
$[42^21^4]/n^{(7)}$	−540	1350	3150	−27000	210	−630	·	−630	1680	5670	−31500	420	−2940
$[42^31^2]/n^{(6)}$	75	−615	−600	9900	·	·	210	210	−630	−1260	9450	·	420
$[42^4]/n^{(5)}$	·	75	·	−900	·	·	·	·	·	·	·	·	420
$[431^5]/n^{(7)}$	−240	240	1200	−7200	−35	35	−35	−280	280	1890	−9240	560	−560
$[4321^3]/n^{(6)}$	180	−420	−1200	10800	·	−35	70	315	−595	−2520	14700	·	560
$[432^21]/n^{(5)}$	·	180	150	−3600	·	·	−35	·	315	315	−4200	·	·
$[43^21^2]/n^{(5)}$	−10	10	50	−600	·	·	·	·	−35	315	280	·	·
$[43^22]/n^{(4)}$	·	−10	·	300	·	·	·	·	35	·	·	·	·
$[4^21^4]/n^{(6)}$	30	−30	−150	900	·	·	·	·	·	−210	350	−35	35
$[4^221^2]/n^{(5)}$	−15	45	75	−900	·	·	·	·	·	210	−1050	·	35
$[4^22^2]/n^{(4)}$	·	−15	·	225	·	·	·	·	·	·	·	·	−35
$[4^231]/n^{(4)}$	·	·	·	·	·	·	·	·	·	210	−1050	·	·
$[4^3]/n^{(3)}$	·	·	·	·	·	·	·	·	·	−35	175	·	·
$[51^7]/n^{(8)}$	36	−36	−264	1440	42	−42	42	84	−84	−252	1728	−336	336
$[521^5]/n^{(7)}$	−72	108	720	−4320	−21	63	−105	−168	252	630	−5544	336	−672
$[52^21^3]/n^{(6)}$	18	−90	−450	3240	·	−21	84	63	−231	−378	5040	·	336
$[52^31]/n^{(5)}$	·	18	30	−360	·	·	−21	·	63	·	−378	·	336
$[531^4]/n^{(6)}$	24	−24	−240	1440	·	·	−21	·	·	63	−1260	−56	56
$[5321^2]/n^{(5)}$	·	24	180	−1440	·	·	·	42	−42	84	1680	·	−56
$[532^2]/n^{(4)}$	·	·	·	·	·	·	·	·	−21	·	420	·	·
$[53^21]/n^{(4)}$	·	·	−10	120	·	·	·	·	·	·	140	·	·
$[541^3]/n^{(5)}$	−6	6	60	−360	·	·	·	·	·	42	−420	·	·
$[5421]/n^{(4)}$	·	−6	−15	180	·	·	·	·	·	−21	315	·	−35
$[543]/n^{(3)}$	·	·	·	·	·	·	·	·	·	·	315	·	·
$[5^21^2]/n^{(4)}$	·	·	−6	36	·	·	·	·	·	·	−35	·	·
$[5^22]/n^{(3)}$	·	·	·	·	·	·	·	·	·	·	42	·	·
$[61^6]/n^{(7)}$	−6	6	24	−240	−7	7	−7	−14	14	42	−168	56	−56
$[621^4]/n^{(6)}$	12	−18	−60	720	·	−7	14	21	−35	−84	420	−28	84
$[62^21^2]/n^{(5)}$	−3	15	30	−540	·	·	−7	·	21	21	−210	·	−28
$[62^3]/n^{(4)}$	·	−3	·	60	·	·	·	·	·	·	·	·	−28
$[631^3]/n^{(5)}$	−4	4	20	−240	·	·	·	·	−7	28	−140	·	·
$[6321]/n^{(4)}$	−4	4	−10	240	·	·	·	·	·	7	70	·	·
$[63^2]/n^{(3)}$	·	·	·	−20	·	·	·	·	·	−7	70	·	·
$[641^2]/n^{(4)}$	1	−1	−5	60	·	·	·	·	·	·	35	·	·
$[642]/n^{(3)}$	1	1	·	−30	·	·	·	·	·	−7	35	·	·
$[651]/n^{(3)}$	5	·	1	−12	·	·	·	·	·	·	−7	·	·
$[6^2]/n^{(2)}$	30	30	12	1	·	·	·	·	·	·	·	·	·
$[71^5]/n^{(6)}$	·	·	·	·	1	−1	1	2	−2	−6	24	−8	8
$[721^3]/n^{(5)}$	·	·	·	·	1	1	−2	−3	5	12	−60	·	8
$[72^21]/n^{(4)}$	·	·	·	·	1	2	1	·	·	−3	30	·	·
$[731^2]/n^{(4)}$	·	·	·	·	1	3	·	1	−1	−1	20	·	·
$[732]/n^{(3)}$	·	·	·	·	1	4	3	1	1	·	20	·	·
$[741]/n^{(3)}$	7	·	·	·	1	6	3	·	·	1	−10	·	·
$[75]/n^{(2)}$	35	·	7	·	1	10	15	4	·	5	1	·	·
$[81^4]/n^{(5)}$	·	·	·	·	8	·	·	10	10	10	5	1	−1
$[821^2]/n^{(4)}$	·	·	·	·	8	8	·	·	·	·	·	1	1
$[82^2]/n^{(3)}$	·	·	·	·	8	16	8	·	8	·	·	1	2
$[831]/n^{(3)}$	·	·	·	·	8	24	·	8	8	·	·	1	2
$[84]/n^{(2)}$	28	28	·	·	8	48	24	32	·	8	·	·	6
$[91^3]/n^{(4)}$	·	·	·	·	36	36	·	·	·	·	·	9	9
$[921]/n^{(3)}$	·	·	·	·	36	72	36	·	·	·	·	9	27
$[93]/n^{(2)}$	·	·	·	·	36	144	108	36	36	·	·	9	9
$[10,1^2]/n^{(3)}$	210	·	·	·	120	360	·	·	·	·	·	·	27
$[10,2]/n^{(2)}$	210	210	·	·	120	480	360	120	120	·	·	45	45
$[11,1]/n^{(2)}$	2310	·	462	·	330	1980	990	1320	·	330	·	45	90
$[12]/n$	13860	13860	5544	462	792	7920	11880	7920	7920	3960	792	495	2970

Table 2·2 (cont.)

Weight 12 (vi)	k_{82^2}	k_{831}	k_{84}	k_{91^3}	k_{921}	k_{93}	$k_{10,1^2}$	$k_{10,2}$	$k_{11,1}$	k_{12}
$[1^{12}]/n^{(12)}$	−5040	−10080	30240	40320	−40320	80640	−362880	362880	3628800	−39916800
$[2\,1^{10}]/n^{(11)}$	30240	55440	−181440	−181440	221760	−483840	1814400	−2177280	−19958400	239500800
$[2^2 1^8]/n^{(10)}$	−70560	−110880	408240	272160	−453600	1088640	−3175200	4989600	39916800	−538876800
$[2^3 1^6]/n^{(9)}$	80640	95760	−423360	−151120	423360	−1118880	2268000	−5443200	−34927200	558352000
$[2^4 1^4]/n^{(8)}$	−45990	−31500	200340	22680	−173880	498960	−567000	2835000	12474000	−261954000
$[2^5 1^2]/n^{(7)}$	11340	1890	−37800	·	22680	−68040	·	−589680	−1247400	44906400
$[2^6]/n^{(6)}$	−630	·	1890	·	·	·	·	22680	·	−1247400
$[3\,1^9]/n^{(10)}$	−6720	−18480	60480	60480	−60480	161280	−604800	604800	6652800	−79833600
$[3\,2\,1^7]/n^{(9)}$	26880	67200	−241920	−151200	211680	−665280	1814400	−2419200	−23284800	319334400
$[3\,2^2 1^5]/n^{(8)}$	−38640	−75600	312480	90720	−241920	907200	−1512000	3326400	24948000	−419126400
$[3\,2^3 1^3]/n^{(7)}$	23520	25200	−141120	−7560	98280	−43480	302400	−1814400	−8316000	199584000
$[3\,2^4 1]/n^{(6)}$	−5040	−630	17640	·	−7560	45360	302400	·	415800	−24948000
$[3^2 1^6]/n^{(8)}$	−1680	−10080	36960	20160	−20160	100800	−252000	252000	3326400	−46569600
$[3^2 2\,1^4]/n^{(7)}$	3920	19600	−77280	−15120	35280	−241920	302400	−554400	−5544000	99792000
$[3^2 2^2 1^2]/n^{(6)}$	−2800	−6720	31920	·	−15120	136080	−37800	340200	1663200	−49896000
$[3^2 2^3]/n^{(5)}$	560	·	−1680	·	·	−7560	−37800	·	·	3326400
$[3^3 1^3]/n^{(6)}$	·	−1680	6720	560	−560	21280	−16800	16800	369600	−7392000
$[3^3 2\,1]/n^{(5)}$	·	560	−2240	·	560	−16800	·	−16800	−92400	4435200
$[3^4]/n^{(4)}$	·	·	·	·	·	560	·	·	·	−92400
$[4\,1^8]/n^{(9)}$	1680	3360	−15120	−15120	15120	−30240	151200	−151200	−1663200	19958400
$[4\,2\,1^6]/n^{(8)}$	−5880	−10080	55440	30240	−45360	105840	−378000	529200	4989600	−69854400
$[4\,2^2 1^4]/n^{(7)}$	7140	8400	−63000	−11340	41580	−113400	226800	−604800	−4158000	74844000
$[4\,2^3 1^2]/n^{(6)}$	−3360	−1260	22680	·	−11340	34020	−18900	245700	831600	−24948000
$[4\,2^4]/n^{(5)}$	420	·	−1890	·	·	·	−18900	·	·	1247400
$[4\,3\,1^5]/n^{(7)}$	560	2800	−16800	−7560	7560	−30240	100800	−100800	−1386000	19958400
$[4\,3\,2\,1^3]/n^{(6)}$	−1120	−4200	30240	2520	−10080	57960	−75600	176400	1663200	−33264000
$[4\,3\,2^2 1]/n^{(5)}$	560	420	−8400	·	2520	−18900	·	−75600	−207900	9979200
$[4\,3^2 1^2]/n^{(5)}$	·	560	−3920	·	·	−7560	4200	−4200	−138600	3326400
$[4\,3^2 2]/n^{(4)}$	·	·	560	·	·	2520	2520	4200	·	−831600
$[4^2 1^4]/n^{(6)}$	−35	−70	1890	630	−630	1260	−9450	9450	138600	−2079000
$[4^2 2\,1^2]/n^{(5)}$	70	105	−2940	·	630	−1890	3150	−12600	−103950	2494800
$[4^2 2^2]/n^{(4)}$	−35	·	525	·	·	·	3150	·	·	−311850
$[4^2 3\,1]/n^{(4)}$	·	−35	700	·	·	630	·	·	11550	−415800
$[4^3]/n^{(3)}$	·	·	−35	·	·	·	·	·	·	11550
$[5\,1^7]/n^{(8)}$	−336	−672	2016	3024	−3024	6048	−30240	30240	332640	−3991680
$[5\,2\,1^5]/n^{(7)}$	1008	1680	−6048	−4536	7560	−18144	60480	−90720	−831600	11975040
$[5\,2^2 1^3]/n^{(6)}$	−1008	−1008	5040	756	−5292	15120	−22680	83160	498960	−9979200
$[5\,2^3 1]/n^{(5)}$	336	·	−1008	·	756	−2268	·	−22680	−41580	1995840
$[5\,3\,1^4]/n^{(6)}$	−56	−448	1680	1008	−1008	5040	−15120	15120	221760	−3326400
$[5\,3\,2\,1^2]/n^{(5)}$	112	504	−2016	·	1008	−7560	5040	−20160	−166320	3991680
$[5\,3\,2^2]/n^{(4)}$	−56	·	168	·	·	756	·	5040	·	−498960
$[5\,3^2 1]/n^{(4)}$	·	−56	224	·	·	1008	·	·	9240	−332640
$[5\,4\,1^3]/n^{(5)}$	·	·	−336	−126	126	−252	2520	−2520	−41580	665280
$[5\,4\,2\,1]/n^{(4)}$	·	·	336	·	−126	378	·	2520	13860	−498960
$[5\,4\,3]/n^{(3)}$	·	·	−56	·	·	−126	·	·	·	55440
$[5^2 1^2]/n^{(4)}$	·	·	·	·	·	·	−126	126	2772	−49896
$[5^2 2]/n^{(3)}$	·	·	·	·	·	·	·	−126	·	16632
$[6\,1^6]/n^{(7)}$	56	112	−336	−504	504	−1008	5040	−5040	−55440	665280
$[6\,2\,1^4]/n^{(6)}$	−140	−224	840	504	−1008	2520	−7560	12600	110880	−1663200
$[6\,2^2 1^2]/n^{(5)}$	112	84	−504	·	504	−1512	1260	−8820	−41580	997920
$[6\,2^3]/n^{(4)}$	−28	·	84	·	·	84	·	1260	·	−83160
$[6\,3\,1^3]/n^{(5)}$	·	56	−224	−84	84	−672	1680	−1680	−27720	443520
$[6\,3\,2\,1]/n^{(4)}$	·	−28	112	·	−84	756	·	1680	9240	−332640
$[6\,3^2]/n^{(3)}$	·	·	·	·	·	−84	·	·	·	18480
$[6\,4\,1^2]/n^{(4)}$	·	·	56	·	·	·	−210	210	4620	−83160
$[6\,4\,2]/n^{(3)}$	·	·	−28	·	·	·	·	−210	·	27720
$[6\,5\,1]/n^{(3)}$	·	·	·	·	·	·	·	·	−462	11088
$[6^2]/n^{(2)}$	·	·	·	·	·	·	·	·	·	−426
$[7\,1^5]/n^{(6)}$	−8	−16	48	72	−72	144	−720	720	7920	−95040
$[7\,2\,1^3]/n^{(5)}$	16	24	−96	−36	108	−288	720	−1440	−11880	190080
$[7\,2^2 1]/n^{(4)}$	−8	·	24	·	−36	108	·	720	1980	−71280
$[7\,3\,1^2]/n^{(4)}$	·	−8	32	·	·	72	−120	120	2640	−47520
$[7\,3\,2]/n^{(3)}$	·	·	·	·	·	−36	·	−120	·	15840
$[7\,4\,1]/n^{(3)}$	·	·	−8	·	·	·	·	·	−330	7920
$[7\,5]/n^{(2)}$	·	·	·	·	·	·	·	·	·	−792
$[8\,1^4]/n^{(5)}$	1	2	−6	−9	9	−18	90	−90	−990	11880
$[8\,2\,1^2]/n^{(4)}$	−2	−3	12	·	−9	27	−45	135	990	−17820
$[8\,2^2]/n^{(3)}$	1	·	−3	·	·	·	·	−45	·	2970
$[8\,3\,1]/n^{(3)}$	·	1	−4	·	·	−9	·	·	−165	3960
$[8\,4]/n^{(2)}$	3	4	1	·	·	·	·	·	·	−495
$[9\,1^3]/n^{(4)}$	·	·	·	1	−1	2	−10	10	110	−1320
$[9\,2\,1]/n^{(3)}$	·	·	·	1	1	−3	·	−10	−55	1320
$[9\,3]/n^{(2)}$	·	9	·	1	3	1	·	·	·	−220
$[10,1^2]/n^{(3)}$	·	·	·	10	10	·	1	−1	−11	132
$[10,2]/n^{(2)}$	45	·	·	10	10	·	·	1	·	−66
$[11,1]/n^{(2)}$	·	165	·	55	55	·	1	·	1	−12
$[12]/n$	1485	1980	495	220	660	220	66	66	12	1

13-2

Table 2·3. *Powers and products of k-statistics (and polykay statistics) in terms of polykay statistics (up to and including weight 8)*

For any given product of k-statistics (and polykay statistics) the table expresses this product as a linear sum of polykays. Thus for example

$$k_{21}k_{11} = \frac{1}{n^{(2)}}\left[2k_{32} + k_{311}(-2+2n) + k_{221}(-4+2n) + k_{2111}(-n+n^2)\right].$$

The dividing factor is the term outside the polykay sum, while the coefficients of n^0, n^1, etc. are the multiplying factor of the particular polykay. The arrangement of the table in this way allows the expansion of any power or product which it is necessary to make in order to take expectations, to be written down immediately in a suitable form. For the K's the same relations hold with the substitution of N for n.

Product of k-statistics	Dividing factor	Polykay	n^0	n^1	n^2	n^3	n^4
k_1^2	n	k_2	1	.	.	.	.
		k_{11}	.	1	.	.	.
$k_2 k_1$	n	k_3	1	.	.	.	.
		k_{21}	.	1	.	.	.
k_1^3	n^2	k_3	1	.	.	.	.
		k_{21}	.	3	.	.	.
		k_{111}	.	.	1	.	.
$k_{11}k_1$	n	k_{21}	2	.	.	.	.
		k_{111}	.	1	.	.	.
$k_3 k_1$	n	k_4	1	.	.	.	.
		k_{31}	.	1	.	.	.
k_2^2	$n^{(2)}$	k_4	-1	1	.	.	.
		k_{22}	.	1	1	.	.
$k_2 k_1^2$	n^2	k_4	1	.	.	.	.
		k_{31}	.	2	.	.	.
		k_{22}	.	1	.	.	.
		k_{211}	.	.	1	.	.
k_1^4	n^3	k_4	1	.	.	.	.
		k_{31}	.	4	.	.	.
		k_{22}	.	3	.	.	.
		k_{211}	.	.	6	.	.
		k_{1111}	.	.	.	1	.
$k_{21}k_1$	n	k_{31}	1	.	.	.	.
		k_{22}	1	.	.	.	.
		k_{211}	.	1	.	.	.
$k_2 k_{11}$	$n^{(2)}$	k_{31}	-2	2	.	.	.
		k_{22}	-2	.	.	.	.
		k_{211}	.	-1	1	.	.
$k_{11}k_1^2$	n^2	k_{31}	2	.	.	.	.
		k_{22}	2	.	.	.	.
		k_{211}	.	5	.	.	.
		k_{1111}	.	.	1	.	.
$k_{111}k_1$	n	k_{211}	3	.	.	.	.
		k_{1111}	.	1	.	.	.
k_{11}^2	$n^{(2)}$	k_{22}	2	.	.	.	.
		k_{211}	-4	4	.	.	.
		k_{1111}	.	-1	1	.	.
$k_4 k_1$	n	k_5	1	.	.	.	.
		k_{41}	.	1	.	.	.
$k_3 k_2$	$n^{(2)}$	k_5	-1	1	.	.	.
		k_{32}	.	5	1	.	.
$k_3 k_1^2$	n^2	k_5	1	.	.	.	.
		k_{41}	.	2	.	.	.
		k_{32}	.	1	.	.	.
		k_{311}	.	.	1	.	.
$k_2^2 k_1$	$nn^{(2)}$	k_5	-1	1	.	.	.
		k_{41}	.	-1	1	.	.
		k_{32}	.	2	2	.	.
		k_{221}	.	.	1	1	.
$k_2 k_1^3$	n^3	k_5	1	.	.	.	.
		k_{41}	.	3	.	.	.
		k_{32}	.	4	.	.	.
		k_{311}	.	.	3	.	.
		k_{221}	.	.	3	.	.
		k_{2111}	.	.	.	1	.
k_1^5	n^4	k_5	1	.	.	.	.
		k_{41}	.	5	.	.	.
		k_{32}	.	10	.	.	.
		k_{311}	.	.	10	.	.
		k_{221}	.	.	15	.	.
		k_{2111}	.	.	.	10	.
		k_{11111}	.	.	.	.	1
$k_{31}k_1$	n	k_{41}	1	.	.	.	.
		k_{32}	1	.	.	.	.
		k_{311}	.	1	.	.	.

Product of k-statistics	Dividing factor	Polykay	n^0	n^1	n^2	n^3
$k_3 k_{11}$	$n^{(2)}$	k_{41}	-2	2	.	.
		k_{32}	-6	.	.	.
		k_{311}	.	-1	1	.
$k_{22}k_1$	n	k_{32}	2	.	.	.
		k_{221}	.	1	.	.
$k_{21}k_2$	$n^{(2)}$	k_{41}	-1	1	.	.
		k_{32}	-3	1	.	.
		k_{221}	.	1	1	.
$k_{21}k_1^2$	n^2	k_{41}	1	.	.	.
		k_{32}	3	.	.	.
		k_{311}	.	2	.	.
		k_{221}	.	3	.	.
		k_{2111}	.	.	1	.
$k_2 k_{11}k_1$	$nn^{(2)}$	k_{41}	-2	2	.	.
		k_{32}	-6	2	.	.
		k_{311}	.	-3	3	.
		k_{221}	.	-4	2	.
		k_{2111}	.	.	-1	1
$k_{11}k_1^3$	n^3	k_{41}	2	.	.	.
		k_{32}	6	.	.	.
		k_{311}	.	7	.	.
		k_{221}	.	12	.	.
		k_{2111}	.	.	9	.
		k_{11111}	.	.	.	1
$k_{211}k_1$	n	k_{311}	1	.	.	.
		k_{221}	2	.	.	.
		k_{2111}	.	1	.	.
$k_{21}k_{11}$	$n^{(2)}$	k_{32}	2	.	.	.
		k_{311}	-2	2	.	.
		k_{221}	-4	2	.	.
		k_{2111}	.	-1	1	.
$k_2 k_{111}$	$n^{(2)}$	k_{311}	-3	3	.	.
		k_{221}	-6	.	.	.
		k_{2111}	.	-1	1	.
$k_{111}k_1^2$	n^2	k_{311}	3	.	.	.
		k_{221}	6	.	.	.
		k_{2111}	.	7	.	.
		k_{11111}	.	.	1	.
$k_{11}^2 k_1$	$nn^{(2)}$	k_{32}	4	.	.	.
		k_{311}	-4	4	.	.
		k_{221}	-8	10	.	.
		k_{2111}	.	-8	8	.
		k_{11111}	.	.	-1	1
$k_{1111}k_1$	n	k_{2111}	4	.	.	.
		k_{11111}	.	1	.	.
$k_{111}k_{11}$	$n^{(2)}$	k_{221}	6	.	.	.
		k_{2111}	-6	6	.	.
		k_{11111}	.	-1	1	.
$k_5 k_1$	n	k_6	1	.	.	.
		k_{51}	.	1	.	.
$k_4 k_2$	$n^{(2)}$	k_6	-1	1	.	.
		k_{42}	.	7	1	.
		k_{33}	.	6	.	.
k_3^2	$n^{(3)}$	k_6	2	-3	1	.
		k_{42}	.	-18	9	.
		k_{33}	.	-16	6	1
		k_{222}	.	.	6	.
$k_4 k_1^2$	n^2	k_6	1	.	.	.
		k_{51}	.	2	.	.
		k_{42}	.	1	.	.
		k_{411}	.	.	1	.
$k_3 k_2 k_1$	$nn^{(2)}$	k_6	-1	1	.	.
		k_{51}	.	-1	1	.
		k_{42}	.	5	1	.
		k_{33}	.	5	1	.
		k_{321}	.	.	5	1

Table 2·3 (*cont.*)

Product of k-statistics	Dividing factor	Polykay	n^0	n^1	n^2	n^3	n^4	n^5
k_2^3	$(n^{(2)})^2$	k_6	1	−2	1	.	.	.
		k_{42}	.	−9	6	3	.	.
		k_{33}	.	−8	4	.	.	.
		k_{222}	.	.	3	4	1	.
$k_3 k_1^3$	n^3	k_6	1	.	.	.	.	.
		k_{51}	.	3	.	.	.	.
		k_{42}	.	3	.	.	.	.
		k_{33}	.	1	.	.	.	.
		k_{411}	.	.	3	.	.	.
		k_{321}	.	.	3	.	.	.
		k_{3111}	.	.	.	1	.	.
$k_2^2 k_1^2$	$n^2 n^{(2)}$	k_6	−1	1	.	.	.	.
		k_{51}	.	−2	2	.	.	.
		k_{42}	.	1	3	.	.	.
		k_{33}	.	2	2	.	.	.
		k_{411}	.	.	−1	1	.	.
		k_{321}	.	.	4	4	.	.
		k_{222}	.	.	1	1	.	.
		k_{2211}	.	.	.	1	1	.
$k_2 k_1^4$	n^4	k_6	1	.	.	.	.	.
		k_{51}	.	4	.	.	.	.
		k_{42}	.	7	.	.	.	.
		k_{33}	.	4	.	.	.	.
		k_{411}	.	.	6	.	.	.
		k_{321}	.	.	16	.	.	.
		k_{222}	.	.	3	.	.	.
		k_{3111}	.	.	.	4	.	.
		k_{2211}	.	.	.	6	.	.
		k_{21111}	.	.	.	.	1	.
k_1^6	n^5	k_6	1	.	.	.	.	.
		k_{51}	.	6	.	.	.	.
		k_{42}	.	15	.	.	.	.
		k_{33}	.	10	.	.	.	.
		k_{411}	.	.	15	.	.	.
		k_{321}	.	.	60	.	.	.
		k_{222}	.	.	15	.	.	.
		k_{3111}	.	.	.	20	.	.
		k_{2211}	.	.	.	45	.	.
		k_{21111}	.	.	.	.	15	.
		k_{111111}	.	.	.	.	.	1
$k_{41} k_1$	n	k_{51}	1	.	.	.	.	.
		k_{42}	1	.	.	.	.	.
		k_{411}	.	1	.	.	.	.
$k_4 k_{11}$	$n^{(2)}$	k_{51}	−2	2	.	.	.	.
		k_{42}	−8	.	.	.	.	.
		k_{33}	−6	.	.	.	.	.
		k_{411}	.	−1	1	.	.	.
$k_{32} k_1$	n	k_{42}	1	.	.	.	.	.
		k_{33}	1	.	.	.	.	.
		k_{321}	.	1	.	.	.	.
$k_{31} k_2$	$n^{(2)}$	k_{51}	−1	1	.	.	.	.
		k_{42}	−2	.	.	.	.	.
		k_{33}	−1	1	.	.	.	.
		k_{321}	.	5	1	.	.	.
$k_3 k_{21}$	$n^{(3)}$	k_{51}	2	−3	1	.	.	.
		k_{42}	8	−6	1	.	.	.
		k_{33}	6	−3	.	.	.	.
		k_{321}	.	10	3	1	.	.
		k_{222}	.	−6	.	.	.	.
$k_{22} k_2$	$n^{(2)}$	k_{42}	−2	2	.	.	.	.
		k_{33}	−2	.	.	.	.	.
		k_{222}	.	3	1	.	.	.
$k_{31} k_1^2$	n^2	k_{51}	1	.	.	.	.	.
		k_{42}	2	.	.	.	.	.
		k_{33}	1	.	.	.	.	.
		k_{411}	.	2	.	.	.	.
		k_{321}	.	3	.	.	.	.
		k_{3111}	.	.	1	.	.	.

Product of k-statistics	Dividing factor	Polykay	n^0	n^1	n^2	n^3	n^4
$k_3 k_{11} k_1$	$n n^{(2)}$	k_{51}	−2	2	.	.	.
		k_{42}	−8	2	.	.	.
		k_{33}	−6	.	.	.	.
		k_{411}	.	−3	3	.	.
		k_{321}	.	−8	2	.	.
		k_{3111}	.	.	−1	1	.
$k_{22} k_1^2$	n^2	k_{42}	2	.	.	.	.
		k_{33}	2	.	.	.	.
		k_{321}	.	4	.	.	.
		k_{222}	.	1	.	.	.
		k_{2211}	.	.	1	.	.
$k_{21} k_2 k_1$	$n n^{(2)}$	k_{51}	−1	1	.	.	.
		k_{42}	−4	2	.	.	.
		k_{33}	−3	1	.	.	.
		k_{411}	.	−1	1	.	.
		k_{321}	.	−1	3	.	.
		k_{222}	.	1	1	.	.
		k_{2211}	.	.	1	1	.
$k_2^2 k_{11}$	$(n^{(2)})^2$	k_{51}	2	−4	2	.	.
		k_{42}	.	−8	.	.	.
		k_{33}	6	−4	2	.	.
		k_{411}	.	1	−2	1	.
		k_{321}	.	−4	.	4	.
		k_{222}	.	−4	.	.	.
		k_{2211}	.	.	−1	.	1
$k_{21} k_1^3$	n^3	k_{51}	1	.	.	.	.
		k_{42}	4	.	.	.	.
		k_{33}	3	.	.	.	.
		k_{411}	.	3	.	.	.
		k_{321}	.	13	.	.	.
		k_{222}	.	3	.	.	.
		k_{3111}	.	.	3	.	.
		k_{2211}	.	.	6	.	.
		k_{21111}	.	.	.	1	.
$k_2 k_{11} k_1^2$	$n^2 n^{(2)}$	k_{51}	−2	2	.	.	.
		k_{42}	−8	4	.	.	.
		k_{33}	−6	2	.	.	.
		k_{411}	.	−5	5	.	.
		k_{321}	.	−20	12	.	.
		k_{222}	.	−4	2	.	.
		k_{3111}	.	.	−4	4	.
		k_{2211}	.	.	−7	5	.
		k_{21111}	.	.	.	−1	1
$k_{11} k_1^4$	n^4	k_{51}	2	.	.	.	.
		k_{42}	8	.	.	.	.
		k_{33}	6	.	.	.	.
		k_{411}	.	9	.	.	.
		k_{321}	.	44	.	.	.
		k_{222}	.	12	.	.	.
		k_{3111}	.	.	16	.	.
		k_{2211}	.	.	39	.	.
		k_{21111}	.	.	.	14	.
		k_{111111}	.	.	.	.	1
$k_{311} k_1$	n	k_{411}	1	.	.	.	.
		k_{321}	2	.	.	.	.
		k_{3111}	.	1	.	.	.
$k_{31} k_{11}$	$n^{(2)}$	k_{42}	2	.	.	.	.
		k_{411}	−2	2	.	.	.
		k_{321}	−8	2	.	.	.
		k_{3111}	.	−1	1	.	.
$k_{221} k_1$	n	k_{321}	2	.	.	.	.
		k_{222}	1	.	.	.	.
		k_{2211}	.	1	.	.	.
$k_{22} k_{11}$	$n^{(2)}$	k_{33}	2	.	.	.	.
		k_{321}	−4	4	.	.	.
		k_{222}	−4	.	.	.	.
		k_{2211}	.	−1	1	.	.

Table 2·3 (*cont.*)

Product of k-statistics	Dividing factor	Polykay	n^0	n^1	n^2	n^3	n^4
$k_3 k_{111}$	$n^{(3)}$	k_{411}	6	−9	3	.	.
		k_{321}	36	−18	.	.	.
		k_{222}	12	.	.	.	.
		k_{3111}	.	2	−3	1	.
$k_{211} k_2$	$n^{(2)}$	k_{411}	−1	1	.	.	.
		k_{321}	−6	2	.	.	.
		k_{222}	−2	.	.	.	.
		k_{2211}	.	1	1	.	.
k_{21}^2	$n^{(3)}$	k_{42}	−2	1	.	.	.
		k_{33}	−2	1	.	.	.
		k_{411}	2	−3	1	.	.
		k_{321}	12	−10	2	.	.
		k_{222}	4	−1	1	.	.
		k_{2211}	.	−2	−1	1	.
$k_{211} k_1^2$	n^2	k_{411}	1	.	.	.	.
		k_{321}	6	.	.	.	.
		k_{222}	2	.	.	.	.
		k_{3111}	.	2	.	.	.
		k_{2211}	.	5	.	.	.
		k_{21111}	.	.	1	.	.
$k_{21} k_{11} k_1$	$nn^{(2)}$	k_{42}	2	.	.	.	.
		k_{33}	2	.	.	.	.
		k_{411}	−2	2	.	.	.
		k_{321}	−12	10	.	.	.
		k_{222}	−4	2	.	.	.
		k_{3111}	.	−3	3	.	.
		k_{2211}	.	−7	5	.	.
		k_{21111}	.	.	−1	1	.
$k_2 k_{111} k_1$	$nn^{(2)}$	k_{411}	−3	3	.	.	.
		k_{321}	−18	6	.	.	.
		k_{222}	−6	.	.	.	.
		k_{3111}	.	−4	4	.	.
		k_{2211}	.	−9	3	.	.
		k_{21111}	.	.	−1	1	.
$k_2 k_{11}^2$	$(n^{(2)})^2$	k_{42}	.	4	.	.	.
		k_{33}	−4	.	.	.	.
		k_{411}	4	−8	4	.	.
		k_{321}	.	−24	8	.	.
		k_{222}	8	−2	2	.	.
		k_{3111}	.	4	−8	4	.
		k_{2211}	.	8	−12	4	.
		k_{21111}	.	.	1	−2	1
$k_{111} k_1^3$	n^3	k_{411}	3	.	.	.	.
		k_{321}	18	.	.	.	.
		k_{222}	6	.	.	.	.
		k_{3111}	.	10	.	.	.
		k_{2211}	.	27	.	.	.
		k_{21111}	.	.	12	.	.
		k_{111111}	.	.	.	1	.
$k_{11}^2 k_1^2$	$n^2 n^{(2)}$	k_{42}	4	.	.	.	.
		k_{33}	4	.	.	.	.
		k_{411}	−4	4	.	.	.
		k_{321}	−24	32	.	.	.
		k_{222}	−8	10	.	.	.
		k_{3111}	.	−12	12	.	.
		k_{2211}	.	−32	34	.	.
		k_{21111}	.	.	−13	13	.
		k_{111111}	.	.	.	−1	1
$k_{2111} k_1$	n	k_{3111}	1	.	.	.	.
		k_{2211}	3	.	.	.	.
		k_{21111}	.	1	.	.	.
$k_{211} k_{11}$	$n^{(2)}$	k_{321}	4	.	.	.	.
		k_{222}	2	.	.	.	.
		k_{3111}	−2	2	.	.	.
		k_{2211}	−6	4	.	.	.
		k_{21111}	.	−1	1	.	.
$k_{21} k_{111}$	$n^{(3)}$	k_{321}	−12	6	.	.	.
		k_{222}	−6	.	.	.	.
		k_{3111}	6	−9	3	.	.
		k_{2211}	18	−15	3	.	.
		k_{21111}	.	2	−3	1	.
$k_2 k_{1111}$	$n^{(2)}$	k_{3111}	−4	4	.	.	.
		k_{2211}	−12	.	.	.	.
		k_{21111}	.	−1	1	.	.
$k_{1111} k_1^2$	n^2	k_{3111}	4	.	.	.	.
		k_{2211}	12	.	.	.	.
		k_{21111}	.	9	.	.	.
		k_{111111}	.	.	1	.	.
$k_{111} k_{11} k_1$	$nn^{(2)}$	k_{321}	12	.	.	.	.
		k_{222}	6	.	.	.	.
		k_{3111}	−6	6	.	.	.
		k_{2211}	−18	24	.	.	.
		k_{21111}	.	−11	11	.	.
		k_{111111}	.	.	−1	1	.
k_{11}^3	$(n^{(2)})^2$	k_{33}	4	.	.	.	.
		k_{321}	.	24	.	.	.
		k_{222}	−16	8	.	.	.
		k_{3111}	8	−16	8	.	.
		k_{2211}	.	−24	30	.	.
		k_{21111}	.	12	−24	12	.
		k_{111111}	.	.	1	−2	1
$k_{11111} k_1$	n	k_{21111}	5	.	.	.	.
		k_{111111}	.	1	.	.	.
$k_{1111} k_{11}$	$n^{(2)}$	k_{2211}	12	.	.	.	.
		k_{21111}	−8	8	.	.	.
		k_{111111}	.	−1	1	.	.
k_{111}^2	$n^{(3)}$	k_{222}	6	.	.	.	.
		k_{2211}	−36	18	.	.	.
		k_{21111}	18	−27	9	.	.
		k_{111111}	.	2	−3	1	.
$k_6 k_1$	n	k_7	1	.	.	.	.
		k_{61}	.	1	.	.	.
$k_5 k_2$	$n^{(2)}$	k_7	−1	1	.	.	.
		k_{52}	.	9	1	.	.
		k_{43}	.	20	.	.	.
$k_4 k_3$	$n^{(3)}$	k_7	2	−3	1	.	.
		k_{52}	.	−24	12	.	.
		k_{43}	.	−58	27	1	.
		k_{322}	.	.	36	.	.
$k_5 k_1^2$	n^2	k_7	1	.	.	.	.
		k_{61}	.	2	.	.	.
		k_{52}	.	1	.	.	.
		k_{511}	.	.	1	.	.
$k_4 k_2 k_1$	$nn^{(2)}$	k_7	−1	1	.	.	.
		k_{61}	.	−1	1	.	.
		k_{52}	.	7	1	.	.
		k_{43}	.	19	1	.	.
		k_{421}	.	.	7	1	.
		k_{331}	.	.	6	.	.
$k_3^2 k_1$	$nn^{(3)}$	k_7	2	−3	1	.	.
		k_{61}	.	2	−3	1	.
		k_{52}	.	−18	9	.	.
		k_{43}	.	−50	21	2	.
		k_{421}	.	.	−18	9	.
		k_{331}	.	.	−16	6	1
		k_{322}	.	.	18	.	.
		k_{2221}	.	.	.	6	.
$k_3 k_2^2$	$(n^{(2)})^2$	k_7	1	−2	1	.	.
		k_{52}	.	−14	12	2	.
		k_{43}	.	−35	22	1	.
		k_{322}	.	.	35	12	1
$k_4 k_1^3$	n^3	k_7	1	.	.	.	.
		k_{61}	.	3	.	.	.
		k_{52}	.	3	.	.	.
		k_{43}	.	1	.	.	.
		k_{511}	.	.	3	.	.
		k_{421}	.	.	3	.	.
		k_{4111}	.	.	.	1	.

Table 2·3 (cont.)

Product of k-statistics	Dividing factor	Polykay	Coefficients of						
			n^0	n^1	n^2	n^3	n^4	n^5	n^6
$k_3 k_2 k_1^2$	$n^2 n^{(2)}$	k_7	-1	1	2	.	.	.	.
		k_{61}	.	-2	2	.	.	.	.
		k_{52}	.	4	2	.	.	.	.
		k_{43}	.	15	3	.	.	.	.
		k_{511}	.	.	-1	1	.	.	.
		k_{421}	.	.	10	2	.	.	.
		k_{331}	.	.	10	2	.	.	.
		k_{322}	.	.	5	1	.	.	.
		k_{3211}	.	.	.	5	1	.	.
$k_2^3 k_1$	$n(n^{(2)})^2$	k_7	1	-2	1	.	.	.	.
		k_{61}	.	1	-2	1	.	.	.
		k_{52}	.	-9	6	3	.	.	.
		k_{43}	.	-25	14	3	.	.	.
		k_{421}	.	.	-9	6	3	.	.
		k_{331}	.	.	-8	4	3	.	.
		k_{322}	.	.	9	12	3	.	.
		k_{2221}	.	.	.	3	4	1	.
$k_3 k_1^4$	n^4	k_7	1	.	.	.	.	.	.
		k_{61}	.	4	.	.	.	.	.
		k_{52}	.	6	.	.	.	.	.
		k_{43}	.	5	.	.	.	.	.
		k_{511}	.	.	6	.	.	.	.
		k_{421}	.	.	12	.	.	.	.
		k_{331}	.	.	4	.	.	.	.
		k_{322}	.	.	3	.	.	.	.
		k_{4111}	.	.	.	4	.	.	.
		k_{3211}	.	.	.	6	.	.	.
		k_{31111}	.	.	.	.	1	.	.
$k_2^2 k_1^3$	$n^3 n^{(2)}$	k_7	-1	1	.	.	.	.	.
		k_{61}	.	-3	3	.	.	.	.
		k_{52}	.	-1	5	.	.	.	.
		k_{43}	.	5	7	.	.	.	.
		k_{511}	.	.	-3	3	.	.	.
		k_{421}	.	.	3	9	.	.	.
		k_{331}	.	.	6	6	.	.	.
		k_{322}	.	.	7	7	.	.	.
		k_{4111}	.	.	.	-1	1	.	.
		k_{3211}	.	.	.	6	6	.	.
		k_{2221}	.	.	.	3	3	.	.
		k_{22111}	.	.	.	.	1	1	.
$k_2 k_1^5$	n^5	k_7	1	.	.	.	.	.	.
		k_{61}	.	5	.	.	.	.	.
		k_{52}	.	11	.	.	.	.	.
		k_{43}	.	15	.	.	.	.	.
		k_{511}	.	.	10	.	.	.	.
		k_{421}	.	.	35	.	.	.	.
		k_{331}	.	.	20	.	.	.	.
		k_{322}	.	.	25	.	.	.	.
		k_{4111}	.	.	.	10	.	.	.
		k_{3211}	.	.	.	40	.	.	.
		k_{2221}	.	.	.	15	.	.	.
		k_{31111}	.	.	.	.	5	.	.
		k_{22111}	.	.	.	.	10	.	.
		k_{211111}	.	.	.	.	.	1	.
k_1^7	n^6	k_7	1	.	.	.	.	.	.
		k_{61}	.	7	.	.	.	.	.
		k_{52}	.	21	.	.	.	.	.
		k_{43}	.	35	.	.	.	.	.
		k_{511}	.	.	21	.	.	.	.
		k_{421}	.	.	105	.	.	.	.
		k_{331}	.	.	70	.	.	.	.
		k_{322}	.	.	105	.	.	.	.
		k_{4111}	.	.	.	35	.	.	.
		k_{3211}	.	.	.	210	.	.	.
		k_{2221}	.	.	.	105	.	.	.
		k_{31111}	.	.	.	.	35	.	.
		k_{22111}	.	.	.	.	105	.	.
		k_{211111}	.	.	.	.	.	21	.
		$k_{1111111}$	.	.	.	.	.	.	1
$k_7 k_1$	n	k_8	1	.	.	.	.	.	.
		k_{71}	.	1	.	.	.	.	.
$k_6 k_2$	$n^{(2)}$	k_8	-1	1	.	.	.	.	.
		k_{62}	.	11	1	.	.	.	.
		k_{53}	.	30	.	.	.	.	.
		k_{44}	.	20	.	.	.	.	.

Product of k-statistics	Dividing factor	Polykay	Coefficients of					
			n^0	n^1	n^2	n^3	n^4	n^5
$k_5 k_3$	$n^{(2)}$	k_8	2	-3	1	.	.	.
		k_{62}	.	-30	15	.	.	.
		k_{53}	.	-88	42	1	.	.
		k_{44}	.	-60	30	.	.	.
		k_{422}	.	.	60	.	.	.
		k_{332}	.	.	90	.	.	.
k_4^2	$n^{(4)}$	k_8	-6	11	-6	1	.	.
		k_{62}	.	96	-80	16	.	.
		k_{53}	.	288	-240	48	.	.
		k_{44}	.	198	-159	28	1	.
		k_{422}	.	.	-216	72	.	.
		k_{332}	.	.	-432	144	.	.
		k_{2222}	.	.	24	24	.	.
$k_6 k_1^2$	n^2	k_8	1	.	.	.	.	.
		k_{71}	.	2	.	.	.	.
		k_{62}	.	1	.	.	.	.
		k_{611}	.	.	1	.	.	.
$k_5 k_2 k_1$	$n n^{(2)}$	k_8	-1	1	.	.	.	.
		k_{71}	.	-1	1	.	.	.
		k_{62}	.	9	1	.	.	.
		k_{53}	.	29	1	.	.	.
		k_{44}	.	20	.	.	.	.
		k_{521}	.	.	9	1	.	.
		k_{431}	.	.	20	.	.	.
$k_4 k_3 k_1$	$n n^{(3)}$	k_8	2	-3	1	.	.	.
		k_{71}	.	2	-3	1	.	.
		k_{62}	.	-24	12	.	.	.
		k_{53}	.	-82	39	1	.	.
		k_{44}	.	-58	27	1	.	.
		k_{521}	.	.	-24	12	.	.
		k_{431}	.	.	-58	27	1	.
		k_{422}	.	.	36	.	.	.
		k_{332}	.	.	72	.	.	.
		k_{3221}	.	.	.	36	.	.
$k_4 k_2^2$	$(n^{(2)})^2$	k_8	1	-2	1	.	.	.
		k_{62}	.	-18	16	2	.	.
		k_{53}	.	-56	40	.	.	.
		k_{44}	.	-39	26	1	.	.
		k_{422}	.	.	63	16	1	.
		k_{332}	.	.	108	12	.	.
$k_3^2 k_2$	$n^{(2)} n^{(3)}$	k_8	-2	5	-4	1	.	.
		k_{62}	.	40	-58	17	1	.
		k_{53}	.	128	-152	40	2	.
		k_{44}	.	90	-99	27	.	.
		k_{422}	.	.	-180	81	9	.
		k_{332}	.	.	-320	104	17	1
		k_{2222}	.	.	.	30	6	.
$k_5 k_1^3$	n^3	k_8	1	.	.	.	.	.
		k_{71}	.	3	.	.	.	.
		k_{62}	.	3	.	.	.	.
		k_{53}	.	1	.	.	.	.
		k_{611}	.	.	3	.	.	.
		k_{521}	.	.	3	.	.	.
		k_{5111}	.	.	.	1	.	.
$k_4 k_2 k_1^2$	$n^2 n^{(2)}$	k_8	-1	1	.	.	.	.
		k_{71}	.	-2	2	.	.	.
		k_{62}	.	6	2	.	.	.
		k_{53}	.	26	2	.	.	.
		k_{44}	.	19	1	.	.	.
		k_{611}	.	.	-1	1	.	.
		k_{521}	.	.	14	2	.	.
		k_{431}	.	.	38	2	.	.
		k_{422}	.	.	7	1	.	.
		k_{332}	.	.	6	.	.	.
		k_{4211}	.	.	.	7	1	.
		k_{3311}	.	.	.	6	.	.

Table 2·3 (*cont.*)

Left table

Product of k-statistics	Dividing factor	Polykay	n^0	n^1	n^2	n^3	n^4	n^5	n^6
$k_3^2 k_1^2$	$n^2 n^{(2)}$	k_8	2	-3	1	.	.	.	.
		k_{71}	.	4	-6	2	.	.	.
		k_{62}	.	-16	6	1	.	.	.
		k_{53}	.	-68	30	2	.	.	.
		k_{44}	.	-50	21	2	.	.	.
		k_{611}	.	.	2	-3	1	.	.
		k_{521}	.	.	-36	18	.	.	.
		k_{431}	.	.	-100	42	4	.	.
		k_{422}	.	.	.	9	.	.	.
		k_{332}	.	.	20	6	1	.	.
		k_{4211}	.	.	.	-18	9	.	.
		k_{331}	.	.	.	-16	6	1	.
		k_{3221}	.	.	.	36	.	.	.
		k_{2222}	.	.	.	6	.	.	.
		k_{22211}	.	.	.	.	6	.	.
$k_3 k_2^2 k_1$	$n(n^{(2)})^2$	k_8	1	-2	1	.	.	.	.
		k_{71}	.	1	-2	1	.	.	.
		k_{62}	.	-14	12	2	.	.	.
		k_{53}	.	-49	34	3	.	.	.
		k_{44}	.	-35	22	1	.	.	.
		k_{521}	.	.	-14	12	2	.	.
		k_{431}	.	.	-35	22	1	.	.
		k_{422}	.	.	35	12	1	.	.
		k_{332}	.	.	70	24	2	.	.
		k_{3221}	.	.	.	35	12	1	.
k_2^4	$(n^{(2)})^3$	k_8	-1	3	-3	1	.	.	.
		k_{62}	.	20	-36	12	4	.	.
		k_{53}	.	64	-96	32	.	.	.
		k_{44}	.	45	-63	23	3	.	.
		k_{422}	.	.	-90	42	42	6	.
		k_{332}	.	.	-160	48	16	.	.
		k_{2222}	.	.	.	15	23	9	1
$k_4 k_1^4$	n^4	k_8	1	.	.	.	.	.	.
		k_{71}	.	4	.	.	.	.	.
		k_{62}	.	6	.	.	.	.	.
		k_{53}	.	4	.	.	.	.	.
		k_{44}	.	1	.	.	.	.	.
		k_{611}	.	.	6	.	.	.	.
		k_{521}	.	.	12	.	.	.	.
		k_{431}	.	.	4	.	.	.	.
		k_{422}	.	.	3	.	.	.	.
		k_{5111}	.	.	.	4	.	.	.
		k_{4211}	.	.	.	6	.	.	.
		k_{41111}	.	.	.	.	1	.	.
$k_3 k_2 k_1^3$	$n^3 n^{(2)}$	k_8	-1	1	.	.	.	.	.
		k_{71}	.	-3	3	.	.	.	.
		k_{62}	.	2	4	.	.	.	.
		k_{53}	.	19	5	.	.	.	.
		k_{44}	.	15	3	.	.	.	.
		k_{611}	.	.	-3	3	.	.	.
		k_{521}	.	.	12	6	.	.	.
		k_{431}	.	.	45	9	.	.	.
		k_{422}	.	.	15	3	.	.	.
		k_{332}	.	.	20	4	.	.	.
		k_{5111}	.	.	.	-1	1	.	.
		k_{4211}	.	.	.	15	3	.	.
		k_{3311}	.	.	.	15	3	.	.
		k_{3221}	.	.	.	15	3	.	.
		k_{32111}	.	.	.	.	5	1	.
$k_2^3 k_1^2$	$n^2(n^{(2)})^2$	k_8	1	-2	1	.	.	.	.
		k_{71}	.	2	-4	2	.	.	.
		k_{62}	.	-8	4	4	.	.	.
		k_{53}	.	-34	20	6	.	.	.
		k_{44}	.	-25	14	3	.	.	.
		k_{611}	.	.	1	-2	1	.	.
		k_{521}	.	.	-18	12	6	.	.
		k_{431}	.	.	-50	28	6	.	.
		k_{422}	.	.	.	18	6	.	.
		k_{332}	.	.	10	28	6	.	.
		k_{4211}	.	.	.	-9	6	3	.
		k_{3311}	.	.	.	-8	4	.	.
		k_{3221}	.	.	.	18	24	6	.
		k_{2222}	.	.	.	3	4	1	.
		k_{22211}	.	.	.	.	3	4	1

Right table

Product of k-statistics	Dividing factor	Polykay	n^0	n^1	n^2	n^3	n^4	n^5	n^6	n^7
$k_3 k_1^5$	n^5	k_8	1	.	.	.	.	.	.	.
		k_{71}	.	5	.	.	.	.	.	.
		k_{62}	.	10	.	.	.	.	.	.
		k_{53}	.	11	.	.	.	.	.	.
		k_{44}	.	5	.	.	.	.	.	.
		k_{611}	.	.	10	.	.	.	.	.
		k_{521}	.	.	30	.	.	.	.	.
		k_{431}	.	.	25	.	.	.	.	.
		k_{422}	.	.	15	.	.	.	.	.
		k_{332}	.	.	10	.	.	.	.	.
		k_{5111}	.	.	.	10	.	.	.	.
		k_{4211}	.	.	.	30	.	.	.	.
		k_{3311}	.	.	.	10	.	.	.	.
		k_{3221}	.	.	.	15	.	.	.	.
		k_{41111}	.	.	.	.	5	.	.	.
		k_{32111}	.	.	.	.	10	.	.	.
		k_{311111}	.	.	.	.	.	1	.	.
$k_2^2 k_1^4$	$n^4 n^{(2)}$	k_8	-1	1	.	.	.	.	.	.
		k_{71}	.	-4	4	.	.	.	.	.
		k_{62}	.	-4	8	.	.	.	.	.
		k_{53}	.	4	12	.	.	.	.	.
		k_{44}	.	5	7	.	.	.	.	.
		k_{611}	.	.	-6	6	.	.	.	.
		k_{521}	.	.	-4	20	.	.	.	.
		k_{431}	.	.	20	28	.	.	.	.
		k_{422}	.	.	10	16	.	.	.	.
		k_{332}	.	.	20	20	.	.	.	.
		k_{5111}	.	.	.	-4	4	.	.	.
		k_{4211}	.	.	.	6	18	.	.	.
		k_{3311}	.	.	.	12	12	.	.	.
		k_{3221}	.	.	.	28	28	.	.	.
		k_{2222}	.	.	.	3	3	.	.	.
		k_{41111}	.	.	.	.	-1	1	.	.
		k_{32111}	.	.	.	.	8	8	.	.
		k_{22211}	.	.	.	.	6	6	.	.
		k_{221111}	.	.	.	.	.	1	1	.
$k_2 k_1^6$	n^6	k_8	1	.	.	.	.	.	.	.
		k_{71}	.	6	.	.	.	.	.	.
		k_{62}	.	16	.	.	.	.	.	.
		k_{53}	.	26	.	.	.	.	.	.
		k_{44}	.	15	.	.	.	.	.	.
		k_{611}	.	.	15	.	.	.	.	.
		k_{521}	.	.	66	.	.	.	.	.
		k_{431}	.	.	90	.	.	.	.	.
		k_{422}	.	.	60	.	.	.	.	.
		k_{332}	.	.	70	.	.	.	.	.
		k_{511}	.	.	.	20	.	.	.	.
		k_{4211}	.	.	.	105	.	.	.	.
		k_{3311}	.	.	.	60	.	.	.	.
		k_{3221}	.	.	.	150	.	.	.	.
		k_{2222}	.	.	.	15	.	.	.	.
		k_{41111}	.	.	.	.	15	.	.	.
		k_{32111}	.	.	.	.	80	.	.	.
		k_{22211}	.	.	.	.	45	.	.	.
		k_{311111}	.	.	.	.	.	6	.	.
		k_{221111}	.	.	.	.	.	15	.	.
		$k_{2111111}$	.	.	.	.	.	.	1	.
k_1^8	n^7	k_8	1	.	.	.	.	.	.	.
		k_{71}	.	8	.	.	.	.	.	.
		k_{62}	.	28	.	.	.	.	.	.
		k_{53}	.	56	.	.	.	.	.	.
		k_{44}	.	35	.	.	.	.	.	.
		k_{611}	.	.	28	.	.	.	.	.
		k_{521}	.	.	168	.	.	.	.	.
		k_{431}	.	.	280	.	.	.	.	.
		k_{422}	.	.	210	.	.	.	.	.
		k_{332}	.	.	280	.	.	.	.	.
		k_{5111}	.	.	.	56	.	.	.	.
		k_{4211}	.	.	.	420	.	.	.	.
		k_{3311}	.	.	.	280	.	.	.	.
		k_{3221}	.	.	.	840	.	.	.	.
		k_{2222}	.	.	.	105	.	.	.	.
		k_{41111}	.	.	.	.	70	.	.	.
		k_{32111}	.	.	.	.	560	.	.	.
		k_{22211}	.	.	.	.	420	.	.	.
		k_{311111}	.	.	.	.	.	56	.	.
		k_{221111}	.	.	.	.	.	210	.	.
		$k_{2111111}$	.	.	.	.	.	.	28	.
		$k_{11111111}$	.	.	.	.	.	.	.	1

The following abbreviations are used throughout this table:

$$n_0 = n^{-1}; \quad n_1 = (n+1)^{-1}; \quad n_2 = (n+5)^{-1}; \quad n_3 = (n^2+15n-4)^{-1};$$

$$n_4 = (n^3+42n^2+119n-42)^{-1}; \quad n_5 = (n^4+98n^3+659n^2-518n+120)^{-1}.$$

Thus

$$k_{32} = n_2(n-1)k_3k_2 - n_0n_2(n-1)k_5$$

is

$$k_{32} = \frac{n-1}{n+5}k_3k_2 - \frac{n-1}{n(n+5)}k_5.$$

The expressions for the poly-K's in terms of the single K's are the same as those given with the exception of the substitution of N for n.

2nd order

$$k_{11} = k_1^2 - n_0k_2.$$

3rd order

$$k_{21} = k_2k_1 - n_0k_3,$$
$$k_{111} = k_1^3 - 3n_0k_2k_1 + 2n_0^2k_3.$$

4th order

$$k_{31} = k_3k_1 - n_0k_4,$$
$$k_{22} = n_1(n-1)k_2^2 - n_0n_1(n-1)k_4,$$
$$k_{211} = k_2k_1^2 - n_0n_1(n-1)k_2^2 - 2n_0k_3k_1 + 2n_0n_1k_4,$$
$$k_{1111} = k_1^4 - 6n_0k_2k_1^2 + 3n_0^2n_1(n-1)k_2^2 + 8n_0^2k_3k_1 - 6n_0^2n_1k_4.$$

5th order

$$k_{41} = k_4k_1 - n_0k_5,$$
$$k_{32} = n_2(n-1)k_3k_2 - n_0n_2(n-1)k_5,$$
$$k_{311} = k_3k_1^2 - n_0n_2(n-1)k_3k_2 - 2n_0k_4k_1 + 2n_0^2n_2(n+2)k_5,$$
$$k_{221} = n_1(n-1)k_2^2k_1 - 2n_0n_2(n-1)k_3k_2 - n_0n_1(n-1)k_4k_1 + 2n_0^2n_2(n-1)k_5,$$
$$k_{2111} = k_2k_1^3 - 3n_0n_1(n-1)k_2^2k_1 - 3n_0k_3k_1^2 + 5n_0^2n_2(n-1)k_3k_2 + 6n_0n_1k_4k_1 - 6n_0^3n_2k_5,$$
$$k_{11111} = k_1^5 - 10n_0k_2k_1^3 + 15n_0^2n_1(n-1)k_2^2k_1 + 20n_0^2k_3k_1^2 - 20n_0^3n_2(n-1)k_3k_2 - 30n_0^2n_1k_4k_1 + 24n_0^3n_2k_5.$$

6th order

$$k_{51} = k_5k_1 - n_0k_6,$$
$$k_{42} = 36n_1^2n_3(n-1)^2k_2^3 - 6n_0n_1n_3(n-1)(n-2)(n+3)k_3^2 + n_1^2n_3(n-1)(n-2)(n+5)(n+7)k_4k_2$$
$$\quad - n_0n_1n_3(n-1)^2(n+4)k_6,$$
$$k_{411} = k_4k_1^2 - 2n_0k_5k_1 - 36n_0n_1^2n_3(n-1)^2k_2^3 - n_0n_1^2n_3(n-1)(n-2)(n+5)(n+7)k_4k_2$$
$$\quad + 6n_0^2n_1n_3(n-1)(n-2)(n+3)k_3^2 + 2n_0n_1n_3(n^2+9n+2)k_6,$$
$$k_{33} = n_0n_1n_3(n-1)(n-2)(n+3)(n+7)k_3^2 - 9n_1^2n_3(n-1)(n^2-9)k_4k_2 - 6n_1^2n_3(n-1)^2(n+7)k_2^3$$
$$\quad - n_0n_1n_3(n-1)(n^2-n+4)k_6,$$
$$k_{321} = n_2(n-1)k_3k_2k_1 + 6n_0n_1n_3(n-1)^2k_2^3 - n_0^2n_3(n-1)(n-2)(n+3)k_3^2$$
$$\quad - n_0n_1n_3(n-1)(n^2+11)k_4k_2 - n_0n_2(n-1)k_5k_1 + 2n_0n_3(n-1)k_6,$$
$$k_{3111} = k_3k_1^3 - 3n_0n_2(n-1)k_3k_2k_1 - 12n_0^2n_1^2n_3(n-1)^2(n-2)k_2^3$$
$$\quad + 2n_0^3n_1n_3(n-1)(n-2)^2(n+3)k_3^2 + 3n_0^2n_1^2n_3(n-1)^2(n^2+5n+16)k_4k_2$$
$$\quad - 3n_0k_4k_1^2 + 6n_0^2n_2(n+2)k_5k_1 - 6n_0n_1n_3(n+3)k_6,$$
$$k_{222} = n_0n_1^2n_3(n-1)^2(n-2)(n^2+15n+2)k_2^3 + 2n_0^2n_1n_3(n-1)(n-2)(7n+1)k_3^2$$
$$\quad - 3n_0n_1^2n_3(n-1)(n-2)(n^2+10n+1)k_4k_2 + 2n_0n_1n_3(n-1)(n-2)k_6,$$
$$k_{2211} = n_1(n-1)k_2^2k_1^2 - 4n_0n_2(n-1)k_3k_2k_1 - n_0^2n_1^2n_3(n-1)^2(n^3+13n^2-16n+8)k_2^3$$
$$\quad + 2n_0^3n_1n_3(n-1)^2(n-2)^2k_3^2 - n_0n_1(n-1)k_4k_1^2 + n_0^2n_1^2n_3(n-1)(5n^3+26n^2-35n+16)k_4k_2$$
$$\quad + 4n_0^2n_2(n-1)k_5k_1 - 6n_0n_1n_3(n-1)k_6,$$

$$k_{21111} = k_2 k_1^4 - 6n_0 n_1(n-1) k_2^2 k_1^2 + 3n_0^2 n_1^2 n_3(n-1)^2 (n^2+13n-12) k_2^3 - 4n_0 k_3 k_1^3$$
$$+ 20n_0^2 n_2(n-1) k_3 k_2 k_1 - 8n_0^3 n_1 n_3(n-1) (n-2)^2 k_3^2 + 12n_0 n_1 k_4 k_1^2$$
$$- 18n_0^2 n_1^2 n_3(n-1) (n^2+5n-4) k_4 k_2 - 24n_0^2 n_2 k_5 k_1 + 24n_0 n_1 n_3 k_6,$$

$$k_{111111} = k_1^6 - 15n_0 k_2 k_1^4 + 45n_0^2 n_1(n-1) k_2^2 k_1^2 - 15n_0^3 n_1^2 n_3(n-1)^2 (n^2+13n-12) k_2^3$$
$$+ 40n_0^2 k_3 k_1^3 - 120n_0^3 n_2(n-1) k_3 k_2 k_1 + 40n_0^4 n_1 n_3(n-1) (n-2)^2 k_3^2$$
$$- 90n_0^2 n_1 k_4 k_1^2 + 90n_0^3 n_1^2 n_3(n-1) (n^2+5n-4) k_4 k_2 + 144n_0^3 n_2 k_5 k_1 - 120n_0^2 n_1 n_3 k_6.$$

7th order

$$k_{61} = k_6 k_1 - n_0 k_7,$$

$$k_{52} = n_2 n_4(n-1) (n+11) (n^2+27n-70) k_5 k_2 - 20n_1 n_4(n-1) (n-2) (n+7) k_4 k_3$$
$$+ 720 n_1 n_2 n_4 n(n-1)^2 k_3 k_2^2 - n_0 n_4(n-1) (n^2+13n-18) k_7,$$

$$k_{511} = k_5 k_1^2 - 2n_0 k_6 k_1 - n_0 n_2 n_4(n-1) (n+11) (n^2+27n-70) k_5 k_2$$
$$+ 20n_0 n_1 n_4(n-1) (n-2) (n+7) k_4 k_3 - 720 n_1 n_2 n_4(n-1)^2 k_3 k_2^2 + 20n_0^2 n_4(n+2) (n^2+25n-6) k_7,$$

$$k_{43} = n_1 n_4(n-1) (n-2) (n+7) (n+9) k_4 k_3 - 36n_1 n_2 n_4 n(n-1)^2 (n+9) k_3 k_2^2$$
$$- 12n_2 n_4(n-1) (n-4) (n+7) k_5 k_2 - n_0 n_4(n-1) (n^2+n+6) k_7,$$

$$k_{421} = n_1^2 n_3(n-1) (n-2) (n+5) (n+7) k_4 k_2 k_1 - 6n_0 n_1 n_3(n-1) (n-2) (n+3) k_3^2 k_1$$
$$+ 36n_1^2 n_3(n-1)^2 k_2^3 k_1 - n_0 n_1 n_3(n-1)^2 (n+4) k_6 k_1 - n_0 n_2 n_4(n-1) (n^3+26n^2+191n-434) k_5 k_2$$
$$- n_0 n_1 n_4(n-1) (n-2) (n-11) (n+7) k_4 k_3 + 36n_1 n_2 n_4(n-1)^2 (n-11) k_3 k_2^2$$
$$+ 2n_0^2 n_4(n-1) (n^2+7n-6) k_7,$$

$$k_{4111} = k_4 k_1^3 - 3n_0 k_5 k_1^2 - 108 n_0 n_1^2 n_3(n-1)^2 k_2^3 k_1 + 18n_0^2 n_1 n_3(n-1) (n-2) (n+3) k_3^2 k_1$$
$$- 3n_0 n_1^2 n_3(n-1) (n-2) (n+5) (n+7) k_4 k_2 k_1 + 2n_0^2 n_1 n_4(n-1) (n-2) (n-21) (n+7) k_4 k_3$$
$$- 72n_0 n_1 n_2 n_4(n-1)^2 (n-21) k_3 k_2^2 + 3n_0^2 n_2 n_4(n-1) (n^3+30n^2+203n-546) k_5 k_2$$
$$+ 6n_0 n_1 n_3(n^2+9n+2) k_6 k_1 - 6n_0^2 n_4(n^2+13n+6) k_7,$$

$$k_{331} = n_0 n_1 n_3(n-1) (n-2) (n+3) (n+7) k_3^2 k_1 + 72n_1 n_2 n_4(n-1)^2 (n+9) k_3 k_2^2 - 6n_1^2 n_3(n-1)^2 (n+7) k_2^3 k_1$$
$$- 9n_1^2 n_3(n-1) (n-3) (n+3) k_4 k_2 k_1 - 2n_0 n_1 n_4(n-1) (n-2) (n+7) (n+9) k_4 k_3$$
$$+ 24n_0 n_2 n_4(n-1) (n-4) (n+7) k_5 k_2 - n_0 n_1 n_3(n-1) (n^2-n+4) k_6 k_1$$
$$+ 2n_0^2 n_4(n-1) (n^2+n+6) k_7,$$

$$k_{322} = n_1 n_2 n_4(n-1)^2 (n-2) (n^2+38n+21) k_3 k_2^2 - n_0 n_1 n_4(n-1) (n-2) (n^2-14n-7) k_4 k_3$$
$$- 2n_0 n_2 n_4(n-1) (n-2) (n^2+28n+7) k_5 k_2 + 2n_0 n_4(n-1) (n-2) k_7,$$

$$k_{3211} = n_2(n-1) k_3 k_2 k_1^2 - 2n_0^2 n_3(n-1) (n-2) (n+3) k_3^2 k_1 - 2n_0 n_1 n_3(n-1) (n^2+11) k_4 k_2 k_1$$
$$+ 12n_0 n_1 n_3(n-1)^2 k_2^3 k_1 - n_0 n_1 n_2 n_4(n-1)^2 (n^3+36n^2+53n+210) k_3 k_2^2$$
$$+ 2n_0^2 n_1 n_4(n-1) (n-2) (2n^2+7n+21) k_4 k_3 - n_0 n_2(n-1) k_5 k_1^2$$
$$+ 3n_0 n_2 n_4(n-1) (n^3+18n^2+7n+70) k_5 k_2 + 4n_0 n_3(n-1) k_6 k_1 - 6n_0^2 n_4(n-1) (n+2) k_7,$$

$$k_{31111} = k_3 k_1^4 - 4n_0 k_4 k_1^3 - 6n_0 n_2(n-1) k_3 k_2 k_1^2 - 48n_0^2 n_1^2 n_3(n-1)^2 (n-2) k_2^3 k_1$$
$$+ 8n_0^3 n_1 n_3(n-1) (n-2)^2 (n+3) k_3^2 k_1 + 3n_0^2 n_1 n_2 n_4(n-1)^2 (n^3+36n^2+77n-294) k_3 k_2^2$$
$$+ 12n_0^2 n_1^2 n_3(n-1)^2 (n^2+5n+16) k_4 k_2 k_1 - 14n_0^3 n_1 n_4(n-1) (n-2) (n-3) (n+4) k_4 k_3$$
$$+ 12n_0^2 n_2(n+2) k_5 k_1^2 - 12n_0^3 n_2 n_4(n-1) (n^3+21n^2+56n-84) k_5 k_2$$
$$- 24n_0 n_1 n_3(n+3) k_6 k_1 + 24n_0^2 n_4(n+4) k_7,$$

$$k_{2221} = n_0 n_1^2 n_3(n-1)^2 (n-2) (n^2+15n+2) k_2^3 k_1 - 3n_0 n_1 n_2 n_4(n-1)^2 (n-2) (n^2+38n+21) k_3 k_2^2$$
$$+ 2n_0^2 n_1 n_3(n-1) (n-2) (7n+1) k_3^2 k_1 - 3n_0 n_1^2 n_3(n-1) (n-2) (n^2+10n+1) k_4 k_2 k_1$$
$$+ 3n_0^2 n_1 n_4(n-1) (n-2) (n^2-14n-7) k_4 k_3 + 6n_0^2 n_2 n_4(n-1) (n-2) (n^2+28n+7) k_5 k_2$$
$$+ 2n_0 n_1 n_3(n-1) (n-2) k_6 k_1 - 6n_0^2 n_4(n-1) (n-2) k_7,$$

$$k_{22111} = n_1(n-1) k_2^2 k_1^3 - 6n_0 n_2(n-1) k_3 k_2 k_1^2 - n_0 n_1(n-1) k_4 k_1^3 + 6n_0^2 n_2(n-1) k_5 k_1^2$$
$$- 3n_0^2 n_1^2 n_3(n-1)^2 (n^3+13n^2-16n+8) k_2^3 k_1 + 8n_0^3 n_1^2 n_2 n_4(n-1)^2 (n^3+36n^2-28n+21) k_3 k_2^2$$
$$+ 6n_0^3 n_1 n_3(n-1)^2 (n-2)^2 k_3^2 k_1 + 3n_0^2 n_1^2 n_3(n-1) (5n^3+26n^2-35n+16) k_4 k_2 k_1$$
$$- 14n_0^3 n_1 n_4(n-1)^2 (n-2) (n-3) k_4 k_3 - 6n_0^3 n_2 n_4(n-1) (3n^3+70n^2-91n+42) k_5 k_2$$
$$- 18n_0 n_1 n_3(n-1) k_6 k_1 + 24n_0^2 n_4(n-1) k_7,$$

$$k_{211111} = k_2 k_1^5 - 5n_0 k_3 k_1^4 - 10n_0 n_1(n-1) k_2^2 k_1^3 + 15n_0^2 n_1^2 n_3(n-1)^2 (n^2+13n-12) k_2^3 k_1$$
$$+ 50n_0^2 n_2(n-1) k_3 k_2 k_1^2 - 40n_0^3 n_1 n_3(n-1) (n-2)^2 k_3^2 k_1 + 20n_0 n_1 k_4 k_1^3 + 120n_0 n_3 k_6 k_1$$

$$-90n_0^2 n_1^2 n_3(n-1)(n^2+5n-4)k_4 k_2 k_1 -60n_0^2 n_2 k_5 k_1^2 +84n_0^3 n_2 n_4(n-1)(n^2+23n-18)k_5 k_2$$
$$+70n_0^3 n_1 n_4(n-1)(n-2)(n-3)k_4 k_3 -35n_0^3 n_1 n_2 n_4(n-1)^2(n^3+36n^2-19n-6)k_3 k_2^2$$
$$-120n_0^2 n_4 k_7,$$

$$k_{1111111} = k_1^7 -21n_0 k_2 k_1^5 +105n_0^2 n_1(n-1)k_2^2 k_1^3 +70n_0^2 k_3 k_1^4 -420n_0^3 n_2(n-1)k_3 k_2 k_1^2$$
$$-105n_0^3 n_1^2 n_3(n-1)^2(n^2+13n-12)k_2^3 k_1 +210n_0^4 n_1 n_2 n_4(n-1)^2(n^3+36n^2-19n-6)k_3 k_2^2$$
$$+280n_0^4 n_1 n_3(n-1)(n-2)^2 k_3^2 k_1 -210n_0^2 n_1 k_4 k_1^3 +630n_0^3 n_1^2 n_3(n-1)(n^2+5n-4)k_4 k_2 k_1$$
$$-420n_0^4 n_1 n_4(n-1)(n-2)(n-3)k_4 k_3 -504n_0^4 n_2 n_4(n-1)(n^2-23n-18)k_5 k_2$$
$$+504n_0^3 n_2 k_5 k_1^2 -840n_0^3 n_1 n_3 k_6 k_1 +720n_0^3 n_4 k_7.$$

8th order

$k_{71} \quad = k_7 k_1 - n_0 k_8,$

$k_{62} \quad = 60n_0^2 n_1^2 n_2 n_3 n_5(n-1)^2(n-2)(93n^6+930n^5+226n^4-7552n^3-5479n^2+7582n-1560)k_3^2 k_2$
$\qquad + 120n_0 n_1^3 n_3 n_5(n-1)^3(4n^5-183n^4-578n^3+2571n^2-1766n+312)k_2^4$
$\qquad + 180n_0 n_1^3 n_3 n_5(n-1)^2(18n^6-41n^5+565n^4+1295n^3-5223n^2+3186n-520)k_4 k_2^2$
$\qquad - 20n_0 n_1^3 n_5(n-1)(n-2)(n-3)(n^4+17n^3-49n^2-227n+78)k_4^2$
$\qquad - 30n_0^2 n_1^2 n_2 n_5(n-1)(n-2)(n^6+20n^5+200n^4+274n^3-945n^2-726n+312)k_5 k_3$
$\qquad + n_1^2 n_3 n_5(n-1)(n^7+104n^6+1978n^5+440n^4-14591n^3-35344n^2+74532n-18480)k_6 k_2$
$\qquad - n_0 n_1^2 n_5(n-1)^2(n^4+40n^3+77n^2-82n-120)k_8,$

$k_{611} \quad = k_6 k_1^2 -2n_0 k_7 k_1 -60n_0^3 n_1^2 n_2 n_3 n_5(n-1)^2(n-2)(93n^6+930n^5+226n^4-7552n^3-5479n^2$
$\qquad +7582n-1560)k_3^2 k_2 -120n_0^2 n_1^3 n_3 n_5(n-1)^3(4n^5-183n^4-578n^3+2571n^2-1766n+312)k_2^4$
$\qquad -180n_0^2 n_1^3 n_3 n_5(n-1)^2(18n^6-41n^5+565n^4+1295n^3-5223n^2+3186n-520)k_4 k_2^2$
$\qquad +20n_0^2 n_1^3 n_5(n-1)(n-2)(n-3)(n^4+17n^3-49n^2-227n+78)k_4^2$
$\qquad +30n_0^3 n_1^2 n_2 n_5(n-1)(n-2)(n^6+20n^5+200n^4+274n^3-945n^2-726n+312)k_5 k_3$
$\qquad -n_0 n_1^2 n_3 n_5(n-1)(n^7+104n^6+1978n^5+440n^4-14591n^3-35344n^2+74532n-18480)k_6 k_2$
$\qquad +2n_0 n_1 n_5(n^4+68n^3+359n^2-8n-60)k_8,$

$k_{53} \quad = 180n_0 n_1^3 n_3 n_5(n-1)^3(7n^5+87n^4+119n^3-1695n^2+1362n-264)k_2^4$
$\qquad -90n_0^2 n_1^2 n_2 n_3 n_5(n-1)^2(n-2)(n^7+5n^6-136n^5-1554n^4-6131n^3-2763n^2$
$\qquad +5754n-1320)k_3^2 k_2$
$\qquad -30n_0 n_1^2 n_3 n_5(n-1)^2(2n^7+17n^6+530n^5+2918n^4+2246n^3-31519n^2+22278n-39660)k_4 k_2^2$
$\qquad -30n_0 n_1^3 n_5(n-1)(n-2)(n-3)(n+11)(n^3+6n^2+15n-6)k_4^2$
$\qquad +n_0^2 n_1^2 n_2 n_5(n-1)(n-2)(n^7+63n^6+868n^5+3306n^4-2795n^3-38865n^2$
$\qquad -21690n+11880)k_5 k_3$
$\qquad -15n_1^2 n_3 n_5(n-1)(n^6+14n^5+86n^4-812n^3-1503n^2+5502n-1560)k_6 k_2$
$\qquad -n_0 n_1^2 n_5(n-1)(n^5+11n^4-3n^3+85n^2-22n-120)k_8,$

$k_{521} \quad = n_2 n_4(n-1)(n+11)(n^2+27n-70)k_5 k_2 k_1 -20n_1 n_4(n-1)(n-2)(n+7)k_4 k_3 k_1$
$\qquad +720n_1 n_2 n_4 n(n-1)^2 k_3 k_2^2 k_1$
$\qquad -n_0 n_4(n-1)(n^2+13n-18)k_7 k_1 -60n_0^2 n_1^2 n_3 n_5(n-1)^3(29n^4-134n^3-665n^2+722n-168)k_2^4$
$\qquad +30n_0^3 n_1 n_2 n_3 n_5(n-1)^2(n-2)(3n^6-174n^5-2094n^4-3020n^3-269n^2+2938n-840)k_3^2 k_2$
$\qquad +30n_0^2 n_1^2 n_3 n_5(n-1)^2(2n^6-93n^5+869n^4-1341n^3-4183n^2+4002n-840)k_4 k_2^2$
$\qquad +10n_0^2 n_1^2 n_5(n-1)(n-2)(n-3)(5n^3+80n^2+65n-42)k_4^2$
$\qquad -n_0^3 n_2 n_5(n-1)(n-2)(n^5+31n^4+205n^3-3135n^2-4950n-2520)k_5 k_3$
$\qquad -n_0 n_1 n_3 n_5(n-1)(n^6+88n^5+1680n^4-2530n^3+119n^2-12918n+4920)k_6 k_2$
$\qquad +2n_0 n_5(n-1)(n^2+23n-30)k_8,$

$k_{5111} \quad = k_5 k_1^3 -3n_0 k_6 k_1^2 -3n_0 n_2 n_4(n-1)(n+11)(n^2+27n-70)k_5 k_2 k_1$
$\qquad +60n_0 n_1 n_4(n-1)(n-2)(n+7)k_4 k_3 k_1 -2160n_1 n_2 n_4(n-1)^2 k_3 k_2^2 k_1 +6n_0^2 n_4(n+2)$
$\qquad \times (n^2+25n-6)k_7 k_1$
$\qquad +360n_0^2 n_1^2 n_3 n_5(n-1)^3(11n^5-96n^4-459n^3+876n^2-404n+48)k_2^4$
$\qquad -180n_0^4 n_1^2 n_2 n_3 n_5(n-1)^2(n-2)(n^7-88n^6-1066n^5-1780n^4+1421n^3+2716n^2-1828n$
$\qquad +240)k_3^2 k_2$

$$-60n_0^3 n_1^3 n_3 n_5 (n-1)^2 (2n^7 - 145n^6 + 899n^5 - 2167n^4 - 9409n^3 + 15488n^2 - 6396n + 720) k_4 k_2^2$$
$$-120n_0^3 n_1^3 n_5 (n-1)(n-2)(n-3)(n^4 + 17n^3 + 16n^2 - 34n + 6) k_4^2$$
$$+2n_0^4 n_1 n_2 n_5 (n-1)(n-2)(n^6 + 17n^5 - 49n^4 - 5645n^3 - 9480n^2 + 13140n - 2160) k_5 k_3$$
$$+3n_0^2 n_1^2 n_3 n_5 (n-1)(n^7 + 94n^6 + 1838n^5 - 420n^4 - 6471n^3 - 20314n_2 + 19512n - 2880) k_6 k_2$$
$$-6n_0 n_1 n_5 (n^3 + 38n^2 + 99n - 18) k_8,$$

$$k_{44} = -24n_0 n_1^3 n_3 n_5 (n-1)^3 (n^6 + 44n^5 + 331n^4 + 392n^3 - 12440n^2 + 10544n - 2112) k_2^4$$
$$-48n_0^2 n_1^2 n_2 n_3 n_5 (n-1)^2 (n-2)(3n^7 + 108n^6 + 1036n^5 + 4054n^4 + 11709n^3 + 4478n^2$$
$$- 11068n + 2640) k_3^2 k_2$$
$$-72n_0 n_1^3 n_3 n_5 (n-1)^2 (n^7 + 9n^6 - 317n^5 - 885n^4 - 276n^3 + 12916n^2 - 9608n + 1760) k_4 k_2^2$$
$$+n_0 n_1^3 n_5 (n-1)(n-2)(n-3)(n+11)(n^4 + 62n^3 + 221n^2 + 448n - 192) k_4^2$$
$$-48n_0^2 n_1^2 n_2 n_5 (n-1)(n-2)(n^6 + 14n^5 + 26n^4 - 152n^3 - 867n^2 - 438n + 264) k_5 k_3$$
$$-16n_1^2 n_3 n_5 (n-1)(n^6 + 8n^5 - 182n^4 + 554n^3 + 1141n^2 - 5242n + 1560) k_6 k_2$$
$$-n_0 n_1^2 n_5 (n-1)(n^5 + 3n^4 + 37n^3 - 51n^2 + 34n + 120) k_8,$$

$$k_{431} = n_1 n_4 (n-1)(n-2)(n+7)(n+9) k_4 k_3 k_1 - 36n_1 n_2 n_4 n(n-1)^2 (n+9) k_3 k_2^2 k_1$$
$$-12n_2 n_4 (n-1)(n-4)(n+7) k_5 k_2 k_1 - n_0 n_4 (n-1)(n^2 + n + 6) k_7 k_1$$
$$+12n_0^2 n_1^2 n_3 n_5 (n-1)^3 (2n^5 - 19n^4 - 624n^3 - 377n^2 + 922n - 264) k_2^4$$
$$+6n_0^3 n_1 n_2 n_3 n_5 (n-1)^2 (n-2)(39n^6 + 900n^5 + 5348n^4 + 3774n^3 - 2067n^2 - 3554n + 1320) k_3^2 k_2$$
$$+6n_0^2 n_1^2 n_3 n_5 (n-1)^2 (22n^6 + 171n^5 - 1325n^4 + 5295n^3 + 2623n^2 - 5226n + 1320) k_4 k_2^2$$
$$-n_0^2 n_1^2 n_5 (n-1)(n-2)(n-3)(n+11)(n^3 + 31n^2 + 10n - 12) k_4^2$$
$$-n_0^3 n_1 n_2 n_5 (n-1)(n-2)(n^6 + 14n^5 + 182n^4 + 1876n^3 + 2625n^2 + 126n - 792) k_5 k_3$$
$$+n_0 n_1 n_3 n_5 (n-1)(31n^5 + 307n^4 - 1929n^3 - 1387n^2 - 2902n + 1560) k_6 k_2$$
$$+2n_0 n_5 (n-1)(n+2)(n+3) k_8,$$

$$k_{4211} = n_1^2 n_3 (n-1)(n-2)(n+5)(n+7) k_4 k_2 k_1^2 - 6n_0 n_1 n_3 (n-1)(n-2)(n+3) k_3^2 k_1^2$$
$$+36n_1^2 n_3 (n-1)^2 k_2^3 k_1^2 - n_0 n_1 n_3 (n-1)^2 (n+4) k_6 k_1^2 + 72n_1 n_2 n_4 (n-1)^2 (n-11) k_3 k_2^2 k_1$$
$$-2n_0 n_1 n_4 (n-1)(n-2)(n+7)(n-11) k_4 k_3 k_1 - 2n_0 n_2 n_4 (n-1)(n^3 + 26n^2 + 191n - 434) k_5 k_2 k_1$$
$$+4n_0^2 n_4 (n-1)(n^2 + 7n - 6) k_7 k_1 - 24n_0^3 n_1^3 n_3 n_5 (n-1)^4 (5n^5 + 202n^4 + 1353n^3$$
$$- 2344n^2 + 1300n - 240) k_2^4$$
$$+12n_0^4 n_1^2 n_2 n_3 n_5 (n-1)^2 (n-2)(n^8 + 57n^7 + 724n^6 + 2092n^5 - 6785n^4 - 3809n^3$$
$$+ 11820n^2 - 6740n + 1200) k_3^2 k_2$$
$$-n_0^3 n_1^3 n_3 n_5 (n-1)^2 (n^9 + 100n^8 + 1176n^7 - 450n^6 - 13701n^5 - 80850n^4 + 26164n^3$$
$$- 247840n^2 + 99120n - 14400) k_4 k_2^2$$
$$+2n_0^3 n_1^3 n_5 (n-1)(n-2)(n-3)(n^5 + 18n^4 + 49n^3 + 432n^2 - 440n + 120) k_4^2$$
$$+2n_0^4 n_1 n_2 n_5 (n-1)(n-2)(n^6 - 23n^5 - 297n^4 - 661n^3 + 3632n^2 - 2940n + 720) k_5 k_3$$
$$+3n_0^2 n_1^2 n_3 n_5 (n-1)^2 (n^6 + 75n^5 + 861n^4 + 117n^3 - 3766n^2 + 3624n - 960) k_6 k_2$$
$$-6n_0 n_1 n_5 (n-1)(n^2 + 11n - 6) k_8,$$

$$k_{41111} = k_4 k_1^4 - 4n_0 k_5 k_1^3 - 216n_0 n_1^2 n_3 (n-1)^2 k_2^3 k_1^2 + 36n_0^2 n_1 n_3 (n-1)(n-2)(n+3) k_3^2 k_1^2$$
$$-6n_0 n_1^2 n_3 (n-1)(n-2)(n+5)(n+7) k_4 k_2 k_1^2 + 8n_0^2 n_1 n_4 (n-1)(n-2)(n+7)(n+21) k_4 k_3 k_1$$
$$-288n_0 n_1 n_2 n_4 (n-1)^2 (n-21) k_3 k_2^2 k_1 + 12n_0^2 n_2 n_4 (n-1)(n^3 + 30n^2 + 203n - 546) k_5 k_2 k_1$$
$$+12n_0 n_1 n_3 (n^2 + 9n + 2) k_6 k_1^2 - 24n_0^2 n_4 (n^2 + 13n + 6) k_7 k_1$$
$$+72n_0^3 n_1^3 n_3 n_5 (n-1)^3 (5n^5 + 142n^4 + 1631n^3 - 1402n^2 - 736n + 480) k_2^4$$
$$-36n_0^4 n_1^2 n_2 n_3 n_5 (n-1)^2 (n-2)(n^7 + 52n^6 + 1164n^5 + 7422n^4 + 2115n^3 - 10914n^2$$
$$- 1760n + 2400) k_3^2 k_2$$
$$+3n_0^3 n_1^3 n_3 n_5 (n-1)^2 (n^8 + 100n^7 + 1216n^6 - 3350n^5 + 4279n^4 - 124190n^3$$
$$+ 72984n^2 + 61920n - 28800) k_4 k_2^2$$
$$-6n_0^3 n_1^3 n_5 (n-1)(n-2)(n-3)(n^4 - 2n^3 - 291n^2 + 112n + 240) k_4^2$$
$$-8n_0^4 n_1 n_2 n_5 (n-1)(n-2)(n^5 - 13n^4 - 235n^3 - 1907n^2 + 354n + 1080) k_5 k_3$$
$$-12n_0^2 n_1^2 n_3 n_5 (n-1)(n^6 + 79n^5 + 1049n^4 - 663n^3 - 4530n^2 + 464n + 1440) k_6 k_2$$
$$+24n_0 n_1 n_5 (n^2 + 17n + 12) k_8,$$

$$k_{332} = -6n_0^2 n_1^3 n_3 n_5 (n-1)^2 (n-2) (n^6 + 97n^5 + 1027n^4 + 3483n^3 - 3596n^2 + 620n + 48) k_2^4$$
$$+ n_0^3 n_1^2 n_2 n_3 n_5 (n-1)^2 (n-2) (n^9 + 103n^8 + 1360n^7 + 4102n_6 - 15011n_5 - 105929n^4 - 59478n^3$$
$$+ 103164n^2 - 19032n - 1440) k_3^2 k_2 - 3n_0^2 n_1^3 n_3 n_5 (n-1)^2 (n-2) (3n^7 + 230n^6 - 12n^5$$
$$- 6910n^4 - 22683n^3 + 19640n^2 - 2988n - 240) k_4 k_2^2 + 2n_0^2 n_1^3 n_5 (n-1) (n-2) (n-3)$$
$$\times (31n^4 + 212n^3 + 491n^2 - 182n - 12) k_4^2 - 2n_0^3 n_1 n_2 n_5 (n-1) (n-2) (n^6 + 50n^5 + 242n^4$$
$$- 344n^3 - 3387n^2 + 1062n + 72) k_5 k_3 - n_0 n_1^2 n_3 n_5 (n-1) (n-2) (n^6 + 58n^5 - 294n^4$$
$$- 1136n^3 + 7133n^2 - 2154n - 120) k_6 k_2 + 2n_0 n_1 n_5 (n-1) (n-2) (n^2 - n + 6) k_8,$$

$$k_{3311} = n_0 n_1 n_3 (n-1) (n-2) (n+3) (n+7) k_3^2 k_1^2 + 144 n_1 n_2 n_4 (n-1)^2 (n+9) k_3 k_2^2 k_1$$
$$- 6n_1^2 n_3 (n-1)^2 (n+7) k_2^3 k_1^2 - 9n_1^2 n_3 (n-1) (n-3) (n+3) k_4 k_2 k_1^2$$
$$- 4n_0 n_1 n_4 (n-1) (n-2) (n+7) (n+9) k_4 k_3 k_1 + 48 n_0 n_2 n_4 (n-1) (n-4) (n+7) k_5 k_2 k_1$$
$$- n_0 n_1 n_3 (n-1) (n^2 - n + 4) k_6 k_1^2 + 4n_0^2 n_4 (n-1) (n^2 + n + 6) k_7 k_1$$
$$+ 6n_0^3 n_1^3 n_3 n_5 (n-1)^3 (n^7 + 87n^6 + 901n^5 + 4001n^4 - 6558n^3 + 5632n^2 - 3824n + 960) k_2^4$$
$$- n_0^4 n_1^2 n_2 n_3 n_5 (n-1)^2 (n-2) (n^9 + 103n^8 + 1828n^7 + 15370n^6 + 59965n^5 + 3535n^4 - 38994n^3$$
$$+ 35712n^2 - 45840n + 1400) k_3^2 k_2 + 3n_0^3 n_1^3 n_3 n_5 (n-1)^2 (3n^8 + 136n^7 - 1244n^6 - 2270n^5$$
$$- 24743n^4 + 33334n^3 - 31856n^2 + 21360n - 4800) k_4 k_2^2$$
$$+ 2n_0^3 n_1^3 n_5 (n-1) (n-2) (n-3) (n^5 + 12n^4 + 181n^3 - 42n^2 + 148n - 120) k_4^2$$
$$+ 4n_0^4 n_1 n_2 n_5 (n-1) (n-2) (n^6 + 32n^5 + 212n^4 + 766n^3 - 381n^2 + 594n - 360) k_5 k_3$$
$$+ n_0^2 n_1^2 n_3 n_5 (n-1) (n^7 - 6n^6 - 1086n^5 + 4968n^4 + 11493n^3 - 7842n^2 + 6872n - 2880) k_6 k_2$$
$$- 6n_0^2 n_1 n_5 (n-1) (n^3 - 5n^2 + 10n - 8) k_8,$$

$$k_{3221} = n_1 n_2 n_4 (n-1)^2 (n^2 + 38n + 21) k_3 k_2^2 k_1 + 2n_0 n_4 (n-1) (n-2) k_7 k_1 - n_0 n_1 n_4 (n-1) (n-2)$$
$$\times (n^2 - 14n - 7) k_4 k_3 k_1 - 2n_0 n_2 n_4 (n-1) (n-2) (n^2 + 28n + 7) k_5 k_2 k_1$$
$$+ 12n_0^2 n_1^2 n_3 n_5 (n-1)^3 (n-2) (n^4 + 88n^3 + 367n^2 - 448n + 112) k_2^4$$
$$- 2n_0^3 n_1 n_2 n_3 n_5 (n-1)^2 (n-2) (n^7 + 96n^6 + 760n^5 - 1254n^4 - 11033n^3 - 4842n^2$$
$$+ 12432n - 3360) k_3^2 k_2$$
$$- n_0^2 n_1^2 n_3 n_5 (n-1)^2 (n-2) (n^6 + 83n^5 - 275n^4 + 3845n^3 + 14554n^2 - 15088n + 3360) k_4 k_2^2$$
$$+ n_0^2 n_1^2 n_5 (n-1) (n-2) (n-3) (n^3 - 82n^2 - 211n + 112) k_4^2$$
$$+ 4n_0^3 n_2 n_5 (n-1) (n-2) (n^4 + 26n^3 - 37n^2 - 374n + 168) k_5 k_3 + 4n_0 n_1 n_3 n_5 (n-1) (n-2)$$
$$\times (n^4 + 70n^3 + 99n^2 + 350n - 160) k_6 k_2 - 6n_0 n_5 (n-1) (n-2) k_8,$$

$$k_{32111} = n_2 (n-1) k_3 k_2 k_1^3 - 3n_0^2 n_3 (n-1) (n-2) (n+3) k_3^2 k_1^2 - 3n_0 n_1 n_3 (n-1) (n^2 + 11) k_4 k_2 k_1^2$$
$$+ 18 n_0 n_1 n_3 (n-1)^2 k_2^3 k_1^2 - 3n_0 n_1 n_2 n_4 (n-1)^2 (n^3 + 36n^2 + 53n + 210) k_3 k_2^2 k_1$$
$$+ 6n_0^2 n_1 n_4 (n-1) (n-2) (2n^2 + 7n + 21) k_4 k_3 k_1 - n_0 n_2 (n-1) k_5 k_1^3$$
$$+ 9n_0^2 n_2 n_4 (n-1) (n^3 + 18n^2 + 7n + 70) k_5 k_2 k_1 + 6n_0 n_3 (n-1) k_6 k_1^2 - 18 n_0^2 n_4 (n-1) (n+2) k_7 k_1$$
$$- 6n_0^3 n_1^3 n_3 n_5 (n-1)^3 (5n^6 + 415n^5 + 1221n^4 - 4567n^3 + 7534n^2 - 5808n + 1440) k_2^4$$
$$+ n_0^4 n_1^2 n_2 n_3 n_5 (n-1)^2 (n-2) (5n^8 + 479n^7 + 4568n^6 + 4706n^5 - 14287n^4 + 21455n^3 + 37074n^2$$
$$- 69840n + 21600) k_3^2 k_2 + 3n_0^3 n_1^3 n_3 n_5 (n-1)^3 (n^8 + 85n^7 + 16n^6 + 373n^5 + 5209n^4$$
$$- 27095n^3 + 46614n^2 - 32880n + 7200) k_4 k_2^2 - 6n_0^2 n_1^3 n_5 (n-1)^2 (n-2) (n-3)$$
$$\times (n^3 - 16n^2 - 37n + 60) k_4^2$$
$$- 2n_0^4 n_1 n_2 n_5 (n-1) (n-2) (7n^5 + 149n^4 + 83n^3 - 773n^2 + 2046n - 1080) k_5 k_3$$
$$- 12 n_0^2 n_1^2 n_3 n_5 (n-1) (n^6 + 64n^5 + 124n^4 + 302n^3 - 485n^2 + 834n - 360) k_6 k_2$$
$$+ 24 n_0^3 n_1 n_5 (n-1)^2 (n^2 + 2n + 2) k_8,$$

$$k_{311111} = k_3 k_1^5 - 5n_0 k_4 k_1^4 - 10 n_0 n_2 (n-1) k_3 k_2 k_1^3 - 120 n_0^2 n_1^2 n_3 (n-1)^2 (n-2) k_2^3 k_1^2$$
$$+ 20 n_0^3 n_1 n_3 (n-1) (n-2)^2 (n+3) k_3^2 k_1^2 + 15 n_0^2 n_1 n_2 n_4 (n-1)^2 (n^3 + 36n^2 + 77n - 294) k_3 k_2^2 k_1$$
$$+ 30 n_0^2 n_1^2 n_3 (n-1)^2 (n^2 + 5n + 16) k_4 k_2 k_1^2 - 70 n_0^3 n_1 n_4 (n-1) (n-2) (n-3) (n+4) k_4 k_3 k_1$$
$$+ 20 n_0^2 n_2 (n+2) k_5 k_1^3 - 60 n_0^3 n_2 n_4 (n-1) (n^3 + 21n^2 + 56n - 84) k_5 k_2 k_1 - 60 n_0 n_1 n_3 (n+3) k_6 k_1^2$$
$$+ 120 n_0^2 n_4 (n+4) k_7 k_1 + 120 n_0^3 n_1^3 n_3 n_5 (n-1)^3 (n^5 + 80n^4 + 159n^3 - 1892n^2 + 2348n - 720) k_2^4$$
$$- 20 n_0^4 n_1^2 n_2 n_3 n_5 (n-1)^2 (n-2) (n^7 + 94n^6 + 820n^5 - 1154n^4 - 16217n^3 + 484n^2 + 27060n$$
$$- 10800) k_3^2 k_2 - 15 n_0^3 n_1^3 n_3 n_5 (n-1)^2 (n^7 + 88n^6 + 256n^5 + 2314n^4 + 5023n^3 - 46514n^2$$

$$+ 51888n - 13920) \, k_4 k_2^2 + 30n_0^3 n_1^3 n_5 (n-1)(n-2)(n-3)(n^3 - 14n^2 - 75n + 100) \, k_4^2$$
$$+ 64n_0^2 n_1 n_2 n_5 (n-1)(n-2)(n^4 + 17n^3 - 19n^2 - 335n + 300) \, k_5 k_3 + 60n_0^2 n_1^2 n_3 n_5 (n-1)$$
$$\times (n^5 + 67n^4 + 309n^3 + 109n^2 - 1296n + 760) \, k_6 k_2 - 24n_0^4 n_1 n_5 (5n^4 + 17n^3 + 8n^2 - 8n + 8) \, k_8,$$

$$k_{2222} = n_0^2 n_1^3 n_3 n_5 (n-1)^3 (n-2)(n-3)(n^6 + 112n^5 + 2035n^4 + 8980n^3 - 8000n^2 + 448n + 384) \, k_2^4$$
$$+ 8n_0^3 n_1^2 n_2 n_3 n_5 (n-1)^2 (n-2)(n-3)(7n^6 + 619n^5 + 5173n^4 + 3709n^3 - 5060n^2$$
$$+ 352n + 240) \, k_3^2 k_2 - 6n_0^2 n_1^3 n_3 n_5 (n-1)^2 (n-2)(n-3)(n^6 + 103n^5 + 1235n^4 + 4825n^3 - 3556n^2$$
$$+ 112n + 160) \, k_4 k_2^2 + 3n_0^2 n_1^3 n_5 (n-1)(n-2)(n-3)(n^4 + 37n^3 + 451n^2 - 97n - 32) \, k_4^2$$
$$- 96n_0^3 n_1 n_2 n_5 (n-1)(n-2)(n-3)(3n^3 + 29n^2 - 6n - 2) \, k_5 k_3$$
$$+ 8n_0 n_1^2 n_3 n_5 (n-1)(n-2)(n-3)(n^4 + 84n^3 + 379n^2 - 84n - 20) \, k_6 k_2$$
$$- 6n_0 n_1 n_5 (n-1)(n-2)(n-3) \, k_8,$$

$$k_{22211} = n_0 n_1^2 n_3 (n-1)^2 (n-2)(n^2 + 15n + 2) \, k_2^3 k_1^2 - 6n_0 n_1 n_2 n_4 (n-1)^2 (n-2)(n^2 + 38n + 21) \, k_3 k_2^2 k_1$$
$$+ 2n_0^2 n_1 n_3 (n-1)(n-2)(7n+1) \, k_3^2 k_1^2 - 3n_0 n_1^2 n_3 (n-1)(n-2)(n^2 + 10n + 1) \, k_4 k_2 k_1^2$$
$$+ 6n_0^2 n_1 n_4 (n-1)(n-2)(n^2 - 14n - 7) \, k_4 k_3 k_1 + 12n_0^2 n_2 n_4 (n-1)(n-2)(n^2 + 28n + 7) \, k_5 k_2 k_1$$
$$+ 2n_0 n_1 n_3 (n-1)(n-2) \, k_6 k_1^2 - 12n_0^2 n_4 (n-1)(n-2) \, k_7 k_1 - n_0^3 n_1^3 n_3 n_5 (n-1)^3 (n-2)$$
$$\times (n^7 + 109n^6 + 1735n^5 + 6079n^4 - 18560n^3 + 21532n^2 - 13056n + 2880) \, k_2^4$$
$$+ 2n_0^4 n_1^2 n_2 n_3 n_5 (n-1)^2 (n-2)(3n^8 + 263n^7 + 176n^6 - 14746n^5 + 10379n^4 + 17123n^3$$
$$- 39358n^2 + 30480n - 7200) \, k_3^2 k_2 + 3n_0^3 n_1^3 n_3 n_5 (n-1)^2 (n-2)(3n^7 + 284n^6 + 1660n^5$$
$$+ 5810n^4 - 17663n^3 + 21026n^2 - 12080n + 2400) \, k_4 k_2^2 - 6n_0^3 n_1^3 n_5 (n-1)^2 (n-2)$$
$$\times (n-3)(n^3 - 21n^2 + 58n - 40) \, k_4^2 - 12n_0^4 n_1 n_2 n_5 (n-1)(n-2)(n^5 + 3n^4 - 171n^3 + 333n^2$$
$$- 334n + 120) \, k_5 k_3 - 4n_0^2 n_1^2 n_3 n_5 (n-1)(n-2)(5n^5 + 375n^4 + 761n^3 - 1095n^2 + 1034n$$
$$- 360) \, k_6 k_2 + 24n_0 n_1 n_5 (n-1)(n-2) \, k_8,$$

$$k_{221111} = n_1 (n-1) \, k_2^2 k_1^4 - 8n_0 n_2 (n-1) \, k_3 k_2 k_1^3 - n_0 n_1 (n-1) \, k_4 k_1^4 + 8n_0^2 n_2 (n-1) \, k_5 k_1^3$$
$$- 6n_0^2 n_1^2 n_3 (n-1)^2 (n^3 + 13n^2 - 16n + 8) \, k_2^3 k_1^2 + 32n_0^2 n_1 n_2 n_4 (n-1)^2 (n^3 + 36n^2 - 28n + 21) \, k_3 k_2^2 k_1$$
$$+ 12n_0^3 n_1 n_3 (n-1)^2 (n-2)^2 \, k_3^2 k_1^2 + 6n_0^2 n_1^2 n_3 (n-1)(5n^3 + 26n^2 - 35n + 16) \, k_4 k_2 k_1^2$$
$$- 56n_0^3 n_1 n_4 (n-1)^2 (n-2)(n-3) \, k_4 k_3 k_1 - 24n_0^3 n_2 n_4 (n-1)(3n^3 + 70n^2 - 91n + 42) \, k_5 k_2 k_1$$
$$- 36n_0 n_1 n_3 (n-1) \, k_6 k_1^2 + 96n_0^2 n_4 (n-1) \, k_7 k_1 + 3n_0^3 n_1^3 n_3 n_5 (n-1)^3 (n^7 + 107n^6$$
$$+ 1537n^5 + 4269n^4 - 25834n^3 - 40384n^2 - 25984n + 5760) \, k_2^4$$
$$- 4n_0^4 n_1^2 n_2 n_3 n_5 (n-1)^2 (n-2)(7n^7 + 634n^6 + 2548n^5 - 19766n^4 + 8425n^3 + 36412n^2$$
$$- 40500n + 10800) \, k_3^2 k_2 - 3n_0^3 n_1^3 n_3 n_5 (n-1)^2 (11n^7 + 1004n^6 + 3308n^5 + 14930n^4$$
$$- 77431n^3 + 114866n^2 - 69168n + 13920) \, k_4 k_2^2 + 30n_0^3 n_1^3 n_5 (n-1)^2 (n-2)(n-3)$$
$$\times (n^2 - 19n + 20) \, k_4^2$$
$$+ 64n_0^4 n_1 n_2 n_5 (n-1)^2 (n-2)(n^3 + 12n^2 - 79n + 60) \, k_5 k_3 + 12n_0^2 n_1^2 n_3 n_5 (n-1)(7n^5 + 493n^4$$
$$+ 259n^3 - 2013n^2 + 2254n - 760) \, k_6 k_2 - 24n_0^4 n_1 n_5 (n-1)(5n^3 - 4n^2 - 4) \, k_8,$$

$$k_{2111111} = k_2 k_1^6 - 6n_0 k_3 k_1^5 - 15n_0 n_1 (n-1) \, k_2^2 k_1^4 + 45n_0^2 n_1^2 n_3 (n-1)^2 (n^2 + 13n - 12) \, k_2^3 k_1^2$$
$$+ 100n_0^2 n_2 (n-1) \, k_3 k_2 k_1^3 - 120n_0^3 n_1 n_3 (n-1)(n-2) \, k_3^2 k_1^2 + 30n_0 n_1 k_4 k_1^4 + 360n_0 n_1 n_3 k_6 k_1^2$$
$$- 270n_0^2 n_1^2 n_3 (n-1)(n^2 + 5n - 4) \, k_4 k_2 k_1^2 - 120n_0^2 n_2 k_5 k_1^3 + 504n_0^3 n_2 n_4 (n-1)$$
$$\times (n^2 + 23n - 18) \, k_5 k_2 k_1 + 420n_0^3 n_1 n_4 (n-1)(n-2)(n-3) \, k_4 k_3 k_1 - 210n_0^3 n_1 n_2 n_4 (n-1)^2$$
$$\times (n^3 + 36n^2 - 19n - 6) \, k_3 k_2^2 k_1 - 720n_0^2 n_4 k_7 k_1 - 15n_0^3 n_1^2 n_3 n_5 (n-1)^3 (n^6 + 107n^5 + 1545n^4$$
$$+ 4909n^3 - 24562n^2 + 25248n - 7200) \, k_2^4 + 160n_0^4 n_1^2 n_2 n_3 n_5 (n-1)^2 (n-2)(n^6 + 91n^5$$
$$+ 421n^4 - 2615n^3 - 974n^2 + 4612n - 1680) \, k_3^2 k_2 + 180n_0^3 n_1^3 n_3 n_5 (n-1)^2 (n^6 + 91n^5$$
$$+ 297n^4 + 1437n^3 - 6034n^2 + 5696n - 1440) \, k_4 k_2^2 - 180n_0^3 n_1^3 n_5 (n-1)(n-2)(n-3)$$
$$\times (n^2 - 19n + 20) \, k_4^2 - 384n_0^4 n_1 n_2 n_5 (n-1)(n-2)(n^3 + 12n^2 - 79n + 60) \, k_5 k_3$$
$$- 480n_0^2 n_1^2 n_3 n_5 (n-1)(n^4 + 70n^3 + 71n^2 - 238n + 120) \, k_6 k_2 + 48n_0^5 n_1 n_5$$
$$\times (15n^4 - 14n^3 + 14n^2 - 14n + 14) \, k_8,$$

$$k_{11111111} = k_1^8 - 28n_0 k_2 k_1^6 + 210n_0^2 n_1 (n-1) \, k_2^2 k_1^4 + 112n_0^2 k_3 k_1^5 - 1120n_0^3 n_2 (n-1) \, k_3 k_2 k_1^3$$
$$- 420n_0^3 n_1^2 n_3 (n-1)^2 (n^2 + 13n - 12) \, k_2^3 k_1^2 + 1680n_0^4 n_1 n_2 n_4 (n-1)^2 (n^3 + 36n^2 - 19n - 6) \, k_3 k_2^2 k_1$$
$$+ 1120n_0^4 n_1 n_3 (n-1)(n-2)^2 \, k_3^2 k_1^2 - 420n_0^2 n_1 k_4 k_1^4 + 2520n_0^3 n_1^2 n_3 (n-1)(n^2 + 5n - 4) \, k_4 k_2 k_1^2$$

$$- 3360 n_0^4 n_1 n_4 (n-1)(n-2)(n-3) k_4 k_3 k_1 + 1344 n_0^3 n_2 k_5 k_1^3 - 4032 n_0^4 n_2 n_4 (n-1)$$
$$\times (n^2 + 23n - 18) k_5 k_2 k_1$$
$$- 3360 n_0^2 n_1 n_3 k_6 k_1^2 + 5760 n_0^3 n_4 k_7 k_1 + 105 n_0^4 n_1^3 n_3 n_5 (n-1)^3 (n^6 + 107 n^5 + 1545 n^4 + 4909 n^3$$
$$- 24562 n^2 + 25248 n - 7200) k_2^4 - 1120 n_0^5 n_1^2 n_3 n_5 (n-1)^2 (n-2)(n^6 + 91 n^5 + 421 n^4$$
$$- 2615 n^3 - 974 n^2 + 4612 n - 1680) k_3^2 k_2 - 1260 n_0^4 n_1^3 n_3 n_5 (n-1)^2 (n^6 + 91 n^5 + 297 n^4$$
$$+ 1437 n^3 - 6034 n^2 + 5696 n - 1440) k_4 k_2^2 + 1260 n_0^4 n_1^3 n_5 (n-1)(n-2)(n-3)(n^2 - 19n + 20) k_4^2$$
$$+ 2688 n_0^5 n_1 n_2 n_5 (n-1)(n-2)(n^3 + 12 n^2 - 79 n + 60) k_5 k_3 + 3360 n_0^3 n_1^2 n_3 n_5 (n-1)$$
$$\times (n^4 + 70 n^3 + 71 n^2 - 238 n + 120) k_6 k_2 - 336 n_0^6 n_1 n_5 (15 n^4 - 14 n^3 + 14 n^2 - 14 n + 14) k_8.$$

Table 3·1. *Cumulants of k-statistics in terms of the population cumulants*

This table differs from others previously published in that all terms have been reduced to a common denominator. Thus instead of the familiar

$$\kappa(2^2) = \mathscr{E}(k_2 - \kappa_2)^2 = \frac{\kappa_4}{n} + \frac{2\kappa_2^2}{n-1}$$

we have, from the table,

$$\kappa(2^2) = \frac{1}{n^{(2)}}\{\kappa_4(n-1) + 2\kappa_2^2 . n\}.$$

To derive the sampling cumulants when the parent population is normal (and therefore $\kappa_r = 0$, $r > 2$) will require the cancellation of factors, which is not necessary in the expressions set out by Fisher and by Kendall. On the other hand, for any other population we believe this form of expression to be preferred.

Sampling cumulants with unit parts, i.e. those relating to k_1, have not been tabulated. The rules whereby these may be derived have been set out in the Introduction (p. 28). The deduction of the sampling moments from the sampling cumulants is given on p. 19 of the Introduction.

Table 3·1 (*cont.*)

Cumulants of k-statistics	Dividing factor	Product of cumulants	Coefficients of						
			n^0	n^1	n^2	n^3	n^4	n^5	n^6
$\kappa(2^2)$	$n^{(2)}$	κ_4	-1	1	.	.	.	.	.
		κ_2^2	.	2	.	.	.	.	.
$\kappa(2^3)$	$(n^{(2)})^2$	κ_6	1	-2	1	.	.	.	.
		$\kappa_4 \kappa_2$	.	-12	12	.	.	.	.
		κ_3^2	.	-8	4	.	.	.	.
		κ_2^3	.	.	8	.	.	.	.
$\kappa(2^4)$	$(n^{(2)})^3$	κ_8	-1	3	-3	1	.	.	.
		$\kappa_6 \kappa_2$	.	24	-48	24	.	.	.
		$\kappa_5 \kappa_3$	.	64	-96	32	.	.	.
		κ_4^2	.	48	-72	32	.	.	.
		$\kappa_4 \kappa_2^2$	.	.	-144	144	.	.	.
		$\kappa_3^2 \kappa_2$	.	.	-192	96	.	.	.
		κ_2^4	.	.	.	48	.	.	.
$\kappa(2^5)$	$(n^{(2)})^4$	κ_{10}	1	-4	6	-4	1	.	.
		$\kappa_8 \kappa_2$	.	-40	120	-120	40	.	.
		$\kappa_7 \kappa_3$	.	-160	400	-320	80	.	.
		$\kappa_6 \kappa_4$	.	-360	840	-680	200	.	.
		κ_5^2	.	-224	496	-384	96	.	.
		$\kappa_6 \kappa_2^2$	.	.	480	-960	480	.	.
		$\kappa_5 \kappa_3 \kappa_2$	.	.	2560	-3840	1280	.	.
		$\kappa_4^2 \kappa_2$	.	.	1920	-2880	1280	.	.
		$\kappa_4 \kappa_3^2$	.	.	2880	-3360	960	.	.
		$\kappa_4 \kappa_2^3$	.	.	.	-1920	1920	.	.
		$\kappa_3^2 \kappa_2^2$	.	.	.	-3840	1920	.	.
		κ_2^5	.	.	.	.	384	.	.
$\kappa(2^6)$	$(n^{(2)})^5$	κ_{12}	-1	5	-10	10	-5	1	.
		$\kappa_{10} \kappa_2$	.	60	-240	360	-240	60	.
		$\kappa_9 \kappa_3$	.	320	-1120	1440	-800	160	.
		$\kappa_8 \kappa_4$	.	960	-3120	3840	-2160	480	.
		$\kappa_7 \kappa_5$	.	1728	-5280	6240	-3360	672	.
		κ_6^2	.	1060	-3200	3800	-2080	452	.
		$\kappa_8 \kappa_2^2$	.	.	-1200	3600	-3600	1200	.
		$\kappa_7 \kappa_3 \kappa_2$	.	.	-9600	24000	-19200	4800	.
		$\kappa_6 \kappa_4 \kappa_2$	.	.	-21600	50400	-40800	12000	.
		$\kappa_6 \kappa_3^2$	.	.	-16960	35360	-23360	4960	.
		$\kappa_5^2 \kappa_2$	.	.	-13440	29760	-23040	5760	.
		$\kappa_5 \kappa_4 \kappa_3$	.	.	-61440	119040	-78720	17280	.
		κ_4^3	.	.	-14880	28320	-19680	5280	.
		$\kappa_6 \kappa_2^3$	.	.	.	9600	-19200	9600	.
		$\kappa_5 \kappa_3 \kappa_2^2$	.	.	.	76800	-115200	38400	.
		$\kappa_4^2 \kappa_2^2$	.	.	.	57600	-86400	38400	.
		$\kappa_4 \kappa_3^2 \kappa_2$	.	.	.	172800	-201600	57600	.
		κ_3^4	.	.	.	23040	-21120	4800	.
		$\kappa_4 \kappa_2^4$	.	.	.	.	-28800	28800	.
		$\kappa_3^2 \kappa_2^3$	.	.	.	.	-76800	38400	.
		κ_2^6	.	.	.	.	.	3840	.
$\kappa(3^2)$	$n^{(2)}$	κ_6	2	-3	1	.	.	.	.
		$\kappa_4 \kappa_2$	.	-18	9	.	.	.	.
		κ_3^2	.	-18	9	.	.	.	.
		κ_2^3	.	.	6	.	.	.	.
$\kappa(3^3)$	$(n^{(3)})^2$	κ_9	4	-12	13	-6	1	.	.
		$\kappa_7 \kappa_2$	.	-108	216	-135	27	.	.
		$\kappa_6 \kappa_3$	.	-432	756	-432	81	.	.
		$\kappa_5 \kappa_4$	.	-756	1188	-621	108	.	.
		$\kappa_5 \kappa_2^2$	.	.	756	-810	216	.	.
		$\kappa_4 \kappa_3 \kappa_2$	.	.	3888	-3564	810	.	.
		κ_3^3	.	.	1224	-1080	252	.	.
		$\kappa_3 \kappa_2^3$	.	.	.	-1296	540	.	.
$\kappa(3^4)$	$(n^{(3)})^3$	κ_{12}	8	-36	66	-63	33	-9	1
		$\kappa_{10} \kappa_2$	.	-432	1512	-2052	1350	-432	54
		$\kappa_9 \kappa_3$	.	-2592	8208	-10152	6156	-1836	216
		$\kappa_8 \kappa_4$	.	-7560	21924	-25218	14391	-4077	459
		$\kappa_7 \kappa_5$	.	-13824	38016	-42336	23760	-6696	756
		κ_6^2	.	-8424	22788	-25218	14175	-4023	459
		$\kappa_8 \kappa_2^2$	.	.	7560	-19116	17442	-6885	999
		$\kappa_7 \kappa_3 \kappa_2$	.	.	69984	-156816	128952	-46332	6156
		$\kappa_6 \kappa_4 \kappa_2$	.	.	151632	-310392	238788	-81810	10530
		$\kappa_6 \kappa_3^2$	.	.	138240	-276048	207360	-69660	8856
		$\kappa_5^2 \kappa_2$	.	.	96768	-191808	144720	-49248	6372
		$\kappa_5 \kappa_4 \kappa_3$	.	.	489888	-903312	635688	-201852	24300
		κ_4^3	.	.	113400	-199260	133974	-40581	4671
		$\kappa_6 \kappa_2^3$	.	.	.	-50544	82080	-43740	7668
		$\kappa_5 \kappa_3 \kappa_2^2$	.	.	.	-489888	689472	-324648	51192
		$\kappa_4^2 \kappa_2^2$	.	.	.	-342144	452952	-202176	30618
		$\kappa_4 \kappa_3^2 \kappa_2$	.	.	.	-1244160	1566864	-668736	96228
		κ_3^4	.	.	.	-190512	230040	-96228	14094
		$\kappa_4 \kappa_2^4$	.	.	.	.	114696	-100116	22356
		$\kappa_3^2 \kappa_2^3$	.	.	.	.	414720	-330480	66744
		κ_2^6	.	.	.	.	.	-7776	3240

Table 3·1 (cont.)

Cumulants of k-statistics	Dividing factor	Product of cumulants	Coefficients of n^0	n^1	n^2	n^3	n^4	n^5	n^6
$\kappa(4^2)$	$n^{(4)}$	κ_8	-6	11	-6	1	·	·	·
		$\kappa_6\kappa_2$	·	96	-80	16	·	·	·
		$\kappa_5\kappa_3$	·	288	-240	48	·	·	·
		κ_4^2	·	204	-170	34	·	·	·
		$\kappa_4\kappa_2^2$	·	·	-216	72	·	·	·
		$\kappa_3^2\kappa_2$	·	·	-432	144	·	·	·
		κ_2^4	·	·	24	24	·	·	·
$\kappa(4^3)$	$(n^{(4)})^2$	κ_{12}	36	-132	193	-144	58	-12	1
		$\kappa_{10}\kappa_2$	·	-1728	4608	-4656	2256	-528	48
		$\kappa_9\kappa_3$	·	-9792	23808	-22544	10416	-2352	208
		$\kappa_8\kappa_4$	·	-28080	64512	-58380	26004	-5700	492
		$\kappa_7\kappa_5$	·	-50112	111168	-97584	42336	-9072	768
		κ_6^2	·	-30240	66384	-57720	24828	-5280	444
		$\kappa_8\kappa_2^2$	·	·	24624	-42984	27576	-7704	792
		$\kappa_7\kappa_3\kappa_2$	·	·	212544	-346464	209376	-55584	5472
		$\kappa_6\kappa_4\kappa_2$	·	·	451872	-702576	405456	-103056	9744
		$\kappa_6\kappa_3^2$	·	·	391392	-594000	338112	-85392	8064
		$\kappa_5^2\kappa_2$	·	·	285120	-436320	247680	-61920	5760
		$\kappa_5\kappa_4\kappa_3$	·	·	1378944	-2018304	1113984	-274176	25344
		κ_4^3	·	·	320832	-460848	251336	-61472	5672
		$\kappa_6\kappa_2^3$	·	·	-1728	-107424	120672	-44640	5472
		$\kappa_5\kappa_3\kappa_2^2$	·	·	-20736	-1029888	1071360	-366336	41472
		$\kappa_4^2\kappa_2^2$	·	·	-15552	-710208	719280	-239328	26352
		$\kappa_4\kappa_3^2\kappa_2$	·	·	-62208	-2436480	2449440	-810432	88992
		κ_3^4	·	·	-10368	-369216	348192	-110592	11808
		$\kappa_4\kappa_2^4$	·	·	·	44928	139680	-97344	15264
		$\kappa_3^2\kappa_2^3$	·	·	·	107136	547776	-338688	50112
		κ_2^6	·	·	·	3456	-5184	-6912	1728
$\kappa(5^2)$	$n^{(5)}$	κ_{10}	24	-50	35	-10	1	·	·
		$\kappa_8\kappa_2$	·	-600	650	-225	25	·	·
		$\kappa_7\kappa_3$	·	-2400	2600	-900	100	·	·
		$\kappa_6\kappa_4$	·	-4800	5200	-1800	200	·	·
		κ_5^2	·	-3000	3250	-1125	125	·	·
		$\kappa_6\kappa_2^2$	·	·	2400	-1400	200	·	·
		$\kappa_5\kappa_3\kappa_2$	·	·	14400	-8400	1200	·	·
		$\kappa_4^2\kappa_2$	·	·	10200	-5950	850	·	·
		$\kappa_4\kappa_3^2$	·	·	18000	-10500	1500	·	·
		$\kappa_4\kappa_2^3$	·	·	-2400	-1800	600	·	·
		$\kappa_3^2\kappa_2^2$	·	·	-7200	-5400	1800	·	·
		κ_2^5	·	·	·	600	120	·	·
$\kappa(6^2)$	$n^{(6)}$	κ_{12}	-120	274	-225	85	-15	1	·
		$\kappa_{10}\kappa_2$	·	4320	-5544	2556	-504	36	·
		$\kappa_9\kappa_3$	·	21600	-27720	12780	-2520	180	·
		$\kappa_8\kappa_4$	·	55800	-71610	33015	-6510	465	·
		$\kappa_7\kappa_5$	·	93600	-120120	55380	-10920	780	·
		κ_6^2	·	55320	-70994	32731	-6454	461	·
		$\kappa_8\kappa_2^2$	·	·	-27000	21150	-5400	450	·
		$\kappa_7\kappa_3\kappa_2$	·	·	-216000	169200	-43200	3600	·
		$\kappa_6\kappa_4\kappa_2$	·	·	-432000	338400	-86400	7200	·
		$\kappa_6\kappa_3^2$	·	·	-378000	296100	-75600	6300	·
		$\kappa_5^2\kappa_2$	·	·	-270000	211500	-54000	4500	·
		$\kappa_5\kappa_4\kappa_3$	·	·	-1296000	1015200	-259200	21600	·
		κ_4^3	·	·	-297000	232650	-59400	4950	·
		$\kappa_6\kappa_2^3$	·	·	48000	26400	-19200	2400	·
		$\kappa_5\kappa_3\kappa_2^2$	·	·	432000	237600	-172800	21600	·
		$\kappa_4^2\kappa_2^2$	·	·	306000	168300	-122400	15300	·
		$\kappa_4\kappa_3^2\kappa_2$	·	·	1080000	594000	-432000	54000	·
		κ_3^4	·	·	162000	89100	-64800	8100	·
		$\kappa_4\kappa_2^4$	·	·	·	·	-135000	5400	·
		$\kappa_3^2\kappa_2^3$	·	·	·	·	-540000	21600	·
		κ_2^6	·	·	-2880	7920	11520	720	·
$\kappa(32)$	$n^{(2)}$	κ_5	-1	1	·	·	·	·	·
		$\kappa_3\kappa_2$	·	6	·	·	·	·	·
$\kappa(42)$	$n^{(2)}$	κ_6	-1	1	·	·	·	·	·
		$\kappa_4\kappa_2$	·	8	·	·	·	·	·
		κ_3^2	·	6	·	·	·	·	·
$\kappa(52)$	$n^{(2)}$	κ_7	-1	1	·	·	·	·	·
		$\kappa_5\kappa_2$	·	10	·	·	·	·	·
		$\kappa_4\kappa_3$	·	20	·	·	·	·	·
$\kappa(62)$	$n^{(2)}$	κ_8	-1	1	·	·	·	·	·
		$\kappa_6\kappa_2$	·	12	·	·	·	·	·
		$\kappa_5\kappa_3$	·	30	·	·	·	·	·
		κ_4^2	·	20	·	·	·	·	·
$\kappa(72)$	$n^{(2)}$	κ_9	-1	1	·	·	·	·	·
		$\kappa_7\kappa_2$	·	14	·	·	·	·	·
		$\kappa_6\kappa_3$	·	42	·	·	·	·	·
		$\kappa_5\kappa_4$	·	70	·	·	·	·	·

Table 3·1 (*cont.*)

Left half

Cumulants of k-statistics	Dividing factor	Product of cumulants	n^0	n^1	n^2	n^3
$\kappa(82)$	$n^{(2)}$	κ_{10}	-1	1	·	·
		$\kappa_8\kappa_2$	·	16	·	·
		$\kappa_7\kappa_3$	·	56	·	·
		$\kappa_6\kappa_4$	·	112	·	·
		κ_5^2	·	70	·	·
$\kappa(43)$	$n^{(3)}$	κ_7	2	-3	1	·
		$\kappa_5\kappa_2$	·	-24	12	·
		$\kappa_4\kappa_3$	·	-60	30	·
		$\kappa_3\kappa_2^2$	·	·	36	·
$\kappa(53)$	$n^{(3)}$	κ_8	2	-3	1	·
		$\kappa_6\kappa_2$	·	-30	15	·
		$\kappa_5\kappa_3$	·	-90	45	·
		κ_4^2	·	-60	30	·
		$\kappa_4\kappa_2^2$	·	·	60	·
		$\kappa_3^2\kappa_2$	·	·	90	·
$\kappa(63)$	$n^{(3)}$	κ_9	2	-3	1	·
		$\kappa_7\kappa_2$	·	-36	18	·
		$\kappa_6\kappa_3$	·	-126	63	·
		$\kappa_5\kappa_4$	·	-210	105	·
		$\kappa_5\kappa_2^2$	·	·	90	·
		$\kappa_4\kappa_3\kappa_2$	·	·	360	·
		κ_3^3	·	·	90	·
$\kappa(73)$	$n^{(3)}$	κ_{10}	2	-3	1	·
		$\kappa_8\kappa_2$	·	-42	21	·
		$\kappa_7\kappa_3$	·	-168	84	·
		$\kappa_6\kappa_4$	·	-336	168	·
		κ_5^2	·	-210	105	·
		$\kappa_6\kappa_2^2$	·	·	126	·
		$\kappa_5\kappa_3\kappa_2$	·	·	630	·
		$\kappa_4^2\kappa_2$	·	·	420	·
		$\kappa_4\kappa_3^2$	·	·	630	·
$\kappa(54)$	$n^{(4)}$	κ_9	-6	11	-6	1
		$\kappa_7\kappa_2$	·	120	-100	20
		$\kappa_6\kappa_3$	·	420	-350	70
		$\kappa_5\kappa_4$	·	720	-600	120
		$\kappa_5\kappa_2^2$	·	·	-360	120
		$\kappa_4\kappa_3\kappa_2$	·	·	-1800	600
		κ_3^3	·	·	-540	180
		$\kappa_3\kappa_2^3$	·	·	240	240
$\kappa(64)$	$n^{(4)}$	κ_{10}	-6	11	-6	1
		$\kappa_8\kappa_2$	·	144	-120	24
		$\kappa_7\kappa_3$	·	576	-480	96
		$\kappa_6\kappa_4$	·	1164	-970	194
		κ_5^2	·	720	-600	120
		$\kappa_6\kappa_2^2$	·	·	-540	180
		$\kappa_5\kappa_3\kappa_2$	·	·	-3240	1080
		$\kappa_4^2\kappa_2$	·	·	-2160	720
		$\kappa_4\kappa_3^2$	·	·	-3780	1260
		$\kappa_4\kappa_2^3$	·	·	480	480
		$\kappa_3^2\kappa_2^2$	·	·	1080	1080
$\kappa(32^2)$	$(n^{(2)})^2$	κ_7	1	-2	1	·
		$\kappa_5\kappa_2$	·	-16	16	·
		$\kappa_4\kappa_3$	·	-36	24	·
		$\kappa_3\kappa_2^2$	·	·	48	·
$\kappa(42^2)$	$(n^{(2)})^2$	κ_8	1	-2	1	·
		$\kappa_6\kappa_2$	·	-20	20	·
		$\kappa_5\kappa_3$	·	-56	40	·
		κ_4^2	·	-40	28	·
		$\kappa_4\kappa_2^2$	·	·	80	·
		$\kappa_3^2\kappa_2$	·	·	120	·
$\kappa(52^2)$	$(n^{(2)})^2$	κ_9	1	-2	1	·
		$\kappa_7\kappa_2$	·	-24	24	·
		$\kappa_6\kappa_3$	·	-80	60	·
		$\kappa_5\kappa_4$	·	-140	100	·
		$\kappa_5\kappa_2^2$	·	·	120	·
		$\kappa_4\kappa_3\kappa_2$	·	·	480	·
		κ_3^3	·	·	120	·
$\kappa(62^2)$	$(n^{(2)})^2$	κ_{10}	1	-2	1	·
		$\kappa_8\kappa_2$	·	-28	28	·
		$\kappa_7\kappa_3$	·	-108	84	·
		$\kappa_6\kappa_4$	·	-224	164	·
		κ_5^2	·	-140	100	·
		$\kappa_6\kappa_2^2$	·	·	168	·
		$\kappa_5\kappa_3\kappa_2$	·	·	840	·
		$\kappa_4^2\kappa_2$	·	·	560	·
		$\kappa_4\kappa_3^2$	·	·	840	·

Right half

Cumulants of k-statistics	Dividing factor	Product of cumulants	n^0	n^1	n^2	n^3	n^4
$\kappa(3^2 2)$	$n^{(2)}n^{(3)}$	κ_8	-2	5	-4	1	·
		$\kappa_6\kappa_2$	·	42	-63	21	·
		$\kappa_5\kappa_3$	·	132	-162	48	·
		κ_4^2	·	90	-99	27	·
		$\kappa_4\kappa_2^2$	·	·	-198	108	·
		$\kappa_3^2\kappa_2$	·	·	-360	162	·
		κ_2^4	·	·	·	36	·
$\kappa(432)$	$n^{(2)}n^{(3)}$	κ_9	-2	5	-4	1	·
		$\kappa_7\kappa_2$	·	52	-78	26	·
		$\kappa_6\kappa_3$	·	192	-240	72	·
		$\kappa_5\kappa_4$	·	340	-390	110	·
		$\kappa_5\kappa_2^2$	·	·	-324	180	·
		$\kappa_4\kappa_3\kappa_2$	·	·	-1536	732	·
		κ_3^3	·	·	-432	180	·
		$\kappa_3\kappa_2^3$	·	·	·	360	·
$\kappa(532)$	$n^{(2)}n^{(3)}$	κ_{10}	-2	5	-4	1	·
		$\kappa_8\kappa_2$	·	62	-93	31	·
		$\kappa_7\kappa_3$	·	262	-333	101	·
		$\kappa_6\kappa_4$	·	550	-645	185	·
		κ_5^2	·	350	-405	5	·
		$\kappa_6\kappa_2^2$	·	·	-480	270	·
		$\kappa_5\kappa_3\kappa_2$	·	·	-2760	1350	·
		$\kappa_4^2\kappa_2$	·	·	-1860	900	·
		$\kappa_4\kappa_3^2$	·	·	-3090	1350	·
		$\kappa_4\kappa_2^3$	·	·	·	720	·
		$\kappa_3^2\kappa_2^2$	·	·	·	1620	·
$\kappa(4^2 2)$	$n^{(2)}n^{(4)}$	κ_{10}	6	-17	17	-7	1
		$\kappa_8\kappa_2$	·	-192	352	-192	32
		$\kappa_7\kappa_3$	·	-1776	2104	-816	104
		$\kappa_6\kappa_4$	·	-1752	2636	-1272	196
		κ_5^2	·	-1104	1616	-764	116
		$\kappa_6\kappa_2^2$	·	·	1560	-1408	296
		$\kappa_5\kappa_3\kappa_2$	·	·	9216	-7680	1536
		$\kappa_4^2\kappa_2$	·	·	6480	-5184	1008
		$\kappa_4\kappa_3^2$	·	·	10800	-8136	1512
		$\kappa_4\kappa_2^3$	·	·	-96	-2592	960
		$\kappa_3^2\kappa_2^2$	·	·	-288	-7632	2448
		κ_2^5	·	·	·	192	192
$\kappa(43^2)$	$(n^{(3)})$	κ_{10}	4	-12	13	-6	1
		$\kappa_8\kappa_2$	·	-132	264	-165	33
		$\kappa_7\kappa_3$	·	-600	1056	-606	114
		$\kappa_6\kappa_4$	·	-1284	2064	-1101	195
		κ_5^2	·	-816	1272	-660	114
		$\kappa_6\kappa_2^2$	·	·	1188	-1278	342
		$\kappa_5\kappa_3\kappa_2$	·	·	7488	-7056	1656
		$\kappa_4^2\kappa_2$	·	·	5184	-4644	1026
		$\kappa_4\kappa_3^2$	·	·	9288	-7992	1782
		$\kappa_4\kappa_2^3$	·	·	·	-2880	1224
		$\kappa_3^2\kappa_2^2$	·	·	·	-7560	2916
		κ_2^5	·	·	·	·	216
$\kappa(32^3)$	$(n^{(2)})^3$	κ_9	-1	3	-3	1	·
		$\kappa_7\kappa_2$	·	30	-60	30	·
		$\kappa_6\kappa_3$	·	106	-168	62	·
		$\kappa_5\kappa_4$	·	192	-276	108	·
		$\kappa_5\kappa_2^2$	·	·	-240	240	·
		$\kappa_4\kappa_3\kappa_2$	·	·	-1080	720	·
		κ_3^3	·	·	-288	120	·
		$\kappa_3\kappa_2^3$	·	·	·	480	·
$\kappa(42^3)$	$(n^{(2)})^3$	κ_{10}	-1	3	-3	1	·
		$\kappa_8\kappa_2$	·	36	-72	36	·
		$\kappa_7\kappa_3$	·	148	-240	92	·
		$\kappa_6\kappa_4$	·	324	-480	188	·
		κ_5^2	·	204	-288	108	·
		$\kappa_6\kappa_2^2$	·	·	-360	360	·
		$\kappa_5\kappa_3\kappa_2$	·	·	-2016	1440	·
		$\kappa_4^2\kappa_2$	·	·	-1440	1008	·
		$\kappa_4\kappa_3^2$	·	·	-2280	1176	·
		$\kappa_4\kappa_2^3$	·	·	·	960	·
		$\kappa_3^2\kappa_2^2$	·	·	·	2160	·
$\kappa(3^2 2^2)$	$(n^{(2)})^2 n^{(3)}$	κ_{10}	2	-7	9	-5	1
		$\kappa_8\kappa_2$	·	-74	185	-148	37
		$\kappa_7\kappa_3$	·	-324	690	-468	102
		$\kappa_6\kappa_4$	·	-702	1347	-864	183
		κ_5^2	·	-452	842	-544	118
		$\kappa_6\kappa_2^2$	·	·	786	-1188	402
		$\kappa_5\kappa_3\kappa_2$	·	·	4848	-5904	1704
		$\kappa_4^2\kappa_2$	·	·	3348	-3708	1044
		$\kappa_4\kappa_3^2$	·	·	5760	-5580	1368
		$\kappa_4\kappa_2^3$	·	·	·	-1656	1008
		$\kappa_3^2\kappa_2^2$	·	·	·	-6336	2736
		κ_2^5	·	·	·	·	288

14·2

Table 3·2. *Sampling cumulants of bivariate k-statistics in terms of bivariate (population) cumulants*

The coefficients of the different products of population cumulants have been reduced to a single common denominator. Thus, from the table, for example

$$\kappa\binom{11}{11} = \frac{1}{n^{(2)}}\{\kappa_{22}(n-1) + \kappa_{20}\kappa_{02}.n + \kappa_{11}^2.n\}.$$

	Dividing factor	Product of cumulants	n^0	n^1	n^2	n^3
$\kappa\binom{22}{00}$	$n^{(2)}$	κ_{40}	-1	1	.	.
		κ_{20}^2	.	2	.	.
$\kappa\binom{21}{01}$	$n^{(2)}$	κ_{31}	-1	1	.	.
		$\kappa_{20}\kappa_{11}$	.	2	.	.
$\kappa\binom{20}{02}$	$n^{(2)}$	κ_{22}	-1	1	.	.
		κ_{11}^2	.	2	.	.
$\kappa\binom{11}{11}$	$n^{(2)}$	κ_{22}	-1	1	.	.
		$\kappa_{20}\kappa_{02}$	.	1	.	.
		κ_{11}^2	.	1	.	.
$\kappa\binom{222}{000}$	$(n^{(2)})^2$	κ_{60}	1	-2	1	.
		$\kappa_{40}\kappa_{20}$	.	-12	12	.
		κ_{30}^2	.	-8	4	.
		κ_{20}^3	.	.	8	.
$\kappa\binom{221}{001}$	$(n^{(2)})^2$	κ_{51}	1	-2	1	.
		$\kappa_{40}\kappa_{11}$	.	-4	4	.
		$\kappa_{31}\kappa_{20}$	.	-8	8	.
		$\kappa_{30}\kappa_{21}$	.	-8	4	.
		$\kappa_{20}^2\kappa_{11}$	.	.	8	.
$\kappa\binom{220}{002}$	$(n^{(2)})^2$	κ_{42}	1	-2	1	.
		$\kappa_{31}\kappa_{11}$	.	-8	8	.
		$\kappa_{22}\kappa_{20}$	.	-4	4	.
		κ_{21}^2	.	-8	4	.
		$\kappa_{20}\kappa_{11}^2$	.	.	8	.
$\kappa\binom{211}{011}$	$(n^{(2)})^2$	κ_{42}	1	-2	1	.
		$\kappa_{40}\kappa_{02}$	.	-1	1	.
		$\kappa_{31}\kappa_{11}$	.	-6	6	.
		$\kappa_{30}\kappa_{12}$	.	-4	2	.
		$\kappa_{22}\kappa_{20}$	.	-5	5	.
		κ_{21}^2	.	-4	2	.
		$\kappa_{20}^2\kappa_{02}$	.	.	2	.
		$\kappa_{20}\kappa_{11}^2$	.	.	6	.
$\kappa\binom{201}{021}$	$(n^{(2)})^2$	κ_{33}	1	-2	1	.
		$\kappa_{31}\kappa_{02}$	.	-2	2	.
		$\kappa_{22}\kappa_{11}$	.	-8	8	.
		$\kappa_{21}\kappa_{12}$	.	-8	4	.
		$\kappa_{20}\kappa_{13}$	.	-2	2	.
		$\kappa_{20}\kappa_{11}\kappa_{02}$	.	.	4	.
		κ_{11}^3	.	.	4	.
$\kappa\binom{111}{111}$	$(n^{(2)})^2$	κ_{33}	1	-2	1	.
		$\kappa_{31}\kappa_{02}$	.	-3	3	.
		$\kappa_{30}\kappa_{03}$	.	-2	1	.
		$\kappa_{22}\kappa_{11}$	.	-6	6	.
		$\kappa_{21}\kappa_{12}$	.	-6	3	.
		$\kappa_{20}\kappa_{13}$	.	-3	3	.
		$\kappa_{20}\kappa_{11}\kappa_{02}$	.	.	6	.
		κ_{11}^3	.	.	2	.
$\kappa\binom{2222}{0000}$	$(n^{(2)})^3$	κ_{80}	-1	3	-3	1
		$\kappa_{60}\kappa_{20}$	.	24	-48	24
		$\kappa_{50}\kappa_{30}$	.	64	-96	32
		κ_{40}^2	.	48	-72	32
		$\kappa_{40}\kappa_{20}^2$	.	.	-144	144
		$\kappa_{30}^2\kappa_{20}$	.	.	-192	96
		κ_{20}^4	.	.	.	48
$\kappa\binom{2221}{0001}$	$(n^{(2)})^3$	κ_{71}	-1	3	-3	1
		$\kappa_{60}\kappa_{11}$	.	6	-12	6
		$\kappa_{51}\kappa_{20}$	.	18	-36	18
		$\kappa_{50}\kappa_{21}$	.	24	-36	12
		$\kappa_{41}\kappa_{30}$	.	40	-60	20
		$\kappa_{40}\kappa_{31}$	.	48	-72	32
		$\kappa_{40}\kappa_{20}\kappa_{11}$	.	.	-72	72
		$\kappa_{31}\kappa_{20}^2$	.	.	-72	72
		$\kappa_{30}^2\kappa_{11}$	.	.	-48	24
		$\kappa_{30}\kappa_{21}\kappa_{20}$	.	.	-144	72
		$\kappa_{20}^3\kappa_{11}$	.	.	.	48

	Dividing factor	Product of cumulants	n^0	n^1	n^2	n^3
$\kappa\binom{2220}{0002}$	$(n^{(2)})^3$	κ_{62}	-1	3	-3	1
		$\kappa_{51}\kappa_{11}$	.	12	-24	12
		$\kappa_{42}\kappa_{20}$	.	12	-24	12
		$\kappa_{41}\kappa_{21}$	.	48	-72	24
		$\kappa_{40}\kappa_{22}$	.	12	-24	12
		$\kappa_{32}\kappa_{30}$	.	16	-24	8
		κ_{31}^2	.	36	-48	20
		$\kappa_{40}\kappa_{11}^2$	.	.	-24	24
		$\kappa_{31}\kappa_{20}\kappa_{11}$	.	.	-96	96
		$\kappa_{30}\kappa_{21}\kappa_{11}$	.	.	-96	48
		$\kappa_{22}\kappa_{20}^2$	.	.	-24	24
		$\kappa_{21}^2\kappa_{20}$	.	.	-96	48
		$\kappa_{20}^2\kappa_{11}^2$	.	.	.	48
$\kappa\binom{2211}{0011}$	$(n^{(2)})^3$	κ_{62}	-1	3	-3	1
		$\kappa_{60}\kappa_{02}$	.	1	-2	1
		$\kappa_{51}\kappa_{11}$	.	10	-20	10
		$\kappa_{50}\kappa_{12}$	.	8	-12	4
		$\kappa_{42}\kappa_{20}$	.	13	-26	13
		$\kappa_{41}\kappa_{21}$	.	32	-48	16
		$\kappa_{40}\kappa_{22}$	.	22	-32	14
		$\kappa_{32}\kappa_{30}$	.	24	-36	12
		κ_{31}^2	.	26	-40	18
		$\kappa_{40}\kappa_{20}\kappa_{02}$	.	.	-12	12
		$\kappa_{40}\kappa_{11}^2$	.	.	-20	20
		$\kappa_{31}\kappa_{20}\kappa_{11}$	.	.	-80	80
		$\kappa_{30}^2\kappa_{02}$	.	.	-8	4
		$\kappa_{30}\kappa_{21}\kappa_{11}$	.	.	-80	40
		$\kappa_{30}\kappa_{20}\kappa_{12}$	.	.	-48	24
		$\kappa_{22}\kappa_{20}^2$	.	.	-32	32
		$\kappa_{21}^2\kappa_{20}$	.	.	-56	28
		$\kappa_{20}^2\kappa_{02}$	.	.	.	8
		$\kappa_{20}^2\kappa_{11}^2$	.	.	.	40
$\kappa\binom{2210}{0012}$	$(n^{(2)})^3$	κ_{53}	-1	3	-3	1
		$\kappa_{51}\kappa_{02}$	.	2	-4	2
		$\kappa_{42}\kappa_{11}$	.	14	-28	14
		$\kappa_{41}\kappa_{12}$	.	16	-24	8
		$\kappa_{40}\kappa_{13}$	.	4	-8	4
		$\kappa_{33}\kappa_{20}$	.	8	-16	8
		$\kappa_{32}\kappa_{21}$	.	40	-60	20
		$\kappa_{31}\kappa_{22}$	.	44	-64	28
		$\kappa_{30}\kappa_{23}$	.	8	-12	4
		$\kappa_{40}\kappa_{11}\kappa_{02}$	.	.	-8	8
		$\kappa_{31}\kappa_{20}\kappa_{02}$	.	.	-16	16
		$\kappa_{31}\kappa_{11}^2$	.	.	-48	48
		$\kappa_{30}\kappa_{21}\kappa_{02}$	.	.	-16	8
		$\kappa_{30}\kappa_{12}\kappa_{11}$	.	.	-32	16
		$\kappa_{22}\kappa_{20}\kappa_{11}$	.	.	-64	64
		$\kappa_{21}\kappa_{20}\kappa_{12}$	.	.	-64	32
		$\kappa_{21}^2\kappa_{11}$	.	.	-80	40
		$\kappa_{20}^2\kappa_{13}$	.	.	-8	8
		$\kappa_{20}^2\kappa_{11}\kappa_{02}$	.	.	.	16
		$\kappa_{20}\kappa_{11}^3$	.	.	.	32
$\kappa\binom{2111}{0111}$	$(n^{(2)})^3$	κ_{53}	-1	3	-3	1
		$\kappa_{51}\kappa_{02}$	.	3	-6	3
		$\kappa_{50}\kappa_{03}$	.	2	-3	1
		$\kappa_{42}\kappa_{11}$	.	12	-24	12
		$\kappa_{41}\kappa_{12}$	.	18	-27	9
		$\kappa_{40}\kappa_{13}$	.	9	-12	5
		$\kappa_{33}\kappa_{20}$	.	9	-18	9
		$\kappa_{32}\kappa_{21}$	.	30	-45	15
		$\kappa_{31}\kappa_{22}$	.	39	-60	27
		$\kappa_{30}\kappa_{23}$	.	14	-21	7
		$\kappa_{40}\kappa_{11}\kappa_{02}$	.	.	-12	12
		$\kappa_{31}\kappa_{20}\kappa_{02}$	.	.	-24	24
		$\kappa_{31}\kappa_{11}^2$	.	.	-36	36
		$\kappa_{30}\kappa_{21}\kappa_{02}$	.	.	-24	12
		$\kappa_{30}\kappa_{20}\kappa_{03}$	.	.	-12	6
		$\kappa_{30}\kappa_{12}\kappa_{11}$	.	.	-48	24
		$\kappa_{22}\kappa_{20}\kappa_{11}$	.	.	-60	60
		$\kappa_{21}^2\kappa_{11}$	.	.	-48	24
		$\kappa_{21}\kappa_{20}\kappa_{12}$	.	.	-60	30
		$\kappa_{20}^2\kappa_{13}$	.	.	-12	12
		$\kappa_{20}^2\kappa_{11}\kappa_{02}$	.	.	.	24
		$\kappa_{20}\kappa_{11}^3$	.	.	.	24

Table 3·2 (*cont.*)

Left table

$\kappa\binom{\cdots}{\cdots}$	Dividing factor	Product of cumulants	n^0	n^1	n^2	n^3
$\kappa\binom{2200}{0022}$	$(n^{(2)})^3$	κ_{44}	-1	3	-3	1
		$\kappa_{42}\kappa_{02}$	·	4	-8	4
		$\kappa_{33}\kappa_{11}$	·	16	-32	16
		$\kappa_{32}\kappa_{12}$	·	32	-48	16
		$\kappa_{31}\kappa_{13}$	·	16	-32	16
		$\kappa_{24}\kappa_{20}$	·	4	-8	4
		$\kappa_{23}\kappa_{21}$	·	32	-48	16
		κ_{22}^2	·	32	-40	16
		$\kappa_{31}\kappa_{11}\kappa_{02}$	·	·	-32	32
		$\kappa_{22}\kappa_{20}\kappa_{02}$	·	·	-16	16
		$\kappa_{22}\kappa_{11}^2$	·	·	-64	64
		$\kappa_{21}^2\kappa_{02}$	·	·	-32	16
		$\kappa_{21}\kappa_{12}\kappa_{11}$	·	·	-128	64
		$\kappa_{20}\kappa_{12}$	·	·	-32	16
		$\kappa_{20}\kappa_{13}\kappa_{11}$	·	·	-32	32
		$\kappa_{20}\kappa_{11}^2\kappa_{02}$	·	·	·	32
		κ_{11}^4	·	·	·	16
$\kappa\binom{2011}{0211}$	$(n^{(2)})^3$	κ_{44}	-1	3	-3	1
		$\kappa_{42}\kappa_{02}$	·	5	-10	5
		$\kappa_{41}\kappa_{03}$	·	4	-6	2
		$\kappa_{40}\kappa_{04}$	·	1	-2	1
		$\kappa_{33}\kappa_{11}$	·	14	-28	14
		$\kappa_{32}\kappa_{12}$	·	28	-42	14
		$\kappa_{31}\kappa_{13}$	·	22	-32	14
		$\kappa_{30}\kappa_{14}$	·	4	-6	2
		$\kappa_{24}\kappa_{20}$	·	5	-10	5
		$\kappa_{23}\kappa_{21}$	·	28	-42	14
		κ_{22}^2	·	25	-38	17
		$\kappa_{40}\kappa_{02}^2$	·	·	-2	2
		$\kappa_{31}\kappa_{11}\kappa_{02}$	·	·	-32	32
		$\kappa_{30}\kappa_{12}\kappa_{02}$	·	·	-16	8
		$\kappa_{30}\kappa_{11}\kappa_{03}$	·	·	-8	4
		$\kappa_{22}\kappa_{20}\kappa_{02}$	·	·	-24	24
		$\kappa_{22}\kappa_{11}^2$	·	·	-52	52
		$\kappa_{21}^2\kappa_{02}$	·	·	-24	12
		$\kappa_{31}\kappa_{20}\kappa_{03}$	·	·	-16	8
		$\kappa_{21}\kappa_{12}\kappa_{11}$	·	·	-104	52
		$\kappa_{20}^2\kappa_{04}$	·	·	-2	2
		$\kappa_{20}\kappa_{13}\kappa_{11}$	·	·	-32	32
		$\kappa_{20}\kappa_{12}^2$	·	·	-24	12
		$\kappa_{20}^2\kappa_{02}^2$	·	·	·	4
		$\kappa_{20}\kappa_{11}^2\kappa_{02}$	·	·	·	32
		κ_{11}^4	·	·	·	12
$\kappa\binom{1111}{1111}$	$(n^{(2)})^3$	κ_{44}	-1	3	-3	1
		$\kappa_{42}\kappa_{02}$	·	6	-12	6
		$\kappa_{41}\kappa_{03}$	·	8	-12	4
		$\kappa_{40}\kappa_{04}$	·	3	-3	1
		$\kappa_{33}\kappa_{11}$	·	12	-24	12
		$\kappa_{32}\kappa_{12}$	·	24	-36	12
		$\kappa_{31}\kappa_{13}$	·	24	-36	16
		$\kappa_{30}\kappa_{14}$	·	8	-12	4
		$\kappa_{24}\kappa_{20}$	·	6	-12	6
		$\kappa_{23}\kappa_{21}$	·	24	-36	12
		κ_{22}^2	·	21	-33	15
		$\kappa_{40}\kappa_{02}^2$	·	·	-3	3
		$\kappa_{31}\kappa_{11}\kappa_{02}$	·	·	-36	36
		$\kappa_{30}\kappa_{12}\kappa_{02}$	·	·	-24	12
		$\kappa_{30}\kappa_{11}\kappa_{03}$	·	·	-24	12
		$\kappa_{22}\kappa_{20}\kappa_{02}$	·	·	-30	30
		$\kappa_{22}\kappa_{11}^2$	·	·	-36	36
		$\kappa_{21}^2\kappa_{02}$	·	·	-24	12
		$\kappa_{21}\kappa_{20}\kappa_{03}$	·	·	-24	12
		$\kappa_{21}\kappa_{12}\kappa_{11}$	·	·	-72	36
		$\kappa_{20}^2\kappa_{04}$	·	·	-3	3
		$\kappa_{20}\kappa_{13}\kappa_{11}$	·	·	-36	36
		$\kappa_{20}\kappa_{12}^2$	·	·	-24	12
		$\kappa_{20}^2\kappa_{02}^2$	·	·	·	6
		$\kappa_{20}\kappa_{11}^2\kappa_{02}$	·	·	·	36
		κ_{11}^4	·	·	·	6
$\kappa\binom{33}{00}$	$n^{(3)}$	κ_{60}	2	-3	1	·
		$\kappa_{40}\kappa_{20}$	·	-18	9	·
		κ_{30}^2	·	-18	9	·
		κ_{20}^3	·	·	6	·
$\kappa\binom{32}{01}$	$n^{(3)}$	κ_{51}	2	-3	1	·
		$\kappa_{40}\kappa_{11}$	·	-6	3	·
		$\kappa_{31}\kappa_{20}$	·	-12	6	·
		$\kappa_{30}\kappa_{21}$	·	-18	9	·
		$\kappa_{20}^2\kappa_{11}$	·	·	6	·

Right table

$\kappa\binom{\cdots}{\cdots}$	Dividing factor	Product of cumulants	n^0	n^1	n^2	n^3	n^4
$\kappa\binom{31}{02}$	$n^{(3)}$	κ_{42}	2	-3	1	·	·
		$\kappa_{31}\kappa_{11}$	·	-12	6	·	·
		$\kappa_{30}\kappa_{12}$	·	-6	3	·	·
		$\kappa_{22}\kappa_{20}$	·	-6	3	·	·
		κ_{21}^2	·	-12	6	·	·
		$\kappa_{20}\kappa_{11}^2$	·	·	6	·	·
$\kappa\binom{22}{11}$	$n^{(3)}$	κ_{42}	2	-3	1	·	·
		$\kappa_{40}\kappa_{02}$	·	-2	1	·	·
		$\kappa_{31}\kappa_{11}$	·	-8	4	·	·
		$\kappa_{30}\kappa_{12}$	·	-8	4	·	·
		$\kappa_{22}\kappa_{20}$	·	-8	4	·	·
		κ_{21}^2	·	-10	5	·	·
		$\kappa_{20}^2\kappa_{02}$	·	·	2	·	·
		$\kappa_{20}\kappa_{11}^2$	·	·	4	·	·
$\kappa\binom{30}{03}$	$n^{(3)}$	κ_{33}	2	-3	1	·	·
		$\kappa_{22}\kappa_{11}$	·	-18	9	·	·
		$\kappa_{21}\kappa_{12}$	·	-18	9	·	·
		κ_{11}^3	·	·	6	·	·
$\kappa\binom{21}{12}$	$n^{(3)}$	κ_{33}	2	-3	1	·	·
		$\kappa_{31}\kappa_{02}$	·	-4	2	·	·
		$\kappa_{30}\kappa_{03}$	·	-2	1	·	·
		$\kappa_{22}\kappa_{11}$	·	-10	5	·	·
		$\kappa_{21}\kappa_{12}$	·	-16	8	·	·
		$\kappa_{20}\kappa_{13}$	·	-4	2	·	·
		$\kappa_{20}\kappa_{11}\kappa_{02}$	·	·	4	·	·
		κ_{11}^3	·	·	2	·	·
$\kappa\binom{333}{000}$	$(n^{(3)})^2$	κ_{90}	4	-12	13	-6	1
		$\kappa_{70}\kappa_{20}$	·	-108	216	-135	27
		$\kappa_{60}\kappa_{30}$	·	-432	756	-432	81
		$\kappa_{50}\kappa_{40}$	·	-756	1188	-621	108
		$\kappa_{50}\kappa_{20}^2$	·	·	756	-810	216
		$\kappa_{40}\kappa_{30}\kappa_{20}$	·	·	3888	-3564	810
		κ_{30}^3	·	·	1224	-1080	252
		$\kappa_{30}\kappa_{20}^3$	·	·	·	-1296	540
$\kappa\binom{332}{001}$	$(n^{(3)})^2$	κ_{81}	4	-12	13	-6	1
		$\kappa_{70}\kappa_{11}$	·	-24	48	-30	6
		$\kappa_{61}\kappa_{20}$	·	-84	168	-105	21
		$\kappa_{60}\kappa_{21}$	·	-144	252	-144	27
		$\kappa_{51}\kappa_{30}$	·	-288	504	-288	54
		$\kappa_{50}\kappa_{31}$	·	-336	528	-276	48
		$\kappa_{41}\kappa_{40}$	·	-420	660	-345	60
		$\kappa_{50}\kappa_{20}\kappa_{11}$	·	·	336	-360	96
		$\kappa_{41}\kappa_{20}^2$	·	·	420	-450	120
		$\kappa_{40}\kappa_{30}\kappa_{11}$	·	·	864	-792	180
		$\kappa_{40}\kappa_{21}\kappa_{20}$	·	·	1296	-1188	270
		$\kappa_{31}\kappa_{30}\kappa_{20}$	·	·	1728	-1584	360
		$\kappa_{30}^2\kappa_{21}$	·	·	1224	-1080	252
		$\kappa_{30}\kappa_{20}^2\kappa_{11}$	·	·	·	-864	360
		$\kappa_{21}\kappa_{20}^3$	·	·	·	-432	180
$\kappa\binom{331}{002}$	$(n^{(3)})^2$	κ_{72}	4	-12	13	-6	1
		$\kappa_{61}\kappa_{11}$	·	-48	96	-60	12
		$\kappa_{60}\kappa_{12}$	·	-24	48	-30	6
		$\kappa_{52}\kappa_{20}$	·	-60	120	-75	15
		$\kappa_{51}\kappa_{21}$	·	-240	408	-228	42
		$\kappa_{50}\kappa_{22}$	·	-96	156	-84	15
		$\kappa_{42}\kappa_{30}$	·	-168	300	-174	33
		$\kappa_{41}\kappa_{31}$	·	-480	744	-384	66
		$\kappa_{40}\kappa_{32}$	·	-180	288	-153	27
		$\kappa_{50}\kappa_{11}^3$	·	·	96	-108	30
		$\kappa_{41}\kappa_{20}\kappa_{11}$	·	·	480	-504	132
		$\kappa_{40}\kappa_{21}\kappa_{11}$	·	·	720	-648	144
		$\kappa_{40}\kappa_{20}\kappa_{12}$	·	·	216	-216	54
		$\kappa_{32}\kappa_{20}^2$	·	·	180	-198	54
		$\kappa_{31}\kappa_{30}\kappa_{11}$	·	·	1008	-936	216
		$\kappa_{31}\kappa_{21}\kappa_{20}$	·	·	1440	-1296	288
		$\kappa_{30}^2\kappa_{12}$	·	·	216	-216	54
		$\kappa_{30}\kappa_{22}\kappa_{20}$	·	·	504	-468	108
		$\kappa_{30}\kappa_{21}^2$	·	·	1008	-864	198
		$\kappa_{30}\kappa_{20}^2\kappa_{11}^2$	·	·	·	-504	216
		$\kappa_{21}^2\kappa_{20}\kappa_{11}$	·	·	·	-720	288
		$\kappa_{20}^3\kappa_{12}$	·	·	·	-72	36

Table 3·2 (cont.)

Left table

	Dividing factor	Product of cumulants	n^0	n^1	n^2	n^3	n^4
$\kappa\binom{322}{011}$	$(n^{(2)})^2$	κ_{72}	4	−12	13	−6	1
		$\kappa_{70}\kappa_{02}$	·	−4	8	−5	1
		$\kappa_{61}\kappa_{11}$	·	−40	80	−50	10
		$\kappa_{60}\kappa_{12}$	·	−40	68	−38	7
		$\kappa_{52}\kappa_{20}$	·	−64	128	−80	16
		$\kappa_{51}\kappa_{21}$	·	−208	368	−212	40
		$\kappa_{50}\kappa_{22}$	·	−136	212	−110	19
		$\kappa_{42}\kappa_{30}$	·	−184	320	−182	34
		$\kappa_{41}\kappa_{31}$	·	−400	632	−332	58
		$\kappa_{40}\kappa_{32}$	·	−220	344	−179	31
		$\kappa_{50}\kappa_{20}\kappa_{02}$	·	·	56	−60	16
		$\kappa_{50}\kappa_{11}^2$	·	·	80	−84	22
		$\kappa_{41}\kappa_{20}\kappa_{11}$	·	·	400	−432	116
		$\kappa_{40}\kappa_{30}\kappa_{02}$	·	·	144	−132	30
		$\kappa_{40}\kappa_{21}\kappa_{11}$	·	·	624	−576	132
		$\kappa_{40}\kappa_{20}\kappa_{12}$	·	·	360	−324	72
		$\kappa_{32}\kappa_{20}^2$	·	·	220	−234	62
		$\kappa_{31}\kappa_{30}\kappa_{11}$	·	·	816	−744	168
		$\kappa_{31}\kappa_{21}\kappa_{20}$	·	·	1248	−1152	264
		$\kappa_{30}^2\kappa_{12}$	·	·	336	−288	66
		$\kappa_{30}\kappa_{22}\kappa_{20}$	·	·	696	−636	144
		$\kappa_{30}\kappa_{31}^2$	·	·	888	−792	186
		$\kappa_{30}\kappa_{20}\kappa_{11}^2$	·	·	·	−408	168
		$\kappa_{30}\kappa_{20}^2\kappa_{02}$	·	·	·	−144	60
		$\kappa_{31}\kappa_{20}^2\kappa_{11}$	·	·	·	−624	264
		$\kappa_{20}^3\kappa_{12}$	·	·	·	−120	48
$\kappa\binom{222}{111}$	$(n^{(3)})^2$	κ_{63}	4	−12	13	−6	1
		$\kappa_{61}\kappa_{02}$	·	−12	24	−15	3
		$\kappa_{60}\kappa_{03}$	·	−8	12	−6	1
		$\kappa_{52}\kappa_{11}$	·	−48	96	−60	12
		$\kappa_{51}\kappa_{12}$	·	−96	168	−96	18
		$\kappa_{50}\kappa_{13}$	·	−48	72	−36	6
		$\kappa_{43}\kappa_{20}$	·	−48	96	−60	12
		$\kappa_{42}\kappa_{21}$	·	−216	384	−222	42
		$\kappa_{41}\kappa_{22}$	·	−264	420	−222	39
		$\kappa_{40}\kappa_{23}$	·	−108	168	−87	15
		$\kappa_{33}\kappa_{30}$	·	−112	192	−108	20
		$\kappa_{32}\kappa_{31}$	·	−336	528	−276	48
		$\kappa_{50}\kappa_{11}\kappa_{02}$	·	·	48	−48	12
		$\kappa_{41}\kappa_{20}\kappa_{02}$	·	·	120	−132	36
		$\kappa_{41}\kappa_{11}^2$	·	·	144	−156	42
		$\kappa_{40}\kappa_{21}\kappa_{02}$	·	·	192	−180	42
		$\kappa_{40}\kappa_{20}\kappa_{03}$	·	·	72	−60	12
		$\kappa_{40}\kappa_{12}\kappa_{11}$	·	·	288	−264	60
		$\kappa_{32}\kappa_{20}\kappa_{11}$	·	·	336	−360	96
		$\kappa_{31}\kappa_{30}\kappa_{02}$	·	·	240	−216	48
		$\kappa_{31}\kappa_{21}\kappa_{11}$	·	·	912	−840	192
		$\kappa_{31}\kappa_{20}\kappa_{12}$	·	·	576	−528	120
		$\kappa_{30}^2\kappa_{03}$	·	·	64	−48	10
		$\kappa_{30}\kappa_{22}\kappa_{11}$	·	·	528	−480	108
		$\kappa_{30}\kappa_{21}\kappa_{12}$	·	·	816	−720	168
		$\kappa_{30}\kappa_{20}\kappa_{13}$	·	·	240	−216	48
		$\kappa_{23}\kappa_{20}^2$	·	·	108	−114	30
		$\kappa_{22}\kappa_{21}\kappa_{20}$	·	·	840	−780	180
		κ_{21}^3	·	·	344	−312	74
		$\kappa_{30}\kappa_{20}\kappa_{11}\kappa_{02}$	·	·	·	−240	96
		$\kappa_{30}\kappa_{11}^3$	·	·	·	−96	40
		$\kappa_{21}\kappa_{20}^2\kappa_{02}$	·	·	·	−192	84
		$\kappa_{21}\kappa_{20}\kappa_{11}^2$	·	·	·	−456	192
		$\kappa_{20}^3\kappa_{03}$	·	·	·	−24	8
		$\kappa_{20}^2\kappa_{12}\kappa_{11}$	·	·	·	−288	120
$\kappa\binom{330}{003}$	$(n^{(3)})^2$	κ_{63}	4	−12	13	−6	1
		$\kappa_{52}\kappa_{11}$	·	−72	144	−90	18
		$\kappa_{51}\kappa_{12}$	·	−72	144	−90	18
		$\kappa_{43}\kappa_{20}$	·	−36	72	−45	9
		$\kappa_{42}\kappa_{21}$	·	−288	468	−252	45
		$\kappa_{41}\kappa_{22}$	·	−288	468	−252	45
		$\kappa_{40}\kappa_{23}$	·	−36	72	−45	9
		$\kappa_{33}\kappa_{30}$	·	−72	144	−90	18
		$\kappa_{32}\kappa_{31}$	·	−432	648	−324	54
		$\kappa_{41}\kappa_{11}^2$	·	·	288	−324	90
		$\kappa_{40}\kappa_{12}\kappa_{11}$	·	·	216	−216	54
		$\kappa_{32}\kappa_{20}\kappa_{11}$	·	·	432	−432	108
		$\kappa_{31}\kappa_{21}\kappa_{11}$	·	·	1728	−1512	324
		$\kappa_{31}\kappa_{20}\kappa_{12}$	·	·	432	−432	108
		$\kappa_{30}\kappa_{22}\kappa_{11}$	·	·	648	−648	162
		$\kappa_{30}\kappa_{21}\kappa_{12}$	·	·	648	−648	162
		$\kappa_{23}\kappa_{20}^2$	·	·	36	−54	18
		$\kappa_{22}\kappa_{21}\kappa_{20}$	·	·	864	−756	162
		κ_{21}^3	·	·	576	−432	90
		$\kappa_{30}\kappa_{11}^3$	·	·	·	−216	108
		$\kappa_{21}\kappa_{20}\kappa_{11}^2$	·	·	·	−864	324
		$\kappa_{20}^2\kappa_{12}\kappa_{11}$	·	·	·	−216	108

Right table

	Dividing factor	Product of cumulants	n^0	n^1	n^2	n^3	n^4
$\kappa\binom{321}{012}$	$(n^{(3)})^2$	κ_{63}	4	−12	13	−6	1
		$\kappa_{61}\kappa_{02}$	·	−8	16	−10	2
		$\kappa_{60}\kappa_{03}$	·	−4	8	−5	1
		$\kappa_{52}\kappa_{11}$	·	−56	112	−70	14
		$\kappa_{51}\kappa_{12}$	·	−92	160	−91	17
		$\kappa_{50}\kappa_{13}$	·	−32	52	−28	5
		$\kappa_{43}\kappa_{20}$	·	−44	88	−55	11
		$\kappa_{42}\kappa_{21}$	·	−236	412	−235	44
		$\kappa_{41}\kappa_{22}$	·	−272	424	−220	38
		$\kappa_{40}\kappa_{23}$	·	−84	132	−69	12
		$\kappa_{33}\kappa_{30}$	·	−100	176	−101	19
		$\kappa_{32}\kappa_{31}$	·	−368	580	−304	53
		$\kappa_{50}\kappa_{11}\kappa_{02}$	·	·	32	−36	10
		$\kappa_{41}\kappa_{20}\kappa_{02}$	·	·	80	−84	22
		$\kappa_{41}\kappa_{11}^2$	·	·	192	−204	54
		$\kappa_{40}\kappa_{21}\kappa_{02}$	·	·	120	−108	24
		$\kappa_{40}\kappa_{20}\kappa_{03}$	·	·	36	−36	9
		$\kappa_{40}\kappa_{12}\kappa_{11}$	·	·	276	−252	57
		$\kappa_{32}\kappa_{20}\kappa_{11}$	·	·	368	−396	106
		$\kappa_{31}\kappa_{30}\kappa_{02}$	·	·	168	−156	36
		$\kappa_{31}\kappa_{21}\kappa_{11}$	·	·	1176	−1080	246
		$\kappa_{31}\kappa_{20}\kappa_{12}$	·	·	552	−504	114
		$\kappa_{30}^2\kappa_{03}$	·	·	36	−36	9
		$\kappa_{30}\kappa_{22}\kappa_{11}$	·	·	564	−516	117
		$\kappa_{30}\kappa_{21}\kappa_{12}$	·	·	780	−684	159
		$\kappa_{30}\kappa_{20}\kappa_{13}$	·	·	168	−156	36
		$\kappa_{23}\kappa_{20}^2$	·	·	84	−90	24
		$\kappa_{22}\kappa_{21}\kappa_{20}$	·	·	828	−756	171
		κ_{21}^3	·	·	408	−360	84
		$\kappa_{30}\kappa_{20}\kappa_{11}\kappa_{02}$	·	·	·	−168	72
		$\kappa_{30}\kappa_{11}^3$	·	·	·	−132	54
		$\kappa_{21}\kappa_{20}^2\kappa_{02}$	·	·	·	−120	48
		$\kappa_{21}\kappa_{20}\kappa_{11}^2$	·	·	·	−588	246
		$\kappa_{20}^3\kappa_{03}$	·	·	·	−12	6
		$\kappa_{20}^2\kappa_{12}\kappa_{11}$	·	·	·	−276	114
$\kappa\binom{221}{112}$	$(n^{(3)})^2$	κ_{54}	4	−12	13	−6	1
		$\kappa_{52}\kappa_{02}$	·	−20	40	−25	5
		$\kappa_{51}\kappa_{03}$	·	−24	40	−22	4
		$\kappa_{50}\kappa_{04}$	·	−8	12	−6	1
		$\kappa_{43}\kappa_{11}$	·	−56	112	−70	14
		$\kappa_{42}\kappa_{12}$	·	−152	268	−154	29
		$\kappa_{41}\kappa_{13}$	·	−128	200	−104	18
		$\kappa_{40}\kappa_{14}$	·	−36	56	−29	5
		$\kappa_{34}\kappa_{20}$	·	−32	64	−40	8
		$\kappa_{33}\kappa_{21}$	·	−200	352	−202	38
		$\kappa_{32}\kappa_{22}$	·	−344	544	−286	50
		$\kappa_{31}\kappa_{23}$	·	−240	376	−196	34
		$\kappa_{30}\kappa_{24}$	·	−56	96	−54	10
		$\kappa_{50}\kappa_{02}^2$	·	·	8	−8	2
		$\kappa_{41}\kappa_{11}\kappa_{02}$	·	·	128	−136	36
		$\kappa_{40}\kappa_{12}\kappa_{02}$	·	·	112	−104	24
		$\kappa_{40}\kappa_{11}\kappa_{03}$	·	·	72	−64	14
		$\kappa_{32}\kappa_{20}\kappa_{02}$	·	·	136	−148	40
		$\kappa_{32}\kappa_{11}^2$	·	·	208	−224	60
		$\kappa_{31}\kappa_{21}\kappa_{02}$	·	·	400	−368	84
		$\kappa_{31}\kappa_{20}\kappa_{03}$	·	·	144	−128	28
		$\kappa_{31}\kappa_{12}\kappa_{11}$	·	·	688	−632	144
		$\kappa_{30}\kappa_{22}\kappa_{02}$	·	·	208	−188	42
		$\kappa_{30}\kappa_{21}\kappa_{03}$	·	·	200	−168	38
		$\kappa_{30}\kappa_{20}\kappa_{04}$	·	·	40	−36	8
		$\kappa_{30}\kappa_{13}\kappa_{11}$	·	·	256	−232	52
		$\kappa_{30}\kappa_{12}^2$	·	·	272	−240	56
		$\kappa_{23}\kappa_{20}\kappa_{11}$	·	·	240	−256	68
		$\kappa_{22}\kappa_{21}\kappa_{11}$	·	·	1000	−920	210
		$\kappa_{22}\kappa_{20}\kappa_{12}$	·	·	568	−524	120
		$\kappa_{21}^2\kappa_{12}$	·	·	752	−672	158
		$\kappa_{21}\kappa_{20}\kappa_{13}$	·	·	400	−368	84
		$\kappa_{20}^2\kappa_{14}$	·	·	36	−38	10
		$\kappa_{30}\kappa_{20}\kappa_{02}^2$	·	·	·	−40	16
		$\kappa_{30}\kappa_{11}^2\kappa_{02}$	·	·	·	−128	52
		$\kappa_{21}\kappa_{20}\kappa_{11}\kappa_{02}$	·	·	·	−400	168
		$\kappa_{21}\kappa_{11}^3$	·	·	·	−200	84
		$\kappa_{20}^2\kappa_{12}\kappa_{02}$	·	·	·	−112	48
		$\kappa_{20}^2\kappa_{11}\kappa_{03}$	·	·	·	−72	28
		$\kappa_{20}\kappa_{11}^2\kappa_{12}$	·	·	·	−344	144

Table 3·2 (*cont.*)

Left table:

	Dividing factor	Product of cumulants	Coefficients of				
			n^0	n^1	n^2	n^3	n^4
$\kappa\binom{311}{022}$	$(n^{(3)})^2$	κ_{54}	4	-12	13	-6	1
		$\kappa_{52}\kappa_{02}$	·	-16	32	-20	4
		$\kappa_{51}\kappa_{03}$	·	-16	32	-20	4
		$\kappa_{50}\kappa_{04}$	·	-4	8	-5	1
		$\kappa_{43}\kappa_{11}$	·	-64	128	-80	16
		$\kappa_{42}\kappa_{12}$	·	-160	272	-152	28
		$\kappa_{41}\kappa_{13}$	·	-112	176	-92	16
		$\kappa_{40}\kappa_{14}$	·	-24	36	-18	3
		$\kappa_{34}\kappa_{20}$	·	-28	56	-35	7
		$\kappa_{33}\kappa_{21}$	·	-208	368	-212	40
		$\kappa_{32}\kappa_{22}$	·	-376	584	-302	52
		$\kappa_{31}\kappa_{23}$	·	-240	384	-204	36
		$\kappa_{30}\kappa_{24}$	·	-48	84	-48	9
		$\kappa_{50}\kappa_{02}^2$	·	·	4	-6	2
		$\kappa_{41}\kappa_{11}\kappa_{02}$	·	·	112	-120	32
		$\kappa_{40}\kappa_{12}\kappa_{02}$	·	·	72	-60	12
		$\kappa_{40}\kappa_{11}\kappa_{03}$	·	·	48	-48	12
		$\kappa_{32}\kappa_{20}\kappa_{02}$	·	·	104	-108	28
		$\kappa_{32}\kappa_{11}^2$	·	·	272	-288	76
		$\kappa_{31}\kappa_{21}\kappa_{02}$	·	·	336	-312	72
		$\kappa_{31}\kappa_{20}\kappa_{03}$	·	·	96	-96	24
		$\kappa_{31}\kappa_{12}\kappa_{11}$	·	·	816	-744	168
		$\kappa_{30}\kappa_{22}\kappa_{02}$	·	·	168	-156	36
		$\kappa_{30}\kappa_{21}\kappa_{03}$	·	·	144	-144	36
		$\kappa_{30}\kappa_{20}\kappa_{04}$	·	·	24	-24	6
		$\kappa_{30}\kappa_{13}\kappa_{11}$	·	·	240	-216	48
		$\kappa_{30}\kappa_{12}^2$	·	·	264	-216	48
		$\kappa_{23}\kappa_{20}\kappa_{11}$	·	·	240	-264	72
		$\kappa_{22}\kappa_{21}\kappa_{11}$	·	·	1200	-1104	252
		$\kappa_{22}\kappa_{20}\kappa_{12}$	·	·	552	-492	108
		$\kappa_{21}^2\kappa_{12}$	·	·	816	-720	168
		$\kappa_{21}\kappa_{20}\kappa_{13}$	·	·	336	-312	72
		$\kappa_{20}^2\kappa_{14}$	·	·	24	-24	6
		$\kappa_{30}\kappa_{20}\kappa_{02}^2$	·	·	·	-24	12
		$\kappa_{30}\kappa_{11}^2\kappa_{02}$	·	·	·	-120	48
		$\kappa_{21}\kappa_{20}\kappa_{11}\kappa_{02}$	·	·	·	-336	144
		$\kappa_{21}\kappa_{11}^3$	·	·	·	-288	120
		$\kappa_{20}^2\kappa_{12}\kappa_{02}$	·	·	·	-72	24
		$\kappa_{20}^2\kappa_{11}\kappa_{03}$	·	·	·	-48	24
		$\kappa_{20}\kappa_{12}\kappa_{11}^2$	·	·	·	-408	168
$\kappa\binom{302}{031}$	$(n^{(3)})^2$	κ_{54}	4	-12	13	-6	1
		$\kappa_{52}\kappa_{02}$	·	-12	24	-15	3
		$\kappa_{51}\kappa_{03}$	·	-12	24	-15	3
		$\kappa_{43}\kappa_{11}$	·	-72	144	-90	18
		$\kappa_{42}\kappa_{12}$	·	-156	276	-159	30
		$\kappa_{41}\kappa_{13}$	·	-96	156	-84	15
		$\kappa_{40}\kappa_{14}$	·	-12	24	-15	3
		$\kappa_{34}\kappa_{20}$	·	-24	48	-30	6
		$\kappa_{33}\kappa_{21}$	·	-228	384	-213	39
		$\kappa_{32}\kappa_{22}$	·	-408	636	-330	57
		$\kappa_{31}\kappa_{23}$	·	-240	372	-192	33
		$\kappa_{30}\kappa_{24}$	·	-36	72	-45	9
		$\kappa_{41}\kappa_{11}\kappa_{02}$	·	·	96	-108	30
		$\kappa_{40}\kappa_{12}\kappa_{02}$	·	·	36	-36	9
		$\kappa_{40}\kappa_{11}\kappa_{03}$	·	·	36	-36	9
		$\kappa_{32}\kappa_{20}\kappa_{02}$	·	·	72	-72	18
		$\kappa_{32}\kappa_{11}^2$	·	·	336	-360	96
		$\kappa_{31}\kappa_{21}\kappa_{02}$	·	·	288	-252	54
		$\kappa_{31}\kappa_{20}\kappa_{03}$	·	·	72	-72	18
		$\kappa_{31}\kappa_{12}\kappa_{11}$	·	·	864	-792	180
		$\kappa_{30}\kappa_{22}\kappa_{02}$	·	·	108	-108	27
		$\kappa_{30}\kappa_{21}\kappa_{03}$	·	·	108	-108	27
		$\kappa_{30}\kappa_{13}\kappa_{11}$	·	·	216	-216	54
		$\kappa_{30}\kappa_{12}^2$	·	·	216	-216	54
		$\kappa_{23}\kappa_{20}\kappa_{11}$	·	·	240	-252	66
		$\kappa_{22}\kappa_{21}\kappa_{11}$	·	·	1476	-1332	297
		$\kappa_{22}\kappa_{20}\kappa_{12}$	·	·	504	-468	108
		$\kappa_{21}^2\kappa_{12}$	·	·	900	-756	171
		$\kappa_{21}\kappa_{20}\kappa_{13}$	·	·	288	-252	54
		$\kappa_{20}^2\kappa_{14}$	·	·	12	-18	6
		$\kappa_{30}\kappa_{11}^2\kappa_{02}$	·	·	·	-108	54
		$\kappa_{21}\kappa_{20}\kappa_{11}\kappa_{02}$	·	·	·	-288	108
		$\kappa_{21}\kappa_{11}^3$	·	·	·	-396	162
		$\kappa_{20}^2\kappa_{12}\kappa_{02}$	·	·	·	-36	18
		$\kappa_{20}^2\kappa_{11}\kappa_{03}$	·	·	·	-36	18
		$\kappa_{20}\kappa_{12}\kappa_{11}^2$	·	·	·	-432	180

Right table:

	Dividing factor	Product of cumulants	Coefficients of	
			n^0	n^1
$\kappa\binom{32}{00}$	$n^{(2)}$	κ_{50}	-1	1
		$\kappa_{30}\kappa_{20}$	·	6
$\kappa\binom{31}{01}$	$n^{(2)}$	κ_{41}	-1	1
		$\kappa_{30}\kappa_{11}$	·	3
		$\kappa_{21}\kappa_{20}$	·	3
$\kappa\binom{22}{10}$	$n^{(2)}$	κ_{41}	-1	1
		$\kappa_{30}\kappa_{11}$	·	2
		$\kappa_{21}\kappa_{20}$	·	4
$\kappa\binom{30}{02}$	$n^{(2)}$	κ_{32}	-1	1
		$\kappa_{21}\kappa_{11}$	·	6
$\kappa\binom{21}{11}$	$n^{(2)}$	κ_{32}	-1	1
		$\kappa_{30}\kappa_{02}$	·	1
		$\kappa_{21}\kappa_{11}$	·	3
		$\kappa_{20}\kappa_{12}$	·	2
$\kappa\binom{12}{20}$	$n^{(2)}$	κ_{32}	-1	1
		$\kappa_{21}\kappa_{11}$	·	4
		$\kappa_{20}\kappa_{12}$	·	2

Table 3·3. *Moments of sampling moments when the population is finite and of size N*

Given a sample $x_1, x_2, \ldots, x_n$, from a finite population, let

$$m_a = \frac{1}{n} \sum_{i=1}^{n} (x_i - m_1')^a$$

where m_1' is the sample mean.

If the population consists of N quantities $u_1, u_2, \ldots, u_N$, let

$$\mu_t = \frac{1}{N} \sum_{j=1}^{N} (u_j - \mu_1')^t$$

where μ_1' is the population mean.

Let
$$M(2^r 1^s) = \mathscr{E}_N[(m_2 - \mathscr{E}_N(m_2))^r (m_1' - \mathscr{E}_N(m_1'))^s]$$

and for brevity denote
$$e_v = n^{(v)}/N^{(v)}.$$

The table gives the expansion of M in terms of products and powers of μ. For example

$$M(21^2) = Nn^{-4}\mu_4[e_1(n-1) - e_2(3n-7) + e_3(2n-12) + 6e_4]$$
$$+ N^2 n^{-4}\mu_2^2[e_2(n-3) - e_3(n-6) - 3e_4].$$

From the point of view of numerical computation this form of expansion is possibly the easiest. The expansions are only given for $a = 1, 2$, since these are the ones commonly called for.

Table 3·3

	Multiplying factor	Coefficients of						
		e_1	e_2	e_3	e_4	e_5	e_6	e_7
$M(2)$	$Nn^{-2}\mu_2$	$n-1$	1					
$M(1^2)$	$Nn^{-2}\mu_2$	1	-1					
$M(21)$	$Nn^{-3}\mu_3$	$n-1$	$-(n-3)$	-2				
$M(1^3)$	$Nn^{-3}\mu_3$	1	-3	2				
$M(2^2)$	$Nn^{-4}\mu_4$	$(n-1)^2$	$-(n^2-6n+7)$	$-(4n-12)$	-6			
	$N^2n^{-4}\mu_2^2$	.	n^2-2n+3	$2n-6$	3			
$M(21^2)$	$Nn^{-4}\mu_4$	$n-1$	$-(3n-7)$	$2n-12$	6			
	$N^2n^{-4}\mu_2^2$	.	$n-3$	$-(n-6)$	-3			
$M(1^4)$	$Nn^{-4}\mu_4$	1	-7	12	-6			
	$N^2n^{-4}\mu_2^2$	.	3	-6	3			
$M(2^2 1)$	$Nn^{-5}\mu_5$	$(n-1)^2$	$-(3n^2-14n+15)$	$2(n^2-12n+25)$	$12(n-5)$	24		
	$N^2n^{-5}\mu_3\mu_2$	.	$2(n^2-4n+5)$	$-2(n^2-9n+20)$	$-10(n-5)$	-20		
$M(21^3)$	$Nn^{-5}\mu_5$	$n-1$	$-(7n-15)$	$12n-50$	$-6(n-10)$	-24		
	$N^2n^{-5}\mu_3\mu_2$	.	$4n-10$	$-(9n-40)$	$5(n-10)$	20		
$M(1^5)$	$Nn^{-5}\mu_5$	1	-15	50	-60	24		
	$N^2n^{-5}\mu_3\mu_2$	.	10	-40	50	-20		
$M(2^3)$	$Nn^{-6}\mu_6$	$(n-1)^3$	$-(3n^3-21n^2+45n-31)$	$2n^3-36n^2+150n-180$	$18n^2-180n+390$	$24n-360$	120	
	$N^2n^{-6}\mu_4\mu_2$	.	$3n^3-9n^2+21n-15$	$-(3n^3-24n^2+93n-120)$	$-(15n^2-126n+285)$	$-(54n-270)$	-90	
	$N^2n^{-6}\mu_3^2$	.	$-(6n^2-12n+10)$	$12n^2-48n+60$	$-(6n^2-60n+130)$	$-(24n-120)$	-40	
	$N^3n^{-6}\mu_2^3$	.	.	$n^3-3n^2+9n-15$	$3n^2-18n+45$	$9n-45$	15	
$M(2^2 1^2)$	$Nn^{-6}\mu_6$	$(n-1)^2$	$-(7n^2-30n+31)$	$12n^2-100n+180$	$-(6n^2-120n+390)$	$-(48n-360)$	-120	
	$N^2n^{-6}\mu_4\mu_2$	.	$3n^2-14n+15$	$-(8n^2-62n+120)$	$5n^2-84n+285$	$36n-270$	90	
	$N^2n^{-6}\mu_3^2$	.	$2n^2-8n+10$	$-(4n^2-32n+60)$	$2n^2-40n+130$	$16n-120$	40	
	$N^3n^{-6}\mu_2^3$	.	.	$n^2-6n+15$	$-(n^2-12n+45)$	$-(6n-45)$	-15	
$M(21^4)$	$Nn^{-6}\mu_6$	$(n-1)$	$-(15n-31)$	$50n-180$	$-(60n-390)$	$24n-360$	120	
	$N^2n^{-6}\mu_4\mu_2$	.	$7n-15$	$-(31n-120)$	$42n-285$	$-(18n-270)$	-90	
	$N^2n^{-6}\mu_3^2$	.	$4n-10$	$-(16n-60)$	$20n-130$	$-(8n-120)$	-40	
	$N^3n^{-6}\mu_2^3$	.	.	$3n-15$	$-(6n-45)$	$3n-45$	15	
$M(1^6)$	$Nn^{-6}\mu_6$	1	-31	180	-390	360	-120	
	$N^2n^{-6}\mu_4\mu_2$	.	15	-120	285	-270	90	
	$N^2n^{-6}\mu_3^2$	.	10	-60	130	-120	40	
	$N^3n^{-6}\mu_2^3$	.	.	15	-45	45	-15	
$M(2^3 1)$	$Nn^{-7}\mu_7$	$(n-1)^3$	$-(7n^3-45n^2+93n-63)$	$12n^3-150n^2+540n-602$	$-(6n^3-180n^2+1170n-2100)$	$-(72n^2-1080n+3360)$	$-(360n-2520)$	-720
	$N^2n^{-7}\mu_5\mu_2$	.	$3n^3-15n^2+33n-21$	$-(9n^3-78n^2+285n-336)$	$6n^3-117n^2+720n-1365$	$54n^2-720n+2310$	$252n-1764$	504
	$N^2n^{-7}\mu_4\mu_3$	.	$3n^2-21n^2+45n-35$	$-(6n^3-84n^2+300n-350)$	$3n^3-105n^2+675n-1225$	$42n^2-630n+1960$	$210n-1470$	420
	$N^3n^{-7}\mu_3\mu_2^2$	.	.	$3n^3-21n^2+75n-105$	$-(3n^3-45n^2+225n-525)$	$-(24n^2-285n+945)$	$-(105n-735)$	-210

Table 3·3 (cont.)

	Multiplying factor	Coefficients of			
		e_1	e_2	e_3	e_4
$M(2^2 1^3)$	$Nn^{-7}\mu_7$	$(n-1)^2$	$-(15n^2-62n+63)$	$50n^2-360n+602$	$-(60n^2-780n+2100)$
	$N^2n^{-7}\mu_5\mu_2$	.	$5n^2-22n+21$	$-(26n^2-190n+336)$	$39n^2-480n+1365$
	$N^2n^{-7}\mu_4\mu_3$	.	$7n^2-30n+35$	$-(28n^2-200n+350)$	$35n^2-450n+1225$
	$N^3n^{-7}\mu_3\mu_2^2$	.	.	$7n^2-50n+105$	$-(15n^2-170n+525)$
$M(21^5)$	$Nn^{-7}\mu_7$	$(n-1)$	$-(31n-63)$	$180n-602$	$-(390n-2100)$
	$N^2n^{-7}\mu_5\mu_2$	.	$11n-21$	$-(95n-336)$	$240n-1365$
	$N^2n^{-7}\mu_4\mu_3$	.	$15n-35$	$-(100n-350)$	$225n-1225$
	$N^3n^{-7}\mu_3\mu_2^2$	.	.	$25n-105$	$-(85n-525)$
$M(1^7)$	$Nn^{-7}\mu_7$	1	-63	602	-2100
	$N^2n^{-7}\mu_5\mu_2$	.	21	-336	1365
	$N^2n^{-7}\mu_4\mu_3$	.	35	-350	1225
	$N^3n^{-7}\mu_3\mu_2^2$	.		105	-525
$M(2^4)$	$Nn^{-8}\mu_8$	$(n-1)^4$	$-(7n^4-60n^3+186n^2-252n+127)$	$12n^4-200n^3+1080n^2-2408n+1932$	$-(6n^4-240n^3+2340n^2-8400n+10206)$
	$N^2n^{-8}\mu_6\mu_2$	.	$4n^4-16n^3+48n^2-64n+28$	$-(12n^4-100n^3+468n^2-1084n+896)$	$8n^4-164n^3+1284n^2-4620n+5908$
	$N^2n^{-8}\mu_5\mu_3$	.	$-(24n^3-72n^2+104n-56)$	$96n^3-528n^2+1184n-1008$	$-(120n^3-1224n^2+4360n-5432)$
	$N^2n^{-8}\mu_4^2$	.	$3n^4-12n^3+42n^2-60n+35$	$-(6n^4-48n^3+264n^2-600n+490)$	$3n^4-60n^3+582n^2-2100n+2555$
	$N^3n^{-8}\mu_4\mu_2^2$	.	.	$6n^4-24n^3+96n^2-240n+210$	$-(6n^4-60n^3+384n^2-1380n+1890)$
	$N^3n^{-8}\mu_3^2\mu_2$	.	.	$-(24n^3-120n^2+280n-280)$	$48n^3-408n^2+1440n-1960$
	$N^4n^{-8}\mu_2^4$	.	.	.	$n^4-4n^3+18n^2-60n+105$
$M(2^3 1^2)$	$Nn^{-8}\mu_8$	$(n-1)^3$	$-(15n^3-93n^2+189n-127)$	$50n^3-540n^2+1806n-1932$	$-(60n^3-1170n^2+6300n-10206)$
	$N^2n^{-8}\mu_6\mu_2$	.	$4n^3-24n^2+48n-28$	$-(25n^3-234n^2+813n-896)$	$41n^3-642n^2+3465n-5908$
	$N^2n^{-8}\mu_5\mu_3$	.	$6n^3-36n^2+78n-56$	$-(24n^3-264n^2+888n-1008)$	$30n^3-612n^2+3270n-5432$
	$N^2n^{-8}\mu_4^2$	.	$3n^3-21n^2+45n-35$	$-(12n^3-132n^2+450n-490)$	$15n^3-291n^2+1575n-2555$
	$N^3n^{-8}\mu_4\mu_2^2$	.	.	$6n^3-48n^2+180n-210$	$-(15n^3-192n^2+1035n-1890)$
	$N^3n^{-8}\mu_3^2\mu_2$	.	.	$6n^3-60n^2+210n-280$	$-(12n^3-204n^2+1080n-1960)$
	$N^4n^{-8}\mu_2^4$	.	.	.	$n^3-9n^2+45n-105$
$M(2^2 1^4)$	$Nn^{-8}\mu_8$	$(n-1)^2$	$-(31n^2-126n+127)$	$180n^2-1204n+1932$	$-(390n^2-4200n+10206)$
	$N^2n^{-8}\mu_6\mu_2$	.	$8n^2-32n+28$	$-(78n^2-542n+896)$	$214n^2-2310n+5908$
	$N^2n^{-8}\mu_5\mu_3$	.	$12n^2-52n+56$	$-(88n^2-592n+1008)$	$204n^2-2180n+5432$
	$N^2n^{-8}\mu_4^2$	.	$7n^2-30n+35$	$-(44n^2-300n+490)$	$97n^2-1050n+2555$
	$N^3n^{-8}\mu_4\mu_2^2$	.	.	$16n^2-120n+210$	$-(64n^2-690n+1890)$
	$N^3n^{-8}\mu_3^2\mu_2$	.	.	$20n^2-140n+280$	$-(68n^2-720n+1960)$
	$N^4n^{-8}\mu_2^4$	.	.	.	$3n^2-30n+105$
$M(21^6)$	$Nn^{-8}\mu_8$	$(n-1)$	$-(63n-127)$	$602n-1932$	$-(2100n-10206)$
	$N^2n^{-8}\mu_6\mu_2$	.	$16n-28$	$-(271n-896)$	$1155n-5908$
	$N^2n^{-8}\mu_5\mu_3$	.	$26n-56$	$-(296n-1008)$	$1090n-5432$
	$N^2n^{-8}\mu_4^2$	.	$15n-35$	$-(150n-490)$	$525n-2555$
	$N^3n^{-8}\mu_4\mu_2^2$	.	.	$60n-210$	$-(345n-1890)$
	$N^3n^{-8}\mu_3^2\mu_2$	.	.	$70n-280$	$-(360n-1960)$
	$N^4n^{-8}\mu_2^4$	.	.	.	$15n-105$
$M(1^8)$	$Nn^{-8}\mu_8$	1	-127	1932	-10206
	$N^2n^{-8}\mu_6\mu_2$	.	28	-896	5908
	$N^2n^{-8}\mu_5\mu_3$	.	56	-1008	5432
	$N^2n^{-8}\mu_4^2$	.	35	-490	2555
	$N^3n^{-8}\mu_4\mu_2^2$	.	.	210	-1890
	$N^3n^{-8}\mu_3^2\mu_2$	.	.	280	-1960
	$N^4n^{-8}\mu_2$	.	.	.	105

(For coefficients of e_5, e_6, e_7 and e_8, see next page.)

Table 3·3 (cont.)

	Multiplying factor	Coefficients of			
		e_5	e_6	e_7	e_8
$M(2^2 1^3)$	$Nn^{-7}\mu_7$	$24n^2 - 720n + 3360$	$240n - 2520$	720	
	$N^2n^{-7}\mu_5\mu_2$	$-(18n^2 - 480n + 2310)$	$-(168n - 1760)$	-504	
	$N^3n^{-7}\mu_4\mu_3$	$-(14n^2 - 420n + 1960)$	$-(140n - 1470)$	-420	
	$N^3n^{-7}\mu_3\mu_2^2$	$8n^2 - 190n + 945$	$70n - 735$	210	
$M(21^5)$	$Nn^{-7}\mu_7$	$360n - 3360$	$-(120n - 2520)$	-720	
	$N^2n^{-7}\mu_5\mu_2$	$-(240n - 2310)$	$84n - 1764$	504	
	$N^3n^{-7}\mu_4\mu_3$	$-(210n - 1960)$	$70n - 1470$	420	
	$N^3n^{-7}\mu_3\mu_2^2$	$95n - 945$	$-(35n - 735)$	-210	
$M(1^7)$	$Nn^{-7}\mu_7$	3360	-2520	720	
	$N^2n^{-7}\mu_5\mu_2$	-2310	1764	-504	
	$N^3n^{-7}\mu_4\mu_3$	-1960	1470	-420	
	$N^3n^{-7}\mu_3\mu_2^2$	945	-735	210	
$M(2^4)$	$Nn^{-8}\mu_8$	$-(96n^3 - 2160n^2 + 13440n - 25200)$	$-(720n^2 - 10080n + 31920)$	$-(2880n - 20160)$	-5040
	$N^2n^{-8}\mu_6\mu_2$	$80n^3 - 1368n^2 + 8160n - 15960$	$504n^2 - 6480n + 21000$	$1920n - 13440$	3360
	$N^3n^{-8}\mu_5\mu_3$	$48n^3 - 1152n^2 + 7120n - 13440$	$384n^2 - 5376n + 17024$	$1536n - 10752$	2688
	$N^3n^{-8}\mu_4^2$	$24n^3 - 540n^2 + 3360n - 6300$	$180n^2 - 2520n + 7980$	$720n - 5040$	1260
	$N^3n^{-8}\mu_4\mu_2^2$	$-(36n^3 - 486n^2 + 2760n - 5670)$	$-(198n^2 - 2340n + 7770)$	$-(720n - 5040)$	-1260
	$N^3n^{-8}\mu_3^2\mu_2$	$-(24n^3 - 456n^2 + 2680n - 5320)$	$-(168n^2 - 2160n + 7000)$	$-(640n - 4480)$	-1120
	$N^4n^{-8}\mu_2^4$	$4n^3 - 36n^2 + 180n - 420$	$18n^2 - 180n + 630$	$60n - 420$	105
$M(2^3 1^2)$	$Nn^{-8}\mu_8$	$24n^3 - 1080n^2 + 10080n - 25200$	$360n^2 - 7560n + 31920$	$2160n - 20160$	5040
	$N^2n^{-8}\mu_6\mu_2$	$-(20n^3 - 684n^2 + 6120n - 15960)$	$-(250n^2 - 4860n + 21000)$	$-(1440n - 13440)$	-3360
	$N^2n^{-8}\mu_5\mu_3$	$-(12n^3 - 576n^2 + 5340n - 13440)$	$-(192n^2 - 4032n + 17024)$	$-(1152n - 10752)$	-2688
	$N^2n^{-8}\mu_4^2$	$-(6n^3 - 270n^2 + 2520n - 6300)$	$-(90n^2 - 1890n + 7980)$	$-(540n - 5040)$	-1260
	$N^3n^{-8}\mu_4\mu_2^2$	$9n^3 - 243n^2 + 2070n - 5670$	$99n^2 - 1755n + 7770$	$540n - 5040$	1260
	$N^3n^{-8}\mu_3^2\mu_2$	$6n^3 - 228n^2 + 2010n - 5320$	$84n^2 - 1620n + 7000$	$480n - 4480$	1120
	$N^4n^{-8}\mu_2^4$	$-(n^3 - 18n^2 + 135n - 420)$	$-(9n^2 - 135n + 630)$	$-(45n - 420)$	-105
$M(2^2 1^4)$	$Nn^{-8}\mu_8$	$360n^2 - 6720n + 25200$	$-(120n^2 - 5040n + 31920)$	$-(1440n - 20160)$	-5040
	$N^2n^{-8}\mu_6\mu_2$	$-(228n^2 - 4080n + 15960)$	$84n^2 - 3240n + 21000$	$960n - 13440$	3360
	$N^2n^{-8}\mu_5\mu_3$	$-(192n^2 - 3560n + 13440)$	$64n^2 - 2688n + 17024$	$768n - 10752$	2688
	$N^2n^{-8}\mu_4^2$	$-(90n^2 - 1680n + 6300)$	$30n^2 - 1260n + 7980$	$360n - 5040$	1260
	$N^3n^{-8}\mu_4\mu_2^2$	$81n^2 - 1380n + 5670$	$-(33n^2 - 1170n + 7770)$	$-(360n - 5040)$	-1260
	$N^3n^{-8}\mu_3^2\mu_2$	$76n^2 - 1340n + 5320$	$-(28n^2 - 1080n + 7000)$	$-(320n - 4480)$	-1120
	$N^4n^{-8}\mu_2^4$	$-(6n^2 - 90n + 420)$	$3n^2 - 90n + 630$	$30n - 420$	105
$M(21^6)$	$Nn^{-8}\mu_8$	$3360n - 25200$	$-(2520n - 31920)$	$720n - 20160$	5040
	$N^2n^{-8}\mu_6\mu_2$	$-(2040n - 15960)$	$1620n - 21000$	$-(480n - 13440)$	-3360
	$N^2n^{-8}\mu_5\mu_3$	$-(1780n - 13440)$	$1344n - 17024$	$-(384n - 10752)$	-2688
	$N^2n^{-8}\mu_4^2$	$-(840n - 6300)$	$630n - 7980$	$-(180n - 5040)$	-1260
	$N^3n^{-8}\mu_4\mu_2^2$	$690n - 5670$	$-(585n - 7770)$	$180n - 5040$	1260
	$N^3n^{-8}\mu_3^2\mu_2$	$670n - 5320$	$-(540n - 7000)$	$160n - 4480$	1120
	$N^4n^{-8}\mu_2^4$	$-(45n - 420)$	$45n - 630$	$-(15n - 420)$	-105
$M(1^8)$	$Nn^{-8}\mu_8$	25200	-31920	20160	-5040
	$N^2n^{-8}\mu_6\mu_2$	-15960	21000	-13440	3360
	$N^2n^{-8}\mu_5\mu_3$	-13440	17024	-10752	2688
	$N^2n^{-8}\mu_4^2$	-6300	7980	-5040	1260
	$N^3n^{-8}\mu_4\mu_2^2$	5670	-7770	5040	-1260
	$N^3n^{-8}\mu_3^2\mu_2$	5320	-7000	4480	-1120
	$N^4n^{-8}\mu_2^4$	-420	630	-420	105

Table 4·1. *Number of r-partitions of an integer n*; $n = 1(1)35$, $r = 1(1)n$

Suppose for example $n = 8$. From the table the number of 3-partitions of 8 is five. (These actually are 611, 521, 431, 422, 332.) On many occasions it is useful to know how many r-partitions there are in order to be certain that all terms in an expansion have been enumerated.

$r \backslash n$	1	2	3	4	5	6	7	8	9	10	11	12	13	14	15	16	17	18	19	20	21	22	23	24	25	26	27	28	29	30	31	32	33	34	35
1	1	1	1	1	1	1	1	1	1	1	1	1	1	1	1	1	1	1	1	1	1	1	1	1	1	1	1	1	1	1	1	1	1	1	1
2		1	1	2	2	3	3	4	4	5	5	6	6	7	7	8	8	9	9	10	10	11	11	12	12	13	13	14	14	15	15	16	16	17	17
3			1	1	2	3	4	5	7	8	10	12	14	16	19	21	24	27	30	33	37	40	44	48	52	56	61	65	70	75	80	85	91	96	102
4				1	1	2	3	5	6	9	11	15	18	23	27	34	39	47	54	64	72	84	94	108	120	136	150	169	185	206	225	249	270	297	321
5					1	1	2	3	5	7	10	13	18	23	30	37	47	57	70	84	101	119	141	164	192	221	255	291	333	377	427	480	540	603	674
6						1	1	2	3	5	7	11	14	20	26	35	44	58	71	90	110	136	163	199	235	282	331	391	454	532	612	709	811	931	1057
7							1	1	2	3	5	7	11	15	21	28	38	49	65	82	105	131	164	201	248	300	364	436	522	618	733	860	1009	1175	1367
8								1	1	2	3	5	7	11	15	22	29	40	52	70	89	116	146	186	230	288	352	434	525	638	764	919	1090	1297	1527
9									1	1	2	3	5	7	11	15	22	30	41	54	73	94	123	157	201	252	318	393	488	598	732	887	1076	1291	1549
10										1	1	2	3	5	7	11	15	22	30	42	55	75	97	128	164	212	267	340	423	530	653	807	984	1204	1455
11											1	1	2	3	5	7	11	15	22	30	42	56	76	99	131	169	219	278	355	445	560	695	863	1060	1303
12												1	1	2	3	5	7	11	15	22	30	42	56	77	100	133	172	224	285	366	460	582	725	905	1116
13													1	1	2	3	5	7	11	15	22	30	42	56	77	101	134	174	227	290	373	471	597	747	935
14														1	1	2	3	5	7	11	15	22	30	42	56	77	101	135	175	229	293	378	478	608	762
15															1	1	2	3	5	7	11	15	22	30	42	56	77	101	135	176	230	295	381	483	615
16																1	1	2	3	5	7	11	15	22	30	42	56	77	101	135	176	231	296	383	486
17																	1	1	2	3	5	7	11	15	22	30	42	56	77	101	135	176	231	297	384
18																		1	1	2	3	5	7	11	15	22	30	42	56	77	101	135	176	231	297
19																			1	1	2	3	5	7	11	15	22	30	42	56	77	101	135	176	231
20																				1	1	2	3	5	7	11	15	22	30	42	56	77	101	135	176
21																					1	1	2	3	5	7	11	15	22	30	42	56	77	101	135
22																						1	1	2	3	5	7	11	15	22	30	42	56	77	101
23																							1	1	2	3	5	7	11	15	22	30	42	56	77
24																								1	1	2	3	5	7	11	15	22	30	42	56
25																									1	1	2	3	5	7	11	15	22	30	42
26																										1	1	2	3	5	7	11	15	22	30
27																											1	1	2	3	5	7	11	15	22
28																												1	1	2	3	5	7	11	15
29																													1	1	2	3	5	7	11
30																														1	1	2	3	5	7
31																															1	1	2	3	5
32																																1	1	2	3
33																																	1	1	2
34																																		1	1
35																																			1

Table 4·2. *Number of bipartitions into r biparts of the number mn*;
$m = 1(1)16,\ n \le m,\ m+n \le 16,\ r = 1(1)m$

For example, given the bipartite number (22) the partitions may be written

(22) (20, 02) (20, 01, 01) (10, 10, 01, 01)
(21, 01) (10, 10, 02)
(12, 10) (11, 10, 01)
(11, 11)

Thus there is 1 one-partition, 4 two-partitions, 3 three-partitions, and 1 four-partition as may be verified from the table.

Table 1 (bipartite number (mm); entries give the number of partitions into r biparts, $r = 1 \ldots 11$)

(mm)	r=1	r=2	r=3	r=4	r=5	r=6	r=7	r=8	r=9	r=10	r=11
(10)	1										
(20)	1	1									
(11)	1	1									
(30)	1	1	1								
(21)	1	2	1								
(40)	1	2	2	1							
(31)	1	3	2	1							
(22)	1	4	3	1							
(50)	1	2	2	1	1						
(41)	1	4	4	2	1						
(32)	1	5	6	3	1						
(60)	1	3	3	3	1	1					
(51)	1	5	6	4	2	1					
(42)	1	7	10	7	3	1					
(33)	1	7	11	8	3	1					
(70)	1	3	4	4	3	2	1				
(61)	1	6	9	7	4	2	1				
(52)	1	8	15	12	7	3	1				
(43)	1	9	18	16	9	3	1				
(80)	1	4	5	5	3	2	1	1			
(71)	1	7	12	13	11	7	3	1			
(62)	1	10	21	21	16	8	4	1			
(53)	1	13	28	28	21	10	3	1			
(44)	1	12	26	32	21	10	3	1			
(90)	1	4	6	6	5	3	2	1	1		
(81)	1	8	16	18	16	11	6	3	1		
(72)	1	11	28	31	23	13	7	3	1		
(63)	1	13	37	45	34	23	13	9	3		
(54)	1	14	45	57	45	34	19	7	1		
(10,0)	1	5	8	10	7	5	3	2	1	1	
(91)	1	9	20	23	18	12	7	4	2	1	
(82)	1	13	36	46	34	24	13	7	3	1	
(73)	1	15	48	68	57	36	19	9	3	1	
(64)	1	17	58	86	75	48	24	10	3	1	
(55)	1	17	60	90	80	52	25	10	3	1	
(11,0)	1	5	8	11	10	7	5	3	2	1	1
(10,1)	1	10	25	31	27	19	12	7	4	2	1
(92)	1	14	45	63	57	39	24	13	7	2	1
(83)	1	17	62	97	92	63	37	19	7	3	1
(74)	1	19	76	126	125	87	50	24	9	3	1
(65)	1	20	83	142	143	102	58	26	10	3	1

Table 2 (bipartite number (mm); entries give the number of partitions into r biparts, $r = 1 \ldots 16$)

(mm)	r=1	r=2	r=3	r=4	r=5	r=6	r=7	r=8	r=9	r=10	r=11	r=12	r=13	r=14	r=15	r=16
(12,0)	1	6	12	15	13	11	7	5	3	2	1	1				
(11,1)	1	11	30	41	38	29	19	12	7	4	2	1				
(10,2)	1	16	55	86	83	63	40	24	13	7	3	1				
(93)	1	19	77	134	138	105	65	37	19	9	3	1				
(84)	1	22	97	181	196	152	93	51	24	10	3	1				
(75)	1	23	108	210	234	185	114	60	26	10	3	1				
(66)	1	24	114	224	251	200	124	64	27	10	3	1				
(13,0)	1	6	14	18	18	14	11	7	5	3	2	1	1			
(12,1)	1	12	36	53	53	42	30	19	12	7	4	2	1			
(11,2)	1	17	66	111	118	94	65	40	24	13	7	3	1			
(10,3)	1	21	94	179	202	164	111	66	37	19	9	3	1			
(94)	1	24	120	246	294	245	165	95	51	24	10	3	1			
(85)	1	26	138	297	367	314	212	120	61	26	10	3	1			
(76)	1	27	148	327	410	354	242	136	66	27	10	3	1			
(77)	1	31	192	479	664	629	461	284	148	68	27	10	3	1		
(14,0)	1	7	16	23	23	20	15	11	7	5	3	2	1	1		
(13,1)	1	13	42	67	71	60	44	30	19	12	7	4	2	1		
(12,2)	1	19	78	144	162	139	100	66	40	24	13	7	3	1		
(11,3)	1	23	112	233	283	248	177	113	66	37	19	9	3	1		
(10,4)	1	27	146	329	424	384	274	171	96	51	24	10	3	1		
(95)	1	29	170	406	545	507	366	225	122	61	24	10	3	1		
(86)	1	31	188	463	636	600	438	269	142	67	27	10	3	1		
(87)	1	35	244	673	1028	1054	832	545	311	154	69	27	10	3	1	
(15,0)	1	7	19	27	30	26	21	15	11	7	5	3	2	1	1	
(14,1)	1	14	49	83	94	83	64	45	30	19	12	7	4	2	1	
(13,2)	1	20	91	179	218	196	150	102	66	40	24	13	7	3	1	
(12,3)	1	25	133	297	389	362	275	183	114	66	37	19	9	3	1	
(11,4)	1	29	174	424	593	572	438	287	173	96	51	24	10	3	1	
(10,5)	1	32	207	538	786	783	607	395	231	123	61	26	10	3	1	
(96)	1	32	207	538	786	783	607	395	231	123	61	26	10	3	1	
(88)	1	40	309	947	1582	1763	1486	1040	620	338	160	70	27	10	3	1
(16,0)	1	8	21	34	37	35	28	22	15	11	7	5	3	2	1	1
(15,1)	1	15	56	102	121	113	90	66	45	30	19	12	7	4	2	1
(14,2)	1	22	105	223	286	274	217	156	103	66	40	24	13	7	3	1
(13,3)	1	27	154	372	519	514	410	288	185	114	66	37	19	7	3	1
(12,4)	1	32	205	542	808	834	673	468	288	174	96	51	24	9	3	1
(11,5)	1	35	246	696	1093	1168	959	664	408	233	123	61	26	10	3	1
(10,6)	1	38	281	832	1351	1477	1232	856	519	288	145	67	27	10	3	1
(97)	1	39	300	914	1517	1682	1415	988	597	324	156	69	27	10	3	1

220

Table 5·1. *Numerical values of the combinatorial coefficient,* ^nC_r; $n = 1(1)\,25$. $r = 0(1)\,[\tfrac{1}{2}n + 1]$

The numerical values of the combinatorial coefficient

$$^nC_r = \frac{n!}{r!\,(n-r)!}$$

are tabulated. Because $^nC_r = {}^nC_{n-r}$ it is necessary to print only one half of the distribution.

n \ r	0	1	2	3	4	5	6	7	8	9	10	11	12	13
25	1	25	300	2300	12650	53130	177100	480700	1081575	2042975	3268760	4457400	5200300	5200300
24	1	24	276	2024	10626	42504	134596	346104	735471	1307504	1961256	2496144	2704156	2496144
23	1	23	253	1771	8855	33649	100947	245157	490314	817190	1144066	1352078	1352078	...
22	1	22	231	1540	7315	26334	74613	170544	319770	497420	646646	705432	646646	...
21	1	21	210	1330	5985	20349	54264	116280	203490	293930	352716	352716	...	
20	1	20	190	1140	4845	15504	38760	77520	125970	167960	184756	167960	...	
19	1	19	171	969	3876	11628	27132	50388	75582	92378	92378	...		
18	1	18	153	816	3060	8568	18564	31824	43758	48620	43758	...		
17	1	17	136	680	2380	6188	12376	19448	24310	24310	...			
16	1	16	120	560	1820	4368	8008	11440	12870	11440	...			
15	1	15	105	455	1365	3003	5005	6435	6435	...				
14	1	14	91	364	1001	2002	3003	3432	3003	...				
13	1	13	78	286	715	1287	1716	1716	...					
12	1	12	66	220	495	792	924	792	...					
11	1	11	55	165	330	462	462	...						
10	1	10	45	120	210	252	210	...						
9	1	9	36	84	126	126	...							
8	1	8	28	56	70	56	...							
7	1	7	21	35	35	...								
6	1	6	15	20	15	...								
5	1	5	10	10										
4	1	4	6	4	...									
3	1	3	3	...										
2	1	2	1											
1	1	1												

Tables 5.2. *Generalized Bernoulli Polynomials and Generalized Bernoulli Numbers*

The generalized Bernoulli polynomials $B_n^{(m)}(x)$ are generated by

$$\frac{t^m e^{xt}}{(e^t - 1)^m} = \sum_{n=0}^{\infty} B_n^{(m)}(x)\frac{t^n}{n!} \quad (|t| < 2\pi).$$

The values of $B_n^{(m)}(x)$ when $m = 1$ are given in Table 5·2·1 and when $x = 0$ in Table 5·2·2.

Table 5·2·1. *Bernoulli polynomials $B_n^{(1)}(x) \equiv B_n(x)$ as polynomials in x; $n = 0(1)12$*

When $m = 1$ we have the quantities equivalent to those commonly referred to as Bernoulli polynomials. Thus, for example,

$$B_4^{(1)}(x) \equiv B_4(x) = \tfrac{1}{30}(-1 + 30x^2 - 60x^3 + 30x^4).$$

When, in addition, $x = 0$ the table gives the Bernoulli numbers. We adopt Nörlund's nomenclature so that $B_{2a+1} = 0$, $a = 1, 2, \dots$.

B_n	Dividing factor	Coefficients of												
		x^0	x^1	x^2	x^3	x^4	x^5	x^6	x^7	x^8	x^9	x^{10}	x^{11}	x^{12}
$B_0(x)$	1	1												
$B_1(x)$	2	-1	2											
$B_2(x)$	6	1	-6	6										
$B_3(x)$	2	.	1	-3	2									
$B_4(x)$	30	-1	.	30	-60	30								
$B_5(x)$	6	.	-1	.	10	-15	6							
$B_6(x)$	42	1	.	-21	.	105	-126	42						
$B_7(x)$	6	.	1	.	-7	.	21	-21	6					
$B_8(x)$	30	-1	.	20	.	-70	.	140	-120	30				
$B_9(x)$	10	.	-3	.	20	.	-42	.	60	-45	10			
$B_{10}(x)$	66	5	.	-99	.	330	.	-462	.	495	-330	66		
$B_{11}(x)$	6	.	5	.	-33	.	66	.	-66	.	55	-33	6	
$B_{12}(x)$	2730	-691	.	13650	.	-45045	.	60060	.	-45045	.	30030	-16380	2730

Table 5·2·2. *Bernoulli polynomials $B_n^{(m)}$ as polynomials in m; $n = 1(1)12$*

As an example $\qquad B_4^{(m)} = \tfrac{1}{240}(2m + 5m^2 - 30m^3 + 15m^4).$

When, in addition, $m = 1$ these again reduce to the Bernoulli numbers.

$B_n^{(m)}$	Dividing factor	Coefficients of											
		m^1	m^2	m^3	m^4	m^5	m^6	m^7	m^8	m^9	m^{10}	m^{11}	m^{12}
$B_1^{(m)}$	2	-1											
$B_2^{(m)}$	12	-1	3										
$B_3^{(m)}$	8	.	1	-1									
$B_4^{(m)}$	240	2	5	-30	15								
$B_5^{(m)}$	96	.	-2	-5	10	-3							
$B_6^{(m)}$	4032	-16	-42	91	315	-315	63						
$B_7^{(m)}$	1152	.	16	42	-7	-105	63	-9					
$B_8^{(m)}$	34560	144	404	-540	-2345	-840	3150	-1260	135				
$B_9^{(m)}$	7680	.	-144	-404	-100	665	448	-630	180	-15			
$B_{10}^{(m)}$	101376	-768	-2288	2068	11792	8195	-8085	-8778	6930	-1485	99		
$B_{11}^{(m)}$	18432	.	768	2288	1100	-2904	-3179	847	1914	-990	165	-9	
$B_{12}^{(m)}$	50319360	1061376	332758	-2037672	-15875860	-14444430	5760755	12882870	315315	-5495490	2027025	-270270	12285

Table 5·3·1. *Difference quotients of zero or Stirling's Numbers of the second kind*

The function tabulated is

$$\frac{\Delta^r O^s}{r!} = {}^s C_r B^{(-r)}_{s-r} \quad (r \leqslant s)$$

for $s = 1(1)25$, $r = 1(1)s$.

r \ s	1	2	3	4	5	6	7	8
1	1	1	1	1	1	1	1	1
2		1	3	7	15	31	63	127
3			1	6	25	90	301	966
4				1	10	65	350	1701
5					1	15	140	1050
6						1	21	266
7							1	28
8								1

r \ s	9	10	11	12	13
1	1	1	1	1	1
2	255	511	1023	2047	4095
3	3025	9330	28501	86526	261625
4	7770	34105	145750	611501	2532530
5	6951	42525	246730	1379400	7508501
6	2646	22827	179487	1323652	9321312
7	462	5880	63987	627396	5715424
8	36	750	11880	159027	1899612
9	1	45	1155	22275	359502
10		1	55	1705	39325
11			1	66	2431
12				1	78
13					1

Table 5·3·1 (cont.)

r \ s	14	15	16	17	18
1	1	1	1	1	1
2	8191	16383	32767	65535	131071
3	788970	2375101	7141686	21457825	64439010
4	10391745	42355950	171798901	694337290	2798806985
5	40075035	210766920	1096190550	5652751651	28958095545
6	63436373	420693273	2734926558	17505749898	110687251039
7	49329280	408741333	3281882604	25708104786	197462483400
8	20912320	216627840	2141764053	20415995028	189036065010
9	5135130	67128490	820784250	9528822303	106175395755
10	752752	12662650	193754990	2758334150	37112163803
11	66066	1479478	28936908	512060978	8391004908
12	3367	106470	2757118	62022324	1256328866
13	91	4550	165620	4910178	125854638
14	1	105	6020	249900	8408778
15		1	120	7820	367200
16			1	136	9996
17				1	153
18					1

r \ s	19	20	21	22
1	1	1	1	1
2	262143	524287	1048575	2097151
3	193448101	580606446	1742343625	5228079450
4	11259666950	45232115901	181509070050	727778623825
5	147589284710	749206090500	3791262568401	19137821912055
6	693081601779	4306078895384	26585679462804	163305339345225
7	1492924634839	11143554045652	82310957214948	602762379967440
8	1709751003480	15170932662679	132511015347084	1142399079991620
9	1144614626805	12011282644725	123272476465204	1241963303533920
10	477297033785	5917584964655	71187132291275	835143799377954
11	129413217791	1900842429486	26826851689001	366282500870286
12	2346951330	411016663391	6833042031078	108823356061937
13	292439160	61068660380	1204909218331	2249686186481
14	243577530	6302524580	149304004500	3295165281331
15	13916778	452329200	13087462580	345615943200
16	527136	22350954	809944464	26046574004
17	12597	741285	34952799	1404142047
18	171	15675	1023435	53374629
19	1	190	19285	1389850
20		1	210	23485
21			1	231
22				1

Table 5·3·1 (cont.)

r \ s	23	24	25
1	1	1	1
2	4194303	8388607	16777215
3	1586335501	47063200806	141197991025
4	2916342574750	11681056634501	46771289738810
5	96416888184100	485000783495250	2436684974110751
6	998969857983405	6090236036084530	37026417000002430
7	4382641999117305	31677463851804540	227832482998716310
8	9741955019900400	82318282158320505	690223721118368580
9	12320068811796900	120622574326072500	1167921451092973005
10	9593401297313460	108254081784931500	1203163392175387500
11	4864251308951100	63100165695775560	802355904438462660
12	1672162773613530	24930204592313460	362262620803537080
13	401282560352190	6888836058192000	114485073348809460
14	68629175807115	1362091021651800	25958110361317200
15	8479404429331	195820242247080	4299394655358000
16	762361127264	20677182465555	526655161695960
17	49916988803	1610949936915	48063331393110
18	2364885369	92484925445	3275678594925
19	79781779	3880739170	166218969675
20	1859550	116972779	6220194750
21	28336	2454606	168519505
22	253	33902	3200450
23	1	276	40250
24		1	300
25			1

Table 5·3·2. *Differentials of zero or Stirling's Numbers of the first kind*

The function tabulated is

$$|S_m^p| = \frac{1}{p!}\left(\frac{d^p}{dx^p}(x^{[m]})\right)_{x=0} = {}^{m-1}C_{p-1}\,B_{m-p}^{(m)}(-1)^{m-p},$$

for $m = 1(1)20$, $p = 1(1)m$.

p \ m	1	2	3	4	5	6	7	8	9	10
1	1	1	2	6	24	120	720	5040	40320	362880
2		1	3	11	50	274	1764	13068	109584	1026576
3			1	6	35	225	1624	13132	118124	1172700
4				1	10	85	735	6769	67284	723680
5					1	15	175	1960	22449	269325
6						1	21	322	4536	63273
7							1	28	546	9450
8								1	36	870
9									1	45
10										1

p \ m	11	12	13	14	15
1	3628800	39916800	479001600	6227020800	87178291200
2	10628640	120543840	1486442880	19802759040	283465647360
3	12753576	150917976	1931559552	26596717056	392156797824
4	8409500	105258076	1414014888	20313753096	310989260400
5	3416930	45995730	657206836	9957703756	159721605680
6	902055	13339535	206070150	3336118786	5663366760
7	157773	2637558	44990231	790943153	14409322928
8	18150	357423	6926634	135036473	2681453775
9	1320	32670	749463	16669653	368411615
10	55	1925	55770	1474473	37312275
11	1	66	2717	91091	2749747
12		1	78	3731	143325
13			1	91	5005
14				1	105
15					1

Table 5·3·2 (*cont.*)

p \ m	16	17	18	19	20
1	1307674368000	20922789888000	355687428096000	6402373705728000	121645100408832000
2	4339163001600	70734282393600	1223405590579200	22376988058521600	431565146817638400
3	6165817614720	102992244837120	1821602444624640	34012249593822720	6686009730341153280
4	5056995703824	87077748875904	1583313975727488	30321254007719424	610116075740491776
5	2706813345600	48366009233424	909299905844112	17950712280921504	371384787345228000
6	1009672107080	18861567058880	369012649234384	7551527592063024	161429736530118960
7	272803210680	537453477960	110228466184200	2353125040549984	52260903362512720
8	54631129553	1146901283528	24871845297936	557921681547048	12953636989943896
9	8207628000	185953177553	4308105301929	102417740732658	2503858755467550
10	928095740	23057159840	577924894833	14710753408923	381922055502195
11	78558480	2185031420	6020269380	1661573386473	4628064775191 0
12	4899622	156952432	4853222764	147560703732	4465226757381
13	218400	8394022	299650806	10246937272	342252511900
14	6580	323680	13896582	549789282	2069293630
15	120	8500	468180	22323822	973941900
16	1	136	10812	662796	34916946
17		1	153	13566	920550
18			1	171	16815
19				1	190
20					1

Table 5·4·1. *Differences of reciprocals of unity (integers)*

The function tabulated is

$$A_{rn} = \{(n+1)!\}^r \left| \frac{1}{n!} \Delta^n \left(\frac{1}{1^r}\right) \right|$$

for $r = 1(1)10$ and some values of n.

r n	1	2	3	4	5
1	1	3	7	15	31
2	1	11	85	575	3661
3	1	50	1660	46760	1217776
4	1	274	48076	6998824	929081776
5	1	1764	1942416	1744835904	1413470290176
6	1	13068	104587344	673781602752	3878864920694016
7	1	109584	7245893376	381495483224064	17810567950611972096
8	1	1026576	628308907776	303443622431870976	
9	1	10628640	66687811660800		
10	1	120543840	8506654697548800		

r n	6	7	8	9	10
1	63	127	255	511	1023
2	22631	137845	83375	5019421	30174551
3	30480800	747497920	18139003520	437786795776	
4	117550462624	1450866102976	1765130436471424		
5	1083688832185344	806595068762689536			
6	21006340945438768128				

Table 5·4·2. *Differences of reciprocals of unity (decimals)*

The function tabulated is $\Delta^n(1/1^r)$ for $r = 1(1)20$, $n = 1(1)20$.

n \ r	1	2	3	4	5
1	− 0·5000000000	− 0·7500000000	− 0·8750000000	− 0·9375000000	− 0·9687500000
2	+ ·3333333333	+ ·6111111111	+ ·7870370370	+ ·8873456790	+ ·9416152263
3	− ·2500000000	− ·5208333333	− ·7204861111	− ·8456307870	− ·9176191165
4	+ ·2000000000	+ ·4566666667	+ ·6677222222	+ ·8100490741	+ ·8961051080
5	− ·1666666667	− ·4083333333	− ·6244907407	− ·7791226852	− ·8766080376
6	+ 0·1428571429	+ 0·3704081633	+ 0·5881932297	+ 0·7518470487	+ 0·8587850391
7	− ·1250000000	− ·3397321429	− ·5571355938	− ·7275081168	− ·8423754239
8	+ ·1111111111	+ ·3143298060	+ ·5301571730	+ ·7055802342	+ ·8271759583
9	− ·1000000000	− ·2928968254	− ·5064311382	− ·6856653246	− ·8130248950
10	+ ·0909090909	+ ·2745343041	+ ·4853496078	+ ·6674548049	+ ·7997912504
11	− 0·0833333333	− 0·2586008899	− 0·4664538813	− 0·6507047279	− 0·7873673735
12	+ ·0769230769	+ ·2446256735	+ ·4493901730	+ ·6352189929	+ ·7756636519
13	− ·0714285714	− ·2322544519	− ·4338804787	− ·6208376705	− ·7646046533
14	+ ·0666666667	+ ·2212152662	+ ·4197027978	+ ·6074286790	+ ·7541262550
15	− ·0625000000	− ·2112955621	− ·4066773456	− ·5948817206	− ·7441734716
16	+ 0·0588235294	+ 0·2023266190	+ 0·3946567146	+ 0·5831037791	+ 0·7346987838
17	− ·0555555556	− ·1941726710	− ·3835187122	− ·5720157198	− ·7256608358
18	+ ·0526315789	+ ·1867231398	+ ·3731610505	+ ·5615496846	+ ·7170234068
19	− ·0500000000	− ·1798869829	− ·3634973471	− ·5516470677	− ·7087545898
20	+ ·0476190476	+ ·1735885098	+ ·3544540691	+ ·5422569249	+ ·7008261296

n \ r	6	7	8	9	10
1	− 0·9843750000	− 0·9921875000	− 0·9960937500	− 0·9980468750	− 0·9990234375
2	+ ·9701217421	+ ·9848322474	+ ·9923399158	+ ·9961445553	+ ·9980638101
3	− ·9569960857	− ·9778732070	− ·9887232386	− ·9942892261	− ·9971201641
4	+ ·9448178902	+ ·9712621436	+ ·9852310196	+ ·9924775848	+ ·9961916482
5	− ·9334495814	− ·9649600499	− ·9818525246	− ·9907067414	− ·9952774971
6	+ 0·9227832182	+ 0·9589347882	+ 0·9785785623	+ 0·9889741444	+ 0·9943770181
7	− ·9127322439	− ·9531594702	− ·9754011758	− ·9872775233	− ·9934895813
8	+ ·9032259900	+ ·9476113057	+ ·9723134124	+ ·9856148443	+ ·9926146105
9	− ·8942058805	− ·9422707632	− ·9693091475	− ·9839842747	− ·9917515769
10	+ ·8856227323	+ ·9371209422	+ ·9663829470	+ ·9823841540	+ ·9908999930
11	− 0·8774347857	− 0·9321470958	− 0·9635299594	− 0·9808129711	− 0·9900594079
12	+ ·8696062370	+ ·9273362605	+ ·9607458288	+ ·9792693448	+ ·9892294030
13	− ·8621061238	− ·9226769651	− ·9580266242	− ·9777520076	− ·9884095891
14	+ ·8549074659	+ ·9181589984	+ ·9553687825	+ ·9762597926	+ ·9875996026
15	− ·8479865913	− ·9137732230	− ·9527690600	− ·9747916218	− ·9867991038
16	+ 0·8413226026	+ 0·9095114218	+ 0·9502244931	+ 0·9733464966	+ 0·9860077740
17	− ·8348969489	− ·9053661733	− ·9477323642	− ·9719234892	− ·9852253137
18	+ ·8286930782	+ ·9013307472	+ ·9452901738	+ ·9705217358	+ ·9844514412
19	− ·8226961538	− ·8973990176	− ·9428956160	− ·9691404298	− ·9836858906
20	+ ·8168928193	+ ·8935653891	+ ·9405465576	+ ·9677788168	+ ·9829284109

Table 5·4·2 (cont.)

r n	11	12	13	14	15
1	− 0·9995117188	− 0·9997558594	− 0·9998779297	− 0·9999389648	− 0·9999694824
2	+ ·9990290825	+ ·9995136004	+ ·9997564866	+ ·9998781388	+ ·9999390345
3	− ·9985518529	− ·9992731635	− ·9996356558	− ·9998175180	− ·9999086554
4	+ ·9980798120	+ ·9990344932	+ ·9995154233	+ ·9997570991	+ ·9998783441
5	− ·9976127595	− ·9987975376	− ·9993957757	− ·9996968785	− ·9998480999
6	+ 0·9971505107	+ 0·9985622481	+ 0·9992767003	+ 0·9996368531	+ 0·9998179218
7	− ·9966928946	− ·9983285789	− ·9991581851	− ·9995770196	− ·9997878090
8	+ ·9962397519	+ ·9980964870	+ ·9990402187	+ ·9995173750	+ ·9997577608
9	− ·9957909344	− ·9978659317	− ·9989227900	− ·9994579165	− ·9997277763
10	+ ·9953463034	+ ·9976368746	+ ·9988058886	+ ·9993986413	+ ·9996978550
11	− 0·9949057287	− 0·9974092791	− 0·9986895045	− 0·9993395465	− 0·9996679959
12	+ ·9944690883	+ ·9971831106	+ ·9985736280	+ ·9992806297	+ ·9996381985
13	− ·9940362669	− ·9969583360	− ·9984582500	− ·9992218883	− ·9996084621
14	+ ·9936071560	+ ·9967349240	+ ·9983433616	+ ·9991633199	+ ·9995787859
15	− ·9931816527	− ·9965128446	− ·9982289543	− ·9991049220	− ·9995491694
16	+ 0·9927596598	+ 0·9962920690	+ 0·9981150199	+ 0·9990466925	+ 0·9995196120
17	− ·9923410851	− ·9960725699	− ·9980015504	− ·9989886290	− ·9994901129
18	+ ·9919258406	+ ·9958543210	+ ·9978885384	+ ·9989307295	+ ·9994606717
19	− ·9915138431	− ·9956372971	− ·9977759763	− ·9988729919	− ·9994312877
20	+ ·9911050130	+ ·9954214740	+ ·9976638571	+ ·9988154140	+ ·9994019604

r n	16	17	18	19	20
1	− 0·9999847412	− 0·9999923706	− 0·9999961853	− 0·9999980927	− 0·9999990463
2	+ ·9999695057	+ ·9999847490	+ ·9999923732	+ ·9999961862	+ ·9999980929
3	− ·9999542931	− ·9999771350	− ·9999885636	− ·9999942805	− ·9999971398
4	+ ·9999391033	+ ·9999695287	+ ·9999847566	+ ·9999923758	+ ·9999961870
5	− ·9999239361	− ·9999619299	− ·9999809522	− ·9999904718	− ·9999952345
6	+ 0·9999087912	+ 0·9999543386	+ 0·9999771502	+ 0·9999885687	+ 0·9999942822
7	− ·9998936684	− ·9999467549	− ·9999733508	− ·9999866665	− ·9999933303
8	+ ·9998785675	+ ·9999391785	+ ·9999695539	+ ·9999847651	+ ·9999923786
9	− ·9998634884	− ·9999316095	− ·9999657595	− ·9999828645	− ·9999914272
10	+ ·9998484308	+ ·9999240478	+ ·9999619675	+ ·9999809648	+ ·9999904761
11	− 0·9998333946	− 0·9999164934	− 0·9999581780	− 0·9999790659	− 0·9999895252
12	+ ·9998183795	+ ·9999089461	+ ·9999543909	+ ·9999771678	+ ·9999885746
13	− ·9998033854	− ·9999014061	− ·9999506063	− ·9999752706	− ·9999876244
14	+ ·9997884121	+ ·9998938732	+ ·9999468241	+ ·9999733741	+ ·9999866743
15	− ·9997734594	− ·9998863473	− ·9999430443	− ·9999714785	− ·9999857246
16	+ 0·9997585272	+ 0·9998788285	+ 0·9999392669	+ 0·9999695837	+ 0·9999847751
17	− ·9997436153	− ·9998713166	− ·9999354919	− ·9999676897	− ·9999838259
18	+ ·9997287236	+ ·9998638117	+ ·9999317192	+ ·9999657965	+ ·9999828770
19	− ·9997138518	− ·9998563137	− ·9999279489	− ·9999639042	− ·9999819284
20	+ ·9996989998	+ ·9998488226	+ ·9999241810	+ ·9999620126	+ ·9999809800

Table 5·5. *Polykays of the first N natural numbers (up to and including weight 8)*

In most rank problems some simplification is achieved by treating the situation as one in which there is sampling without replacement from a finite population of (say) the first N natural numbers. For rank moment problems this means that the polykay techniques may usefully be employed. It is possible to calculate the polykays in terms of N as is done in this table. Thus, reading horizontally, for example

$$K_{31^2} = \tfrac{1}{120}\{N^4 + 3N^3 + 3N^2 + N\}.$$

Note that K_3, K_5, K_{32}, K_7 K_{52}, K_{43} and K_{32^2} are all zero.

Polykay	Dividing factor	Coefficients of								
		N^8	N^7	N^6	N^5	N^4	N^3	N^2	N^1	N^0
K_1	2	.	.	.	.	.	.	.	1	1
K_2	12	.	.	.	.	.	.	1	1	.
K_{1^2}	12	.	.	.	.	.	.	3	5	2
K_{21}	24	.	.	.	.	.	1	2	1	.
K_{1^3}	8	.	.	.	.	.	1	2	1	.
K_4	−120	.	.	.	.	1	2	1	.	.
K_{31}	120	.	.	.	.	.	1	2	1	.
K_{2^2}	720	.	.	.	.	5	6	−5	−6	.
K_{21^2}	720	.	.	.	.	15	40	33	8	.
K_{1^4}	240	.	.	.	.	15	30	5	−18	−8
K_{41}	−240	.	.	.	1	3	3	1	.	.
K_{31^2}	120	.	.	.	.	1	3	3	1	.
K_{2^21}	1440	.	.	.	15	30	5	−18	−8	.
K_{21^3}	480	.	.	.	5	15	13	1	−2	.
K_{1^5}	96	.	.	.	3	5	−5	−13	−6	.
K_6	504	.	.	2	6	5	.	−1	.	.
K_{51}	−504	.	.	.	2	6	5	.	−1	.
K_{42}	−10080	.	.	7	5	−27	−25	20	20	.
K_{41^2}	−10080	.	.	21	77	81	−9	−54	−20	.
K_{3^2}	−5040	.	.	.	6	15	4	−15	−10	.
K_{321}	10080	.	.	.	7	17	3	−17	−10	.
K_{2^3}	60480	.	.	35	21	−157	−141	122	120	.
K_{31^3}	10080	.	.	.	63	231	299	153	22	.
$K_{2^21^2}$	60480	.	.	105	301	147	−257	−252	−44	.
K_{21^4}	20160	.	.	105	315	161	−275	−322	−96	.
K_{1^6}	4032	.	.	63	63	−315	−539	−74	236	96

Table 5·5 (*cont.*)

Polykay	Dividing factor	Coefficients of								
		N^8	N^7	N^6	N^5	N^4	N^3	N^2	N^1	N^0
K_{61}	1008	.	2	8	11	5	-1	-1	.	.
K_{51^2}	-504	.	.	2	8	11	5	-1	-1	.
K_{421}	-20160	.	7	12	-22	-52	-5	40	20	.
K_{3^21}	-10080	.	.	6	21	19	-11	-25	-10	.
K_{41^3}	-6720	.	7	28	18	-68	-133	-88	-20	.
K_{321^2}	10080	.	.	7	24	20	-14	-27	-10	.
K_{2^31}	120960	.	35	56	-136	-298	-19	242	120	.
K_{31^4}	5040	.	.	21	84	110	32	-35	-20	.
$K_{2^21^3}$	40320	.	35	112	56	-138	-131	26	40	.
K_{21^5}	8064	.	21	56	-28	-210	-201	-38	16	.
K_{1^7}	1152	.	9	.	-84	-98	91	194	80	.
K_8	-720	3	12	14	.	-7	.	2	.	.
K_{71}	720	.	3	12	14	.	-7	.	2	.
K_{62}	30240	10	4	-71	-53	145	133	-84	-84	.
K_{53}	5040	.	3	7	-7	-28	-10	21	41	.
K_{4^2}	100800	7	-20	-62	180	503	120	-448	-280	.
K_{61^2}	30240	30	140	155	-183	-475	-175	146	84	.
K_{521}	-30240	.	10	22	-29	-95	-23	73	42	.
K_{431}	-100800	.	7	40	78	40	-57	-80	-28	.
K_{42^2}	-604800	35	-48	-380	42	1225	678	-880	-672	.
K_{3^22}	-302400	.	30	63	-103	-325	-95	262	168	.
K_{51^3}	-10080	.	30	63	-103	-325	-95	262	168	.
K_{421^2}	-604800	105	250	-344	-1410	-785	960	1024	200	.
$K_{3^21^2}$	-302400	.	90	333	297	-245	-455	-88	68	.
K_{32^21}	604800	.	35	72	-128	-370	-75	298	168	.
K_{2^4}	3628800	175	-140	-1754	-176	-5335	3340	-3756	-3024	.
K_{321^3}	604800	.	35	1290	1778	-380	-1765	-910	-48	.
K_{41^4}	-201600	105	420	-160	2256	6145	4830	1480	64	.
$K_{2^21^2}$	3628800	525	1190	-1690	-6052	-2275	4670	3440	192	.
K_{31^5}	120960	.	315	1260	1200	-1418	-3419	-2146	-400	.
$K_{2^21^4}$	241920	105	336	14	-940	-795	444	676	160	.
K_{21^6}	241920	315	630	-1890	-6160	-3865	3226	4288	1152	.
K_{1^8}	34560	135	-180	-1890	-840	6055	8140	884	-3088	-1152

Table 5·6·1. *Asymmetric sums of products of natural numbers; sum A_1*

$$A_1 = \sum_{i=1}^{n-1} \sum_{j=i+1}^{n} i^r j^s \quad \text{for} \quad r+s = 1(1)8.$$

For example $\displaystyle\sum_{i<j} ij^3 = \frac{1}{120}\{160(n+1)^{(3)} + 285(n+1)^{(4)} + 108(n+1)^{(5)} + 10(n+1)^{(6)}\}.$

	Dividing factor	Coefficients of							
		$(n+1)^{(3)}$	$(n+1)^{(4)}$	$(n+1)^{(5)}$	$(n+1)^{(6)}$	$(n+1)^{(7)}$	$(n+1)^{(8)}$	$(n+1)^{(9)}$	$(n+1)^{(10)}$
i	6	1	.	.	.	.	.	.	.
j	3	1	.	.	.	.	.	.	.
i^2	12	2	1	.	.	.	.	.	.
ij	24	8	3	.	.	.	.	.	.
j^2	12	8	3	.	.	.	.	.	.
i^3	60	10	15	3	.	.	.	.	.
i^2j	120	40	45	8	.	.	.	.	.
ij^2	120	80	75	12	.	.	.	.	.
j^3	60	80	75	12	.	.	.	.	.
i^4	60	10	35	18	2	.	.	.	.
i^3j	120	40	105	48	5	.	.	.	.
i^2j^2	360	240	495	204	20	.	.	.	.
ij^3	120	160	285	108	10	.	.	.	.
j^4	60	160	285	108	10	.	.	.	.
i^5	84	14	105	105	28	2	.	.	.
i^4j	840	280	1575	1400	350	24	.	.	.
i^3j^2	840	560	2415	1932	455	30	.	.	.
i^2j^3	840	1120	3885	2828	630	40	.	.	.
ij^4	168	448	1365	924	196	12	.	.	.
j^5	84	448	1365	924	196	12	.	.	.
i^6	168	28	434	756	364	60	3	.	.
i^5j	336	112	1302	2016	910	144	7	.	.
i^4j^2	2520	1680	14805	20580	8750	1332	63	.	.
i^3j^3	672	896	6132	7728	3108	456	21	.	.
i^2j^4	504	1344	7497	8652	3304	468	21	.	.
ij^5	336	1792	8862	9576	3500	480	21	.	.
j^6	168	1792	8862	9576	3500	480	21	.	.
i^7	2520	420	13230	37926	29400	8400	945	35	.
i^6j	5040	1680	39690	101136	73500	20160	2205	80	.
i^5j^2	15120	10080	179550	409752	280350	73980	7875	280	.
i^4j^3	5040	6720	91350	188328	121800	30984	3213	112	.
i^3j^4	5040	13440	143010	268632	164850	40500	4095	140	.
i^2j^5	15120	80640	704970	1218168	711900	169200	16695	560	.
ij^6	5040	53760	418950	680904	382200	88200	8505	280	.
j^7	2520	53760	418950	680904	382200	88200	8505	280	.
i^8	180	30	1905	8694	10206	4500	855	70	2
i^7j	720	240	11430	46368	51030	21600	3990	320	9
i^6j^2	5040	3360	120330	436296	451780	183960	33075	2600	72
i^5j^3	1440	1920	52020	169776	166020	65160	11415	880	24
i^4j^4	1800	4800	99675	295140	273550	103680	17715	1340	36
i^3j^5	1440	7680	125460	339696	299460	109800	18315	1360	36
i^2j^6	720	7680	103590	258888	217840	77400	12615	920	24
ij^7	360	7680	92655	218484	177030	61200	9765	700	18
j^8	180	7680	92655	218484	177030	61200	9765	700	18

Table 5·6·2. *Asymmetric sums of products of natural numbers; sum A_2*

$$A_2 = \sum_{i=1}^{n-2} \sum_{j=i+1}^{n-1} \sum_{l=j+1}^{n} i^r j^s l^t \quad \text{for} \quad w = r+s+t = 1(1)5,$$

where all possible 3-partitions of w (including zero) are used for r, s and t. For example

$$\sum_{i<j<l} i l^3 = \tfrac{1}{840} \{945(n+1)^{(4)} + 1036(n+1)^{(5)} + 280(n+1)^{(6)} + 20(n+1)^{(7)}\}.$$

Table 5·6·2 (*cont.*)

	Dividing factor	Coefficients of				
		$(n+1)^{(4)}$	$(n+1)^{(5)}$	$(n+1)^{(6)}$	$(n+1)^{(7)}$	$(n+1)^{(8)}$
i	24	1	.	.	.	.
j	12	1	.	.	.	.
l	8	1	.	.	.	.
i^2	120	5	2	.	.	.
j^2	60	10	3	.	.	.
l^2	40	15	4	.	.	.
ij	120	10	3	.	.	.
il	120	15	4	.	.	.
jl	60	15	4	.	.	.
i^3	120	5	6	1	.	.
j^3	60	20	15	3	.	.
l^3	120	135	84	10	.	.
i^2j	360	30	27	4	.	.
ij^2	120	20	15	2	.	.
i^2l	360	45	36	5	.	.
il^2	360	135	84	10	.	.
j^2l	24	12	8	1	.	.
jl^2	180	135	84	10	.	.
ijl	48	12	8	1	.	.
i^4	840	35	98	42	4	.
j^4	420	280	399	126	10	.
l^4	280	945	1036	280	20	.
i^3j	840	70	147	56	5	.
ij^3	840	280	399	126	10	.
i^3l	840	105	196	70	6	.
il^3	840	945	1036	280	20	.
j^3l	840	840	1064	315	24	.
jl^3	420	945	1036	280	20	.
i^2jl	5040	1260	1848	595	48	.
ij^2l	1680	840	1064	315	24	.
ijl^2	5040	3780	4368	1225	90	.
i^2j^2	2520	420	693	238	20	.
i^2l^2	504	189	252	77	6	.
j^2l^2	2520	3780	4368	1225	90	.
i^5	336	14	84	70	16	1
j^5	168	224	546	308	56	3
l^5	336	3402	5880	2716	432	21
i^4j	2520	210	945	700	150	9
ij^4	336	224	546	308	56	3
i^4l	5040	630	2520	1750	360	21
il^4	1008	3402	5880	2716	432	21
j^4l	240	480	1040	550	96	5
jl^4	504	3402	5880	2716	432	21
i^3j^2	3360	560	1932	1288	260	15
i^2j^3	2520	840	2331	1414	270	15
i^3l^2	10080	3780	10416	6160	1152	63
i^2l^3	10080	11340	23184	11900	2040	105
j^3l^2	2520	7560	14784	7385	1242	63
j^2l^3	3360	15120	27328	13020	2120	105
i^3jl	6720	1680	5152	3220	624	35
ij^3l	480	480	1040	550	96	5
ijl^3	6720	15120	27328	13020	2120	105
i^2j^2l	5040	2520	6216	3535	648	35
i^2jl^2	240	180	400	215	38	2
ij^2l^2	5040	7560	14784	7385	1242	63

Table 5·7. *Distribution of rank sums (Wilcoxon's statistic)*;

$$r = 4(1)20, \quad r_1 = 2(1)[\tfrac{1}{2}r]$$

Given a set of ranks $r_2 = r - r_1$ chosen without replacement from the first r natural numbers, the frequency function of their sum, S, is given by the coefficients of successive powers of h in the expansion of

$$\prod_{i=1}^{r_2} \left(\frac{h^i - h^{r+1}}{1 - h^i} \right).$$

Since the distribution of S must start from $\tfrac{1}{2}r_2(r_2 + 1)$ it is convenient to tabulate the function

$$U = r_1 r_2 + \tfrac{1}{2}r_2(r_2 + 1) - S.$$

Further, since the distribution of U is symmetrical about its mean, namely $\tfrac{1}{2}r_1 r_2$, approximately one half only of the frequency distribution is tabulated, the mid-point about which there is symmetry being given in bold type.

Table 5·7

r	4	5	6	7	8	9	10	11	12	13	14	15	16	17	18	19	20
r_1	2	2	2	2	2	2	2	2	2	2	2	2	2	2	2	2	2
U																	
0	1	1	1	1	1	1	1	1	1	1	1	1	1	1	1	1	1
1	1	1	1	1	1	1	1	1	1	1	1	1	1	1	1	1	1
2	**2**	2	2	2	2	2	2	2	2	2	2	2	2	2	2	2	2
3	1	**2**	2	2	2	2	2	2	2	2	2	2	2	2	2	2	2
4	...	2	**3**	3	3	3	3	3	3	3	3	3	3	3	3	3	3
5		...	2	**3**	3	3	3	3	3	3	3	3	3	3	3	3	3
6			...	3	**4**	4	4	4	4	4	4	4	4	4	4	4	4
7				...	3	**4**	4	4	4	4	4	4	4	4	4	4	4
8					...	4	**5**	5	5	5	5	5	5	5	5	5	5
9						...	4	**5**	5	5	5	5	5	5	5	5	5
10							...	5	**6**	6	6	6	6	6	6	6	6
11								...	5	**6**	6	6	6	6	6	6	6
12									...	6	**7**	7	7	7	7	7	7
13										...	6	**7**	7	7	7	7	7
14											...	7	**8**	8	8	8	8
15												...	7	**8**	8	8	8
16													...	8	**9**	9	9
17														...	8	**9**	9
18															...	9	**10**
19																...	9
																	...

Table 5·7 (cont.)

U \\ r	6	7	8	9	10	11	12	13	14	15	16	17	18	19	20
r_1	3	3	3	3	3	3	3	3	3	3	3	3	3	3	3
0	1	1	1	1	1	1	1	1	1	1	1	1	1	1	1
1	1	1	1	1	1	1	1	1	1	1	1	1	1	1	1
2	2	2	2	2	2	2	2	2	2	2	2	2	2	2	2
3	3	3	3	3	3	3	3	3	3	3	3	3	3	3	3
4	**3**	4	4	4	4	4	4	4	4	4	4	4	4	4	4
5	**3**	4	5	5	5	5	5	5	5	5	5	5	5	5	5
6	3	**5**	6	7	7	7	7	7	7	7	7	7	7	7	7
7	...	4	**6**	7	8	8	8	8	8	8	8	8	8	8	8
8		...	**6**	8	9	10	10	10	10	10	10	10	10	10	10
9			6	**8**	10	11	12	12	12	12	12	12	12	12	12
10			...	8	**10**	12	13	14	14	14	14	14	14	14	14
11				...	**10**	12	14	15	16	16	16	16	16	16	16
12					10	**13**	15	17	18	19	19	19	19	19	19
13					...	12	**15**	17	19	20	21	21	21	21	21
14						...	**15**	18	20	22	23	24	24	24	24
15							15	**18**	21	23	25	26	27	27	27
16							...	18	**21**	24	26	28	29	30	30
17								...	**21**	24	27	29	31	32	33
18									21	**25**	28	31	33	35	36
19									...	24	**28**	31	34	36	38
20										...	**28**	32	35	38	40
21											28	**32**	36	39	42
22											...	32	**36**	40	43
23												...	**36**	40	44
24													36	**41**	45
25													...	40	**45**
26														...	**45**
27															45
															...

Table 5·7 (*cont.*)

U	8	9	10	11	12	13	14	15	16	17	18	19	20	10	11	12	13	14	15	16	17	18	19	20
r_1	4	4	4	4	4	4	4	4	4	4	4	4	4	5	5	5	5	5	5	5	5	5	5	5
0	1	1	1	1	1	1	1	1	1	1	1	1	1	1	1	1	1	1	1	1	1	1	1	1
1	1	1	1	1	1	1	1	1	1	1	1	1	1	1	1	1	1	1	1	1	1	1	1	1
2	2	2	2	2	2	2	2	2	2	2	2	2	2	2	2	2	2	2	2	2	2	2	2	2
3	3	3	3	3	3	3	3	3	3	3	3	3	3	3	3	3	3	3	3	3	3	3	3	3
4	5	5	5	5	5	5	5	5	5	5	5	5	5	5	5	5	5	5	5	5	5	5	5	5
5	5	6	6	6	6	6	6	6	6	6	6	6	6	7	7	7	7	7	7	7	7	7	7	7
6	7	8	9	9	9	9	9	9	9	9	9	9	9	9	10	10	10	10	10	10	10	10	10	10
7	7	9	10	11	11	11	11	11	11	11	11	11	11	11	12	13	13	13	13	13	13	13	13	13
8	8	11	13	14	15	15	15	15	15	15	15	15	15	14	16	17	18	18	18	18	18	18	18	18
9	7	11	14	16	17	18	18	18	18	18	18	18	18	16	19	21	22	23	23	23	23	23	23	23
10	...	12	16	19	21	22	23	23	23	23	23	23	23	18	23	26	28	29	30	30	30	30	30	30
11		11	16	20	23	25	26	27	27	27	27	27	27	19	25	30	33	35	36	37	37	37	37	37
12		...	18	23	27	30	32	33	34	34	34	34	34	20	29	35	40	43	45	46	47	47	47	47
13			16	23	28	32	35	37	38	39	39	39	39	20	30	39	45	50	53	55	56	57	57	57
14			...	24	31	36	40	43	45	46	47	47	47	19	32	43	52	58	63	66	68	69	70	70
15				23	31	38	43	47	50	52	53	54	54	...	32	46	57	66	72	77	80	82	83	84
16				...	33	41	48	53	57	60	62	63	64		32	48	63	74	83	89	94	97	99	100
17					31	41	49	56	61	65	68	70	71		...	49	66	81	92	101	107	112	115	117
18					...	43	53	61	68	73	77	80	82			49	70	88	103	114	123	129	134	137
19						41	53	63	71	78	83	87	90			48	71	93	111	126	137	146	152	157
20						...	55	67	77	85	92	97	101			...	73	98	121	139	154	165	174	180
21							53	67	79	89	97	104	109				71	101	127	150	168	183	194	203
22							...	69	83	95	105	113	120				...	102	134	161	184	202	217	228
23								67	83	97	109	119	127					102	137	170	197	220	238	253
24								...	86	102	116	128	138					101	141	178	212	239	262	280
25									83	102	118	132	144					...	141	184	222	256	283	306
26									...	104	123	139	153						141	188	233	272	306	333
27										102	123	142	158						...	190	240	286	325	359
28										...	126	147	166							190	247	299	346	385
29											123	147	168							188	249	309	362	409
30											...	150	174							...	252	317	379	433
31												147	174								249	322	390	453
32												...	177								...	325	402	472
33													174									325	408	488
34													...									322	414	501
35																						...	414	511
36																							414	518
37																							...	521
38																								521
39																								518

Table 5·7 (cont.)

U	r=12 r₁=6	r=13 r₁=6	r=14 r₁=6	r=15 r₁=6	r=16 r₁=6	r=17 r₁=6	r=18 r₁=6	r=19 r₁=6	r=20 r₁=6	r=14 r₁=7	r=15 r₁=7	r=16 r₁=7	r=17 r₁=7	r=18 r₁=7	r=19 r₁=7	r=20 r₁=7
0	1	1	1	1	1	1	1	1	1	1	1	1	1	1	1	1
1	1	1	1	1	1	1	1	1	1	1	1	1	1	1	1	1
2	2	2	2	2	2	2	2	2	2	2	2	2	2	2	2	2
3	3	3	3	3	3	3	3	3	3	3	3	3	3	3	3	3
4	5	5	5	5	5	5	5	5	5	5	5	5	5	5	5	5
5	7	7	7	7	7	7	7	7	7	7	7	7	7	7	7	7
6	11	11	11	11	11	11	11	11	11	11	11	11	11	11	11	11
7	13	14	14	14	14	14	14	14	14	15	15	15	15	15	15	15
8	18	19	20	20	20	20	20	20	20	20	21	21	21	21	21	21
9	22	24	25	26	26	26	26	26	26	26	27	28	28	28	28	28
10	28	31	33	34	35	35	35	35	35	34	36	37	38	38	38	38
11	32	37	40	42	43	44	44	44	44	42	45	47	48	49	49	49
12	39	46	51	54	56	57	58	58	58	53	58	61	63	64	65	65
13	42	52	59	64	67	69	70	71	71	63	70	75	78	80	81	82
14	48	61	71	78	83	86	88	89	90	75	86	93	98	101	103	104
15	51	68	81	91	98	103	106	108	109	87	101	112	119	124	127	129
16	55	76	94	107	117	124	129	132	134	100	120	134	145	152	157	160
17	55	81	103	121	134	144	151	156	159	112	137	157	171	182	189	194
18	58	88	116	139	157	170	180	187	192	125	158	184	204	218	229	236
19	55	90	123	152	175	193	206	216	223	136	176	210	236	256	270	281
20	...	94	134	169	199	222	240	253	263	146	197	239	274	300	320	334
21		94	139	182	218	248	271	289	302	155	214	268	311	346	372	392
22		94	146	196	241	278	308	331	349	162	233	297	353	397	432	458
23		...	147	205	258	304	341	371	394	166	247	325	392	449	493	528
24			151	217	280	335	382	419	449	169	263	354	437	506	564	608
25			147	221	293	359	415	462	499	169	272	379	477	563	633	691
26			...	227	310	387	455	512	559	166	282	403	520	623	711	782
27				227	319	408	488	557	614	...	285	424	558	682	788	877
28				227	330	431	525	607	677		289	441	598	742	871	979
29				...	332	446	553	650	733		285	454	629	799	950	1082
30					338	464	587	699	798		...	464	663	856	1036	1192
31					332	471	608	737	852			468	709	908	1114	1301
32					...	480	634	780	914			464	722	957	1197	1413
33						480	648	813	965			...	734	1000	1271	1524
34						480	664	847	1021				734	1038	1346	1635
35						...	668	870	1064				734	1069	1410	1741
36							676	896	1113				...	1093	1475	1846
37							668	907	1145					1109	1524	1943
38							...	920	1182					1117	1572	2034
39								920	1203					1117	1605	2117
40								920	1226					1109	1634	2191
41								...	1232					...	1646	2253
42									1242						1656	2306
43									1232						1646	2345
44									...						...	2371
45																2385
46																2371

Table 5·7 (*cont.*)

r r_1 U	16 8	17 8	18 8	19 8	20 8	18 9	19 9	20 9	20 10
0	1	1	1	1	1	1	1	1	1
1	1	1	1	1	1	1	1	1	1
2	2	2	2	2	2	2	2	2	2
3	3	3	3	3	3	3	3	3	3
4	5	5	5	5	5	5	5	5	5
5	7	7	7	7	7	7	7	7	7
6	11	11	11	11	11	11	11	11	11
7	15	15	15	15	15	15	15	15	15
8	22	22	22	22	22	22	22	22	22
9	28	29	29	29	29	30	30	30	30
10	38	39	40	40	40	40	41	41	42
11	48	50	51	52	52	52	53	54	54
12	63	66	68	69	70	69	71	72	73
13	77	82	85	87	88	87	90	92	93
14	97	104	109	112	114	111	116	119	121
15	116	127	134	139	142	138	145	150	152
16	141	156	167	174	179	171	182	189	193
17	164	185	200	211	218	207	222	233	237
18	194	222	243	258	269	251	273	288	295
19	221	258	286	307	322	297	326	348	356
20	255	302	340	368	389	352	392	421	433
21	284	345	393	431	459	411	462	502	515
22	319	394	457	506	544	476	544	596	615
23	348	441	519	583	632	545	630	699	720
24	383	495	593	673	738	622	731	818	847
25	409	543	662	763	844	699	833	945	978
26	440	597	742	866	969	782	949	1088	1131
27	461	645	816	968	1095	867	1067	1241	1289
28	486	696	900	1082	1239	954	1197	1408	1470
29	499	738	974	1192	1381	1040	1326	1584	1652
30	515	783	1057	1313	1542	1128	1468	1775	1860
31	519	816	1127	1427	1697	1210	1603	1971	2065
32	**526**	851	1204	1550	1870	1292	1749	2180	2293
33	519	873	1265	1662	2034	1368	1887	2393	2517
34	...	894	1331	1780	2212	1437	2030	2613	2761
35		902	1379	1885	2378	1499	2161	2834	2994
36		**910**	1430	1993	2557	1555	2297	3060	3246
37		902	1460	2083	2716	1598	2414	3280	3481
38		...	1492	2174	2885	1632	2532	3500	3729
39			1502	2244	3032	1656	2630	3712	3956
40			**1514**	2313	3184	**1667**	2724	3916	4192
41			1502	2358	3308	**1667**	2794	4107	4397
42			...	2400	3436	1656	2860	4287	4609
43				2417	3531	...	2897	4447	4784
44				**2430**	3627		2929	4591	4959
45				2417	3688		**2934**	4714	5095
46				...	3746		2929	4814	5226
47					3768		...	4890	5311
48					**3788**			4943	5392
49					3768			**4968**	**5424**
50					...			4968	5424
51								4943	5392

Table 5·8. *Distribution of $C_{n,S}$ required for Kendall's τ and Mann's T;*

$$n = 1(1)16, \quad S = 0(1)\left[\tfrac{1}{4}n(n-1)\right]$$

The quantities tabulated are the coefficients of x^S in the expansion of

$$\prod_{j=1}^{n}\left(\frac{1-x^j}{1-x}\right) = \sum_{S=1}^{M} C_{n,S}\,x^S,$$

where $M = \tfrac{1}{2}n(n-1)$. The p.d.f. of S is

$$p(S) = C_{n,S}/n!, \quad 0 \leqslant S \leqslant \tfrac{1}{2}n(n-1).$$

Each column in the table is symmetrical about the central value or values (shown in bold type).

Kendall's τ is

$$\tau = -1 + 4S/\{n(n-1)\}.$$

Mann's T is just S.

The sum of the frequencies of S for given n is $n!$.

S \ n	1	2	3	4	5	6	7	8	9	10	11
0	1	1	1	1	1	1	1	1	1	1	1
1		1	2	3	4	5	6	7	8	9	10
2			2	5	9	14	20	27	35	44	54
3			1	**6**	15	29	49	76	111	155	209
4				5	20	49	98	174	285	440	649
5				...	**22**	71	169	343	628	1068	1717
6					20	90	259	602	1230	2298	4015
7					...	**101**	359	961	2191	4489	8504
8						**101**	455	1415	3606	8095	16599
9						90	531	1940	5545	13640	30239
10						...	**573**	2493	8031	21670	51909
11							**573**	3017	11021	32683	84591
12							531	3450	14395	47043	131625
13							...	3736	17957	64889	196470
14								**3836**	21450	86054	282369
15								3736	24584	110010	391939
16								...	27073	135853	526724
17									28675	162337	686763
18									**29228**	187959	870233
19									28675	211089	1073227
20									...	230131	1289718
21										243694	1511742
22										**250749**	1729808
23										**250749**	1933514
24										243694	2112319
25										...	2256396
26											2357475
27											**2409581**
28											**2409581**
29											2357475

Table 5·8 (*cont.*)

S \ n	12	13	14	15	16
0	1	1	1	1	1
1	11	12	13	14	15
2	65	77	90	104	119
3	274	351	441	545	664
4	923	1274	1715	2260	2924
5	2640	3914	5629	7889	10813
6	6655	10569	16198	24087	34900
7	15159	25728	41926	66013	100913
8	31758	57486	99412	165425	266338
9	61997	119483	218895	384320	650658
10	113906	233389	452284	836604	1487262
11	198497	431886	884170	1720774	3208036
12	330121	762007	1646177	3366951	6574987
13	526581	1288587	2934764	6301715	12876702
14	808896	2097472	5032235	11333950	24210652
15	1200626	3298033	8330256	19664205	43874857
16	1726701	5024460	13354639	33018831	76893687
17	2411747	7435284	20789572	53808313	130701986
18	3277965	10710609	31498907	85306779	216008661
19	4342688	15046642	46541635	131846699	347854815
20	5615807	20647290	67178356	199019426	546871981
21	7097310	27712842	94865470	293868698	840732790
22	8775209	36426054	131234038	425060810	1265769513
23	10624132	46936280	178050835	603012233	1868715733
24	12604826	59342609	237160055	839953393	2708503701
25	14664752	73677240	310405409	1149906518	3858025899
26	16739858	89890517	399533919	1548556267	5405745562
27	18757500	107839121	506084453	2052994543	7457019331
28	20640359	127278852	631265833	2681325612	10134977992
29	22311069	147863220	775831020	3452124397	13580800674
30	23697232	169148705	939955265	4383749406	17953216130
31	24736324	190607064	1123127045	5493521812	23427073737
32	25380120	211644496	1324060932	6796793172	30190848078
33	**25598186**	231626875	1540641165	8305935430	38442975195
34	25380120	249909685	1769903560	10029297355	48386965771
35	...	265870800	2008061518	11970180517	60225299589
36		278943900	2250579364	14125894411	74152174574
37		288650143	2492293227	16486953600	90345259476
38		294625748	2727576366	19036479131	108956677797
39		**296643390**	2950542516	21749861592	130103527156
40		294625748	3155277747	24594733930	153858307693
41		...	3336088769	27531288780	180239689955
42			3487753817	30512958144	209204091832
43			3605761397	33487453708	240638550997
44			3686522377	36398145065	274355370450
45			**3727542188**	39185731988	310088978041
46			**3727542188**	41790147131	347495375766
47			3686522377	44152608576	386154462530
48			...	46217728808	425575398166
49				47935579065	465205041801
50				49263606316	504439350762
51				50168304699	542637474944
52				**50626553988**	579138134521
53				**50626553988**	613277734909
54				50168304699	644409560477
55				...	671923305201
56					695264150336
57					713950590364
58					727590240796
59					735892934219
60					**738680521142**

Table 6·1. *Distribution of the number of multiple runs.*
Sequence length 6(1)12, colours 2(1)9

These tables 6·1·2–6·1·9 show the distribution of the number of runs T in a k-colour sequence, $k = 2(1)9$, for sequence length r and colour specification represented by all possible k-partitions $r_1, r_2, \ldots, r_k$ of r. The probability distributions of T are obtained by dividing the tabled frequencies by $r!/\prod_i r_i!$.

Table 6·1·2. *Two-colour runs*

r	r_1	r_2	$^rC_{r_1}$	Values of T										
				2	3	4	5	6	7	8	9	10	11	12
6	3	3	20	2	4	8	4	2						
	4	2	15	2	4	6	3	.						
	5	1	6	2	4	.	.	.						
7	4	3	35	2	5	12	9	6	1					
	5	2	21	2	5	8	6	.	.					
	6	1	7	2	5	.	.	.	.					
8	4	4	70	2	6	18	18	18	6	2				
	5	3	56	2	6	16	16	12	4	.				
	6	2	28	2	6	10	10	.	.	.				
	7	1	8	2	6	.	.	.	.	.				
9	5	4	126	2	7	24	30	36	18	8	1			
	6	3	84	2	7	20	25	20	10	.	.			
	7	2	36	2	7	12	15	.	.	.	.			
	8	1	9	2	7	.	.	.	.	.	.			
10	5	5	252	2	8	32	48	72	48	32	8	2		
	6	4	210	2	8	30	45	60	40	20	5	.		
	7	3	120	2	8	24	36	30	20	.	.	.		
	8	2	45	2	8	14	21	.	.	.	.	.		
	9	1	10	2	8	.	.	.	.	.	.	.		
11	6	5	462	2	9	40	70	120	100	80	30	10	1	
	7	4	330	2	9	36	63	90	75	40	15	.	.	
	8	3	165	2	9	28	49	42	35	.	.	.	.	
	9	2	55	2	9	16	28	.	.	.	.	.	.	
	10	1	11	2	9	.	.	.	.	.	.	.	.	
12	6	6	924	2	10	50	100	200	200	200	100	50	10	2
	7	5	792	2	10	48	96	180	180	160	80	30	6	.
	8	4	495	2	10	42	84	126	126	70	35	.	.	.
	9	3	220	2	10	32	64	56	56	.	.	.	.	.
	10	2	66	2	10	18	36	.	.	.	.	.	.	.
	11	1	12	2	10	.	.	.	.	.	.	.	.	.

Table 6·1·3. *Three-colour runs*

r	r_1	r_2	r_3	$\dfrac{r!}{\Pi r_i!}$	Values of T									
					3	4	5	6	7	8	9	10	11	12
6	2	2	2	90	6	18	36	30						
	3	2	1	60	6	18	26	10						
	4	1	1	30	6	18	6	.						
7	3	2	2	210	6	24	62	80	38					
	3	3	1	140	6	24	52	40	18					
	4	2	1	105	6	24	42	30	3					
	5	1	1	42	6	24	12	.	.					
8	3	3	2	560	6	30	100	180	170	74				
	4	2	2	420	6	30	90	150	120	24				
	4	3	1	280	6	30	80	90	60	14				
	5	2	1	168	6	30	60	60	12	.				
	6	1	1	56	6	30	20	.	.	.				
9	3	3	3	1680	6	36	150	360	510	444	174			
	4	3	2	1260	6	36	140	310	405	284	79			
	5	2	2	756	6	36	120	240	252	96	6			
	4	4	1	630	6	36	120	180	180	84	24			
	5	3	1	504	6	36	110	160	132	56	4			
	6	2	1	252	6	36	80	100	30	.	.			
	7	1	1	72	6	36	30	.	.	.	.			
10	4	3	3	4200	6	42	202	580	1050	1234	838	248		
	4	4	2	3150	6	42	192	510	870	894	498	138		
	5	3	2	2520	6	42	182	470	752	692	332	44		
	5	4	1	1260	6	42	162	300	372	252	108	18		
	6	2	2	1260	6	42	152	350	440	240	30	.		
	6	3	1	840	6	42	142	250	240	140	20	.		
	7	2	1	360	6	42	102	150	60	.	.	.		
	8	1	1	90	6	42	42	.	.	.	.	.		
11	4	4	3	11550	6	48	266	900	2010	3064	3012	1764	480	
	5	3	3	9240	6	48	256	840	1802	2568	2340	1168	212	
	5	4	2	6930	6	48	246	750	1527	1968	1548	702	135	
	6	3	2	4620	6	48	226	660	1220	1360	870	220	10	
	5	5	1	2772	6	48	216	480	744	672	432	144	30	
	6	4	1	2310	6	48	206	450	645	560	300	90	5	
	7	2	2	1980	6	48	186	480	690	480	90	.	.	
	7	3	1	1320	6	48	176	360	390	280	60	.	.	
	8	2	1	495	6	48	126	210	105	.	.	.	.	
	9	1	1	110	6	48	56	.	.	.	.	.	.	
12	4	4	4	34650	6	54	342	1350	3618	6894	9036	7938	4320	1092
	5	4	3	27720	6	54	332	1270	3300	5974	7388	5982	2826	588
	6	3	3	18480	6	54	312	1140	2778	4570	5060	3360	1100	100
	5	5	2	16632	6	54	312	1080	2592	4104	4272	2880	1110	222
	6	4	2	13860	6	54	302	1030	2388	3620	3550	2130	710	70
	7	3	2	7920	6	54	272	880	1818	2350	1820	660	60	.
	6	5	1	5544	6	54	272	700	1320	1400	1120	540	110	22
	7	4	1	3960	6	54	252	630	1008	1050	660	270	30	.
	8	2	2	2970	6	54	222	630	1008	840	210	.	.	.
	8	3	1	1980	6	54	212	490	588	490	140	.	.	.
	9	2	1	660	6	54	152	280	168	.	.	.	.	.
	10	1	1	132	6	54	72	.	.	.	.	.	.	.

Table 6·1·4. *Four-colour runs*

r	r_1	r_2	r_3	r_4	$\dfrac{r!}{\Pi r_i!}$	Values of T								
						4	5	6	7	8	9	10	11	12
6	2	2	1	1	180	24	72	84						
	3	1	1	1	120	24	72	24						
7	2	2	2	1	630	24	108	252	246					
	3	2	1	1	420	24	108	192	96					
	4	1	1	1	210	24	108	72	6					
8	2	2	2	2	2520	24	144	504	984	864				
	3	2	2	1	1680	24	144	444	684	384				
	3	3	1	1	1120	24	144	384	384	184				
	4	2	1	1	840	24	144	324	294	54				
9	3	2	2	2	7560	24	180	780	2010	2880	1686			
	3	3	2	1	5040	24	180	720	1560	1720	836			
	4	2	2	1	3780	24	180	660	1320	1260	336			
	4	3	1	1	2520	24	180	600	870	660	186			
	5	2	1	1	1512	24	180	480	600	216	12			
10	3	3	2	2	25200	24	216	1140	3720	7480	8416	4204		
	3	3	3	1	16800	24	216	1080	3120	5160	5016	2184		
	4	2	2	2	18900	24	216	1080	3330	6210	6066	1974		
	4	3	2	1	12600	24	216	1020	2730	4170	3366	1074		
	5	2	2	1	7560	24	216	900	2160	2736	1368	156		
	4	4	1	1	6300	24	216	900	1740	1980	1116	324		
	5	3	1	1	5040	24	216	840	1560	1536	768	96		
11	3	3	3	2	92400	24	252	1584	6360	16680	27756	27408	12336	
	4	3	2	2	69300	24	252	1524	5820	14400	22056	18708	6516	
	4	3	3	1	46200	24	252	1464	5070	10720	14256	10848	3566	
	5	2	2	2	41580	24	252	1404	4950	11016	14184	8364	1386	
	4	4	2	1	34650	24	252	1404	4350	9000	10656	6768	2016	
	5	3	2	1	27720	24	252	1344	4200	7896	8484	4704	816	
	5	4	1	1	13860	24	252	1224	2910	4176	3384	1584	306	
	6	2	2	1	13860	24	252	1164	3210	4920	3480	780	30	
	6	3	1	1	9240	24	252	1104	2460	2920	1980	480	20	
12	3	3	3	3	369600	24	288	2112	10176	33360	74016	109632	98688	41304
	4	3	3	2	277200	24	288	2052	9486	29590	61916	83952	66638	23254
	4	4	2	2	207900	24	288	1992	8796	26100	51216	62892	43128	13464
	5	3	2	2	166320	24	288	1932	8316	23856	44520	50484	30468	6432
	4	4	3	1	138600	24	288	1932	7896	20580	35616	39312	25428	7524
	5	3	3	1	110880	24	288	1872	7416	18616	30320	31104	17528	3712
	5	4	2	1	83160	24	288	1812	6726	15966	23820	21384	10818	2322
	6	2	2	2	83160	24	288	1752	6876	17460	27120	22080	7020	540
	6	3	2	1	55440	24	288	1692	5976	13060	17120	12780	4160	340
	5	5	1	1	33264	24	288	1632	4656	8352	9024	6336	2448	504
	6	4	1	1	27720	24	288	1572	4386	7410	7620	4680	1590	150

Table 6·1·5. *Five-colour runs*

r	r_1	r_2	r_3	r_4	r_5	$\dfrac{r!}{\Pi r_i!}$	Values of T							
							5	6	7	8	9	10	11	12
5	1	1	1	1	1	120	120							
6	2	1	1	1	1	360	120	240						
7	2	2	1	1	1	1260	120	480	660					
	3	1	1	1	1	840	120	480	240					
8	2	2	2	1	1	5040	120	720	1980	2220				
	3	2	1	1	1	3360	120	720	1560	960				
	4	1	1	1	1	1680	120	720	720	120				
9	2	2	2	2	1	22680	120	960	3960	8880	8760			
	3	2	2	1	1	15120	120	960	3540	6360	4140			
	3	3	1	1	1	10080	120	960	3120	3840	2040			
	4	2	1	1	1	7560	120	960	2700	3000	780			
	5	1	1	1	1	3024	120	960	1440	480	24			
10	2	2	2	2	2	113400	120	1200	6600	22200	43800	39480		
	3	2	2	2	1	75600	120	1200	6180	18420	29940	19740		
	3	3	2	1	1	50400	120	1200	5760	14640	18600	10080		
	4	2	2	1	1	37800	120	1200	5340	12540	13980	4620		
	4	3	1	1	1	25200	120	1200	4920	8760	7680	2520		
	5	2	1	1	1	15120	120	1200	4080	6240	3144	336		
11	3	2	2	2	2	415800	120	1440	9480	39360	103680	157920	103800	
	3	3	2	2	1	277200	120	1440	9060	34320	78480	99120	54660	
	3	3	3	1	1	184800	120	1440	8640	29280	55800	60480	29040	
	4	2	2	2	1	207900	120	1440	8640	30960	65880	73080	27780	
	4	3	2	1	1	138600	120	1440	8220	25920	45720	42000	15180	
	4	4	1	1	1	69300	120	1440	7380	17520	23040	15120	4680	
	5	2	2	1	1	83160	120	1440	7380	20880	31104	19152	3084	
	5	3	1	1	1	55440	120	1440	6960	15840	18504	10752	1824	
	6	2	1	1	1	27720	120	1440	5700	10800	7920	1680	60	
12	3	3	2	2	2	1663200	120	1680	13020	65100	216720	464520	579180	322860
	3	3	3	2	1	1108800	120	1680	12600	58800	175560	327600	357000	175440
	4	2	2	2	2	1247400	120	1680	12600	60480	191520	383040	422520	175440
	4	3	2	2	1	831600	120	1680	12180	54180	152880	263760	249900	96900
	4	3	3	1	1	554400	120	1680	11760	47880	116760	174720	147840	53640
	4	4	2	1	1	415800	120	1680	11340	43260	99120	133560	95760	30960
	5	2	2	2	1	498960	120	1680	11340	46620	118944	174888	120372	24996
	5	3	2	1	1	332640	120	1680	10920	40320	87864	108528	68712	14496
	5	4	1	1	1	166320	120	1680	10080	29400	49224	46368	24192	5256
	6	2	2	1	1	166320	120	1680	9660	31500	57120	49560	15540	1140
	6	3	1	1	1	110880	120	1680	9240	25200	36120	28560	9240	720

Table 6·1·6. *Six-colour runs*

r	r_1	r_2	r_3	r_4	r_5	r_6	$\dfrac{r!}{\Pi r_i!}$	Values of T						
								6	7	8	9	10	11	12
6	1	1	1	1	1	1	720	720						
7	2	1	1	1	1	1	2520	720	1800					
8	3	1	1	1	1	1	6720	720	3600	2400				
	2	2	1	1	1	1	10080	720	3600	5760				
9	2	2	2	1	1	1	45360	720	5400	17280	21960			
	3	2	1	1	1	1	30240	720	5400	13920	10200			
	4	1	1	1	1	1	15120	720	5400	7200	1800			
10	2	2	2	2	1	1	226800	720	7200	34560	87840	96480		
	3	2	2	1	1	1	151200	720	7200	31200	64320	47760		
	3	3	1	1	1	1	100800	720	7200	27840	40800	24240		
	4	2	1	1	1	1	75600	720	7200	24480	32400	10800		
	5	1	1	1	1	1	30240	720	7200	14400	7200	720		
11	2	2	2	2	2	1	1247400	720	9000	57600	219600	482400	478080	
	3	2	2	2	1	1	831600	720	9000	54240	184320	336240	247080	
	3	3	2	1	1	1	554400	720	9000	50880	149040	215280	129480	
	4	2	2	1	1	1	415800	720	9000	47520	128880	164880	64800	
	4	3	1	1	1	1	277200	720	9000	44160	93600	94320	35400	
	5	2	1	1	1	1	166320	720	9000	37440	68400	43920	6840	
	6	1	1	1	1	1	55440	720	9000	24000	18000	3600	120	
12	2	2	2	2	2	2	7484400	720	10800	86400	439200	1447200	2868480	2631600
	3	2	2	2	2	1	4989600	720	10800	83040	392160	1154880	1944480	1403520
	3	3	2	2	1	1	3326400	720	10800	79680	345120	887760	1247280	755040
	3	3	3	1	1	1	2217600	720	10800	76320	298080	645840	776880	408960
	4	2	2	2	1	1	2494800	720	10800	76320	313200	751680	935640	406440
	4	3	2	1	1	1	1663200	720	10800	72960	266160	534960	553440	224160
	4	4	1	1	1	1	831600	720	10800	66240	187200	282960	212400	71280
	5	2	2	1	1	1	997920	720	10800	66240	217440	373680	272880	56160
	5	3	1	1	1	1	665280	720	10800	62880	170400	232560	155280	32640
	6	2	1	1	1	1	332640	720	10800	52800	120000	111600	34320	2400
	7	1	1	1	1	1	95040	720	10800	36000	36000	10800	720	0

247

Table 6·1·7. *Seven-colour runs*

r	r_1	r_2	r_3	r_4	r_5	r_6	r_7	$\dfrac{r!}{\Pi r_i!}$	Values of T					
									7	8	9	10	11	12
7	1	1	1	1	1	1	1	5040	5040					
8	2	1	1	1	1	1	1	20160	5040	15120				
9	2	2	1	1	1	1	1	90720	5040	30240	55440			
	3	1	1	1	1	1	1	60480	5040	30240	25200			
10	2	2	2	1	1	1	1	453600	5040	45360	166320	236880		
	3	2	1	1	1	1	1	302400	5040	45360	136080	115920		
	4	1	1	1	1	1	1	151200	5040	45360	75600	25200		
11	2	2	2	2	1	1	1	2494800	5040	60480	332640	947520	1149120	
	3	2	2	1	1	1	1	1663200	5040	60480	302400	705600	589680	
	3	3	1	1	1	1	1	1108800	5040	60480	272160	463680	307440	
	4	2	1	1	1	1	1	831600	5040	60480	241920	372960	151200	
	5	1	1	1	1	1	1	332640	5040	60480	151200	100800	15120	
12	2	2	2	2	2	1	1	14968800	5040	75600	554400	2368800	5745600	6219360
	3	2	2	2	1	1	1	9979200	5040	75600	524160	2005920	4067280	3301200
	3	3	2	1	1	1	1	6652800	5040	75600	493920	1643040	2666160	1769040
	4	2	2	1	1	1	1	4989600	5040	75600	463680	1431360	2071440	942480
	4	3	1	1	1	1	1	3326400	5040	75600	433440	1068480	1224720	519120
	5	2	1	1	1	1	1	1995840	5040	75600	372960	796320	619920	126000
	6	1	1	1	1	1	1	665280	5040	75600	252000	252000	75600	5040

Table 6·1·8. *Eight-colour runs*

r	r_1	r_2	r_3	r_4	r_5	r_6	r_7	r_8	$\dfrac{r!}{\Pi r_i!}$	Values of T				
										8	9	10	11	12
8	1	1	1	1	1	1	1	1	40320	40320				
9	2	1	1	1	1	1	1	1	181440	40320	141120			
10	2	2	1	1	1	1	1	1	907200	40320	282240	584640		
	3	1	1	1	1	1	1	1	604800	40320	282240	282240		
11	2	2	2	1	1	1	1	1	4989600	40320	423360	1753920	2772000	
	3	2	1	1	1	1	1	1	3326400	40320	423360	1451520	1411200	
	4	1	1	1	1	1	1	1	1663200	40320	423360	846720	352800	
12	2	2	2	2	1	1	1	1	29937600	40320	564480	3507840	11088000	14736960
	3	2	2	1	1	1	1	1	19958400	40320	564480	3205440	8366400	7781760
	3	3	1	1	1	1	1	1	13305600	40320	564480	2903040	5644800	4152960
	4	2	1	1	1	1	1	1	9979200	40320	564480	2600640	4586400	2187360
	5	1	1	1	1	1	1	1	3991680	40320	564480	1693440	1411200	282240

Table 6·1·9. *Nine-colour runs*

r	r_1	r_2	r_3	r_4	r_5	r_6	r_7	r_8	r_9	$\dfrac{r!}{\Pi r_i!}$	Values of T			
											9	10	11	12
9	1	1	1	1	1	1	1	1	1	362880	362880			
10	2	1	1	1	1	1	1	1	1	1814400	362880	1451520		
11	2	2	1	1	1	1	1	1	1	9979200	362880	2903040	6713280	
	3	1	1	1	1	1	1	1	1	6652800	362880	2903040	3386880	
12	2	2	2	1	1	1	1	1	1	59875200	362880	4354560	20139840	35017920
	3	2	1	1	1	1	1	1	1	39916800	362880	4354560	16813440	18385920
	4	1	1	1	1	1	1	1	1	19958400	362880	4354560	10160640	5080320

Table 6·2. *Maximum runs coefficients*

Given a sequence of length r, the maximum runs coefficients are tabulated for $r = 3(1)15$ and for many colour specifications of the sequence, the specifications being written for convenience in the partition notation. The coefficient is the number of permutations with the maximum number of runs in a sequence of given length and given colour specification divided by factorials representing the repetition of numbers in the colour specification of the sequence. Thus if $N(T\text{max.})$ is the number of permutations with the maximum number of runs, in a sequence of length 14 and of colour specification represented by $3^3 1^5$, it is the quantity $N(T\text{max.})/(3!\,5!)$ which is tabulated.

Table 6·2

Group 1

r	Partition	Coefficient
3	21	1
	1^3	1
4	2^2	1
	21^2	3
	1^4	1
5	32	1
	31^2	1
	$2^2 1$	6
	21^3	6
	1^5	1
6	3^2	1
	321	10
	2^3	5
	31^3	4
	$2^2 1^2$	21
	21^4	10
	1^6	1
7	43	1
	421	3
	$3^2 1$	9
	32^2	19
	41^3	1
	321^2	48
	$2^3 1$	41
	31^4	10
	$2^2 1^3$	55
	21^5	15
	1^7	1
8	4^2	1
	431	14
	42^2	12
	$3^2 2$	37
	421^2	27
	$3^2 1^2$	46
	$32^2 1$	192
	2^4	36

Group 2

r	Partition	Coefficient
8	41^4	5
	321^3	160
	$2^3 1^2$	185
	31^5	20
	$2^2 1^4$	120
	21^6	21
	1^8	1
9	54	1
	531	4
	52^2	3
	$4^2 1$	12
	432	79
	3^3	29
	521^2	6
	431^2	93
	$42^2 1$	168
	$3^2 21$	418
	32^3	281
	51^4	1
	421^3	130
	$3^2 1^3$	170
	$32^2 1^2$	1035
	$2^4 1$	365
	41^5	15
	321^4	425
	$2^3 1^3$	610
	31^6	35
	$2^2 1^5$	231
	21^7	28
	1^9	1
10	5^2	1
	541	18
	532	44
	$4^2 2$	69
	43^2	124
	531^2	48
	$52^2 1$	78
	$4^2 1^2$	81
	4321	1074
	42^3	329
	$3^3 1$	364
	$3^2 2^2$	1051

Group 3

r	Partition	Coefficient
10	521^3	56
	431^3	420
	$42^2 1^2$	1155
	$3^2 21^2$	2520
	$32^3 1$	3290
	2^5	329
	51^5	6
	421^4	450
	$3^2 1^4$	505
	$32^2 1^3$	3980
	$2^4 1^2$	2010
	41^6	35
	321^5	966
	$2^3 1^4$	1645
	31^7	392
	$2^2 1^6$	2436
	21^8	36
	1^{10}	1
11	65	1
	641	5
	632	10
	$5^2 1$	15
	542	135
	53^2	106
	$4^2 3$	240
	631^2	10
	$62^2 1$	15
	541^2	153
	5321	816
	52^3	231
	$4^2 21$	1008
	$43^2 1$	1783
	432^2	3258
	$3^3 2$	2056
	621^3	10
	531^3	304
	$52^2 1^2$	771
	$4^2 1^3$	390
	4321^2	7590
	$42^3 1$	4630
	$3^3 1^2$	2420
	$3^2 2^2 1$	13665
	32^4	4325
	61^5	1
	521^4	285
	431^4	1475
	$42^2 1^3$	5400
	$3^2 21^3$	10790
	$32^3 1^2$	20590
	$2^5 1$	3984

Group 4

r	Partition	Coefficient
11	51^6	21
	421^5	1260
	$3^2 1^5$	1281
	$32^2 1^4$	12285
	$2^4 1^3$	7980
	41^7	70
	321^6	1960
	$2^3 1^5$	3850
	31^8	840
	$2^2 1^7$	666
	21^9	45
	1^{11}	1
12	66	1
	651	22
	642	70
	63^2	50
	$5^2 2$	111
	543	588
	4^3	182
	641^2	75
	6321	340
	62^3	90
	$5^2 1^2$	126
	5421	2322
	$53^2 1$	1856
	532^2	3216
	$4^2 31$	3762
	$4^2 2^2$	3366
	$43^2 2$	11627
	3^4	1721
	621^4	100
	531^4	1360
	$52^2 1^3$	6040
	$4^2 1^4$	1485
	4321^3	37360
	$42^3 1^2$	33870
	$3^3 1^3$	11360
	$3^2 2^2 1^2$	94380
	$32^4 1$	58480
	2^6	3655
	61^6	7
	521^5	1050
	431^5	4326
	$42^2 1^4$	19635
	$3^2 21^4$	36855
	$32^3 1^3$	91700
	$2^5 1^2$	25914

Group 5

r	Partition	Coefficient
12	51^7	56
	421^6	3038
	$3^2 1^6$	2884
	$32^2 1^5$	32424
	$2^4 1^4$	25585
	41^8	126
	321^7	3648
	$2^3 1^6$	8106
	31^9	120
	$2^2 1^8$	1035
	21^{10}	55
	1^{12}	1
13	76	1
	751	6
	742	15
	73^2	10
	$6^2 1$	18
	652	206
	643	445
	$5^2 3$	513
	54^2	834
	651^2	228
	6421	1620
	$63^2 1$	1200
	632^2	1975
	$5^2 21$	1998
	5431	10470
	542^2	9090
	$53^2 2$	14590
	$4^3 1$	3086
	$4^2 32$	27048
	43^3	15131
	731^3	20
	$72^2 1^3$	45
	641^3	590
	6321^2	4290
	$62^3 1$	2340
	$5^2 1^3$	750
	5421^2	20502
	$53^2 1^2$	16446
	$532^2 1$	57042
	52^4	8236
	$4^2 31^2$	30810
	$4^2 2^2 1$	54540
	$43^2 21$	184470
	432^3	109650
	$3^4 1$	26485
	$3^3 2^2$	95285

Group 6

r	Partition	Coefficient
13	721^4	15
	631^4	775
	$62^2 1^3$	2530
	541^4	3855
	5321^3	42860
	$52^3 1^2$	37110
	$4^2 21^3$	41520
	$43^2 1^3$	70430
	$432^2 1^2$	377400
	$42^4 1$	112635
	$3^3 21^2$	219810
	$3^2 2^3 1$	398030
	32^5	72430
	71^6	1
	621^5	546
	531^5	4830
	$52^2 1^4$	20895
	$4^2 1^5$	4746
	4321^4	144480
	$42^3 1^3$	172760
	$3^3 1^4$	42315
	$3^2 2^2 1^3$	460950
	$32^4 1^2$	420630
	$2^6 1$	51499
	61^7	28
	521^6	3136
	431^6	11074
	$42^2 1^5$	59682
	$3^2 21^5$	106932
	$32^3 1^4$	326270
	$2^5 1^3$	240548
	51^8	126
	421^7	6552
	$3^2 1^7$	5916
	$32^2 1^6$	76062
	$2^4 1^5$	70371
	41^9	210
	321^8	6345
	$2^3 1^7$	15720
	31^{10}	165
	$2^2 1^9$	1540
	21^{11}	66
	1^{13}	1

Table 6·2 (*cont.*)

Column group 1

r	Partition	Coefficient
14	77	1
	761	26
	752	102
	743	190
	$6^2 2$	163
	653	1150
	64^2	850
	$5^2 4$	1614
	751^2	108
	7421	630
	$73^2 1$	440
	732^2	690
	$6^2 1^2$	181
	6521	4296
	6431	9610
	$63^2 2$	12180
	642^2	8055
	$5^2 31$	9756
	$5^2 2^2$	8379
	$54^2 1$	15624
	5432	85758
	53^3	22808
	$4^3 2$	24159
	$4^2 3^2$	58951
	741^3	220
	7321^2	1440
	$72^3 1$	750
	651^3	1584
	6421^2	17730
	$63^2 1^2$	13380
	$632^2 1$	44610
	62^4	6285
	$5^2 21^2$	18756
	5431^2	96228
	$542^2 1$	166158
	$53^2 21$	265488
	532^3	152966
	$4^3 1^2$	27245
	$4^2 321$	465660
	$4^2 2^3$	135885
	$43^3 1$	255740
	$43^2 2^2$	674355
	$3^4 2$	188150

Column group 2

r	Partition	Coefficient
14	731^4	250
	$72^2 1^3$	9360
	641^4	3250
	6321^3	32600
	$62^3 1^2$	27080
	$5^2 1^4$	3501
	5421^3	124440
	$53^2 1^3$	99460
	$532^2 1^2$	515940
	$52^4 1$	148710
	$4^2 31^3$	175760
	$4^2 2^2 1^2$	462195
	$43^2 21^2$	1533180
	$432^3 1$	1803460
	42^5	159480
	$3^4 1^2$	214635
	$3^3 2^2 1$	1524660
	$3^2 2^4$	679120
	721^5	162
	631^5	3570
	$62^2 1^4$	14805
	541^5	13986
	5321^4	193410
	$52^3 1^3$	223020
	$4^2 21^4$	174825
	$43^2 1^4$	290745
	$432^2 1^3$	205506
	$42^4 1^2$	910035
	$3^4 21^3$	1162980
	$3^2 2^3 1^2$	6229020
	$32^5 1$	1116360
	2^7	47844
	71^7	8
	621^6	2156
	531^6	14504
	$52^2 1^5$	75264
	$4^2 1^6$	13237
	4321^5	468804
	$42^3 1^4$	693070
	$3^3 1^5$	133224
	$3^2 2^2 1^4$	1787730
	$32^4 1^3$	2141020
	$2^6 1^2$	386407
	61^8	84
	521^7	8064
	431^7	25488
	$42^2 1^6$	158508
	$3^2 21^6$	274008
	$32^3 1^5$	986916
	$2^5 1^4$	446544
	51^9	252
	421^8	12960
	$3^2 1^8$	11265
	$32^2 1^7$	162720
	$2^4 1^6$	172305

Column group 3

r	Partition	Coefficient
14	41^{10}	330
	321^9	10450
	$2^3 1^8$	28545
	31^{11}	220
	$2^2 1^{10}$	2211
	21^{12}	78
	1^{14}	1
15	87	1
	861	7
	852	21
	843	35
	$7^2 1$	21
	762	292
	753	806
	74^2	555
	$6^2 3$	941
	654	4315
	5^3	1198
	851^2	21
	8421	105
	$83^2 1$	70
	832^2	105
	761^2	318
	7521	2832
	7431	5610
	742^2	4530
	$73^2 2$	6520
	$6^2 21$	3493
	6531	24572
	652^2	20691
	$64^2 1$	18255
	6432	43860
	63^3	23870
	$5^2 41$	32421
	$5^2 32$	86352
	$54^2 2$	135777
	543^2	213937
	$4^3 3$	115631

Column group 4

r	Partition	Coefficient
15	841^3	35
	8321^2	210
	$82^3 1$	105
	751^3	1016
	7421^2	9690
	$73^2 1^2$	6960
	$732^2 1$	22350
	72^4	2985
	$6^2 1^3$	1285
	6521^2	45678
	6431^2	103470
	$642^2 1$	173820
	$63^2 21$	264750
	632^3	147960
	$5^2 31^2$	96108
	$5^2 2^2 1$	164043
	$54^2 1^2$	151401
	54321	1637124
	542^3	468084
	$53^3 1$	431824
	$53^2 2^2$	1112571
	$4^3 21$	446585
	$4^2 3^2 1$	1074540
	$4^2 32^2$	1860045
	$43^3 2$	2005690
	3^5	1633861
	831^4	35
	$82^2 1^3$	105
	741^4	1725
	7321^3	15860
	$72^3 1^2$	12690
	651^4	8405
	6421^3	127700
	$63^2 1^3$	97170
	$632^2 1^2$	488490
	$62^4 1$	136295
	$5^2 21^3$	121866
	5431^3	609860
	$542^2 1^2$	1570410
	$53^2 21^2$	2490630
	$532^3 1$	2856020
	52^5	245799
	$4^3 1^3$	167035
	$4^2 321^2$	4185750
	$4^2 2^3 1$	2421765
	$43^3 1^2$	2261430
	$43^2 2^2 1$	1181526
	432^4	3436130
	$3^4 21$	3229875
	$3^3 2^3$	3774730

Column group 5

r	Partition	Coefficient
15	821^5	21
	731^5	1686
	$72^2 1^4$	6735
	641^5	14070
	6321^4	179130
	$62^3 1^3$	199780
	$5^2 1^5$	13587
	5421^4	585690
	$53^2 1^4$	464835
	$532^2 1^3$	3199980
	$52^4 1^2$	1377705
	$4^2 31^4$	785505
	$4^2 2^2 1^3$	2729895
	$43^2 21^3$	8898960
	$432^3 1^2$	15547980
	$42^5 1$	2722125
	$3^4 1^3$	1219715
	$3^3 2^2 1^2$	12848850
	$3^2 2^4 1$	11309130
	32^6	1330994
	81^7	1
	721^6	952
	631^6	13132
	$62^2 1^5$	65856
	541^6	43708
	5321^5	717360
	$52^3 1^4$	1029980
	$4^2 21^5$	615909
	$43^2 1^5$	1006194
	$432^2 1^4$	8801520
	$42^4 1^3$	5143565
	$3^3 21^4$	4864440
	$3^2 2^3 1^3$	17155180
	$32^5 1^2$	9102324
	$2^7 1$	769159
	71^8	36
	621^7	6888
	531^7	38304
	$52^2 1^6$	231084
	$4^2 1^7$	33111
	4321^6	1329804
	$42^3 1^5$	2334654
	$3^3 1^6$	368704
	$3^2 2^2 1^5$	5861646
	$32^4 1^4$	8655780
	$2^6 1^3$	2052309
	61^9	210
	521^8	18522
	431^8	53865
	$42^2 1^7$	378810
	$3^2 21^7$	636540
	$32^3 1^6$	2637180
	$2^5 1^5$	1410003

Column group 6

r	Partition	Coefficient
15	51^{10}	462
	421^9	23925
	$3^2 1^9$	20185
	$32^2 1^8$	323235
	$2^4 1^7$	384945
	41^{11}	495
	321^{10}	16456
	$2^3 1^9$	49115
	31^{12}	286
	$2^2 1^{11}$	33891
	21^{13}	91
	1^{15}	1

Table 6·3·1. *Number of ring runs (Whitworth) for different colour specifications*

The supposition is that the r elements have arisen through a free sampling procedure.

6·3·1 (*a*), $r = 2(1)12$. The colour specification is represented by all possible two-partitions of r. The probabilities are obtained by dividing the number of permutations by the appropriate multinomial term.

Multinomial term	r	Partition	Runs					
			2	4	6	8	10	12
2	2	(1^2)	2					
3	3	(21)	3					
4	4	(31)	4					
6	4	(2^2)	4	2				
5	5	(41)	5					
10	5	(32)	5	5				
6	6	(51)	6					
15	6	(42)	6	9				
20	6	(3^2)	6	12	2			
7	7	(61)	7					
21	7	(52)	7	14				
35	7	(43)	7	21	7			
8	8	(71)	8					
28	8	(62)	8	20				
56	8	(53)	8	32	16			
70	8	(4^2)	8	36	24	2		
9	9	(81)	9					
36	9	(72)	9	27				
84	9	(63)	9	45	30			
126	9	(54)	9	54	54	9		
10	10	(91)	10					
45	10	(82)	10	35				
120	10	(73)	10	60	50			
210	10	(64)	10	75	100	25		
252	10	(5^2)	10	80	120	40	2	
11	11	$(10,1)$	11					
55	11	(92)	11	44				
165	11	(83)	11	77	77			
330	11	(74)	11	99	165	55		
462	11	(65)	11	110	220	110	11	
12	12	$(11,1)$	12					
66	12	$(10,2)$	12	54				
220	12	(93)	12	96	112			
495	12	(84)	12	126	252	105		
792	12	(75)	12	144	360	240	36	
924	12	(6^2)	12	150	400	300	60	2

6·3·1 (b), r = 3(1)12. The colour specification is represented by all possible three-partitions of r. The probabilities are obtained by dividing the number of permutations by the appropriate multinomial term.

Multinomial term	r	Partition	Runs									
			3	4	5	6	7	8	9	10	11	12
6	3	(1^3)	6									
12	4	(21^2)	8	4								
20	5	(31^2)	10	10								
30	5	(2^21)	10	10	10							
30	6	(41^2)	12	18								
60	6	(321)	12	18	24	6						
90	6	(2^3)	12	18	36	24						
42	7	(51^2)	14	28								
105	7	(421)	14	28	42	21						
140	7	(3^21)	14	28	56	28	14					
210	7	(32^2)	14	28	70	70	28					
56	8	(61^2)	16	40								
168	8	(521)	16	40	64	48						
280	8	(431)	16	40	96	72	48	8				
420	8	(42^2)	16	40	112	144	96	12				
560	8	(3^22)	16	40	128	176	144	56				
72	9	(71^2)	18	54								
252	9	(621)	18	54	90	90						
504	9	(531)	18	54	144	144	108	36				
756	9	(52^2)	18	54	162	252	216	54				
630	9	(4^21)	18	54	162	162	162	54	18			
1260	9	(432)	18	54	198	333	378	225	54			
1680	9	(3^3)	18	54	216	396	486	378	132			
90	10	(81^2)	20	70								
360	10	(721)	20	70	120	150						
840	10	(631)	20	70	200	250	200	100				
1260	10	(62^2)	20	70	220	400	400	150				
1260	10	(541)	20	70	240	300	360	180	80	10		
2520	10	(532)	20	70	280	550	760	580	240	20		
4200	10	(43^2)	20	70	320	700	1100	1130	680	180		
3150	10	(4^22)	20	70	300	600	900	780	380	100		
110	11	(91^2)	22	88								
495	11	(821)	22	88	154	231						
1320	11	(731)	22	88	264	396	330	220				
1980	11	(72^2)	22	88	286	594	660	330				
2310	11	(641)	22	88	330	495	660	440	220	55		
4620	11	(632)	22	88	374	836	1320	1210	660	110		
2772	11	(5^21)	22	88	352	528	792	528	352	88	22	
6930	11	(542)	22	88	418	957	1716	1848	1276	517	88	
9240	11	(53^2)	22	88	440	1100	2046	2508	2024	880	132	
11550	11	(4^23)	22	88	462	1188	2310	3058	2662	1408	352	
132	12	$(10,1^2)$	24	108								
660	12	(921)	24	108	192	336						
1980	12	(831)	24	108	336	588	504	420				
2970	12	(82^2)	24	108	360	840	1008	630				
3960	12	(741)	24	108	432	756	1080	900	480	180		
7920	12	(732)	24	108	480	1200	2088	2220	1440	360		
5544	12	(651)	24	108	480	840	1440	1200	960	360	120	12
13860	12	(642)	24	108	552	1416	2880	3630	3120	1620	480	30
18480	12	(63^2)	24	108	576	1608	3384	4740	4640	2640	720	40
16632	12	(5^22)	24	108	576	1488	3168	4176	3840	2304	792	156
27720	12	(543)	24	108	624	1812	4104	6396	7008	5076	2160	408
34650	12	(4^3)	24	108	648	1944	4536	7506	8712	6912	3456	804

6·3·1 (c), r = 4(1)12. The colour specification is represented by all possible four-partitions of r. The probabilities are obtained by dividing the number of permutations by the appropriate multinomial term.

Multinomial term	r	Partition	Runs								
			4	5	6	7	8	9	10	11	12
24	4	(1^4)	24								
60	5	(21^3)	30	30							
120	6	(31^3)	36	72	12						
180	6	(2^21^2)	36	72	72						
210	7	(41^3)	42	126	42						
420	7	(321^2)	42	126	182	70					
630	7	(2^31)	42	126	252	210					
336	8	(51^3)	48	192	96						
840	8	(421^2)	48	192	336	240	24				
1120	8	(3^21^2)	48	192	416	320	144				
1680	8	(32^21)	48	192	496	640	304				
2520	8	(2^4)	48	192	576	960	744				
504	9	(61^3)	54	270	180						
1512	9	(521^2)	54	270	540	540	108				
2520	9	(431^2)	54	270	720	810	540	126			
3780	9	(42^21)	54	270	810	1350	1080	216			
5040	9	(3^221)	54	270	900	1620	1530	666			
7560	9	(32^3)	54	270	990	2160	2700	1386			
720	10	(71^3)	60	360	300						
2520	10	(621^2)	60	360	800	1000	300				
5040	10	(531^2)	60	360	1100	1600	1320	560	40		
7560	10	(52^21)	60	360	1200	2400	2520	960	60		
6300	10	(4^21^2)	60	360	1200	1800	1800	840	240		
12600	10	(4321)	60	360	1400	3100	4050	2840	790		
18900	10	(42^3)	60	360	1500	3900	6300	5340	1440		
16800	10	(3^31)	60	360	1500	3600	5100	4440	1740		
25200	10	(3^22^2)	60	360	1600	4400	7700	7640	3440		
990	11	(81^3)	66	462	462						
3960	11	(721^2)	66	462	1122	1650	660				
9240	11	(631^2)	66	462	1562	2750	2640	1540	220		
13860	11	(62^21)	66	462	1672	3850	4840	2640	330		
13860	11	(541^2)	66	462	1782	3300	4092	2772	1188	198	
27720	11	(5321)	66	462	2002	5170	8272	7612	3652	484	
41580	11	(52^3)	66	462	2112	6270	12012	13332	6534	792	
34650	11	(4^221)	66	462	2112	5610	9570	9834	5478	1518	
46200	11	(43^21)	66	462	2222	6380	11550	13574	9218	2728	
69300	11	(432^2)	66	462	2332	7480	16060	22604	16258	5038	
92400	11	(3^32)	66	462	2442	8250	18810	27654	24618	10098	
1320	12	(91^3)	72	576	672						
5940	12	(821^2)	72	576	1512	2520	1260				
15840	12	(731^2)	72	576	2112	4320	4680	3360	720		
23760	12	(72^21)	72	576	2232	5760	8280	5760	1080		
27720	12	(641^2)	72	576	2472	5400	7740	6720	3600	1080	60
55440	12	(6321)	72	576	2712	7920	14640	16320	10440	2640	120
83160	12	(62^3)	72	576	2832	9360	20340	27120	18360	4320	180
33264	12	(5^21^2)	72	576	2592	5760	8928	8064	5184	1728	360
83160	12	(5421)	72	576	2952	9000	18324	23616	18576	8424	1620
110880	12	(53^21)	72	576	3072	10080	21624	30816	28080	14016	2544
166320	12	(532^2)	72	576	3192	11520	28584	46656	46584	24768	4368
138600	12	(4^231)	72	576	3192	10800	24120	36768	36144	21168	5760
207900	12	(4^22^2)	72	576	3312	12240	31500	54288	59184	36288	10440
277200	12	(43^22)	72	576	3432	13320	36060	66528	80784	58008	18420
369600	12	(3^4)	72	576	3552	14400	41040	80448	107424	88128	33960

Table 6·3·2. *Number of ring runs (randomization)*

All distinguishable circular permutations of the given r elements are treated as a randomization set. When the numbers of the different colours in the ring have no common factor, Tables 6·3·1 should be used with the frequencies divided by r. When there is a common factor the present table is appropriate.

(a) Two-colour runs

Total	r	Partition	Runs					
			2	4	6	8	10	12
2	4	(2^2)	1	1				
3	6	(42)	1	2				
4	6	(3^2)	1	2	1			
4	8	(62)	1	3				
10	8	(4^2)	1	5	3	1		
10	9	(63)	1	5	4			
5	10	(82)	1	4				
22	10	(64)	1	8	10	3		
26	10	(5^2)	1	8	12	4	1	
6	12	$(10, 2)$	1	5				
19	12	(93)	1	8	10			
43	12	(84)	1	11	21	10		
80	12	(6^2)	1	13	34	26	5	1

(b) Three-colour runs

Total	r	Partition	Runs									
			3	4	5	6	7	8	9	10	11	12
16	6	(2^3)	2	3	6	5						
54	8	(42^2)	2	5	14	19	12	2				
188	9	(3^3)	2	6	24	44	54	42	16			
128	10	(62^2)	2	7	22	41	40	16				
318	10	(4^22)	2	7	30	61	90	79	38	11		
250	12	(82^2)	2	9	30	71	84	54				
1160	12	(642)	2	9	46	119	240	304	260	137	40	3
1542	12	(63^2)	2	9	48	134	282	395	388	220	60	4
2896	12	(4^3)	2	9	54	163	378	627	726	579	288	70

(c) Four-colour runs

Total	r	Partition	Runs								
			4	5	6	7	8	9	10	11	12
318	8	(2^4)	6	24	72	120	96				
1896	10	(42^3)	6	36	150	390	633	534	147		
6940	12	(62^3)	6	48	236	780	1698	2260	1536	360	16
17340	12	(4^22^2)	6	48	276	1020	2628	4524	4938	3024	876
30804	12	(3^4)	6	48	296	1200	3420	6704	8952	7344	2834

A *composition* (used here in the technical sense) of a number N, as opposed to a partition, is any collection of integers whose sum is N but in which the order of the parts of the partition is important. The number of *compositions* of a number N into s parts, the largest part being equal to m is tabulated for $N = 1(1)18$, $s = 1(1)N$, $m = 1(1)N$. This function is denoted by $g^*(N, s, m)$.

Thus, for example, the number of compositions of 8 into 4 parts, the largest of which is 4, is, from the table, 12.

Or, alternatively, given r balls, r_1 of one colour and r_2 the total number of the remaining colours, the number of ways in which the r balls may be arranged so that a longest run of the r_1 colour is exactly l is $g^*(r+1, r_2+1, l+1)$. For example, given 7 balls, 4 white, 2 red and 1 black, the number of arrangements which will give a longest white run of 3 is $g^*(8, 4, 4)$ or, from the table, 12.

$N = 1$ (columns $s = 1$)

m \ s	1
1	1

$N = 2$ (columns $s = 1 \ 2$)

m \ s	1	2
1	.	1
2	1	.

$N = 3$ (columns $s = 1 \ 2 \ 3$)

m \ s	1	2	3
1	.	.	1
2	.	2	.
3	1	.	.

$N = 4$ (columns $s = 1 \ 2 \ 3 \ 4$)

m \ s	1	2	3	4
1	.	.	.	1
2	.	1	3	.
3	.	2	.	.
4	1	.	.	.

$N = 5$ (columns $s = 1 \ 2 \ 3 \ 4 \ 5$)

m \ s	1	2	3	4	5
1	.	.	.	.	1
2	.	.	3	4	.
3	.	2	3	.	.
4	.	2	.	.	.
5	1	.	.	.	.

$N = 6$ (columns $s = 1 \ 2 \ 3 \ 4 \ 5 \ 6$)

m \ s	1	2	3	4	5	6
1	.	.	.	.	.	1
2	.	.	1	6	5	.
3	.	1	6	4	.	.
4	.	2	3	.	.	.
5	.	2	.	.	.	.
6	1	.	.	.	.	.

$N = 7$ (columns $s = 1 \ 2 \ 3 \ 4 \ 5 \ 6 \ 7$)

m \ s	1	2	3	4	5	6	7
1	.	.	.	.	.	.	1
2	.	.	.	4	10	6	.
3	.	.	6	12	5	.	.
4	.	2	6	4	.	.	.
5	.	2	3	.	.	.	.
6	.	2	.	.	.	.	.
7	1	.	.	.	.	.	.

$N = 8$ (columns $s = 1 \ 2 \ 3 \ 4 \ 5 \ 6 \ 7 \ 8$)

m \ s	1	2	3	4	5	6	7	8
1	.	.	.	.	.	.	.	1
2	.	.	.	1	10	15	7	.
3	.	.	3	18	20	6	.	.
4	.	1	9	12	5	.	.	.
5	.	2	6	4	.	.	.	.
6	.	2	3	.	.	.	.	.
7	.	2	.	.	.	.	.	.
8	1	.	.	.	.	.	.	.

$N = 9$ (columns $s = 1 \ 2 \ 3 \ 4 \ 5 \ 6 \ 7 \ 8 \ 9$)

m \ s	1	2	3	4	5	6	7	8	9
1	.	.	.	.	.	.	.	.	1
2	.	.	.	.	5	20	21	8	.
3	.	.	1	16	40	30	7	.	.
4	.	.	9	24	20	6	.	.	.
5	.	2	9	12	5	.	.	.	.
6	.	2	6	4	.	.	.	.	.
7	.	2	3	.	.	.	.	.	.
8	.	2	.	.	.	.	.	.	.
9	1	.	.	.	.	.	.	.	.

$N = 10$ (columns $s = 1 \ldots 10$)

m \ s	1	2	3	4	5	6	7	8	9	10
1	.	.	.	.	.	.	.	.	.	1
2	.	.	.	.	1	15	35	28	9	.
3	.	.	.	10	50	75	42	8	.	.
4	.	.	6	34	50	30	7	.	.	.
5	.	1	12	24	20	6	.	.	.	.
6	.	2	9	12	5	.	.	.	.	.
7	.	2	6	4	.	.	.	.	.	.
8	.	2	3	.	.	.	.	.	.	.
9	.	2	.	.	.	.	.	.	.	.
10	1	.	.	.	.	.	.	.	.	.

$N = 11$ (columns $s = 1 \ldots 11$)

m \ s	1	2	3	4	5	6	7	8	9	10	11
1	.	.	.	.	.	.	.	.	.	.	1
2	.	.	.	.	.	6	35	56	36	10	.
3	.	.	.	4	45	120	126	56	9	.	.
4	.	.	3	36	90	90	42	8	.	.	.
5	.	.	12	40	50	30	7	.	.	.	.
6	.	2	12	24	20	6	.	.	.	.	.
7	.	2	9	12	5	.	.	.	.	.	.
8	.	2	6	4	.	.	.	.	.	.	.
9	.	2	3	.	.	.	.	.	.	.	.
10	.	2	.	.	.	.	.	.	.	.	.
11	1	.	.	.	.	.	.	.	.	.	.

$N = 12$ (columns $s = 1 \ldots 12$)

m \ s	1	2	3	4	5	6	7	8	9	10	11	12
1	.	.	.	.	.	.	.	.	.	.	.	1
2	.	.	.	.	.	1	21	70	84	45	11	.
3	.	.	.	1	30	140	245	196	72	10	.	.
4	.	.	1	30	125	195	147	56	9	.	.	.
5	.	.	9	54	100	90	42	8	.	.	.	.
6	.	1	15	40	50	30	7	.	.	.	.	.
7	.	2	12	24	20	6	.	.	.	.	.	.
8	.	2	9	12	5	.	.	.	.	.	.	.
9	.	2	6	4	.	.	.	.	.	.	.	.
10	.	2	3	.	.	.	.	.	.	.	.	.
11	.	2	.	.	.	.	.	.	.	.	.	.
12	1	.	.	.	.	.	.	.	.	.	.	.

$N = 13$ (columns $s = 1 \ldots 13$)

m \ s	1	2	3	4	5	6	7	8	9	10	11	12	13
1	.	.	.	.	.	.	.	.	.	.	.	.	1
2	.	.	.	.	.	.	7	56	126	120	55	12	.
3	.	.	.	.	15	126	350	448	288	90	11	.	.
4	.	.	.	20	140	330	371	224	72	10	.	.	.
5	.	.	6	60	165	210	147	56	9	.	.	.	.
6	.	.	15	60	100	90	42	8	.	.	.	.	.
7	.	2	15	40	50	30	7	.	.	.	.	.	.
8	.	2	12	24	20	6	.	.	.	.	.	.	.
9	.	2	9	12	5	.	.	.	.	.	.	.	.
10	.	2	6	4	.	.	.	.	.	.	.	.	.
11	.	2	3	.	.	.	.	.	.	.	.	.	.
12	.	2	.	.	.	.	.	.	.	.	.	.	.
13	1	.	.	.	.	.	.	.	.	.	.	.	.

$N = 14$ (columns $s = 1 \ldots 14$)

m \ s	1	2	3	4	5	6	7	8	9	10	11	12	13	14
1	.	.	.	.	.	.	.	.	.	.	.	.	.	1
2	.	.	.	.	.	.	1	28	126	210	165	66	13	.
3	.	.	.	.	5	90	392	756	756	405	110	12	.	.
4	.	.	.	10	130	456	735	644	324	90	11	.	.	.
5	.	.	3	58	230	405	392	224	72	10	.	.	.	.
6	.	.	12	78	175	210	147	56	9	.	.	.	.	.
7	.	1	18	60	100	90	42	8	.	.	.	.	.	.
8	.	2	15	40	50	30	7	.	.	.	.	.	.	.
9	.	2	12	24	20	6	.	.	.	.	.	.	.	.
10	.	2	9	12	5	.	.	.	.	.	.	.	.	.
11	.	2	6	4	.	.	.	.	.	.	.	.	.	.
12	.	2	3	.	.	.	.	.	.	.	.	.	.	.
13	.	2	.	.	.	.	.	.	.	.	.	.	.	.
14	1	.	.	.	.	.	.	.	.	.	.	.	.	.

$N = 15$ (columns $s = 1 \ldots 15$)

m \ s	1	2	3	4	5	6	7	8	9	10	11	12	13	14	15
1	.	.	.	.	.	.	.	.	.	.	.	.	.	.	1
2	.	.	.	.	.	.	.	8	84	252	330	220	78	14	.
3	.	.	.	.	1	50	357	1008	1470	1200	550	132	13	.	.
4	.	.	.	4	100	530	1197	1456	1044	450	110	12	.	.	.
5	.	.	1	48	280	666	861	672	324	90	11	.	.	.	.
6	.	.	9	88	270	420	392	224	72	10	.	.	.	.	.
7	.	.	18	84	175	210	147	56	9	.	.	.	.	.	.
8	.	2	18	60	100	90	42	8	.	.	.	.	.	.	.
9	.	2	15	40	50	30	7	.	.	.	.	.	.	.	.
10	.	2	12	24	20	6	.	.	.	.	.	.	.	.	.
11	.	2	9	12	5	.	.	.	.	.	.	.	.	.	.
12	.	2	6	4	.	.	.	.	.	.	.	.	.	.	.
13	.	2	3	.	.	.	.	.	.	.	.	.	.	.	.
14	.	2	.	.	.	.	.	.	.	.	.	.	.	.	.
15	1	.	.	.	.	.	.	.	.	.	.	.	.	.	.

Table 6·4 (*cont.*)

N = 16

m \ s	1	2	3	4	5	6	7	8	9	10	11	12	13	14	15	16
1																1
2								1	36	210	462	495	286	91	15	
3						21	266	1106	2268	2640	1815	726	156	14		
4				1	65	525	1652	2716	2646	1605	605	132	13			
5				34	300	960	1617	1652	1080	450	110	12				
6			6	90	370	741	882	672	324	90	11					
7			15	106	280	420	392	224	72	10						
8		1	21	84	175	210	147	56	9							
9		2	18	60	100	90	42	8								
10		2	15	40	50	30	7									
11		2	12	24	20	6										
12		2	9	12	5											
13		2	6	4												
14		2	3													
15		2														
16	1															

N = 17

m \ s	1	2	3	4	5	6	7	8	9	10	11	12	13	14	15	16	17
1																	1
2									9	120	462	792	715	364	105	16	
3						6	161	1016	2898	4620	4455	2640	936	182	15		
4					35	450	1967	4312	5544	4500	2365	792	156	14			
5				20	285	1230	2667	3472	2934	1650	605	132	13				
6			3	84	460	1170	1743	1680	1080	450	110	12					
7			12	130	410	756	882	672	324	90	11						
8			21	112	280	420	392	224	72	10							
9		2	21	84	175	210	147	56	9								
10		2	18	60	100	90	42	8									
11		2	15	40	50	30	7										
12		2	12	24	20	6											
13		2	9	12	5												
14		2	6	4													
15		2	3														
16		2															
17	1																

N = 18

m \ s	1	2	3	4	5	6	7	8	9	10	11	12	13	14	15	16	17	18
1																		1
2									1	45	330	924	1287	1001	455	120	17	
3						1	77	784	3138	6720	8712	7150	3718	1183	210	16		
4					15	335	2051	5944	9912	10440	7260	3366	1014	182	15			
5				10	240	1415	3927	6412	6804	4905	2420	792	156	14				
6			1	70	525	1680	3087	3668	2970	1650	605	132	13					
7			9	126	550	1245	1764	1680	1080	450	110	12						
8			18	138	420	756	882	672	324	90	11							
9		1	24	112	280	420	392	224	72	10								
10		2	21	84	175	210	147	56	9									
11		2	18	60	100	90	42	8										
12		2	15	40	50	30	7											
13		2	12	24	20	6												
14		2	9	12	5													
15		2	6	4														
16		2	3															
17		2																
18	1																	

Table 6·5. *Longest run of two colours. Sequence length* $r = 10(1)15, 20$

For two different colours, r_1 and r_2 in number, with $r_1 + r_2 = r$, the table gives the distribution of the longest run, of length g. The probabilities are obtained by dividing the number tabled by $^rC_{r_1}$.

r	r_1	r_2	$^rC_{r_1}$	19	18	17	16	15	14	13	12	11	10	9	8	7	6	5	4	3	2	1
10	9	1	10	.	.	.	.	.	.	.	.	.	.	2	2	2	2	2	.	.	.	.
	8	2	45	.	.	.	.	.	.	.	.	.	.	.	3	6	9	12	12	3	.	.
	7	3	120	.	.	.	.	.	.	.	.	.	.	.	.	4	12	24	40	36	4	.
	6	4	210	.	.	.	.	.	.	.	.	.	.	.	.	.	5	20	51	91	43	.
	5	5	252	.	.	.	.	.	.	.	.	.	.	.	.	.	.	10	48	110	82	2
11	10	1	11	.	.	.	.	.	.	.	.	.	2	2	2	2	2	1	.	.	.	.
	9	2	55	.	.	.	.	.	.	.	.	.	.	3	6	9	12	15	9	1	.	.
	8	3	165	.	.	.	.	.	.	.	.	.	.	.	4	12	24	40	54	30	1	.
	7	4	330	.	.	.	.	.	.	.	.	.	.	.	.	5	20	50	100	107	48	.
	6	5	462	.	.	.	.	.	.	.	.	.	.	.	.	.	6	33	106	204	112	1
12	11	1	12	.	.	.	.	.	.	.	.	2	2	2	2	2	2	.	.	.	.	.
	10	2	66	.	.	.	.	.	.	.	.	.	3	6	9	12	15	15	6	.	.	.
	9	3	220	.	.	.	.	.	.	.	.	.	.	4	12	24	40	60	60	20	.	.
	8	4	495	.	.	.	.	.	.	.	.	.	.	.	5	20	50	100	165	140	15	.
	7	5	792	.	.	.	.	.	.	.	.	.	.	.	.	6	30	92	220	330	114	.
	6	6	924	.	.	.	.	.	.	.	.	.	.	.	.	.	12	72	224	408	206	2
13	12	1	13	.	.	.	.	.	.	.	2	2	2	2	2	2	1	.	.	.	.	.
	11	2	78	.	.	.	.	.	.	.	.	3	6	9	12	15	18	12	3	.	.	.
	10	3	286	.	.	.	.	.	.	.	.	.	4	12	24	40	60	78	58	10	.	.
	9	4	715	.	.	.	.	.	.	.	.	.	.	5	20	50	100	175	230	130	5	.
	8	5	1287	.	.	.	.	.	.	.	.	.	.	.	6	30	90	211	410	453	87	.
	7	6	1716	.	.	.	.	.	.	.	.	.	.	.	.	7	46	177	466	735	284	1
14	13	1	14	.	.	.	.	.	.	2	2	2	2	2	2	2	.	.	.	.	.	.
	12	2	91	.	.	.	.	.	.	.	3	6	9	12	15	18	18	9	1	.	.	.
	11	3	364	.	.	.	.	.	.	.	.	4	12	24	40	60	84	88	48	4	.	.
	10	4	1001	.	.	.	.	.	.	.	.	.	5	20	50	100	175	270	280	100	1	.
	9	5	2002	.	.	.	.	.	.	.	.	.	.	6	30	90	210	420	668	528	50	.
	8	6	3003	.	.	.	.	.	.	.	.	.	.	.	7	42	150	418	919	1167	300	.
	7	7	3432	.	.	.	.	.	.	.	.	.	.	.	.	14	100	378	968	1454	516	2
15	14	1	15	.	.	.	.	.	2	2	2	2	2	2	2	1	.	.	.	.	.	.
	13	2	105	.	.	.	.	.	.	3	6	9	12	15	18	21	15	6	.	.	.	.
	12	3	455	.	.	.	.	.	.	.	4	12	24	40	60	84	106	90	34	1	.	.
	11	4	1365	.	.	.	.	.	.	.	.	5	20	50	100	175	280	370	300	65	.	.
	10	5	3003	.	.	.	.	.	.	.	.	.	6	30	90	210	420	741	960	525	21	.
	9	6	5005	.	.	.	.	.	.	.	.	.	.	7	42	147	394	900	1655	1614	246	.
	8	7	6435	.	.	.	.	.	.	.	.	.	.	.	8	61	270	846	1954	2576	719	1
20	19	1	20	2	2	2	2	2	2	2	2	2	2	.	.	.	.	.	.	.	.	.
	18	2	190	.	3	6	9	12	15	18	21	24	27	27	18	9	1	.	.	.	.	.
	17	3	1140	.	.	4	12	24	40	60	84	112	144	180	196	196	168	96	20	.	.	.
	16	4	4845	.	.	.	5	20	50	100	175	280	420	600	815	960	880	470	70	.	.	.
	15	5	15504	.	.	.	.	6	30	90	210	420	756	1260	1980	2880	3540	3086	1100	56	.	.
	14	6	38760	.	.	.	.	.	7	42	147	392	882	1764	3234	5523	8442	10192	7007	1127	1	.
	13	7	77520	.	.	.	.	.	.	8	56	224	672	1680	3696	7392	13516	20936	21672	7556	112	.
	12	8	125970	.	.	.	.	.	.	.	9	72	324	1080	2973	7180	15708	30124	42164	24945	1391	.
	11	9	167960	.	.	.	.	.	.	.	.	10	90	456	1736	5464	14784	34030	58406	47640	5344	.
	10	10	184756	.	.	.	.	.	.	.	.	.	20	208	1140	4464	13180	34570	64058	58290	8194	2

Table 7·1. *Number of permutations of the first N natural numbers with s positive successive differences*; N = 1(1)16

For given N the distribution of s is symmetrical about $\frac{1}{2}(N-1)$ so that one half only of the distribution is tabulated. The probability is obtained by dividing the appropriate number of permutations by N!.

s \ N	1	2	3	4	5	6	7	8	9	10	11	12	13	14	15	16
0	1	1	1	1	1	1	1	1	1	1	1	1	1	1	1	1
1		1	4	11	26	57	120	247	502	1013	2036	4083	8178	16369	32752	65519
2			1	11	66	302	1191	4293	14608	47840	152637	478271	1479726	4537314	13824739	41932745
3				...	26	302	2416	15619	88234	455192	2203488	10187685	45533450	198410786	848090912	3572085255
4					...	...	1191	15619	156190	1310354	9738114	66318474	423281535	2571742175	15041229521	85383238549
5								...	88234	1310354	15724248	162512286	1505621508	12843262863	102776998928	782115518299
6										...	9738114	162512286	2275172004	27971176092	311387598411	3207483178157
7												...	1505621508	27971176092	447538817472	6382798925475
8														...	311387598411	6382798925475

Table 7·2·1. *One-half the number of permutations of the first N natural numbers giving rise to t sign runs*; N = 2(1)15

Given the first N natural numbers arranged in a random order and differenced, there will be N−1 successive differences, some positive and some negative, in sign. The number of runs of positive and negative signs is t. One half the number of permutations (out of a total of N!), which have t runs, is tabulated. The probability of a given arrangement is (2 × Number tabulated/N!).

t \ N	2	3	4	5	6	7	8	9	10	11	12	13	14	15
1	1	1	1	1	1	1	1	1	1	1	1	1	1	1
2		2	6	14	30	62	126	254	510	1022	2046	4094	8190	16382
3			5	29	118	418	1383	4407	13736	42236	128761	390385	1179354	3554454
4				16	150	926	4788	22548	100530	433162	1825296	7577120	31130190	126969558
5					61	841	7311	51663	325446	1910706	10715506	58258210	309958755	1623847695
6						272	5166	59982	553410	4474002	33264396	233794460	1579900140	10379490060
7							1385	34649	517496	6031076	60719066	555855290	4766383420	39004154260
8								7936	252750	4717222	67695936	829327600	9149098140	93784852940
9									50521	1995181	45484301	787626845	11526573255	150635656095
10										353792	16962726	463683670	9529400790	164092645110
11											2702765	154624469	4991117034	120067381974
12												22368256	1505035350	56630860638
13													199360981	15583997521
14														1903757312

Table 7·2·2. *Number of permutations of the first N natural numbers with t_u runs up*; N = 2(1)15

Given N! is the number of permutations of the first N natural numbers and T_t is the number of permutations in which there are t_u runs up in these numbers, the probability of such an arrangement is $T_t/N!$.

t_u \ N	2	3	4	5	6	7	8	9	10	11	12	13	14	15
0	1	1	1	1	1	1	1	1	1	1	1	1	1	1
1	1	5	18	58	179	543	1636	4916	14757	44281	132854	398574	1195735	3587219
2			5	61	479	3111	18270	101166	540242	2819266	14494859	73802835	373398489	1881341265
3					61	1385	19028	206276	1949762	16889786	137963364	1081702420	8236142455	61386982075
4							1385	50521	1073517	17460701	241595239	3002137335	34591152955	377209516235
5									50521	2702765	82112518	1866618654	35576491869	598888328289
6											2702765	199360981	8200548715	248913100771
7													199360981	19391512145

Table 7·3. *Number of permutations, T_p, of the first N natural numbers with p peaks;*
$$N = 2(1)15$$

Given a sequence of the first N natural numbers, $x_1, x_2, ..., x_N$ arranged in a random order, there will be a 'peak' at the ith position if

$$x_{i-1} < x_i > x_{i+1}.$$

The probability that there will be just p peaks is $T_p/N!$.

p \ N	2	3	4	5	6	7	8	9	10	11	12	13	14	15
0	2	4	8	16	32	64	128	256	512	1024	2048	4096	8192	16384
1		2	16	88	416	1824	7680	31616	128512	518656	2084864	8361984	33497088	134094848
2				16	272	2880	24576	185856	1304832	8728576	56520704	357888000	2230947840	13754155008
3						272	7936	137216	1841152	21253376	222398464	2174832640	20261765120	182172651520
4								7936	353792	9061376	175627264	2868264960	41731645440	559148810240
5										353792	22368256	795300864	21016670208	460858269696
6												22368256	1903757312	89702612992
7														1903757312

Table 7·4·1. *Number of permutations T_l, of the first N natural numbers, in which the longest ascending run is of length l_u; $N = 2(1)18$*

Given a sequence of the first N natural numbers $x_1, x_2, ..., x_N$ arranged in a random order, an ascending run of length l_u will occur, for example, when

$$x_{i-1} > x_i < x_{i+1} < ... < x_{i+l_u-1} > x_{i+l_u}.$$

The probability that the *longest* ascending run is of length l_u is $T_l/N!$.

l_u \ N	1	2	3	4	5	6	7	8	9	10	11	12	13
1	1	1	1	1	1	1	1	1	1	1	1	1	1
2		1	4	16	69	348	2016	13357	99376	822040	7477161	74207208	797771520
3			1	6	41	293	2309	19975	189324	1960041	21993884	266361634	3465832370
4				1	8	67	602	5811	60875	690729	8457285	111323149	1569068565
5					1	10	99	1024	11304	133669	1695429	23023811	333840443
6						1	12	137	1602	19710	257400	3574957	52785901
7							1	14	181	2360	32010	456720	6881160
8								1	16	231	3322	49236	761904
9									1	18	287	4512	72540
10										1	20	349	5954
11											1	22	417
12												1	24
13													1

Table 7·4·1 (cont.)

l_u \ N	14	15	16	17	18
1	1	1	1	1	1
2	9236662345	114579019468	1516103040832	21314681315997	317288088082404
3	48245601976	715756932697	11277786883706	188135296650845	3313338641688673
4	23592426102	377105857043	6387313185590	114303481217895	2155348564851616
5	5153118154	84426592621	1463941342191	26793750988542	516319125748337
6	827242933	13730434111	240806565782	4445225178694 6	86585391630673
7	109546009	1841298059	32629877967	608572228291	11923667699474
8	12372360	211170960	3788091451	71356438043	1409672722481
9	1209936	21064680	383685120	7315701120	145957544981
10	103194	1845480	34288800	663848640	1406178240
11	7672	142590	2721600	53749920	1102187520
12	491	9690	192240	3900480	81591840
13	26	571	12032	253776	5454144
14	1	28	657	14722	328950
15		1	30	749	17784
16			1	32	847
17				1	34
18					1

Table 7·4·2. *Number of permutations, T_l^*, of the first N natural numbers in which the longest run up or down is of length l; N = 2(1)14*

The run 'down' or descending run is defined similarly to the ascending run of the previous table. The probability that the longest run ascending or descending is of length l is given by $T_l^*/N!$

l \ N	2	3	4	5	6	7	8	9	10	11	12	13	14
2	2	4	10	32	122	544	2770	15872	101042	707584	5405530	44736512	398721962
3		2	12	70	442	3108	24216	208586	1972904	20373338	228346522	2763259364	35927135944
4			2	16	134	1164	10982	112354	1245676	14909340	191916532	2646066034	38932027996
5				2	20	198	2048	22468	264538	3340962	45173518	652197968	10024549190
6					2	24	274	3204	39420	514296	7137818	105318770	1649428474
7						2	28	362	4720	64020	913440	13760472	219040274
8							2	32	462	6644	98472	1523808	24744720
9								2	36	574	9024	145080	2419872
10									2	40	698	11908	206388
11										2	44	834	15344
12											2	48	982
13												2	52
14													2
15													

Table 7·5·1. *Number of permutations, T_c, of the first N natural numbers with c_u consecutive ascending pairs*; $N = 1(1)16$

A random sequence of the first N natural numbers is supposed. For illustration let $N = 7$ and let the sequence be

$$3 \;\; 2 \;\; 1 \;\; 4 \;\; 5 \;\; 7 \;\; 6.$$

This sequence has one ascending pair of numbers and three descending pairs. The probability of an arrangement of the N numbers containing c_u consecutive ascending pairs is $T_c/N!$

c_u \ N	1	2	3	4	5	6	7	8	9	10	11	12	13	14	15	16
0	1	1	3	11	53	309	2119	16687	148329	1468457	16019531	190899411	2467007773	34361893981	513137616783	8178130767479
1		1	2	9	44	265	1854	14833	133496	1334961	14684570	176214841	2290792932	32071101049	481066515734	7697064251745
2			1	3	18	110	795	6489	59332	600732	6674805	80765135	1057289046	14890154058	224497707343	3607998868005
3				1	4	30	220	1855	17304	177996	2002440	24474285	323060540	4581585866	69487385604	1122488536715
4					1	5	45	385	3710	38934	444990	5506710	73422855	1049946755	16035550531	260577696015
5						1	6	63	616	6678	77868	978978	13216104	190899423	2939850914	48106651593
6							1	7	84	924	11130	142758	1957956	28634892	445431987	7349627285
7								1	8	108	1320	17490	244728	3636204	57269784	954497115
8									1	9	135	1815	26235	397683	6363357	107380845
9										1	10	165	2420	37895	618618	10605595
10											1	11	198	3146	53053	927927
11												1	12	234	4004	72345
12													1	13	273	5005
13														1	14	315
14															1	15
15																1

Table 7·5·2. *One-half the number of permutations, T_c^*, in which there are c consecutive ascending and descending pairs*; $N = 2(1)16$

The probability of a sequence of the first N natural numbers containing c consecutive ascending and descending pairs is given by $T_c^*/N!$

c \ N	2	3	4	5	6	7	8	9	10	11	12	13	14	15	16
0			1	7	45	323	2621	23811	239653	2648395	31889517	415641779	5830753109	87601592187	1403439027805
1	1	2	5	20	115	790	6217	55160	545135	5938490	70686805	912660508	12702694075	189579135710	3019908731105
2		1	5	24	128	835	6423	56410	554306	6016077	71426225	920484892	12793635300	190730117959	3035659077083
3			1	8	60	444	3599	32484	325322	3582600	43029621	559774736	7841128936	117668021988	1883347579515
4				1	11	113	1099	11060	118484	1366134	16970322	226574211	3240161105	49453685911	802790789101
5					1	14	183	2224	27152	342684	4543314	63698088	946271907	14890232426	247843374361
6						1	17	270	3950	57090	837234	12672240	199784640	3295943235	57008766755
7							1	20	374	6408	107370	1803312	30868440	543965592	9927502779
8								1	23	495	9729	185769	3526263	67606449	1322489055
9									1	26	633	14044	301345	6390738	135988283
10										1	29	788	19484	464437	10895833
11											1	32	960	26180	686665
12												1	35	1149	34263
13													1	38	1355
14														1	41
15															1

Table 7·6·1. *Number of permutations, T_w, of the first N natural numbers with w_u runs of consecutive pairs up; $N = 2(1)16$*

A random sequence of the first N natural numbers is supposed. For illustration let this be $N = 7$ with the sequence

$$3 \ 2 \ 1 \ 4 \ 5 \ 7 \ 6.$$

Here there is one run of consecutive pairs up, and two runs of consecutive pairs down. The probability that in a sequence of N there will be w_u runs of consecutive pairs up is $T_w/N!$.

w_u \ N	2	3	4	5	6	7	8	9	10	11	12	13	14	15	16
0	1	3	11	53	309	2119	16687	148329	1468457	16019531	190899411	2467007773	34361893981	513137616783	8178130767479
1	1	3	12	56	321	2175	17008	150504	1485465	16170035	192384876	2483177808	34554278857	515620794591	8212685046336
2			1	11	87	693	5934	55674	572650	6429470	78366855	1031378445	14583751161	220562730171	3553474001452
3					3	53	680	8064	96370	1200070	15778800	220047400	3257228485	51125192475	849388162448
4							11	309	5805	95575	1516785	24206055	396475975	6733084365	119143997490
5									53	2119	54564	1186632	24097899	479209437	9567687136
6											309	16687	562723	15591849	393815324
7													2119	148329	6333648
8															16687

Table 7·6·2. *One-half the number of permutations, T_w^*, of the first N natural numbers with w runs of consecutive pairs up and down; $N = 2(1)16$*

The probability that in a random sequence of the first N natural numbers there will be w runs of consecutive pairs up and down is given by $2T_w^*/N!$.

w \ N	1	2	3	4	5	6	7	8	9	10	11	12	13	14	15	16
0	1			1	7	45	323	2621	23811	239653	2648395	31889517	415641779	5830753109	87601592187	1403439027805
1		1	3	8	28	143	933	7150	62310	607445	6545935	77232740	989893248	13692587323	203271723033	3223180454138
2				3	25	155	1005	7488	64164	619986	6646750	78161249	999473835	13801761213	204631472475	3241541125110
3						17	259	2770	27978	294602	3331790	40682144	535206440	7557750635	114101726625	1834757172082
4								131	3177	51433	740375	10495759	152731785	2320039735	37026077865	622039472960
5										1281	45155	1024252	19832856	364000521	6640162083	123218209230
6												15139	730457	22043447	551171457	12703420042
7														209617	13258275	512736710
8																3325923

Table 8·1. *The solution of the equation* $\exp(-a) + ka = 1$ *for a, given* $0 < k < 1$

k	a	k	a	k	a	k	a	k	a	k	a
0·050	20·0000000	0·100	9·9995458	0·150	6·6581095	0·200	4·9651142	0·250	3·9206904		
·051	19·6078431	·101	9·9004937	·151	6·6136295	·201	4·9395120	·251	3·9037180		
·052	19·2307691	·102	9·8033798	·152	6·5697225	·202	4·9141487	·252	3·8868671		
·053	18·8679244	·103	9·7081477	·153	6·5263771	·203	4·8890207	·253	3·8701363		
·054	18·5185184	·104	9·6147429	·154	6·4835823	·204	4·8641246	·254	3·8535241		
0·055	18·1818180	0·105	9·5231129	0·155	6·4413272	0·205	4·8394568	0·255	3·8370292		
·056	17·8571425	·106	9·4332073	·156	6·3996012	·206	4·8150140	·256	3·8206502		
·057	17·5438592	·107	9·3449775	·157	6·3583940	·207	4·7907929	·257	3·8043857		
·058	17·2413787	·108	9·2583768	·158	6·3176956	·208	4·7667902	·258	3·7882344		
·059	16·9491518	·109	9·1733599	·159	6·2774962	·209	4·7430026	·259	3·7721949		
0·060	16·6666657	0·110	9·0898836	0·160	6·2377864	0·210	4·7194272	0·260	3·7562660		
·061	16·3934414	·111	9·0079060	·161	6·1985568	·211	4·6960608	·261	3·7404463		
·062	16·1290307	·112	8·9273866	·162	6·1597983	·212	4·6729003	·262	3·7247347		
·063	15·8730138	·113	8·8482865	·163	6·1215021	·213	4·6499430	·263	3·7091299		
·064	15·6249974	·114	8·7705681	·164	6·0836596	·214	4·6271857	·264	3·6936307		
0·065	15·3846122	0·115	8·6941952	0·165	6·0462625	0·215	4·6046258	0·265	3·6782358		
·066	15·1515112	·116	8·6191326	·166	6·0093024	·216	4·5822605	·266	3·6629441		
·067	14·9253682	·117	8·5453466	·167	5·9727714	·217	4·5600869	·267	3·6477545		
·068	14·7058763	·118	8·4728044	·168	5·9366617	·218	4·5381024	·268	3·6326657		
·069	14·4927463	·119	8·4014745	·169	5·9009657	·219	4·5163044	·269	3·6176767		
0·070	14·2857054	0·120	8·3313262	0·170	5·8656758	0·220	4·4946903	0·270	3·6027863		
·071	14·0844963	·121	8·2623301	·171	5·8307850	·221	4·4732575	·271	3·5879935		
·072	13·8888760	·122	8·1944575	·172	5·7962859	·222	4·4520036	·272	3·5732972		
·073	13·6986147	·123	8·1276809	·173	5·7621718	·223	4·4309262	·273	3·5586962		
·074	13·5134952	·124	8·0619734	·174	5·7284359	·224	4·4100228	·274	3·5441897		
0·075	13·3333117	0·125	7·9973091	0·175	5·6950714	0·225	4·3892910	0·275	3·5297765		
·076	13·1578693	·126	7·9336629	·176	5·6620721	·226	4·3687286	·276	3·5154556		
·077	12·9869832	·127	7·8710106	·177	5·6294315	·227	4·3483333	·277	3·5012261		
·078	12·8204781	·128	7·8093287	·178	5·5971436	·228	4·3281028	·278	3·4870869		
·079	12·6581876	·129	7·7485942	·179	5·5652022	·229	4·3080351	·279	3·4730370		
0·080	12·4999534	0·130	7·6887851	0·180	5·5336015	0·230	4·2881278	0·280	3·4590756		
·081	12·3456253	·131	7·6298800	·181	5·5023357	·231	4·2683789	·281	3·4452017		
·082	12·1950603	·132	7·5718581	·182	5·4713992	·232	4·2487864	·282	3·4314143		
·083	12·0481222	·133	7·5146991	·183	5·4407865	·233	4·2293482	·283	3·4177125		
·084	11·9046814	·134	7·4583837	·184	5·4104922	·234	4·2100622	·284	3·4040955		
0·085	11·7646144	0·135	7·4028927	0·185	5·3805110	0·235	4·1909266	0·285	3·3905624		
·086	11·6278033	·136	7·3482078	·186	5·3508378	·236	4·1719393	·286	3·3771122		
·087	11·4941358	·137	7·2943110	·187	5·3214675	·237	4·1530985	·287	3·3637441		
·088	11·3635044	·138	7·2411851	·188	5·2923952	·238	4·1344024	·288	3·3504572		
·089	11·2358068	·139	7·1888131	·189	5·2636161	·239	4·1158490	·289	3·3372508		
0·090	11·1109450	0·140	7·1371786	0·190	5·2351255	0·240	4·0974366	0·290	3·3241240		
·091	10·9888254	·141	7·0862658	·191	5·2069187	·241	4·0791634	·291	3·3110759		
·092	10·8693583	·142	7·0360593	·192	5·1789912	·242	4·0610277	·292	3·2981058		
·093	10·7524581	·143	6·9865438	·193	5·1513386	·243	4·0430277	·293	3·2852129		
·094	10·6380427	·144	6·9377049	·194	5·1239566	·244	4·0251617	·294	3·2723964		
0·095	10·5260334	0·145	6·8895283	0·195	5·0968408	0·245	4·0074282	0·295	3·2596555		
·096	10·4163548	·146	6·8420002	·196	5·0699872	·246	3·9898254	·296	3·2469895		
·097	10·3089347	·147	6·7951072	·197	5·0433917	·247	3·9723518	·297	3·2343976		
·098	10·2037037	·148	6·7488362	·198	5·0170503	·248	3·9550058	·298	3·2218790		
·099	10·1005954	·149	6·7031744	·199	4·9909591	·249	3·9377858	·299	3·2094331		
0·100	9·9995458	0·150	6·6581095	0·200	4·9651142	0·250	3·9206904	0·300	3·1970591		

N.B. for $k < 0.05$, take $a = 1/k$.

Table 8·1 (*cont.*)

k	a	k	a	k	a	k	a	k	a	k	a
0·300	3·1970591	0·350	2·6566127	0·400	2·2316119	0·450	1·8847348	0·500	1·5936243		
·301	3·1847564	·351	2·6471411	·401	2·2240016	·451	1·8784240	·501	1·5882633		
·302	3·1725241	·352	2·6377143	·402	2·2164218	·452	1·8721351	·502	1·5829188		
·303	3·1603617	·353	2·6283321	·403	2·2088724	·453	1·8658679	·503	1·5775904		
·304	3·1482684	·354	2·6189939	·404	2·2013532	·454	1·8596224	·504	1·5722783		
0·305	3·1362436	0·355	2·6096996	0·405	2·1938638	0·455	1·8533984	0·505	1·5669823		
·306	3·1242866	·356	2·6004487	·406	2·1864042	·456	1·8471958	·506	1·5617022		
·307	3·1123968	·357	2·5912409	·407	2·1789740	·457	1·8410145	·507	1·5564381		
·308	3·1005735	·358	2·5820758	·408	2·1715732	·458	1·8348542	·508	1·5511898		
·309	3·0888160	·359	2·5729530	·409	2·1642015	·459	1·8287149	·509	1·5459572		
0·310	3·0771238	0·360	2·5638723	0·410	2·1568586	0·460	1·8225965	0·510	1·5407403		
·311	3·0654961	·361	2·5548333	·411	2·1495444	·461	1·8164989	·511	1·5355389		
·312	3·0539325	·362	2·5458356	·412	2·1422588	·462	1·8104218	·512	1·5303531		
·313	3·0424323	·363	2·5368790	·413	2·1350014	·463	1·8043652	·513	1·5251826		
·314	3·0309949	·364	2·5279631	·414	2·1277721	·464	1·7983290	·514	1·5200275		
0·315	3·0196198	0·365	2·5190875	0·415	2·1205707	0·465	1·7923131	0·515	1·5148876		
·316	3·0083062	·366	2·5102520	·416	2·1133971	·466	1·7863172	·516	1·5097628		
·317	2·9970537	·367	2·5014563	·417	2·1062510	·467	1·7803413	·517	1·5046532		
·318	2·9858617	·368	2·4927000	·418	2·0991322	·468	1·7743854	·518	1·4995585		
·319	2·9747297	·369	2·4839828	·419	2·0920406	·469	1·7684491	·519	1·4944788		
0·320	2·9636570	0·370	2·4753044	0·420	2·0849759	0·470	1·7625325	0·520	1·4894139		
·321	2·9526432	·371	2·4666645	·421	2·0779381	·471	1·7566355	·521	1·4843637		
·322	2·9416876	·372	2·4580629	·422	2·0709268	·472	1·7507578	·522	1·4793282		
·323	2·9307898	·373	2·4494991	·423	2·0639420	·473	1·7448995	·523	1·4743074		
·324	2·9199493	·374	2·4409730	·424	2·0569834	·474	1·7390603	·524	1·4693011		
0·325	2·9091655	0·375	2·4324843	0·425	2·0500510	0·475	1·7332402	0·525	1·4643092		
·326	2·8984379	·376	2·4240326	·426	2·0431444	·476	1·7274391	·526	1·4593317		
·327	2·8877660	·377	2·4156176	·427	2·0362635	·477	1·7216568	·527	1·4543685		
·328	2·8771494	·378	2·4072391	·428	2·0294083	·478	1·7158932	·528	1·4494195		
·329	2·8665874	·379	2·3988969	·429	2·0225784	·479	1·7101483	·529	1·4444847		
0·330	2·8560797	0·380	2·3905906	0·430	2·0157738	0·480	1·7044219	0·530	1·4395640		
·331	2·8456257	·381	2·3823199	·431	2·0089942	·481	1·6987139	·531	1·4346573		
·332	2·8352249	·382	2·3740846	·432	2·0022395	·482	1·6930242	·532	1·4297645		
·333	2·8248770	·383	2·3658845	·433	1·9955095	·483	1·6873527	·533	1·4248856		
·334	2·8145814	·384	2·3577192	·434	1·9888042	·484	1·6816993	·534	1·4200205		
0·335	2·8043377	0·385	2·3495885	0·435	1·9821232	0·485	1·6760639	0·535	1·4151691		
·336	2·7941455	·386	2·3414922	·436	1·9754665	·486	1·6704464	·536	1·4103313		
·337	2·7840042	·387	2·3334300	·437	1·9688339	·487	1·6648467	·537	1·4055072		
·338	2·7739134	·388	2·3254016	·438	1·9622253	·488	1·6592647	·538	1·4006965		
·339	2·7638728	·389	2·3174068	·439	1·9556405	·489	1·6537002	·539	1·3958994		
0·340	2·7538818	0·390	2·3094453	0·440	1·9490792	0·490	1·6481533	0·540	1·3911155		
·341	2·7439401	·391	2·3015170	·441	1·9425415	·491	1·6426237	·541	1·3863450		
·342	2·7340472	·392	2·2936214	·442	1·9360271	·492	1·6371114	·542	1·3815878		
·343	2·7242027	·393	2·2857585	·443	1·9295360	·493	1·6316164	·543	1·3768437		
·344	2·7144062	·394	2·2779280	·444	1·9230678	·494	1·6261384	·544	1·3721128		
0·345	2·7046573	0·395	2·2701297	0·445	1·9166226	0·495	1·6206774	0·545	1·3673948		
·346	2·6949555	·396	2·2623632	·446	1·9102001	·496	1·6152333	·546	1·3626899		
·347	2·6853006	·397	2·2546284	·447	1·9038002	·497	1·6098061	·547	1·3579979		
·348	2·6756921	·398	2·2469251	·448	1·8974228	·498	1·6043955	·548	1·3533187		
·349	2·6661296	·399	2·2392530	·449	1·8910677	·499	1·5990016	·549	1·3486523		
0·350	2·6566127	0·400	2·2316119	0·450	1·8847348	0·500	1·5936243	0·550	1·3439987		

Table 8·1 (*cont.*)

k	a	k	a	k	a	k	a	k	a
0·550	1·3439987	0·600	1·1262612	0·650	0·9336939	0·700	0·7614337	0·750	0·6058600
·551	1·3393577	·601	1·1221820	·651	·9300635	·701	·7581691	·751	·6028986
·552	1·3347293	·602	1·1181127	·652	·9264412	·702	·7549111	·752	·5999427
·553	1·3301135	·603	1·1140533	·653	·9228268	·703	·7516598	·753	·5969924
·554	1·3255101	·604	1·1100038	·654	·9192204	·704	·7484150	·754	·5940475
0·555	1·3209191	0·605	1·1059641	0·655	0·9156220	0·705	0·7451767	0·755	0·5911081
·556	1·3163405	·606	1·1019342	·656	·9120315	·706	·7419450	·756	·5881742
·557	1·3117742	·607	1·0979141	·657	·9084489	·707	·7387198	·757	·5852457
·558	1·3072201	·608	1·0939037	·658	·9048742	·708	·7355010	·758	·5823227
·559	1·3026782	·609	1·0899029	·659	·9013073	·709	·7322887	·759	·5794050
0·560	1·2981485	0·610	1·0859117	0·660	0·8977481	0·710	0·7290829	0·760	0·5764927
·561	1·2936307	·611	1·0819301	·661	·8941968	·711	·7258834	·761	·5735858
·562	1·2891250	·612	1·0779581	·662	·8906531	·712	·7226903	·762	·5706842
·563	1·2846312	·613	1·0739955	·663	·8871172	·713	·7195036	·763	·5677880
·564	1·2801493	·614	1·0700424	·664	·8835889	·714	·7163232	·764	·5648970
0·565	1·2756792	0·615	1·0660987	0·665	0·8800683	0·715	0·7131491	0·765	0·5620114
·566	1·2712209	·616	1·0621643	·666	·8765553	·716	·7099813	·766	·5591310
·567	1·2667743	·617	1·0582393	·667	·8730498	·717	·7068198	·767	·5562558
·568	1·2623394	·618	1·0543236	·668	·8695519	·718	·7036645	·768	·5533859
·569	1·2579161	·619	1·0504171	·669	·8660616	·719	·7005154	·769	·5505213
0·570	1·2535043	0·620	1·0465198	0·670	0·8625787	0·720	0·6973725	0·770	0·5476618
·571	1·2491040	·621	1·0426317	·671	·8591032	·721	·6942358	·771	·5448075
·572	1·2447151	·622	1·0387527	·672	·8556352	·722	·6911053	·772	·5419584
·573	1·2403377	·623	1·0348828	·673	·8521746	·723	·6879808	·773	·5391144
·574	1·2359715	·624	1·0310219	·674	·8487214	·724	·6848625	·774	·5362756
0·575	1·2316167	0·625	1·0271701	0·675	0·8452755	0·725	0·6817503	0·775	0·5334419
·576	1·2272730	·626	1·0233272	·676	·8418370	·726	·6786441	·776	·5306132
·577	1·2229405	·627	1·0194933	·677	·8384057	·727	·6755440	·777	·5277897
·578	1·2186192	·628	1·0156683	·678	·8349816	·728	·6724499	·778	·5249712
·579	1·2143089	·629	1·0118521	·679	·8315648	·729	·6693618	·779	·5221578
0·580	1·2100097	0·630	1·0080447	0·680	0·8281552	0·730	0·6662796	0·780	0·5193494
·581	1·2057214	·631	1·0042461	·681	·8247528	·731	·6632035	·781	·5165460
·582	1·2014440	·632	1·0004563	·682	·8213575	·732	·6601332	·782	·5137476
·583	1·1971774	·633	0·9966752	·683	·8179694	·733	·6570689	·783	·5109542
·584	1·1929217	·634	0·9929027	·684	·8145883	·734	·6540104	·784	·5081658
0·585	1·1886768	0·635	0·9891389	0·685	0·8112143	0·735	0·6509579	0·785	0·5053823
·586	1·1844426	·636	·9853837	·686	·8078473	·736	·6479112	·786	·5026037
·587	1·1802190	·637	·9816371	·687	·8044874	·737	·6448703	·787	·4998301
·588	1·1760060	·638	·9778989	·688	·8011344	·738	·6418352	·788	·4970613
·589	1·1718036	·639	·9741693	·689	·7977884	·739	·6388059	·789	·4942975
0·590	1·1676118	0·640	0·9704481	0·690	0·7944493	0·740	0·6357824	0·790	0·4915385
·591	1·1634304	·641	·9667354	·691	·7911171	·741	·6327646	·791	·4887843
·592	1·1592594	·642	·9630310	·692	·7877918	·742	·6297526	·792	·4860350
·593	1·1550988	·643	·9593350	·693	·7844733	·743	·6267462	·793	·4832905
·594	1·1509485	·644	·9556473	·694	·7811617	·744	·6237456	·794	·4805508
0·595	1·1468085	0·645	0·9519679	0·695	0·7778568	0·745	0·6207506	0·795	0·4778159
·596	1·1426788	·646	·9482967	·696	·7745587	·746	·6177612	·796	·4750858
·597	1·1385592	·647	·9446338	·697	·7712674	·747	·6147775	·797	·4723604
·598	1·1344498	·648	·9409790	·698	·7679828	·748	·6117994	·798	·4696398
·599	1·1303505	·649	·9373324	·699	·7647049	·749	·6088269	·799	·4669239
0·600	1·1262612	0·650	0·9336939	0·700	0·7614337	0·750	0·6058600	0·800	0·4642128

Table 8·1 (*cont.*)

k	a	k	a	k	a	k	a
0·800	0·4642128	0·850	0·3343447	0·900	0·2145557	0·950	0·1034788
·801	·4615063	·851	·3318552	·901	·2122529	·951	·1013381
·802	·4588045	·852	·3293697	·902	·2099535	·952	·0992005
·803	·4561073	·853	·3268882	·903	·2076576	·953	·0970658
·804	·4534148	·854	·3244107	·904	·2053652	·954	·0949342
0·805	0·4507270	0·855	0·3219371	0·905	0·2030762	0·955	0·0928056
·806	·4480438	·856	·3194675	·906	·2007906	·956	·0906799
·807	·4453652	·857	·3170019	·907	·1985084	·957	·0885573
·808	·4426911	·858	·3145402	·908	·1962296	·958	·0864376
·809	·4400217	·859	·3120824	·909	·1939542	·959	·0843209
0·810	0·4373568	0·860	0·3096286	0·910	0·1916824	0·960	0·0822071
·811	·4346965	·861	·3071786	·911	·1894137	·961	·0800963
·812	·4320407	·862	·3047325	·912	·1871484	·962	·0779885
·813	·4293894	·863	·3022903	·913	·1848866	·963	·0758836
·814	·4267426	·864	·2998520	·914	·1826281	·964	·0737816
0·815	0·4241004	0·865	0·2974176	0·915	0·1803729	0·965	0·0716825
·816	·4214626	·866	·2949869	·916	·1781211	·966	·0695864
·817	·4188292	·867	·2925601	·917	·1758725	·967	·0674932
·818	·4162003	·868	·2901372	·918	·1736274	·968	·0654028
·819	·4135759	·869	·2877180	·919	·1713855	·969	·0633154
0·820	0·4109558	0·870	0·2853027	0·920	0·1691469	0·970	0·0612308
·821	·4083402	·871	·2828911	·921	·1669116	·971	·0591492
·822	·4057290	·872	·2804833	·922	·1646795	·972	·0570704
·823	·4031222	·873	·2780793	·923	·1624508	·973	·0549944
·824	·4005197	·874	·2756790	·924	·1602253	·974	·0529213
0·825	0·3979215	0·875	0·2732825	0·925	0·1580031	0·975	0·0508511
·826	·3953278	·876	·2708897	·926	·1557841	·976	·0487837
·827	·3927383	·877	·2685007	·927	·1535683	·977	·0467191
·828	·3901532	·878	·2661153	·928	·1513558	·978	·0446574
·829	·3875723	·879	·2637337	·929	·1491465	·979	·0425985
0·830	0·3849957	0·880	0·2613557	0·930	0·1469404	0·980	0·0405424
·831	·3824235	·881	·2589815	·931	·1447374	·981	·0384891
·832	·3798554	·882	·2566109	·932	·1425377	·982	·0364386
·833	·3772916	·883	·2542439	·933	·1403412	·983	·0343909
·834	·3747321	·884	·2518806	·934	·1381478	·984	·0323460
0·835	0·3721768	0·885	0·2495210	0·935	0·1359576	0·985	0·0303038
·836	·3696257	·886	·2471649	·936	·1337706	·986	·0282644
·837	·3670787	·887	·2448125	·937	·1315867	·987	·0262278
·838	·3645360	·888	·2424637	·938	·1294060	·988	·0241939
·839	·3619974	·889	·2401185	·939	·1272283	·989	·0221628
0·840	0·3594630	0·890	0·2377769	0·940	0·1250538	0·990	0·0201345
·841	·3569327	·891	·2354389	·941	·1228825	·991	·0181088
·842	·3544066	·892	·2331044	·942	·1207142	·992	·0160859
·843	·3518846	·893	·2307735	·943	·1185490	·993	·0140657
·844	·3493667	·894	·2284461	·944	·1163869	·994	·0120482
0·845	0·3468528	0·895	0·2261222	0·945	0·1142279	0·995	0·0100335
·846	·3443431	·896	·2238019	·946	·1120720	·996	·0080214
·847	·3418374	·897	·2214851	·947	·1099191	·997	·0060120
·848	·3393358	·898	·2191718	·948	·1077693	·998	·0040053
·849	·3368383	·899	·2168620	·949	·1056226	·999	·0020013
0·850	0·3343447	0·900	0·2145557	0·950	0·1034788	1·000	0·0000000

Table 8·2. *The solution of the equation* $\exp(b) - b/(1-p) = 1$ *for b, given* $0 < p < 1$

p	b	p	b	p	b	p	b	p	b
0·000	0·0000000	0·050	0·1017243	0·100	0·2071465	0·150	0·3166873	0·200	0·4308422
·001	·0020007	·051	·1037953	·101	·2092954	·151	·3189233	·201	·4331762
·002	·0040027	·052	·1058676	·102	·2114460	·152	·3211613	·202	·4355124
·003	·0060060	·053	·1079415	·103	·2135983	·153	·3234011	·203	·4378506
·004	·0080107	·054	·1100169	·104	·2157522	·154	·3256427	·204	·4401908
0·005	0·0100167	0·055	0·1120937	0·105	0·2179077	0·155	0·3278862	0·205	0·4425332
·006	·0120241	·056	·1141721	·106	·2200649	·156	·3301316	·206	·4448777
·007	·0140328	·057	·1162519	·107	·2222238	·157	·3323788	·207	·4472243
·008	·0160429	·058	·1183332	·108	·2243844	·158	·3346279	·208	·4495731
·009	·0180543	·059	·1204161	·109	·2265466	·159	·3368789	·209	·4519239
0·010	0·0200671	0·060	0·1225004	0·110	0·2287105	0·160	0·3391317	0·210	0·4542769
·011	·0220813	·061	·1245863	·111	·2308761	·161	·3413865	·211	·4566319
·012	·0240968	·062	·1266737	·112	·2330434	·162	·3436431	·212	·4589892
·013	·0261137	·063	·1287625	·113	·2352123	·163	·3459017	·213	·4613485
·014	·0281319	·064	·1308529	·114	·2373830	·164	·3481621	·214	·4637101
0·015	0·0301515	0·065	0·1329449	0·115	0·2395553	0·165	0·3504245	0·215	0·4660737
·016	·0321725	·066	·1350383	·116	·2417294	·166	·3526887	·216	·4684395
·017	·0341949	·067	·1371333	·117	·2439051	·167	·3549549	·217	·4708075
·018	·0362186	·068	·1392298	·118	·2460826	·168	·3572229	·218	·4731777
·019	·0382438	·069	·1413278	·119	·2482618	·169	·3594929	·219	·4755500
0·020	0·0402703	0·070	0·1434274	0·120	0·2504427	0·170	0·3617649	0·220	0·4779245
·021	·0422982	·071	·1455285	·121	·2526253	·171	·3640387	·221	·4803012
·022	·0443275	·072	·1476312	·122	·2548096	·172	·3663145	·222	·4826801
·023	·0463582	·073	·1497354	·123	·2569957	·173	·3685923	·223	·4850611
·024	·0483903	·074	·1518411	·124	·2591835	·174	·3708720	·224	·4874444
0·025	0·0504237	0·075	0·1539485	0·125	0·2613730	0·175	0·3731536	0·225	0·4898299
·026	·0524586	·076	·1560573	·126	·2635643	·176	·3754372	·226	·4922176
·027	·0544949	·077	·1581678	·127	·2657573	·177	·3777227	·227	·4946075
·028	·0565326	·078	·1602798	·128	·2679521	·178	·3800102	·228	·4969996
·029	·0585717	·079	·1623933	·129	·2701486	·179	·3822997	·229	·4993940
0·030	0·0606123	0·080	0·1645085	0·130	0·2723469	0·180	0·3845912	0·230	0·5017906
·031	·0626542	·081	·1666252	·131	·2745469	·181	·3868846	·231	·5041894
·032	·0646976	·082	·1687435	·132	·2767487	·182	·3891800	·232	·5065905
·033	·0667424	·083	·1708633	·133	·2789523	·183	·3914774	·233	·5089938
·034	·0687886	·084	·1729848	·134	·2811576	·184	·3937768	·234	·5113994
0·035	0·0708362	0·085	0·1751079	0·135	0·2833647	0·185	0·3960782	0·235	0·5138073
·036	·0728853	·086	·1772325	·136	·2855736	·186	·3983816	·236	·5162174
·037	·0749358	·087	·1793587	·137	·2877843	·187	·4006870	·237	·5186298
·038	·0769878	·088	·1814866	·138	·2899967	·188	·4029944	·238	·5210445
·039	·0790411	·089	·1836160	·139	·2922110	·189	·4053039	·239	·5234614
0·040	0·0810960	0·090	0·1857470	0·140	0·2944271	0·190	0·4076153	0·240	0·5258807
·041	·0831523	·091	·1878797	·141	·2966449	·191	·4099288	·241	·5283023
·042	·0852100	·092	·1900139	·142	·2988646	·192	·4122443	·242	·5307261
·043	·0872692	·093	·1921498	·143	·3010860	·193	·4145619	·243	·5331523
·044	·0893298	·094	·1942873	·144	·3033093	·194	·4168815	·244	·5355808
0·045	0·0913919	0·095	0·1964264	0·145	0·3055344	0·195	0·4192031	0·245	0·5380116
·046	·0934554	·096	·1985672	·146	·3077613	·196	·4215268	·246	·5404447
·047	·0955205	·097	·2007096	·147	·3099900	·197	·4238526	·247	·5428802
·048	·0975870	·098	·2028536	·148	·3122206	·198	·4261804	·248	·5453180
·049	·0996549	·099	·2049992	·149	·3144530	·199	·4285103	·249	·5477582
0·050	0·1017243	0·100	0·2071465	0·150	0·3166873	0·200	0·4308422	0·250	0·5502007

Table 8·2 (*cont.*)

p	b	p	b	p	b	p	b	p	b
0·250	0·5502007	0·300	0·6754716	0·350	0·8075177	0·400	0·9474049	0·450	1·0964715
·251	·5526456	·301	·6780429	·351	·8102346	·401	·9502912	·451	1·0995575
·252	·5550929	·302	·6806170	·352	·8129547	·402	·9531813	·452	1·1026478
·253	·5575425	·303	·6831938	·353	·8156780	·403	·9560750	·453	1·1057425
·254	·5599945	·304	·6857733	·354	·8184043	·404	·9589724	·454	1·1088416
0·255	0·5624489	0·305	0·6883555	0·355	·8211339	0·405	0·9618735	0·455	1·1119451
·256	·5649057	·306	·6909405	·356	·8238666	·406	·9647784	·456	1·1150530
·257	·5673648	·307	·6935282	·357	·8266025	·407	·9676870	·457	1·1181653
·258	·5698264	·308	·6961187	·358	·8293416	·408	·9705993	·458	1·1212821
·259	·5722904	·309	·6987120	·359	·8320840	·409	·9735154	·459	1·1244033
0·260	0·5747568	0·310	0·7013080	0·360	0·8348295	0·410	0·9764352	0·460	1·1275291
·261	·5772256	·311	·7039068	·361	·8375782	·411	·9793588	·461	1·1306593
·262	·5796969	·312	·7065084	·362	·8403302	·412	·9822863	·462	1·1337941
·263	·5821706	·313	·7091128	·363	·8430854	·413	·9852175	·463	1·1369333
·264	·5846467	·314	·7117200	·364	·8458439	·414	·9881526	·464	1·1400772
0·265	0·5871253	0·315	0·7143300	0·365	0·8486057	0·415	0·9910914	0·465	1·1432255
·266	·5896063	·316	·7169428	·366	·8513707	·416	·9940342	·466	1·1463785
·267	·5920898	·317	·7195585	·367	·8541390	·417	·9969808	·467	1·1495361
·268	·5945758	·318	·7221770	·368	·8569106	·418	·9999312	·468	1·1526983
·269	·5970643	·319	·7247983	·369	·8596855	·419	1·0028856	·469	1·1558651
0·270	0·5995552	0·320	0·7274225	0·370	0·8624637	0·420	1·0058438	0·470	1·1590366
·271	·6020486	·321	·7300496	·371	·8652453	·421	1·0088060	·471	1·1622127
·272	·6045445	·322	·7326795	·372	·8680301	·422	1·0117721	·472	1·1653935
·273	·6070429	·323	·7353123	·373	·8708184	·423	1·0147421	·473	1·1685790
·274	·6095439	·324	·7379480	·374	·8736099	·424	1·0177160	·474	1·1717692
0·275	0·6120473	0·325	0·7405866	0·375	0·8764049	0·425	1·0206940	0·475	1·1749642
·276	·6145533	·326	·7432281	·376	·8792032	·426	1·0236759	·476	1·1781639
·277	·6170618	·327	·7458725	·377	·8820049	·427	1·0266618	·477	1·1813684
·278	·6195728	·328	·7485198	·378	·8848101	·428	1·0296517	·478	1·1845777
·279	·6220864	·329	·7511701	·379	·8876186	·429	1·0326456	·479	1·1877918
0·280	0·6246025	0·330	0·7538232	0·380	0·8904305	0·430	1·0356436	0·480	1·1910107
·281	·6271212	·331	·7564794	·381	·8932459	·431	1·0386456	·481	1·1942345
·282	·6296424	·332	·7591385	·382	·8960647	·432	1·0416516	·482	1·1974631
·283	·6321662	·333	·7618005	·383	·8988870	·433	1·0446618	·483	1·2006966
·284	·6346926	·334	·7644656	·384	·9017127	·434	1·0476760	·484	1·2039350
0·285	0·6372216	0·335	0·7671336	0·385	0·9045420	0·435	1·0506944	0·485	1·2071784
·286	·6397532	·336	·7698046	·386	·9073747	·436	1·0537168	·486	1·2104266
·287	·6422874	·337	·7724786	·387	·9102109	·437	1·0567434	·487	1·2136798
·288	·6448242	·338	·7751556	·388	·9130506	·438	1·0597741	·488	1·2169380
·289	·6473636	·339	·7778356	·389	·9158938	·439	1·0628090	·489	1·2202012
0·290	0·6499056	0·340	0·7805187	0·390	0·9187405	0·440	1·0658481	0·490	1·2234694
·291	·6524502	·341	·7832048	·391	·9215908	·441	1·0688913	·491	1·2267426
·292	·6549975	·342	·7858939	·392	·9244447	·442	1·0719388	·492	1·2300209
·293	·6575474	·343	·7885861	·393	·9273021	·443	1·0749904	·493	1·2333042
·294	·6601000	·344	·7912814	·394	·9301631	·444	1·0780463	·494	1·2365926
0·295	0·6626553	0·345	0·7939797	0·395	0·9330277	0·445	1·0811065	0·495	1·2398862
·296	·6652132	·346	·7966811	·396	·9358959	·446	1·0841709	·496	1·2431848
·297	·6677737	·347	·7993856	·397	·9387677	·447	1·0872396	·497	1·2464887
·298	·6703370	·348	·8020932	·398	·9416431	·448	1·0903126	·498	1·2497976
·299	·6729030	·349	·8048039	·399	·9445222	·449	1·0933899	·499	1·2531118
0·300	0·6754716	0·350	0·8075177	0·400	0·9474049	0·450	1·0964715	0·500	1·2564312

Table 8·2 (*cont.*)

p	b	p	b	p	b	p	b	p	b
0·500	1·2564312	0·550	1·4295288	0·600	1·6187881	0·650	1·8284236	0·700	2·0645683
·501	1·2597558	·551	1·4331439	·601	1·6227645	·651	1·8328620	·701	2·0696186
·502	1·2630857	·552	1·4367654	·602	1·6267490	·652	1·8373109	·702	2·0746833
·503	1·2664208	·553	1·4403934	·603	1·6307417	·653	1·8417704	·703	2·0797623
·504	1·2697613	·554	1·4440279	·604	1·6347426	·654	1·8462407	·704	2·0848558
0·505	1·2731070	0·555	1·4476690	0·605	1·6387518	0·655	1·8507216	0·705	2·0899639
·506	1·2764581	·556	1·4513166	·606	1·6427692	·656	1·8552133	·706	2·0950866
·507	1·2798145	·557	1·4549709	·607	1·6467950	·657	1·8597160	·707	2·1002241
·508	1·2831763	·558	1·4586317	·608	1·6508291	·658	1·8642295	·708	2·1053765
·509	1·2865435	·559	1·4622992	·609	1·6548717	·659	1·8687540	·709	2·1105438
0·510	1·2899161	0·560	1·4659734	0·610	1·6589227	0·660	1·8732896	0·710	2·1157261
·511	1·2932942	·561	1·4696543	·611	1·6629822	·661	1·8778362	·711	2·1209237
·512	1·2966777	·562	1·4733420	·612	1·6670502	·662	1·8823941	·712	2·1261364
·513	1·3000667	·563	1·4770364	·613	1·6711269	·663	1·8869632	·713	2·1313646
·514	1·3034613	·564	1·4807376	·614	1·6752121	·664	1·8915436	·714	2·1366082
0·515	1·3068613	0·565	1·4844457	0·615	1·6793061	0·665	1·8961353	0·715	2·1418674
·516	1·3102669	·566	1·4881606	·616	1·6834087	·666	1·9007385	·716	2·1471422
·517	1·3136781	·567	1·4918825	·617	1·6875201	·667	1·9053532	·717	2·1524329
·518	1·3170948	·568	1·4956112	·618	1·6916404	·668	1·9099795	·718	2·1577395
·519	1·3205172	·569	1·4993469	·619	1·6957694	·669	1·9146174	·719	2·1630620
0·520	1·3239452	0·570	1·5030896	0·620	1·6999074	0·670	1·9192670	0·720	2·1684007
·521	1·3273789	·571	1·5068394	·621	1·7040543	·671	1·9239284	·721	2·1737556
·522	1·3308183	·572	1·5105962	·622	1·7082102	·672	1·9286017	·722	2·1791269
·523	1·3342634	·573	1·5143600	·623	1·7123751	·673	1·9332868	·723	2·1845146
·524	1·3377142	·574	1·5181310	·624	1·7165491	·674	1·9379840	·724	2·1899190
0·525	1·3411708	0·575	1·5219092	0·625	1·7207323	0·675	1·9426932	0·725	2·1953400
·526	1·3446332	·576	1·5256945	·626	1·7249246	·676	1·9474146	·726	2·2007778
·527	1·3481013	·577	1·5294871	·627	1·7291261	·677	1·9521481	·727	2·2062325
·528	1·3515753	·578	1·5332869	·628	1·7333370	·678	1·9568939	·728	2·2117043
·529	1·3550552	·579	1·5370940	·629	1·7375571	·679	1·9616521	·729	2·2171933
0·530	1·3585409	0·580	1·5409084	0·630	1·7417866	0·680	1·9664228	0·730	2·2226996
·531	1·3620326	·581	1·5447302	·631	1·7460255	·681	1·9712059	·731	2·2282234
·532	1·3655301	·582	1·5485593	·632	1·7502739	·682	1·9760016	·732	2·2337646
·533	1·3690336	·583	1·5523959	·633	1·7545318	·683	1·9808100	·733	2·2393236
·534	1·3725431	·584	1·5562399	·634	1·7587993	·684	1·9856311	·734	2·2449004
0·535	1·3760586	0·585	1·5600915	0·635	1·7630764	0·685	1·9904651	0·735	2·2504951
·536	1·3795801	·586	1·5639505	·636	1·7673632	·686	1·9953119	·736	2·2561079
·537	1·3831077	·587	1·5678171	·637	1·7716597	·687	2·0001717	·737	2·2617389
·538	1·3866414	·588	1·5716914	·638	1·7759660	·688	2·0050446	·738	2·2673883
·539	1·3901811	·589	1·5755732	·639	1·7802821	·689	2·0099306	·739	2·2730561
0·540	1·3937270	0·590	1·5794627	0·640	1·7846081	0·690	2·0148298	0·740	2·2787425
·541	1·3972790	·591	1·5833600	·641	1·7889440	·691	2·0197424	·741	2·2844478
·542	1·4008373	·592	1·5872650	·642	1·7932899	·692	2·0246683	·742	2·2901719
·543	1·4044017	·593	1·5911777	·643	1·7976459	·693	2·0296077	·743	2·2959151
·544	1·4079724	·594	1·5950983	·644	1·8020120	·694	2·0345607	·744	2·3016774
0·545	1·4115493	0·595	1·5990267	0·645	1·8063882	0·695	2·0395273	0·745	2·3074591
·546	1·4151325	·596	1·6029631	·646	1·8107746	·696	2·0445077	·746	2·3132603
·547	1·4187220	·597	1·6069073	·647	1·8151713	·697	2·0495019	·747	2·3190811
·548	1·4223179	·598	1·6108596	·648	1·8195784	·698	2·0545100	·748	2·3249218
·549	1·4259202	·599	1·6148198	·649	1·8239958	·699	2·0595321	·749	2·3307823
0·550	1·4295288	0·600	1·6187881	0·650	1·8284236	0·700	2·0645683	0·750	2·3366630

Table 8·2 (*cont.*)

p	b	p	b	p	b	p	b	p	b
0·750	2·3366630	0·800	2·6603991	0·850	3·0649236	0·900	3·6149504	0·950	4·5139125
·751	2·3425639	·801	2·6675649	·851	3·0741763	·901	3·6283267	·951	4·5394849
·752	2·3484852	·802	2·6747623	·852	3·0834847	·902	3·6418272	·952	4·5655524
·753	2·3544271	·803	2·6819917	·853	3·0928496	·903	3·6554544	·953	4·5921358
·754	2·3603898	·804	2·6892533	·854	3·1022717	·904	3·6692110	·954	4·6192568
0·755	2·3663733	0·805	2·6965476	0·855	3·1117519	0·905	3·6830995	0·955	4·6469386
·756	2·3723780	·806	2·7038747	·856	3·1212908	·906	3·6971227	·956	4·6752063
·757	2·3784038	·807	2·7112352	·857	3·1308893	·907	3·7112834	·957	4·7040862
·758	2·3844511	·808	2·7186292	·858	3·1405482	·908	3·7255846	·958	4·7336067
·759	2·3905200	·809	2·7260571	·859	3·1502684	·909	3·7400293	·959	4·7637983
0·760	2·3966106	0·810	2·7335194	0·860	3·1600507	0·910	3·7546205	0·960	4·7946936
·761	2·4027231	·811	2·7410162	·861	3·1698959	·911	3·7693615	·961	4·8263277
·762	2·4088578	·812	2·7485481	·862	3·1798050	·912	3·7842557	·962	4·8587385
·763	2·4150148	·813	2·7561153	·863	3·1897789	·913	3·7993063	·963	4·8919667
·764	2·4211942	·814	2·7637183	·864	3·1998185	·914	3·8145171	·964	4·9260566
0·765	2·4273964	0·815	2·7713573	0·865	3·2099247	0·915	3·8298917	0·965	4·9610560
·766	2·4336214	·816	2·7790329	·866	3·2200985	·916	3·8454338	·966	4·9970169
·767	2·4398694	·817	2·7867453	·867	3·2303410	·917	3·8611475	·967	5·0339958
·768	2·4461407	·818	2·7944950	·868	3·2406530	·918	3·8770368	·968	5·0720546
·769	2·4524354	·819	2·8022824	·869	3·2510358	·919	3·8931059	·969	5·1112609
0·770	2·4587538	0·820	2·8101079	0·870	3·2614903	0·920	3·9093593	0·970	5·1516890
·771	2·4650960	·821	2·8179719	·871	3·2720175	·921	3·9258014	·971	5·1934204
·772	2·4714622	·822	2·8258748	·872	3·2826187	·922	3·9424371	·972	5·2365456
·773	2·4778527	·823	2·8338170	·873	3·2932950	·923	3·9592711	·973	5·2811645
·774	2·4842677	·824	2·8417991	·874	3·3040475	·924	3·9763087	·974	5·3273885
0·775	2·4907073	0·825	2·8498213	0·875	3·3148774	0·925	3·9935551	0·975	5·3753417
·776	2·4971718	·826	2·8578843	·876	3·3257859	·926	4·0110158	·976	5·4251636
·777	2·5036615	·827	2·8659884	·877	3·3367743	·927	4·0286965	·977	5·4770112
·778	2·5101764	·828	2·8741341	·878	3·3478438	·928	4·0466032	·978	5·5310624
·779	2·5167169	·829	2·8823219	·879	3·3589958	·929	4·0647422	·979	5·5875196
0·780	2·5232832	0·830	2·8905522	0·880	3·3702316	0·930	4·0831199	0·980	5·6466149
·781	2·5298754	·831	2·8988256	·881	3·3815526	·931	4·1017430	·981	5·7086156
·782	2·5364940	·832	2·9071426	·882	3·3929602	·932	4·1206186	·982	5·7738322
·783	2·5431389	·833	2·9155036	·883	3·4044558	·933	4·1397540	·983	5·8426281
·784	2·5498106	·834	2·9239092	·884	3·4160410	·934	4·1591569	·984	5·9154323
0·785	2·5565093	0·835	2·9323599	0·885	3·4277171	0·935	4·1788355	0·985	5·9927565
·786	2·5632351	·836	2·9408563	·886	3·4394859	·936	4·1987979	·986	6·0752175
·787	2·5699884	·837	2·9493988	·887	3·4513488	·937	4·2190532	·987	6·1635688
·788	2·5767694	·838	2·9579881	·888	3·4633076	·938	4·2396103	·988	6·2587436
·789	2·5835784	·839	2·9666248	·889	3·4753639	·939	4·2604792	·989	6·3619174
0·790	2·5904156	0·840	2·9753093	0·890	3·4875195	0·940	4·2816697	0·990	6·4746004
·791	2·5972812	·841	2·9840423	·891	3·4997761	·941	4·3031927	·991	6·5987782
·792	2·6041756	·842	2·9928244	·892	3·5121355	·942	4·3250592	·992	6·7371353
·793	2·6110990	·843	3·0016561	·893	3·5245997	·943	4·3472810	·993	6·8934287
·794	2·6180518	·844	3·0105382	·894	3·5371705	·944	4·3698706	·994	7·0731496
0·795	2·6250340	0·845	3·0194712	0·895	3·5498500	0·945	4·3928409	0·995	7·2847924
·796	2·6320462	·846	3·0284559	·896	3·5626401	·946	4·4162056	·996	7·5425516
·797	2·6390885	·847	3·0374927	·897	3·5755431	·947	4·4399794	·997	7·8729578
·798	2·6461612	·848	3·0465825	·898	3·5885609	·948	4·4641775	·998	8·3353540
·799	2·6532646	·849	3·0557259	·899	3·6016960	·949	4·4888162	·999	9·1181296
0·800	2·6603991	0·850	3·0649236	0·900	3·6149504	0·950	4·5139125	1·000	∞

Table 9. *Partition coefficients for the reversion of series*

(i) Coefficient:
$$M_r(P) = \frac{(n+w+r-1)!}{\pi_1!\dots\pi_\lambda!\{(p_1+1)!\}^{\pi_1}\dots\{(p_\lambda+1)!\}^{\pi_\lambda}},$$

where
$$n = \sum_{i=1}^{\lambda} p_i\pi_i, \quad w = \sum_{i=1}^{\lambda}\pi_i \quad\text{and}\quad P = (p_1^{\pi_1},\dots,p_\lambda^{\pi_\lambda}).$$

(ii) McMahon's coefficient:
$$M_r'(P) = \frac{r(n+w+r-1)!}{(n+r)!\,\pi_1!\dots\pi_\lambda!},$$

where n and w are defined as above.

n	w	Partition	Coefficient (i) $r=1$	Multipliers for coefficient (i) $r=2$	$r=3$	$r=4$	Coefficient (ii) $r=1$	$r=2$	$r=3$	$r=4$
1	1	1	1	3	12	60	1	2	3	4
2	1	2	1	4	20	120	1	2	3	4
	2	1^2	3	5	30	210	2	5	9	14
3	1	3	1	5	30	210	1	2	3	4
	2	21	10	6	42	336	5	12	21	32
	3	1^3	15	7	56	504	5	14	28	48
4	1	4	1	6	42	336	1	2	3	4
	2	31	15	7	56	504	6	14	24	36
	2	2^2	10	7	56	504	3	7	12	18
	3	21^2	105	8	72	720	21	56	108	180
	4	1^4	105	9	90	990	14	42	90	165
5	1	5	1	7	56	504	1	2	3	4
	2	41	21	8	72	720	7	16	27	40
	2	32	35	8	72	720	7	16	27	40
	3	31^2	210	9	90	990	28	72	135	220
	3	2^21	280	9	90	990	28	72	135	220
	4	21^3	1260	10	110	1320	84	240	495	880
	5	1^5	945	11	132	1716	42	132	297	572
6	1	6	1	8	72	720	1	2	3	4
	2	51	28	9	90	990	8	18	30	44
	2	42	56	9	90	990	8	18	30	44
	3	41^2	378	10	110	1320	36	90	165	264
	2	3^2	35	9	90	990	4	9	15	22
	3	321	1260	10	110	1320	72	180	330	528
	4	31^3	3150	11	132	1716	120	330	660	1144
	3	2^3	280	10	110	1320	12	30	55	88
	4	2^21^2	6300	11	132	1716	180	495	990	1716
	5	21^4	17325	12	156	2184	330	990	2145	4004
	6	1^6	10395	13	182	2730	132	429	1001	2002
7	1	7	1	9	90	990	1	2	3	4
	2	61	36	10	110	1320	9	20	33	48
	2	52	84	10	110	1320	9	20	33	48
	3	51^2	630	11	132	1716	45	110	198	312
	2	43	126	10	110	1320	9	20	33	48
	3	421	2520	11	132	1716	90	220	396	624
	4	41^3	6930	12	156	2184	165	440	858	1456
	3	3^21	1575	11	132	1716	45	110	198	312
	3	32^2	2100	11	132	1716	45	110	198	312
	4	321^2	34650	12	156	2184	495	1320	2574	4368

Table 9 (*cont.*)

n	w	Partition	Coefficient (i)	Multipliers for coefficient (i)			Coefficient (ii)			
			$r = 1$	$r = 2$	$r = 3$	$r = 4$	$r = 1$	$r = 2$	$r = 3$	$r = 4$
7	5	31^4	51975	13	182	2730	495	1430	3003	5460
	4	2^31	15400	12	156	2184	165	440	858	1456
	5	2^21^3	138600	13	182	2730	990	2860	6006	10920
	6	21^5	270270	14	210	3360	1287	4004	9009	17472
	7	1^7	135135	15	240	4080	429	1430	3432	7072
8	1	8	1	10	110	1320	1	2	3	4
	2	71	45	11	132	1716	10	22	36	52
	2	62	120	11	132	1716	10	22	36	52
	3	61^2	990	12	156	2184	55	132	234	364
	2	53	210	11	132	1716	10	22	36	52
	3	521	4620	12	156	2184	110	264	468	728
	4	51^3	13860	13	182	2730	220	572	1092	1820
	2	4^2	126	11	132	1716	5	11	18	26
	3	431	6930	12	156	2184	110	264	468	728
	3	42^2	4620	12	156	2184	55	132	234	364
	4	421^2	83160	13	182	2730	660	1716	3276	5460
	5	41^4	135135	14	210	3360	715	2002	4095	7280
	3	3^22	5775	12	156	2184	55	132	234	364
	4	3^21^2	51975	13	182	2730	330	858	1638	2730
	4	32^21	138600	13	182	2730	660	1716	3276	5460
	5	321^3	900900	14	210	3360	2860	8008	16380	29120
	6	31^5	945945	15	240	4080	2002	6006	13104	24752
	4	2^4	15400	13	182	2730	55	143	273	455
	5	2^31^2	600600	14	210	3360	1430	4004	8190	14560
	6	2^21^4	3153150	15	240	4080	5005	15015	32760	61880
	7	21^6	4729725	16	272	4896	5005	16016	37128	74256
	8	1^8	2027025	17	306	5814	1430	4862	11934	25194
9	1	9	1	11	132	1716	1	2	3	4
	2	81	55	12	156	2184	11	24	39	56
	2	72	165	12	156	2184	11	24	39	56
	3	71^2	1485	13	182	2730	66	156	273	420
	2	63	330	12	156	2184	11	24	39	56
	3	621	7920	13	182	2730	132	312	546	840
	4	61^3	25740	14	210	3360	286	728	1365	2240
	2	54	462	12	156	2184	11	24	39	56
	3	531	13860	13	182	2730	132	312	546	840
	3	52^2	9240	13	182	2730	66	156	273	420
	4	521^2	180180	14	210	3360	858	2184	4095	6720
	5	51^4	315315	15	240	4080	1001	2730	5460	9520
	3	4^21	8316	13	182	2730	66	156	273	420
	3	432	27720	13	182	2730	132	312	546	840
	4	431^2	270270	14	210	3360	858	2184	4095	6720
	4	42^21	360360	14	210	3360	858	2184	4095	6720
	5	421^3	2522520	15	240	4080	4004	10920	21840	38080
	6	41^5	2837835	16	272	4896	3003	8736	18564	34272
	3	3^3	5775	13	182	2730	22	52	91	140
	4	3^221	450450	14	210	3360	858	2184	4095	6720
	5	3^21^3	1576575	15	240	4080	2002	5460	10920	18040
	4	32^3	200200	14	210	3360	286	728	1365	2240
	5	32^21^2	6306300	15	240	4080	6006	16380	32760	57120
	6	321^4	23648625	16	272	4896	15015	43680	92820	171360

Table 9 (*cont.*)

n	w	Partition	Coefficient (i) $r=1$	Multipliers for coefficient (i) $r=2$	$r=3$	$r=4$	Coefficient (ii) $r=1$	$r=2$	$r=3$	$r=4$
9	7	31^6	18918900	17	306	5814	27132	24752	55692	108528
	5	2^41	1401400	15	240	4080	1001	2730	5460	9520
	6	2^31^3	21021000	16	272	4896	10010	29120	61880	114240
	7	2^21^5	75675600	17	306	5814	24024	74256	167076	325584
	8	21^7	91891800	18	342	6840	19448	63648	151164	310080
	9	1^9	34459425	19	380	7980	4862	16796	41990	90440
10	1	10	1	12	156	2184	1	2	3	4
	2	91	66	13	182	2730	12	26	42	60
	2	82	220	13	182	2730	12	26	42	60
	3	81^2	2145	14	210	3360	78	182	315	480
	2	73	495	13	182	2730	12	26	42	60
	3	721	12870	14	210	3360	156	364	630	960
	4	71^3	45045	15	240	4080	364	910	1680	2720
	2	64	792	13	182	2730	12	26	42	60
	3	631	25740	14	210	3360	156	364	630	960
	3	62^2	17160	14	210	3360	78	182	315	480
	4	621^2	360360	15	240	4080	1092	2730	5040	8160
	5	61^4	675675	16	272	4896	1365	3640	7140	12240
	2	5^2	462	13	182	2730	6	13	21	30
	3	541	36036	14	210	3360	156	364	630	960
	3	532	60060	14	210	3360	156	364	630	960
	4	531^2	630630	15	240	4080	1092	2730	5040	8160
	4	52^21	840840	15	240	4080	1092	2730	5040	8160
	5	521^3	6306300	16	272	4896	5460	14560	28560	48960
	6	51^5	7567560	17	306	5814	4368	12376	25704	46512
	3	4^22	36036	14	210	3360	78	182	315	480
	4	4^21^2	378378	15	240	4080	546	1365	2520	4080
	3	43^2	45045	14	210	3360	78	182	315	480
	4	4321	2522520	15	240	4080	2184	5460	10080	16320
	5	431^3	9459450	16	272	4896	5460	14560	28560	48960
	4	42^3	560560	15	240	4080	364	910	1680	2720
	5	42^21^2	18918900	16	272	4896	8190	21840	42840	73440
	6	421^4	75675600	17	306	5814	21840	61880	128520	232560
	7	41^6	64324260	18	342	6840	12376	37128	81396	155040
	4	3^31	525525	15	240	4080	364	910	1680	2720
	4	3^22^2	9459450	15	240	4080	546	1365	2520	4080
	5	3^221^2	23648625	16	272	4896	8190	21840	42840	73440
	6	3^21^4	47297250	17	306	5814	10920	30940	64260	116280
	5	32^31	21021000	16	272	4896	5460	14560	28560	48960
	6	32^21^3	252252000	17	306	5814	43680	123760	257040	465120
	7	321^5	643242600	18	342	6840	74256	222768	488376	930240
	8	31^7	413513100	19	380	7980	31824	100776	232560	465120
	5	2^5	1401400	16	272	4896	273	728	1428	2448
	6	2^41^2	84084000	17	306	5814	10920	30940	64260	116280
	7	2^31^4	714714000	18	342	6840	61880	185640	406980	775200
	8	2^21^6	1929727800	19	380	7980	111384	352716	813960	1627920
	9	21^8	1964187225	20	420	9240	75582	251940	610470	1279080
	10	1^{10}	654729075	21	462	10626	16796	58786	149226	326876

APPENDIX

ADDITIONAL BIBLIOGRAPHY COMPILED BY D. E. BARTON

The purpose of the Introduction to these tables has been to present them in the simplest and clearest way possible, keeping the number of references quoted to the minimum necessary for that purpose. It is thought, however, that the following additional bibliography may be helpful to table-users who wish to go more deeply into some particular area of the wide field which the tables cover.

Given below is a numbered list of 59 publications which contain matter in one or more of the following categories:

(T) previous tables and tables used in checking;

(t) more extensive and derived tables;

(a) further applications;

(s) sources of formulae and theory used in the Introduction.

The key gives the serial number and category letter, or letters, under the relevant Part headings.

References secondary to and cited in those which are either given below or on pp. 63–4 above are not systematically listed.

Key

Part I A: 25(s), 47(T). *Part I B:* 9(t), 19(t), 20(t), 33(s). *Part I D:* 5(a), 34(t).

Part II A: 29(s), 34(t). *Part II B:* 58(T), 59(t). *Part II C:* 15(T), 29(s), 30(s), 32(s), 34(t), 44(a, s). *Part II D:* 4(t), 6(t), 32(s), 53(t).

Part IV B: 5(a), 9(T, t), 19(T), 23(T, t), 34(t), 40(T, t).

Part V A: 41(s). *Part V C:* 3(a, s), 7(s), 8(s), 14(a, s), 21(t), 22(t), 27(s), 35(s), 36(t), 38(s). *Part V G:* 1(s), 12(s), 17(s), 22(T), 24(T, s), 54(s). *Part V H:* 14(s), 22(T), 31(s), 46(s).

Part VI A: 14(s), 22(T). *Part VI B:* 10(s), 11(t), 37(s). *Part VI C:* 22(T). *Part VI D & E:* 39(t), 49(t).

Part VII A & B: 2(t, a), 8(s), 43(T, s). *Part VII C:* 8(t), 22(t). *Part VII D:* 14(s), 42(T), 57(s). *Part VII E & F:* 14(s), 28(T), 55(s), 56(s).

Part VIII: 13(T), 18(a), 26(s), 50(a), 51(T).

Part IX: 16(s), 48(s), 52(T).

REFERENCES

1. BARCROFT, D. (1886). On partitions. *Johns Hopk. Univ. Circ.* **5**, 64.
2. BARTON, D. E. (1953). On Neyman's smooth test of goodness-of-fit and its power with respect to a particular system of alternatives. *Skand. Aktuar Tidskr.* **36**, 24–63.
3. BARTON, D. E. (1957). The modality of Neyman's contagious distribution of type A. *Trab. Estadistica*, **8**, 13–22.
4. BARTON, D. E. & DAVID, F. N. (1961). The central sampling moments of the mean in samples from a finite population (Aty's formulae and Madow's central limit). *Biometrika*, **48**, 199–201.
5. BARTON, D. E. & DAVID, F. N. (1962). Randomization bases for multivariate tests I. The bivariate case; randomness on N points in a plane. (Read at 33rd Session of I.S.I., Paris, 1961.) *Bull. Int. Statist. Inst.* **37**, (1), 158–9, (2), 455–67.
6. BARTON, D. E., DAVID, F. N. & FIX, E. (1960). Polykays of the natural numbers. *Biometrika*, **47**, 53–9.
7. BARTON, D. E. & MALLOWS, C. L. (1961). The randomization bases of the problem of the amalgamation of weighted means. *J.R. Statist. Soc.* B, **23**, 423–33.
8. BARTON, D. E. & MALLOWS, C. L. (1965). Some aspects of the random sequence. (Shortened and revised version of paper read at 2nd European Congress of I.M.S., Copenhagen, 1963.) *Ann. Math. Statist.* **36**, 236–60.
9. BENNETT, J. H. (1956). Partitions in more than one dimension. *J.R. Statist. Soc.* B, **18**, 104–12.
10. BIZLEY, M. T. L. (1963). A problem in permutations. *Amer. Math. Mon.* **70**, 727–30.
11. BIZLEY, M. T. L. & JOSEPH, A. N. (1960). The two pack matching problem. *J.R. Statist. Soc.* B, **22**, 114–30.
12. CAUCHY, A. (1843). Mémoire sur les fonctions dont plusieurs valeurs sont liées entre elles par une équation linéaire, et sur diverses transformations de produits composés d'un nombre indéfini de facteurs. *Comptes Rendues*, **17**, 523–31.
13. COHEN, A. C. (1960). Estimating parameters of the truncated Poisson. *Biometrics*, **16**, 203–11.
14. DAVID, F. N. & BARTON, D. E. (1962). *Combinatorial Chance*. London: Charles Griffin and Co.
15. DRESSEL, P. L. (1940). Statistical semivariants and their estimates with particular emphasis on their relation to algebraic invariants. *Ann. Math. Statist.* **11**, 33–57.
16. ERDÈLYI, A. (1960). *Asymptotic Expansions.* §2.4: Laplace's Method. New York: Dover Publishing Co.
17. FESTINGER, L. (1946). The significance of the difference between means without reference to the frequency distribution function. *Psychometrika*, **11**, 97–105.
18. FISHER, R. A. (1930). Distribution of gene ratios for rare mutations. *Proc. Roy. Soc. Edinb.* **50**, 205–20.
19. FISHER, R. A. (1947). Theory of linkage in polysomic inheritance. *Phil. Trans.* B, **233**, 55–87.
20. FISHER, R. A. (1950). A class of enumerations of importance in genetics. *Proc. Roy. Soc.* B, **136**, 509–20.
21. FLETCHER, A., MILLER, J. C. P., ROSENHEAD, L. & COMRIE, L. J. (1962). *An Index of Mathematical Tables* (2nd edition). Oxford: Blackwell Scientific Publications Ltd.
22. GREENWOOD, J. ARTHUR & HARTLEY, H. O. (1962). *Guide to Tables in Mathematical Statistics.* Princeton University Press.
23. GUPTA, H. (1951). A generalization of the partition function. *Proc. Nat. Inst. Sci. India*, **17**, 231–8.
24. HALDANE, J. B. S. & SMITH, C. A. B. (1947). A simple exact test for birth order effect. *Ann. Eugen., Lond.*, **14**, 117–24.
25. HAMMOND, J. (1882). On the calculation of symmetric functions. (1883). On the use of certain differential operators in the theory of equations. *Proc. Lond. Math. Soc.* **13**, 79–84; **14**, 119–29.
26. IRWIN, J. O. (1959). On estimation of the mean of a Poisson distribution from a sample with the zero class missing. *Biometrics*, **15**, 324–6.
27. JORDAN, C. (1950). *Calculus of Finite Differences* (2nd edition). New York: Chelsea Publishing Co.
28. KAPLANSKY, I. (1945). The asymptotic distribution of runs of consecutive elements. *Ann. Math. Statist.* **16**, 200–3.
29. KENDALL, M. G. (1940). Derivation of multivariate sampling formulae from univariate formulae by symbolic operations. *Ann. Eugen., Lond.*, **10**, 392–402.
30. KENDALL, M. G. (1952). Moment statistics in samples from a finite population. *Biometrika*, **39**, 14–16.
31. KENDALL, M. G. (1962). *Rank Correlation Methods* (3rd edition). London: Charles Griffin and Co.
32. KENDALL, M. G. & STUART, A. (1958). *The Advanced Theory of Statistics*, **1**, chapters 12 and 13. London: Charles Griffin and Co.
33. MCMAHON, P. A. (1920, 1955). *Introduction to Combinatory Analysis*. Cambridge: University Press. Also in *Famous Problems and other Monographs*, by Klein *et al.* (1955). New York: Chelsea Publishing Co.

References

34. MIKHAIL, N. N. (1965). Ph.D. Thesis, London University. [Containing tables of trivariate symmetric function relations.]
35. MILNE-THOMPSON, L. M. (1933). *Calculus of Finite Differences.* London: MacMillan and Co.
36. MITRINOVIĆ, D. S. (1959–61). *Pub. de la Fac. Electro. Tech. (Math. Phys.), Univ. Belgrade,* nos. 23, 34, 43 and 60. (With R. S. Mitrinović.)
37. MOOD, A. M. (1940). The distribution theory of runs. *Ann. Math. Statist.* **11**, 367–92.
38. MOSER, L. & WYMAN, M. (1957). Stirling numbers of the second kind. *Duke Math. J.* **25**, 29–43.
39. MOSTELLER, F. (1941). Note on an application of runs to quality control charts. *Ann. Math. Statist.* **12**, 228–32.
40. NANDA, V. S. (1951). Partition theory and thermodynamics of multi-dimensional oscillator assemblies. *Proc. Camb. Phil. Soc.* **47**, 591–601.
41. NÖRLUND, N. E. (1922). Mèmoire sur les Polynômes de Bernoulli. *Acta Math.* **43**, 121–96.
42. OLMSTEAD, P. S. (1946). Distribution of sample arrangements for runs up and down. *Ann. Math. Statist.* **17**, 24–33.
43. RIORDAN, J. (1958). *Introduction to Combinatorial Analysis.* New York: John Wiley and Sons.
44. ROBSON, D. S. (1957*a*). Application of multivariate polykays to the theory of unbiased ratio type estimation. *J. Amer. Statist. Ass.* **52**, 511–22.
45. ROBSON, D. S. (1957*b*). *Applications of Multivariate Polykays to Genetic Covariance Component Analysis.* Ithaca: Cornell University.
46. RODRIGUES, O. (1839). Note sur les inversions, ou dérangements produits dans les permutations. *J. Math. Pures Appl.* **4**, 236–40.
47. ROE, JOSEPHINE R. (1931). *Interfunctional Expressibility Tables of Symmetric Functions.* Published by Syracuse University and the author (at Cambridge, Mass.). (See also *Interfunctional Expressibility Problems of Symmetric Functions,* published by the author, Cambridge, Mass.) These two publications are commonly listed as one work.
48. SIDDIQUI, M. M. (1962). Approximations to the moments of the sample median. *Ann. Math. Statist.* **33**, 157–64.
49. TAKASHIMA, M. (1955). Tables for testing randomness by means of length of runs. *Bull. Math. Statist.* **6**, 17–23.
50. TANNER, J. C. (1951). Delay to pedestrians crossing a road. *Biometrika,* **38**, 383–92.
51. THIJSEN, W. P. (1950). Over de geïtereerde machtsverscheffing. *Simon Stevin,* **27**, 177–92.
52. VAN ORSTRAND, C. E. (1910). Reversion of power series. *Phil. Mag.* **6**/**19**, 366–76.
53. WEIBULL, M. (1959). Moments of the difference between means in two samples from a finite population. *Skand. Aktuar Tidskr.* **42**, 36–60.
54. WILCOXON, F. (1945). Individual comparisons by ranking methods. *Biometrics Bull.* **1**, 80–3.
55. WOLFOWITZ, J. (1942). Additive partition functions and a class of statistical hypotheses. *Ann. Math. Statist.* **13**, 247–79.
56. WOLFOWITZ, J. (1944*a*). Note on runs of consecutive elements. *Ann. Math. Statist.* **15**, 97–8.
57. WOLFOWITZ, J. (1944*b*). Asymptotic distribution of runs up and down. *Ann. Math. Statist.* **15**, 163–72.
58. ZIAUD-DIN, M. (1958). On the expression of the k-statistic k_{11} in terms of power sums and sample moments. *Bull. Int. Statist. Inst.* **36** (3), 102–6.
59. ZIAUD-DIN, M. & AHMAD, M. (1960). On the expression of the k-statistic k_{12} in terms of power sums and sample moments. *Bull. Int. Statist. Inst.* **38**, 635–40.